Methods in Enzymology

Volume 205

METALLOBIOCHEMISTRY

Part B

Metallothionein and Related Molecules

METHODS IN ENZYMOLOGY

EDITORS-IN-CHIEF

John N. Abelson   Melvin I. Simon

DIVISION OF BIOLOGY
CALIFORNIA INSTITUTE OF TECHNOLOGY
PASADENA, CALIFORNIA

Methods in Enzymology

Volume 205

Metallobiochemistry

Part B

Metallothionein and Related Molecules

EDITED BY

James F. Riordan

Bert L. Vallee

CENTER FOR BIOCHEMICAL AND BIOPHYSICAL SCIENCES AND MEDICINE
HARVARD MEDICAL SCHOOL
BOSTON, MASSACHUSETTS

ACADEMIC PRESS, INC.
Harcourt Brace Jovanovich, Publishers
San Diego New York Boston
London Sydney Tokyo Toronto

This book is printed on acid-free paper. ♾

Academic Press, Inc.
San Diego, California 92101

United Kingdom Edition published by
ACADEMIC PRESS LIMITED
24-28 Oval Road, London NW1 7DX

Library of Congress Catalog Card Number: 54-9110

ISBN 0-12-182106-4 (alk. paper)

PRINTED IN THE UNITED STATES OF AMERICA
91 92 93 94 9 8 7 6 5 4 3 2 1

Table of Contents

Section I. Introduction

Section II. Isolation of Metallothioneins

Section III. Quantification in Tissues and Body Fluids

Section IV. Isolation and Purification of Metallothioneins

Section V. Chemical Characterization of Metallothioneins

Section VI. Physicochemical Characterization of Metallothioneins

Section VII. Induction of Metallothioneins

Section VIII. Chapter Related to Section IV: Isolation and Purification of Metallothioneins

Section IX. Overview

Contributors to Volume 205

Article numbers are in parentheses following the names of contributors.
Affiliations listed are current.

YASUNOBU AOKI (16), *Environmental Health Sciences Division, National Institute for Environmental Studies, Tsukuba, Ibaraki 305, Japan*

D. BANERJEE (14), *Department of Pathology, Health Sciences Centre, University of Western Ontario, London, Ontario N6A 5Cl, Canada*

WERNER R. BERNHARD (50), *Institut für Pflanzenbiologie, Universität Zürich, CH-8008 Zürich, Switzerland*

FRANK O. BRADY (63), *Department of Biochemistry and Molecular Biology, University of South Dakota School of Medicine, Vermillion, South Dakota 57069*

IAN BREMNER (4, 10, 66), *Rowett Research Institute, Bucksburn, Aberdeen AB2 9SB, Scotland*

M. GEORGE CHERIAN (3, 12, 13, 14), *Department of Pathology, Health Sciences Centre, University of Western Ontario, London, Ontario N6A 5C1, Canada*

MARY J. CISMOWSKI (37), *Department of Molecular Biology, Scripps Research Institute, La Jolla, California 92037*

ROBERT J. COUSINS (19, 65), *Center for Nutritional Sciences, University of Florida, Gainesville, Florida 32611*

MARTA CZUPRYN (8, 47), *Center for Biochemical and Biophysical Sciences and Medicine, Harvard Medical School, Boston, Massachusetts 02115*

DAVID L. EATON (13), *Department of Environmental Health, University of Washington, Seattle, Washington 98195*

M. EBADI (43), *Departments of Pharmacology and Neurology, University of Nebraska College of Medicine, Omaha, Nebraska 68198*

M. E. ELMES (15), *Department of Pathology, University of Wales College of Medicine, Cardiff CF4 4XN, Wales*

KENNETH H. FALCHUK (8, 47), *Center for Biochemical and Biophysical Sciences and Medicine, and the Department of Medicine, Brigham and Women's Hospital, Harvard Medical School, Boston, Massachusetts 02115*

BRUCE A. FOWLER (30, 67), *Program in Toxicology, The University of Maryland, Baltimore, Maryland 21201*

ROBIN E. GANDLEY (67), *Program in Toxicology, University of Maryland, Baltimore, Maryland 21201*

JUSTINE S. GARVEY (20), *Division of Biology, California Institute of Technology, Pasadena, California 91125*

ZBIGNIEW GASYNA (62), *Department of Chemistry, University of Virginia, Charlottesville, Virginia 22901*

JUDITH A. GLAVEN (67), *Program in Toxicology, University of Maryland, Baltimore, Maryland 21201*

ERWIN GRILL (39), *Institut für Pflanzenwissenschaften, ETH-Zentrum, CH-8092 Zürich, Switzerland*

HANS-JÜRGEN HARTMANN (31), *Anorganische Biochemie, Physiologisch-Chemisches Institut, Universität Tübingen, D-7400 Tübingen 1, Germany*

Y. HAYASHI (40, 41), *Institute for Developmental Research, Aichi Prefectural Colony, Kasugai, Aichi 480-03, Japan*

P. C. HUANG (37), *Department of Biochemistry, The Johns Hopkins University, School of Hygiene and Public Health, Baltimore, Maryland 21205*

PETER E. HUNZIKER (27, 45, 48, 49, 53), *Biochemisches Institut, Universität Zürich, CH-8057 Zürich, Switzerland*

M. ISOBE (41), *School of Agriculture, Nagoya University, Nagoya, Aichi 464-01, Japan*

B. JASANI (15), *Department of Pathology,*

University of Wales College of Medicine, Cardiff CF4 4XN, Wales

JEREMIAS H. R. KÄGI (69), *Biochemisches Institut, Universität Zürich, CH-8057 Zürich, Switzerland*

M. KAWABATA (41), *Institute for Developmental Research, Aichi Prefectural Colony, Kasugai, Aichi 480-03, Japan*

MASAMI KIMURA (17, 21, 35, 44), *Department of Molecular Biology, Keio University School of Medicine, Shinjuku-ku Shinanomachi, Tokyo 160, Japan*

CURTIS D. KLAASSEN (22, 64), *Department of Pharmacology, Toxicology, and Therapeutics, University of Kansas Medical Center, Kansas City, Kansas 66103*

DOMINIK KLEIN (9), *Institute of Toxicology, GSF Research Center, D-8042 Neuherberg, Germany*

SHINZI KOIZUMI (17), *Central Institute of Experimental Animals, National Institute of Industrial Health, Kawasaki 214, Japan*

YUTAKA KOJIMA (2, 48), *Department of Environmental Medicine, Graduate School of Environmental Science, Hokkaido University, Sapporo 606, Japan*

SUSAN K. KREZOSKI (36), *Department of Chemistry, University of Wisconsin—Milwaukee, Milwaukee, Wisconsin 53201*

JAMES E. LAIB (56), *Department of Chemistry, University of Wisconsin—Milwaukee, Milwaukee, Wisconsin 53201*

LOIS D. LEHMAN-MCKEEMAN (22), *Miami Valley Laboratories, Procter & Gamble Company, Cincinnati, Ohio 45239*

KONRAD LERCH (32), *Givaudan Research Company, CH-8600 Dübendorf, Switzerland*

LIH-YUAN LIN (11), *Institute of Radiation Biology, National Tsing Hua University, Hsinchu, Taiwan, Republic of China*

JIE LIU (64), *Department of Pharmacology, Toxicology, and Therapeutics, University of Kansas Medical Center, Kansas City, Kansas 66103*

CHARLES C. MCCORMICK (11), *Division of Nutritional Sciences, Cornell University, Ithaca, New York 14853*

RAJESH K. MEHRA (10), *Division of Hematology-Oncology, University of Utah Medical Center, Salt Lake City, Utah 84132*

KIM MELNICK (56), *Department of Chemistry, University of Wisconsin—Milwaukee, Milwaukee, Wisconsin 53201*

ANATOL G. MORELL (34), *Department of Medicine, Albert Einstein College of Medicine, Bronx, New York 10461*

N. MUTOH (40, 41), *Institute for Developmental Research, Aichi Prefectural Colony, Kasugai, Aichi 480-03, Japan*

C. W. NAKAGAWA (41), *Institute for Developmental Research, Aichi Prefectural Colony, Kasugai, Aichi 480-03, Japan*

KATSUYUKI NAKAJIMA (21, 44), *Japan Immunoresearch Laboratory, 17-5 Takasaki, Gunma, Japan*

MONICA NORDBERG (28), *Department of Environmental Hygiene, Karolinska Institute, S-104 01 Stockholm, Sweden*

ROBERT W. OLAFSON (24, 33), *Department of Biochemistry and Microbiology, University of Victoria, Victoria, British Columbia V8W 3P6, Canada*

PER-ERIC OLSSON (24, 26), *Department of Medical Nutrition, Huddinge University Hospital, F60 NOVUM, Huddinge, Sweden*

NORIKO OTAKI (21, 44), *Departments of Occupational Disease and Experimental Toxicology, National Institute of Industrial Health, 6-21-1 Nagao, Kawasaki, Kanagawa, Japan*

FUMINORI OTSUKA (17), *Department of Environmental Toxicology, Faculty of Pharmaceutical Sciences, Teikyo University, Sagamiko, Kanagawa 199-01, Japan*

DAVID H. PETERING (36, 46, 56, 57), *Department of Chemistry, University of Wisconsin—Milwaukee, Milwaukee, Wisconsin 53201*

DONALD J. PLOCKE (68), *Department of Biology, Boston College, Chestnut Hill, Massachusetts 02167*

WILFRIED E. RAUSER (38), *Department of*

Botany, University of Guelph, Guelph, Ontario N1G 2W1, Canada

MARK P. RICHARDS (25), *Livestock and Poultry Sciences Institute, Nonruminant Animal Nutrition Laboratory, U. S. Department of Agriculture, Agricultural Research Service, Beltsville, Maryland 20705*

A. H. ROBBINS (58), *Miles Research Center, West Haven, Connecticut 06516*

G. ROESIJADI (30), *Chesapeake Biological Laboratory, Center for Environmental and Estuarine Studies, The University of Maryland System, Solomons, Maryland 20688*

M. MERAL SAVAS (46, 56), *Department of Chemistry, University of Wisconsin—Milwaukee, Milwaukee, Wisconsin 53201*

ANDREAS SCHÄFFER (61), *CIBA-GEIGY Ltd., Agricultural Division, CH-4002 Basel, Switzerland*

A. M. SCHEUHAMMER (12), *Environment Canada, Canadian Wildlife Service, Hull, Quebec K1A OH3, Canada*

JOSEPH J. SCHROEDER (65), *Department of Biochemistry, Emory University School of Medicine, Atlanta, Georgia 30322*

ZAHIR A. SHAIKH (18), *Department of Pharmacology and Toxicology, University of Rhode Island, Kingston, Rhode Island 02881*

C. FRANK SHAW III (36, 46, 56, 57), *Department of Chemistry, University of Wisconsin—Milwaukee, Milwaukee, Wisconsin 53201*

IRMIN STERNLIEB (34), *Department of Medicine and Liver Research Center, Albert Einstein College of Medicine, Bronx, New York 10461*

MARTIN J. STILLMAN (62), *Department of Chemistry, University of Western Ontario, London, Ontario N6A 5B7, Canada*

RICHARD J. STOCKERT (34), *Departments of Medicine and Biochemistry, and Liver Research Center, Albert Einstein College of Medicine, Bronx, New York 10461*

C. D. STOUT (58), *Department of Molecular Biology, Research Institute of Scripps Clinic, La Jolla, California 92037*

KARL H. SUMMER (9), *Institute of Toxicology, GSF Research Center, D-8042 Neuherberg, Germany*

KAZUO T. SUZUKI (16, 23, 29), *Environmental Health Sciences Division, National Institute for Environmental Studies, Tsukuba, Ibaraki 305, Japan*

KEIJI SUZUKI (21, 44), *College of Medical Care and Technology, Gunma University, Maebashi, Gunma, Japan*

DOUGLAS M. TEMPLETON (3), *Department of Clinical Biochemistry, University of Toronto, Toronto, Ontario M5G 1L5, Canada*

BERT L. VALLEE (1), *Center for Biochemical and Biophysical Sciences and Medicine, Harvard Medical School, Boston, Massachusetts 02115*

MILAN VAŠÁK (5, 6, 7, 54, 60), *Biochemisches Institut, Universität Zürich, CH-8057 Zürich, Switzerland*

ULRICH WESER (31), *Anorganische Biochemie, Physiologisch-Chemisches Institut, Universität Tübingen, D-7400 Tübingen 1, Germany*

P. D. WHANGER (42), *Department of Agricultural Chemistry, Oregon State University, Corvallis, Oregon 97331*

DENNIS R. WINGE (52, 55), *Departments of Medicine and Biochemistry, University of Utah Medical Center, Salt Lake City, Utah 84132*

ERNST-LUDWIG WINNACKER (39), *Genzentrum der Universität München, Am Klopferspitz, D-8033 Martinsried, Germany*

KURT WÜTHRICH (59), *Institut für Molekularbiologie und Biophysik, Eidgenössische Technische Hochschule-Hönggerberg, CH-8093 Zürich, Switzerland*

JIN ZENG (51), *Biochemisches Institut, Universität Zürich, CH-8057 Zürich, Switzerland*

MEINHART H. ZENK (39), *Lehrstuhl für Pharmazeutische Biologie, Universität München, D-8000 München 2, Germany*

Preface

This Metallobiochemistry volume of *Methods of Enzymology* is devoted to metallothionein and related proteins because of the tremendous upsurge of interest in this area, particularly in its biological function. While this function has remained enigmatic for more than three decades, its elucidation may well be imminent.

The detection of metallothionein was the consequence of a search for a cadmium protein. Spectroscopy revealed that it contained significant amounts of both cadmium and zinc as well as small amounts of copper and iron. Much as it was clearly a cadmium-containing protein, its zinc content was sufficiently impressive to lead to the descriptive term "metallothionein," which has proved to be an enduring epithet. It is and has remained the only protein known to contain cadmium in its native state. Since cadmium has been thought to be toxic, a steadily increasing concern with environmental protection focused special attention of toxicologists on this protein. Much research effort has continued to be devoted to this aspect of metallothionein chemistry and biology.

As the significance of its zinc content became more apparent, the quest for a role for metallothionein in zinc metabolism began to develop, particularly in relation to zinc metalloenzymes. Among those, the enzymes involved in nucleic acid metabolism seemed to provide a plausible rationale for the nutritional essentiality of zinc and even suggested a role for zinc in gene expression. In the past few years the world of "zinc fingers" began to flourish, and the similarities in zinc binding among some of these transcription factors, other DNA-binding proteins, and metallothionein have suggested a functional relationship between them and the postulate of a zinc regulatory process.

Difficulties in defining both the protein itself and the conditions for investigating its properties had greatly confused the metallothionein picture in the past. The state of that knowledge is now sufficiently secure so that it is appropriate to compile in this volume definitions of terms and standard operating procedures for the isolation, quantitation, and chemical and physical characterization of metallothioneins. We hope that those in the field and those wishing to enter it will find these contributions timely and useful in establishing the baselines for communication.

In assembling this volume, we have been extremely fortunate in having had the cooperation and support of a distinguished group of contributing authors. In particular, we would like to acknowledge the important editorial assistance of Jeremias H. R. Kägi, whose collaboration, cooperation, and friendship we have enjoyed ever since the discovery of metallothionein in 1957. His advice and counsel were and remain invaluable. Finally, we thank the staff of Academic Press for their continued support and assistance. Our interaction with them has been a truly pleasant experience.

JAMES F. RIORDAN
BERT L. VALLEE

METHODS IN ENZYMOLOGY

VOLUME XXVII. Enzyme Structure (Part D)
Edited by C. H. W. HIRS AND SERGE N. TIMASHEFF

VOLUME XXVIII. Complex Carbohydrates (Part B)
Edited by VICTOR GINSBURG

VOLUME XXIX. Nucleic Acids and Protein Synthesis (Part E)
Edited by LAWRENCE GROSSMAN AND KIVIE MOLDAVE

VOLUME XXX. Nucleic Acids and Protein Synthesis (Part F)
Edited by KIVIE MOLDAVE AND LAWRENCE GROSSMAN

VOLUME XXXI. Biomembranes (Part A)
Edited by SIDNEY FLEISCHER AND LESTER PACKER

VOLUME XXXII. Biomembranes (Part B)
Edited by SIDNEY FLEISCHER AND LESTER PACKER

VOLUME XXXIII. Cumulative Subject Index Volumes I–XXX
Edited by MARTHA G. DENNIS AND EDWARD A. DENNIS

VOLUME XXXIV. Affinity Techniques (Enzyme Purification: Part B)
Edited by WILLIAM B. JAKOBY AND MEIR WILCHEK

VOLUME XXXV. Lipids (Part B)
Edited by JOHN M. LOWENSTEIN

VOLUME XXXVI. Hormone Action (Part A: Steroid Hormones)
Edited by BERT W. O'MALLEY AND JOEL G. HARDMAN

VOLUME XXXVII. Hormone Action (Part B: Peptide Hormones)
Edited by BERT W. O'MALLEY AND JOEL G. HARDMAN

VOLUME XXXVIII. Hormone Action (Part C: Cyclic Nucleotides)
Edited by JOEL G. HARDMAN AND BERT W. O'MALLEY

VOLUME XXXIX. Hormone Action (Part D: Isolated Cells, Tissues, and Organ Systems)
Edited by JOEL G. HARDMAN AND BERT W. O'MALLEY

VOLUME XL. Hormone Action (Part E: Nuclear Structure and Function)
Edited by BERT W. O'MALLEY AND JOEL G. HARDMAN

VOLUME XLI. Carbohydrate Metabolism (Part B)
Edited by W. A. WOOD

VOLUME XLII. Carbohydrate Metabolism (Part C)
Edited by W. A. WOOD

VOLUME XLIII. Antibiotics
Edited by JOHN H. HASH

VOLUME XLIV. Immobilized Enzymes
Edited by KLAUS MOSBACH

VOLUME XLV. Proteolytic Enzymes (Part B)
Edited by LASZLO LORAND

VOLUME XLVI. Affinity Labeling
Edited by WILLIAM B. JAKOBY AND MEIR WILCHEK

VOLUME XLVII. Enzyme Structure (Part E)
Edited by C. H. W. HIRS AND SERGE N. TIMASHEFF

VOLUME XLVIII. Enzyme Structure (Part F)
Edited by C. H. W. HIRS AND SERGE N. TIMASHEFF

VOLUME XLIX. Enzyme Structure (Part G)
Edited by C. H. W. HIRS AND SERGE N. TIMASHEFF

VOLUME L. Complex Carbohydrates (Part C)
Edited by VICTOR GINSBURG

VOLUME LI. Purine and Pyrimidine Nucleotide Metabolism
Edited by PATRICIA A. HOFFEE AND MARY ELLEN JONES

VOLUME LII. Biomembranes (Part C: Biological Oxidations)
Edited by SIDNEY FLEISCHER AND LESTER PACKER

VOLUME LIII. Biomembranes (Part D: Biological Oxidations)
Edited by SIDNEY FLEISCHER AND LESTER PACKER

VOLUME 80. Proteolytic Enzymes (Part C)
Edited by LASZLO LORAND

VOLUME 81. Biomembranes (Part H: Visual Pigments and Purple Membranes, I)
Edited by LESTER PACKER

VOLUME 82. Structural and Contractile Proteins (Part A: Extracellular Matrix)
Edited by LEON W. CUNNINGHAM AND DIXIE W. FREDERIKSEN

VOLUME 83. Complex Carbohydrates (Part D)
Edited by VICTOR GINSBURG

VOLUME 84. Immunochemical Techniques (Part D: Selected Immunoassays)
Edited by JOHN J. LANGONE AND HELEN VAN VUNAKIS

VOLUME 85. Structural and Contractile Proteins (Part B: The Contractile Apparatus and the Cytoskeleton)
Edited by DIXIE W. FREDERIKSEN AND LEON W. CUNNINGHAM

VOLUME 86. Prostaglandins and Arachidonate Metabolites
Edited by WILLIAM E. M. LANDS AND WILLIAM L. SMITH

VOLUME 87. Enzyme Kinetics and Mechanism (Part C: Intermediates, Stereochemistry, and Rate Studies)
Edited by DANIEL L. PURICH

VOLUME 88. Biomembranes (Part I: Visual Pigments and Purple Membranes, II)
Edited by LESTER PACKER

VOLUME 89. Carbohydrate Metabolism (Part D)
Edited by WILLIS A. WOOD

VOLUME 90. Carbohydrate Metabolism (Part E)
Edited by WILLIS A. WOOD

VOLUME 91. Enzyme Structure (Part I)
Edited by C. H. W. HIRS AND SERGE N. TIMASHEFF

VOLUME 92. Immunochemical Techniques (Part E: Monoclonal Antibodies and General Immunoassay Methods)
Edited by JOHN J. LANGONE AND HELEN VAN VUNAKIS

VOLUME 93. Immunochemical Techniques (Part F: Conventional Antibodies, Fc Receptors, and Cytotoxicity)
Edited by JOHN J. LANGONE AND HELEN VAN VUNAKIS

VOLUME 94. Polyamines
Edited by HERBERT TABOR AND CELIA WHITE TABOR

VOLUME 95. Cumulative Subject Index Volumes 61–74, 76–80
Edited by EDWARD A. DENNIS AND MARTHA G. DENNIS

VOLUME 96. Biomembranes [Part J: Membrane Biogenesis: Assembly and Targeting (General Methods; Eukaryotes)]
Edited by SIDNEY FLEISCHER AND BECCA FLEISCHER

VOLUME 97. Biomembranes [Part K: Membrane Biogenesis: Assembly and Targeting (Prokaryotes, Mitochcndria, and Chloroplasts)]
Edited by SIDNEY FLEISCHER AND BECCA FLEISCHER

VOLUME 98. Biomembranes [Part L: Membrane Biogenesis (Processing and Recycling)]
Edited by SIDNEY FLEISCHER AND BECCA FLEISCHER

VOLUME 99. Hormone Action (Part F: Protein Kinases)
Edited by JACKIE D. CORBIN AND JOEL G. HARDMAN

VOLUME 100. Recombinant DNA (Part B)
Edited by RAY WU, LAWRENCE GROSSMAN, AND KIVIE MOLDAVE

VOLUME 101. Recombinant DNA (Part C)
Edited by RAY WU, LAWRENCE GROSSMAN, AND KIVIE MOLDAVE

VOLUME 102. Hormone Action (Part G: Calmodulin and Calcium-Binding Proteins)
Edited by ANTHONY R. MEANS AND BERT W. O'MALLEY

VOLUME 103. Hormone Action (Part H: Neuroendocrine Peptides)
Edited by P. MICHAEL CONN

VOLUME 104. Enzyme Purification and Related Techniques (Part C)
Edited by WILLIAM B. JAKOBY

VOLUME 105. Oxygen Radicals in Biological Systems
Edited by LESTER PACKER

VOLUME 106. Posttranslational Modifications (Part A)
Edited by FINN WOLD AND KIVIE MOLDAVE

VOLUME 107. Posttranslational Modifications (Part B)
Edited by FINN WOLD AND KIVIE MOLDAVE

VOLUME 108. Immunochemical Techniques (Part G: Separation and Characterization of Lymphoid Cells)
Edited by GIOVANNI DI SABATO, JOHN J. LANGONE, AND HELEN VAN VUNAKIS

VOLUME 109. Hormone Action (Part I: Peptide Hormones)
Edited by LUTZ BIRNBAUMER AND BERT W. O'MALLEY

VOLUME 110. Steroids and Isoprenoids (Part A)
Edited by JOHN H. LAW AND HANS C. RILLING

VOLUME 111. Steroids and Isoprenoids (Part B)
Edited by JOHN H. LAW AND HANS C. RILLING

VOLUME 112. Drug and Enzyme Targeting (Part A)
Edited by KENNETH J. WIDDER AND RALPH GREEN

VOLUME 113. Glutamate, Glutamine, Glutathione, and Related Compounds
Edited by ALTON MEISTER

VOLUME 114. Diffraction Methods for Biological Macromolecules (Part A)
Edited by HAROLD W. WYCKOFF, C. H. W. HIRS, AND SERGE N. TIMASHEFF

VOLUME 115. Diffraction Methods for Biological Macromolecules (Part B)
Edited by HAROLD W. WYCKOFF, C. H. W. HIRS, AND SERGE N. TIMASHEFF

VOLUME 116. Immunochemical Techniques (Part H: Effectors and Mediators of Lymphoid Cell Functions)
Edited by GIOVANNI DI SABATO, JOHN J. LANGONE, AND HELEN VAN VUNAKIS

VOLUME 117. Enzyme Structure (Part J)
Edited by C. H. W. HIRS AND SERGE N. TIMASHEFF

VOLUME 118. Plant Molecular Biology
Edited by ARTHUR WEISSBACH AND HERBERT WEISSBACH

VOLUME 119. Interferons (Part C)
Edited by SIDNEY PESTKA

VOLUME 120. Cumulative Subject Index Volumes 81–94, 96–101

VOLUME 121. Immunochemical Techniques (Part I: Hybridoma Technology and Monoclonal Antibodies)
Edited by JOHN J. LANGONE AND HELEN VAN VUNAKIS

VOLUME 122. Vitamins and Coenzymes (Part G)
Edited by FRANK CHYTIL AND DONALD B. MCCORMICK

VOLUME 123. Vitamins and Coenzymes (Part H)
Edited by FRANK CHYTIL AND DONALD B. MCCORMICK

VOLUME 124. Hormone Action (Part J: Neuroendocrine Peptides)
Edited by P. MICHAEL CONN

VOLUME 125. Biomembranes (Part M: Transport in Bacteria, Mitochondria, and Chloroplasts: General Approaches and Transport Systems)
Edited by SIDNEY FLEISCHER AND BECCA FLEISCHER

VOLUME 126. Biomembranes (Part N: Transport in Bacteria, Mitochondria, and Chloroplasts: Protonmotive Force)
Edited by SIDNEY FLEISCHER AND BECCA FLEISCHER

VOLUME 127. Biomembranes (Part O: Protons and Water: Structure and Translocation)
Edited by LESTER PACKER

VOLUME 128. Plasma Lipoproteins (Part A: Preparation, Structure, and Molecular Biology)
Edited by JERE P. SEGREST AND JOHN J. ALBERS

VOLUME 129. Plasma Lipoproteins (Part B: Characterization, Cell Biology, and Metabolism)
Edited by JOHN J. ALBERS AND JERE P. SEGREST

VOLUME 130. Enzyme Structure (Part K)
Edited by C. H. W. HIRS AND SERGE N. TIMASHEFF

VOLUME 131. Enzyme Structure (Part L)
Edited by C. H. W. HIRS AND SERGE N. TIMASHEFF

VOLUME 132. Immunochemical Techniques (Part J: Phagocytosis and Cell-Mediated Cytotoxicity)
Edited by GIOVANNI DI SABATO AND JOHANNES EVERSE

VOLUME 133. Bioluminescence and Chemiluminescence (Part B)
Edited by MARLENE DELUCA AND WILLIAM D. MCELROY

VOLUME 134. Structural and Contractile Proteins (Part C: The Contractile Apparatus and the Cytoskeleton)
Edited by RICHARD B. VALLEE

VOLUME 135. Immobilized Enzymes and Cells (Part B)
Edited by KLAUS MOSBACH

VOLUME 136. Immobilized Enzymes and Cells (Part C)
Edited by KLAUS MOSBACH

VOLUME 137. Immobilized Enzymes and Cells (Part D)
Edited by KLAUS MOSBACH

VOLUME 138. Complex Carbohydrates (Part E)
Edited by VICTOR GINSBURG

VOLUME 139. Cellular Regulators (Part A: Calcium- and Calmodulin-Binding Proteins)
Edited by ANTHONY R. MEANS AND P. MICHAEL CONN

VOLUME 154. Recombinant DNA (Part E)
Edited by RAY WU AND LAWRENCE GROSSMAN

VOLUME 155. Recombinant DNA (Part F)
Edited by RAY WU

VOLUME 156. Biomembranes (Part P: ATP-Driven Pumps and Related Transport: The Na,K-Pump)
Edited by SIDNEY FLEISCHER AND BECCA FLEISCHER

VOLUME 157. Biomembranes (Part Q: ATP-Driven Pumps and Related Transport: Calcium, Proton, and Potassium Pumps)
Edited by SIDNEY FLEISCHER AND BECCA FLEISCHER

VOLUME 158. Metalloproteins (Part A)
Edited by JAMES F. RIORDAN AND BERT L. VALLEE

VOLUME 159. Initiation and Termination of Cyclic Nucleotide Action
Edited by JACKIE D. CORBIN AND ROGER A. JOHNSON

VOLUME 160. Biomass (Part A: Cellulose and Hemicellulose)
Edited by WILLIS A. WOOD AND SCOTT T. KELLOGG

VOLUME 161. Biomass (Part B: Lignin, Pectin, and Chitin)
Edited by WILLIS A. WOOD AND SCOTT T. KELLOGG

VOLUME 162. Immunochemical Techniques (Part L: Chemotaxis and Inflammation)
Edited by GIOVANNI DI SABATO

VOLUME 163. Immunochemical Techniques (Part M: Chemotaxis and Inflammation)
Edited by GIOVANNI DI SABATO

VOLUME 164. Ribosomes
Edited by HARRY F. NOLLER, JR. AND KIVIE MOLDAVE

VOLUME 165. Microbial Toxins: Tools for Enzymology
Edited by SIDNEY HARSHMAN

VOLUME 166. Branched-Chain Amino Acids
Edited by ROBERT HARRIS AND JOHN R. SOKATCH

VOLUME 179. Complex Carbohydrates (Part F)
Edited by VICTOR GINSBURG

VOLUME 180. RNA Processing (Part A: General Methods)
Edited by JAMES E. DAHLBERG AND JOHN N. ABELSON

VOLUME 181. RNA Processing (Part B: Specific Methods)
Edited by JAMES E. DAHLBERG AND JOHN N. ABELSON

VOLUME 182. Guide to Protein Purification
Edited by MURRAY P. DEUTSCHER

VOLUME 183. Molecular Evolution: Computer Analysis of Protein and Nucleic Acid Sequences
Edited by RUSSELL F. DOOLITTLE

VOLUME 184. Avidin-Biotin Technology
Edited by MEIR WILCHEK AND EDWARD A. BAYER

VOLUME 185. Gene Expression Technology
Edited by DAVID V. GOEDDEL

VOLUME 186. Oxygen Radicals in Biological Systems (Part B: Oxygen Radicals and Antioxidents)
Edited by LESTER PACKER AND ALEXANDER N. GLAZER

VOLUME 187. Arachidonate Related Lipid Mediators
Edited by ROBERT C. MURPHY AND FRANK A. FITZPATRICK

VOLUME 188. Hydrocarbons and Methylotrophy
Edited by MARY E. LIDSTROM

VOLUME 189. Retinoids (Part A: Molecular and Metabolic Aspects)
Edited by LESTER PACKER

VOLUME 190. Retinoids (Part B: Cell Differentiation and Clinical Applications)
Edited by LESTER PACKER

VOLUME 191. Biomembranes (Part V: Cellular and Subcellular Transport: Epithelial Cells)
Edited by SIDNEY FLEISCHER AND BECCA FLEISCHER

VOLUME 192. Biomembranes (Part W: Cellular and Subcellular Transport: Epithelial Cells)
Edited by SIDNEY FLEISCHER AND BECCA FLEISCHER

VOLUME 193. Mass Spectrometry
Edited by JAMES A. MCCLOSKEY

VOLUME 194. Guide to Yeast Genetics and Molecular Biology
Edited by CHRISTINE GUTHRIE AND GERALD R. FINK

VOLUME 195. Adenylyl Cyclase, G Proteins, and Guanylyl Cyclase
Edited by ROGER A. JOHNSON AND JACKIE D. CORBIN

VOLUME 196. Molecular Motors and the Cytoskeleton
Edited by RICHARD B. VALLEE

VOLUME 197. Phospholipases
Edited by EDWARD A. DENNIS

VOLUME 198. Peptide Growth Factors (Part C)
Edited by DAVID BARNES, J. P. MATHER, AND GORDON H. SATO

VOLUME 199. Cumulative Subject Index Volumes 168–174, 176–194 (in preparation)

VOLUME 200. Protein Phosphorylation (Part A: Protein Kinases: Assays, Purification, Antibodies, Functional Analysis, Cloning, and Expression)
Edited by TONY HUNTER AND BARTHOLOMEW M. SEFTON

VOLUME 201. Protein Phosphorylation (Part B: Analysis of Protein Phosphorylation, Protein Kinase Inhibitors, and Protein Phosphatases)
Edited by TONY HUNTER AND BARTHOLOMEW M. SEFTON

VOLUME 202. Molecular Design and Modeling: Concepts and Applications (Part A: Proteins, Peptides, and Enzymes)
Edited by JOHN J. LANGONE

VOLUME 203. Molecular Design and Modeling: Concepts and Applications (Part B: Antibodies and Antigens, Nucleic Acids, Polysaccharides, and Drugs)
Edited by JOHN J. LANGONE

VOLUME 204. Bacterial Genetic Systems
Edited by JEFFREY H. MILLER

VOLUME 205. Metallobiochemistry (Part B: Metallothionein and Related Molecules)
Edited by JAMES F. RIORDAN AND BERT L. VALLEE

VOLUME 206. Cytochrome P450
Edited by MICHAEL R. WATERMAN AND ERIC F. JOHNSON

VOLUME 207. Ion Channels (in preparation)
Edited by BERNARDO RUDY AND LINDA IVERSON

VOLUME 208. Protein–DNA Interactions (in preparation)
Edited by ROBERT T. SAUER

VOLUME 209. Phospholipid Biosynthesis (in preparation)
Edited by EDWARD A. DENNIS AND DENNIS E. VANCE

VOLUME 210. Numerical Computer Methods (in preparation)
Edited by LUDWIG BRAND AND MICHAEL L. JOHNSON

Section I

Introduction

[1] Introduction to Metallothionein

By BERT L. VALLEE

Zinc is a component of many proteins and, as such, it is involved in virtually all aspects of metabolism. It has been found to be an integral component of nearly 300 enzymes in different species of all phyla, indispensable to their functions, which encompass the synthesis and/or degradation of all major metabolites. Advances in the isolation and characterization of enzymes and analysis of metals were basic to the rapid growth of the field. The capacity to measure picograms of zinc in nanograms of proteins is now widespread, and such analyses of biological matter have ceased to be a problem.[1,2]

About a decade ago, the role of zinc in gene expression began to attract interest[3-6] and this field has recently gained wide attention. At least 200 DNA-binding proteins are now known that are thought—although not shown—to contain zinc, which is essential for their function.[7] In addition to its roles in catalysis and gene expression, zinc stabilizes the structure of proteins and nucleic acids, preserves the integrity of subcellular organelles, participates in transport processes, and plays important roles in viral and immune phenomena.[8] Its nutritional essentiality has focused attention on the pathology and clinical consequences of both its deficiency and toxicity.[1,9]

There is one other, widely occurring zinc protein that has served as a source of fascination and frustration for biochemists for almost 40 years. Metallothionein was discovered contemporaneously with the first of the zinc metalloenzymes,[10] but the lack of a readily measurable biological

[1] B. L. Vallee, *BioFactors* **1,** 31 (1988).

[2] J. F. Riordan and B. L. Vallee, this series, Vol. 158, p. 3.

[3] B. L. Vallee, *Experientia* **33,** 600 (1977).

[4] B. L. Vallee, *in* "Biological Aspects of Inorganic Chemistry" (D. Dolphin, ed.), p. 37. Wiley, New York, 1977.

[5] B. L. Vallee and K. H. Falchuk, *Philos. Trans. R. Soc. London, Ser. B* **294,** 185 (1981).

[6] J. S. Hanas, D. J. Hazuda, D. F. Bogenhagen, F. Y.-H. Wu, and C.-W. Wu, *J. Biol. Chem.* **258,** 14120 (1983).

[7] B. L. Vallee and D. S. Auld, *in* "Matrix Metalloproteinases and Inhibitors" (H. Birkedal-Hansen, Z. Werb, H. Welgus, and H. E. Van Wart, eds.), in press. Gustav Fischer Verlag, Stuttgart, 1991.

[8] B. L. Vallee and D. S. Auld, *Proc. Natl. Acad. Sci. U.S.A.* **87,** 220 (1990).

[9] B. L. Vallee, *in* "Zinc Enzymes" (I. Bertini, C. Luchinat, W. Maret, and M. Zeppezauer, eds.), p. 1. Birkhäuser, Boston, 1986.

[10] M. Margoshes and B. L. Vallee, *J. Am. Chem. Soc.* **79,** 4813 (1957).

activity coupled with the need for sensitive methods for metal analysis stifled interest in the protein. Its existence was virtually unnoticed for a very long time and those few who were drawn by its siren call are to be commended for their vision and steadfastness.

Two features of metallothionein were sufficient to sustain interest in it as a chemical and biological curiosity during the early years. One was its highly unusual amino acid composition—no aromatic or heterocyclic amino acids while one-third of its residues are cysteines—and the other was its extraordinary metal content—7 g-atoms of metal per mole.[11] That most of this metal seemed to be cadmium only added to the attraction. Moreover, it was a small protein of just 60 or so amino acids, readily amenable to structural analysis. Another characteristic that eventually captured the attention of molecular geneticists was the inducibility of metallothionein in response to heavy metals *in vivo* and to a great variety of metabolites such as glucocorticoids, glucagon, interleukin 1, catecholamines, progesterone, estrogen, ethanol, and interferon, among others, thus making it a convenient target for studying the regulation of gene expression.

Although metallothionein was first isolated from equine kidney cortex, it has been found throughout the animal kingdom.[12] Indeed, plants and fungi contain cadmium(II) thiolate oligopeptides, which are nontranslationally biosynthesized, or metal-binding polypeptides only distantly related to equine renal metallothionein.[13] Thus, these molecules appear to be important for a great many species and yet, as has been stated so often, it has not been possible to assign them a specific biological role. Among the functions considered early are detoxification of cadmium and other heavy metals, regulation of zinc and copper metabolism, and providing zinc for newly synthesized apoenzymes.[14,15] In addition, it was suggested to function in the expression of the genetic message.[16–18] Perhaps complicating the situation a bit more, the occurrence of different isoforms of metallothionein has suggested to some that they may have different biological functions. Long ago thionein, the apoprotein, rather than the metalloprotein, was thought to play a cardinal metabolic role. If this were found to be true

[11] Y. Kojima, C. Berger, B. L. Vallee, and J. H. R. Kägi, *Proc. Natl. Acad. Sci. U.S.A.* **73,** 3413 (1976).

[12] D. Hamer, *Annu. Rev. Biochem.* **55,** 913 (1986).

[13] W. E. Rauser, *Annu. Rev. Biochem.* **59,** 61 (1990).

[14] J. H. R. Kägi and B. L. Vallee, *J. Biol. Chem.* **235,** 3460 (1960).

[15] J. H. R. Kägi and B. L. Vallee, *J. Biol. Chem.* **236,** 2435 (1961).

[16] B. L. Vallee, *Experientia, Suppl.* **34,** p. 19 (1979).

[17] B. L. Vallee, *Experientia, Suppl.* **52,** p. 5 (1987).

[18] J. H. R. Kägi and A. Schäffer, *Biochemistry* **27,** 8509 (1988).

then zinc, cadmium, copper, and mercury might inhibit an essential biological function. The reported increase in the production of the protein in instances of heavy metal poisoning could then be interpreted to be an attempt to overcome this inhibition.[16]

Some new perspective on metallothionein may be gained from considerations of zinc coordination, function, and structure of zinc enzymes and DNA-binding proteins. A sufficient number of three-dimensional structures of zinc enzymes have now been established to constitute standards of reference. In all of these the catalytic zinc atom is coordinated to three amino acid side chains of the protein, leaving a fourth coordination position to be filled by water which, in turn, is hydrogen bonded to or activated by other protein side chain(s). The nature of the particular ligands varies from one enzyme to another, with histidine being the most common binding residue along with carboxylates and thiols. The objective of zinc binding seems to be to poise the atom for catalysis.[7,8,19,20]

An examination of the protein and gene sequence databanks and literature has revealed structural relationships between zinc enzyme families related by common ancestry. In many, the zinc is coordinated to two, closely spaced ligands (separated by 1–3 amino acids) and a third that is separated from the first 2 by from 20 to 120 amino acids and completes the coordination sphere. In the zinc-dependent alcohol dehydrogenases, the first two ligands are separated by a longer, 20-residue sequence, with the third being 106 residues away.[7,8,19,20]

In contrast to active-site zinc atoms, structural zinc atoms have four protein ligands. There are too few examples to allow generalizations, but these ligands tend to be closely arrayed in the linear sequence of the protein. Structural zinc atoms do not have an open coordination site containing water. Instead, they are believed to contribute to conformational stability. In both instances known, the four ligands are all cysteine sulfurs.[7,8,20]

What appears to be emerging from these investigations is a realization that zinc coordination is critical to its ability to assume different biological roles. The chemistry of zinc does not seem exceptional. It forms a rather stable divalent oxidation state and can exist in both metal–hydrate and metal–hydroxide forms near neutral pH. However, it does have a particularly flexible coordination sphere, and can assume multiple coordination geometries, and this may be its most distinguishing feature. It can adapt to constraints imposed by biological macromolecules and thereby acquire or provide chemical characteristics important for biological function.

[19] B. L. Vallee and D. S. Auld, *FEBS Lett.* **257,** 138 (1989).
[20] B. L. Vallee and D. S. Auld, *Biochemistry* **29,** 5647 (1990).

This aspect of zinc is dramatically illustrated in more recent studies of the functionally heterogeneous group of DNA-binding proteins.[20] These have been inferred to be zinc-binding proteins based largely on their primary sequences in relation to the zinc-containing *Xenopus* transcription factor IIIA. Patterns of conserved cysteine and histidine residues separated by variable numbers of amino acids have been formulated into a popular model known as a "zinc finger." This model has been adopted for TFIIIA and other proteins with similarly spaced Cys_2-His_2 sequences. However, it does not pertain to DNA-binding proteins in which cysteines constitute the only zinc ligands.

Nuclear magnetic resonance (NMR) structures for the rat glucocorticoid receptor or the fungal transcription factor GAL4 reveal that both proteins contain two zinc-binding sites. In GAL4, the two zincs bind to six cysteines to form a zinc cluster reminiscent of the metal-binding sites of metallothionein.[21] In the glucocorticoid receptor, each zinc is bound to four cysteines and the DNA recognition site is a helical segment located between the two zinc atoms.[22] Thus, there are at least three types of zinc-binding sites in the DNA-binding domains of these proteins, giving rise to three types of DNA recognition sites. In each case the zinc is coordinated to four ligands, but the structures of their zinc complexes differ significantly. There is no catalytic activity associated with these zinc atoms. They seem to serve an important structural role by generating and/or stabilizing zinc "fingers," zinc "clusters," and zinc "twists." Significantly, it is the ability to acquire or accept a range of coordination geometries that seemingly underlies the utilization of zinc for this purpose.[23]

Within this context, it is now possible to reexamine the possible function of metallothionein. The cell has chosen to utilize zinc for at least two important purposes. It places it at the active site of a broad variety of enzymes, which in each instance supplies the metal with three appropriately spaced ligands while reserving a fourth coordination site to bind an activated water that will participate in catalysis. On the other hand, it exploits the coordination flexibility of zinc by incorporating it into structural sites of some enzymes, where it can confer conformational stability, and more frequently into DNA-binding proteins, where it is likely instrumental in generating a specific recognition site.

Where does all of this zinc come from? How is it stored within the cell?

[21] T. Pan and J. E. Coleman, *Proc. Natl. Acad. Sci. U.S.A.* **87,** 2077 (1990).

[22] T. Hard, E. Kellenbach, R. Boelens, B. A. Maler, K. Dahlman, L. P. Freedman, J. Carlstedt-Duke, K. R. Yamamoto, J.-A. Gustafsson, and R. Kaptein, *Science* **249,** 157 (1990).

[23] B. L. Vallee, J. E. Coleman, and D. S. Auld, *Proc. Natl. Acad. Sci. U.S.A.* **88,** 999 (1991).

How is it distributed among the various potential recipients? Assuming that extremely few copies of these recipients may be present per cell, how is the stability of the complexes established and regulated?

It has been suggested that metallothionein provides metal where and when needed for whatever role.[16,17] A process would be needed to achieve this that would not run contrary to the high stability of the metal mercaptides. Metallothionein could be a quick storage and delivery system pressed into service when zinc absolutely, positively has to function and to be there at a specific time. Thionein could be stored so that its oxidoreductive SH groups are available for that particular potential.

On the delivery side, this calls for an extraordinary system, one which not only has to make certain that the zinc atoms get to where they are supposed to but which also must ensure, protect, and defend against all intruders, including heavy metal imposters, oxidizers, and alkylating agents. It must anticipate both short-term and long-term needs and rapidly respond to environmental changes. It must be able to bind zinc tenaciously but release it under appropriate circumstances. For all of these reasons, the molecule of choice cannot have a single, unique biological function but must be adaptable.

Biology has clearly availed itself of zinc and its coordination for specific biological objectives. The zinc clusters of metallothionein seem to go beyond this, representing an example of "inorganic natural products chemistry"; a zinc cluster structure had never been observed in nonbiological zinc chemistry prior to the discovery of metallothionein. In fact, its zinc thiolate cluster is a structural motif unique to biology, perhaps indicative of its function. Ultimately, such zinc cluster structures may, in part, be characteristic and synonymous with functions yet to emerge—much as is the case for catalytically active zinc atoms in enzymes. The zinc thiolate cluster of the transcription factor GAL4 may be the first of its kind in DNA-binding proteins.[20,21]

Considering all of this new insight, it seems reasonable to conclude that metallothionein is the keystone that has been endowed with all the necessary attributes to safeguard and regulate zinc homeostasis.

[2] Definitions and Nomenclature of Metallothioneins

By YUTAKA KOJIMA

Introduction

Recommendations concerning the nomenclature of metallothionein were adopted by the Committee on the Nomenclature of Metallothionein appointed at the General Discussion Session of the Second International Meeting on Metallothionein and Other Low-Molecular-Weight Metal-Binding Proteins in 1985.[1] They supersede the original recommendations made by the plenum of the First International Meeting on Metallothionein and Other Low-Molecular-Weight Metal-Binding Proteins in 1978.[2]

Definition of Metallothionein

Historically, the term metallothionein was introduced to designate the cadmium-, zinc-, and copper-containing sulfur-rich protein from equine renal cortex.[3] This protein was characterized as follows[4-6]:

1. low molecular weight
2. high metal content
3. characteristic amino acid composition (high cysteine content, no aromatic amino acids nor histidine)
4. unique amino acid sequence (characteristic distribution of cysteinyl residues such as Cys-X-Cys)
5. spectroscopic features characteristic of tetrahedral metal–thiolate (–mercaptide) complexes
6. metal–thiolate clusters

Prompted by the conspicuousness of these features, the designation metallothionein is now used as a generic term for a variety of similar metal–thiolate polypeptides. Accordingly, the following definition has been adopted[1]: *Polypeptides resembling equine renal metallothionein in several of their features can be designated as "metallothionein."*

[1] B. A. Fowler, C. E. Hildebrand, Y. Kojima, and M. Webb, *Experientia, Suppl.* **52,** 19 (1987).

[2] M. Nordberg and Y. Kojima, eds., *Experientia, Suppl.* **34,** 41 (1979).

[3] J. H. R. Kägi and B. L. Vallee, *J. Biol. Chem.* **235,** 3460 (1960).

[4] J. H. R. Kägi, S. R. Himmelhoch, P. D. Whanger, J. L. Bethune, and B. L. Vallee, *J. Biol. Chem.* **249,** 3537 (1974).

[5] Y. Kojima, C. Berger, B. L. Vallee, and J. H. R. Kägi, *Proc. Natl. Acad. Sci. U.S.A.* **73,** 3413 (1976).

[6] M. Vašák, *J. Am. Chem. Soc.* **102,** 3953 (1980).

On the basis of structural characteristics, the "metallothioneins" are subdivided into classes[7,8]:

class I: Polypeptides with locations of cysteine closely related to those in equine renal metallothionein

class II: Polypeptides with locations of cysteine only distantly related to those in equine renal metallothionein, such as yeast metallothionein[9]

class III: Atypical, nontranslationally synthesized metal–thiolate polypeptides, such as cadystin,[10] phytometallothionein, phytochelatin, and homophytochelatin[11–14]

Nomenclature of Metallothionein

The term metallothionein should be used for all polypeptides fitting the above definition. This designation should be employed in publication titles, running titles, and as a key word for computer filing. The metal-free form (apoprotein) may be designated either as apometallothionein or thionein.[15] More specific terms, such as cadmium-metallothionein (or cadmium-thionein) or zinc-metallothionein (or zinc-thionein), are appropriate for metallothioneins that contain only one metal. The metal prefixed to the word "thionein" should not be used to designate the metal employed for metallothionein induction.

The molar metal content can be specified by a subscript, i.e., Cd_7-metallothionein or Zn_7-metallothionein. When the metallothionein contains more than one metal, for example, 5.3 mol cadmium and 1.7 mol zinc/mol, terms such as Cd,Zn-metallothionein, Cd,Zn-thionein, $Cd_{5.3},Zn_{1.7}$-metallothionein, and $Cd_{5.3},Zn_{1.7}$-thionein are recommended. The sequence in which the metals are listed should indicate the order of their

[7] J. H. R. Kägi, M. Vašák, K. Lerch, D. E. O. Gilg, P. Hunziker, W. R. Bernhard, and M. Good, *Environ. Health Perspect.* **54,** 93 (1984).

[8] J. H. R. Kägi and Y. Kojima, *Experientia, Suppl.* **52,** 25 (1987).

[9] D. R. Winge, K. B. Nielson, W. R. Gray, and D. H. Hamer, *J. Biol. Chem.* **260,** 14464 (1985).

[10] Y. Hayashi, C. W. Nakagawa, and A. Murasugi, *Environ. Health Perspect.* **65,** 13 (1986).

[11] W. R. Bernhard and J. H. R. Kägi, *Experientia, Suppl.* **52,** 309 (1987).

[12] E. Grill, E. L. Winnacker, and M. H. Zenk, *Proc. Natl. Acad. Sci. U.S.A.* **84,** 439 (1987).

[13] W. E. Rauser, *Annu. Rev. Biochem.* **59,** 61 (1990).

[14] N. J. Robinson and P. J. Jackson, *Physiol. Plant.* **67,** 499 (1986).

[15] The nomenclature recommendations of 1985[1] included the following additional sentence: "Only 'apometallothionein' should be employed in publication titles, running titles and as a key word for computer filing." Actually, "thionein" is now also used in publication titles, running titles, and as a key word for computer filing.

relative amounts. In contexts where specification of the metal composition is not available or is of no special interest, the term metallothionein should be used.

Nomenclature of Multiple Forms of Metallothionein

The nomenclature adopted is based on the recommendation of the "Nomenclature of Multiple Forms of Enzymes," IUPAC-IUB Commission on Biochemical Nomenclature (CBN).[16] The terms multiple forms of metallothionein or isometallothioneins should be used as broad terms covering all metallothioneins occurring naturally in a single species. These terms should apply only to forms of metallothionein arising from genetically determined differences in primary structure and not to forms that differ only in metal composition or are derived by modification of the same primary sequence. Different isometallothioneins should be defined by the term metallothionein followed by an arabic numeral.[17] Where complex isometallothionein families occur, the major groups should be designated by numbers and the subforms by letters (metallothionein-1a, metallothionein-1b, metallothionein-1c; metallothionein-2a, metallothionein-2b, etc.).

Nomenclature of Metallothionein Genes

The nomenclature, in general, follows that of multiple forms of metallothionein. For cases in which the isolation and sequencing of an isometallothionein-encoding gene precede the identification, isolation, and sequencing of the protein, the gene will be designated as metallothionein followed by an arabic numeral and a letter specifying, if feasible, the structural correspondence of the amino acid sequence (derived by decoding the nucleic acid sequence) to a known isometallothionein. Isometallothionein-encoding, but not expressed, sequences lacking introns or essential control sequences flanking the metallothionein-encoding regions will be designated as ψ-metallothionein followed, if feasible, by an arabic numeral and a letter denoting the homology of the decoded amino acid sequence with a known isometallothionein.

[16] IUPAC-IUB Commission on Biochemical Nomenclature (CBN), *J. Biol. Chem.* **252,** 5939 (1977).

[17] Conventionally, charge-separable isometallothioneins are numbered according to the sequence in which they elute from anion exchangers, i.e., metallothionein-1, metallothionein-2. In mammals, the marked difference between the metallothionein-1 and metallothionein-2 isoforms is the presence of an acidic amino acid residue in position 10 or 11 in the standard sequence of metallothionein-2 in place of a neutral residue in metallothionein-1.[8]

[3] Toxicological Significance of Metallothionein

By DOUGLAS M. TEMPLETON and M. GEORGE CHERIAN

From the time of its isolation and characterization as a cadmium-binding protein, a number of functions have been proposed for metallothionein (MT), including the detoxification of such metals.[1] A reduction in the toxicity of metal ions following the induction of MT in animals has long been recognized. Comparative resistance to metals of cultured cells with elevated levels of MT has also been described frequently. In an earlier review, Webb noted that these effects have been demonstrated repeatedly in various systems.[2] Nevertheless, they continue to be documented both *in vivo* and *in vitro*. More recently, the ability of MT to modulate the toxicity of a number of xenobiotics, including chemotherapeutic alkylating agents, and the toxicity of radiation and chemicals that act as free radical generators, has also gained prominence. The present chapter focuses mainly on additional information concerning metal resistance that has appeared since the review by Webb.[2] For a more thorough discussion of the earlier documentation of MT-related metal resistance, the reader is referred to that source.[2] Here we also incorporate newer information on drug resistance that allows a more critical discussion of molecular mechanisms. Some aspects of this latter subject have been discussed recently.[3,4] The somewhat paradoxical toxicity of MT under some circumstances is also considered here.

Although the protective effects of MT have been studied extensively both in animal experiments and in cell culture systems, detoxification is by no means universally accepted as a primary function of MT.[5] Bremner and Beattie[6] have suggested that such a role is probably adventitious and reflects the chemical similarities of cadmium and zinc. This argument can be extended to other metals and electrophiles by noting that the behavior of MT is dominated by the chemistry of the thiol group. Furthermore, the toxicity of Cd-MT, and in particular the nephrotoxicity of extracellular Cd-MT, has been demonstrated.[7,8] This too has been interpreted as casting

[1] J. H. R. Kägi and B. L. Vallee, *J. Biol. Chem.* **235,** 3460 (1960).
[2] M. Webb, *Experientia, Suppl.* **52,** 109 (1987).
[3] A. Basu and J. S. Lazo, *Toxicol. Lett.* **50,** 123 (1990).
[4] M. P. Hacker, *Toxicol. Lett.* **50,** 121 (1990).
[5] M. Karin, *Cell (Cambridge, Mass.)* **41,** 9 (1985).
[6] I. Bremner and J. H. Beattie, *Annu. Rev. Nutr.* **10,** 63 (1990).
[7] M. G. Cherian, R. A. Goyer, and L. Delaquerriere-Richardson, *Toxicol. Appl. Pharmacol.* **38,** 399 (1976).
[8] K. S. Squibb and B. A. Fowler, *Environ. Health Perspect.* **54,** 31 (1984).

doubt on the importance of MT synthesis in detoxification.[9] The synthesis of hepatic MT exclusively in free polysomes is indicative of a nonsecretory protein destined for an intracellular location.[10] Together with the absence of significant biliary or urinary excretion, this suggests that although MT may afford short-term protection, a long-term role in reducing the body burden of toxins such as cadmium is doubtful. On the other hand, the obvious protective outcome of MT induction in some cases (including the development of metal-resistant cell lines), the biochemical parallels between MT and glutathione, and the occurrence of MT-like molecules necessary for tolerance to essential elements in microorganisms (e.g., the copper-binding *CUP1* gene product in yeast) have led many authors to propose detoxification as a major function of mammalian MT.[11] This debate between functional and adventitious detoxification has been useful insofar as it has focused attention on requisite proof of causation. Nevertheless, the true function of this protein remains to be decided.

Toxic Effects of Metallothionein

Whereas induction of MT allows most cells to sequester intracellular Cd in a nontoxic form, at least temporarily, extracellular Cd-MT is a potent nephrotoxin targeting the proximal tubule. This has been demonstrated by intravenous injection of Cd-MT into rats[7,8] and in primary cultures of renal epithelial cells.[12] An intravenous dose of Cd-MT producing a concentration of 10 μg Cd/g renal cortex was sufficient to cause renal damage in mice[13] and rats,[7] in contrast to an estimated critical concentration of about 200 μg/g after chronic exposure to environmental cadmium[14] and a value of 100–200 μg/g in rats exposed repeatedly to $CdCl_2$ for up to several weeks.[15] While not directly comparable, these studies support the conclusion arrived at from *in vitro* studies[16] that Cd-MT is a more potent nephrotoxin than cadmium salts. Repeated exposure to Cd-MT results in a dose-dependent induction of renal MT, however, and raises the subsequent threshold for Cd-MT toxicity.[13] Thus intracellular MT appears to be protective during cadmium exposure whereas extracellular MT is harmful.

[9] J. H. R. Kägi and A. Schäffer, *Biochemistry* **27,** 8509 (1988).
[10] M. G. Cherian, S. Yu, and C. M. Redman, *Can. J. Biochem.* **59,** 301 (1981).
[11] D. H. Hamer, *Annu. Rev. Biochem.* **55,** 913 (1986).
[12] M. G. Cherian, *in* "Biological Roles of Metallothionein" (E. C. Foulkes, ed.), p. 193. Elsevier, North-Holland, Amsterdam, 1982.
[13] T. Maitani, F. E. Cuppage, and C. D. Klaassen, *Fundam. Appl. Toxicol.* **10,** 98 (1988).
[14] L. Friberg, *Environ. Health Perspect.* **54,** 1 (1984).
[15] C. Tohyama, N. Sugihira, and H. Saito, *J. Toxicol. Environ. Health* **22,** 255 (1987).
[16] M. G. Cherian, *In Vitro Cell. Dev. Biol.* **21,** 505 (1985).

The mechanism of Cd-MT nephrotoxicity is not well understood. Apothionein itself is not toxic, implicating cadmium as the cause of renal damage.[8] It is generally held that uptake of MT and subsequent release of cadmium, perhaps in acidified lysosomal compartments, allows "free Cd^{2+}" to act as an intracellular toxin. Such a mechanism would be supported by demonstration of higher local concentrations of non-protein-bound cadmium, and perhaps different compartmentalization after Cd-MT injection. Cadmium released under these circumstances may interfere with the formation of secondary lysosomes.[17] A single dose of 0.25 to 1 mg Cd/kg given as Cd-MT at 8 to 14 days of gestation in the rat is teratogenic, apparently due to maternal nephrotoxicity.[18] Cadmium can be sequestered in maternal tissues by bismuth subnitrate ($BiONO_3$) pretreatment that results in induction of maternal hepatic and renal MT, thereby decreasing accumulation of cadmium in the decidua and abolishing its teratogenicity.[19]

The toxicity of MT to other cell types has been less often studied. Interactions of MT with the immune system are complex. Thus, Cd-MT, Zn-MT, and Cd,Zn-MT all induce lymphocyte proliferation *in vitro,* whereas Hg- and Cu-MT do not. In fact, the latter species inhibit lipopolysaccharide- and concanavalin A-stimulated lymphocyte proliferation.[20] In contrast to the renal epithelium, hepatocytes are more sensitive to the toxic effects of $CdCl_2$ than to those of Cd-MT, although zinc pretreatment offers a protective effect only against $CdCl_2$.[21] The somewhat unique sensitivity of the proximal tubule to Cd-MT suggests that long-term transport of small amounts of MT from other sites may underlie the slow development of renal compromise as the clinical manisfestation of chronic cadmium intoxication.

Protective Effects of Metallothionein

Cadmium and Copper Resistance in Cultured Cells

The metal-binding properties of cytosolic MT account for resistance of cultured cells to metals that are sequestered by the protein. Thus it has been demonstrated, using numerous cell lines, that elevations in MT correlate with the tolerated concentration of metals, such as copper and cad-

[17] B. A. Fowler, P. L. Goering, and K. S. Squibb, *Experientia, Suppl.* **52,** 661 (1987).
[18] M. Webb, D. Holt, N. Brown, and G. C. Hard, *Arch. Toxicol.* **61,** 457 (1988).
[19] I. Naruse and Y. Hayashi, *Teratology* **40,** 459 (1989).
[20] M. A. Lynes, J. S. Garvey, and D. A. Lawrence, *Mol. Immunol.* **27,** 211 (1990).
[21] L. E. Sendelbach, W. M. Bracken, and C. D. Klaassen, *Toxicology* **55,** 83 (1989).

mium, that induce and bind to MT. In general, these experiments rely on demonstrating an increased expression of MT, either following pretreatment with less toxic inducing agents such as zinc, bismuth, or cortisol, or by selecting for metal-resistant cells that are subsequently shown to have undergone *MT* gene amplification.

Direct implication for a causative role of MT in metal resistance has come from studies of copper sensitivity. A series of copper-resistant hepatoma cell lines have been reported that demonstrate a level of resistance proportional to the cellular concentration of Cu-MT.[22] Yeasts require the *CUP1* gene, encoding a metallothionein-like protein, to utilize copper and avoid copper-induced damage.[23-25] Deletion of the copper-binding protein gene renders yeasts sensitive to copper, and introduction of a plasmid construct containing a mammalian *MT* gene confers both copper and cadmium resistance in a copy-dependent manner.[26] Yeasts with a normal *CUP1* gene autoregulate its transcription and display copper resistance, whereas mutants that constitutively express the gene also acquire cadmium resistance, but not resistance to a number of other metals.[27] Mammalian *MT* genes can be expressed in *CUP1*-deficient yeasts, are autoregulated, and confer copper resistance.[28] Taken together, these observations satisfy Koch's postulates for a causative role of MT in cellular resistance to copper. Thus, the resistant phenotype is consistently associated with MT, is modified by removal of MT, and is restored by reintroduction of the *MT* gene. By inference, cross-resistance to cadmium is also a consequence of MT accumulation.

Primary rat hepatocytes have been added to the list of cells showing increased cadmium tolerance following pretreatment with zinc, concomitant with transfer of most cytosolic cadmium to the MT pool.[29,30] No net increase in the accumulation of cadmium by these cells was noted.[30] Mouse LMTK cells become resistant to cadmium with a spontaneous mutation rate of about 1×10^6/cell/generation, consistent with a one-hit event.[31] Interestingly, because this single event results in increased expres-

[22] J. H. Freedman and J. Peisach, *Biochim. Biophys. Acta* **992,** 145 (1989).

[23] D. J. Thiele, *Mol. Cell. Biol.* **8,** 2745 (1988).

[24] J. M. Huibregtse, D. R. Engelke, and D. J. Thiele, *Proc. Natl. Acad. Sci. U.S.A.* **86**(1), 65 (1989).

[25] D. J. Ecker, T. R. Butt, and S. T. Crooke, *Met. Ions Biol. Syst.* **25,** 147 (1989).

[26] P. C. Huang and L. Y. Lin, *in* "Biologically Active Proteins and Peptides" (S. H. Chiou, K. T. Wang, and S. H. Wu, eds.), p. 101. Academia Sinica Press, 1989.

[27] D. J. Ecker, T. R. Butt, E. J. Sternberg, M. P. Neeper, C. Debouck, J. A. Gorman, and S. T. Crooke, *J. Biol. Chem.* **261,** 16895 (1986).

[28] D. J. Thiele, M. J. Walling, and D. H. Hamer, *Science* **231,** 854 (1986).

[29] J. M. Frazier and W. S. Din, *Experientia, Suppl.* **52,** 619 (1987).

[30] J. Liu, W. C. Kershaw, and C. D. Klaassen, *In Vitro Cell. Dev. Biol.* **26**(1), 75 (1990).

[31] A. Chopra, J. Thibodeau, Y. C. Tam, C. Marengo, M. Mbikay, and J.-P. Thirion, *J. Cell. Physiol.* **142**(2), 316 (1990).

sion of both mouse *mt1* and *mt2* genes, the authors speculate that a common regulatory pathway, e.g., a trans-acting factor, may be involved. Rat hepatocytes grown in the presence of 5-azacytidine (AZA-CR) increase their expression of MT.[32] Probably incorporation of AZA-CR into DNA leads to hypomethylation of the *MT* gene(s).[33] In these circumstances, the increased MT expression parallels an increased cadmium resistance.[32] Deoxy-AZA-CR is also effective, and synergistic with cadmium in evoking the resistance response, due to a shortening of induction time in the presence of cadmium.[34]

Chinese hamster V79 cells, rendered especially sensitive to cadmium by depletion of glutathione following treatment with buthionine sulfoximine, showed a decrease in cadmium toxicity after exposure to zinc.[35] This effect required a lag period for induction of MT, and again paralleled the synthesis of the protein. There is mounting evidence that glutathione serves as a first line of defense against cadmium cytotoxicity prior to induction of MT, and then protein thiols become more important than that of the tripeptide as scavengers of the metal.[36,37]

Copper and Cadmium Resistance in Animals

Decreased copper and cadmium toxicity attributable to MT induction is frequently demonstrated in animal studies, supporting the toxicological relevance of the above observations in cultured cells. Copper-loaded rats given zinc by subcutaneous injection further elevate their hepatic MT and show an increase in the proportion of hepatic copper bound to MT.[38] Metallothionein has been demonstrated immunohistochemically in the hepatocytes of two patients with Wilson's disease, and copper-saturated MT was isolated from their livers.[39] Hepatic damage is abolished by sequestration of copper in this form, and it has been suggested that this may in part explain the beneficial effects of zinc administration in cases of Wilson's disease.[38,40] It has further been suggested that hepatic damage occurs in Wilson's disease when MT biosynthesis is compromised or when copper levels overwhelm the binding capacity of MT.[41] Rats receiving

[32] M. P. Waalkes, M. J. Wilson, and L. A. Poirier, *Toxicol. Appl. Pharmacol.* **81,** 250 (1985).
[33] S. J. Compare and R. D. Palmiter, *Cell (Cambridge, Mass.)* **25,** 233 (1981).
[34] M. P. Waalkes, M. S. Miller, M. J. Wilson, R. M. Bare, and A. E. McDowell, *Chem.-Biol. Interact.* **66,** 189 (1988).
[35] T. Ochi, F. Otsuka, K. Takahashi, and M. Ohsawa, *Chem.-Biol. Interact.* **65,** 1 (1988).
[36] R. K. Singhal, M. E. Anderson, and A. Meister, *FASEB J.* **1,** 220 (1987).
[37] C. A. M. Suzuki and M. G. Cherian, *Toxicol. Appl. Pharmacol.* **98,** 544 (1989).
[38] D.-Y. Lee, G. J. Brewer, and Y. Wang, *J. Lab. Clin. Med.* **114**(6), 639 (1989).
[39] N. O. Nartey, J. V. Frei, and M. G. Cherian, *Lab. Invest.* **57,** 397 (1987).
[40] G. J. Brewer, V. Yuzbasayan-Gurkan, and D.-Y. Lee, *J. Lab. Clin. Med.* **114,** 633 (1989).
[41] M. E. Elmes and B. Jasani, *Lancet* **Oct. 10, 2 (8563),** 866 (1987).

nonlethal doses of cadmium are better able to tolerate subsequent exposures,[42] and liver slices prepared from rats pretreated with zinc to induce Zn-MT leak decreased amounts of cytosolic enzymes, compared to untreated controls, when exposed to cadmium.[43] Nevertheless, factors other than MT are clearly important in cadmium resistance in rodents. Thus, Huang and co-workers have shown that three mouse strains with altered sensitivity to cadmium are indistinguishable from two wild-type strains with respect to the rate or level of induction of hepatic MT.[44] Cadmium-resistant *(v;bw)* and cadmium-sensitive (Austin) strains of *Drosophila* were compared with respect to their abilities to induce MT in response to cadmium.[45] In the *v;bw* strain, even small increases in MT were associated with markedly increased cadmium resistance, whereas in the Austin flies even large increases in MT had little effect on the degree of cadmium sensitivity. The authors conclude that differences in MT level alone are insufficient to account for cadmium resistance.

Resistance to Other Metals

Platinum. Much interest surrounds the potential ability of MT to afford protection against platinum when the latter is administered as the chemotherapeutic agent *cis*-dichlorodiammine-Pt(II) (*cis*-DDP). Several earlier studies on platinum resistance in cultured cells produced some ambiguities regarding the ability of platinum to bind and (or) induce MT.[46–49] Thus the ability of MT to modify the toxicity of *cis*-DDP remains without a clear mechanistic basis. Two observations may help to clarify the situation: (1) MT appears to contain a nonspecific Pt(II)-binding site at the N-terminal Met residue that is occupied at low concentrations of platinum.[50] At higher concentrations, the thiol groups become involved in S–Pt bonds expected to have strong covalent character. The cluster structure of the protein is

[42] M. Webb and R. D. Vershoyle, *Biochem. Pharmacol.* **25,** 673 (1976).

[43] U. Wormser and S. Ben-Zakine, *Arch. Toxicol., Suppl.* **13,** 316 (1989).

[44] P. C. Huang, S. Morris, J. Dinman, R. Pine, and B. Smith, *Experientia, Suppl.* **52,** 439 (1987).

[45] H. J. Gill, D. L. Nida, D. A. Dean, M. W. England, and K. B. Jacobson, *Toxicology* **56,** 315 (1989).

[46] A. Bakka, L. Endresen, B. B. S. Johnsen, P. D. Edminson, and H. E. Rugstad, *Toxicol. Appl. Pharmacol.* **61,** 215 (1981).

[47] R. P. Sharma and L. R. Edwards, *Biochem. Pharmacol.* **32,** 2665 (1983).

[48] A. J. Zelazowski, J. S. Garvey, and J. D. Hoeschele, *Arch. Biochem. Biophys.* **229,** 246 (1984).

[49] C. L. Litterst, F. Bertolero, and J. Uozumi, *in* "Biochemical Mechanisms of Platinum Antitumor Drugs" (D. C. H. McBrien and T. F. Slater, eds.), p. 227. IRL Press, Oxford, 1986.

[50] J. Bongers, J. U. Bell, and D. E. Richardson, *J. Inorg. Biochem.* **34,** 55 (1988).

disrupted and oligomerization can result. Variable degrees of oxidative cross-linking during this process probably accounts for the poorly defined stoichiometries of MT-bound Pt(II).[51] (2) Induction of MT is probably not a property of *cis*-DDP itself, but hydrolysis of the Pt–Cl bonds to give *cis*-diamminediaquo-$Pt(II)^{2+}$ results in more potent induction of MT in the liver and kidney of mice.[52] Induction of MT in cultured choriocarcinoma cells was increased when refrigerated aqueous solutions of *cis*-DDP were allowed to stand for several weeks.[53] Limited hydrolysis to this putative primary inducer *in vivo* could explain the modest and variable inductions reported by others. The resulting Cu,Zn-MT binds Pt to a limited extent,[52] and so the significance of these events in *cis*-DDP toxicity remains uncertain.

Yeast strains constitutively expressing *CUP1* are resistant to both cadmium and copper, but not to platinum.[27] Nevertheless, several tumor cell lines with acquired *cis*-DDP resistance were found to overexpress MT, and maintenance of high MT levels either by cadmium exposure or transfection with human MT-2a gave a similar degree of resistance.[54] In each case, cross-resistance to several alkylating agents was also conferred, and the degree of resistance correlated with the cellular content of MT. Administration to mice over 5 days of sufficient bismuth subnitrate to induce MT resulted in partial protection of renal (blood urea nitrogen; BUN) and bone marrow (leukocyte count) functions when *cis*-DDP was given subsequently,[55] A corresponding reduction in the cardiotoxicity of adriamycin in these experiments has been shown more convincingly elsewhere by the same authors to correlate with MT accumulation (see below), and they have also found that the renal level of bismuth-induced MT is inversely proportional to the degree of *cis*-DDP-induced nephrotoxicity.[56] Because MT was not induced in transplanted tumors, bismuth did not affect the antitumor activity of *cis*-DDP, suggesting a promising chemotherapeutic strategy.[56] Chinese hamster ovary (CHO) cells selected by use of cadmium and zinc to overexpress MT were resistant to *cis*-DDP. The proportion of cytosolic MT-bound ^{195m}Pt increased to 17% as compared to 4% in the unselected parent line.[57] However, selecting with *cis*-DDP itself produced

[51] K. B. Nielson, C. L. Atkin, and D. R. Winge, *J. Biol. Chem.* **260,** 5342 (1985).

[52] P. G. Farnworth, B. L. Hillcoat, and I. A. G. Roos, *Chem.-Biol. Interact.* **69,** 319 (1989).

[53] C. Harford and B. Sarkar, *J. Mol. Toxicol.* **2,** 67 (1989).

[54] S. L. Kelley, A. Basu, B. A. Teicher, M. P. Hacker, D. H. Hamer, and J. S. Lazo, *Science* **241,** 1813 (1988).

[55] M. Satoh, A. Naganuma, and N. Imura, *Cancer Chemother. Pharmacol.* **21,** 176 (1988).

[56] A. Naganuma, M. Satoh, and N. Imura, *Cancer Res.* **47,** 983 (1987).

[57] P. A. Andrews, M. P. Murphy, and S. P. Howell, *Cancer Chemother. Pharmacol.* **19,** 149 (1987).

resistant lines that did not have increased MT levels.[57] A *cis*-DDP-resistant murine leukemia cell, L1210-PPD, was compared with its nonresistant L1210 parent and actually found to have less MT, make less MT mRNA on induction with zinc, and be more sensitive to cadmium.[58] Schilder *et al.* found that MT was expressed in only two of six clones derived from a human ovarian carcinoma cell line by selection for increasing degrees of *cis*-DDP resistance.[59] Although MT was inducible by cadmium in all clones, only one that constitutively expressed the protein was capable of induction in response to *cis*-DDP. Several cell lines derived from patients with ovarian tumors refractory to chemotherapy showed variable levels of MT expression, including no measurable message in some, that were generally unresponsive to *cis*-DDP.[59] Nor did cadmium-resistant mouse C127 fibroblasts transfected with a human MT-2a construct show any increased resistance to *cis*-DDP in a clonogenic assay.[59] Recently, Suzuki and Cherian[60] have found that injection of *cis*-DDP into rats in divided doses fails to induce MT or alter the pattern of *cis*-DDP-related nephropathy. Preinduction of MT with zinc was also without effect on the subsequent response of the rats to *cis*-DDP. These latter studies[57-60] demonstrate that overexpression of MT is neither necessary nor sufficient to confer *cis*-DDP resistance on cultured cells.

Mercury. Inorganic mercury is a relatively potent inducer of MT,[9] and can apparently substitute for zinc and cadmium in seven-metal thionein clusters.[51] However, the organ distribution of Hg(II) is not altered by cadmium pretreatment, and only a few percent of the renal burden of Hg(II) is bound to MT. Thus Webb has argued that the impact of MT induction on mercury nephrotoxicity is probably small.[2] Induction of Zn-MT by estradiol increases the amount of Hg-MT in the renal cortex of rats, and treatment of amniotic fluid cells with estradiol and mercury together increases the cellular uptake of mercury.[61] Nevertheless, a decrease in mercury-induced renal damage was reported to occur in these studies, despite the increased mercury content of the cortical cells and unchanged mercury levels in the liver. That the amount of Hg-MT can be increased while decreasing renal damage suggests that sequestration of the metal as the thionein complex is protective to the kidney.

Lead. Rats given lead acetate induce Pb- and Zn-MTs, the lead protein

[58] P. Farnworth, B. Hillcoat, and I. Roos, *Cancer Chemother. Pharmacol.* **25,** 411 (1990).

[59] R. J. Schilder, L. Hall, A. Monks, L. M. Handel, A. J. Fornace, Jr., R. F. Ozols, A. T. Fojo, and T. C. Hamilton, *Int. J. Cancer* **45,** 416 (1990).

[60] C. A. M. Suzuki and M. G. Cherian, *Toxicology* **64,** 113 (1990).

[61] S. Nishiyama, T. Taguchi, and S. Onosaka, *Biochem. Pharmacol.* **36,** 3387 (1987).

reaching maximal levels at 6 hr.[62,63] CHO Cd^r cells having a 50- to 100-fold gene amplification for MT were found to accumulate Zn-MT with mRNA levels peaking at 4 hr after exposure to lead.[64] The response is independent of glucocorticoid induction of MT, and occurs in adrenalectomized rats.[65] Because induction was maximal at acutely toxic doses, it was suggested that impending lead-induced cell death might be responsible for MT synthesis, lead being therefore a gratuitous inducer of genes transcribed late before cell death.[64] Lead inhibits the cytosolic zinc-dependent enzyme porphobilinogen synthase (δ-aminolevulinic acid dehydratase), but induction of MT in rats with zinc prior to sacrifice reactivates the enzyme in tissue homogenates.[66] An increasing cadmium content in the MT reduces the effect. Presumably the Zn-MT sequesters lead and donates zinc to the enzyme.

Gold. Failure of chrysotherapy with auranofin or sodium aurothiomalate is commonly due to development of resistance to the antimitogenic effect of the Au-containing drugs. Human epithelial cells in culture develop auranofin resistance with an elevation of MT and a decrease in cytosolic gold.[67] Cells made resistant to either auranofin or sodium aurothiomalate show cross-resistance to the other compound, although MT synthesis appears to be much more important for auranofin resistance,[68] raising the likelihood of multiple mechanisms.

Protection Against Alkylating Agents and Radicals

Based on the kinetic lability of group IIB metals in MT[9] and the chemistry of the nucleophilic thiol group, it is reasonable to expect that MT will be reactive toward electrophiles, including many alkylating agents and radical species. Differential reactivity of bridging and terminal thiolates[69] may afford a means of limiting reaction with electrophiles to a repairable degree. Metallothionein has been shown to be an extremely potent sacrificial scavenger of HO· that may be renewed by reaction with

[62] H. Ikebuchi, R. Teshima, K. Suzuki, T. Terao, and Y. Yamane, *Biochem. J.* **233,** 541 (1986).

[63] H. Ikebuchi, R. Teshima, K. Suzuki, J.-I. Sasada, T. Terao, and Y. Yamane, *Biochem. Biophys. Res. Commun.* **136,** 535 (1986).

[64] S.-J. Rhee and P. C. Huang, *Chem.-Biol. Interact.* **72,** 347 (1989).

[65] K. Arizono, T. Ito, M. Yamaguchi, and T. Ariyoshi, *Eisei Kagaku* **28,** 94 (1982).

[66] P. L. Goering and B. A. Fowler, *Biochem. J.* **245,** 339 (1987).

[67] A. Glennås, P. E. Hunziker, J. S. Garvey, J. H. R. Kägi, and H. E. Rugstad, *Biochem. Pharmacol.* **35,** 2033 (1986).

[68] H. E. Rugstad and A. Glennås, *Scand. J. Rheumatol.* **17,** 175 (1988).

[69] D. M. Templeton, P. A. W. Dean, and M. G. Cherian, *Biochem. J.* **234,** 685 (1986).

glutathione.[70] At 13 μM, MT was as effective as 10 mM glutathione at protecting calf thymus DNA from damage by HO· generated *in situ.*[71] Chinese hamster V79 cells rendered tolerant to cadmium and having elevated MT also showed relative resistance to oxidative stress (H_2O_2 and $O_2^{\cdot-}$).[72] While concomitant elevations in catalase and superoxide dismutase (SOD) were ruled out, the mechanism remains unknown and could involve radical scavenging or reduction of the radical by MT-mediated hydrogen donation. Elimination of Fenton-mediated radical damage through iron chelation or exchange of iron with MT has also been suggested as a possible mechanism.[72] Participation in repair mechanisms, e.g., as a regulatory element, or as a zinc donor or cofactor, must also be considered.[73]

Induction of Zn-MT decreased lipid peroxidation in rat liver, and the induction and protection were synergistic with respect to zinc and stress [starvation, restraint, or dimethyl sulfoxide (DMSO) administration].[74] However, hepatic MT increases under each of these stressful conditions, which themselves promote lipid peroxidation, lending credence to the view that MT is turned on as a first-line protective mechanism under conditions that favor oxygen radical formation. This is further supported by the induction of MT by interleukin 1 and interferon γ,[5] potent activators of the respiratory burst in neutrophils and macrophages. Nevertheless, the attribution of these effects to thionein remains problematic. The hepatotoxin CCl_4 damages tissue through mechanisms probably involving carbon-centered radicals. Preinduction of MT with zinc protects against CCl_4-induced damage in exposed rats[75] and in liver slices derived from them.[43] Although metal ions are released from MT by CCl_4-derived radicals *in vitro,* this appears to be due to radical-mediated oxidation of protein thiols rather than to reactions that produce covalent adducts, as no such adducts were found with $^{14}CCl_4$.[76] A protective role of the released metal, in contrast to the thionein itself, must therefore be considered. In support of this mechanism, it was shown that although Zn- and Cd,Zn-MTs were effective in decreasing lipid peroxidation in erythrocyte ghosts treated with a xanthine oxidase radical generating system, Cd(II) and Zn(II) *alone* were equally

[70] P. J. Thornalley and M. Vasak, *Biochim. Biophys. Acta* **827,** 36 (1985).
[71] J. Abel and N. de Ruiter, *Toxicol. Lett.* **47,** 191 (1989).
[72] A. C. Mello-Fihlo, L. S. Chubatsu, and R. Meneghinin, *Biochem. J.* **256,** 475 (1988).
[73] B. Kaina, H. Lohrer, M. Karin, and P. Herrlich, *Proc. Natl. Acad. Sci. U.S.A.* **87**(7), 2710 (1990).
[74] J. Hidalgo, L. Campmany, M. Borras, J. S. Garvey, and A. Armario, *Am. J. Physiol.* **255,** E518 (1988).
[75] P. L. Goering and C. D. Klaassen, *Toxicol. Appl. Pharmacol.* **74,** 299 (1984).
[76] Z. E. Suntres and E. M. K. Lui, *Biochem. Pharmacol.* **39,** 833 (1990).

effective.[77] The authors postulate that the metals interfere with the redox cycling of iron, and, when supplied as their thionein complexes, may simply be released from the protein oxidation of the thiols by reactive oxygen. Additional mechanisms of suppression of lipid peroxidation by the divalent ions could be postulated.

Adriamycin is cardiotoxic in mice and produces products of lipid peroxidation in the myocardium. The preinduction of MT by $BiONO_3$ decreases this effect, and also reduces the apparent marrow toxicity of the drug, under conditions that also decrease *cis*-DDP nephrotoxicity and marrow suppression.[55] Zinc, cadmium, cobalt, or mercury can be substituted as the inducing agent, and a decrease in adriamycin-related myocardial lipid peroxidation is seen in each case. The decrease parallels the level of cardiac (but not hepatic) MT.[78] Pretreatment with bismuth is necessary to protect mice against lethal doses of adriamycin, its coadministration with the drug having no effect.[79] Under these circumstances, bismuth in heart is mostly associated with Bi,Zn-MT, and the adriamycin-induced production of conjugated dienes and malonyldialdehyde in heart is negatively correlated with the concentration of MT in that tissue.[79] In an elegant series of experiments, Webber and co-workers[80] made human prostatic carcinoma cells resistant to cadmium by adaptation. In the presence of cadmium, these cells (Cd^r-ind) had 3.5 times the basal rate of MT synthesis compared with unadapted cells that were exposed to cadmium (Cd^s-ind). When Cd^r cells were deinduced by removal of cadmium (Cd^r-deind), they had lower levels of MT than Cd^s-ind. Adriamycin sensitivity increased in the order Cd^r-ind < Cd^s-ind < Cd^r-deind, with decreasing MT. These experiments show, however, that it is the overexpression of MT, and not the adaptation to cadmium per se, that accounts for the adriamycin resistance. Adriamycin appears to inhibit MT synthesis in these cells. It was suggested that a useful therapeutic strategy might be to give a second dose of adriamycin after the level of cellular MT was decreased.[80]

Human epithelial and murine fibrosarcoma cells selected for growth in the presence of 100 μM $CdCl_2$ demonstrated cross-resistance to chlorambucil and its glucocorticoid ester prednimustine, as well as to *cis*-DDP.[81] Most platinum and 20–40% of intracellular chlorambucil appeared to be associated with MT on gel-permeation chromatography. When a variety of

[77] J. P. Thomas, G. J. Bachowski, and A. W. Girotti, *Biochim. Biophys. Acta* **884,** 448 (1986).
[78] M. Satoh, A. Naganuma, and N. Imura, *Toxicology* **53,** 231 (1988).
[79] A. Naganuma, M. Satoh, and N. Imura, *Jpn. J. Cancer Res.* **79,** 406 (1988).
[80] M. M. Webber, S. M. Maseehur Rehman, and G. T. James, *Cancer Res.* **48,** 4503 (1988).
[81] L. Endresen and H. E. Rugstad, *Experientia, Suppl.* **52,** 595 (1987).

human carcinoma cell lines was made resistant to *cis*-DDP by exposure to $CdCl_2$ or *cis*-DDP, cross-resistance to the alkylating agents melphalan and chlorambucil was associated with the overexpression of MT.[54] To rule out the involvement of other consequences of metal exposure, such as the induction of stress proteins, cells were transfected with human MT-2a. These cells acquired resistance to *cis*-DDP, melphalan, and chlorambucil. Little or no change in resistance to doxorubicin, bleomycin, 5-fluorouracil, or vincristine was seen in either the metal-treated or transfected cells, and the resistance may therefore show some selectivity for sulfhydryl-reactive agents.[54] CHO cells transfected with human MT-2a similarly acquired cross-resistance to cadmium, monofunctional nitrosourea and nitrosoguanidine derivatives, and mitomycin C, but not to methylmethane sulfonate, an ethyl nitrosourea, bleomycin, or ionizing radiation.[73,82] Control transfectants lacking a functional coding region demonstrated that the resistance was due to the coding region of hMT-2a. There was no decrease in the initial formation of DNA methylation products caused by the alkylating agents, however, arguing against a straightforward scavenging of the agents by thiol groups.[73]

Protection Against Radiation

Pretreatment with $BiONO_3$ increased the MT content of bone marrow in mice while decreasing marrow suppression and lethality caused by ^{60}Co γ radiation.[83] In this study no effect was found on the sensitivity of implanted tumors to radiation, and no MT induction was observed in the tumor tissues. HeLa cells, B16 cells derived from a murine melanoma, and murine fibrosarcoma WHFIB cells were all exposed to ^{60}Co irradiation.[84] Modest improvements in survival of HeLa and WHFIB cells, but not B16 cells, were observed at intermediate doses of radiation when the cells were pretreated with cadmium. This effect was enhanced when other inducers of MT (dexamethasone and serum factors) were included along with cadmium. However, cellular MT was not measured in this study. In mice pretreated with cadmium or manganese, survival after irradiation correlated with hepatic MT levels,[85] which may of course simply be an indicator of the level of exposure to the metal. In contrast to these experiments, which provide circumstantial evidence for a role of MT in protection

[82] H. Lohrer and T. Robson, *Carcinogenesis* **10,** 2279 (1989).

[83] M. Satoh, N. Miura, A. Naganuma, N. Matsuzaki, E. Kawamura, and N. Imura, *Eur. J. Cancer Clin. Oncol.* **25,** 1727 (1989).

[84] M. J. Rennan and P. I. Dowman, *Radiat. Res.* **120,** 442 (1989).

[85] J. Matsubara, Y. Tajima, A. Ikeda, T. Kinoshita, and T. Shimoyama, *Pharmacol. Ther.* **39,** 331 (1988).

against radiation damage, others have been unable to demonstrate such an effect directly. Although ionizing radiation increased MT 20-fold in the cadmium-resistant wild-type CHO- and BC11-hMT-2a transfectants described above, no protection against ionizing radiation was observed.[82] These transfectants were also as sensitive as their nontransfected counterparts to bleomycin, an agent with a free radical-based chemotherapeutic action.[82] Nor did the transfectants and controls differ in their response to γ irradiation under either anoxic or oxygenated conditions, with or without additional zinc- or cadmium-induced thionein, or under conditions of glutathione depletion.[73]

Because about 90% of the DNA damage induced by ionizing radiation is due to hydroxyl radicals produced by the radiolysis of water,[86] efficient HO · scavengers such as MT might be expected to afford a degree of protection against the effects of radiation. Protective agents could, in principle, also act at a later stage by facilitating the repair of DNA damaged either by HO · or by the effects of direct energy absorption. The extreme reactivity and consequent short survival of HO · ensure that its effects are local. Therefore, it would seem that to be effective, a protective agent would have to occur in a nuclear location in close proximity to DNA. Although MT is primarily a cytosolic protein, it has been localized to the nucleus shortly after induction by zinc and cadmium salts[87,88] and in the perinatal period.[89,90] Perhaps differences in intra- vs extranuclear location at various times or following different modes of induction can explain different degrees of protection from radiation. Interestingly, in the studies of Rennan and Dowman,[84] protection of HeLa and WHFIB cells against ^{60}Co radiation was achieved with 10-hr exposures to cadmium, but not with 2 or 18 hr, when the exposures immediately preceded irradiation. This intermediate time would be expected to be closer to the time of onset of induction of MT, even though a greater accumulation of the protein should occur by 18 hr.

Summary of Protective Effects

The correlation between cadmium resistance in cultured cells and binding of cadmium to increased levels of MT is strong and has been amply demonstrated. Evidence that a similar mechanism operates *in vivo* is com-

[86] T. Alper, "Cellular Radiobiology." Cambridge Univ. Press, Cambridge, England, 1979.
[87] D. Banerjee, S. Onosaka, and M. G. Cherian, *Toxicology* **24,** 94 (1982).
[88] K. G. Danielson, S. Ohi, and P. C. Huang, *Proc. Natl. Acad. Sci. U.S.A.* **79,** 2301 (1982).
[89] M. Panemangelore, D. Banerjee, S. Onosaka, and M. G. Cherian, *Dev. Biol.* **97,** 95 (1983).
[90] D. M. Templeton, D. Banerjee, and M. G. Cherian, *Can. J. Biochem. Cell Biol.* **63,** 16 (1985).

pelling. An involvement of MT in modifying the toxicity of copper, particularly in yeast, is also clear. In these cases, sequestration of the metal as its corresponding thionein complex is the important event. However, in other circumstances the protective role of MT is less clear. Overexpression of MT is associated with resistance of cultured cells to *cis*-DDP, but MT expression is neither necessary nor sufficient for resistance to this agent. Different mechanisms of resistance to gold compounds probably operate, and involve MT to different degrees, if at all. Transfection of eukaryotic cells with *MT* genes confers resistance to some (e.g., adriamycin, melphalan, chlorambucil, prednimustine, monofunctional nitrosourea derivatives, nitrosoguanidine derivatives, mitomycin C) but not other (e.g., doxorubicin, bleomycin, 5-fluorouracil, vincristine, methylmethane sulfonate, ethyl nitrosourea) cytotoxic agents. Although MT is an excellent scavenger of hydroxyl radicals *in vivo,* the ability of MT-inducing agents to protect biological systems from radiation and other radical-mediated damage ranges from significant to nil.

When experiments designed to demonstrate resistance to any of these agents are based on selection for MT overexpression or on its induction, other effects of the selecting or inducing agent cannot be ruled out. These could include activation/induction of other stress response or repair systems, changes in cellular metabolic and transport processes, and delays in cell cycling. Even when effects in the recipient cell occur with expression of a transfected *MT* gene, mechanisms could include other unknown gene-regulatory properties of MT, changes in metal ion-dependent metabolic pathways, effects of metals released from MT by the toxic agent, or binding of electrophiles (including metal ions) by thionein.

The evolutionary pressures giving rise to MT may have selected for the unique chemistry of a polythiol, and metal ions may have been a solution to the problem of keeping proximate thiols in a reduced state. Alternatively, thionein may have been a particularly successful means of managing a number of essential metals. Regardless of which viewpoint is more appealing, the presence of both thiols and labile metal centers can be expected to modify cellular responses to a variety of organic and inorganic substances. It is less important to assign detoxification as a *function* of MT than it is to understand the mechanisms by which metal toxicity can be modulated and multidrug resistance manifested.

[4] Nutritional and Physiologic Significance of Metallothionein

By IAN BREMNER

Introduction

Metallothionein (MT) and metals are closely linked. The protein binds 7–10 g atoms of metal/mol, appears only to occur with its full complement of metal atoms, and its synthesis is induced by many metals by a regulated process involving increased gene transcription. It is therefore not surprising that the long list of functions proposed for MT includes the control of metal metabolism. However, there is some uncertainty as to its precise role in the handling of metals, as there have been suggestions that it is involved in such diverse processes as the control of their absorption, tissue uptake, transport, storage, and detoxification.

Synthesis of MT can also be induced by many physiological and nutritional factors, including starvation and imposition of various types of physical or inflammatory stress. This has implied that the protein could have other physiological roles, such as in the acute phase response, the scavenging of free radicals, the regulation of cell differentiation, and the storage of sulfur. However, as the list of proposed functions of MT grows, it becomes increasingly difficult to believe that any one protein, even one with such unique properties, could be so versatile. It seems more likely therefore that MT has some relatively basic functions, consistent with its highly conserved structure, the existence of an MT "housekeeping" gene, and the ease with which its synthesis can be induced by a plethora of metals, hormones, and related factors.

Nutritional Factors Affecting Metallothionein Production

Metals

Although MT binds to and its synthesis is induced by many metals, copper and zinc are the only ones of nutritional importance. The others, including cadmium and mercury, are nonessential metals the cytotoxic effects of which appear to be reduced by binding to MT. Such detoxification of heavy metals represents an important role for MT, although it could be an adventitious consequence of the ability of these metals to induce and bind to a protein that is primarily concerned with the metabolism of zinc and copper.

Metallothionein has been isolated as a major zinc- and copper-binding protein from many tissues, such as liver, kidneys, intestine, and pancreas.[1] Indeed, as immunological techniques for the detection and measurement of MT have improved, it has been found in most tissues, including thymus, bone marrow, brain, and reproductive organs. It is located mainly in the cell cytosol but can also occur in the nucleus in amounts that depend on the tissue metal concentration and the stage of development. For example, the intranuclear localization of MT is particularly evident in the hepatocytes of fetal and neonatal rats but decreases with age, so that at weaning the protein is localized mainly in the cytosol.[2] Appreciable amounts of copper-MT also occur in the lysosomes and other particulate fractions of copper-loaded liver from Bedlington terriers[3] and pigs.[4] In the latter case, over 80% of the hepatic copper may be present as MT, distributed evenly between the cytosol and particulate fractions.

Close relationships between tissue MT and metal content have been demonstrated in humans and in domestic animals receiving normal diets. Thus liver MT and zinc concentrations are closely related in human, sheep, and calf liver, whereas in pigs of normal zinc status the best correlation is between liver MT and copper concentrations.[5] However, it has been necessary in most experiments with rats to inject the copper or zinc or feed diets that are severely deficient in or contain excessive amounts of the metals before such correlations are seen. For example, in rats given severely zinc-deficient diets, liver and intestinal MT concentrations are rapidly reduced to nondetectable levels, whereas injection of zinc or feeding diets with very high zinc contents greatly increases tissue MT levels such that most of the additional tissue zinc is bound to MT.[6,7] Injection of copper induces liver and to a lesser extent kidney MT synthesis in rats[8] whereas feeding high-copper diets increases MT levels in kidneys but does not greatly affect those in the liver.[9] Such experiments have provided valuable insight into the factors that control MT production but have not necessarily helped in the elucidation of the nutritional and physiological significance of the protein.

It is only in recent years that serious attempts have been made to

[1] J. H. R. Kägi and Y. Kojima, *Experientia, Suppl.* **52,** 1 (1987).
[2] M. Panemangalore, D. Banerjee, S. Onosaka, and M. G. Cherian, *Dev. Biol.* **97,** 95 (1983).
[3] G. F. Johnson, A. G. Morell, R. J. Stockert, and I. Sternlieb, *Hepatology* **1,** 243 (1981).
[4] R. K. Mehra and I. Bremner, *Biochem. J.* **219,** 539 (1984).
[5] I. Bremner, *Experientia, Suppl.* **52,** 81 (1987).
[6] P. Menard, C. C. McCormick, and R. J. Cousins, *J. Nutr.* **111,** 1358 (1981).
[7] M. P. Richards and R. J. Cousins, *J. Nutr.* **106,** 1591 (1976).
[8] I. Bremner, W. G. Hoekstra, N. T. Davies, and B. W. Young, *Biochem. J.* **174,** 883 (1978).
[9] I. Bremner, R. K. Mehra, J. N. Morrison, and A. M. Wood, *Biochem. J.* **235,** 735 (1986).

establish the effects of moderate and nutritionally relevant changes in dietary zinc and copper content on tissue MT concentrations. For example, liver, kidney, and pancreas MT-1 concentrations decrease within a few days of rats being given a diet containing only 3 or 6 mg zinc/kg instead of the requirement level of 12 mg/kg.[10] Similarly, kidney MT levels decrease in line with the decrease in kidney copper concentrations in rats made copper deficient by feeding iron-supplemented diets,[11] but dietary copper deficiency does not affect liver MT concentrations.[10] In other studies in which rats were given diets with 5, 30, or 180 mg zinc/kg and 1, 6, or 36 mg copper/kg, kidney concentrations of MT and of MT mRNA were proportional to the dietary zinc intake.[12] However, hepatic concentrations of MT and MT mRNA were not greatly affected by the dietary copper and zinc content, and MT mRNA levels in the intestine increased only at the highest zinc and lowest copper intakes. It was noted that this tissue-specific response in MT production occurred primarily in the organs of absorption and excretion since this could imply that MT plays a role in the control of these processes. It seems that a primary determinant of whether a change in dietary intake of zinc affects tissue MT synthesis is the tissue zinc concentration. Only when this increases above a critical basal level does interaction occur with the promoter region in the MT gene with increased transcription of MT mRNA.

The influence of changes in dietary zinc intake on tissue MT content is also evident in the maternal–fetal complex. Hepatic MT levels are often greatly elevated in fetal and neonatal animals, with the nature of the bound metal and the gestational age at which maximum concentrations are found depending on species.[5] In hamsters and rats, for example, the main metals associated with MT in fetal liver are copper and zinc, respectively.[13] When maternal rats are given diets of low zinc content during pregnancy and lactation, liver MT and MT mRNA concentrations in the pups are reduced in line with the reduced tissue zinc contents.[14–16] In contrast, maternal copper or iron deficiency does not affect hepatic MT levels in the pups, emphasizing the fact that copper is often a less potent inducer of hepatic MT synthesis than is zinc.

Although iron can bind to MT *in vitro* it does not do so *in vivo,* and changes in iron status do not have a major effect on MT production.

[10] I. Bremner, J. N. Morrison, A. M. Wood, and J. R. Arthur, *J. Nutr.* **117,** 1595 (1987).
[11] R. K. Mehra and I. Bremner, *Biochem. J.* **213,** 459 (1983).
[12] T. L. Blalock, M. A. Dunn, and R. J. Cousins, *J. Nutr.* **118,** 222 (1988).
[13] A. Bakka and M. Webb, *Biochem. Pharmacol.* **30,** 721 (1981).
[14] K. R. Gallant and M. G. Cherian, *Biochem. Cell Biol.* **64,** 8 (1986).
[15] K. R. Gallant and M. G. Cherian, *J. Nutr.* **117,** 709 (1987).
[16] J. N. Morrison and I. Bremner, *J. Nutr.* **117,** 1588 (1987).

Increased hepatic synthesis of zinc MT can occur at very high dietary iron intakes although this is probably a stress response.[17] Kidney MT levels may decrease in iron-loaded[11] and iron-deficient rats,[18] but this results from induced copper deficiency and anorexia, respectively. Iron deficiency does not consistently affect MT levels in the liver but increases those in the red blood cells because of the increased production of MT-rich reticulocytes.[18]

Other Nutrients

No systematic study has been made of the effects of other changes in nutritional status on MT production. Reduction of food intake increases liver MT concentrations, probably because of the influence of glucagon and other "stress factors" on MT synthesis.[5] Protein deficiency also increases liver MT concentrations, even though liver zinc concentrations are decreased, but its effects on kidney MT concentrations are variable and depend on the degree of protein deprivation.[19] Because of the high cysteine content of MT there has been some interest in the effects of dietary sulfur supply on MT production. Surprisingly, liver MTs were increased in rats given sulfydryl-deficient diets, probably because of the reduction in their food intake, indicating that sulfur is not a limiting factor for MT synthesis.[20]

Metallothionein in the Control of Metal Absorption

The presence of MT in most cell types suggests that it plays a general role in the handling of metals. Nevertheless there have been claims that it plays a specific role in some tissues, such as in the control of metal absorption in the gastrointestinal tract. This suggestion was based on the existence of an inverse relationship between the efficiency of zinc absorption and intestinal MT concentrations.[6,7,21] In zinc-deficient animals, which absorb zinc with high efficiency, little MT is present in the intestinal mucosa to limit zinc transfer to the plasma. Conversely, in zinc-loaded animals, where homeostatic control of zinc metabolism results in reduced zinc absorption, zinc is apparently incorporated into intestinal MT with concomitant reduction in transfer of zinc into the plasma. This attractive

[17] C. C. McCormick, *Proc. Soc. Exp. Biol. Med.* **176,** 392 (1984).

[18] A. Robertson, J. N. Morrison, A. M. Wood, and I. Bremner, *J. Nutr.* **119,** 439 (1988).

[19] I. Bremner, *in* "Essential and Toxic Trace Elements in Human Health and Disease" (A. S. Prasad, ed.), in press. Wiley, New York, 1990.

[20] L. E. Sendelbach, C. A. White, S. Howell, Z. Gregusa, and C. D. Klaassen, *Toxicol. Appl. Pharmacol.* **102,** 259 (1990).

[21] R. J. Cousins, *Physiol. Rev.* **65,** 238 (1985).

hypothesis was originally based on experiments that involved parenteral injection of rats with zinc and are therefore not necessarily of nutritional relevance. Other criticisms of this hypothesis include the failure to take cognizance of specific activity changes in the endogenous mucosal zinc pool that would affect estimates of zinc absorption.[22] Moreover, no clear inverse relationship was found between mucosal MT concentrations and the efficiency of zinc absorption in mice.[22,23] Indeed, it has even been claimed that MT facilitates rather than inhibits zinc absorption. However, studies on intestinally perfused rats indicate that the mucosal buffer role of MT and the opposing view of its facilitating role in zinc absorption are not necessarily mutually exclusive views of its function in the intestine.[24]

Nevertheless, when dietary zinc intakes in rats are varied over a nutritionally relevant range, where changes in the efficiency of zinc absorption have been recorded, no major changes in mucosal MT concentrations have been found.[25] Only with very high dietary zinc intakes do MT concentrations increase, possibly because it is only then that cellular zinc concentrations increase significantly. It is possible also that MT is then involved in the control of zinc excretion, as has been implied by some immunocytochemical investigations.[26] When dietary zinc and copper intakes by rats were varied over a limited range in one investigation, significant increases in MT and MT mRNA levels were detected only in the group receiving both a high zinc (180 mg/kg diet) and low copper (1 mg/kg diet) intake, indicating that *MT* gene expression was highest in that group.[12] However, the treatments had no effect on zinc or copper absorption. The importance of MT as a regulator of zinc absorption is still therefore far from clear.

Similarly, there is no evidence that changes in dietary copper supply over a physiologically relevant range have any effect on mucosal MT concentrations, indicating that MT is not part of the homeostatic control mechanism regulating copper absorption.[12,27] Nevertheless, reduced copper absorption has been reported in circumstances where mucosal MT concentrations are elevated, such as in brindled mice, which have a genetic abnormality that limits copper absorption.[28] This may reflect a defect in the efflux of copper from the mucosal cells and the stimulation of *MT* gene transcription by the copper that accumulates in the cell. Similarly the decrease in the efficiency of copper absorption in animals given high-zinc

[22] B. C. Starcher, J. G. Glauber, and J. G. Madaras, *J. Nutr.* **110,** 1391 (1980).
[23] P. R. Flanagan, J. Haist, and L. S. Valberg, *J. Nutr.* **113,** 962 (1983).
[24] J. E. Hoadley, A. S. Leinart, and R. J. Cousins, *J. Nutr.* **118,** 497 (1988).
[25] A. C. Hall, B. W. Young, and I. Bremner, *J. Inorg. Biochem.* **11,** 57 (1979).
[26] H. Nishimura, N. Nishimura, and C. Tohyama, *J. Histochem. Cytochem.* **37,** 715 (1989).
[27] P. Oestreicher and R. J. Cousins, *J. Nutr.* **115,** 159 (1985).
[28] I. J. Crane and D. M. Hunt, *Chem.-Biol. Interact.* **45,** 113 (1983).

diets has been attributed to the induction, by the excess zinc, of mucosal MT synthesis, with preferential binding of copper to the protein and inhibition of its transfer into the plasma.[25] This therefore provides a plausible explanation at a molecular level for the zinc–copper interaction. However, it has yet to be proved that copper associated with mucosal MT is unavailable to the animal and is excreted on desquamation of the intestinal cells, as is commonly assumed. Moreover, the effects of zinc on copper absorption and mucosal copper-MT accumulation have only been demonstrated in the rat at very high zinc intakes and have not been substantiated at more physiological levels.[27] Nevertheless, the protective effect of zinc against Wilson's disease is commonly attributed to induction of MT synthesis in the intestinal mucosa.

Function of Metallothionein in Control of Metal Metabolism

The finding that MT production is stimulated when tissue zinc and copper concentrations are increased has indicated that it functions in the cellular detoxification or storage of the metals, especially as cytotoxic effects are more commonly encountered when the metals are bound in other forms. If MT does have a storage function it is relatively transient insofar as the protein has a short half-life and the metals are rapidly released when exposure to the metals is reduced. The elevated concentrations of MT in the liver of fetal and neonatal animals have also been regarded as evidence that MT acts as a storage reserve for copper and zinc in later life, although they could also reflect the immaturity of biliary excretory mechanisms for copper in neonates.

Unfortunately little is known of the fate of metals after binding to MT. It has been reported that direct transfer of zinc and copper can occur from MT to apoenzymes such as alkaline phosphatase and superoxide dismutase, implying that MT might be involved in regulation of the activation of these and other enzymes.[29–31] However, such transfer has only been demonstrated *in vitro* and there is as yet no evidence that it occurs *in vivo*. Indeed, transfer of copper even *in vitro* occurs only under oxidizing conditions. Nevertheless, the kinetics of zinc transfer from MT in Ehrlich cells do indicate that these are rate-limiting ligand substitution processes that do not involve degradation of the protein.[30]

Another possibility is that MT acts as a metal transport protein in the movement of metal between tissues. There is good evidence for such a role

[29] B. L. Geller and D. R. Winge, *Arch. Biochem. Biophys.* **213,** 109 (1982).
[30] S. K. Krezoski, J. Villalobos, C. F. Shaw, and D. H. Petering, *Biochem. J.* **255,** 483 (1988).
[31] A. O. Udom and F. O. Brady, *Biochem. J.* **187,** 329 (1980).

in the transport of cadmium from the liver to the kidneys, since cadmium-MT has been detected in the plasma of cadmium-exposed subjects and animals and parenterally administered cadmium-MT is selectively absorbed by the kidneys.[32] Similarly, elevated copper-MT concentrations occur in the plasma of copper-loaded animals, although this may partly reflect leakage of the protein because of copper-induced liver damage.[9] However, such secretion appears to make only a limited contribution to the loss of copper-MT from the liver.

It seems likely that the fate of metals in MT depends on the particular needs of the animal and that the copper and zinc can be used for metabolic processes, deposited in new tissue, or excreted if not required. Such a concept is consistent with MT acting in the homeostatic regulation of copper and zinc metabolism. Autoregulation of *MT* genes serves to keep intracellular concentrations of potentially toxic free Zn^{2+} and Cu^{2+} at a low level and to bind the excess metal in a nontoxic form.[33] This metal may be released on degradation of the protein or in some cases may accumulate in lysosomes or other organelles in the form of insoluble aggregates of copper-MT. Alternatively the protein may be secreted in intact form from the cell or it may donate its metal to apoenzymes by ligand exchange reactions. In essence, therefore, MT acts as a metal buffer that establishes steady-state kinetics for intracellular Cu^{2+} and Zn^{2+} levels. The autoregulation of *MT* genes involves binding of incoming metal ions to a small pool of thionein. However, if the influx of metal ions is excessive, they then interact with other proteins that bind to the *MT* genes and induce synthesis of additional thionein, which then binds the excess metal ions. The subsequent reduction in the concentration of free metal ions reduces the stimulus for increased thionein production, which then returns to basal levels.

Physiological Factors Affecting Metallothionein Synthesis

The above hypothesis explains many aspects of the link between MT synthesis and metal load. However, it does not explain why MT synthesis in the liver and in some other tissues can be induced by restriction of food intake, bacterial infection, and by imposition of many types of physical and inflammatory stress (Table I).[5] Such induction is mediated by a range of factors that interact directly or indirectly with regulatory elements on the *MT* gene and increase gene transcription. Since imposition of stress increases circulating levels of glucocorticoids they have been proposed as

[32] M. Nordberg and G. F. Nordberg, *Experientia, Suppl.* **52,** 669 (1987).
[33] D. H. Hamer, *Annu. Rev. Biochem.* **55,** 913 (1986).

TABLE I
PATHOPHYSIOLOGICAL FACTORS THAT INDUCE MT SYNTHESIS

Carbon tetrachloride	Interleukin 1
Catecholamines	Irradiation
Endotoxin	Oxidative challenge
Glucagon	Physical stress
Glucocorticoids	Starvation
Infection	Streptozotocin
Inflammation	Tumor necrosis factor
Interferon	

inducing agents for MT synthesis.[34] Injection of rats with dexamethasone or its addition to cell cultures results in increased MT synthesis and glucocorticoid regulatory elements have been identified on the *MT* gene. However, there is some doubt as to the physiological importance of the glucocorticoids in stress-induced synthesis of MT. Thus chronic administration of adrenocorticotropic hormone (ACTH) to rats had no effect on basal MT concentrations in the liver and suppressed the increase caused by restraint stress.[35] Moreover, adrenalectomy or treatment with a glucocorticoid receptor blocker increased basal and stress-induced MT levels. It was claimed that glucocorticoids may on occasion have a permissive role in mobilizing MT from tissues to serum and that in physiological concentrations corticosterone has an inhibitory role in the maintenance of hepatic MT levels.[35]

An alternative possibility is that stress-induced MT synthesis is caused by catecholamines, which have been shown to increase hepatic MT concentrations in rats.[36] Adrenergic blockade in male rats decreased liver MT induction by sham adrenalectomy and exogenous catecholamines but adrenoceptor blocks were ineffective in another study with female rats.[37-39] These results are not consistent with the relative paucity of β receptors in male rat liver. Cyclic AMP may mediate the induction of MT synthesis by epinephrine and glucagon because analogs of this second messenger also increase liver MT levels.[21,38,40] The effects of dexamethasone and cyclic

[34] L. J. Hager and R. D. Palmiter, *Nature (London)* **291,** 340 (1981).
[35] J. Hidalgo, M. Giralt, J. S. Garvey, and A. Armario, *Am. J. Physiol.* **254,** E71 (1988).
[36] F. O. Brady and B. S. Helvig, *Am. J. Physiol.* **247,** E318 (1984).
[37] F. O. Brady, *Life Sci.* **28,** 1647 (1981).
[38] F. O. Brady, B. S. Helvig, A. E. Funk, and S. H. Garrett, *Experientia, Suppl.* **52,** 555 (1987).
[39] J. Hidalgo, M. Giralt, J. S. Garvey, and A. Armario, *Horm. Metab. Res.* **20,** 530 (1988).

AMP analogs on *MT* gene transcription are additive, suggesting that the two regulatory pathways are independent.

Bacterial infection also increases hepatic MT synthesis while decreasing serum zinc concentrations, which are characteristic of an acute phase response.[41] These effects can be mimicked by administration of endotoxin, although the lipopolysaccharide is not a primary inducer of MT synthesis. Moreover, studies on transgenic mice carrying an *MT*–thymidine kinase fusion gene showed that the endotoxin-related promoter site for *MT* gene transcription is independent of the site used by glucocorticoids.[42]

Bacterial infection and other types of stress cause the release of the cytokine interleukin 1 (IL-1) from monocytes and activated macrophages. This, like tumor necrosis factor and other cytokines, induces hepatic MT synthesis in rats.[43,44] IL-1 also increases MT levels in bone marrow and thymus. This tissue-specific regulation of *MT* gene expression appears to be responsible for the changes in zinc metabolism and particularly the increased zinc uptake by these same tissues. It has been suggested that this satisfies the increased demand in IL-1 treated rats for zinc for hematopoietic cell production.

Another explanation for the induction of MT synthesis by injection involves the release of a macrophage-derived heat-stable protein factor, which is distinct from all other known inducers of MT, including IL-1.[45] This factor, which is produced by macrophages in response to endotoxin exposure, stimulates MT synthesis and zinc accumulation by Chang cells in culture.[46] Although endotoxin itself does not induce MT synthesis in the human B cell line RPM1 1788, the products obtained when peripheral mononuclear cells and spleen cells are stimulated by endotoxin or concanavalin do induce the protein.[47,48] Glucagon is another possible mediator

[40] M. A. Dunn and R. J. Cousins, *Am. J. Physiol.* **256,** E420 (1989).

[41] P. Z. Sobocinski, W. J. Canterbury, C. A. Mapes, and R. E. Dinterman, *Am. J. Physiol.* **234,** E399 (1978).

[42] D. M. Durnam, J. S. Hoffman, C. J. Quaife, E. P. Benditt, H. Y. Chen, and R. D. Palmiter, *Proc. Natl. Acad. Sci. U.S.A.* **81,** 1053 (1984).

[43] R. J. Cousins and A. S. Leinart, *FASEB J.* **2,** 2884 (1988).

[44] K. L. Huber and R. J. Cousins, *J. Nutr.* **118,** 1570 (1988).

[45] Y. Iijima, T. Takahashi, T. Fukushima, S. Abe, Y. Itano, and F. Kosaka, *Toxicol. Appl. Pharmacol.* **89,** 135 (1987).

[46] T. Fukushima, Y. Iijima, and F. Kosaka, *Biochem. Biophys. Res. Commun.* **152,** 874 (1988).

[47] S. Abe, M. Matsumi, M. Tsukioki, S. Mizukawa, T. Takahishi, I. Iijima, Y. Itano, and F. Kosaka, *Experientia, Suppl.* **52,** 587 (1987).

[48] J. Oberbarnscheidt, P. Kind, J. Abel, and E. Gleichmann, *Res. Commun. Chem. Pathol. Pharmacol.* **60,** 211 (1988).

for IL-1- or endotoxin-induced MT synthesis, since increased MT levels occur in glucagon-treated rats.[49] However, serum zinc levels are unaffected.

The induction of MT synthesis by "stress" factors is therefore extremely complex. It may be mediated by several factors that directly or indirectly induce increased gene transcription. A common feature of the responses is increased cellular uptake and, *in vivo,* increased hepatic uptake of zinc.[43] In most instances this results in reduced serum zinc concentrations as part of an acute phase response. It can only be assumed that this redistribution of zinc benefits the host defense process.

Metallothionein as a Free Radical Scavenger

Another suggestion is that MT plays a role in the scavenging of free radicals, the production of which is often stimulated in "stress" conditions. Thus, treatments such as administration with IL-1 or interferon or exposure to X rays or increased oxygen tension cause increased production of oxygen free radicals and also induce MT synthesis.[5] Production of these radicals by neutrophils and macrophages is part of the inflammatory response, designed to destroy bacteria and virus-infected cells. However, these radicals can have deleterious effects on DNA and cell membranes unless they are scavenged by some antioxidant system, such as vitamin E, glutathione, or glutathione peroxidase. Considerable interest has been shown in the possible role of MT as another free radical scavenger.[50] Zinc-MT has been shown to scavenge hydroxyl radicals *in vitro* and to be more effective than glutathione in preventing DNA degradation by hydroxyl radicals.[51] However, zinc-MT was less effective than glutathione in inhibiting lipid peroxidation of microsomal membranes and copper-MT tended to promote free radical-mediated damage to the membranes.[52] Aerobic radiolysis of an MT solution induces metal loss and thiolate oxidation. Damage by hydroxyl radicals, which involves the metal thiolate clusters, can be reversed by lowering the pH and adding glutathione and metals.[50]

The concomitant increase in lipid peroxidation and in hepatic MT levels in rats after restriction of food and water intake and their further enhancement in the presence of dimethyl sulfoxide have suggested that

[49] P. Z. Sobocinski and W. J. Canterbury, *Ann. N.Y. Acad. Sci.* **210,** 354 (1982).

[50] P. J. Thornalley and M. Vašák, *Biochim. Biophys. Acta* **827,** 36 (1985).

[51] J. Abel and N. de Ruiter, *Toxicol. Lett.* **47,** 191 (1989).

[52] J. R. Arthur, I. Bremner, P. C. Morrice, and C. F. Mills, *Free Radical Res. Commun.* **4,** 15 (1987).

these factors are linked.[53] However, the enhancement in MT levels caused by deprivation stress was less evident when the rats were supplemented with vitamin E, which could imply competition between MT and the vitamin in free radical scavenging. Cadmium-tolerant cells, which have increased ability to synthesize MT, tend also to have enhanced ability to resist oxidative stress.[54] Addition of zinc to hepatocyte culture also results in increased ability to withstand chemically induced free radical damage.[55] This could reflect increased cellular MT levels, although it may also be due to a stabilizing effect of zinc on membranes or effects of zinc on cytochrome P-450 or glutathione peroxidase. It has also been suggested that the cardiac toxicity of the anti-tumor drug, adriamycin, which results from extensive lipid peroxidation, can be reduced by pretreatment with metals that induce heart MT synthesis.[56]

There is therefore strong circumstantial evidence that MT plays a role in the scavenging of free radicals but definitive proof has yet to be obtained. It is noteworthy that although MT synthesis is induced in different types of oxidant stress, it does not prevent the oxidative damage to tissues. Moreover, high levels of MT do not appear to suppress further MT synthesis in response to the generation of free radicals. Thus preinduction by zinc of liver MT levels does not prevent the increase in MT synthesis caused by immobilization stress.[53]

[53] J. Hidalgo, L. Campmany, M. Borras, J. S. Garvey, and A. Armario, *Am. J. Physiol.* **255,** E518 (1988).

[54] A. C. Mello-Filho, L. S. Chubatsu, and R. Meneghini, *Biochem. J.* **256,** 475 (1988).

[55] D. E. Coppen, D. E. Richardson, and R. J. Cousins, *Proc. Soc. Exp. Biol. Med.* **189,** 100 (1988).

[56] M. Satoh, A. Naganuma, and I. Nobumasa, *Toxicology* **53,** 231 (1988).

Section II

Isolation of Metallothioneins

[5] Large-Scale Preparation of Metallothionein: Biological Sources

By MILAN VAŠÁK

In animals, metallothionein (MT) is most abundant in parenchymatous tissues, the highest concentrations being found in kidney and liver. However, their occurrence and biosynthesis have been documented also in many other tissues and cell types. The absolute amounts present in different species and tissues are highly variable, reflecting effects of differences in age, state of development, dietary regime, and other, not yet fully identified factors.[1] Thus, in human and equine kidney and liver, which are rich natural sources of MT, concentration can vary over a factor of 10 or more (Table I[2,3]). The biosynthesis of MT can be increased by heavy metal administration to laboratory animals, i.e., rat and rabbit, yielding about 10 mg of protein/g wet weight of tissue (Table I). Cadmium is the best MT inducer, followed by zinc. Liver is the most commonly used source of MT, because the MT isoforms are well characterized and good yields are obtained.

Metallothionein Induction by Cadmium and Zinc

In male New Zealand White rabbits or Sprague-Dawley rats, the biosynthesis of MT is increased by subcutaneous injection of a chloride salt of the appropriate metal. The original procedure of Kimura for cadmium injection into rabbits (21 injections over the time period of 7 weeks)[4] may now be shortened to 9 injections, 3 times per week, of sterile filtered $CdCl_2$ (0.1 *M*) in 0.15 *M* NaCl at a dose of 1 mg/kg body weight. While an even shorter time protocol (six injections) results in approximately 10% lower yield of MT, the use of higher levels of Cd dosing (2 mg/kg body weight) or longer induction periods (more than 3 weeks) resulted in increased rabbit mortalities. For rats, a dose of 1 mg/kg body weight injected on the first day, increased to 3 mg/kg body weight on three subsequent days, is used. In spite of the cadmium induction, Cd_7-MT has never been isolated from mammalian tissues. The highest cadmium content obtained was about 5 mol Cd/mol of protein, the remainder being zinc.

[1] J. H. R. Kägi and Y. Kojima, *Experientia, Suppl.* **52,** 25 (1987).
[2] J. H. R. Kägi, S. R. Himmelhoch, P. D. Whanger, J. L. Bethune, and B. L. Vallee, *J. Biol. Chem.* **249,** 3537 (1974).
[3] M. Sutter and M. Vašák, unpublished (1984).
[4] M. Kimura, N. Otaki, and I. Imano, *Experientia, Suppl.* **34,** 163 (1979).

TABLE I
BIOLOGICAL SOURCES OF METALLOTHIONEIN

Tissue	Concentration [mg MT/g (wet weight)][a]	Metal induction[b]	Ref.
Human liver	0.1–15	—	2
Horse liver	0.1–10	—	2
Horse kidney[c]	0.5–10	—	2
Rabbit liver	13	Cd	3
Rat liver	10	Cd	3

[a] Both electrophoretically different isoforms MT-1 and MT-2 are considered.
[b] Metal induction as described below.
[c] Isolated from kidney cortex.

The induction of MT by zinc is the method of choice if a metal-homogeneous Zn-MT form is required. Since zinc is far less toxic than cadmium, much larger amounts of this metal must be administered. For rabbits, a procedure similar to that described for cadmium induction (see above) is employed, the only exception being a dose of 10 mg Zn/kg body weight. The yield of Zn-MT is usually 20–30% lower than that of cadmium-induced Zn,Cd-MT (Table I). For rats, no established protocol for a large-scale preparation of Zn-MT exists. In all induction protocols the animals were sacrificed on the day following the last injection and the livers were stored at -70°. It should be noted, moreover, that metal-homogeneous Zn-MT can also be generated from Zn,Cd-MT by the method of metal reconstitution as described elsewhere in this volume.[5]

Comments

Because of the large variation of MT concentrations in tissues without metal induction (Table I), it is advisable to estimate their MT content prior to a large-scale protein isolation. For this purpose the first preparative step, up to and including the second alcohol precipitation step and pellet lyophilization, as described later in this volume,[6] should be performed on 25 g of tissue. The crude fraction is subsequently taken up in a known volume (about 2 ml) of 20 m*M* Tris-HCl buffer, pH 8.6, and the zinc, cadmium, and copper content determined by flame atomic absorption

[5] M. Vašák, this volume [54].
[6] M. Vašák, this volume [6].

spectroscopy.[7] At this preparative stage the metal ions are predominantly bound to MT; therefore, the determined metal content is a reasonable measure of the MT concentration present. Although liver MTs from adults usually contain almost exclusively zinc ions, the presence of large amounts of copper can sometimes be encountered. A special isolation procedure is required to purify the oxygen-sensitive Cu(I)-MT.[8]

[7] M. Czupryn and K. H. Falchuk, this volume [47].
[8] R. J. Stockert, A. G. Morell, and I. Sternlieb, this volume [34].

[6] Standard Isolation Procedure for Metallothionein

By MILAN VAŠÁK

As in the preparation of any metalloprotein, the presence of adventitious metal ions should be kept to a minimum. All solutions should be passed through a Chelex 100 column (Bio-Rad, Richmond, CA) to remove metal contaminants and labware rendered metal free by acid washing.[1] The large-scale isolation and purification of metallothionein (MT), a cytosolic protein, involves alcohol precipitation followed by size-exclusion and anion-exchange chromatography. This procedure is used to isolate MT from human, rat, and rabbit livers[2] and is basically similar to that published by Kägi *et al.* for human liver, and horse liver and kidney MT.[3,4] All operations when not otherwise indicated are carried out at 4°.

Metallothionein Isolation from Rabbit Liver

Step 1. Preparation of Crude Fraction. Fresh of half-frozen liver from rabbits in which the MT concentration has been increased by zinc or cadmium induction,[5] are cut into small pieces and homogenized in portions of 250 g in a 1-liter Waring blendor (30 sec) with 500 ml of 20 m*M* Tris-HCl buffer at 4°. The homogenate is placed in a 4-liter beaker immersed in a water/ice bath and equipped with an efficient stirring arrangement. An equal volume of a 96% ethanol/chloroform solution (1.05 : 0.08,

[1] B. Holmquist, this series, Vol. 158, p. 6.
[2] M. Sutter and M. Vašák, unpublished (1984).
[3] J. H. R. Kägi, S. R. Himmelhoch, P. D. Whanger, J. L. Bethune, and B. L. Vallee, *J. Biol. Chem.* **249**, 3537 (1974).
[4] R. Bühler and J. H. R. Kägi, *FEBS Lett.* **39**, 229 (1974).
[5] M. Vašák, this volume [5].

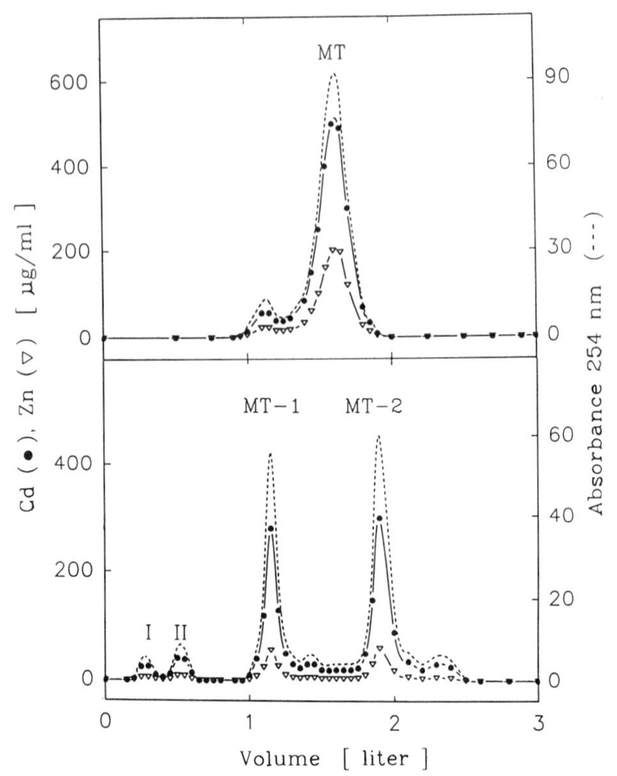

FIG. 1. *Top:* Gel filtration (Sephadex G-50) of the lyophilized crude fraction (5 g) dissolved in a minimum volume of 20 m*M* Tris-HCl, pH 8.6, and eluted with 20 m*M* Tris-HCl/10 m*M* NaCl at pH 8.6. *Bottom:* Ion-exchange chromatography (DEAE-cellulose, Whatman DE-52) equilibrated with 20 m*M* Tris-HCl, pH 8.6. After adsorption, the column is washed with 800 ml of the same buffer. Metallothionein is eluted with a linear gradient of 20–200 m*M* Tris-HCl at pH 8.6 (2 × 1 liter). The chromatographic peaks I and II contain, in addition to zinc and cadmium, a varying amount of copper.

v/v), prechilled to −20°, is slowly added (10 min) to the homogenate with stirring. The resultant precipitate is removed by centrifugation at 3000 *g* for 20 min. The precipitate is discarded, and the supernatant is transferred back to the 4-liter beaker. Three volumes of 96% ethanol (−20°) per volume of supernatant is slowly added with stirring. The resultant suspension is incubated overnight at −20°. The clear supernatant is partially decanted and the remaining white precipitate collected by centrifugation at 3000 *g* for 20 min. Further centrifugation of the pellet at 34,000 *g* for 5 min removes most of the ethanol. The pellet is transferred to a small beaker and solubilized with stirring in 15 ml of 10 m*M* Tris-HCl/25 m*M* NaCl, pH 8.6, for 20 min at room temperature. The insoluble particles are

TABLE I
PURIFICATION OF METALLOTHIONEIN FROM RABBIT LIVER[a]

Fraction	Volume (ml)	Total metal content ($\mu g \times 10^4$) Cadmium	Zinc	Protein (mg)
Crude extract (5 g)	—	9.8	2.1	—
Sephadex G-50 eluate	320	8.5	1.7	—
DEAE-cellulose eluate				
MT-1	250	4.4	0.95	510
MT-2	300	3.3	0.61	365

[a] Metallothionein was induced by administration of cadmium salt as described elsewhere in this volume.[5]

removed by centrifugation at 34,000 *g* for 20 min and the supernatant is lyophilized. The crude fraction (approximately 1.25 g) is stored at −20°.

Step 2. Gel-Filtration Chromatography. The lyophilized crude fraction (5 g) is dissolved in a minimum volume of 20 m*M* Tris-HCl, pH 8.6 (approximately 15 ml) and applied to a Sephadex G-50 column (150 × 5 cm) equilibrated with 20 m*M* Tris-HCl/10 m*M* NaCl at pH 8.6. The protein is eluted with the same buffer at a flow rate of 40 ml/hr. The protein fractions (12 ml) are monitored at 280 and 254 nm and their metal content (zinc, cadmium, and copper) determined by atomic absorption spectroscopy.[6] Low-molecular-weight fractions of a high metal content (K_D between 0.5 and 0.75) are concentrated by ultrafiltration [400 ml Amicon (Danvers, MA) Ultrafiltration cell using YM2 membrane] and the conductivity is adjusted below that of 20 m*M* Tris-HCl, pH 8.6 (see Fig. 1, *top*).

Step 3. Ion-Exchange Chromatography. The protein is adsorbed on an anion-exchange DEAE-cellulose column (Whatman DE-52) (40 × 3 cm) equilibrated with 20 m*M* Tris-HCl, pH 8.6. After adsorption the column is washed with 800 ml of the same buffer. Metallothionein is eluted with a linear gradient of 20–200 m*M* Tris-HCl at pH 8.6 (2 × 1 liter) at a flow rate of 30 ml/hr. The resulting chromatographic profile yields two major peaks of high metal content; these represent metallothionein isoforms MT-1 and MT-2, separated based on their single charge difference (see Fig. 1, *bottom*). The corresponding fractions are concentrated by ultrafiltration. Prior to lyophilization (approximately 50 ml) the salt concentration is reduced to that of 2 m*M* Tris-HCl, pH 8.6 (see Table I).

The same purification steps are used in the isolation of MT from rat liver.

[6] M. Czupryn and K. H. Falchuk, this volume [47].

Purification of Metallothionein from Human Liver and Horse Liver and Kidney

Although the above-described preparative procedure for rabbit liver MT is also applicable for MT isolation from human liver and horse liver and kidney, some changes are required. In the case of horse kidney, the capsule is removed and only the cortex is processed. The human liver MT-2 fractions, after DEAE-cellulose chromatography, contain a coeluting protein, which is characterized by a strong 280-nm absorption and is separated by additional anion-exchange chromatography.[7] The conductivity of the concentrated MT-2 fraction is reduced below that of 15 m*M* Tris-HCl, pH 8.6, and adsorbed on a DEAE BioGel A (Bio-Rad) column (20×2 cm) preequilibrated with 15 m*M* Tris-HCl, pH 8.6. The column is washed with 100 ml of the same buffer before applying a linear gradient of 15–30 m*M* Tris-HCl at pH 8.6 (2×0.5 liter). A flow rate of 16 ml/hr is used.

Comments

Since MT has no aromatic amino acids the protein-bound metal provides a convenient way of monitoring the isolation process. The metal peaks from the chromatographic profile should reflect the protein absorption monitored at 254 nm arising from the bound cadmium and zinc, which have their low-energy charge-transfer bands at 250 and 231 nm, respectively.[8] Large amounts of copper may be encountered in the preparation of human and horse liver MT. The described procedure is unsuitable for the isolation of the air-sensitive Cu(I)-containing MT.

[7] A. Schäffer, private communication (1989).
[8] M. Vašák and J. H. R. Kägi, *Met. Ions Biol. Syst.* **15**, 213 (1983).

[7] Criteria of Purity for Metallothioneins

By MILAN VAŠÁK

Over the years, a number of analytical techniques have been used to analyze the purity of mammalian metallothionein (MT). Initially, MT moves as a low-molecular-weight fraction off a Sephadex G-50 column with a high metal content. Based on its further separation on an anion-exchange chromatography column into two protein peaks, differing by a single charge difference, the protein has been subclassed as MT-1 and MT-2.[1] As described elsewhere in this volume, each subclass constitutes a family of isoproteins that can be separated further by reversed-phase high-performance liquid chromatography (HPLC).[2] In this respect, the MT-2 subclass of various species is considerably more homogeneous than MT-1. Although the analytical techniques described in this chapter have been established as some criteria of purity of the MT-1 and MT-2 subclasses, they are also applicable to an individual MT isoform. Throughout the analytical checks, procedures to control adventitious metal ion concentrations should be followed to ensure that reagents and apparatus are "metal free."[3] The described analytical methods can be applied without difficulties to characterize the zinc- and/or cadmium-containing MTs, if present in milligram amounts.

Protein Quantification. Metallothionein has no aromatic amino acids, the absorption features of which are commonly applied in protein quantification. Due to its highly unusual amino acid composition,[4] protein quantitation by commercially available kits is unsatisfactory. The use of the absorption features of the metal-containing forms is also not feasible, due to possible metal loss in the course of protein isolation and/or variability of metal composition. Although MT has no aromatic amino acids, its concentration may be quantified using the absorption of the apoprotein ($\epsilon_{220} = 48{,}200\ M^{-1}\ \text{cm}^{-1}$).[5] A small aliquot of the sample at neutral pH is quickly added into a quartz cuvette containing 0.1 *M* HCl. The sulfur–metal charge-transfer bands[6] will disappear as the metal is released from the protein structure. There is no significant contribution of the free metal ions to the protein absorption. This rather easy and quick method takes advantage of the tailing of the 190-nm protein band[6] into the region accessible to common spectrophotometers. Care must be taken, as a small inaccuracy in the absorption measurement will lead to a large error in the protein concentration due to the steep slope of the absorption profile. Again, trace copper can lead to large inaccuracies with this method, as it is not released from the protein structure under these conditions. The concentration of MT has been determined in the micromolar range using inductively coupled plasma emission spectrometry (ICP-AES).[7]

[1] M. Nordberg and Y. Kojima, *Experientia, Suppl.* **34,** 41 (1979).
[2] P. Hunziker, this volume [27].
[3] B. Holmquist, this series, Vol. 158, p. 6.
[4] J. H. R. Kägi and A. Schäffer, *Biochemistry* **27,** 8509 (1988).
[5] R. H. O. Bühler and J. H. R. Kägi, *Experientia, Suppl.* **34,** 211 (1979).
[6] M. Vašák and J. H. R. Kägi, *Met. Ions Biol. Syst.* **15,** 213 (1983).
[7] J. Bongers, C. D. Walton, and D. E. Richardson, *Anal. Chem.* **60,** 2683 (1988).

Metal Analysis. The metal content is determined by flame atomic absorption spectroscopy as described.[8]

Sulfhydryl Group Determination. Sulfhydryl groups of the metal-containing or metal-free MT are quantified either with Ellman's reagent DTNB [5,5′-dithiobis(2-nitrobenzoic acid)] in 10 m*M* potassium phosphate buffer, 2 *M* guanidinium hydrochloride, 20 m*M* EDTA, pH 7.5 ($\epsilon_{412} = 13{,}600\ M^{-1}\ cm^{-1}$)[9] or with 2,2′-dithiodipyridine in 0.2 *M* sodium acetate buffer, pH 4.0 ($\epsilon_{343} = 7600\ M^{-1}\ cm^{-1}$)[10] using a small aliquot of the sample. The protein concentration is determined via apo-MT (see above). Although the results of both methods are comparable, with typically 19–21 sulfhydryl groups found in mammalian MTs, the latter method proved to be better suited as the assay is less pH sensitive. The availability of all 20 sulfhydryl groups to modification rules out the presence of intramolecular disulfide bridges, due to metal loss at high pH.

Size-Exclusion Chromatography. The size homogeneity of the MT preparation is checked on a small portion of the protein applied to a Sephadex G-50 column (150 × 1 cm) equilibrated with 20 m*M* Tris-HCl/10 m*M* NaCl at pH 8.6 or to a Superose 12 HR 10/30 column (Pharmacia, Piscataway, NJ) attached to a fast protein liquid chromatography (FPLC) apparatus. Note that in spite of its molecular weight between 6K and 7K, depending on the metal present, the monomeric protein elutes with an apparent molecular weight of 10K[6] due to its elongated shape.

Polyacrylamide Gel Electrophoresis. Nondenaturing polyacrylamide slab gel electrophoresis in 7.5% gels at pH 8.9 on the MT sample (10–20 μg) is used as a check of the protein purity.[11] The electrophoretic mobility of MT-1 and MT-2 differ in that MT-2 moves faster toward the anode than MT-1. The MT-2 isoforms isolated from human, rabbit, and rat carry three negative charges compared with two for the MT-1 isoforms.[4,6] Since the appearance of additional bands in the gel may also be due to the presence of polymeric MT species, a reducing 10–20% sodium dodecyl sulfate polyacrylamide electrophoresis (SDS-PAGE)[12] of an alkylated MT sample should be performed. The sample (10–50 μg) is reduced by incubation with 0.2 *M* dithiothreitol (DTT) at 60° for 15 min and alkylated by adding a fourfold molar excess of iodoacetamide at 60° for 15 min. After adding an equal volume of concentrated reducing sample buffer,[12] the pH is adjusted to neutral using 2 *M* NaOH prior to boiling for 3 min.

[8] M. Czupryn and K. H. Falchuk, this volume [47].

[9] D. McGilvray and J. G. Morris, this series, Vol. 17B, p. 585.

[10] A. O. Pedersen and J. Jacobsen, *Eur. J. Biochem.* **106,** 291 (1980).

[11] B. J. Davis, *Ann. N.Y. Acad. Sci.* **121,** 404 (1964).

[12] U. K. Laemmli, *Nature (London)* **227,** 680 (1970).

Amino Acid Analysis. Amino acid analysis is performed after performic acid oxidation and hydrolysis at 110° for 22 hr.[13]

Acknowledgments

The financial support of the Schweizerischer Nationalfonds throughout this work is gratefully acknowledged.

[13] S. Moore and W. H. Stein, this series, Vol. 6, p. 819.

[8] Isolation of Metallothioneins under Metal-Free Conditions

By KENNETH H. FALCHUK and MARTA CZUPRYN

Introduction

Metallothioneins are molecules characterized by a high metal-to-protein stoichiometry.[1,2] This high metal content has been a useful gauge of the presence of metallothionein in protein mixtures and provides a sensitive and reliable guide to its purification. The amounts and types of metals associated with metallothionein depend on the source of the protein and can be quite varied. In most mammalian tissues, zinc or copper are the most common metal components of metallothionein, whereas in the kidney it is cadmium.[1] However, even when isolated from a single source, metallothionein can be found to contain more than one metal, e.g. zinc and copper in fetal bovine liver[3] or cadmium and zinc in equine kidney cortex.[4] There is also a wide variety of other d^{10} metals, such as mercury, lead, silver, antimony, gallium, bismuth, and gold, that have been reported to be associated with this protein under certain conditions.[1,5] This metal ion heterogeneity of metallothionein has been ascribed to differences in the uptake by and/or supply to an organism of diverse metals or in the metal-binding capability of metallothionein isoforms.[1]

[1] J. H. R. Kägi and A. Schäffer, *Biochemistry* **27,** 8509 (1988).
[2] W. E. Rauser, *Annu. Rev. Biochem.* **59,** 61 (1990).
[3] K. Münger, U. A. Germann, M. Beltramini, D. Niedermann, G. Baitella-Eberle, J. H. R. Kägi, and K. Lerch, *J. Biol. Chem.* **260,** 10032 (1985).
[4] M. Margoshes and B. L. Vallee, *J. Am. Chem. Soc.* **79,** 4813 (1957).
[5] M. Webb, *Experientia, Suppl.* **52,** 109 (1987).

In any metallothionein purification procedure it is necessary to be aware that the metal composition of this protein, as found in intact tissues, can be artificially altered during its isolation by exposure to extraneous metal ions. Extraneous metal ions can be present as contaminants in the solutions used for tissue homogenization and fractionation, in chromatographic media as well as on the surfaces of glassware, storage containers, and other handling equipment used throughout the purification procedure. Under certain conditions, these extraneous metals can be exchanged for metals already bound to the protein.

It is believed that the amount and type of metals bound to metallothionein relate to its physiological role. Therefore, it is essential to avoid adventitious metal contamination throughout the entire purification procedure because these may significantly affect and alter the original composition and/or function of the protein.

This chapter describes methods that reduce the risk of contamination from adventitious metal ions and that can be used to isolate metallothioneins from various sources under metal-controlled conditions.

Preparation of Buffers and Glassware

Water, buffer reagents, and laboratory glassware are common sources of metal ion contamination. Many reagents, even of analytical grade, are frequently contaminated by metal ions, especially zinc, copper, and iron. These sources must be rendered metal free before use to obviate exogenous contamination.

Ultrapure water suitable for isolation of metalloproteins can be obtained from systems, such as a Milli-Q water purification unit (Millipore, Bedford, MA), that combine mixed-bed ion exchange, adsorption of organic compounds, and membrane microfiltration. Such systems supply water with resistivity of 18 MΩ/cm and metal ion content that is practically below the detection limits for most elements analyzed by graphite furnace atomic absorption.

Buffers commonly used in metallothionein preparation, such as Tris, HEPES, and phosphate, should be prepared as concentrated stock solutions. Two methods are suitable for removing trace metal contaminants from these buffers. The first approach involves extraction with diphenylthiocarbazone (dithizone), a complexing agent soluble in organic solvents that reacts with various metals to form organic chelates. The reagent is suitable for extraction of solutions in the pH range 3 to 7.5. Extraction is carried out in a hood to protect against exposure to halogenated solvents. A fresh 0.01% dithizone solution is prepared in pure-grade carbon tetrachloride or chloroform, and 1 ml is mixed with each 10–20 ml of the solution to be extracted. The solutions are mixed in a separatory funnel with a Teflon stopcock by vigorously shaking the funnel for approximately 5 min or until the green color of the organic phase turns red, indicative of the metal–dithizone complex. The organic phase is then removed, replaced by a fresh aliquot of dithizone solution, and the extraction repeated. The procedure is continued with fresh portions of dithizone solution until there is no noticeable change of color. Traces of dithizone remaining in the aqueous phase are removed by adding pure organic solvent (chloroform or carbon tetrachloride) to the solution in the separating funnel and extracting as above. The residual organic solvent is removed from the aqueous phase by carefully applying a vacuum to the top of the separatory funnel for as long as 1–5 hr. During this step, care is taken to prevent implosion of the separatory funnel, intake of dust into the aqueous solutions, and evaporative losses by connecting a water trap to the bottom of the funnel.

The second method uses the cation-exchange resins Chelex 100 or Chelex 20 (Bio-Rad, Richmond, CA). It is recommended for removal of metal ion contaminants when the pH of the buffer to be used is greater than 7.5. The functional groups of these resins, iminodiacetate anions, are capable of chelating a broad spectrum of metal ions. The resins require precycling, i.e., sequential treatment with acid and alkali before use. The amount of the resin to be used (usually 1 g of resin/100 ml of the solution to be extracted) is equilibrated in a sintered glass funnel with two bed volumes of 1 *M* HCl, followed by 5 vol of metal-free water, 2 vol of 1 *M* NaOH, and 5 vol of water. A suitable column and its inlet and outlet attachments are acid cleaned and rinsed with ultrapure water, and filled with the prepared cation exchanger. The solution to be extracted is chromatographed at a flow rate of 4–10 ml/min/cm^2. The first two column volumes are discarded and the subsequent eluate is collected in a metal-free container. The pH of the eluate should be measured and adjusted with metal-free acid, if necessary.

Stock solutions should be prepared in acid-clean glassware (see below) using metal-free water and stored in closed, metal-free containers to prevent exposure to airborne contamination. Prior to use, the solutions are diluted to a final concentration under metal-free conditions. High-quality organic solvents (e.g., Fisher, Pittsburgh, PA; HPLC grade) can be used without additional purification; however, spectrographic analysis of these solvents should be carried out to demonstrate absence of significant amounts of metals. For example, zinc concentration in organic solvent-based solutions should not exceed 1.5×10^{-8} *M*.

Glassware should be rendered metal free by soaking for 24 hr in 6 *N* HNO_3 and washing extensively in metal-free water. Colorless plastic (polypropylene or polystyrene) is generally metal free and does not require additional treatment. The disposable laboratory plasticware, such as pipettes, pipette tips, and tubes, can be used either directly or after rinsing with metal-free water. Reusable plastic items should be cleaned by soaking in a solution containing a chelating agent (e.g., 20 m*M* EDTA) and rinsed extensively with metal-free water. Acid cleaning is not recommended for plastic, because it may change the plastic into an excellent ion exchanger.

Dialysis tubing is rendered metal free by heating at 70° in several changes of metal-free water.

Sources of Metallothionein

Class I metallothioneins have been isolated from the cytosolic fractions of various animal tissues and cells grown in tissue culture.[6] The advantage of using tissue cultures is that the cells can be maintained under controlled conditions and metallothionein formation induced by an excess of zinc, copper, cadmium, and other metals and inducers.

Class II and III metallothioneins have been isolated from fungi,[7] plants,[2] and microorganisms.[8] Unicellular yeast[9] and microalgae[10] are a good source because they are easy to cultivate and show inducible metallothionein synthesis in response to treatment with doses of cadmium and other metals, 10^{-5} *M* or higher. Plant cell suspension cultures[11] also are suitable systems to study metal-binding complexes. Metal-resistant strains of yeast,[9] and plant[11] and animal cell lines,[12] have been selected to overproduce metallothioneins and facilitate protein purification and characterization.

Purification of Metallothionein

Preparation of Tissue and Cell Extracts

Metallothionein can be isolated either from fresh or previously frozen animal or plant tissue. The tissue is homogenized in a buffer (e.g., 10 m*M* Tris-HCl, 25 m*M* potassium phosphate, pH 7.5) made up by diluting a

[6] J. H. R. Kägi and Y. Kojima, *Experientia, Suppl.* **52,** 25 (1987).

[7] K. Lerch, *Experientia, Suppl.* **34,** 173 (1979).

[8] R. W. Olafson, W. D. McCubbin, and C. M. Kay, *Biochem. J.* **251,** 691 (1988).

[9] D. R. Winge, K. B. Nielson, W. R. Gray, and D. H. Hamer, *J. Biol. Chem.* **260,** 14464 (1985).

[10] W. Gekeler, E. Grill, E.-L. Winnacker, and M. H. Zenk, *Arch. Microbiol.* **150,** 197 (1988).

[11] P. J. Jackson, C. J. Unkefer, J. A. Doolen, K. Watt, and N. J. Robinson, *Proc. Natl. Acad. Sci. U.S.A.* **84,** 6619 (1987).

[12] J. H. Freedman and J. Peisach, *Biochim. Biophys. Acta* **992,** 145 (1989).

dithizone-extracted stock solution with metal-free water. This, and the following procedures, should be done under N_2 wherever possible, and all solutions used should contain a reducing agent [1 to 5 mM 2-mercaptoethanol or dithiothreitol (DTT)] to prevent protein oxidation and metal redistribution among other than metallothionein cellular constituents.[13] Particular care must be taken with tissue preparation. Metal-free plastic gloves (powder free) should always be used in handling the tissue, as well as throughout the entire metallothionein isolation procedure, to prevent metal contamination from the skin. Unicellular organisms or cells grown in tissue culture should be washed before homogenization in metal-free water or metal-free isotonic solution to remove growth medium. It is especially important if the medium contains excess metal ion used to induce formation of metallothionein.

Centrifugation of the homogenate to prepare cytosolic fractions should be done using either disposable plastic or acid-cleaned glass centrifuge tubes.

Fractionation of Cell Extracts

In most purification protocols, the tissue or cell extract is first resolved by gel filtration on Sephadex G-50 or G-75 columns and then chromatographed on anion-exchange DEAE- or QAE-based columns.

Both these types of chromatographic media are capable of binding metal ions and are a source of extraneous metal contamination. The metals bound to the matrix material should be removed before use and/or the ability of the columns to bind metals should be decreased. The primary source of interaction with metal ions in Sephadex gels is carboxyl groups. A common procedure suitable to eliminate these active groups from the Sephadex matrix is alkaline reduction of the gel.[14] The swollen gel is taken up in 2.5 vol of water and the pH adjusted to 10–11 with sodium hydroxide. $NaBH_4$ is added to a concentration of 2 g/liter and the slurry heated at 80° for 2 hr with gentle stirring. After cooling the gel is washed extensively with metal-free water until it reaches neutral pH. Such treatment largely decreases the capacity of the Sephadex gel to interact with metal ions.

Ion-exchange media, the function of which depend on binding ionized solutes, obviously cannot be made incapable of interacting with metal ions or their complexes. Heavy metal impurities, therefore, must be removed from these media before use. DEAE- and QAE-based resins are rendered metal free by washing with 0.5 M NaOH and 0.5 M HCl. NaOH and HCl

[13] D. T. Minkel, K. Poulson, S. Wielgus, C. F. Shaw III, and D. M. Petering, *Biochem. J.* **191,** 475 (1980).

[14] B. Lönnerdal and B. Hoffman, *Biol. Trace Elem. Res.* **3,** 301 (1981).

used for this procedure should be of the highest purity (metal-free reagents can be purchased from Johnson Matthey/Aesar, Seabrook, NH). Either batch or column procedure can be used. The medium is allowed to swell in metal-free water and then washed extensively with 0.5 *M* NaOH until free of chloride. The excess NaOH is removed by washing with metal-free water. The ion exchanger is then treated with 0.5 *M* HCl, washed extensively with water, and equilibrated with the appropriate dithizone-extracted eluant buffer.

An alternative procedure, which can be applied for both gel filtration and anion-exchange media, involves removal of contaminating metal ions by elution with a chelating agent. A column procedure is recommended. The column is washed with three column volumes of 2 m*M* 1,10-phenanthroline in metal-free equilibration buffer to remove trace metals from the gel matrix. The column is eluted then with metal-free buffer until the absorbance at 320 nm, characteristic of the presence of 1,10-phenanthroline, is negligible (<0.01 absorbance unit).

Instead of 1,10-phenanthroline, bovine serum albumin (BSA) can be used as a chelating agent. Approximately one-tenth of the bed volume of 1% (w/v) BSA in metal-free equilibration buffer is applied to the column. Bovine serum albumin is eluted and the column regenerated under metal-free conditions suitable for the type of chromatographic medium used.

Each procedure used to remove adventitious metal should be verified by the actual analysis of the metal content of the column eluate by atomic absorption spectroscopy.

If glass columns are to be used for chromatography, they should be acid cleaned as other glassware. Columns made of plastic and Teflon tubings should be soaked overnight in 20 m*M* EDTA, pH 7.5, and rinsed extensively with metal-free water. Column eluate should be collected in disposable plastic test tubes, which should be plugged immediately after the procedure is finished by tight-fitting metal-free stoppers (e.g., Bio-Rad supplies metal-free polypropylene Chemtubes with fitting Chemplugs) or covered by Parafilm to prevent airborne metal contamination. The tubes and stoppers should never be handled with uncovered hands, because they are a good source of metal contamination.

High-performance liquid chromatography (HPLC) techniques, including anion-exchange[15,16] and reversed-phase[17-19] chromatography, have been used to resolve isoforms of mammalian and plant metallothioneins.

[15] L.-Y. Lin, W. C. Lin, and P. C. Huang, *Biochim. Biophys. Acta* **1037,** 248 (1990).
[16] L. D. Lehman and C. D. Klaassen, *Anal. Biochem.* **153,** 305 (1986).
[17] S. Klauser, J. H. R. Kägi, and K. Y. Wilson, *Biochem. J.* **209,** 71 (1983).
[18] M. P. Richards and N. C. Steele, *J. Chromatogr.* **402,** 243 (1987).
[19] R. K. Mehra and D. R. Winge, *Arch. Biochem. Biophys.* **265,** 381 (1988).

To assure metal-free conditions in this type of isolation, the usual care of solutions, labware, and columns should be taken. DEAE-based HPLC columns should be washed with mobile phase B (e.g., 0.5 *M* Nacl, 10m*M* Tris-HCl, pH 8.6) to remove any remaining contaminants, followed by column reequilibration with mobile phase A (e.g., 10 m*M* Tris-HCl, pH 8.6). If necessary the column can be regenerated by several injections (1–2 ml) of 0.1 *M* NaOH or 30% (v/v) acetic acid.

Reversed-phase type HPLC media interact with metals only minimally[20] and do not require any special treatment before being used, except that chromatographic separation is carried out under metal-free conditions. It is recommended, however, that the eluate be checked each time for metal contamination by atomic absorption spectroscopy. Trifluoroacetic acid should not be used as an ion-pairing agent because of disruption of metal–thiolate complexes at the low pH.

[20] K. T. Suzuki, H. Sunaga, Y. Aoki, and M. Yamamura, *J. Chromatogr.* **281,** 159 (1983).

Section III

Quantification in Tissues and Body Fluids

[9] Determination of Metallothionein in Biological Materials

By KARL H. SUMMER and DOMINIK KLEIN

Current methods for the quantification of metallothionein (MT) in biological materials are based either on the direct determination of the protein moiety or the indirect determination via the metal and SH content of MT (Table I). Some forms of MT, such as polymeric forms, hardly can be quantitatively analyzed at present.

Several parameters must be taken into account when deciding which method is most appropriate for the quantification of MT in a particular sample. The most important are outlined in detail below, but basic procedural steps of sample preparation and handling, such as homogenization (Potter–Elvehjem, Ultra-Turrax, sonification), the buffer system (pH, ionic strength), and reductive conditions (2-mercaptoethanol, dithiothreitol, inert gas atmosphere), should also be considered.

Criteria for Selection of Metallothionein Determination Methods

Sensitivity

A major criterion for choosing a particular method is the amount of MT to be determined. As shown by comparative studies,[1-3] immunological methods[4-7] are the most sensitive. As little as 100 pg of MT can be determined using radioimmunological[8] and enzyme-linked immunosorbent (ELISA)[9,10] procedures. The limits of detection with the thiomolybdate,[11] silver staining,[12,13] and cadmium–Chelex assay[14] are 15, 50, and

[1] M. P. Waalkes, J. S. Garvey, and C. D. Klaassen, *Toxicol. Appl. Pharmacol.* **79,** 524 (1985).
[2] H. H. Dieter, L. Müller, J. Abel, and K. H. Summer, *Toxicol. Appl. Pharmacol.* **85,** 380 (1986).
[3] C. N. Nolan and Z. A. Shaikh, *Anal. Biochem.* **154,** 213 (1986).
[4] R. J. Vander Mallie and J. S. Garvey, *J. Biol. Chem.* **254,** 8416 (1979).
[5] F. O. Brady and R. L. Kafka, *Anal. Biochem.* **98,** 89 (1979).
[6] C. Tohyama and Z. A. Shaikh, *Fundam. Appl. Toxicol.* **1,** 1 (1981).
[7] R. K. Mehra and I. Bremner, *Biochem. J.* **213,** 459 (1983).
[8] J. S. Garvey, R. J. Vander Mallie, and C. C. Chang, this series, Vol. 84, p. 121.
[9] D. G. Thomas, H. J. Linton, and J. S. Garvey, *J. Immunol. Methods* **89,** 239 (1986).
[10] J. S. Garvey, D. G. Thomas, H., and J. Linton, *Experentia, Suppl.* **52,** 335 (1987).
[11] D. Klein, R. Bartsch, and K. H. Summer, *Anal. Biochem.* **189,** 35 (1991).
[12] F. Otsuka, S. Koizumi, M. Kimura, and M. Ohsawa, *Anal. Biochem.* **168,** 184 (1988).
[13] L.-Y. Lin, and C. C. McCormick, *Comp. Biochem. Physiol. C: Comp. Pharmacol. Toxicol.* **85C,** 75 (1986).

TABLE I
QUANTIFICATION OF METALLOTHIONEIN IN BIOLOGICAL MATERIALS

Required sample treatment	MT determination	
	Direct	Indirect
Homogenization	Immunological methods RIA ELISA	—
Preparation of cytosolic fraction followed by denaturing high-molecular-weight proteins (heat, TCA, organic solvents)[a]	Immunological methods RIA ELISA	Metal affinity methods Cd saturation Hg saturation Ag saturation SH determination Electrochemistry Photometry
Preparation of cytosolic fraction followed by chromatography (Sephadex G-75, HPLC)	Absorption	Metal determination
Preparation of cytosolic fraction followed by electrophoresis (SDS-PAGE)	Western blot Protein staining	—

[a] TCA, Trichloroacetic acid.

100 ng of MT, respectively. For other methods, e.g., high-performance liquid chromatography (HPLC)/atomic absorption spectroscopy (AAS),[15-17] electrochemical[18,19] and metal affinity assays [cadmium–heme,[2,20,21] mercury–trichloroacetic acid (TCA),[22,23] and silver saturation assay[24]], a minimum amount of about 1 μg MT is required. The latter methods are nevertheless suited to determine basal MT levels in most tissues, but at present quantification of MT in body fluids such as plasma and urine[7,25-27]

[14] R. Bartsch, D. Klein, and K. H. Summer, *Arch. Toxicol.* **64,** 177 (1990).
[15] L. D. Lehman and C. D. Klaassen, *Anal. Biochem.* **153,** 305 (1986).
[16] M. P. Richards, *J. Chromatogr.* **482,** 87 (1989).
[17] K. T. Suzuki, *Anal. Biochem.* **102,** 31 (1980).
[18] R. W. Olafson and R. G. Sim, *Anal. Biochem.* **100,** 343 (1979).
[19] R. W. Olafson, *Experentia, Suppl.* **52,** 329 (1987).
[20] S. Onosaka and M. G. Cherian, *Toxicol. Appl. Pharmacol.* **63,** 270 (1982).
[21] D. L. Eaton and B. F. Toal, *Toxicol. Appl. Pharmacol.* **66,** 134 (1982).
[22] J. R. Piotrowski, W. Bolanowska, and A. Saporta, *Acta Biochem. Pol.* **20,** 207 (1973).
[23] P. B. Lobel and J. F. Payne, *Comp. Biochem. Physiol. C: Comp. Pharmacol. Toxicol.* **86C,** 37 (1987).
[24] A. M. Scheuhammer and M. G. Cherian, *Toxicol. Appl. Pharmacol.* **82,** 417 (1986).
[25] G. F. Nordberg, J. S. Garvey, and C. C. Chang, *Environ. Res.* **28,** 179 (1982).

is possible only with immunological methods. The detection limits are only approximate because the linear range of a particular assay also depends on the type of sample. The linearity of the assay, therefore, must be determined separately for each method and kind of sample.

Nonmetallothionein Proteins

High amounts of non-MT proteins and a high non-MT protein-to-MT ratio may interfere with the quantification of MT. For example, when using immunological methods, antibodies may bind nonspecifically to non-MT proteins. With chromatographic methods, resolution of the separation is adversely affected by high non-MT protein-to-MT ratios. Nonspecific binding of cadmium, mercury, and silver to non-MT proteins is of minor importance for metal-affinity assays, since metal-binding high-molecular-weight proteins are denatured by heat, acid, or organic solvents. However, the results of these assays may be influenced by low-molecular-weight thiols [e.g., cysteine, glutathione (GSH)], which are stable toward heat, acid, and organic solvents, particularly if present at high concentrations.[14,21] SH-containing compounds other than MT additionally can cause adverse effects on MT quantification via its SH groups.

Metallothionein Isoforms and Metals Bound to Metallothionein

The presence of MT isoforms in biological material may affect the results of immunological methods due to the different antigenicity of the isoforms against a particular antibody. This is of special importance in induction experiments leading to different patterns of MT isoforms.[28,29] Thus when assessing total amounts of MT it is necessary to ensure that the antibodies raised against MT do not differ in their specificities toward MT isoforms or, although normally not practicable, the standard MT used must represent the isoform composition of MT in the sample. Clearly, the use of standard MT from other species may also lead to erroneous results.

The metals bound to MT should be considered when selecting a method for quantification, since the metal composition of MT influences its heat stability, sensitivity to oxidation, and the pH-dependent dissocia-

[26] C. C. Chang, R. J. Vander Mallie, and J. S. Garvey, *Toxicol. Appl. Pharmacol.* **55,** 490 (1980).

[27] C. Tohyama, Z. A. Shaikh, K. Nogawa, E. Kobayashi, and R. Honda, *Toxicology* **20,** 289 (1981).

[28] K. T. Suzuki and M. Yamamura, *Biochem. Pharmacol.* **29,** 2407 (1980).

[29] L. D. Lehman-McKeeman, G. K. Andrews, and C. D. Klaassen, *Biochem. J.* **249,** 429 (1988).

tion of metals. For example, heating of the sample should be avoided when determining copper-containing MT. Immunological methods are less likely to be affected by oxidation of MT than the indirect methods (SH determination and metal affinity assays), provided that the epitopes remain unchanged. The metal composition of MT also affects its absorption spectrum,[30] which can have consequences for the quantification of MT by HPLC/photometric analysis.

Furthermore, due to the different affinities of the metals to the protein, the use of metal saturation and SH determination methods is limited. Copper-containing MT, for example, cannot be determined with the cadmium–heme assay.

In general, indirect methods cause only minor problems when measuring MT with different isoforms, whereas the direct immunological methods cause only minor problems when measuring MT with different metal composition. Both types of analysis, however, can result in a loss of information, namely, the isoform pattern (indirect methods) or the metal composition of MT (direct methods). Clearly there is no universal method for the quantification of MT, so that the parallel use of different methods is considered to be most appropriate.

Comment

When several methods are applicable, one should also bear in mind that the various methods require different laboratory skills and procedures. In general, immunological methods are more time consuming and difficult to perform than metal affinity assays. Comparative interpretation of results obtained from different methods seems to be the optimal approach for deciding which of the MT assays is most suitable for a particular sample.

[30] M. Vasak and J. H. R. Kägi, *Met. Ions Biol. Syst.* **15,** 213 (1983).

[10] Assay of Extracellular Metallothionein

By IAN BREMNER and RAJESH K. MEHRA

Introduction

Although metallothionein (MT) is rightly regarded mainly as an intracellular protein located in the cytoplasm and nucleus of the cell, it also occurs in small amounts in extracellular fluids. This was first reported in studies of cadmium-exposed animals and it was suggested that transfer of cadmium from the liver to kidneys involves the secretion of cadmium-MT from the liver and its subsequent uptake by the kidneys.[1] Unfortunately, the concentrations of plasma MT, even in cadmium-exposed animals, are too low for its detection and quantitation by the chromatographic and metal-binding assays that have been developed for MT. It has therefore been essential to develop immunological techniques for the assay of MT in extracellular fluids like plasma, bile, and urine. Using a radioimmunoassay procedure, Garvey and colleagues[2] demonstrated that MT is present in the plasma of normal rats at concentrations of 1 – 2 ng/ml and that the urinary excretion of MT in human subjects provides a measure of industrial exposure to cadmium.[3]

Alternative radioimmunoassays have been developed using ^{109}Cd-labeled tracer but these have generally been less sensitive.[4] Enzyme-linked immunosorbent assays (ELISA) incorporating fluorimetric[5] or peroxidase[6] detection systems have been reported, but they are also less sensitive than the best of the radioimmunoassays and are not always suitable for the analysis of plasma. Monoclonal antibodies have been raised to MT, but their binding properties have not been suitable for the development of sensitive immunoassays. However, application of radioimmunoassays[2,7] for MT has shown that its secretion into plasma and its excretion into bile and urine are affected by many factors, including age, species, nutritional status, liver function, and exposure to metals.[8,9] Degradation products of MT have also been tentatively identified in bile and urine.[10,11] Such studies have provided information on the physiologic roles of MT and have led to the suggestion that assay of MT in urine, blood plasma, or blood cells can be used in the assessment of nutritional status and particularly in the diagnosis of zinc deficiency.[8]

These studies have been carried out using a specific and sensitive radioimmunoassay for rat MT-1 developed in these laboratories.[7]

[1] G. F. Nordberg, *Environ. Physiol. Biochem.* **2,** 7 (1972).
[2] J. S. Garvey, R. J. Vander Maillie, and C. C. Chang, this series, Vol. 84, p. 121.
[3] G. F. Nordberg, J. S. Garvey, and C. C. Chang, *Environ. Res.* **28,** 179 (1982).
[4] C. Tohyama and Z. A. Shaikh, *Fundam. Appl. Toxicol.* **1,** 1 (1981).
[5] D. G. Thomas, H. J. Linton, and J. S. Garvey, *J. Immunol. Methods* **89,** 239 (1986).
[6] A. Grider, K.-J. Kao, P. A. Klein, and R. J. Cousins, *J. Lab. Clin. Med.* **113,** 221 (1989).
[7] R. K. Mehra and I. Bremner, *Biochem. J.* **213,** 459 (1983).
[8] I. Bremner, R. K. Mehra, and M. Sato, *Experientia, Suppl.* **52,** 507 (1987).
[9] J. Hidalgo, M. Giralt, J. S. Garvey, and A. Armario, *Am. J. Physiol.* **254,** E71 (1988).
[10] M. Sato and I. Bremner, *Biochem. J.* **223,** 475 (1984).
[11] M. Sato, Y. Nagai, and I. Bremner, *Toxicology* **56,** 23 (1989).

Purification of Antigen

The two main isoproteins of cadmium-MT, MT-1 and MT-2, are isolated from the livers of rats that have been injected intraperitoneally with 0.5 mg cadmium (as $CdSO_4$)/kg body weight on day 1 followed by five injections of 2 mg Cd/kg body weight on alternative days. The rats are killed 24 hr after the last injection and their livers removed. The MT-1 and MT-2 are purified by a combination of gel filtration on Sephadex G-75 and anion-exchange chromatography on DEAE-Sephadex A-25, as described elsewhere in this volume (see [6]). However, the use of a heat-coagulation step to separate MT from the bulk of other cytosolic proteins, as used by some investigators,[2] is avoided and all procedures are carried out at $<4°$. The purity of the final products is confirmed by slab gel electrophoresis in 7.5% (w/v) polyacrylamide with Tris/glycine buffer, pH 8.9, and determination of their amino acid composition.

Zinc-containing MT, for use as standards in the immunoassay and as the source of iodinated MT, is purified in the same way from rats that are injected intraperitoneally on three successive days with 10 mg zinc (as $ZnSO_4$)/kg body weight. The rats are killed 24 hr after the last injection. Two isoforms of MT, MT-1 and MT-2, are obtained.

Preparation of Antibodies

Because of the high degree of structural homology between MTs from different sources and its low molecular weight, which leads to its rapid clearance from the circulation, MT is not a good immunogen. In order to raise its immunogenicity, it is necessary to convert it into a suitable polymeric form before injection into animals. This has been done by formation of self-polymers by reaction with glutaraldehyde[2] or by polymerization of MT and bovine serum albumin with glutaraldehyde.[4] However, a superior antigenic response is obtained when MT is conjugated with rabbit immunoglobulin G.

Rat liver Cd-MT-1 (6.0 mg) and rabbit immunoglobulin G (12.0 mg) are dissolved in 1.0 ml 0.05 *M* phosphate buffer, pH 7.5, whereupon 10 μl of 9.5% (w/v) glutaraldehyde is added. After 2 hr at room temperature, the mixture is diluted with 1.4 ml 0.05 *M* phosphate buffer, pH 7.5. When an aliquot of the mixture is separated on Sephadex G-75, about 35% of the cadmium and zinc elute at the void volume of the column, suggesting that an equivalent amount of protein has formed polymeric species. The use of the above quantities of reagents and of phosphate buffer, pH 7.5, results in formation of soluble polymers. However, insoluble polymers are obtained at pH 6.2 and with increased amounts of glutaraldehyde.

Although rabbits are normally used for the raising of antibodies, in our experience a better response is obtained using sheep. No systematic attempt has been made to establish the optimum immunization procedure, but the following procedure consistently produces a high titer of antibodies.

A homogeneous emulsion is made of equal volumes of the MT-1/immunoglobulin G conjugate solution (7.5 mg protein/ml) and Freund's complete adjuvant by repeatedly forcing the mixture from a syringe through a fine-bore needle. Sheep are immunized by injection of this emulsion (4 ml) at several intramuscular and subcutaneous/intradermal sites. Booster injections are given after 6, 16, 24, and 27 weeks using intramuscular injection of the emulsions prepared with the conjugate and a 1 : 1 mixture of Freund's complete and incomplete adjuvant.

Blood is collected from the jugular vein 1–3 weeks after each immunization. It is allowed to clot at room temperature, whereupon the antiserum is separated by centrifugation at 1500 *g* for 30 min at 4°. The antiserum is heated at 56° for 30 min to deactivate the complement and sodium azide is added (1 g/liter) as a preservative. The antisera are stored at −20°.

Detection of Antibodies

Although the presence of antibodies can usually be detected by double-diffusion techniques, this is not particularly successful and antisera are routinely screened instead by measuring the specific binding of ^{125}I-labeled zinc-MT-1 to the antibody. Iodination of proteins is often carried out using the chloramine-T reaction, which depends on iodination of tyrosine residues. However, as MT is devoid of tyrosine residues, an alternative reagent must be used, such as the Bolton–Hunter reagent [*N*-succinimidyl-3-(4-hydroxy-5-[^{125}I]iodophenyl) propionate; NSHPP], which reacts with free NH_2 groups.[12] An alternative option is to react MT and nonradioactive NSHPP and subject the product to iodination with chloramine-T. However, in our experience the most stable tracer is obtained by reaction of zinc-MT-1 with iodinated NSHPP.

^{125}I-Labeled NSHPP is obtained from Amersham International (Bucks, UK) in the form of a solution in dry benzene (500 μCi; 2000 Ci/mmol). It is important that this reagent be stored at +1° and not allowed to freeze. The benzene is carefully evaporated in a stream of dry N_2, ensuring that the residue is concentrated in a small area in the bottom of the conical reagent tube, which is then placed in an ice bath. A solution of zinc-MT-1 (5 μg) in 10 μl 0.1 *M* borate buffer, pH 8.5, is then added to the ^{125}I-labeled

[12] A. E. Bolton and W. M. Hunter, *Biochem. J.* **133,** 529 (1973).

NSHPP and the mixture vortexed. The reaction is allowed to proceed at 0° for 60 min, whereupon it is stopped by the addition of 0.5 ml 0.2 *M* glycine/0.1 *M* borate buffer, pH 8.5. After 5 min, the mixture is diluted with 1.0 ml 0.05 *M* Tris-HCl (pH 8.0) containing 1 g gelatin/liter and 1 g sodium azide/liter (gelatin buffer) and then fractionated by gel filtration on Sephadex G-25 (17 × 65 mm), using gelatin buffer as eluant and collecting 1-ml fractions. The ^{125}I-labeled MT-1 is eluted in a sharp peak at the void volume of the column, whereupon the eluant is changed to 2% (v/v) normal sheep serum. The tube containing the maximum amount of ^{125}I-labeled MT-1 (as determined using a γ counter) is diluted with gelatin buffer to a final concentration of 40–80 ng ^{125}I-labeled MT-1/ml (2–5 × 10^6 cpm/ml). In most preparations at least 40% of the ^{125}I-labeled NHSPP is bound to the MT-1. Fractionation of the ^{125}I-labeled MT-1 on Sephadex G-75 confirms that the ^{125}I and zinc elute in the same fractions and at the elution volume expected for MT-1; there is therefore no aggregation of MT-1 after the iodination. Aliquots of diluted tracer are stored at −20° and are stable for up to 5 weeks. After that time, the degree of binding of the tracer to antibody declines to unacceptably low levels.

The ^{125}I-labeled MT-1 is used to screen the sheep sera for antibody content and also in the development of the radioimmunoassay. Doubling serial dilutions of antiserum (1 : 500 to 1 : 64,000) are made in gelatin buffer [0.05 *M* Tris-HCl (pH 8.0) containing 1 g gelatin/liter and 1 g sodium azide/liter] containing 0.2% heated normal sheep serum. The diluted sera (400 μl) are reacted with 100 μl of ^{125}I-labeled MT-1 (10,000 cpm) diluted in gelatin buffer containing 0.05 *M* 2-mercaptoethanol. After 18 hr at 4°, antigen–antibody complexes are precipitated by the addition of 100 μl of an appropriate dilution of donkey antisheep IgG antiserum in gelatin buffer. (The amount of serum required to precipitate all the bound tracer is determined in preliminary experiments; typical dilutions of the donkey anti-sheep IgG antiserum are 1 : 7.) After 18 hr at 4°, the mixture is diluted with 2 ml 0.05 *M* Tris-HCl, pH 8.0, and the precipitate collected by centrifugation at 1500 *g* for 60 min at 4°. The tubes are drained and blotted on tissue paper and the radioactivity measured using a γ counter with an NaI crystal. Corrections for nonspecific binding are made using normal sheep serum in place of antiserum. The titer of antisera increases from about 4000 after the first booster injection to about 150,000 after the fourth booster, although the specific response varies between sheep. The percentage binding of tracer to the antiserum is nearly 80% at a serum dilution of 1 : 2000 and decreases to about 20% at a dilution of 1 : 128,000. The binding of ^{125}I-labeled zinc-MT-1 to the antibody is markedly reduced if 2-mercaptoethanol is omitted from the solution.

Immunoassay Procedure

The antiserum is used in a competitive binding assay to determine MT-1 concentrations. The antiserum is diluted 1:21,000 with gelatin buffer containing 1.66 μl normal sheep serum/ml; this gives an antiserum dilution of 1:35,000 in the final assay mixture and is sufficient to bind 50% of the tracer in the absence of unlabeled antigen. Aliquots of the diluted antiserum (300 μl) are added to 100 μl of zinc-MT-1 standards or of unknown samples in gelatin buffer. The standard solutions contain 100–25,000 pg of MT-1. After incubation at 4° for 40 hr, 100 μl of ^{125}I-labeled zinc-MT-1 in gelatin buffer containing 0.05 *M* 2-mercaptoethanol (1.5 ng/ml; 10,000 cpm) is added. After incubation for a further 7 hr 100 μl of diluted donkey anti-sheep IgG antiserum is added to precipitate the antibody–antigen complex as described above. The precipitate is collected after 16 hr by centrifugation and the tube decanted and counted in a γ counter. Corrections for nonspecific binding are made by using normal sheep serum in place of antiserum in tubes with no added antigen. The data are expressed as $[(b/b_0) \times 100]$ versus log concentration of MT-1, or as logit y versus log concentration: logit $y = \log[(100\ b)/(b_0 - b)]$, where b_0

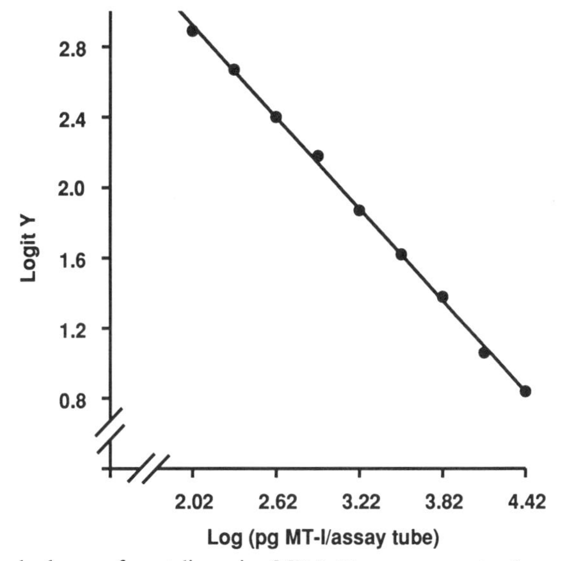

FIG. 1. Standard curve for rat liver zinc-MT-1. Known amounts of pure rat liver zinc-MT-1 (92–12,000 pg) were incubated with serum containing anti-MT-1 for 30 hr at 4°. Approximately 150 pg of ^{125}I-labeled zinc-MT-1 (10,000 cpm) was then added and the incubation was continued for a further 16 hr at 4°.

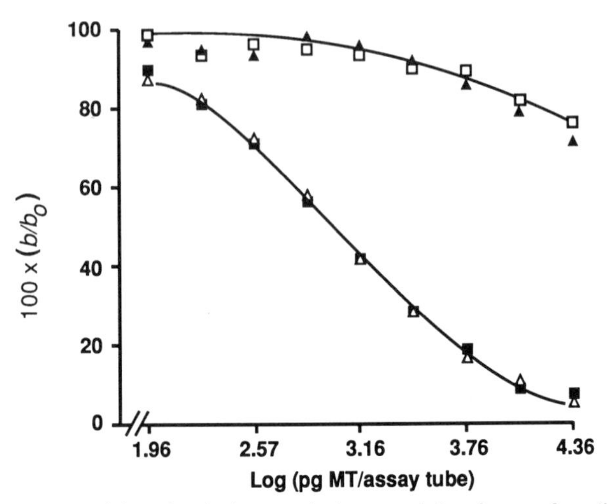

FIG. 2. Cross-reactivity of cadmium- and zinc-containing forms of rat liver MT-1 and MT-2. Various amounts (100–25,000 pg) of rat liver zinc-MT-1 (■), cadmium-MT-1 (△), zinc-MT-2 (□), and cadmium-MT-2 (▲) were incorporated in the inhibition assays with ^{125}I-labeled zinc-MT-1 as tracer and anti-cadmium-MT-1 antiserum.

is radioactivity precipitated in the absence of competing unlabeled antigen and b is radioactivity precipitated in the presence of standard or samples. Both b and b_0 are corrected for nonspecific binding.

A linear plot of logit y against log concentration is obtained over the range 1–250 ng MT-1/ml (Fig. 1). However, the sensitivity of the assay decreases to 2 ng/ml as the tracer ages, in line with the decrease in binding of the tracer to the antibody. When a standard sample of zinc-MT-1 (10 ng/ml) is assayed repeatedly, the intraassay variability is about 4%. The interassay variability is 15%. In all assays, standards and samples are analyzed in triplicate.

Before any assay for MT can be applied in the analysis of unknown samples, it is necessary to establish how effectively all isoforms and metalloforms of the protein cross-react. This is done by using purified MT-1 and MT-2 from the livers of cadmium-, zinc-, and copper-injected rats in competitive binding assays using ^{125}I-labeled zinc-MT-I as tracer (Fig. 2). As has been found in most immunoassay systems, the three metalloforms of MT-1 produce identical inhibition curves. However, the metalloforms of MT-2 are relatively ineffective at inhibiting the binding of the ^{125}I-labeled MT-1 tracer. For example, 200 ng of cadmium-MT-2 is required to produce the same degree of inhibition (50%) as 1.2 ng of any of the metalloforms of MT-1. The assay system is therefore specific for MT-1 but is not affected by the metal bound to the protein. When other species of

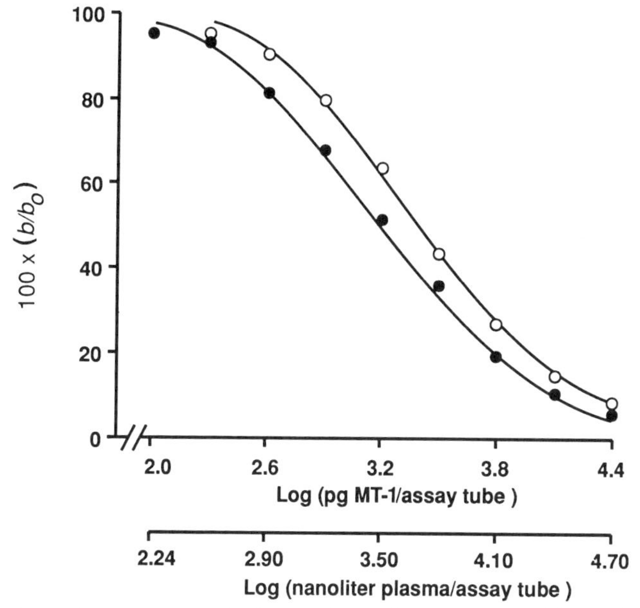

FIG. 3. Comparison of the standard curve for rat liver zinc-MT-1 (●) and the inhibition curve produced by various amounts (0.4–50 μl) of the plasma (○) that was obtained from 1-day-old female rats.

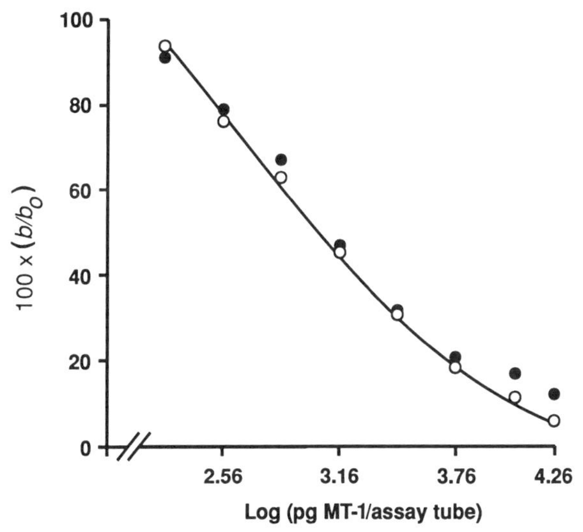

FIG. 4. Influence of rat plasma (●) on the standard curve for MT-1 (○).

MT are checked, only mouse and guinea pig liver MT-1 show complete cross-reaction. There is only poor cross-reaction with human, sheep, cattle, and pig MT-1.

These results are unusual, as similar assay systems developed in other laboratories have generally shown complete cross-reactivity between MT-1 and MT-2 from most mammalian species.[2,4] In these assays, the total amount of MT in the sample is measured whereas only MT-1 is measured in the method described here. The antibody does not show the same specificity when used in an enzyme-linked immunosorbent assay (ELISA)[13] or for immunocytochemical studies.[14] The reason for the unusual response is not known but it might reflect the differences in the immunization procedure.

Assay of Metallothionein in Extracellular Fluids

Before the radioimmunoassay can be used to analyze biological samples, it has to be shown that there are no matrix effects that can lead to erroneous results. If the basal levels of MT-1 in the sample are sufficiently high, a series of doubling dilutions can be carried out and the effects of the diluted samples on the binding of ^{125}I-labeled zinc-MT-1 established. The shape of the inhibition curve (b/b_0 versus log concentration) should be identical to that obtained with zinc-MT-1 standards. Alternatively a range of standard solutions of zinc-MT-1 should be prepared in the extracellular fluid if it is of low basal MT-1 content. The shape of the standard curve should again be identical to that obtained for standards in buffer solutions. Finally, the recoveries of internal standards of zinc-MT-1 added to the relevant samples should be determined. With the assay reported here, recoveries of MT-1 added to plasma are generally 90–110%. The inhibition curve obtained from doubling dilutions of plasma from neonatal rats is parallel to that of authentic MT-1 (Fig. 3). Standard curves for MT-1 prepared in adult rat plasma and in buffer are identical (Fig. 4). There are no matrix effects that interfere with the analysis of plasma. Depending on MT-1 concentrations, up to 100 μl of plasma can be used in the assay. Concentrations of MT-1 are only about 2 ng/ml in adult rats, but are 1000-fold greater in newborn pups.[15] They are also increased in situations where liver MT-1 concentrations are elevated, such as after metal loading

[13] A. Ghaffar, P. J. Aggett, and I. Bremner, *in* "Nutrient Availability: Chemical and Biological Aspects" (D. A. T. Southgate, I. T. Johnson, and G. R. Fenwick, eds.), p. 74. Royal Society of Chemistry, London, 1989.

[14] L. M. Williams, H. Cunningham, A. Ghaffar, G. I. Riddoch, and I. Bremner, *Toxicology* **55,** 307 (1989).

[15] J. N. Morrison and I. Bremner, *J. Nutr.* **117,** 1588 (1987).

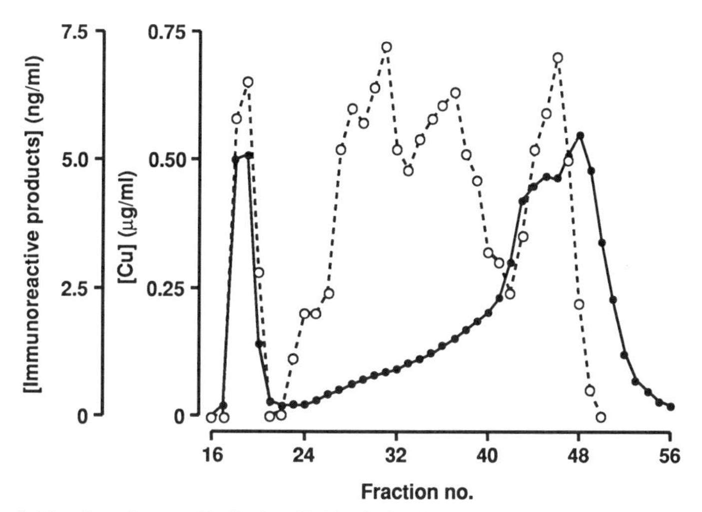

FIG. 5. Fractionation on Sephadex G-50 of bile from copper-loaded rats. The elution profiles for copper (●) in bile and for material that cross-reacts in the MT-1 immunoassay (○) are shown.

or injection with endotoxin. Indeed there is often but not invariably a close correlation between liver and plasma MT-1 concentrations.[9,16] Secretion of MT-1 into plasma is also enhanced if there is any liver damage. It appears that the liver is one of the principal sources of plasma MT-1.

Concentrations of MT-1 in rat urine and bile are generally higher than those in plasma and it is therefore possible to reduce the sample volume to only 10 or 20 μl. When this is done, there is again no evidence of any matrix effects and recoveries of internal MT-1 standards are 90–110%. Typical concentrations of MT-1 in the bile and urine of normal rats are 20 and 80 ng/ml, respectively, but these are also affected by factors such as metal loading or injection of endotoxin.[8] Liver and bile MT-1 concentrations are closely related.

Fractionation of rat urine and bile on Sephadex G-50 reveals the presence of several components that react with the antibody to MT-1 (Fig. 5).[10,11] One of these components appears to be aggregated MT-1, because it elutes at the void volume of the column but dissociates on treatment with 2-mercaptoethanol. Other components are of lower apparent molecular weight than MT-1, indicating that they might be degradation products. Indeed, one of them has the same elution volume as the α and β fragments obtained by partial hydrolysis of MT-1.

[16] M. Sato, R. K. Mehra, and I. Bremner, *J. Nutr.* **114,** 1683 (1984).

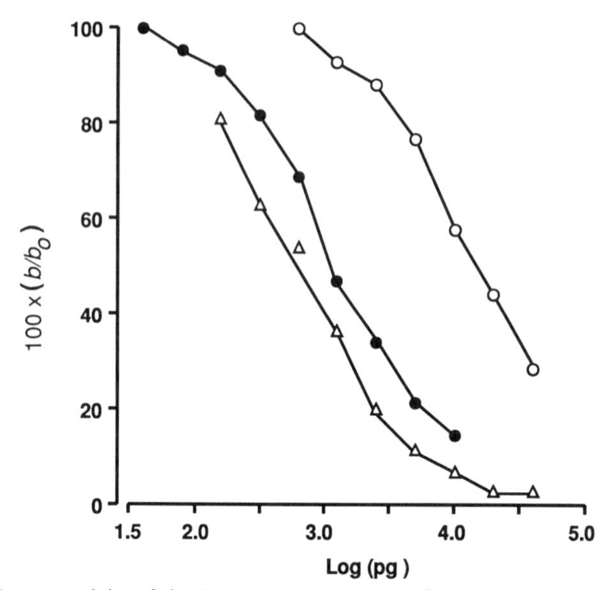

FIG. 6. Cross-reactivity of zinc-MT-1 (●) and the Cu_6 β fragment (△) and Cd_4 α fragment (○) of MT-1 in the immunoassay. b and b_0 refer to the ^{125}I-labeled MT-1 precipitated in the presence and in the absence of competing immunoreactive material. Both are corrected for nonspecific binding.

Winge and Garvey[17] showed that the antigenic determinants of MT are located mainly in two immunologically dominant regions in the N-terminal part of the polypeptide chain (corresponding to the β fragment). When the cross-reactivity of the Cd_4 α and Cu_6 β fragments were determined in the assay described here, only the β fragment cross-reacted completely and yielded an inhibition curve parallel to that obtained with authentic MT-1 (Fig. 6). The amounts of Cu_6 β fragment, zinc-MT-1, and Cd_4 α fragment to produce 50% inhibition of binding of ^{125}I-labeled zinc-MT-1 were 0.8, 1.1 and 18.0 ng respectively, equivalent to a ratio of 0.7 : 1.0 : 16. An important implication of this finding is that the amount of immunoreactive material in bile and, to a lesser extent, urine will be underestimated if the putative degradation products behave like the α fragment.

[17] D. R. Winge and J. S. Garvey, *Proc. Natl. Acad. Sci. U.S.A.* **80,** 2472 (1983).

[11] Quantification and Identification of Metallothioneins by Gel Electrophoresis and Silver Staining

By CHARLES C. MCCORMICK and LIH-YUAN LIN

Introduction

The traditional methods of metallothionein (MT) analysis in biological tissues are largely indirect. Most methods exploit the metal-binding properties of MT and/or its small molecular size. The conventional method of gel filtration combined with metal analysis was the first procedure for MT analysis in tissues.[1-2] It is based the characteristic elution volume of MT on Sephadex G-75 filtration medium and the quantification of metals in the MT elution volume. This method provides a measure of the amount of metal (as MT) per unit of tissue and consequently is an indirect analysis. Despite the potential nonspecific nature of this procedure, it was widely used and continues to be employed. The alternative to this method as originally described by Onosaka *et al.*[3] (cited by Eaton and Toal[4]) and later evaluated by Eaton and Toal[4] is the cadmium/hemoglobin affinity assay. Although this method was potentially more specific, it was nonetheless an indirect assay based on the ability of MT to bind metals, specifically cadmium. The isolation and quantification of MT by high-performance liquid chromatography (HPLC)[5] appeared to be more sensitive and much less time consuming. Again, it is a procedure based largely on metal quantification and thus is also indirect. In lieu of a radioimmunoassay, these procedures sufficed and were responsible for much of the research information obtained on the nature of MT accumulation in tissues.

Several years ago, our laboratory began work to develop a polyacrylamide gel electrophoresis (PAGE) procedure for MT quantification. The aim of this work was to develop an assay that was specific, direct, and quantitative. Our initial work focused on MT in the chick, because in this species (in contrast to mammals) MT exists as a single predominant form. Additional work, as presented in this chapter, suggests that MT isoforms such as those found in rat tissues can also be successfully separated and presumably quantitated by PAGE. In both instances, silver stain enhancement

[1] J. H. R. Kägi and M. Nordberg, *Experientia, Suppl.* **34,** 56 (1979).
[2] Z. A. Shaikh and O. J. Lucis, *Experientia* **27,** 1024 (1971).
[3] S. Onosaka, T. Keiichi, M. Doi, and O. Kunio, *Eisei Kagaku* **24,** 128 (1978).
[4] D. L. Eaton and B. R. Toal, *Toxicol. Appl. Pharmacol.* **66,** 134 (1982).
[5] K. T. Suzuki, *Anal. Biochem.* **102,** 31 (1980).

markedly improves sensitivity and perhaps more importantly provides a convenient means by which MT can be positively identified among heat-stable cytosolic proteins. The latter characteristic has been invaluable, in our laboratory, as a means of identifying MT in tissue extracts and various preparations.

Procedures

Tissue Preparation

Tissue cytosol is prepared from either a 1 : 2 or 1 : 3 (w : v) tissue : buffer homogenate. Tissue is weighed and combined with sucrose–Tris buffer (0.25 *M* sucrose, 10 m*M* Tris-HCl, pH 7.4, 5 m*M* dithiothreitol) and homogenized (Potter–Elvejhem). The homogenate is subjected to high-speed centrifugation (1 hr at 100,000 *g*) at 4°. The supernatant is collected and stored at −20°. We have noted that electrophoresis of freshly prepared cytosol (frozen at −20° for <1 week) appears to be more successful than that of cytosol stored for several weeks.

On the day of electrophoresis, 1 ml of cytosol is incubated in a heating block at 95° for 1 min. The sample is allowed to cool at room temperature for at least 10 min and then centrifuged at approximately 15,000 *g* for 2 min. The clear supernatant is collected and placed on ice until used. Similar heat treatment of pancreatic cytosol results in incomplete removal of heat-denatured proteins. Heat-treated pancreatic cytosol must be subjected to high-speed centrifugation (100,000 *g*) for 15 min at 4°.

Electrophoresis

In general, the electrophoretic system is similar to that described by Davis.[6] We employ a vertical slab gel apparatus (Hoefer, San Francisco, CA, model SE500) with a water-cooling chamber. The separating gel is 14 cm wide × approximately 9 cm high × 0.15 cm thick. These gel dimensions appear to result in minimal band distortion as compared to longer gels, i.e., >9 cm height. The separation gel (20 ml/slab) is formed as a linear gradient of polyacrylamide, 7.5 to 30% (w/v) total acrylamide (T) and 5% (w/v) bisacrylamide (Bis). The separation gel also contains 0.375 *M* Tris, 0.06 *M* HCl, 0.0575% (v/v) *N,N,N′,N′*-tetramethylethylenediamine (TEMED), pH 8.6, and 0.035% (w/v) ammonium persulfate. All ingredients except ammonium persulfate are added to a gradient former (Haake Buchler, Lenexa, KA) that has been prechilled at 4° for 1 hr. While

[6] B. J. Davis, *Ann. N.Y. Acad. Sci.* **121,** 404 (1964).

mixing, the ammonium persulfate is added and the casting begun. The flow of acrylamide from the gradient former is accomplished by gravity at approximately 20 cm above the electrophoretic unit. Care is taken to assure that mixing is accomplished without the formation of air bubbles. After casting is complete, a 3- to 4-mm layer of running buffer (0.005 *M* Tris-HCl, pH 8.3, 0.0384 *M* glycine) is applied to the surface of the acrylamide and the gel is then allowed to polymerize (approximately 30–45 min). Following several rinses with running buffer, the stacking gel [3% T and 20% Bis, containing 0.06 *M* Tris, 0.032 *M* H_3PO_4, 0.575% TEMED, and 0.05% (w/v) riboflavin] is applied and combs (10 wells, each 8 mm wide × 2.8 cm long) inserted. The gel is then exposed to fluorescent light (accomplished by maintaining lights within 5 cm of glass plates for no more then 2.5 hr). The upper and lower buffer reservoirs are filled with running buffer and samples applied. The samples are mixed with 0.5 vol of 1 *M* sucrose containing bromophenol (0.005%, w/v). Electrophoresis is performed at constant current (20 mA/slab) with cooling for approximately 9 hr. The tracking dye reaches the end of the gel in approximately 4 hr.

Staining

Immediately following electrophoresis, gels are removed and placed directly in a staining suspension containing 0.25% (w/v) Coomassie Blue R250, 7% (v/v) acetic acid, and 25% (v/v) methanol. Gels are stained overnight and then destained for at least 24 hr in a circulating destaining apparatus (Bio-Rad, Richmond, CA) containing 45% methanol and 7% acetic acid. At this point, gels can be scanned, especially those containing relatively intense bands of MT. Gels can then be further stained (enhanced) with silver stain by a procedure essentially as described by Merril *et al.*[7] We routinely employ the concentrate solutions provided by Bio-Rad and thus follow the manufacturer's recommended procedure (with slight modification). Gels, previously stained with Coomassie Blue, are placed in 10% ethanol/5% acetic acid for 1 hr as specified by the manufacturer. Gels are then treated with oxidizer for 10 min, washed three times with distilled water, and then placed in 200 ml of silver reagent for 30 min. Finally, gels are washed and placed in developer solution for a total of 6 min, after which time we routinely observe maximum band intensity with little background. Development is stopped by placing gels in 5% acetic acid. Gels can then be stored indefinitely at 4° in plastic bags containing 5% acetic acid.

[7] C. R. Merril, D. Goldman, S. A. Sedman, and M. H. Ebert, *Science* **211,** 1437 (1981).

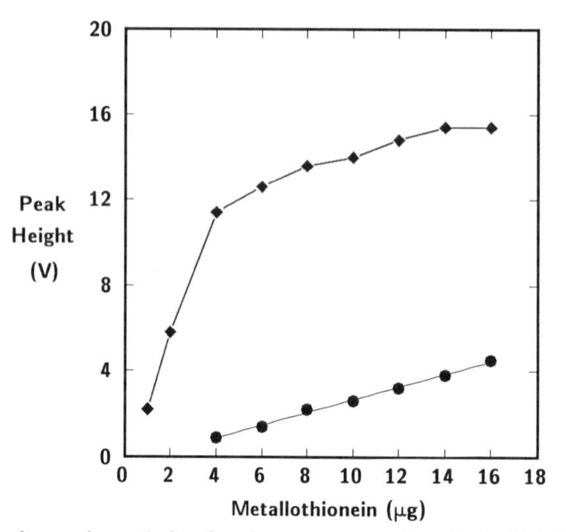

FIG. 1. Densitometric analysis of various amounts of purified chick MT isolated by nondenaturing PAGE. Various amounts of purified chick hepatic MT were added to heat-treated hepatic cytosol from noninduced chicks. The samples were then subjected to nondenaturing PAGE as described in Procedures. The electrophoretic gel was stained initially with Coomassie Blue (●), then subsequently enhanced with silver stain (◆). The lanes were scanned by densitometer and peak height recorded. (Adapted from Ref. 8.)

Quantification

Gels are scanned by densitometer (EC910; E-C Apparatus Corp., St. Petersburg, FL) at a rate of 2.8 cm/min using a 540-nm filter and a 0.3×3 mm slit. The measurement of light absorbance is recorded by computer (Chromlab; Data Translation, Marlboro, MA) as voltage changes during the scan. We have found that the measurement of peak height rather than peak area is more closely correlated with amount of MT. Thus, we routinely measure peak height.

Comments

We have previously reported that the above PAGE procedure is useful for the quantification of MT in liver and pancreatic cytosol of chicks.[8] Figure 1 illustrates typical results when purified chick MT is added to heat-treated hepatic cytosol from control chicks. The presence of cytosolic proteins appears to enhance the uniformity of MT bands represented by

[8] L.-Y. Lin and C. C. McCormick, *Comp. Biochem. Physiol. C: Comp. Pharmacol. Toxicol.* **85C,** 75 (1986).

small amounts (<2 μg) of MT. Diffusion of small amounts of MT is a particular problem, especially for quantification. Therefore, we recommend that in the construction of a standard curve, known amounts of MT be added to cytosol from noninduced animals. Peak height is then recorded as an increase above baseline. The use of cytosol from zinc-depleted animals, i.e., animals fed a zinc-deficient diet for 2 weeks, could be employed to obviate the contribution of endogenous MT. We have indicated that approximately 1 μg of MT can be detected and quantified in tissue cytosol.[8] As described, the procedure is thus capable of detecting approximately 20 μg MT/g tissue. Sato *et al.*[9] have reported approximately 30 μg MT-1/g of liver using a radioimmunoassay. Olafson,[10] employing a polarographic analysis, found 24 μg total MT/g of tissue from control mice. Since even mild induction results in three- to fourfold changes in tissue MT, the PAGE procedure would appear to be suitable for tissue MT analysis.

As mentioned previously, MT in tissues from the chick is predominantly one protein. In mammals, however, and particularly in the rat, MT exists as two distinct isoforms. We show here that nondenaturing PAGE is clearly amenable to the isolation and identification of both MTs in tissues from this species. Principally, we show that silver stain, under the conditions employed, markedly and perhaps specifically enhances MT. Figure 2A illustrates the results of electrophoresis and Coomassie Blue stain of purified, semipurified, and crude preparations of rat MTs. Purified MTs were obtained by DEAE ion-exchange chromatography and subsequently determined to be pure by amino acid compositional analysis. Semipurified samples represent the MT peaks from Sephadex G-75 chromatography of hepatic, pancreatic, and renal cytosol from zinc-treated rats (lanes 5, 6, and 7, respectively). The peaks were concentrated by ultrafiltration (YM2 membrane, Amicon, Danvers, MA) and equal portions of the concentrate applied to the gel. Finally, heat-treated hepatic cytosol from control and zinc-treated rats is also shown (lanes 8 and 9, respectively). The overall results illustrate two important points: (1) that the isoforms can be clearly separated in this electrophoretic system and (2) that silver stain markedly enhances MT. Regarding the latter, MT is substantially more enhanced by silver stain when compared to other low-molecular-weight, heat-stable proteins (compare Fig. 2A and B). This point may be particularly important in identifying the purity of an MT preparation. For example, the sample in lane 2 (Fig. 2A) represents the leading fractions of the MT-1 peak from DEAE chromatography. The major band appears to be a non-

[9] M. Sato, R. K. Mehra, and I. Bremner, *J. Nutr.* **114,** 1683 (1984).
[10] R. W. Olafson, *J. Biol. Chem.* **256,** 1263 (1981).

MT protein by virtue of the lack of silver stain enhancement (lane 2, Fig. 2B); a protein similar to that present in the original Sephadex G-75 fractions (compare to lanes 4 and 5). Others have also reported a significant contaminant in MT-1 preparations present in the leading edge of the MT-1 DEAE peak.[11] We suggest that this is a non-MT protein based on the its lack of silver stain enhancement. Many other experiments conducted in our laboratory indicate that MT is one of the first proteins to enhance with silver stain. Results of PAGE using hepatic cytosol from cadmium-injected rats (Fig. 3A and B) again indicates the apparent sensitivity of MT for silver stain (Fig. 3B). It is especially clear in Fig. 3 that other heat-stable proteins in hepatic cytosol do not enhance when treated with silver stain under the conditions described. These and other results suggest that silver stain may be employed to distinguish MT among heat-stable cytosolic proteins.

One of the essential features of the above procedure is the reduction of other proteins in a tissue cytosol. Heat precipitation was chosen largely due to the heat stability of MT. It is possible to vary the reduction in concen-

[11] R. J. Vander Mallie and J. S. Garvey, *Immunochemistry* **15,** 857 (1978).

FIG. 2. Gel electrophoresis of control and zinc-induced rat liver cytosol and corresponding concentrates of MT peaks from Sephadex G-75 chromatography of hepatic, pancreatic, and renal cytosol. Male rats were given (ip) either zinc (5 mg Zn/kg) or an equivalent volume of saline. Hepatic, pancreatic, and renal cytosols were prepared and chromatographed on Sephadex G-75 (2.5 × 50 cm). The fractions corresponding to the V_e for MT were collected and concentrated by ultrafiltration (YM2, Amicon). Electrophoresis of the concentrates, as well as of heat-treated (80° for 2 min) hepatic cytosol, was conducted as described in Procedures. Lanes: (1) trailing fractions of first DEAE peak (corresponding to MT-1); (2) leading fractions of first DEAE peak (MT-1); (3) second DEAE peak (MT-2); (4) ultrafiltration concentrate of Sephadex G-75 MT peak from hepatic cytosol of saline-treated rats; (5) ultrafiltration concentrate of Sephadex G-75 MT peak from hepatic cytosol of zinc-treated rats; (6) ultrafiltration concentrate of Sephadex G-75 MT peak from pancreatic cytosol of zinc-treated rats; (7) ultrafiltration concentrate of Sephadex G-75 MT peak from renal cytosol of zinc-treated rats; (8) heat-treated hepatic cytosol from saline-treated rats; and (9) heat-treated hepatic cytosol from zinc-treated rats. The gel was first stained with Coomassie Blue (A), then subsequently enhanced with silver stain (B) as described in Procedures.

FIG. 3. Gel electrophoresis of hepatic cytosol from rats given a single injection of cadmium. Rats were given (ip) a single cadmium injection (50 μmol Cd/kg) and killed 1, 2, or 3 days later. Hepatic cytosol was prepared and immediately subjected to nondenaturing PAGE as described. Sixty microliters of cytosol was applied to the gel and electrophoresis was conducted for 10 hr. The gel was first stained with Coomassie Blue (A), then subsequently enhanced with silver stain (B) as described in Procedures. Lanes: (1) 10 μg purified rat liver MT-1; (2) 20 μg purified rat liver MT-2; (3–5) 60 μl heat-treated hepatic cytosol from rats killed 1, 2, and 3 days, respectively, following cadmium administration.

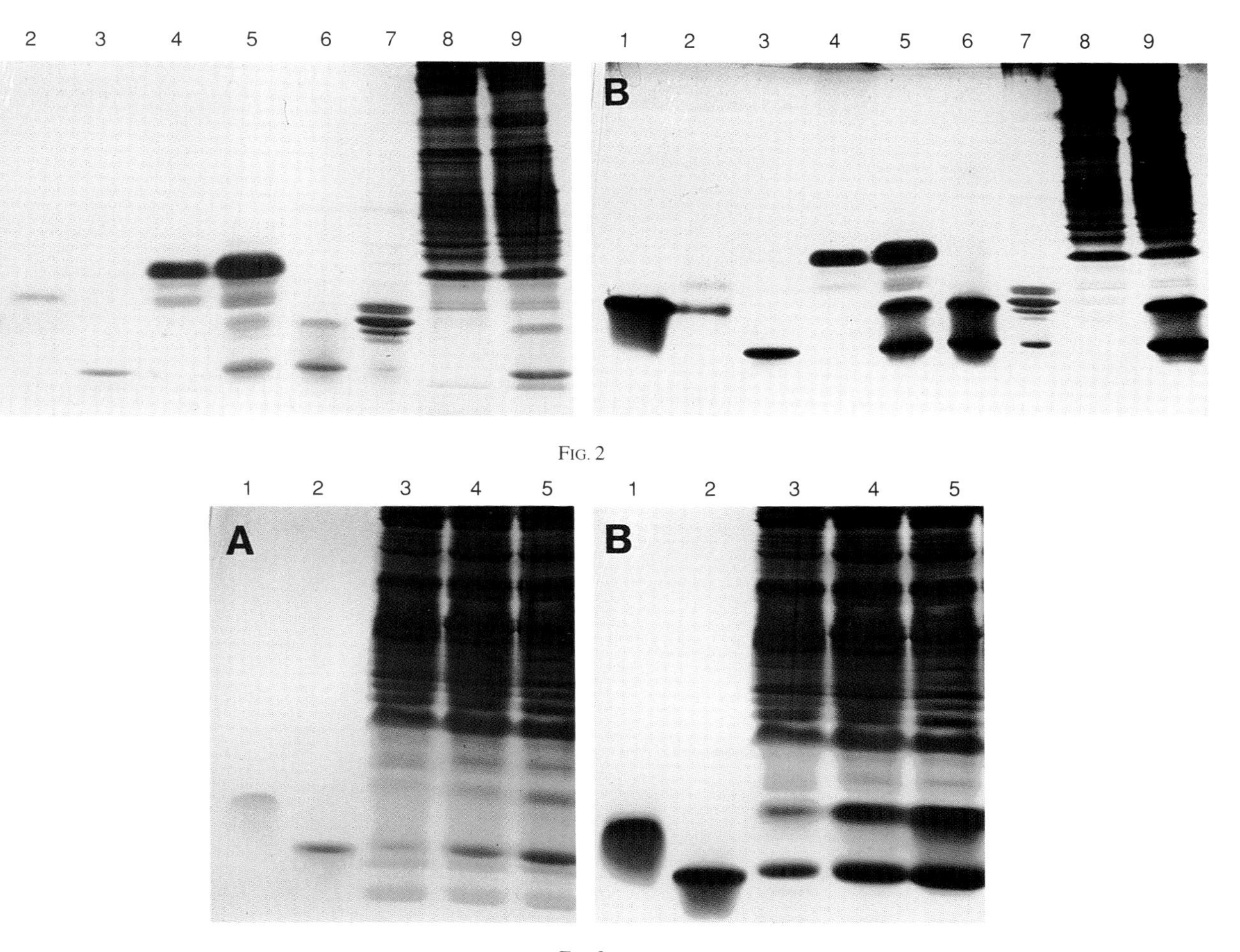

FIG. 2

FIG. 3

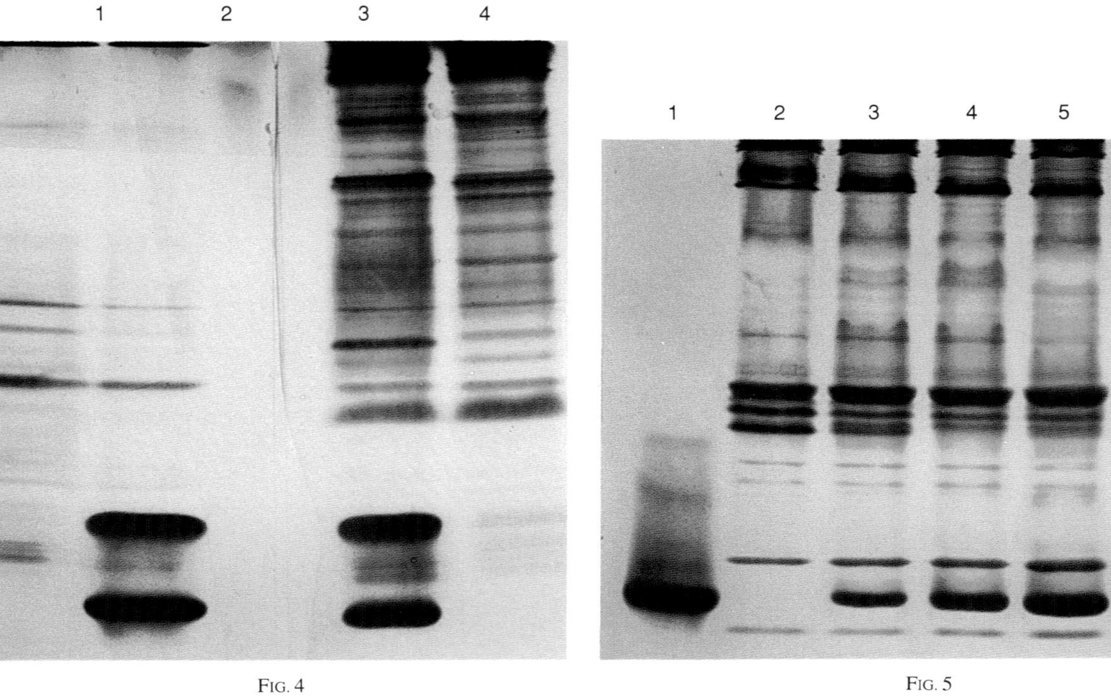

FIG. 4

FIG. 5

tration of other proteins in cytosol by heat treatment. Figure 4 shows the results of heat treatment at two different temperatures, 100 and 80°. The results indicate no loss of MT (as confirmed by densitometry and chromatographic analysis of heat-treated supernatant) at the higher temperatures, but as can be seen, a marked reduction of other cytosolic proteins occurs at the higher temperature. This may be an advantage when low concentrations of MT are anticipated and thus the need to load greater amounts of cytosol. It is possible to reduce or eliminate the loading dye solution and thus increase the amount of heat-treated supernatant applied.

We have also investigated the use of PAGE for MT analysis in tissues other than the liver, e.g., the pancreas. To evaluate MT in pancreatic extracts, a slightly different procedure must be employed. For example, low-speed centrifugation is insufficient to obtain a clear supernatant following heat treatment of pancreatic cytosol. We have employed high-speed centrifugation to accomplish this (100,000 *g* for 15 min). In addition, a different gradient of polyacrylamide (10–20% T) must be used to effectively separate MT from other interfering proteins. Figure 5 shows an electrophoretic gel of pancreatic heat-treated cytosol from chicks given 0, 1, 2, 4, or 8 mg Zn/kg body weight. Under these conditions, MT migrates between other low-molecular-weight proteins and thus can be isolated. Again, silver stain enhancement was effective in localizing (identifying) MT among the other heat-stable proteins.

Finally, we have attempted to use the above PAGE system to isolate and quantify MTs containing copper. At this point, we have not been successful with several MT preparations containing variable amounts of copper. The presence of copper, but not cadmium, appears to consistently affect the mobility of MT (either purified or in a tissue extract) such that the lane is generally a smear. Consequently carboxymethylation may be required to achieve acceptable results. However, under these conditions,

---

FIG. 4. Gel electrophoresis of hepatic cytosol subjected to two temperatures of heat treatment. Hepatic cytosol from control or cadmium-injected rats was heated to either 100 or 80° for 2 min. Following centrifugation, 40 μl of supernatant was subjected to nondenaturing PAGE as described. Lanes: (1) hepatic cytosol treated at 100° from control rats; (2) hepatic cytosol treated at 100° from cadmium-injected rats; (3) hepatic cytosol treated at 80° from cadmium-injected rats; and (4) hepatic cytosol treated at 80° from control rats.

FIG. 5. Gel electrophoresis of heat-treated pancreatic cytosol from chicks given various amounts of zinc. Chicks received either 0, 1, 2, 4, or 8 mg Zn/kg (ip) and were killed 24 hr later. Pancreas cytosol was prepared, heat treated, and subjected to electrophoresis as described in the sections Procedures and Comments. Lanes: (1) 20 μg purified chick MT; (2–6) 40 μl of heat-treated pancreatic cytosol from chicks given 0, 1, 2, 4, or 8 mg Zn/kg, respectively. (Taken with permission from Lin and McCormick.[8])

silver staining may be less effective to enhance MT. Although our laboratory has not investigated this, Otsuka *et al.*[12] have reported that carboxymethylated MTs can be visualized following sodium dodecyl sulfate-PAGE. Separation of the isoforms under these conditions, however, appears to be minimal.

In summary, a nondenaturing PAGE for isolating and identifying metallothioneins has been presented. The method employs gradient PAGE and extended electrophoresis of heat-treated tissue extracts (cytosol). The salient feature of the procedure is the use of Coomassie Blue stain as an initial treatment followed by silver stain enhancement. The latter process appears to specifically identify (enhance) MTs among other heat-stable proteins.

[12] F. Otsuka, S. Koizumi, M. Kimura, and M. Ohsawa, *Anal. Biochem.* **168,** 184 (1988).

[12] Quantification of Metallothionein by Silver Saturation

By A. M. SCHEUHAMMER and M. GEORGE CHERIAN

Introduction

The silver (Ag) saturation assay[1] for the measurement of metallothionein (MT) in tissues was developed as a modification of the cadmium (Cd) saturation assay.[2,3] Its main purpose is to improve analytical accuracy when measuring MT containing a high proportion of copper (Cu) or other metals having a higher binding affinity for MT than cadmium. The silver saturation assay is comparable to the cadmium saturation assay when measuring zinc (Zn)-MT or Cd,Zn-MT concentrations. Like other metal saturation assays, the silver saturation assay is most useful for measuring induced levels of MT in tissues. It is not adequately sensitive for measuring MT in biological fluids such as plasma and urine. This is a simple and rapid method, and a number of samples can be analyzed with good reproducibility in a research laboratory without specialized equipment.

[1] A. M. Scheuhammer and M. G. Cherian, *Toxicol. Appl. Pharmacol.* **82,** 417 (1986).
[2] S. Onosaka, K. Tanaka, M. Doi, and K. Okahara, *Eisei Kagaku* **24,** 128 (1978).
[3] S. Onosaka and M. G. Cherian, *Toxicol. Appl. Pharmacol.* **63,** 270 (1982).

Reagents and Equipment

Prepare the following solutions:

Sucrose (0.25 *M*)

Glycine (0.5 *M*): Adjust the pH to 8.5 with 4 *N* NaOH. (Other buffers that may cause the formation of insoluble silver salts, such as AgCl in Tris-HCl, or any other assay conditions that favor the formation of silver precipitates, must be avoided)

$AgNO_3$ (9.27 m*M*; 1000 μg Ag^+/ml): This stock solution should be stored in a dark bottle covered with aluminum foil to protect from direct light because of the photosensitivity of Ag^+. Working solutions (20 μg Ag^+/ml glycine buffer) can be made weekly and should also be kept in dark bottles

Red blood cell hemolysates according to the procedures of Onosaka and Cherian[3] can be prepared from any mammalian blood samples, including human. Briefly, 10 ml of blood is added to 20 ml of heparinized isotonic KCl solution (1.15%, w/v) and is centrifuged in a Sorvall RC-5 (Sorvall/Du Pont, Newtown, CT) at 500 *g* for 5 min. The pellet is resuspended in 30 ml of 1.15% KCl and centrifuged again at 500 *g* for 5 min. The washing and centrifugation steps are repeated once. The red blood cell pellet is lysed in 10–15 ml of 30 m*M* Tris base buffer, pH 8.0 at room temperature. After 10 min, the lysate is centrifuged at 8000 *g* for 10 min. The pellet is discarded and the supernatant (red blood cell hemolysate) is divided into aliquots (2 ml) and stored frozen at −20° or colder. The hemolysate should be discarded when oxidized as indicated by a dark brown color.

A Tissumizer (Tekmar, Cincinnati, OH), Polytron (Brinkmann, Westbury, NY), or equivalent is required for rapid homogenization of tissue samples.

Sample Preparation

Place 0.2–0.5 g fresh or frozen tissue into an appropriate-sized centrifuge tube and mince into small pieces. Add 4 vol of the sucrose solution, and homogenize at high speed for 60 sec. For human liver and other fibrous tissues a 2-min homogenization may be needed. Centrifuge the homogenate in a refrigerated centrifuge at 18,000 *g* for 20 min. Transfer the resulting supernatant fluid into a 5-ml plastic test tube, seal, and store frozen [preferably in an ultracold (−85°) freezer] until required for MT determinations. These samples can be stored for months without degradation of MT.

Silver Saturation Assay

Principle

The estimation of MT by the silver saturation method exploits the high Ag^+-binding capacity and heat stability of MT as compared to other cytosolic ligands. Addition of excess Ag^+ to tissue supernatant results in displacement of other metals from MT and also in Ag^+ binding to MT and other cytosolic proteins and ligands. The addition of hemoglobin to the sample removes silver from all ligands other than MT. During heating, hemoglobin-bound Ag^+ is precipitated. Repetition of hemoglobin addition and heating completes the removal of excess silver and all other silver-binding ligands, and thus the heated supernatant contains Ag^+ exclusively bound to MT. It is possible to calculate the concentration of MT from the amount of silver in the final supernatant and the known stoichiometry of Ag^+ in MT.

Procedure

1. Dispense an appropriate volume, usually 50–400 μl, of tissue supernatant or sucrose solution (blank) into a glass test tube. The appropriate volume depends on the MT concentration in the tissue. Duplicate aliquots of the sample are used for the assay. Most tissues in experimental animals have basal levels of MT < 50 μg/g (wet wt), whereas after induced synthesis of MT levels are typically several hundred micrograms per gram.
2. Add glycine buffer to make 0.8 ml.
3. Add 0.5 ml of the 20 μg Ag^+/ml solution, and incubate at room temperature for 5 min.
4. Add 100 μl hemolysate and mix.
5. Heat in a boiling water bath for 1.5 min.
6. Centrifuge for 5 min at 1200 *g* at room temperature.
7. Add another 100 μl hemolysate and repeat the heating/centrifugation steps. The final volume of the sample is 1.5 ml.
8. With a Pasteur pipette, transfer the resulting supernatant to plastic Eppendorf micro-test tubes (1.5 ml), and centrifuge for 5 min at 15,000 *g*.
9. Measure the silver concentration of the final supernatant by atomic absorption spectrophotometry (AAS), using an air–acetylene flame.
10. An AAS standard curve for silver is generated using known amounts of silver (0.25, 0.5, 1.0, and 2.0 ppm) in glycine buffer.

Alternatively, radioactive ^{110m}Ag-labeled silver salt can be used and silver concentrations in the final supernatant can be determined by radioactive γ isotope counting instead of atomic absorption spectrometry. The sensitivity of the method can be increased approximately 100-fold by using ^{110m}Ag with high specific activity. Thus this method is very similar to that using radiolabeled ^{109}Cd in the cadmium saturation assay.[4,5] Analysts should be aware that the radioactive ^{110m}Ag isotope is a strong γ-emitting radionuclide with a long half-life and it should be handled with special care. Cross-contamination from sample to sample during radioactivity measurement must be avoided.

Both Ag^+- and Cd^{2+}-binding assays provide comparable results[1,6] for MT estimation using samples from various sources. However, the silver saturation assay is superior to the cadmium-binding assay for measurement of MT containing a high content of copper or any other metal having a higher affinity than cadmium for MT at pH 8.6. In addition, the silver saturation assay is potentially more sensitive than the cadmium-binding assay because MT can bind with more silver atoms than cadmium per mole protein.

Calculation of Metallothionein Concentration

The amount of silver in the final supernatant fraction is proportional to the amount of MT present. Although purified MT can potentially bind up to 20 mol Ag/mol protein,[7] the oxidative formation of intra- or intermolecular disulfide bonds usually results in a decreased availability of reactive thiols. We have found that Zn-MT or Cd-MT purified from rat liver reaches saturation at 17–18 mol Ag/mol protein, and that the calculation of Zn-MT or Cd-MT concentration based on the binding of 17 mol Ag/mol protein gives comparable results to the cadmium saturation assay, which assumes 7 mol Cd/mol protein. Therefore, for the purposes of calculation, 1 μg Ag represents 3.55 μg MT.

The amount of MT in the sample is calculated with the following equation:

$$\mu\text{g MT/g tissue} = \frac{(C_{Ag} - C_{BKG}) \times 3.55 \times V_T \times \text{SDF}}{S_V}$$

where C_{Ag} is the concentration of silver in the final supernatant; C_{BKG}, background reading in the supernatant of the blank (without sample); V_T, total volume in the assay sample; SDF, sample dilution factor, which depends on the weight of the sample and the volume of the homogenizing

[4] D. L. Eaton and B. F. Toal, *Toxicol. Appl. Pharmacol.* **66,** 134 (1982).
[5] D. L. Eaton and M. G. Cherian, this volume [13].
[6] J. Chung, N. O. Nartey, and M. G. Cherian, *Arch. Environ. Health* **41,** 319 (1986).
[7] J. H. R. Kägi and B. L. Vallee, *J. Biol. Chem.* **236,** 2435 (1961).

media and also on the dilution of the sample before taking aliquots for assay; and S_V, sample volume used in the assay.

The method as described using flame AAS estimation of silver has a detection limit of ~5 μg MT/g wet tissue. However, the sensitivity of this method can be increased by using either radioactive ^{110m}Ag or by measuring silver in the final supernatant by graphite furnace atomic absorption spectrometry. Although originally standardized using Zn-MT and Cd-MT purified from rat liver,[1] the silver saturation assay has also been successfully applied to the measurement of hepatic and renal MT in other species, including humans,[6] mice,[8] quail,[9] doves,[10] and horses.[11]

Assay Characteristics and Limitations

The use of Ag^+ has intrinsic value for measurement of MT because it has a higher *in vitro* affinity for the protein thiols than any of the other metals that are commonly found in association with MT. Moreover, Ag^+ can be easily measured either by atomic absorption spectrophotometry (flame or flameless) or by using a radioactive ^{110m}Ag isotope. A number of experimental precautions should be undertaken in this assay because of the physicochemical properties of silver salts. All the experimental conditions that favor the formation of AgCl should be avoided. The silver solutions should be kept in dark bottles to avoid any direct contact with light because of the photosensitivity of silver salts.

During the assay procedure, samples should be mixed well after each reagent addition. In order to assure the saturation of MT with Ag^+, duplicate aliquots of each sample are used. In samples where MT synthesis has been induced with metals, dilution of samples is recommended. Also, human liver samples typically have high concentrations (200 μg/g) of MT[6] and they should be diluted before analysis. Certain commercially available glycine contains high concentrations of Ag^+ and should be analyzed before use. It is also important to prepare Ag^+ standards for atomic absorption spectrophotometry in the same glycine buffer used in the assay.

The incomplete removal of Ag^+-binding ligands other than MT in the samples by hemolysate/heating steps can result in false positive values. However, Sephadex G-75 column fractionation of final rat liver supernatant (after hemolysate addition and heating steps) showed the presence of Ag^+ bound almost exclusively to MT.[1] Although the cadmium-binding

[8] J. Koropatnick, M. Leibbrant, and M. G. Cherian, *Radiat. Res.* **119,** 356 (1989).

[9] A. M. Scheuhammer, *Toxicol. Appl. Pharmacol.* **95,** 153 (1989).

[10] A. M. Scheuhammer and D. M. Templeton, *Toxicology* **60,** 151 (1990).

[11] E. H. Jeffery, R. Noseworthy, and M. G. Cherian, *Comp. Biochem. Physiol. C: Comp. Pharmacol. Toxicol.* **93C,** 327 (1989).

assay gave high MT values for control rat testes, the silver-binding assay gave values that were similar to other tissues.

For accurate measurement of MT by the silver saturation assay, complete removal of hemoglobin by heating is also important. The presence of any red color in the supernatant after heating indicates incomplete removal and these samples should be heated again or discarded. Addition of cysteine or glutathione at 40 μM concentration did not interfere with the assay. Certain metal-chelating agents in high concentration can interfere with the assay. Instead of hemolysate, a Chelex resin (thiol resin) can also be used to remove the excess of Ag^+, but hemolysate addition gave more reliable results than the resin.[1]

All the metal-binding assays for MT are indirect methods based on certain properties of this protein. Since most of these properties of MT have been demonstrated experimentally, MT levels in tissues can be accurately measured by these methods. However, these assays are not sufficiently sensitive to measure MT levels in biological fluids such as plasma, urine, and bile.

[13] Determination of Metallothionein in Tissues by Cadmium–Hemoglobin Affinity Assay

By DAVID L. EATON and M. GEORGE CHERIAN

Introduction

The high affinity of metallothionein (MT) for mercury and cadmium, coupled with good stability at low pH and high temperatures, has served as the basis for the development of several simple, rapid, and inexpensive metal-binding assays to quantitate MT in biological samples. The first such method, described by Piotrowski *et al.*,[1] took advantage of the high affinity of MT for mercury at low pH, and relied on nonspecific binding and/or adsorption of excess mercury to acid-precipitable macromolecules. Onosaka *et al.*[2] first published a method for measuring MT in tissues that depended on the ability of heat-precipitable proteins in red blood cell hemolysates (e.g., hemoglobin) to bind and remove excess cadmium that was not bound to MT. The method was later modified to utilize crude bovine

[1] J. K. Piotrowski, B. Trojanowska, and A. Sapota, *Arch. Toxicol.* **32,** 351 (1974).
[2] S. Onosaka, K. Tanaka, M. Doi, and K. Okahara, *Eisei Kagaku* **24,** 128 (1978).

hemoglobin,[3] and it has been thoroughly evaluated for specificity[4,5] and also compared with other methods for quantitating metallothionein.[6-9]

Method of Assay

Reagents

Buffer: 10 m*M* Tris-HCl, pH 7.4

Cadmium solution: Carrier-free ^{109}Cd, 1 μCi/ml, and 2.0 μg/ml $CdCl_2$ in Tris buffer

Hemoglobin solution: 2% (w/v) crude bovine hemoglobin (type II, Sigma Chemical Co., St. Louis, MO) in Tris buffer

Preparation of Tissues for Assay

Tissues are removed, rinsed in ice-cold buffer, and then homogenized in 4 vol of buffer, with a glass homogenizer and a Teflon pestle. The homogenate is centrifuged at 10,000 *g* for 10 min, and the supernatant fraction is heated for 2 min in a boiling water bath. The heated samples are then centrifuged at 10,000 *g* for 2 min to remove precipitated proteins. As the assay requires only 200 μl of heat-denatured supernatant for a single analysis, this step can be conveniently performed in 1.5-ml microcentrifuge tubes. Heat-denatured supernatant fractions can be stored at $-70°$ for later analysis, if necessary.

Procedure

Two hundred microliters of ^{109}Cd solution (2 μg/ml, 1 μCi/ml) is mixed with 200 μl of sample (heat-denatured supernatant) and allowed to incubate at room temperature for 10 min. Then 100 μl of a 2% bovine hemoglobin solution is added to the tubes, and mixed and heated in a 100° boiling water bath for 2 min. (This assay can be conveniently performed in 1.5-ml microcentrifuge tubes. However, if the caps are closed tightly, pressure buildup during heating will cause the caps to pop off, possibly resulting in cross-sample contamination and distribution of aerosols of ^{109}Cd into room air. To avoid this, a small hole should be placed through

[3] D. L. Eaton and B. F. Toal, *Toxicol. Appl. Pharmacol.* **66,** 134 (1982).

[4] D. L. Eaton and B. F. Toal, *Sci. Total Environ.* **28,** 375 (1983).

[5] M. P. Waalkes, J. S. Garvey, and C. D. Klaassen, *Toxicol. Appl. Pharmacol.* **79,** 524 (1985).

[6] S. Onosaka and M. G. Cherian, *Toxicol. Appl. Pharmacol.* **63,** 270 (1982).

[7] C. V. Nolan and Z. A. Shaikh, *Anal. Biochem.* **154,** 213 (1986).

[8] L. Lehman and C. D. Klaassen, *Anal. Biochem.* **153,** 305 (1986).

[9] H. H. Dieter, L. Muller, J. Abel, and K. H. Summer, *Toxicol. Appl. Pharmacol.* **85,** 380 (1986).

each microfuge cap with a needle.) After 2 min in the boiling water bath, the tubes are placed on ice for several minutes, then centrifuged at 10,000 *g* for 2 min in a microfuge, and another 100 μl aliquot of 2% hemoglobin is added. The heating, cooling, and centrifugation are again repeated. A 500-μl aliquot of the supernatant fraction should be carefully removed from the microfuge tube and transferred to a clean γ-counting tube so as to avoid any carryover of precipitate. The amount of radioactivity in the supernatant fraction is then determined by direct γ counting.

Blank samples with buffer in place of the tissue sample and samples to determine total radioactivity (200 μl of buffer added in place of hemoglobin) should be run with each assay.

Calculations and Definition of Units

The amount of MT (cadmium-binding potential) in each sample is calculated with the following equation:

$$\text{nmol Cd bound/ml} = (\text{Cts}_\text{S} - \text{Cts}_\text{BKG}) \times (17.8/\text{Cts}_\text{T}) \times \text{DF}$$

where Cts_S is the counts per minute in the sample; Cts_BKG, counts per minute in the background sample; Cts_T, counts per minute in the "total counts" sample; DF, dilution factor.

The value of 17.8 is obtained by dividing the grams of cadmium in each sample (0.4 μg) by the atomic weight of cadmium (0.1124 μg/nmol) and the MT sample size (0.2 ml) to yield units of nanomoles of cadmium per milliliter of sample. This numerical value is identical to units of micrograms of MT if one assumes that 7 g-atoms of cadmium is bound per mole MT and that the molecular weight of Cd-MT is 7000.[6]

Assay Characteristics and Limitations

Sensitivity and Reproducibility

Under the standard assay conditions described above, the detection limit is approximately 0.8 μg MT/g of tissue. A final cadmium concentration of 1.0 μg/ml was selected because at this concentration of cadmium about 5% of the total cadmium was bound to MT in tissue homogenates from control animals.[3] As background levels (counts in sample with no MT present) are consistently less than 1.0% of the total cadmium, the 5% of cadmium bound in control animals provides a reliable estimate of basal levels of MT (1.0 to 2.0 μg MT/g tissue) yet allows for measurement of levels 10 to 15 times greater than basal levels without further sample dilution. When only very small amounts of tissue are available (i.e., fetal tissues), or when very low concentrations of MT are present, the concentration of cadmium in the assay procedure may be reduced to 0.1 μg Cd/ml, which results in a 10-fold increase in assay sensitivity, but some loss in reproducibility due to an increase in background counts relative to MT-related counts. Under the standard conditions of the assay, coefficients of variation of replicate analyses range from 1 to 4%, but increase up to 10% when the ratio of background to total counts becomes large.[3] The sensitivity of this method under standard conditions is adequate to accurately determine the levels of MT in tissues of control animals as well as those treated with metals.[10]

When measuring MT in tissues from cadmium-treated animals, tissue samples should be diluted sufficiently to reduce the error introduced by isotope dilution from the endogenous cadmium already complexed with the MT. The error introduced by isotope dilution from unlabeled cadmium present in the MT is minimal if samples are diluted such that less than 50% of added radioactivity (^{109}Cd) is bound to MT (e.g., remains in the supernatant fraction after hemoglobin treatment).[3] Where large amounts of MT are present, the concentration of exogenous cadmium added should be increased to ensure complete saturation of cadmium-binding sites. For example, Chellman *et al.*[11] found it necessary to increase the amount of exogenous cadmium added in the assay from 0.4 to 100 μg when measuring cadmium-binding proteins in the testes. This assay can also be performed with addition of nonradioactively labeled $CdCl_2$ and determination of cadmium concentration in the final supernatant by atomic absorption spectrometry.[2,6,10]

For tissues in which MT concentrations are very low (e.g., plasma), this assay is not adequately sensitive. Radioimmunoassay appears to be the only assay with adequate sensitivity to measure MT in plasma.[5-8]

Specificity

A potential pitfall of any metal-binding assay for MT is the nonspecific binding of metal to non-MT peptides. Analysis of Sephadex G-75 eluate fractions from a 100,000 *g* supernatant fraction of rat liver with the cadmium–hemoglobin affinity assay indicated that more than 99% of total binding of ^{109}Cd occurred in fractions from the MT peak.[3] Comparison of the cadmium-binding assay with quantitation of rat liver MT isoforms by high-performance liquid chromatography–atomic absorption spectrophotometry (HPLC-AAS) yielded nearly identical results.[8]

[10] S. Onosaka and M. G. Cherian, *Toxicology* **23,** 11 (1982).

[11] G. J. Chellman, Z. A. Shaikh, R. B. Baggs, and G. L. Diamond, *Toxicol. Appl. Pharmacol.* **79,** 511 (1985).

Low-molecular-weight thiols, such as endogenous cysteine and glutathione (GSH), or exogenously added thiols such as 2-mercaptoethanol, could bind to cadmium and result in an overestimation of MT. Indeed, final concentrations of cysteine or GSH as low 0.1 m*M* resulted in a two-fold increase in "background" counts,[3] and this could result in an overestimation of MT in samples where MT concentrations are very low. 2-Mercaptoethanol did not appear to have a significant effect on background counts at concentrations of 1 m*M* or less.

In addition to specificity of cadmium binding to MT, the assay is also dependent on complete removal of unbound ^{109}Cd from the supernatant fraction. The complexation of unbound ^{109}Cd with hemoglobin and subsequent heat denaturation appear to be quite effective under most circumstances. However, any factor that interferes with the heat denaturation and precipitation of hemoglobin will result in erroneous estimations of MT. The presence of relatively high concentrations of nonprotein thiols prevents the complete heat denaturation of hemoglobin. For example, a final concentration of GSH of 5 m*M* resulted in nearly complete inhibition of hemoglobin precipitation.[3] Some metals are also capable of preventing the complete heat denaturation of hemoglobin.[12] For example, final concentrations of tin and zinc above 5 and 15 μM, respectively, prevent the complete precipitation of hemoglobin on heating. Fortunately, because hemoglobin is an intense chromophore, it is quite easy to identify samples in which precipitation of hemoglobin is incomplete. The appearance of any red coloration in the supernatant fraction after heat denaturation and centrifugation is indicative of incomplete precipitation, and thus results should be considered suspect. Certain metal-chelating agents, such as ethylenediaminetetraacetic acid (EDTA), can also interfere in the cadmium-binding assay and give false positive results.[13]

Underestimation Due to Metal Competition or Incomplete Cadmium Saturation

Metallothionein complexed with copper, mercury, or silver may not be accurately quantitated by this method, as these metals appear to have a higher affinity for MT than cadmium under the conditions of the assay.[11] Thus, this method should not be used to quantitate MT in tissues obtained from animals with high exposure to these metals. The presence of exogenous metals at concentrations exceeding that of the cadmium used in the assay (e.g., $> 1\ \mu M$) will also result in substantial underestimation of MT

[12] D. L. Eaton, *Toxicol. Appl. Pharmacol.* **78,** 158 (1985).

[13] C. R. Miller, S. Y. Zhu, W. Victery, and R. A. Goyer, *Toxicol. Appl. Pharmacol.* **84,** 584 (1986).

by this method. The addition of 1 μM mercury, silver, or copper resulted in 36.5, 10.6, and 34.8% underestimation of MT, respectively, when the assay was conducted with Zn-MT under standard assay conditions.[11]

Removal of the "free" cadmium on hemoglobin addition and heat precipitation will result in dissociation of the Cd-MT complex to reestablish equilibrium. The net effect is an apparent "stripping" of cadmium from MT after each hemoglobin addition.[3] Consequently, excessive repetition of the hemoglobin addition and heat-precipitation step will result in slight underestimation of the amount of MT in the supernatant fraction. However, use of only a single hemoglobin addition results in a "background" count of nearly 1% of the total, which can be reduced substantially by a second hemoglobin treatment. Thus, we recommend the use of two additions of hemoglobin and heat precipitation in the assay as the optimum balance between a desirable low background and an undesirable removal of bound ^{109}Cd from MT.

[14] Immunohistochemical Localization of Metallothionein

By M. GEORGE CHERIAN and D. BANERJEE

The immunohistochemical localization of metallothionein (MT) has provided new information on its intracellular distribution in various tissues. Studies[1-3] have demonstrated that MT is present in both nucleus and cytoplasm of hepatocytes in livers from fetal and newborn human and rats, whereas it is mainly a cytoplasmic protein in adult liver. Metallothionein has been localized in various tissues of rats[4,5] and also of humans having diseases such as cancer.[6-8] The preparation of a specific antibody to MT is essential for the localization of MT by this technique. This is a simple morphological technique that is widely used for identification and localiza-

[1] D. Banerjee, S. Onasaka, and M. G. Cherian, *Toxicology* **24,** 95 (1982).
[2] M. Panemangalore, D. Banerjee, S. Onasaka, and M. G. Cherian, *Dev. Biol.* **97,** 95 (1983).
[3] N. O. Nartey, D. Banerjee, and M. G. Cherian, *Pathology* **19,** 233 (1987).
[4] K. G. Danielson, S. Ohi, and P. C. Huang, *Proc. Natl. Acad. Sci. U.S.A.* **79,** 2301 (1982).
[5] H. Nishimura, N. Nishimura, and C. Tohyama, *J. Histochem. Cytochem.* **37,** 715 (1989).
[6] N. O. Nartey, J. V. Frei, and M. G. Cherian, *Lab. Invest.* **57,** 397 (1987).
[7] N. O. Nartey, M. G. Cherian, and D. Banerjee, *Am. J. Pathol.* **129,** 177 (1987).
[8] T. E. Kontozoglou, D. Banerjee, and M. G. Cherian, *Virchows Arch. A: Pathol. Anat. Histopathol.* **415,** 545 (1989).

tion of various antigens in tissues. However, it is important to use appropriate controls to rule out any artifacts. The use of uncharacterized antibodies from different sources and the use of inappropriate controls have resulted in the false localization of MT in blood vessels in one report.[9]

Reagents

Primary antibody: A specific antibody to metallothionein (e.g., rabbit antibody to rat MT)
Secondary antibody: A biotin-conjugated antibody that can bind with the primary antibody (e.g., biotinylated goat antibody to rabbit IgG)
Peroxidase-conjugated avidin–biotin complex (ABC): Three of the four biotin-binding sites on the avidin molecule are bound to biotinylated peroxidase, leaving one site free to bind to the biotin on the secondary antibody. Most of the reagents except primary antibody can be obtained from Vector Laboratories, Inc. (Burlingame, CA).
Tris-buffered saline (TBS): Tris buffer (0.05 *M*) with sodium chloride (0.15 *M*) is adjusted to pH 7.6 with hydrochloric acid
3,3′-Diaminobenzidine tetrahydrochloride (DAB; Sigma Co., St. Louis, MO) in hydrogen peroxide. Dissolve 5 mg DAB in 10 ml TBS and add 33 μl of aqueous 3% (v/v) hydrogen peroxide immediately before use
Buffered formalin: Formalin (10%,v/v) is adjusted to pH 7.4 with phosphate buffer (0.1 *M*)

Tissue Preparation

Sections of tissues are immersed in 10% buffered formalin for at least 24 hr. The sections are paraffin embedded and 5-μm sections are cut. They are placed on slides and dried overnight at 37° on a slide warmer.

Principle

The primary antibody reacts with the MT in the cells and the peroxidase-conjugated avidin–biotin complex (ABC) reacts with the biotin on the secondary antibody. The peroxidase enzymes in each ABC complex react with hydrogen peroxide, resulting in the oxidation of the DAB to an insoluble brown precipitate. The reaction sites are visualized as brown staining of the tissue. This staining is proportional to the concentration of MT in the cell. The principle of this technique is shown in Fig. 1. The specificity of the staining depends on the reaction of the primary antibody with MT. However, a number of controls, as described in Control Experi-

[9] M. E. Elmes, J. P. Clarkson, and B. Jasani, *Experientia, Suppl.* **52,** 533 (1987).

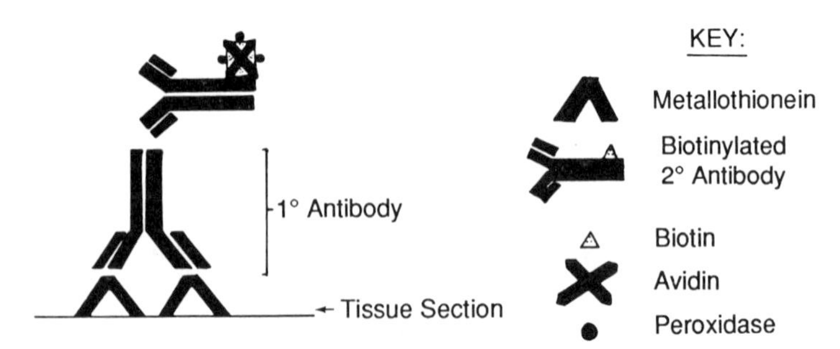

FIG. 1. Description of the principle involved in the immunohistochemical localization of metallothionein in tissues by using the avidin–biotin complex (ABC)/peroxidase method. The primary antibody (1° antibody) is rabbit anti-rat MT and the secondary antibody (2° antibody) is goat anti-rabbit IgG.

ments below, should be used along with the sample to determine the specificity of the staining.

Procedure

All incubations are performed with 200 μl of solution/slide in a humid tray chamber.

1. The sections are deparaffinized by washing the slides in xylene for 3 min with two changes of xylene. This is essential because residual paraffin will interfere with reactions.
2. The slides are rinsed three times in 100% ethanol for 2 min each and then dipped in absolute methanol.
3. The endogenous peroxidase activity in the tissue is destroyed by placing the slides in 3% hydrogen peroxide in 100% methanol for 30 min with stirring at room temperature.
4. The slides are washed twice in TBS for 10 min each at room temperature.
5. In order to block any nonspecific binding the tissue sections in the slide are incubated in 10% normal goat serum in TBS for 60 min at room temperature. The excess serum is drained from slides, one at a time, without allowing the tissue to dry.
6. The slides are incubated in the appropriate dilutions of rabbit anti-rat liver MT (primary antibody) or in the appropriate control solutions overnight at 4° in a humid chamber.
7. The next day, the slides are allowed to warm to room temperature for 60 min. Open the tray slightly to allow air to exchange. From this step

onward, the slides that are treated differently (i.e., in different solutions and/or different dilutions of the same solution) must be washed in separate containers or trays.

8. The slides are washed four times in TBS for 5 min each time at room temperature with stirring.

9. The sections are incubated in biotinylated goat anti-rabbit IgG (secondary antibody) in 10% normal goat serum, diluted at 1:200 with TBS, for 30 min at room temperature.

10. The slides are washed in TBS four times for 5 min each at room temperature.

11. The slides are then incubated in peroxidase-conjugated avidin–biotin complex diluted with TBS. The avidin and biotin are prepared as follows and incubated at room temperature for 30 min before being applied to the slides: Dilute avidin (A) to 1:100 with TBS and mix. Add biotinylated peroxidase enzyme solution (B) in equal volume to A and then mix well. Let stand for 30 min at room temperature. Apply it to the slides and incubate at room temperature for 30 min.

12. The slides are washed in TBS four times for 5 min at room temperature.

13. The staining solution is prepared by dissolving 3,3′-diaminobenzidine tetrahydrochloride (DAB) in TBS to give a 50% (w/v) solution. Immediately before staining, 3% H_2O_2 is added to the DAB solution (33 μl to 10 ml of DAB solution) and quickly mixed.

14. The slides are drained and incubated with the substrate solution (DAB/H_2O_2) for 5–10 min at room temperature. The endpoint is determined with the positive control slide examined under a microscope. After appropriate staining appears on these sections the reaction is stopped by pouring out the DAB/H_2O_2 solution and by washing the slides in running tap water for 10 min. If the reaction is allowed to proceed for more than 10 min, excessive background staining may occur.

15. The slides are counterstained for nuclei using Harris hematoxylin for 30 sec. The nuclei stain blue and the slides are washed in running tap water for 20 min.

16. The slides are dehydrated in graded ethanols and rinsed with 100% ethanol for 2 min. Slides are then cleared in two changes of xylene for 1 and 2 min, respectively, and then mounted in Permount. This procedure is similar to that developed for other antigens using the avidin–biotin peroxidase technique.[10]

For publication of black-and-white photographs, the slides should be photographed before counterstaining. Otherwise it will be difficult to distin-

[10] M. Hsu, L. Raine, and H. Fanger, *Am. J. Clin. Pathol.* **75,** 734 (1981).

guish the counterstain from nuclear staining for MT. After counterstaining, color photographs can be used for publication that will show staining for MT as brown and the counterstain as blue.

Control Experiments

A number of control experiments should be performed for the correct interpretation of MT staining by the avidin–biotin complex (ABC) method. These are especially important when working with a new batch of antibody or tissues that has not been previously stained for MT. The appearance of background staining in tissue sections makes it difficult for accurate interpretation of antigen-specific staining. Several methods have been developed to eliminate or minimize these false positive results.

Negative Controls

1. Hydrogen peroxide: As described in the procedure, diluted hydrogen peroxide (3%) is applied to tissue sections before incubation with the primary antibody to destroy the endogenous peroxidase activity. This step will reduce the background staining due to the presence of endogenous peroxidase activity in tissues.

2. Normal goat serum: Nonspecific interactions may occur between the antibody reagents used and the tissue proteins and receptors. Incubation of tissue sections in normal serum of the species in which the secondary antibody was made (in this procedure, goat) prior to incubation in the primary antibody solution will block the nonmetallothionein binding. In addition, this normal serum is often used as the diluent for the antibody reagents to ensure the binding sites other than MT in the primary antibody stay blocked throughout the procedure.

3. Normal rabbit serum: Nonimmune rabbit serum (NRS) is the serum of the species in which the primary antibody to MT was made. The dilution of NRS used as a control is established by serial dilution in separate experiments. Any staining due to nonspecific binding of reagents should be minimal and if present should be diffused. The dilution of NRS (usually 1:200 to 1:400) found to obtain these results is the minimal dilution at which one can use the primary antibody. If staining of NRS-treated slides is greater than minimal, the problem is often that the concentration of secondary antibody is too high and/or time of incubation in the secondary antibody is too long, or there are natural antibodies in the rabbit serum that cross-react with human antigens. Other ways to decrease the background staining are to increase the concentration of normal goat serum, incubate the sections in the primary antibody/NRS overnight at

4–6°, decrease the staining time in the substrate/chromogen solution, or obtain a more specific secondary antibody. A useful method to find all these parameters in a relatively short time is to set up a series of titrations of concentrations and/or incubation times of the various reagents until the combination that provides optimal results is obtained. This will provide the working dilutions for all the reagents in the procedure. The primary antibody can now be used at the same dilution as or higher than the NRS in the procedure. It is important always to run a tissue section incubated with normal rabbit serum along with the rabbit primary antibody to MT.

4. Anti-MT preabsorbed with MT in 10% NGS/TBS: If the primary rabbit MT antibody is preincubated with MT or its polymer, this preabsorbed MT antibody should not give any positive staining for MT when the tissue sections are incubated with it in the procedure. This is the most appropriate control to identify the specificity of the primary antibody and detect the nonspecific staining. However, it should be pointed out that the binding of MT to its antibody is rather weak and it may require overnight incubations to completely block all the MT-binding sites on the antibody.

Positive Control

In order to ensure that all the reagents used in the procedure work, it is important to run a positive control using tissue sections that had demonstrated positive rabbit anti-rat MT localization with the ABC/DAB method in a previous experiment. This is especially needed when a new batch of the primary antibody or a new tissue is used in the procedure. We use sections of kidney from cadmium-injected rats as a positive control. The concentration of MT in one-half of these kidneys is determined by cadmium or silver binding assay[11,12] and the other half is processed for immunohistochemistry.

Preparation of Antibody to Rat Liver Metallothionein

Metallothionein is isolated from livers of rats that were repeatedly injected subcutaneously with $CdCl_2$ (1 mg/kg) daily for 2 weeks. The steps involved in the purification were separation of cytosol, heating (80° for 2 min), fractionation on a Sephadex G-75 column and chromatography on a DEAE-Sephadex column.[13] Metallothionein^{-2} (MT-2), eluted out at higher ionic strength (0.15 *M*, Tris-HCl buffer), is used for the antibody

[11] A. M. Scheuhammer and M. G. Cherian, *Toxicol. Appl. Pharmacol.* **82,** 417 (1986).

[12] A. M. Scheuhammer and M. G. Cherian, this volume [12].

[13] D. M. Templeton and M. G. Cherian, *Biochem. J.* **221,** 569 (1984).

preparation. Metallothionein-2 is cross-linked with glutaraldehyde to obtain a polymer.

Cross-linking with glutaraldehyde is carried out by addition of a 10-fold excess of glutaraldehyde (5%, v/v; Eastman Chemicals, Rochester, NY) to 0.5–5 mg of MT-2 in 1 ml of 0.1 *M* sodium phosphate buffer, pH 6.2, with stirring at room temperature. After 2 hr lysine is added to 0.1 *M*, and after a further 30 min, a 20-fold excess of sodium borohydride (to MT) is added in five divided portions as the solid. The polymers of MT-2 formed are separated from lower molecular weight species by elution with 10 m*M* Tris-HCl buffer, pH 8.6, from Sephadex G-75 columns. The polymer of MT-2 formed under these experimental conditions had a molecular weight of 58,900. Two milliliters of 1 mg/ml polymerized rat MT-2 was emulsified in 2 ml of Freund's complete adjuvant. Rabbits (male, 2 kg) were injected at multiple sites weekly for 1 month and then once every 2 weeks for 4 months. Two rabbits out of six developed antisera against MT. The antiserum was characterized by immunodiffusion and electrophoresis. The isolated antiserum was frozen at $-80°$ in small aliquots (2 ml). The antiserum cross-reacts with both monomer and polymer of MT from rat, guinea pig, human, and horse.

Characteristics and Limitations of Method

Avidin–biotin complex immunoperoxidase is a simple technique that is extensively used for identification and localization of many antigens. It can be easily standardized in any research laboratory with some experience in morphological techniques. The preparation of an antibody to MT is essential for this technique. The other reagents can be obtained as a standard kit commercially (Vector Laboratories, Burlingame, CA). The principle involved in this technique is described in Fig. 1. The final color reaction is dependent on the specificity and affinity of the primary antibody for MT. The staining method for MT was first developed using the peroxidase–anti-peroxidase (PAP) complex method.[1,4] This was later modified by using avidin–biotin complex (ABC), which gave much better results than using the peroxidase–anti-peroxidase complex method because of increased sensitivity.

The method is sensitive and good staining for MT can be obtained when the concentration of MT in tissues is at least 5 μg/g. When the concentration of MT is lower than this, the dilution of the primary antibody can be reduced to obtain staining. Paraffin-embedded sections stored in pathology laboratories for a number of years have been successfully used for the staining of MT.[7,8] Because MT is heat stable, storage of the tissue section does not affect the antigenicity of MT.

The cross-reactivity of rabbit anti-rat MT antibody with other species has allowed the localization of MT in human[6-8] and horse[14] tissues. The antibodies prepared in other species should be analyzed for their cross-reactivity before undertaking the immunohistochemical localization. As discussed earlier in the section Control Experiments, appropriate controls —both negative and positive control slides—should be run along with the experiment. The specific staining for MT can be thus determined.

[14] E. H. Jeffery, R. Noseworthy, and M. G. Cherian, *Comp. Biochem. Physiol. C: Comp. Pharmacol. Toxicol.* **93c,** 327 (1989).

[15] Immunohistochemical Detection of Metallothionein

By B. JASANI and M. E. ELMES

The object of this chapter is to give an account of the basic concepts and methodology necessary for reliable performance, evaluation, and validation of immunohistochemical (IHC) detection of mammalian metallothionein (MT) under the light microscope. The development of IHC localization of MT at the electron microscope level is still in its infancy[1] and therefore not yet considered ready for inclusion in this volume.

Introduction

The detection of MT in tissue by histological means until recently has been based on techniques capable of disclosing the presence of high concentrations of sulfhydryl (SH) or disulfide bonds (S–S), or group IIb heavy metal ions known to be associated with MT. Such methods, although effective, have been found to be either tedious, expensive, and destructive of the specimen (e.g., X-ray microprobe analysis capable of detecting heavy metals)[2] or relatively nonspecific [e.g., rubeanic acid or orcein-based dyes routinely used for the detection of copper or copper-associated protein (CAP), respectively],[3] or demanding of the use of unusual tissue fixatives (e.g., Bohm fixative for the detection of sulfhydryl groups or disulfide

[1] W. Savino, P. C. Huang, A. Corrigan, S. Berrih, and M. Dardenne, *J. Histochem. Cytochem.* **32,** 942 (1984).
[2] J. Gwyn-Jones and M. E. Elmes, *Scand. J. Gastroenterol. Suppl.* **16** (70) 37 (1981).
[3] C. F. A. Culling, R. T. Allison, and W. T. Barr, "Cellular Pathology Techniques," 4th Ed., pp. 292, 339. Butterworths, London, 1985.

TABLE I
CHARACTERISTICS OF PRIMARY ANTIBODIES USED FOR DETECTION OF METALLOTHIONEIN

Antibody type	Immunogen used	Tissue cross-reactivity	Ref.
Polyclonal rabbit	Rat Cd-MT-1, MT-2	Helly or Bouin's fluid-fixed, paraffin-embedded, Cd-injected rat tissues	*a*, *b*
Polyclonal rabbit	Rat liver MT	Ten percent buffered formalin-fixed, paraffin-embedded rat and human liver, kidney, and/or thyroid	*c*, *d*
Polyclonal rabbit	Rat liver Zn-MT-2	Neutral buffered formalin-fixed, paraffin-embedded rat and human liver and ileum	*e*, *f*
Polyclonal rabbit	Cd, Zn-treated rat liver and kidney MT-1	Ten percent buffered formalin-fixed, paraffin-embedded rat liver, kidney, uterus, and ovaries; guinea pig mammary glands	*g*, *h*
Polyclonal rabbit	Cd-induced rat MT	Formol–saline-fixed, paraffin-embedded, normal, diseased, and copper-loaded human livers	*i*, *j*
Monoclonal mouse	Horse MT-1, MT-2	Formol–saline-fixed, paraffin-embedded, copper-loaded rat liver, kidney, ileum	*k*, *l*, *m*
Polyclonal sheep	Zn-induced rat MT-1	Formol–saline-fixed, paraffin-embedded rat liver, ileum	*n*
Monoclonal rat Antibodies A to I Antibody J	Human MT-2 (MAbs A to I) Rainbow trout MT (MAb J)	Formol–saline-fixed, paraffin-embedded rat liver, ileum; human liver, ileum (MAbs A to J), ovaries (MAb B)	*o*, *p*

[a] S. Ohi, G. Cardenosa, R. Pine, and P. C. Huang, *J. Biol. Chem.* **256,** 2180 (1981).
[b] K. G. Danielson, S. Ohi, and P. C. Huang, *Proc. Natl. Acad. Sci. U.S.A.* **79,** 2301 (1982).
[c] D. Banerjee, S. Onosaka, and M. G. Cherian, *Toxicology* **24,** 95 (1982).
[d] N. O. Nartey, J. V. Frei, and M. G. Cherian, *Lab. Invest.* **57,** 397 (1987).
[e] F. O. Brady and R. L. Kafka, *Anal. Biochem.* **98,** 89 (1989).
[f] J. P. Clarkson, M. E. Elmes, B. Jasani, and M. Webb, *Histochem. J.* **17,** 343 (1985).

bonds associated with MT).[4] These difficulties have led to the development, over the past decade, of the potentially more economical, versatile, and specific immunohistochemical (IHC) method for *in situ* detection of MT.

The success of the IHC approach in the *in situ* detection of the MT is critically dependent on the choice of the primary antibody, the secondary detection system, and the method of tissue substrate preparation employed. These issues are described next, reviewing the available data.

Choice of Primary Antibody and Tissue Substrate

The majority of the antibodies used for IHC work have been those generated primarily for the purpose of developing specific and sensitive immunoassays for the detection of MT in the fluid phase.[5-8] These, as well as the antibodies specifically produced for IHC application,[9,10] have proved eminently effective as primary antibodies for IHC detection of a wide variety of mammalian MT in widely differing tissue substrates (see Table I for a summary). The reason for such a universally versatile character of antibodies to MT stems from the antigenic nature of MT, as shown by the elegant studies of Winge and Garvey (1983)[11] and Kikuchi *et al.* (1988),[12]

[4] A. F. W. Morselt, D. Broekaert, E. J. Jongstra-Spaapen, and J. H. J. Copius-Peereboom-Stegeman, *Arch. Toxicol.* **55,** 155 (1984).

[5] R. S. Vander Mallie and J. S. Garvey, *Immunocytochemistry* **15,** 857 (1978).

[6] F. O. Brady and R. L. Kafka, *Anal. Biochem.* **98,** 89 (1979).

[7] R. K. Mehra and I. Bremner, *Biochem. J.* **213,** 459 (1983).

[8] S. Ohi, G. Cardenosa, R. Pine, and P. C. Huang, *J. Biol. Chem.* **256,** 2180 (1981).

[9] D. Banerjee, S. Onosaka, and M. G. Cherian, *Toxicology* **24,** 95 (1982).

[10] C. Tohyama, H. Nishimura, and N. Nishimura, *Acta Histochem. Cytochem.* **21,** 91 (1988).

[11] D. R. Winge and J. S. Garvey, *Proc. Natl. Acad. Sci. U.S.A.* **80,** 2472 (1983).

[12] Y. Kikuchi, N. Wada, M. Irie, H. Ikekuchi, J.-I. Sawada, T. Terao, S. Nakayama, S. Iguchi, and Y. Okada, *Mol. Immunol.* **25,** 1033 (1988).

---

[g] C. Tohyama, H. Nishimura, and N. Nishimura, *Acta. Histochem. Cytochem.* **21,** 91 (1988).

[h] N. Nishimura, H. Nishimura, and C. Tohyama, *J. Histochem. Cytochem.* **37,** 1601 (1989).

[i] R. S. Vander Mallie and J. S. Garvey, *Immunochemistry* **15,** 857 (1978).

[j] M. E. Elmes, J. P. Clarkson, N. J. Mahy, and B. Jasani, *J. Pathol.* **158,** 131 (1989).

[k] Jasani *et al.* (in preparation).

[l] W. E. Evering, S. Haywood, M. E. Elmes, B. Jasani, and J. Trafford, *J. Pathol.* **160,** 305 (1990).

[m] M. E. Elmes, S. Haywood, and B. Jasani *J. Pathol.* **164,** 83 (1991).

[n] J. P. Clarkson, M. E. Elmes, and B. Jasani (unpublished data), 1991.

[o] See [20] in this volume.

[p] J. P. Clarkson, L. Ungar, M. E. Elmes, and B. Jasani (unpublished data), 1991.

using polyclonal and monoclonal antibodies produced against different species of MT. The MT molecule appears to possess only a single immunodominant antigenic determinant or epitope. The bulk of the activity possessed by this epitope is accounted for by the first five amino acid residues of the N-terminal portion of the molecule (see Fig. 1 for a pictorial representation).

The possession of only a single antigenically active epitope per molecule of MT also accounts for the behavior of the MTs as hapten-like antigens when used as immunogens. Thus, they have been consistently found to require covalent (e.g., glutaraldehyde based) self-polymerization or covalent cross-linking with other proteins to make them optimally immunogenic. It is also, probably for the same reason, why they have proved to be rather unsatisfactory targets for developing the otherwise

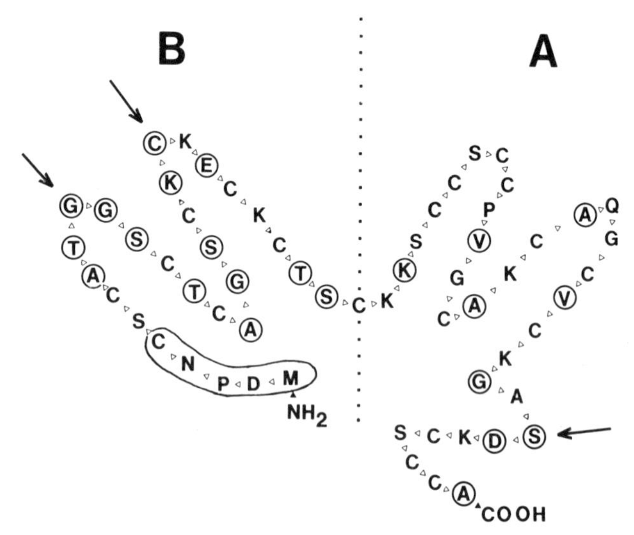

FIG. 1. Pictorial representation of the primary, secondary, and tertiary structures of mammalian MT based on the data reviewed by Hamer [D. Hamer, *Annu. Rev. Biochem.* **55,** 913 (1986)] showing the primary sequence positions of the immunoreactive region (outlined) and the potentially immunogenic sites (arrows) as proposed by the data of Winge and Garvey [D. R. Winge and J. R. Garvey, *Proc. Natl. Acad. Sci. U.S.A.* **80,** 2472 (1983)] and Kikuchi *et al.* [Y. Kikuchi, N. Wada, M. Irie, H. Ikekuchi, J.-I. Sawada, T. Temo, S. Nakayama, S. Iguchi, and Y. Okada, *Mol. Immunol.* **25,** 1033 (1988)]. According to the primary sequence and the X-ray crystallographic data reviewed by Hamer (1986), the mammalian MT consists of 61 amino acid residues (unconserved, indicated by circles) that are folded up into two β sandwich-bearing globular domains (A and B, respectively) separated by a short flexible region (position indicated by the dotted line). The bulk of the immunoreactivity of the mammalian MT appears to be accounted for by the short, N-terminal five-amino acid sequence situated in the B domain.

highly sensitive two-site competitive immunological assay for MT. Furthermore, as discussed later, MT in its monomeric state also appears to be inefficient in its capacity to preabsorb antibodies to MT when used as primary antibody reagents in immunohistochemistry.

It is noteworthy that apart from the obvious structural differences defined by the heavy metal ion composition and the primary sequence, MT also shows potential differences in its secondary and tertiary configurations depending on the type and number of heavy metals bound by it. Furthermore, MT also shows differences according to its separation characteristics on ion-exchange chromatography. Thus MT isoproteins or isoforms encoded by two separate sets of genes are separable into two classes of MT, MT-1 and MT-2, depending on their elution profile positions on DEAE-cellulose column chromatography. However, it seems from the available IHC data that both MT-1 and MT-2 exist in tissue in forms that are indistiguishable from each other simply on the basis of their IHC detection.

To date only one antibody preparation has been cited in the literature as capable of distinguishing between the MT-1 and MT-2 isoforms.[7] This antibody, which is a polyclonal sheep anti-rat MT-2, has been found to achieve this function only in radioimmunoassay (RIA) and not in enzyme-linked immunosorbent assay (ELISA) or IHC detection systems set up for MT (I. Bremner, personal communication, 1991; B. Jasani and M. E. Elmes, unpublished data, 1991).

In summary, in terms of the IHC detection of MT using the available set of primary antibodies, it does not seem to be possible to differentiate MT according to its secondary, tertiary, or charge-dependent structural characteristics. On the contrary, because of the highly conserved primary sequence-based nature of its only epitope, it seems to be quite feasible to immunohistochemically detect any variety of MT species with a single anti-MT antibody in any variety of tissues from many different mammalian species.

Choice of Secondary Detection Systems

The main object of a secondary detection system in IHC detection of antigens is to disclose the sites of primary antibody binding in the tissue substrate. Under the light microscope this is achieved by either direct or indirect attachment to the primary antibody molecules of a fluorogenic or a chromogenic dye system. The direct method, which involves chemical coupling of either a fluorescent dye or an enzyme (capable of depositing a colored dye in the presence of a chromogenic substrate), is rarely used for

TABLE II
SECONDARY INDIRECT ENZYME TECHNIQUES FOR IMMUNOHISTOCHEMICAL DETECTION OF METALLOTHIONEIN

Authors	Method	Original Ref.
a, b	Peroxidase–anti-peroxidase (PAP)	*c*
d	Avidin–biotin complex (ABC)	*e*
f, g	Modified DNP–hapten sandwich staining (DHSS)	*h*
i	DNP–Localization system (DLS)	*j*

[a] K. G. Danielson, S. Ohi, and P. C. Huang, *Proc. Soc. Natl. Acad. Sci. U.S.A.* **79,** 2301 (1982).
[b] N. O. Nartey, J. V. Frei, and M. G. Cherian, *Lab. Invest.* **57,** 397 (1987).
[c] L. A. Sternberger, P. H. Hardy, J. J. Cuculis, and H. G. Meyer, *J. Histochem. Cytochem.* **18,** 315 (1970).
[d] N. Nishimura, H. Nishimura, and C. Tohyama, *J. Histochem. Cytochem.* **37,** 1601 (1989).
[e] S.-M. Hsu, L. Raine, and H. Fanger, *J. Histochem. Cytochem.* **29,** 577 (1981).
[f] J. P. Clarkson, M. E. Elmes, B. Jasani, and M. Webb, *Histochem. J.* **17,** 343 (1985).
[g] M. E. Elmes, J. P. Clarkson, N. J. Mahy, and B. Jasani, *J. Pathol.* **158,** 131 (1989).
[h] B. Jasani, R. E. Edwards, N. D. Thomas, and A. R. Gibbs, *Virchows Arch. Pathol. Anat. Histopathol.* **406,** 441 (1985).
[i] W. E. Evering, S. Haywood, M. E. Elmes, B. Jasani, and J. Trafford, *J. Pathol.* **160,** 305 (1990).
[j] B. Jasani, N. D. Thomas, M. Ludgate, G. R. Newman, and E. D. Williams, *Proc. RMS* **20,** IM7 (1985).

routine IHC work and has not been used at all for the IHC detection of MT.

The indirect method, which represents the most popular choice, involves specific attachment to the tissue-bound primary antibody molecules of a fluorescent dye or a chromogenic enzyme-bearing complex through formation of reversible, noncovalent ligand/receptor-type bonds. Most of the IHC work on MT has been based on the use of the enzyme-based secondary detection systems. Thus only a single group of main workers in the field has used an indirect fluorescent dye-linked secondary antibody system for the detection of MT,[13] while there are several independent groups[9,13–15] that have described the use of the immunoenzyme approach.

The greater popularity of the enzyme-based secondary system derives

[13] K. G. Danielson, S. Ohi, and P. C. Huang, *Proc. Natl. Acad. Sci. U.S.A.* **79,** 2301 (1982).
[14] J. P. Clarkson, M. E. Elmes, B. Jasani, and M. Webb, *Histochem. J.* **17,** 343 (1985).
[15] N. Nishimura, H. Nishimura, and C. Tohyama, *J. Histochem. Cytochem.* **37,** 1601 (1989).

TABLE III
PEROXIDASE–ANTI-PEROXIDASE METHOD

Step	Comment
1. Dewax in series of xylene and absolute ethanol baths	Necessary step for paraffin wax-embedded tissue substrates
2. Inhibit endogenous peroxidase with methanol/H_2O_2	Necessary step in all peroxidase enzyme-based IHC procedures
3. Rehydrate and equilibrate in diluent buffer	Usually phosphate-buffered saline (PBS)
4. Incubate in primary antibody for 0.5 to 2 hr at 17 to 37°; or 16 hr at 4°	Unlabeled polyclonal rabbit antibody diluted 1:20–1:500 in PBS containing 5–20% normal swine serum as blocking agent
5. Wash in PBS three times (1–10 min each time) at 17°	To remove any unbound primary antibody
6. Incubate in secondary bridge antibody 0.5–2 hr at 17°	Polyclonal bivalent anti-rabbit IgG prepared in goat, sheep, or swine; added in excess to allow univalent binding
7. Wash as in step 5	To remove any unbound bridge antibody
8. Incubate in PAP complex 0.5–1 hr at 17°	PAP made up of a stable rabbit anti-peroxidase–peroxidase immune complex designed to bind to free valency of bridge antibody
9. Wash as in step 5	To remove any unbound PAP
10. Incubate in diaminobenzidine (DAB)/H_2O_2 substrate solution	Leads to formation of insoluble polymeric DAB producing amorphous brown color deposit at the site of tissue-bound primary antibody/bridge antibody/PAP immune complexes
11. Counterstain for nuclei, dehydrate in reversed series of alcohol and xylene baths, mount under coverslip using permanent mountant	Counterstain usually hematoxylin; mountant usually Terpene

from its greater sensitivity on the one hand and compatibility with conventional histological techniques on the other. For example, in the latter context it permits the use of both the light microscope optics and the various routinely applicable nuclear counterstains. Its only disadvantage for the newcomer is the diversity of the available indirect immunoenzyme methods in terms of their design and method of application. (see Polak and Van Noorden, 1988,[16] for a comprehensive review). The choice in terms of

[16] J. M. Polak and S. Van Noorden, "An Introduction to Immunocytochemistry: Current Methods and Problems," 1st Ed. Oxford Univ. Press, Oxford, 1987.

TABLE IV
AVIDIN–BIOTIN COMPLEX PROCEDURE

Step	Comment
1–5. As in Table III	Primary antibody dilution may be used as high as 1 : 1000+ because of higher sensitivity of technique
6. Incubate in biotin-labeled anti-rabbit IgG, 0.5–1 hr at 17°	a
7. Wash as in Table III	
8. Incubate in preformed ABC, 0.5–1 hr at 17°	a
9–11. As in Table III	

[a] ABC kit obtainable from Vector Lab., Inc. (Burlingame, CA).

IHC detection of MT as far as published work is concerned has focused on the use of three distinct multilayer varieties of indirect immunoperoxidase methods[17–19] (see Table II for a summary). The basic methodological details of each of these three techniques are outlined for comparison in Tables III to V, respectively.

Despite the differences in their design and method of application, all three methods lead to formation of the same type of insoluble polymeric compound of diaminobenzidine (DAB), which usually varies in its color from light to dark brown when viewed under the light (i.e., bright-field) microscope. A formal comparison of the sensitivities of the peroxidase–anti-peroxidase (PAP),[17] the avidin–biotin complex (ABC),[18] and the dinitrophenyl (DNP)–localization system (DLS)[19–21] methods has not been made so far with respect to the IHC detection of MT. However, in terms of the primary antibody working dilutions stipulated for these techniques, it would appear that the ABC method is more sensitive compared to the PAP procedure, whereas the DLS technique appears more effective than the ABC system.

Evaluation and Validation of Immunohistochemical Staining

Using any one of the above standardized and optimized methods for staining, it is important next to evaluate the staining both in terms of the

[17] L. A. Sternberger, P. H. Hardy, J. J. Cuculis, and H. G. Meyer, *J. Histochem. Cytochem.* **18,** 35 (1979).

[18] S.-M. Hsu, L. Raine, and H. Fanger, *J. Histochem. Cytochem.* **29,** 577 (1981).

[19] B. Jasani, D. Wynford-Thomas, and E. D. Williams, *J. Clin. Pathol.* **34,** 1000 (1981).

[20] B. Jasani, N. D. Thomas, M. Ludgate, G. R. Newman, and E. D. Williams, *Proc. RMS* **20,** IM7 (1985).

TABLE V
DNP-LOCALIZATION SYSTEM

Step	Comment
1-3. As in Table III	
4. Incubate in mouse monoclonal anti-MT at 1:50,000, 16 hr at 4°	*a*
5. Wash as in Table III	
6. Incubate in DNP-labeled anti-rabbit IgG + mouse IgG	*b*
7. Wash as in Table III	
8. Incubate in polyvalent monoclonal IgM anti-DNP bridge antibody, 0.5 hr at 17°	*b*
9. Wash as in Table III	
10. Incubate in DNP-labeled peroxidase conjugate, 0.5 hr at 17°	*b*
11. Wash as in Table III	
12. Incubate DNP-labeled glucose oxidase, 0.5 hr at 17°	*b*
13. Wash as in Table III	
14. Incubate in DAB/glucose solution in PBS, 16 hr at 17°	*b*
15. Counterstain and mount as in Table III	

[a] W. E. Evering, S. Haywood, M. E. Elmes, B. Jasani, and J. Trafford, *J. Pathol.* **160,** 305 (1990).
[b] DNP-localization system kit obtainable from Bioclinical Services, Ltd. (Cardiff, Wales).

range of the tissue elements stained and the quality with which they have been stained. It is also necessary to ensure that the staining obtained is related specifically to MT and not to any form of nonspecific staining due to a spurious binding of either the primary antibody reagent or any one of the secondary detection reagents.

The specificity of the staining can be assessed in several ways. For example, the specificity of the secondary detection system is best ascertained in terms of the result obtained through the application of the staining procedure to the tissue substrate in the absence of the primary antibody reagent.

As for checking the specificity of the primary antibody a purified or synthetically pure preparation of MT is used to preabsorb or preadsorb the antibody activity. The preabsorption is likely to prove consistently more effective if the MT preparation is first self-polymerized, e.g., by its glutaraldehyde cross-linking according to the method originally described by Vander Mallie and Garvey[5] (see Table VI for practical details). The biological specificity is validated by inclusion of tissue specimens known independently to be variably positive or negative for the presence of MT activity as

[21] B. Jasani, R. E. Edwards, N. D. Thomas, and A. R. Gibbs, *Virchows Arch. A: Pathol. Anat. Histopathol.* **406,** 441 (1985).

TABLE VI
GLUTARALDEHYDE CROSS-LINKING OF METALLOTHIONEIN FOR PREABSORPTION OF ANTI-METALLOTHIONEIN PRIMARY ANTIBODY[a]

1. Dissolve 100 μg of freeze-dried purified MT in 10 μl of 0.1 *M* sodium acetate, pH 6.2
2. Add 10 μl of 32 m*M* purified glutaradehyde (Sigma, St. Louis, MO)
3. Incubate at 25° for 1 hr
4. Add 1.0 ml of 0.1 *M* lysine hydrochloride to quench reaction
5. Dialyze overnight against 1 liter of phosphate-buffered saline (PBS, 0.01 *M*, pH 7.2) at 4°
6. Repeat dialysis with at least two more changes of PBS
7. Store material at −70° in small aliquots (e.g., 100 μl)
8. Dilute anti-MT to 10× desired concentration
9. Add 1 vol of 10× anti-MT to 9 vol of cross-linked MT
10. Incubate for 0.5 hr at 37° prior to application to tissue substrate

[a] R. S. Vander Mallie and J. Garvey, *Immunochemistry* **15,** 857 (1978).

well as by assessing the staining according to morphologic identity of the stained and unstained tissue components.

The range and character of the staining due to MT may be assessed in one of the following ways. The character of the staining may be qualified in terms of whether it is amorphous or granular. The distribution of the staining may be characterized in terms of its localization at the individual cell level (e.g., nuclear, cytoplasmic, cell surface, or extracellular, lumenal) or at the total cell population or tissue section level. It may be classified as focal, multifocal, or uniform. It may also prove worthwhile to enumerate the stained cells as either a percentage of the total number of nucleated cell or in terms of their surface density, i.e., number of stained cells profiles per unit area (usually in mm^2) of the tissue section stained.

The intensity of staining relating to the different stained areas may then be assessed to give a complete description of the IHC staining obtained in a given experiment or test. The intensity may be either objectively measured using a microdensitometric approach[22] or estimated on an arbitrary subjective scale of pluses and minuses (see Table VII for an illustration of a commonly used scale).

An illustration of MT-related IHC staining, in terms of the differing character of its distribution and intensity, is presented in Fig. 2 based on observations made of liver and kidney sections taken from rats fed on varying amounts of copper in their diet.[23]

[22] W. Hazelhoff Roelfzema, C. Tohyama, H. Nishimura, N. Nishimura, and A. F. W. Morselt, *Histochemistry* **90,** 365 (1989).

[23] W. Evering, S. Haywood, M. E. Elmes, B. Jasani, and J. Trafford, *J. Pathol.* **160,** 305 (1990).

TABLE VII
ARBITRARY SCALE FOR RECORDING VARIATIONS IN INTENSITY OF IMMUNOHISTOCHEMICAL STAINING

Level of staining	Scale of intensity
Nil	−
Borderline: just visible above background level at highest magnification	+/−
Weak: Just visible above background at lower magnifications	+
Moderate: easily visible at lower magnifications	++
Strong: easily visible at lowest magnification	+++

Having evaluated the IHC staining, it is important to ascertain that the staining is a true reflection of the presence of MT and not simply due to some cross-reactive antigenic species.

The most compelling biochemical proof of the IHC-detected MT identity can be achieved through a parallel *in situ* hybridization study using a complementary nucleic acid probe to MT mRNA on an IHC-positive tissue preparation. Alternatively, a gel electrophoretic/immunoblot study of the IHC-positive tissue can be undertaken to establish the identity of the anti-MT-related reactivity on the basis of its molecular weight profile. However, neither of these approaches appear to have been utilized so far in the case of the IHC-detectable MT.

Although not entirely dependable in terms of specificity, X-ray microprobe analysis aimed at identifying the presence of MT according to its heavy metal composition and content, or histochemical techniques involving the use of rubeanic acid as a stain specific for copper(II) ions,[24] the Sipple method for SH groups or S–S bonds,[25] or the Shikata orcein method[26] as a marker of the copper-associated protein (CAP), which is presumed to represent a polymerized form of MT, may all be used as alternative *in situ* markers of MT. Interestingly, the CAP-associated polymerized MT does not seem to be detectable by anti-MT, unlike the glutaraldehyde self-polymerized MT. It has been postulated by Suzuki and Yamamura (1980)[27] that the CAP represents self-polymerized MT by virtue of intermolecular oxidative coupling of its SH groups. It is therefore

[24] A. G. E. Pearse, "Histochemistry, Theoretical and Applied," 4th Ed., Vol. 2, p. 991. Churchill Livingstone, Edinburgh, 1985.

[25] T. O. Sippel, *Histochem. J.* **5,** 413 (1978).

[26] T. Shikata, T. Uzawa, N. Yoshiwara, T. Akatsuka, and S. Yamazaki, *Jpn. J. Exp. Med.* **44,** 25 (1974).

[27] K. T. Suzuki and M. Yamamura, *Biochem. Pharmacol.* **29,** 689 (1980).

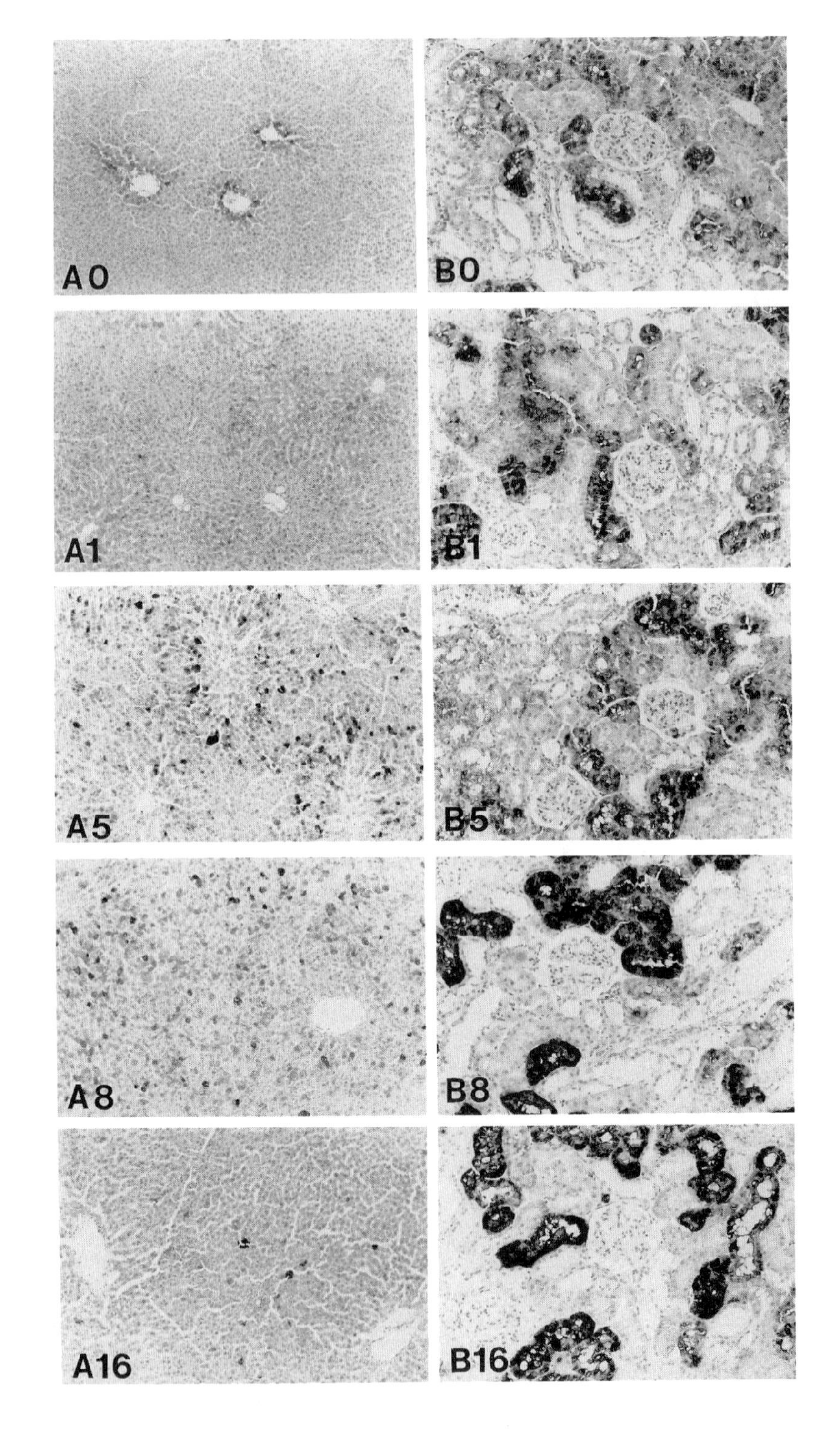

A0
B0
A1
B1
A5
B5
A8
B8
A16
B16

possible that this form of polymerization leads to masking of the N-terminal epitope.

Applications of Immunohistochemistry

Once having established a reliable and validated IHC technique for the detection of MT, there are a number of applications one may undertake. The IHC approach can be used as a powerful approach to the understanding of the distribution and the dynamics of MT[1,13–15,22,23,28] and the associated heavy metals in normal subjects and in animals and humans exposed to heavy metals. More recently the IHC detection of MT has been shown to have a potential for providing a more intimate understanding of the role played by MT in inflammatory[29] and neoplastic diseases.[30,31]

[28] M. E. Elmes, J. P. Clarkson, N. J. Mahy, and B. Jasani, *J. Pathol.* **158,** 131 (1989).
[29] A. G. Douglas-Jones, N. D. Thomas, and B. Jasani. *J. Pathol.* **160,** 170A (1990).
[30] N. Nartey, M. G. Cherian, and D. Banerjee, *Am. J. Pathol.* **129,** 177 (1987).
[31] T. E. Kontozoglou, D. Banerjee, and M. G. Cherian, *Virchows Arch. A: Pathol. Anat. Histopathol.* **415,** 545 (1989).

FIG. 2. Photomicrograph of variation in MT immunostaining due to changes in biochemical MT status induced by high copper intake in rats: Liver (A) and Kidney (B) formalin-fixed, paraffin-embedded sections immunoperoxidase stained with monoclonal antibody (E9) to horse MT (MT-1 and MT-2) in conjunction with the DLS procedure [see W. E. Evering, S. Haywood, M. E. Elmes, B. Jasani, and J. Trafford, *J. Pathol.* **160,** 305 (1990)]. The sections were taken from tissues removed from rats fed on a high copper (1 g/kg) diet for 0, 1, 5, 8, and 16 weeks, respectively. In the liver the baseline immunostaining was limited to a single layer of hepatocytes around the central venules (see A0). At 1 week this pattern became replaced by fairly even but weak immunostaining of the majority of the hepatocytes (A1). At 5 weeks many individually scattered hepatocytes were very strongly positive for MT immunostaining (A5), which persisted to a slightly lower intensity level up to 8 weeks (A8) and then virtually vanished to baseline levels by 16 weeks (A16). This bell-shaped response in the MT immunostaining was accompanied by parallel changes in the hepatic copper as measured by atomic absorption spectrophotometry on separate liver samples taken from the same rats, and the development of copper tolerance (see Evering *et al.*, 1990). The kidney sections (B0, B1, B5, B8, and B16, respectively) showed variation in the MT immunostaining that again paralleled the changes observed in the copper concentration over the same time points. Thus over the first 5 weeks (B1 and B5) there was no significant change in the distribution or the intensity of MT immunostaining from the baseline levels (B0). From 8 weeks up to 16 weeks there was a dramatic increase in the intensity of the immunostaining without any apparent change in its distribution. These results provide a good indication of the fidelity and the overall value of the MT immunostaining obtained using the primary antibody and the secondary detection system used. Hematoxylin counterstain. Magnification: ×210. (Sections for these studies were kindly provided by Dr. Susan Haywood, Department of Veterinary Pathology, University of Liverpool, Liverpool, England.)

[16] Detection of Metallothionein by Western Blotting

By YASUNOBU AOKI and KAZUO T. SUZUKI

A convenient method for detecting metallothionein (MT) in a small sample was developed using the Western blotting technique. The Western blotting technique was originally established to detect a protein specifically by immunological procedures.[1] In this method a protein mixture was subjected to sodium dodecyl sulfate (SDS)/polyacrylamide gel electrophoresis (SDS-PAGE), and the proteins separated on the gel were electrophoretically transferred to the synthetic membrane [a poly(vinylidene fluoride) membrane or a nitrocellulose membrane]. Immobilized proteins on the membrane were detected by specific binding of the radioactive ligand or by immunochemical staining. These procedures have been applied to analyze various proteins with specific reactions or ligands. Calcium-binding proteins were detected after incubating the membrane in a buffer containing ^{45}Ca.[2]

Likewise, MT was shown to be detectable on the membrane by incubation in a buffer containing radioactive cadmium (^{109}Cd)[3] or by immunochemical staining.[4] Under a standard procedure of SDS-PAGE, MT migrated as an irregular broad band on the gel. Some modifications of the procedure allowed MT to be separated on the gel as a compact band.[3] Cadmium-binding proteins other than MT were also identified using the same method.[5]

Preparation of SDS/Polyacrylamide Slab Gel[3,4]

SDS-PAGE is carried out according to the method of Laemmli.[6] Two isoforms of rat cadmium-containing MT (MT-1 and MT-2) are used as standards.[7] Each isoform is dissolved at a concentration of 1 mg/ml in 20 m*M* Tris-HCl buffer (pH 7.5 at 25°) containing 0.9% NaCl and stored at $-70°$ in small portions until used. These MT solutions should be used within a month.

A standard separating gel consists of 15% (w/v) acrylamide, and is $10 \times 14 \times 0.1$ cm in size.

[1] H. Towbin, T. Staehelin, and J. Gordon, *Proc. Natl. Acad. Sci. U.S.A.* **76,** 4350 (1979).
[2] K. Maruyama, T. Mikawa, and S. Ebashi, *J. Biochem. (Tokyo)* **95,** 511 (1984).
[3] Y. Aoki, M. Kunimoto, Y. Shibata, and K. T. Suzuki, *Anal. Biochem.* **157,** 117 (1986).
[4] Y. Aoki, C. Tohyama, and K. T. Suzuki, *J. Biochem. Biophys. Meth.* in press (1991).
[5] Y. Aoki, H. Sunaga, and K. T. Suzuki, *Biochem. J.* **250,** 735 (1988).
[6] U. K. Laemmli, *Nature (London)* **227,** 680 (1970).
[7] The method for purification of MT is described in [29] in this volume.

Separating Gel Solutions

Acrylamide (44%, w/v)/bisacrylamide (0.8%, w/v), 8.5 ml
Tris-HCl buffer (pH 8.8 at 25°), 12.5 ml of a 0.75 *M* solution
Distilled water, 3.1 ml

The solutions listed above are mixed and placed *in vacuo* in a glass bell for several minutes. Gels having different acrylamide concentrations can be obtained by changing the volumes of the acrylamide solution and water. The deaerated solution is mixed gently with 250 μl of 20% (w/v) SDS, 0.63 ml of 10 mg/ml ammonium persulfate, and 25 μl of N,N,N′,N′-tetramethylethylenediamine (TEMED). The solution is pipetted into the sandwich of glass plates, and is overlaid with distilled water. Polymerization takes about 1–2 hr.

Stacking Gel Solutions

Acrylamide (30%, w/v)/bisacrylamide (0.8%, w/v), 2 ml
Tris-HCl (pH 6.8 at 25°), 5 ml of a 0.5 *M* solution
Distilled water, 12.3 ml

The solutions listed above are mixed and deaerated. Add to the mixed solution 200 μl of 20% (w/v) SDS, 0.48 ml of 10 mg/ml ammonium persulfate solution, and 25 μl of TEMED and shake gently. After the liquid is poured from the surface of polymerized separating gel, the surface of gel is rinsed with 1–2 ml of the stacking gel solution. The sandwich of glass plates is filled with the stacking gel solution, and a template comb of sample wells (the width of each well is 5 mm) is inserted into the sandwich to a level of 1.5–2 cm above the top of the separating gel. The polymerized gel is used after allowing it to stand overnight to get reproducible results. After removing the comb slowly, each well is rinsed with a tank buffer [25 m*M* Tris/190 m*M* glycine (pH 8.5) containing 0.1% SDS]. The gel is set in an electrophoresis chamber, which is filled with the tank buffer.

Sample solutions with cadmium-containing MT are combined with 0.5 *M* Tris-HCl buffer (pH 6.8 at 25°), 20% (w/v) SDS, 50% (v/v) glycerol, and 0.2% bromphenol blue to make final concentrations of 125 m*M*, 1%, 10% (v/v), and 0.01% in a final volume of 10–50 μl, respectively. About 1 μg of rat MT-1 or MT-2 is adequate as standard. This sample solution does not contain 2-mercaptoethanol (2-ME), which is an original component of the sample buffer, because MT migrates on the gel as an irregular broad band in the presence of 2-ME. When a sample with zinc-containing MT is separated on the gel, $CdCl_2$ is added to the sample solution at a concentration of 0.25 mg Cd/ml because MT does not migrate as a compact band without $CdCl_2$.

After the sample solutions are heated in a water bath at 70° for 20 min, each sample is underlaid in each well using a microsyringe. The cathode is connected to the upper buffer chamber, and the gel is run first at 22 mA (constant) until bromphenol blue enters into the separating gel and then at 28 mA. The run is stopped when bromphenol blue reaches the bottom of the gel. Electrophoresis is completed in 2–3 hr.

Transfer of Proteins to Membranes (Western Blotting)[3-5]

The gel is carefully removed from the sandwich of glass plates, and is incubated in 100–200 ml of 25 m*M* Tris/190 m*M* glycine (pH 8.5) containing 0.2% (w/v) SDS and 5% (v/v) 2-ME at 37° for 1 hr. Without this 2-ME treatment, MT is not retained on the membrane. But this incubation is not essential to identify heavy metal-binding proteins other than MT. The gel is soaked in 200 ml of a transfer buffer [25 m*M* Tris, 190 m*M* glycine (pH 8.5)/methanol (v/v), 4 : 1] for 20 min to remove excess 2-ME. A poly(vinylidene fluoride) membrane (Durapore membrane; Millipore, Bedford, MA or Clear Blot Membrane-p, Atto Co., Ltd., Tokyo, Japan) is soaked in the transfer buffer after wetting in methanol for 20–30 sec. A nitrocellulose membrane can be used instead, but proteins are retained less efficiently on it than on a poly(vinylidene fluoride) membrane. Because a nitrocellulose membrane is soluble in methanol, wetting with methanol should be omitted. After the gel is moved to a flat plastic tray containing the transfer buffer, the membrane is placed under the gel (taking care not to trap bubbles between the gel and the membrane). A sheet of filter paper (e.g., 3MM; Whatman, Maidstone, England) is placed under the membrane. The gel covered with the membrane and filter paper is taken out from the tray, and excess buffer is absorbed on tissue paper. The other face of the gel, which is not covered with the membrane, is carefully covered with a sheet of filter paper wetted with the transfer buffer. Then the gel is placed between two sheets of fiber pad, and is set in a cassette. The cassette is placed into the transfer chamber (for example, Bio-Rad, Richmond, CA) filled with the transfer buffer. The cassette should be oriented so that the membrane is facing toward the anode side. Chilled water is circulated in a cooling coil. To cool the transfer chamber efficiently, it is placed in an ice-water bath set on a magnetic stirrer and the transfer buffer is stirred gently during transfer.

The proteins separated on the gel are transferred to a poly(vinylidene fluoride) membrane at a constant voltage of 60 V for 30 min followed by 150 V for 2 hr (distance between electrodes, 3.5 cm). Lower voltage (60 V for more than 2 hr; distance between electrodes, 5 cm) is used for the nitrocellulose membrane. After the transfer, the membrane is soaked in

400 ml of 10 m*M* Tris-HCl buffer (pH 7.4 at 25°) for 1–2 hr at 4°. To visualize MT and other proteins on the membrane, the proteins are stained with colloidal gold. The membrane is washed twice in 20 m*M* Tris-HCl buffer (pH 7.5 at 25°) containing 0.9% (w/v) NaCl and 0.05% (v/v) Tween 20 each for 20 min, and then is soaked in colloidal gold total protein stain (Bio-Rad) for more than 2 hr.[8] The stained membrane is washed with distilled water and dried. Proteins on the membrane can also be stained by Amido black. The membrane is stained with 0.1% Amido black in a methanol–acetic acid solution [methanol/acetic acid/distilled water (v/v/v), 5:1:4] for 10 min and is washed well with a methanol–acetic acid solution until the background turns white. However, it is difficult to stain MT by Amido black.

Detection of Metallothionein by Binding of Radioactive Cadmium on Membrane[3]

Two solutions are prepared:

Binding solution: 10 m*M* Tris-HCl buffer (pH 7.4 at 25°) containing 1 μCi/ml $^{109}CdCl_2$, 0.1 m*M* zinc acetate,[9] and 0.1 *M* KCl:

1. Combine 1.25 ml of 0.2 *M* Tris-HCl buffer (pH 7.4 at 25°) [0.39 g of Trizma base (Sigma, St. Louis, MO) and 2.64 g of Trizma-HCl (Sigma) dissolved in 100 ml of distilled water] with 25 μCi of $^{109}CdCl_2$ (carrier free; New England Nuclear, Boston, MA) (volume depends on each lot), 25 μl of 0.1 *M* zinc acetate, and 2.5 ml of 1 *M* KCl
2. Make up to 25 ml with distilled water

If a white precipitate forms in the 0.1 *M* zinc acetate, it can be removed by centrifuging at 1000 *g* for 5 min before use. A binding solution can be stocked for a month.

Washing solution: 10 m*M* Tris-HCl buffer (pH 8.0 at 25°) containing 0.1 *M* KCl:

1. Combine 10 ml of 0.2 *M* Tris-HCl buffer (pH 8.0 at 25°) (1.06 g of Trizma base and 1.78 g of Trizma-HCl are dissolved in 100 ml of distilled water) and 20 ml of 1 *M* KCl
2. Make up to 200 ml with distilled water

The 0.2 *M* Tris-HCl buffer (pH 8.0 at 25°) should be used within a week.

[8] The protocol is available from Bio-Rad Company, Richmond, CA.

[9] Zinc acetate (1 m*M*) is given in an original recipe for the binding solution, but the concentration of zinc is reduced to increase the sensitivity.

The membrane is incubated in a flat plastic tray containing 25 ml of the binding solution for 10 min at room temperature with gentle shaking. Then the membrane is rinsed well twice with 100 ml of the washing solution for 1 min. The membrane is dried on filter paper at room temperature for 3 hr or overnight. The dried membrane is exposed to an X-ray film (X-Omat AR; Kodak, Rochester, NY) using the intensifying screen (Cronex Lighting Plus; Du Pont, Wilmington, DE) at −70°. After expo-

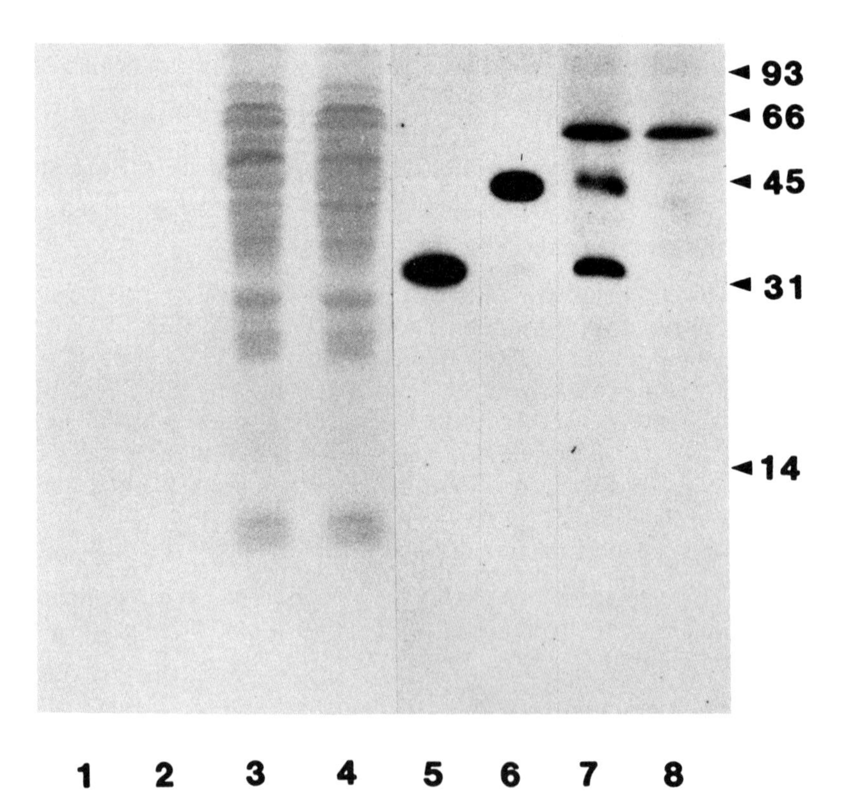

FIG. 1. Detection of MT-1 and MT-2 in rat liver cytosol on a nitrocellulose membrane.[3] Lanes 1–4, bands of proteins stained by Amido black; lanes 5–8, autoradiograms; lanes 1 and 5, rat MT-1 (8.8 μg); lanes 2 and 6, MT-2 (8.8 μg); lanes 3 and 7, a cytosol prepared from cadmium-treated rat liver (50 μg protein); lanes 4 and 8, a cytosol from control rat liver (50 μg protein). It is difficult to stain MTs with Amido black (lanes 1 and 2). The membrane is incubated in a binding solution containing 1 m*M* zinc acetate. Similar results are obtained using poly(vinylidene fluoride).[4] Arrows indicate molecular weight markers: 93, phosphorylase *b* (M_r 93,000); 66, bovine serum albumin (66,000); 45, ovalbumin (45,000); 31, carbonate dehydratase (31,000); 14, lysozyme (14,000).

sure for 36–48 hr, the X-ray film is developed. The limit of detection for each cadmium-containing MT from rat is 0.4 μg. The MT-1 and MT-2 induced in rat liver after cadmium administration are detected by the method described above (Fig. 1).[3] New binding solution may give dirty radioactive spots on the membrane. These spots can be removed by incubating a plain membrane in the binding solution before the membrane that retains proteins is incubated.

The presence of zinc in the binding solution is essential to reduce nonspecific binding of ^{109}Cd to the membrane, as shown in Fig. 2.[3] In the absence of zinc, ^{109}Cd binds to most proteins on the membrane and nonspecific binding of ^{109}Cd to the membrane causes severe interference in

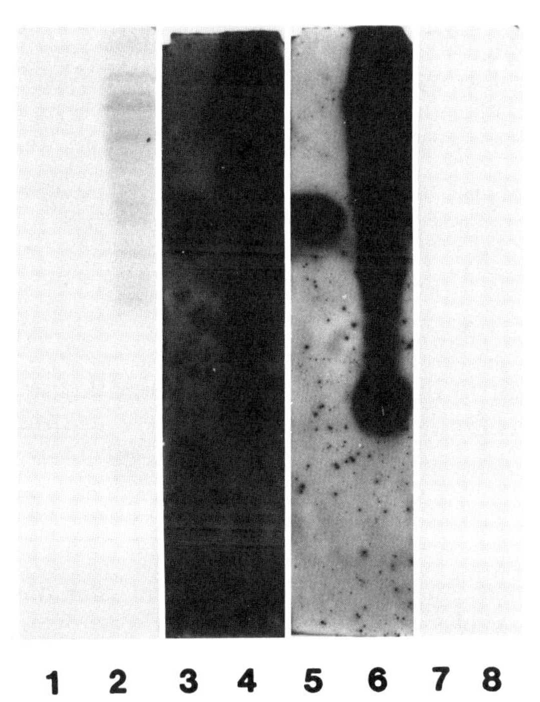

FIG. 2. Effect of divalent metal ions on detection of MT.[3] The nitrocellulose membranes retaining rat MT-2 (8.8 μg) and rat liver cytosol (50 μg protein) are incubated in a binding solution without divalent metal ions (lanes 3 and 4), with 1 m*M* magnesium acetate (lanes 5 and 6), or with 1 m*M* $CuSO_4$ (lanes 7 and 8) instead of zinc acetate. Lanes 1, 3, 5, and 7, MT-2; lanes 2, 4, 6, and 8, rat liver cytosol. Lanes 1 and 2, bands of proteins stained by Amido black; lanes 3–8, autoradiogram.

detecting MT. Similar interference is observed even in the presence of magnesium ion. In the presence of cupric ion, ^{109}Cd does not bind MT.

Detection of Metallothionein on Membranes by the Immunochemical Method[4]

Metallothionein transferred to the membrane can also be detected by immunochemical staining. Rat MT-1 and MT-2 retained on the membrane are stained with anti-rat MT rabbit serum and protein A–colloidal gold conjugate (protein A gold; Bio-Rad). The staining is enhanced with the gold enhancement kit (Bio-Rad) used according to the manufacturer's instructions.[8] The limit of detection for each MT is 0.1 μg.

Applications

Using Western blotting and ^{109}Cd binding the heavy metal-binding protein(s) of an insect (a species of mayfly) was detected.[10] A cadmium-binding protein other than MT was also detected in a rat liver and was identified as ornithine carbamoyltransferase.[5]

[10] Y. Aoki, S. Hatakeyama, N. Kobayashi, Y. Sumi, T. Suzuki, and K. T. Suzuki, *Comp. Biochem. Physiol. C: Comp. Pharmacol. Toxicol.* **93C**, 345 (1989).

[17] Detection of Carboxymethylmetallothionein by Sodium Dodecyl Sulfate-Polyacrylamide Gel Electrophoresis

By MASAMI KIMURA, SHINZI KOIZUMI, and FUMINORI OTSUKA

Several methods, such as gel filtration, metal saturation, high-speed liquid chromatography with atomic absorption spectrophotometry, ion-exchange high-performance liquid chromatography, sodium dodecyl sulfate-polyacrylamide gel electrophoresis (SDS-PAGE), and radioimmunoassay (RIA), have been used to detect metallothioneins (MTs) in research laboratories.

Among them, RIA is the most effective method by which MTs can be directly measured as protein; but it is not easy to obtain the antibody with a constant high titer against MTs. RIA is not always the most convenient method for determining MTs.

Often SDS-PAGE, which is used for the analysis of protein, is employed to detect MTs in the cytosol fraction obtained from tissues and cells and to check their purity. However, because MT is rich in sulfhydryl residues of cysteines, it easily forms inter- and/or intramolecular linkages following the aggregation and configuration changes of the molecules. The electrophoretic patterns of MTs are sometimes obtained as broad, obscure bands in the gel. This can be resolved by blocking the sulfhydryl residues before electrophoresis.

This treatment involves dissociation of metals from the protein, as well as reduction of the sulfhydryl residues of constituent cysteines by denaturants such as sodium dodecyl sulfate or guanidine hydrochloride, and a reducing agent, such as dithiothreitol (DTT) or dithioerythritol, and then carboxymethylation of the residues with iodoacetic acid.

The carboxymethylated MTs are analyzed by SDS-PAGE. The simple electrophoretic patterns can be detected by dye and silver staining. In the case of the labeled MTs, which have a radioactive amino acid such as [^{35}S]cysteine incorporated *in vivo* or *in vitro,* the proteins can be detected by autoradiography.

Carboxymethylation of Protein

Protein solutions (10 μl) are mixed with 5 μl of 0.2 *M* Tris-HCl buffer (pH 8.8) containing 8% (w/v) sodium dodecyl sulfate (SDS), 50% (v/v) glycerol, and 2 μl of 0.2 *M* dithioerythritol (DTE). Then the mixture is boiled for 5 min.[1] Protein solutions can be slightly modified.[2] A portion of sample (7 μl) can be mixed with 10 μl of 100 m*M* Tris-HCl buffer (pH 8.8) containing 60 m*M* DTE, 40 m*M* EDTA, 4% SDS, and 25% glycerol.

After the reaction mixture is cooled, 3 μl of 1 *M* recrystallized iodoacetic acid (IA) solution, which is adjusted to about pH 8 with NaOH just before use, is added to the mixture. Then it is incubated at 50° for 15 min.

The final concentrations of SDS, DTE, and IA are 2%, 20 m*M*, and 150 m*M* in the reaction mixture, respectively. The concentration of DTE is equivalent to that of the sulfhydryl residues contained in about 13 mg of MT/ml. The amount of IA added is 7.5 times that of DTE. Under these conditions, at least 1.6 mg/ml in the final reaction mixture of purified rabbit liver MT-1[3] is completely carboxymethylated.

[1] S. Koizumi, N. Otaki, and M. Kimura, *Ind. Health* **20,** 101 (1982).
[2] F. Otsuka, S. Koizumi, M. Kimura, and M. Ohsawa, *Anal. Chem.* **168** (1988).
[3] M. Kimura, N. Otaki, and M. Imano, *Experientia, Suppl.* **34** (1979).

SDS-Polyacrylamide Gel Electrophoresis

SDS–15% polyacrylamide slab gels (1 mm thick) and electrode buffer are prepared according to the method of Laemmli.[4] The carboxymethylated protein solution (5–20 μl/2.5- or 4.5-mm slot) is mixed with one-tenth volume of 0.1% (w/v) Bromphenol Blue and applied to the gels (10-cm running gels). Electrophoresis is performed at 45–50 V for 12–24 hr until the Bromphenol Blue is about 9 cm from the starting edge of the running gel.

Dye Staining

Method A

Step 1. The gel is fixed in 10% (v/v) trichloroacetic acid/10% (v/v) acetic acid/30% (v/v) methanol for 1 hr.

Step 2. After fixation the gel is soaked in 0.025% (w/v) Coomassie Brilliant Blue/10% (v/v) acetic acid/50% (v/v) methanol with gentle shaking for 1 hr.

Step 3. The gel is destained in 7% acetic acid/25% methanol for several hours.

Method B

Step 1. The gel is soaked three times in 40% methanol containing 15% trichloroacetic acid for 20 min.

Step 2. The gel is stained with 0.25% Coomassie Brilliant Blue in 50% methanol/12.5% trichloroacetic acid for 15 min.

Step 3. The gel is destained three times for 10 min with 5% trichloroacetic acid, three times for 15 min with 40% methanol/10% acetic acid, and twice for 20 min with 10% ethanol/5% acetic acid.

Overnight destaining is not recommended, because the intensity of the MT bands decreases, presumably due to the loss of MTs from the gel.

Method A seems to be easier and simpler than method B, but the gel must be treated for silver staining as in method B.

Silver Staining

Step 1. The gel stained by method B is washed twice for 10 min with water. Then it is soaked in 1% (w/v) sodium thiosulfate for 15 min.

Step 2. After the gel is washed twice for 5 min with water, it is soaked in

[4] U. K. Laemmli, *Nature (London)* **227**, 680 (1970).

0.0032 *M* potassium dichromate containing 0.0032 *N* nitric acid for 10 min.

Step 3. After potassium dichromate is removed from the gel by washing twice for 5 min with water, it is soaked in 0.2% (w/v) silver nitrate for 20 min.

Step 4. The gel is washed for 1 min with water and then with a small amount of developer [0.28 *M* sodium carbonate containing 0.0185% (v/v) formaldehyde] for 30 sec.

Step 5. The gel is soaked in the developer until the desired image is obtained.

Step 6. The development is stopped by soaking the gel in 5% acetic acid for 5 min.

It is very important that all solutions are prepared with double-distilled water and all steps are carried out with gentle agitation at 35° under normal lighting. The period required for silver development to obtain sufficient sensitivity simultaneously yields a relatively high background.

Step 7. Silver-stained bands can be analyzed densitometrically at 420 nm. Before photographing, excess surface deposits of silver are removed with 20-fold diluted photographic fixer.[5]

When MTs are carboxymethylated with iodoacetic acid, the image is unexpectively negative using silver staining, although discrete bands are detected with dye staining. This negative image of carboxymethylated MTs is successfully reversed when the gel is restained after destaining with a photographic fixer such as sodium thiosulfate. The image reversal can be achieved by the pretreatment of the gel with this agent just before the silver-staining procedure. Washing with water in step 2 after the sodium thiosulfate treatment is critical because sodium thiosulfate remaining in the gel forms a soluble complex with silver ions, thereby preventing access of silver ions to MTs. Effective concentration of sodium thiosulfate ranges from 0.05 to 1%. Concentrations of sodium thiosulfate over or under this range greatly decrease the intensity of the MT bands.

The potassium dichromate treatment is necessary for enhancing the intensity of the MT bands. Also, Coomassie Brilliant Blue staining before silver staining is quite effective for intensification of the silver image of MTs.[6,7]

Although the detection limit by Coomassie Brilliant Blue staining is about 50 ng/lane in the gel, the detection limit using silver staining is

[5] R. C. Swutzer III, C. R. Merril, and S. Shifrin, *Anal. Biochem.* **98,** 231 (1979).

[6] M. R. DeMoreno, J. F. Smith, and R. V. Smith, *Anal. Biochem.* **151,** 466 (1985).

[7] S. Irie, M. Sezaki, and Y. Kato, *Anal. Biochem.* **126,** 350 (1982).

approximately 1.5 ng/lane for mouse liver MT-1 and MT-2. Both calibration curves for MTs from densitometric analysis are similar and of good linearity, at least up to 50 ng/lane. The silver-staining method is approximately 30-fold more sensitive than the dye-staining method.

Sample Preparation

MTs are induced in the tissues, as well as in the cultured cells, by heavy metals and hormones. Determination of MTs has often been undertaken with the cytosol fractions obtained from animal tissues and cultured cells treated with various inducers. When intact tissue (e.g., liver) cytosol is used for carboxymethylation, MTs cannot be detected because of the heavy interference throughout the lane. Such interference is hardly eliminated simply by heat treatment (100°, 3 min), although the amount of proteins per lane in the heat-stable fraction of cytosol is about 60 μg protein less than that of the intact cytosol. Dilution of samples after heat treatment can be effective for elimination of such interference when a sufficient amount of MTs is induced in the cytosol; otherwise the metallothioneins can be partially purified by acid treatment.[8] When the heat-stable, 5% tannic acid-insoluble fraction is used for carboxymethylation, protein bands corresponding to MTs emerge with slight interference. Treatment with 2% trichloroacetic acid before the tannic acid treatment can eliminate the interference from other proteins for the MT bands. The heat-stable, 2% trichloroacetic acid-soluble and 5% tannic acid-insoluble fraction is almost completely free from interference. No significant loss of MTs by heat treatment is observed, but the yield of MTs from cytosol after double treatments with trichloroacetic acid and tannic acid is approximately 50%. Therefore, quantitative analysis of MTs in cytosol must be carried out carefully.

Autoradiographic Analysis for SDS-PAGE

The supernatant cytosol fraction obtained from [^{35}S]cysteine-labeled cells, which induce MTs by various inducers, is applied to SDS-PAGE after carboxymethylation with method A.

The gels are soaked in 3 vol of autoradiography enhancer solution (EN^{3}HANCE, New England Nuclear, Boston, MA) for 1 hr after fixation with 10% trichloroacetic acid/10% acetic acid/30% methanol for 1 hr. Then the gels are washed with water for 1 hr; change the water two or three times. All procedures are done with gentle shaking. The gels are dried on filter paper with a slab gel drier. Kodak (Rochester, NY) X-Omat RP film

[8] A. J. Zelazowski and J. K. Piotrowski, *Acta Biochim. Pol.* **24,** 97 (1977).

is exposed to the dried gel sheets at $-70°$ for appropriate times. The developed film is analyzed by a densitometer at 540 nm.

Relative amounts of the labeled MTs can be estimated by this method, since the densitometric concentrations of MT bands on the fluorogram are proportional to the amounts of protein applied to the gel. For example, RK-13 cells, of the established cell line derived from rabbit kidney, are labeled with [^{35}S]cysteine (0.33 μCi/ml, 31 Ci/mmol) for 15 hr after addition of $CdCl_2$. A series of dilutions of the cytosol fraction containing the induced MTs is carboxymethylated and then electrophoresed on a gel. The gel is autoradiographically fluorographed. The densitometric concentrations of the thionein bands are plotted against the amount of input protein. The curve is linear, demonstrating that this method is suitable for the quantitative estimation of labeled thioneins. As MT contains most of the cysteine residues in the molecule, it is very effective to use [^{35}S]cysteine as the incorporated amino acid of the induced thionein. The exposure time as well as the labeling periods can be shortened by using larger amounts of [^{35}S]cysteine with high specific radioactivity.

Small amounts of MTs in several samples can be simultaneously measured on gel by silver staining, as well as by autoradiography, after carboxymethylation following SDS-PAGE. These methods are very convenient and effective for various experiments of MT induction *in vitro* and *in vivo*.

Further, iso-MTs can be separated by this procedure, although it is very difficult to separate them by regular SDS-PAGE. On 20-cm gels the separation is sufficient to identify each of the iso-MTs. Chemical modification of the amino acid residues specific to each iso-MT may result in changes in their mobilities and better separation from each other. However, electrophoresis at a lower current and for a long time is not recommended, because MT-1 and MT-2 are not separated. For good separation the electrophoresis must be performed at 20 mA (constant) for 3.5 hr. In addition, gel composition is also important to separate isoforms; use 17.5% total acrylamide and 2.23% bisacrylamide.

Summary

A sensitive method for detecting metallothioneins (MTs) by using silver staining and autoradiography after sodium dodecyl sulfate-polyacrylamide gel electrophoresis (SDS-PAGE) of carboxymethylated MTs is described. Carboxymethylation of MTs is indispensable because it prevents their aggregation, thereby allowing each of them to appear as a single band using SDS-PAGE. Metallothioneins can be detected with a limit of nanogram levels per lane. This method can be applied to MTs induced in *in vitro* cultured cells and in *in vivo* tissues.

[18] Radioimmunoassay for Metallothionein in Body Fluids and Tissues

By ZAHIR A. SHAIKH

The radioimmunoassay (RIA) of metallothionein (MT) has been in use in our laboratory since the late 1970s.[1] A number of improvements have been made.[2-4] Both single- and double-antibody assays have been developed and applied to the study of MT in biological materials derived from animals as well as humans.[4-22] The single most important advancement made possible with RIA has been the quantitation of MT in human urine. The studies with urine specimens from cadmium-exposed populations have revealed that urinary MT is a biological indicator of the cadmium body burden.[6-8,10,12,16,18,19,21] In fact, there is evidence that in individuals chronically exposed to cadmium, the urinary MT, like other low-molecular-weight plasma proteins is, in addition, a specific indicator of renal tubular dysfunction.[8,12,20,22]

[1] C. Tohyama and Z. A. Shaikh, *Biochem. Biophys. Res. Commun.* **84,** 907 (1978).
[2] C. Tohyama and Z. A. Shaikh, *Fundam. Appl. Toxicol.* **1,** 1 (1981).
[3] C. V. Nolan and Z. A. Shaikh, *Anal. Biochem.* **154,** 213 (1986).
[4] C. V. Nolan and Z. A. Shaikh, *Toxicol. Appl. Pharmacol.* **85,** 135 (1986).
[5] Z. A. Shaikh and C. V. Nolan, *Experientia, Suppl.* **52,** 343 (1987).
[6] C. Tohyama, Z. A. Shaikh, K. Nogawa, E. Kobayashi, and R. Honda, *Toxicology* **20,** 289 (1981).
[7] C. Tohyama, Z. A. Shaikh, K. J. Ellis, and S. H. Cohn, *Toxicology* **22,** 181 (1981).
[8] C. Tohyama, Z. A. Shaikh, K. Nogawa, E. Kobayashi, and R. Honda, *Arch. Toxicol.* **50,** 159 (1982).
[9] Y. H. Lee, Z. A. Shaikh, and C. Tohyama, *Toxicology* **27,** 337 (1983).
[10] Z. A. Shaikh and C. Tohyama, *Environ. Health Perspect.* **54,** 171 (1984).
[11] T. Kido, Z. A. Shaikh, H. Kito, R. Honda, and K. Nogawa, *Toxicology* **65,** 323 (1991).
[12] Z. A. Shaikh, C. Tohyama, and C. V. Nolan, *Arch. Toxicol.* **59,** 360 (1987).
[13] C. V. Nolan and Z. A. Shaikh, *Biol. Trace Elem. Res.* **12,** 419 (1987).
[14] K. J. McVety and Z. A. Shaikh, *Toxicology* **46,** 295 (1987).
[15] M. E. Blazka, C. V. Nolan, and Z. A. Shaikh, *J. Appl. Toxicol.* **8,** 217 (1988).
[16] Z. A. Shaikh, K. M. Harnett, S. A. Perlin, and P. C. Huang, *Experientia* **45,** 146 (1989).
[17] Z. A. Shaikh and P. C. Tewari, *Experientia* **46,** 694 (1990).
[18] Z. A. Shaikh, *in* "Metallothionein in Biology and Medicine" (C. D. Klaassen and K. T. Suzuki, eds.), in press. CRC Press, Boca Raton, Florida, 1991.
[19] Z. A. Shaikh, K. J. Ellis, K. S. Subramanian, and A. Greenberg, *Toxicology* **63,** 53 (1990).
[20] Z. A. Shaikh, T. Kido, H. Kito, R. Honda, and K. Nogawa, *Toxicology* **64,** 59 (1990).
[21] T. Kawada, C. Tohyama, and S. Suzuki, *Int. Arch. Occup. Environ. Health* **62,** 95 (1990).
[22] C. Tohyama, Y. Mitane, E. Kobayashi, H. N. Sugihira, A. Nakano, and H. Saito, *J. Appl. Toxicol.* **8,** 15 (1988).

In this chapter details of the currently used RIA method for determining MT in tissues, as well as in urine, serum, and plasma, will be described. As with any immunoassay, the first and foremost requirement is an antibody with high titer as well as specificity. We have developed antibodies against rat and human MTs in rabbits and against rat MT in goats and sheep. In most cases the antibodies were developed against a mixture of MT-1 and MT-2. Therefore, the antibodies are not specific against either of the isomers and react evenly with both. Since the immunological response varies widely, a sufficient number of animals must be immunized to select the sera with the best titer as well as reactivity with MTs from various origins. Due to its low molecular weight, MT is rapidly removed from circulation by the kidneys. Therefore, in order to increase its plasma half-life, the MT can be either conjugated to another protein or polymerized to form larger, unfilterable aggregates.[2] In the case of MT, the latter approach appears to be more effective. However, one practical problem is that the antibodies recognize the MT polymer as well as the monomer and proper precautions must be taken in sample storage and preparation to avoid obtaining erroneously high values.[3]

Single-Antibody RIA

Reagents

Reagents for MT iodination: 0.1 *M* potassium phosphate buffer, pH 8.6; phosphate buffer containing 0.25% (w/v) gelatin; and phosphate buffer containing 0.2% (v/v) Tween 20. The Bolton–Hunter reagent can be obtained from Amersham Corporation (Arlington Heights, IL) or New England Nuclear (Boston, MA)

BNT buffer: 0.125 *M* boric acid, 0.125 *M* sodium chloride, 1 ml/liter Tween 20, and 0.02% (w/v) sodium azide. The pH is adjusted to 8.6 with 5 *N* hydrochloric acid

BNTB buffer: This buffer is the same as above, except that it also contains 0.04% bovine serum albumin (BSA)

Pansorbin (Calbiochem Corporation, La Jolla, CA): The stock is a 10% suspension of killed *Staphylococcus aureus* cells. The binding capacity of this product for IgG varies between 1 and 3%. A lot with higher binding is desirable. For use in RIA, the Pansorbin stock is diluted 1 : 20 with BNTB buffer

Serum: The antiserum is diluted with BNTB buffer to get at least 20% binding with ^{125}I-labeled MT alone. The preimmune serum is diluted to the same dilution.

MT standard: The stock solution should be kept at 4° at a concentration of at least 500 μg/ml in phosphate or BNT buffer. The working standards (1 – 100 ng/100 μl) are prepared by serial dilution of the stock with BNTB buffer

Procedures

Iodination of Metallothionein. This procedure is similar to the one described in detail by Bolton and Hunter[23] and in our previous publications.[2,3] Briefly, the Bolton-Hunter reagent (containing 1 mCi ^{125}I) is dried under a gentle stream of nitrogen or argon at 4°. The benzene evaporates in about 10 min. To this, 5 μg of rat hepatic Cd-MT-2 in 10 μl of phosphate buffer is added and quickly mixed. The reaction is allowed to proceed overnight at 4°. The next day the contents of the vial are transferred with several washings of the phosphate buffer to a 0.9×15 cm column packed with Sephadex G-25 (fine) and eluted at 13 ml/hr with the phosphate buffer containing gelatin. The fractions (0.8 ml) are collected in polystyrene tubes containing 0.2 ml of phosphate buffer with Tween. A 10-μl aliquot of each fraction is assayed for ^{125}I radioactivity in a γ spectrometer and the first ^{125}I peak, which contains the conjugated MT, is pooled, diluted to 40 ml with phosphate buffer, and stored in 0.5-ml aliquots at −85°. The yield of ^{125}I bound to MT varies between 30 and 40%. ^{125}I-Labeled MT degrades over time and must be repurified just before use. The stock solution of ^{125}I-labeled MT should be discarded after 2 months and a fresh batch prepared as needed.

Purification of ^{125}I-Labeled Metallothionein. A plastic pipette tip (bed volume 1.2 ml) or a glass pipette (bed volume 5.5 ml) is packed with Sephadex G-25. The column is rinsed with BNT buffer. Up to 0.2 ml of the ^{125}I-labeled MT stock is applied to the pipette tip column and up to 1 ml to the glass pipette column. The selection of the column is based on the age of the stock solution and the total amount of ^{125}I-labeled MT needed for the assay. The columns are eluted with a fixed volume of BNT buffer (800 μl for pipette tip column and 3.5 ml for the glass column). The volume of the eluate is adjusted with BNTB buffer to get about 20,000 cpm/100 μl.

Sample Preparation. Urine, serum, and plasma: The frozen specimens are thawed to 4°, mixed thoroughly, and centrifuged at 3000 rpm for 10 min at 4°. The samples are maintained on ice. Heat treatment (see below) is not necessary.

Tissues: The frozen tissue specimens are thawed to 4°. A small piece is cut, weighed, and homogenized in BNTB buffer (5%, w/v). An aliquot (1 ml) is transferred to a polystyrene microcentrifuge tube (Fisher Scien-

[23] A. E. Bolton and W. M. Hunter, *Biochem. J.* **133,** 529 (1973).

tific Company, Pittsburgh, PA), heated at 80° for 10 min, cooled on ice, and centrifuged in a microcentrifuge (Fisher Scientific Company or Eppendorf) at 7000–10,000 *g* for 10 min at 4°. The supernatant is maintained on ice and used within 1 hr. Sample dilution is often necessary; typically the control tissue supernatants are diluted 1 : 1 and those from metal-treated animals are diluted 100 times or more.

Radioimmunoassay Procedure. To 12 × 75 mm conical bottom polystyrene tubes 100 μl of sample or standard solution is added in triplicate, followed by 100 μl of freshly purified ^{125}I-labeled MT and 100 μl diluted antiserum. The tubes are mixed and incubated at 4° for 16–24 hr. At the end of the incubation 200 μl diluted Pansorbin is added to bind the rabbit ^{125}I-labeled MT–IgG complex. After thorough mixing the tubes are incubated at room temperature for 30 min and centrifuged at 3000 rpm for 10 min. The supernatant is aspirated and discarded. The pellet is resuspended in 1.5 ml BNTB buffer, centrifuged, and the supernatant discarded. ^{125}I-Labeled MT activity in the pellet is determined in a γ spectrometer.

For determining nonspecific binding (blank) 100 μl of diluted preimmune serum is used in place of the antiserum and 100 μl of BNTB buffer is used in place of the sample or standard. The maximum binding (B_0) is determined by incubating 100 μl ^{125}I-labeled MT, 100 μl BNTB buffer, and 100 μl antiserum. Incubation times and treatment with Pansorbin remain the same as above. The net binding (B or B_0) is calculated after subtraction of the preimmune blank. For the standard curve the log of MT concentrations and the respective B/B_0 values are plotted. Metallothionein concentration in unknown samples is determined from their B/B_0 values using the standard curve.

Double-Antibody RIA

Reagents. All reagents for this assay are the same as for the single-antibody assay except that Pansorbin is replaced with the anti-IgG serum for the animal species in which the primary antibody is raised. The second antibody can be commercially obtained. Optimum dilution of this antiserum for precipitating the primary antibody is determined and is prepared freshly.

Procedure

The first four steps of the procedure are the same as for the single-antibody assay. After incubation with the primary antibody for 16–24 hr 100 μl of diluted serum containing the second antibody is added. The contents are mixed and incubated at 4° for an additional 24 hr. The tubes

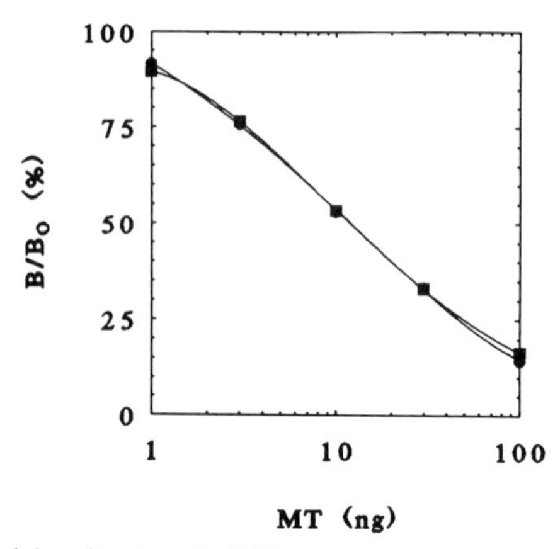

FIG. 1. Cross-reactivity of rat hepatic Cd-MT-1 (●) and Cd-MT-2 (■) with rabbit anti-rat hepatic MT serum in a single-antibody RIA. The ratio of ^{125}I-labeled MT binding to the antibody in the presence (B) and absence (B_0) of added MT is plotted against the MT content in the assay tube. The rat hepatic MTs were induced in response to cadmium and isolated and purified by gel filtration and ion-exchange chromatography as described previously. [Z. A. Shaikh and O. J. Lucis, *Experientia* **27**, 1024 (1971).]

are centrifuged at 3000 rpm for 30 min to sediment the fine particles. The supernatant is aspirated and discarded. The pellet, which is barely visible, is washed with 1.5 ml BNTB buffer, centrifuged again for 30 min, and the supernatant is discarded. ^{125}I-Labeled MT activity remaining in each tube is determined in a γ spectrometer and the results are calculated as described above for the single-antibody assay.

Single-Antibody RIA. In the original assay ammonium sulfate precipitation procedure was used to separate the immunoglobin-bound ^{125}I-labeled MT from the free ^{125}I-labeled MT.[2] This has since been replaced with Pansorbin. Pansorbin (*Staphylococcus aureus* cells) contains protein A, which binds strongly with the IgG. With the use of Pansorbin the nonspecific binding has been reduced from as high as 10% to about 1%. The single-antibody assay is relatively rapid in that results can be obtained within 2 days. In our laboratory this assay has been widely used to determine total MT in serum, plasma, urine, and tissues. Of the 25 rabbit serums tested so far, all reacted equally with both rat MT-1 and MT-2 (Fig. 1). The RIA can easily measure as little as a few nanograms of MT in 100 μl urine, plasma, or serum and in 2.5 mg or less of tissue.

The antisera react differently with MTs from various mammalian species. Some anti-rat MT sera cross-react completely with MTs from species other than rat[2,3] while others show varying degrees of cross-reactivities (Fig. 2). Thus, it is important that, along with the samples, purified MTs from

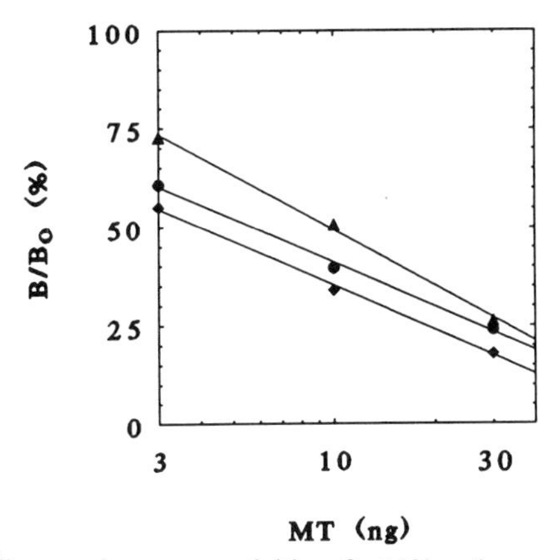

FIG. 2. Species differences in cross-reactivities of rat (●) and mouse (◆) hepatic Cd-MT-2 and human (▲) renal MT-1 with a rabbit anti-rat hepatic MT serum. The proteins were isolated and purified as listed under Fig. 1. Complete cross-reactivity of other rabbit anti-rat MT sera with mouse, rabbit, and human MTs is previously reported.[2-4,6]

the same species are used to generate the standard curve. Obviously, if the antiserum reacts equally with MT from other species, the standard curve for rat MT can be substituted for calculating the MT in body fluids or tissue samples from another species. So far, we have found no antiserum that cross-reacts with nonmammalian species.

As mentioned in the procedure section, tissue samples are heated at 85° to remove the high-molecular-weight proteins and the subcellular particles.

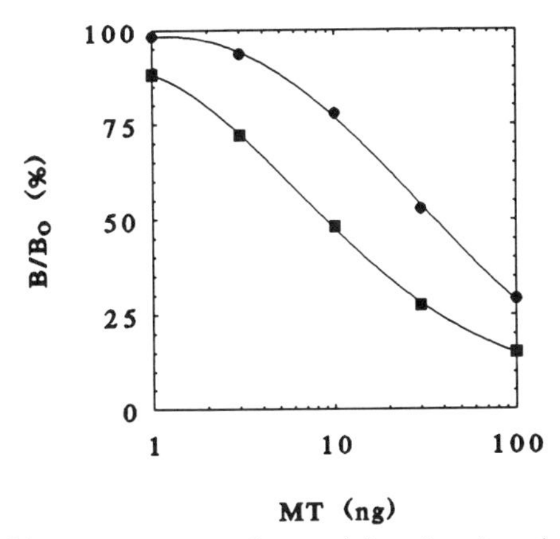

FIG. 3. Effect of heat treatment on the reactivity of rat hepatic Cd-MT-2 with rabbit anti-rat hepatic MT serum. The tubes containing MT standards were kept either at 4° (■) or heated at 80° (●) for 10 min then cooled to 4°, before incubating with the antiserum.

This is helpful in reducing the nonspecific binding (with preimmune serum) to about 1%. It is essential that the appropriate standard MT solutions are also heated at the same temperature as the tissue samples. We have discovered that if the standard curve from unheated MT is used to calculate the MT in supernatants from heat-treated homogenates, considerably higher MT values are obtained. The heated standards generate a curve that is much different from the unheated standards (Fig. 3) and

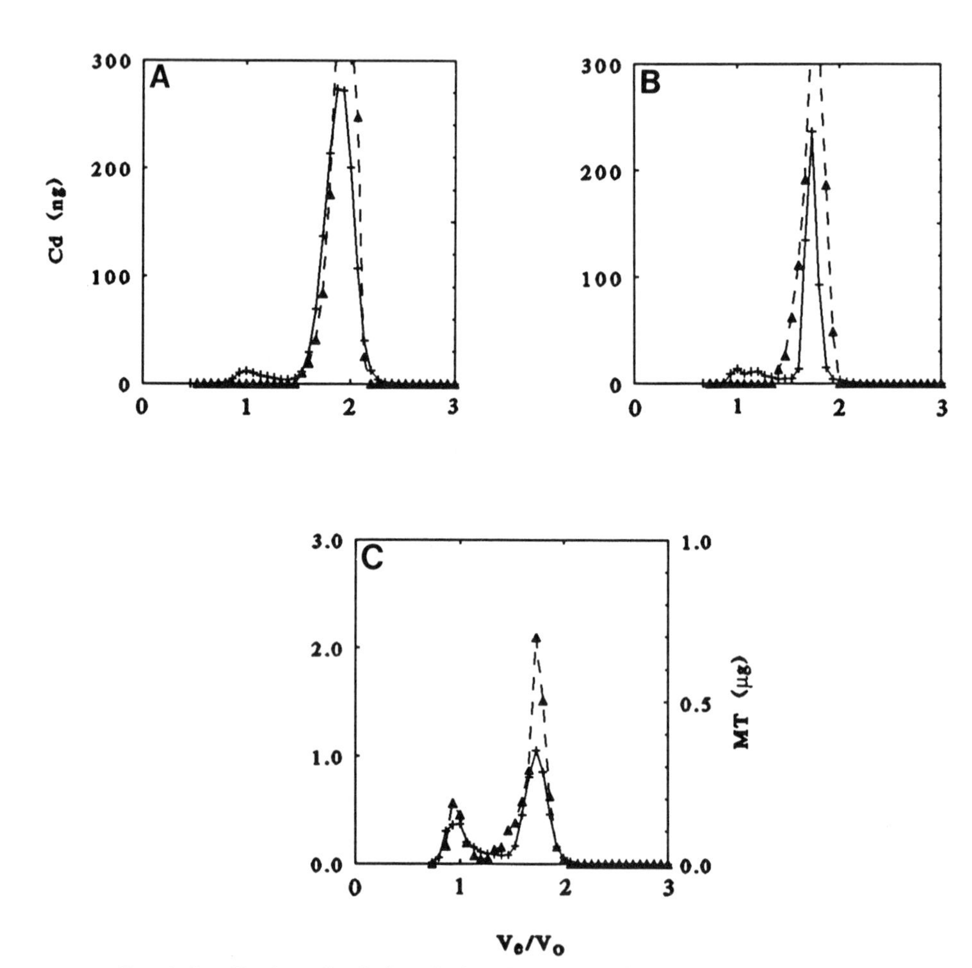

FIG. 4. Localization of cadmium (—) and MT(-▲-) in supernatants of (A) liver, (B) kidney, and (C) testis homogenates of mice injected subcutaneously with 10 μmol $CdCl_2$/kg for 24 hr. The supernatants (1 ml) were applied to 0.9 × 60 cm Sephadex G-75 (superfine) columns and eluted at 2 ml/hr. Metallothionein was analyzed in 1-ml column fractions from heat-treated supernatants by single-antibody RIA. (Modified from Shaikh and Tewari.[17])

TABLE I
METALLOTHIONEIN LEVELS IN MOUSE LIVER DETERMINED BY RADIOIMMUNOASSAY AND CADMIUM SATURATION METHOD

Method	MT (μg/g)[a]	
	Control	Cd injected[b]
RIA	3.5 ± 1.9	240 ± 33
Cd saturation[c]	3.8 ± 3.1	230 ± 41

[a] Values are mean ± SD ($N = 4$).
[b] Male mice of C3H strain were injected subcutaneously with 30 μmol $CdCl_2$/kg body weight and sacrificed 24 hr later.
[c] Method described by D. L. Eaton and B. F. Toal, *Toxicol. Appl. Pharmacol.* **66,** 134 (1982).

provides MT values in samples that are comparable to those obtained by another method.

Figure 4 shows that in supernatants from homogenates of liver, kidney, and testis of mice injected with cadmium, the majority of the cadmium elutes as a single peak with a V_e/V_0 of about 1.8; a smaller cadmium peak is also present in the void volume of the Sephadex G-75 column. When the column fractions from heat-treated supernatants were analyzed for reactivity with the MT antibody, the reactivity was observed only in the low-molecular-weight cadmium-binding (i.e., MT) region of liver and kidney. This demonstrates that in supernatants of heated tissue homogenates the RIA detects only MT and no other higher or lower molecular weight substance. Similar results are obtained for unheated human urine.[6] In the testis, both the high- and low-molecular-weight cadmium-binding proteins react with the antibody. It appears that the low-molecular-weight protein in the testis is MT and that the high-molecular-weight species is an MT polymer.[17] Reactivity of the antibody with MT polymers produced due to oxidation of hepatic and renal MT has been discussed previously.[3]

As cadmium is unable to displace copper from MT the cadmium saturation assay gives lower values in the presence of copper.[24] However, the MT induced in liver by cadmium or zinc contains very little copper. Therefore, this method can be used to compare the hepatic MT values obtained by the RIA. Table I shows MT concentration in livers of control

[24] D. L. Eaton, *Toxicol. Appl. Pharmacol.* **78,** 158 (1985).

TABLE II
METALLOTHIONEIN CONCENTRATION IN RAT SERUM, LIVER, AND KIDNEY[a]

Group	Serum (μg/ml)	Liver (μg/g)	Kidney (μg/g)
Control	0.20 ± 0.01	31.2 ± 3.0	54.9 ± 8.8
Cd exposed	0.24 ± 0.02	583.2 ± 21.6	794.8 ± 45.4

[a] MT was determined by single-antibody RIA in tissues of 4-month-old male Sprague-Dawley rats maintained, for 3 months prior to sacrifice, on rat chow and either deionized water (controls) or on water containing 100 mg Cd/liter. The values of MT are expressed as mean ± SE ($N = 4$).

and cadmium-injected mice determined by the two methods. The two sets of values, at both the low and high end of the scale, were very similar, further confirming that the RIA measures MT only and that for tissues the heated standard is the appropriate choice.

Metallothionein values in tissues of control rats and rats treated with cadmium in the drinking water, determined by the single-antibody RIA, are shown in Table II. The liver MT values in controls were somewhat higher than previously reported from this laboratory.[3,5,13-15] This could be

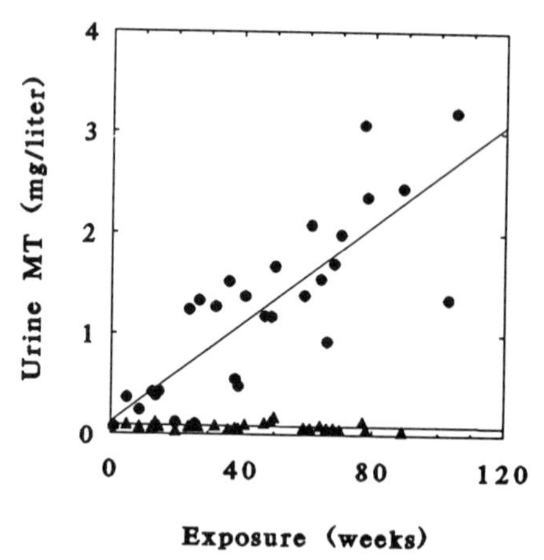

FIG. 5. Metallothionein concentration in urine of cadmium-exposed rats. The animals were given 50 mg Cd/liter drinking water for the duration indicated. Metallothionein was analyzed by single-antibody RIA. (▲), Control; (●), cadmium exposed. (Adapted from Shaikh *et al.*[16])

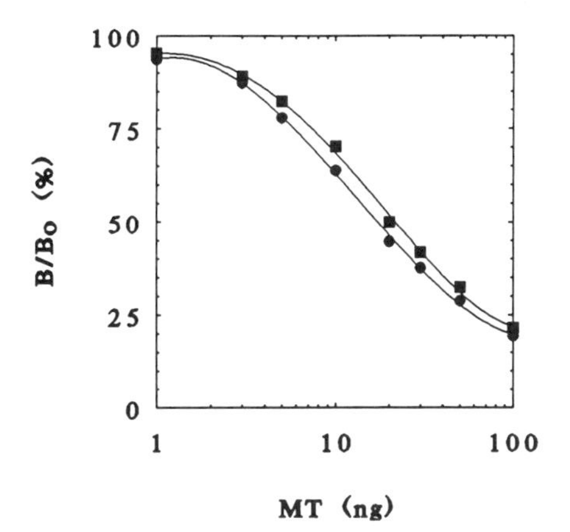

FIG. 6. Cross-reactivity of rat hepatic Cd-MT-1 (●) and Cd-MT-2 (■) with sheep anti-rat hepatic MT serum in a double-antibody RIA. The second antibody was rabbit anti-sheep IgG.

due to a reaction to stress, as only the hepatic MT levels are elevated and the renal MT levels are within the normal range.

Urinary MT levels, determined in another group of control rats and rats kept on cadmium in their drinking water, are depicted in Fig. 5. The values in controls ranged between 41 and 180 μg/liter, with an overall group mean of 87 μg/liter. In cadmium-exposed animals the MT excretion in urine increased linearly during the course of the study to values that were greater than 3 mg/liter. Such increases in urinary MT excretion have also been observed in cadmium-exposed human populations.[6-8,10,12,18-22]

Double-Antibody RIA. This RIA requires a second antibody to precipitate the ^{125}I-labeled MT–IgG complex. The second antibody replaces Pansorbin. This assay takes 3 days to complete as opposed to only 2 days for the single-antibody assay. An advantage of this assay is that the nonspecific binding is always lower and the specific binding higher than with some lots of Pansorbin.

In our laboratory both rabbit and sheep antisera against the rat MT have been used for the double-antibody RIA. The results of MT analysis in tissues and fluids were similar to those with the single-antibody RIA. Double-antibody RIAs for MT have also been described by Vander Mallie and Garvey[25] and Mehra and Bremner.[26] One major difference between our assay and that of the latter group[26] is that their sheep antiserum reacted with MT-1 only, whereas our sheep antiserum, like the rabbit antiserum, reacts about equally with both MT-1 and MT-2 (Fig. 6).

[25] R. J. Vander Mallie and J. S. Garvey, *J. Biol. Chem.* **254,** 8416 (1979).

[26] R. K. Mehra and I. Bremner, *Biochem. J.* **213,** 459 (1983).

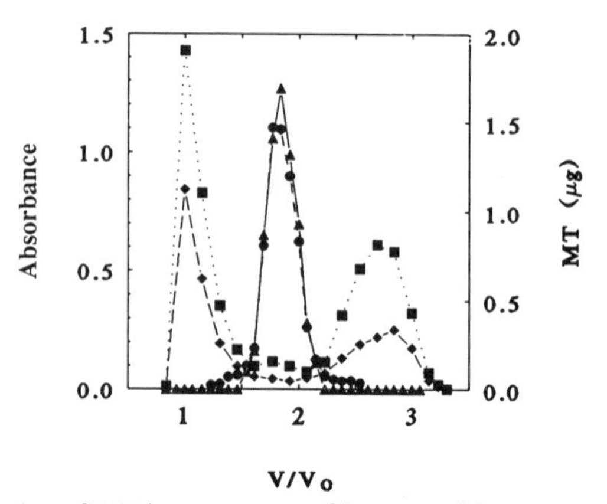

FIG. 7. Localization of MT in supernatant of heat-treated homogenate of liver from a cadmium-treated rat by double-antibody RIA (▲) and cadmium saturation (●) assay. The supernatant (1 ml) was applied to a 0.9 × 60 cm Sephadex G-75 column and eluted with BNTB buffer at 10 ml/hr. Each fraction (1 ml) was analyzed for absorbance at 250 nm and (■) and 280 nm (◆) as well as MT.

Like the single-antibody RIA, the double-antibody RIA can be validated by assay of column fractions following separation of a heated liver supernatant on a Sephadex column. Such data for a liver of cadmium-treated rat are shown in Fig. 7. The Cd–S bond has characteristic absorbance at 250 nm. This helps identify the approximate location of Cd–MT fractions. However, we analyzed all column fractions by both the RIA and the cadmium saturation assay. The latter assay helps confirm the location of the cadmium-binding protein (MT) peak. As can be seen in Fig. 7, the RIA shows reactivity only in the MT region. Also, the values obtained with the two assays are very similar, further confirming the accuracy of the RIA. It should be pointed out, however, that although the cadmium saturation assay gives hepatic MT values that are similar to those obtained by the RIA, it is less sensitive and cannot be used for MT analysis in plasma, serum, or urine. Furthermore, the inability of cadmium to displace copper, silver, and mercury from MT causes the cadmium saturation assay to underestimate MT concentration[24] in tissues with naturally high copper concentration (i.e., kidney) and in situations where MT is induced by metals other than cadmium or zinc.

Acknowledgment

This work was supported by a U.S. Public Health Service Grant No. ES03187.

[19] Measurement of Human Metallothionein by Enzyme-Linked Immunosorbent Assay

By ROBERT J. COUSINS

Introduction

Metallothionein has been a focus of research interest for a variety of reasons. Ours has been the involvement of the protein in certain aspects of zinc metabolism and function. Early studies showed that synthesis of metallothionein is sensitive to dietary zinc intake in liver and intestine.[1,2] Subsequent studies have shown conclusively that expression of the metallothionein gene is sensitive to the dietary zinc supply[3] as well as to specific hormones and mediators of disease (see Refs. 4 and 5 for reviews). Up regulation of expression has been related to cellular zinc accumulation and cytoprotection by undefined mechanisms.[6]

Research directed at the physiologic characteristics of human metallothionein has received little attention. Human metallothionein genes have been cloned[7] and their cell type-specific expression in isolated human cells has been studied.[8] Physical properties of human metallothionein, particularly the microheterogeneity of the metallothionein-1 isoform, have been well characterized.[9] Information about the metallothionein content of human tissues and fluids is scant, limited mainly to measurement in liver and kidney samples obtained at autopsy.[10,11]

Radioimmunoassays (RIAs) for rat metallothionein have been developed. These are sufficiently sensitive to measure plasma and urinary metallothionein in animal studies.[12,13] RIA methods have been applied in limited studies to examine human metallothionein.[14] Following the notion

[1] C. C. McCormick, M. P. Menard, and R. J. Cousins, *Am. J. Physiol.* **240,** E414 (1981).
[2] M. P. Menard, C. C. McCormick, and R. J. Cousins, *J. Nutr.* **111,** 1353 (1981).
[3] T. L. Blalock, M. A. Dunn, and R. J. Cousins, *J. Nutr.* **118,** 222 (1988).
[4] R. J. Cousins, *Physiol. Rev.* **65,** 238 (1985).
[5] M. A. Dunn, T. L. Blalock, and R. J. Cousins, *Proc. Soc. Exp. Biol. Med.* **185,** 107 (1987).
[6] J. J. Schroeder and R. J. Cousins, *Proc. Natl. Acad. Sci. U.S.A.* **87,** 3137 (1990).
[7] R. I. Richards, A. Heguy, and M. Karin, *Cell (Cambridge, Mass.)* **37,** 263 (1984).
[8] U. Varshney, N. Jahroudi, R. Foster, and L. Gedamu, *Mol. Cell. Biol.* **6,** 26 (1986).
[9] J. H. R. Kägi and Y. Kojima, *Experientia, Suppl.* **52,** 34 (1987).
[10] S. Onosaka, K.-S. Min, C. Fukuhara, K. Tanaka, S.-I. Tashiro, I. Shimizu, M. Futura, T. Yasutomi, K. Kobashi, and K.-I. Yamamoto, *Toxicology* **38,** 261 (1986).
[11] J. Chung, N. O. Nartey, and M. G. Cherian, *Arch. Environ. Health* **41,** 319 (1986).
[12] R. K. Mehra and I. Bremner, *Biochem. J.* **213,** 459 (1983).
[13] M. Sato, R. K. Mehra, and I. Bremner, *J. Nutr.* **114,** 1683 (1984).
[14] J. S. Garvey, R. J. Vander Mallie, and C. C. Chang, this series, Vol. 84, p. 121.

that sufficient background exists to explore dietary and physiological aspects of metallothionein in human subjects, we initiated a project to develop an enzyme-linked immunosorbent assay (ELISA) for human metallothionein based on anti-human metallothionein IgG. ELISA technology offers a decrease in assay time compared to the RIA as well as the advantage of not requiring radioisotopes.

Purification of Human Liver Metallothionein

Samples of human liver are obtained as soon as possible after death. Tissue is frequently received frozen. On receipt, all tissues are stored at −80° until used. Metallothionein levels in tissues vary considerably, which is related to subject age and cause of death as well as length of time before the tissue is initially frozen.[15]

Liver samples in 50-g aliquots are combined (1:2, w/v) with 1 m*M* Tris-HCl (pH 8.0, 0.02% NaN_3) and homogenized with a Polytron tissue grinder (Brinkmann Instruments, Westbury, NY). Homogenization is carried out in a laminar flow hood following appropriate aseptic measures. The homogenate is centrifuged at 40,000 *g* for 10 min (4°). The supernatant is heated in a boiling water bath with intermittent stirring until the temperature of the extract is at 100° for 5 min. The supernatant containing cytoplasmic proteins, which are now predominantly heat denatured, is centrifuged at 40,000 *g* (max) for 30 min (4°). The clear supernatant solution is removed and concentrated to 10–15 ml by ultrafiltration (PM10 membrane; Amicon, Danvers, MA) at 40 psi of N_2. About 100 ml of supernatant solution, representing 150 g of liver, can be concentrated for this step. The sample is applied to a 2.5 × 60 cm column of Sephadex G-75 at 4°. Elution is with 1 m*M* Tris-HCl (pH 8.0, 0.02% NaN_3). Zinc in fractions is monitored by atomic absorption spectrophotometry, and fractions composing the metallothionein peak (approximate $V_e/V_0 \approx 2$) are pooled. These fractions are concentrated (YM5 ultrafiltration membrane; Amicon) using several volume exchanges of the column buffer. This impure metallothionein is stored in aliquots at −80° until further purification steps.

Preparation of usable quantities of metallothionein isoforms is achieved with anion-exchange high-performance liquid chromatography (HPLC). Samples containing up to 10 mg protein are injected onto a Mono Q 5/5 anion-exchange column (Pharmacia, Piscataway, NJ). Metallothionein-1 (MT-1) and -2 (MT-2) isoforms are eluted using a step gradient of 10 m*M* Tris-HCl (pH 8.0) for MT-1 and 30 m*M* Tris-HCl (pH

[15] A Grider, K. J. Kao, P. A. Klein, and R. J. Cousins, *J. Lab. Clin. Med.* **113,** 221 (1989).

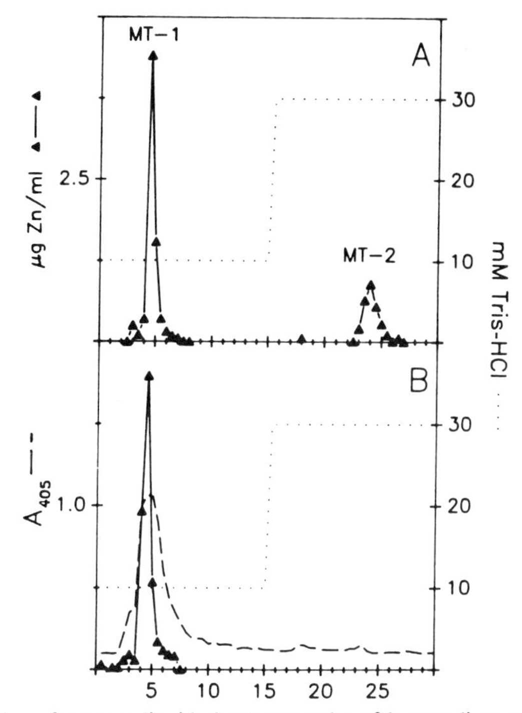

FIG. 1. High-performance liquid chromatography of human liver metallothionein isoforms using a Pharmacia Mono Q anion-exchange column and a step gradient of 10 and 30 m*M* Tris-HCl. Samples were derived from heat-treated liver cytosol followed by partial purification using gel-filtration chromatography. (A) Separation of isoform, metallothionein-1, and metallothionein-2 as measured by elution of zinc-containing fractions. (B) Rechromatography of metallothionein-1 showing immunoreactivity to sheep anti-human metallothionein-1 IgG with the ELISA procedure as measured by absorbance at 405 nm. (Reproduced from Grider *et al.*[15] with kind permission of C. V. Mosby, St. Louis.)

8.0) for MT-2. At a flow rate of 2 ml/min, a complete chromatography cycle can be completed in 1 hr. For this method, the microheterogeneity of MT-1[9] is not considered. A typical separation of the isoforms is shown in Fig. 1.[15] Isoforms are identified by zinc content (atomic absorption) and are separately concentrated by ultrafiltration (YM2 membrane; Amicon) and then stored at −80° as aliquots at a concentration of 100 μg/ml.

Metallothionein protein concentrations are calculated using measurement of protein by the Lowry method.[16] The high cysteine content of this protein appears to enhance chromophore development and produces an atypically high absorbance. The value obtained by Lowry is reduced four-

[16] O. H. Lowry, N. J. Rosebrough, A. L. Farr, and R. J. Randal, *J. Biol. Chem.* **193,** 265 (1951).

fold to obtain the correct metallothionein concentration. This correction factor was verified by zinc content and amino acid analysis.[15] For each laboratory the exact correction factor should be separately obtained. Furthermore, amino acid composition or microsequence data to verify the purification procedure are strongly recommended as an essential step for assay standardization.

Preparation of Sheep Anti-Human Metallothionein Immunoglobulin G

Antisera are generated in male adult sheep using a method similar to that outlined by Mehra and Bremner.[12] The high IgG titers and large pool of IgG produced are attractive features of using this species. Purified human MT-1 or MT-2 is added to purified rat IgG (Sigma, St. Louis, MO) at 1 : 2 (w/w) and dissolved in 1 ml 50 m*M* sodium phosphate buffer, pH 7.4, followed by 3.8 μl of 25% (w/v) glutaraldehyde. Typically, 1–3 mg of purified metallothionein is used. The mixture is incubated for 2 hr at room temperature, then 1.4 ml of 50 m*M* sodium phosphate buffer (pH 7.4) is added and the mixture is placed at 4° overnight. Just prior to injection, the mixture is combined (1 : 1, v/v) with Freund's complete adjuvant. The antigen is administered as multiple intradermal and intramuscular injections. Booster injections, in which Freund's incomplete adjuvant replaces complete adjuvant, are given at roughly 3- to 4-week intervals. After titers are sufficiently high, initially tested by Ouchterlony double diffusion,[17] serum is obtained, frozen, and stored at −80°. Prior to obtaining serum, after a period when there are no immunizations, the sheep receive at least two booster injections.

Sheep anti-human metallothionein IgG is prepared by protein A affinity chromatography. Protein G may provide higher yields of sheep IgG. Sheep serum (10 ml) is diluted with 5–10 vol of binding buffer (10 m*M* Na_2HPO_4, 0.15 *M* NaCl, pH 8.2), and passed through a 0.7 × 15 cm column of protein A agarose (Bethesda Research Laboratories, Gaithersburg, MD) basically as outlined by the manufacturer. The column is washed with binding buffer until absorbance at 280 nm is near 0 (usually 10 bed volumes). IgG is eluted with 0.1 *M* sodium citrate (pH 3.0) in 1-ml fractions into tubes containing 0.57 ml of 1 *M* Tris (Sigma) to buffer the IgG-containing fractions to pH 8.0. Fractions with the maximum absorbance are pooled, dialyzed against phosphate-buffered saline for 24 hr at 4°, and measured for protein concentration. Routinely, the IgG concentration of serum after multiple metallothionein injections is about 125 mg/ml of serum.

[17] J. S. Garvey, N. E. Cremer, and D. H. Sussdorf, "Methods in Immunology, Third Edition," p. 313. Benjamin, Reading, Massachusetts, 1977.

ELISA Procedure

Stock Solutions

Phosphate-buffered saline (PBS), 10× stock per liter: NaCl, 80 g; Na_2HPO_4, 14.3 g; KH_2PO_4, 4 g; NaN_3, 2 g; adjust to pH 7.2 with NaOH, then autoclave [dilute 1 : 10 (v/v) for use]

Substrate buffer (per liter): $NaHCO_3$, 1.34 g; Na_2CO_3, 1.48 g; $MgCl_2$, 0.4 g; adjust to pH 9.6 if necessary

Diluting buffer (per liter): PBS (10× stock solution), 100 ml; bovine serum albumin, 10 g (fatty acid free, Sigma); Tween 20, 5 ml

Washing buffer (per liter): PBS (10× stock solution), 100 ml; Tween 20, 5 ml (used for all washing steps as indicated)

Assay

Microtiter plates (96-well) are used. These may differ among manufacturers, and should be performance evaluated. A human metallothionein standard solution containing 1 μg/ml in PBS with 2-mercaptoethanol (10 μl/ml) is diluted 1 : 10 (v/v) with coating buffer and 100 μl is added per well. After 1 hr at room temperature or overnight at 4°, plates are washed three times. Diluting buffer is added (300 μl/well) for 30 min to block nonspecific binding sites and is then removed. A washing step is optional here. Purified human metallothionein standards or test samples are added (50 μl/well). Serial dilutions made with diluting buffer are prepared in adjacent wells. An appropriate dilution of sheep anti-human metallothionein IgG (usually 1 : 4000 in our studies) is added at 50 μl/well. This is the primary antibody. After 3 hr at room temperature, wash the plate three times and add 100 μl/well of a 1 : 1000 dilution of donkey anti-sheep IgG alkaline phosphatase conjugate (Sigma). This is the secondary antibody. After 30 min, wash the plate three times. An appropriate quantity of *p*-nitrophenyl phosphate (Sigma) in substrate buffer is added (usually 200 μl/well) followed by incubation in an incubator at 37° for 15–60 min. If color is present after 15 min, incubate for only 30 min total. After sufficient color development, measure absorbance at 405 nm in an automatic microtiter plate reader (ELISA spectrophotometer). The effects of human metallothionein coating concentration and sheep anti-human metallothionein IgG dilution on absorbance (405 nm) are shown in Fig. 2.

The original papers should be consulted for additional graphs showing optimization and standardization of the ELISA method.[15,18] Multidispensing pipettes and semiautomatic plate-washing devices are essential basic equipment needed for the assay. Optimal conditions and dilutions must be experimentally determined (Fig. 2). Anti-human metallothionein IgG di-

[18] A. Grider, L. B. Bailey, and R. J. Cousins, *Proc. Natl. Acad. Sci. U.S.A.* **87,** 1259 (1990).

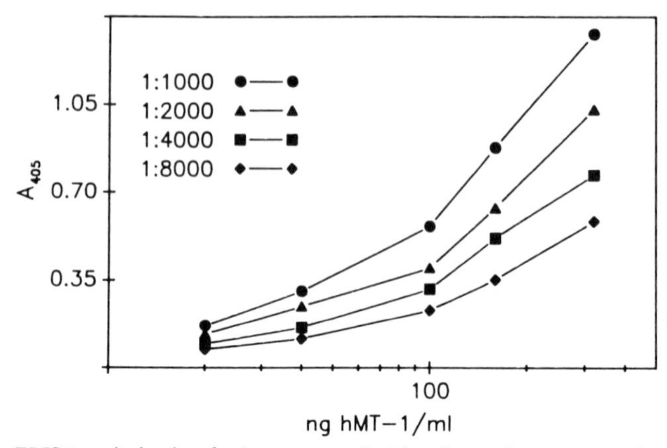

FIG. 2. ELISA optimization for human metallothionein coating concentration and sheep anti-human metallothionein IgG dilution. Absorbance at 405 nm was measured after a 30-min incubation using ELISA at four IgG dilutions and five metallothionein concentrations. Metallothionein concentrations were based on the amino acid analysis. (Reproduced from Grider *et al.*[15] with the kind permission of C. V. Mosby, St. Louis.)

lutions (1000- to 8000-fold) with at least five concentrations (20 to 300 ng/ml) of human metallothionein for plate coating should be used initially. The latter could be a single isoform or a mixture, depending on cross-reactivity. In the ELISA described here, isoforms of human metallothionein cross-react with the sheep anti-human metallothionein IgG. Excellent suggestions for ELISA optimization are available.[19]

Preparation of Samples

Exact methods of preparation must be determined by trial and error. Methods for human liver, serum, and red blood cell lysates are presented here. Use of protease inhibitors has not been extensively investigated in metallothionein research. However, addition of protease inhibitors should be considered in sample preparation.

Liver

Samples of 400 mg (if possible) are homogenized (Polytron) in 4 vol of 10 m*M* Tris-HCl, pH 7.4, and centrifuged at 10,000 *g* for 10 min at 4°. The supernatant is heated in a 100° bath for 5 min, then centrifuged for 30 min at 4°. The sample is then mixed with diluting buffer in an initial dilution of 1 : 100 (v/v) depending on the metallothionein concentration.[15]

[19] J. S. Garvey, D. G. Thomas, and H. J. Linton, *Experientia, Suppl.* **52,** 335 (1987).

Dilutions may have to be less for other tissues where metallothionein concentrations are lower. Tissue samples should be stored at −80° until the homogenates can be prepared, which then should be used immediately for the ELISA.

Red Blood Cells

Erythrocytes can be obtained by standard centrifugation and washing steps with ice-cold PBS or saline or with Percoll (Pharmacia) gradients. The final cell preparation is diluted with ice-cold H_2O at a ratio of 1 : 1.4 (v/v). The components are mixed by immersion and then frozen at −80°. The thawed mixture is considered the erythrocyte lysate.[18]

Serum

Blood is collected and placed at room temperature for 15 min. The clot is removed by centrifugation at 5000 *g* for 10 min at 4°. The serum is removed and the protease inhibitor, phenylmethylsulfonyl fluoride (PMSF), is added at 0.5 mg/ml. Storage is at −80°.

Data Analysis and Calculations

Competing metallothionein concentrations of 150 to 20,000 ng/ml produce changes in percentage binding that are linear when concentrations are on a log scale (Fig. 3). Routinely, metallothionein concentrations of

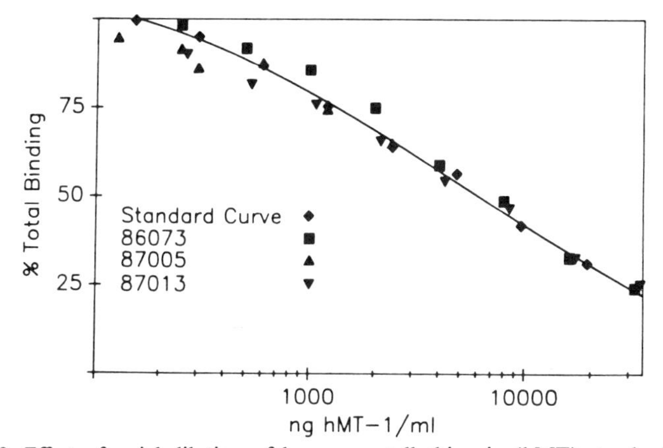

FIG. 3. Effect of serial dilution of human metallothionein (hMT) standard and heat-treated human liver cytosol on percentage total binding in the ELISA. Metallothionein concentrations of unknown samples are determined from a standard curve. (Reproduced from Grider *et al.*[15] with the kind permission of C. V. Mosby, St. Louis.)

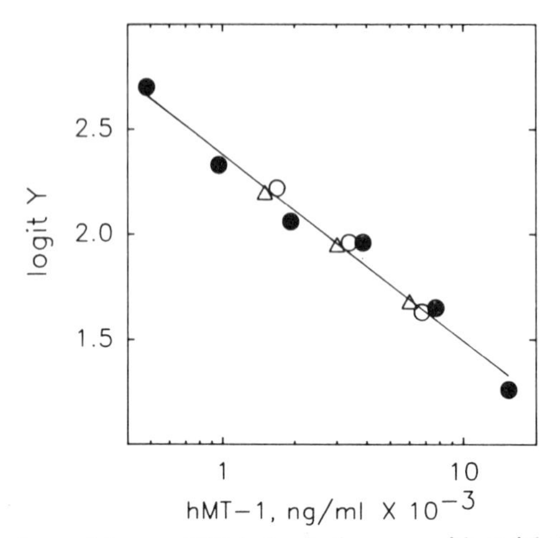

FIG. 4. Comparison of human MT-1 standard curves with serial dilution of human erythrocyte lysates (1.4- to 5.6-fold) from two subjects (Δ, O) vs. known amounts of MT-1 standard (●). Expressed as logit Y ($y = -0.89x + 5.04$, $r = 0.989$). (Reproduced from Grider *et al.*[18] with the kind permission of the National Academy of Sciences.)

unknowns are determined by linear regression after data are transformed to logit Y format.[20] As described here, the ELISA had coefficients of interassay and intraassay variation of 15.4 and 4.2%, respectively. An example of a logit plot showing serial dilutions of human erythrocyte lysate vs purified human metallothionein standard is shown in Fig. 4.

Comparison of Metallothionein ELISA to Other Methods

Usually, immunologically based assays involve measuring extremely small amounts of antigen. Metallothionein is unique in that reliable chemical assays and chromatographically purified protein (estimated by zinc content) can be used to measure the protein in human liver for comparative purposes.[15] These methods were used to compare estimates of metallothionein in samples of human liver derived from a variety of subjects.

The ^{109}Cd-binding method involves heat denaturation of nonmetallothionein proteins, Cd^{2+} binding to metallothionein, and removal of excess Cd^{2+} with hemoglobin.[21] For calculations of the metallothionein content, it is assumed that 6 g-atoms of Zn^{2+} is replaced by Cd^{2+}. In the tissue used for comparative purposes, the ELISA value averaged 4.6 ± 0.4

[20] T. Chard, "An Introduction to Radioimmunoassay and Related Techniques," p. 163. Elsevier North-Holland, New York, 1982.

[21] D. L. Eaton and B. F. Toal, *Toxicol. Appl. Pharmacol.* **66,** 134 (1982).

(SE) times that of the ^{109}Cd-binding method.[15] The reason(s) for this discrepancy is not clear. It could relate to the calculations involved in each assay. The ^{109}Cd-binding method does not require an estimate of tissue protein concentration because it is a calculated value. Furthermore, Cd^{2+} may not replace all of the metals associated with human liver metallothionein, e.g., copper.[22] This would underestimate the concentration. Amino acid analysis was used to standardize the ELISA.[15] Therefore, the value obtained by ELISA may be more correct. Alternatively, it is possible that a contaminating protein may react with the protein to give a higher ELISA value. If the latter is true, it would affect absolute values but not relative values. These are questions that must be addressed to verify accuracy of the ELISA.

A significant aspect of the ELISA is that radioisotopes are not involved. This precludes the need to routinely iodinate purified metallothionein required to conduct the assay. The ELISA is also a more rapid method. However, it does require more IgG and purified human metallothionein preparation, which is time consuming.

Uses of ELISA for Human Metallothionein

Limited data are available on the metallothionein concentration of human tissues. Initial data obtained with this ELISA method as well as the ^{109}Cd-binding method suggest that, in humans, the level of the protein is one to two orders of magnitude greater than in the laboratory rat.[10,11,15] This could be due to differences in sensitivity of the gene to regulation by metals and/or physiologic mediators in the two species.

Liver metallothionein was shown to increase during cancer and infection.[15] This may be expected based on animal studies; nevertheless, it shows that experiments on this protein with animals are applicable to the human biology and pathophysiology of metallothionein. Biopsy-size materials can be used as well. Preliminary data show that human patients with Wilson's disease respond to therapeutic doses of zinc as manifested in increased intestinal metallothionein levels.[23] These micromethods require imaginative means to standardize the data, e.g., very sensitive methods to measure protein concentrations. Fluorometric ELISA methods[19] of greater sensitivity than colorimetric versions may prove useful when sample size is limited.

[22] J. R. Riordan and V. Richards, *J. Biol. Chem.* **255,** 5380 (1980).

[23] V. Yuzbasiyan-Gurkan, A. Grider, T. Nostrant, R. J. Cousins, and G. J. Brewer, unpublished observations.

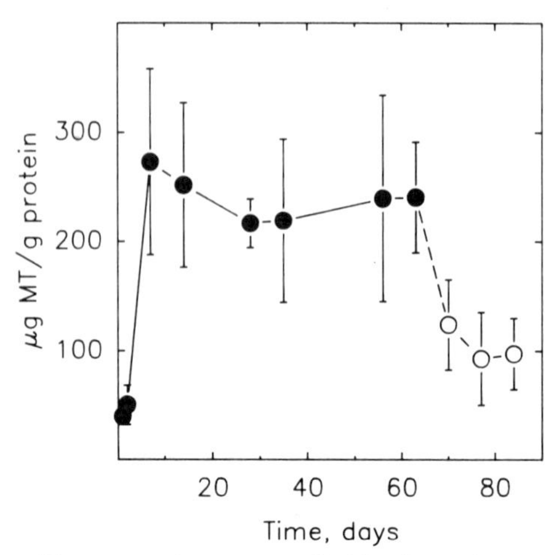

FIG. 5. Response of human erythrocyte metallothionein to zinc supplementation (50 mg Zn/day). ●, 50 mg of zinc supplement per day; ○, supplement withdrawn after day 63. Expressed as micrograms MT per gram of lysate protein (mean ± SE, $n = 3-6$). (Reproduced from Grider *et al.*[18] with the kind permission of the National Academy of Sciences.)

Human erythrocytes contain measurable amounts of metallothionein.[18] This probably reflects synthesis during cellular differentiation in the marrow. These levels of metallothionein are sensitive to dietary restriction or supplementation of zinc. The response of erythrocyte metallothionein to zinc supplementation of human subjects is shown in Fig. 5. Serum metallothionein levels change in these subjects, but not as rapidly. The ELISA described above has not been found effective with plasma as the starting sample. This does not appear to be a handicap with available data in terms of usefulness of the method for studies of zinc status assessment.

Acknowledgments

The methods described here, which were previously published, were supported, in part, by NIH Grant DK 31127 and Boston Family Endowment Funds.

[20] Antibodies to Metallothionein

By JUSTINE S. GARVEY

Introduction

During the past few years, there have been reports of both polyclonal antibodies[1-5] and monoclonal antibodies (MAbs)[6-9] with specificity for metallothioneins (MTs). As the term implies, poly indicates a product of many clones of lymphocytes, whereas mono indicates a product of one clone of lymphocytes. Both polyclonal and monoclonal antibodies have advantages and disadvantages as immunological reagents, and from guidelines[10,11] for antibody production and/or usage a general conclusion is that both sources of antibodies may prove useful if readily available. However, when analysis suggests that a choice be made, then consideration of the most prominent criteria of immunological reactions, specificity and sensitivity, must contribute to a decision favoring but one type, either polyclonal or monoclonal. Antibody specificity is defined as recognition of the antibody-binding site for a unique topographical feature of the antigen, this being a specific antigenic determinant or epitope. An epitope consists of a region of five to seven amino acid residues that may be either sequential (continuous) or conformational (discontinuous segments brought into proximity by secondary and tertiary structure of the molecule). Whereas polyclonal denotes a combination of specificities, monoclonal denotes a single specificity; the use of a panel of MAbs thus provides the capability to blend at will individual specificities of the members of the panel. Sensitivity represents the least amount of antigen that is detectable by the antibody and it should refer to a particular type of assay. Sensitivity is dependent on or correlated with affinity, the latter being a thermodynamic measurement

[1] G. Madapallimatam and J. R. Riordan, *Biochem. Biophys. Res. Commun.* **77,** 1286 (1977).
[2] R. J. Vander Mallie and J. S. Garvey, *Immunochemistry* **15,** 857 (1978).
[3] F. O. Brady and R. L. Kafka, *Anal. Biochem.* **98,** 89 (1979).
[4] C. Tohyama and Z. A. Shaikh, *Fundam. Appl. Toxicol.* **1,** 1 (1981).
[5] R. K. Mehra and I. Bremner, *Biochem. J.* **213,** 459 (1983).
[6] T. Masui, T. Utakoji, and M. Kimura, *Experientia* **39,** 182 (1983).
[7] B. G. Talbot, G. Bilodeau, and J.-P. Thirion, *Mol. Immunol.* **23,** 1133 (1986).
[8] Y. Kikuchi, N. Wada, M. Irio, H. Ikebuchi, J.-I. Sawada, T. Terao, S. Nakayama, S. Iguchi, and Y. Okada, *Mol. Immunol.* **25,** 1033 (1988).
[9] J. S. Garvey, *in* "Metallothionein in Biology and Medicine" (C. D. Klaassen and K. T. Suzuki, eds.), in press. (Proceedings of a U.S.A.-Japan Seminar, University of Hawaii, Honolulu, 10-14 December, 1989), CRC Press, Boca Raton, Florida, 1991.
[10] B. S. Dunbar and E. D. Schwoebel, this series, Vol. 182, p. 663.
[11] B. S. Dunbar and S. M. Skinner, this series, Vol. 182, p. 670.

of the strength of the antibody–antigen reaction and an indication of the closeness of fit of the epitope in the antibody-combining site. The term affinity derives from primary binding experimentally determined by monovalent hapten–anti-hapten attraction. This "intrinsic" affinity is enhanced when the influence of other contributions to binding are considered; the enhanced, "functional" affinity is termed avidity. Avidity increases with the maturation of the immune response; it is dependent on antibody valence, antigen valence, and other nonspecific factors associated with binding. If the class of antibody is immunoglobulin M (IgM) with a valence of 5 or 10 rather than IgG with a valence of 2, avidity is increased; this is also the case when polyclonal binding involving two or more epitopes is compared to monoclonal binding involving a single epitope. Polyclonal antibody is an obvious choice for radioimmunoassay (RIA), in which the required sensitivity involves detection of antigen at the picogram level, and hence an antibody of high avidity. Since the initial aim of the author's laboratory in MT research was to contribute an immunochemical approach to improve on the then-existing lack of both sensitive and specific detection/quantitation of MT, it was the firm conclusion from the previous remarks on avidity that the investigation should begin with the production of a polyclonal antibody. An immediate corollary of that decision was that there should be development of a radioimmunoassay (RIA) to fully exploit the capability of the antibody. Both objectives were successfully attained; the antibody and the RIA were employed regularly in the period from 1978 to the present in the analysis of MTs and candidate MTs in the author's laboratory. Additionally, the antibodies were made available to other laboratories to provide a useful probe for MT in a spectrum of assays with applications in various subdisciplines of the biological sciences. Of the many RIAs performed in the author's laboratory, two are selected as examples for this chapter. One involves the detection and quantitation of MT in normal physiological fluids of humans, the other involves the determination of the principal epitopes of MT with which the polyclonal antibody binds. The first-mentioned demonstrates the exquisite sensitivity and specificity of an antibody probe for quantitation and the second demonstrates its utility for structural analysis. Following optimization of the RIA, it became clear that an ELISA (enzyme-linked immunosorbent assay) would be a valuable supplement to the RIA, and its development is also discussed in the first part of this report on polyclonal antibodies. Moreover, pertinent to the earlier comments on the potential usefulness of MAbs, the second part of this chapter discusses MAbs, an important aspect of their production being the use of the ELISA for screening.

Polyclonal Antibodies

Preparation of Immunogen and Immunization Protocol

The principal procedures that contribute to obtaining an antibody are described in Ref. 12 and in other chapters of this volume. Briefly, they involve the following steps: (1) MT is induced in rats injected subcutaneously with $CdCl_2$; (2) liver MT-1 and MT-2 isoforms are isolated and purified from a tissue homogenate; (3) the bulk removal of liver proteins is achieved by heat treatment, organic solvent, or by a combination of these treatments; (4) chromatography is performed on either a pair of open columns (sizing followed by ion exchange), or alternatively by high-performance liquid chromatography (HPLC), or by combined open and HPLC columns; (5) purity is assessed by both polyacrylamide gel electrophoresis (PAGE) and amino acid compositional analysis. From results of immunization with both monomeric MT and MT that is cross-linked with glutaraldehyde (GDA), the latter is clearly the method of choice. Both soluble and insoluble polymer obtained by varying the conditions of polymerization (see protocols at the end of this section) are effective immunogens when emulsified with Freund's adjuvant and used to immunize rabbits. Injections are made into multiple sites that include subcutaneous, intramuscular, and intraperitoneal routes for both soluble and insoluble polymers; additionally, the intravenous route is used for soluble polymers (Fig. 1). Serum obtained from a preinjection bleeding and from each of the bleedings throughout the complete course of initial and booster injections are so handled as to avoid any alteration that might be caused by contamination, denaturation, etc. An important restriction is that there be no pooling of serum samples. The binding assays that are discussed in the following section are used to monitor the progress, i.e., maturation, of the immune response. They determine the timing of multiple booster injections or whether immunization is to be aborted because of a low level of serum antibody. In obtaining a high avidity response, the immunization protocol involves a time span of several months for injections and additional months for bleedings; this applies only to polymeric MT because the immunization with monomeric MT is aborted at an earlier time. In the author's experience, the overall very low antibody response with monomeric MT generally caused immunization to be aborted after the first booster injection; attention is thus concentrated on immunization with polymeric MT. Moreover, the development of assays involved only anti-

[12] J. S. Garvey, R. J. Vander Mallie, and C. C. Chang, this series, Vol. 84, p. 121.

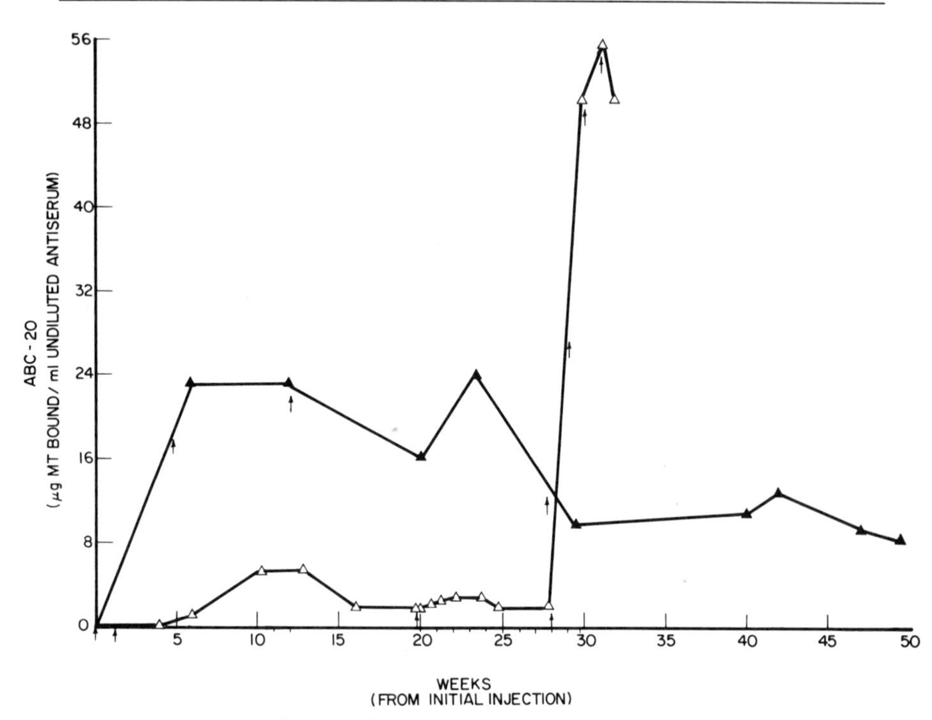

FIG. 1. Antigen-binding capacity (ABC)-20 values for antisera of rabbits 4 and 7 tested, respectively, with ^{125}I-labeled rat MT-1 and ^{125}I-labeled rat MT-2. Complexes, formed with constant amounts of ^{125}I-labeled MT and varying dilutions of the two preimmunization serums and each antiserum, were precipitated at 37.5% saturation with ammonium sulfate, as specified in the described (see text) ABC-20 assay protocol. From a plotted curve of data points obtained for each antiserum, the ABC-20 values are obtained. These ABC-20 values, one for each of the preimmunization serums and one for each antiserum (▲, No. 4; △, No. 7), are plotted here to correspond with the elapsed time from the collection of the blood sample and the beginning of immunization [time (t) = 0]. Each injection of immunogen (indicated by arrows) was insoluble polymer into No. 4 and soluble polymer into No. 7. The detailed immunization of No. 4 is as follows for four injections, each administered into multiple sites: $t = 0$, 1 mg; $t = 4.7$ weeks, 2 mg; $t = 12$ weeks, 3.1 mg; $t = 27.8$ weeks, 1.7 mg. Freund's complete adjuvant (FCA) was used in a 1 : 1 ratio with immunogen for injections 1–3; a mixture of adjuvant composed of nine parts Freund's incomplete adjuvant (FIA) and one part CFA was used in a 1 : 1 ratio with immunogen for injection 4. The number of ABC-20 values shown are fewer than the number of blood collections; this is because the procedure to obtain an ABC-20 value was not performed with every antiserum. The detailed immunization of No. 7 is as follows for seven injections, three administered into multiple sites followed by four given only by the intravenous route: $t = 0$, 4.4 mg; $t = 1.3$ weeks, 4.6 mg; $t = 19.7$ weeks, 4.1 mg; $t = 28$ weeks, 1.2 mg; $t = 29$ weeks, 1.2 mg; $t = 30$ weeks, 1.2 mg; $t = 31$ weeks, 1.2 mg. FCA was used in a 1 : 1 ratio with immunogen for injection 1; a mixture of adjuvant composed of nine parts FIA and one part FCA was used in a 1 : 1 ratio with immunogen for injections 2 and 3; and no adjuvant was used with the immunogen injected

body obtained by immunization with polymeric MT, either insoluble polymer or soluble polymer as discussed below. However, it is important to note that the form of the antigen to which the antibody binds is monomeric MT and therefore only monomeric MT is used in all assays.

Insoluble Polymer. Insoluble aggregation occurs in this procedure because of an extreme excess of GDA. To a solution of MT (1–2 mg/ml) in 0.10 *M* acetate buffer, pH 6.2, is added 2.5% (v/v) GDA to obtain an 80 : 1 molar excess of GDA, based on an assumed molecular weight of 7000 for MT. After a 1-hr ambient temperature incubation, lysine is added to a final concentration of 0.10 *M*. The precipitate is dispersed mechanically by repeated passage through a hypodermic syringe to which a higher gauge needle is attached prior to each successive passage. As the final step prior to use as immunogen, the total reaction mixture of precipitate plus soluble products is dialyzed for 24 hr at 4–8° against borate-buffered saline (BBS): five parts borate buffer to 95 parts saline. One liter of borate buffer is prepared with 6.184 g boric acid, 9.536 g borax, 4.384 g NaCl in distilled water, and the pH is adjusted to 8.4–8.5 with dilute HCl or NaOH. One liter of saline is prepared as 8.5 g NaCl in distilled water.

Soluble Polymer. To a solution of MT (4–5 mg/ml) in 0.10 *M* acetate buffer, pH 6.2, is added 2.5% GDA to obtain a 17 : 1 molar excess of GDA, again based on an assumed molecular weight of 7000 for MT. After a 1-hr incubation at ambient temperature, lysine is added to a final concentration of 0.10 *M*. After another 1-hr incubation at ambient temperature, the reaction mixture is dialyzed against BBS for 24 hr at 4–8°. The soluble polymer of MT obtained as the product of the reaction is used as immunogen.

---

intravenously. The more numerous ABC-20 values shown for No. 7 than shown for No. 4 are important for the discussion of the response. Number 7 shows some responses that are lower than those of No. 4, and this may be attributed to the form of immunogen (No. 4, insoluble polymer; No. 7, soluble polymer). However, there are also ABC-20 responses of No. 7 that exceed those of No. 4. At $t = 50$ weeks, No. 4 was bled out; as noted, the injection at $t = 27.8$ weeks failed to boost No. 4. Additional intravenous injections into No. 7 as a series of three, beginning at $t = 40$ weeks and continued weekly, failed to boost the responses; the ABC-20 values (not shown) continued to decline from the peak ABC-20 value at $t = 31$ weeks. Thus, the injection at $t = 31$ weeks led to a decreased ABC-20 value in serum obtained 5 days later (the last ABC-20 value shown). During the time interval thereafter, the response continued to decline without further injection of immunogen and no booster effect resulted from the resumption of immunization following a 9-week period without immunization. (Modified from Vander Mallie and Garvey[2] with publisher's permission.)

Assays for High Avidity and Cross-Reactivity

For further details on the methodology of this section, see Ref. 13.

Interfacial Assay. Each serum is initially screened by the interfacial precipitation assay, which is performed by pipetting three to four drops of whole serum into each of two 6 × 50 mm test tubes, then overlaying the serum in one tube with MT at a concentration of 1 mg/ml in BBS, the serum in the other (control) tube with the BBS that is used to prepare the MT solution. Antiserums that show a precipitate with MT but no precipitate with their control are further assayed to obtain a relative measure of avidity by a modified Farr ABC (antigen-binding capacity) test as described.

ABC Assay. In brief, this is a primary binding reaction, performed with dilutions of the antiserum that are prepared in BBS containing 10% (v/v) normal rabbit serum (NRS) and with a constant concentration of ^{125}I-labeled MT, 300 ng labeled MT/ml borate buffer, pH 8.4–8.5, that contains 1% NRS. For the labeling of the MT, see the section Iodination of Metallothionein; however, the described iodination is for the use of 1 ng ^{125}I-labeled MT in the RIA. Since 150 ng MT is used in the ABC assay, it is obvious that the specific radioactivity is higher for the RIA than is required in the ABC assay. ^{125}I-Labeled MT is diluted appropriately with nonradioactive MT, but both MTs must have the same immunoreactivity.

The diluted antibody and the antigen, 500 μl each, are mixed and incubated 16–18 hr at 4–8° to allow formation of complexes. The complexes are precipitated with precooled (4–8°) saturated ammonium sulfate (SAS), added as 600 μl to each assay tube. Precipitation of the complexes is allowed to continue for 30 min at 4–8°. The precipitates are collected by centrifugation at 1800 *g* for 30 min at 4–8° and the supernatants decanted. The precipitates are resuspended in precooled 40% SAS and the centrifugation is repeated. The pellets of washed precipitates are counted in a γ counter and the data are plotted as percentage total counts precipitated *vs* log antibody dilution. The dilution of antiserum that binds 20% of the added ^{125}I-labeled MT is used to calculate the antigen-binding capacity (ABC-20) of the antiserum. See Fig. 1 and its legend for the ABC assays of antiserums obtained from rabbit No. 4 (the immunogen was rat MT-1, the polymer was insoluble) and from rabbit No. 7 (the immunogen was rat MT-2, the polymer was soluble). The responses shown by rabbits No. 4 and No. 7 have been replicated generally, although not precisely. This is to be expected from general experience with any immunization protocol that is performed in rabbits. Moreover, it is realistic to immunize three or four

[13] J. S. Garvey, N. E. Cremer, and D. H. Sussdorf, "Methods in Immunology," pp. 79, 275, 301, and 313. Benjamin, Reading, Massachusetts, 1977.

rabbits in order to have assurance that one rabbit produces antibody of very high avidity that can be used as primary antibody reagent for the RIA. The precautions to note with use of the intravenous route of injection are that there is an added risk of anaphylaxis over that occurring when injections are by multiple routes, and the booster effect that is obtained by use of the intravenous route of injection is relatively brief in duration. Thus, monitoring of the response is especially important, in order that the animal may be bled out at the peak of the response for a high volume yield of serum that may provide a reagent of exquisite sensitivity and specificity in the binding of antigen.

Ouchterlony Analysis. An initial examination of the specificity of the antiserums is accomplished by Ouchterlony analysis. This is a precipitation reaction performed on a microscope slide coated with buffered agarose in which uniform wells are punched; a convenient pattern is a group of peripheral wells equidistant from a central well. Antigen solutions, MT, and other test proteins are placed in the peripheral wells, and undiluted antiserum is placed in the central well. After an incubation time that is dependent on various factors, e.g., temperature and concentration of the sample solutions, the diffused antigen and antibody form precipitation lines. Based on this assay, evidence is obtained for a reaction of identity with the immunizing antigen, and for cross-reactivity, complete or partial, of the antibody with antigens other than the immunogen. In the author's laboratory, it is this assay that provided the initial evidence for (1) complete cross-reactivity of the antibody with MT isoforms other than the pure isoform used for immunization, (2) complete cross-reactivity of the antibody with mammalian MTs in addition to rat MTs and particularly including human MT, and (3) complete cross-reactivity of the antibody with native rat MT, the apothionein, and thioneins binding various metals.

Iodination of Metallothionein

In this section are discussed potential pitfalls in the iodination of MT. Rat hepatic MT is iodinated by a method[12] adapted from Bolton and Hunter, i.e., MT is reacted initially with *N*-succinimidyl-3-(4-hydroxyphenyl) propionate (NSHPP) and subsequently ^{125}I is incorporated into MT using the chloramine-T reaction. Briefly, the NSHPP·MT conjugate is prepared using ^{109}Cd-MT. This trace labeling with ^{109}Cd is for the quantitation of MT in NSHPP·MT conjugate, and it is achieved either during the *in vivo* induction of MT or by incubation of the liver supernatant with ^{109}Cd following the bulk removal of liver proteins (see the section Preparation of Immunogen and Immunization Protocol, above). The ^{109}Cd-specific radioactivity is determined for the purified MT of known protein concentration; it is necessary that quantitation be accurate and is preferably derived by more than one method. The conjugation reaction is performed using 10 mol of NSHPP and 1 mol Cd-MT. Following the Sephadex filtration of the conjugation reaction mixture, the MT concentration is determined in the pure, eluted NSHPP·MT conjugate. This is determined from the concentration of ^{109}Cd and calculated from the previously determined specific radioactivity. The conjugate is distributed as a known amount of MT (3–5 μg in 5–10 μl 0.25 *M* phosphate buffer, pH 7.5) into 1.5-ml Eppendorf tubes, and the tubes are tightly capped and stored at −20° for use in iodinations. It is important that the pH of the NSHPP·MT be sufficiently buffered at pH 7.5 prior to the addition of ^{125}I, since a successful iodination requires that the protein not be denatured by the addition of the basic ^{125}I-labeled sodium iodide and that the pH be maintained at 7.5 during the iodination. Delivery of the ^{125}I solution to an investigator's laboratory should be as close to the date of production of the ^{125}I as possible, and it should be free from reducing agents and be specified by the manufacturer as "carrier free and for protein labeling." An example is the product supplied by Amersham Corporation (Arlington Heights, IL) and coded as IMS.30. ^{125}I is used as 0.5 mCi in the iodination of 3–5 μg MT conjugate. Strict adherence to the details of published protocols for iodination, e.g., Ref. 12, that have repeatedly yielded successful radioimmunoassays is recommended. Use of pure chemicals and freshly prepared reagents is obviously important to the success of the iodination. However, in the case of highly reactive chloramine-T, it is wise to purchase the chemical with a recent date of manufacture, to store it refrigerated and in air-tight condition, and to make a fresh replacement at least annually after the initial exposure to air and despite adherence to the already-mentioned recommendations for storage.

With reference to the ^{125}I-labeled MT reagent that is used in the RIA, the MT content is calculated from the known amount of MT in the NSHPP·MT conjugate that is iodinated. Recovery of MT from the Sephadex column that is used to fractionate protein-bound ^{125}I is >90%. It is from evaluation of the eluate profile that fractions of protein-bound ^{125}I are pooled for use as the ^{125}I-labeled MT reagent; however, the final concentration of the collected eluate diluted for use as ^{125}I-labeled MT reagent is 1 ng MT/50 μl and a minimum of 10^5 counts per minute (cpm) by γ counting.

Radioimmunoassay for Detection and Quantitation of Metallothionein

In current use is the following assay adapted from Ref. 12.

Reagents and Equipment. Doubly distilled water or water of equivalent purity is used in the preparation of all reagents.

Tris-gelatin azide buffer (TGA): Tris buffer, 0.05 *M*, is prepared with 0.1% gelatin and 0.02% sodium azide. For the preparation of 1 liter, 4.44 g Trizma hydrochloride, 2.65 g Trizma-base, 1.0 g gelatin, and 0.2 g sodium azide are added to 800 ml doubly distilled H_2O. This solution is heated, stirred magnetically until the chemicals are dissolved, cooled to ambient temperature, transferred to a 1-liter volumetric flask, and adjusted to the liter volume with doubly distilled H_2O. If this solution is prepared properly by avoiding overheating (i.e., warming to about 80° provides sufficient heat to dissolve gelatin), then the pH will be 7.9-8.0 and no pH adjustment is necessary. This buffer is used as diluent throughout the assay and for the collection of precipitates.

Ethylenediaminetetraacetic acid: This reagent (EDTA, molecular weight 292.254) is prepared at 0.125 *M* and pH 7.8. For 100 ml, 3.65 g EDTA or equivalent disodium EDTA is heated with about 75 ml doubly distilled H_2O, with stirring and the addition of NaOH (solution or pellets), to form the sodium salt and to raise the pH to 7.8. After cooling, final adjustments of volume and pH are made.

Primary (specific) antibody reagent: The following solutions are used in the volume relationship shown for 10 ml of the reagent, to be added as 50 μl to each assay tube, but *not* to nonspecifically bound (NSB) assay tubes: EDTA, 0.125 *M*, 4000 μl; polyclonal antibody (rabbit anti-rat liver MT), 1:100 dilution,[13a] 500 μl; normal rabbit serum (NRS), undiluted, 250 μl; Tris-gelatin azide buffer (TGA), 5250 μl.

NSB reagent: This reagent is prepared to include all the solutions used in the preparation of primary (specific) antibody reagent except the polyclonal antibody. Accordingly, the volume relationships for 10 ml of reagent are the following: EDTA, 0.125 *M*, 4000 μl; normal rabbit serum, undiluted, 250 μl; Tris-gelatin azide buffer (TGA), 5750 μl.

Second antibody reagent: This is goat anti-rabbit immunoglobulin added in a 100-μl volume at a dilution[13b] determined by optimization, i.e., to provide complete precipitation of the antibody-antigen complexes formed by the reaction of the primary antibody with MT while using the least amount of reagent to do so.

^{125}I-labeled metallothionein: This is prepared as described above (see *Iodination of Metallothionein*) and is added as a 50-μl volume in the assay.

[13a] The primary antibody is used in the assay in a volume of 50 μl and at that titered dilution that will allow 50% of the maximal precipitable counts to be complexed. This value is based on the total number of precipitable counts taken as 100% when a 1:100 dilution of primary antibody is used, e.g., serums obtained from rabbit No. 4 at about 30 weeks after the initial injection (see Fig. 1).

[13b] The dilution used by the author is goat anti-rabbit IgG serum diluted 1:16 or 1:18 in TGA; it is retitered whenever the serum source is changed or a change is made in either NRS or the primary antibody reagent.

Unlabeled metallothionein: Both the accurately quantitated MT used as competitor in development of the standard curve (SC)[13c] and the competitor of unknown concentration to be quantitated using that standard curve are unlabeled. The unknown competitors may be quantitated as a specified volume of physiological fluid, e.g., serum or urine, or a homogeneous cytosol of known wet tissue concentration.

In regard to species of unlabeled MT, both the competitor for the development of the standard curve and the unknown to be quantitated should be the same, since they should compete for the labeled MT in an immunologically identical manner. Although the labeled MT is preferably identical to the competitor MTs, its being so is not as critical as is the identical nature of the competing MTs.

Assay tubes: 10 × 75 mm, 3 ml, supplied by Sarstedt (Princeton, NJ)

Dilution tubes (for preparation of the competitor MT): 12 × 75 mm, 5 ml, with tight-fitting caps, supplied by Sarstedt

Dilution bottles and/or tubes (for preparation and storage of competitor MTs to be used for development of a standard curve): Volumes vary, but high-quality polypropylene with tight-fitting caps is used

Assay tube racks: Polypropylene, 80-tube capacity, accommodating a tube diameter of 10–13 mm, available from various vendors, e.g., Fisher Scientific (Pittsburgh, PA) and specified as "holder and dryer tube"

Dispensers: Manual delivery is suggested with an Eppendorf repeater pipettor, Combitips differing in dispensing range, and pipette tips, available from various vendors, e.g., PGC Scientific Corporation (Gaithersburg, MD)

γ counter: A prototype is Gamma 4000, supplied by Beckman Instruments (Fullerton, CA), having a 200-tube capacity, with tube holders of size compatible with the described assay tubes

Refrigerated centrifuge: A prototype is a Sorvall superspeed model sup-

[13c] The symbol SC identifies a standard curve developed for quantitation of unknowns. Usually it is derived from a series of doubling dilutions of a known competitor MT, covering the range 40,000 pg MT to 1 pg MT, each response performed in quadruplicate. Critical values such as "maximum bound" (max, no competitor MT) and "nonspecific bound" (NSB, no competitor MT and no primary antibody) are performed in octuplets. The 100% bound value is determined as max − NSB. SC(0) denotes a standard curve developed without addition of a specified amount of a known control fluid, such as serum or urine, or homogeneous cytosol. SC(50), for example, denotes a standard curve developed with the addition in each assay tube of 50 μl of a control fluid of known MT concentration; the fluid to be quantitated will then be assayed in 50-μl amounts. Homogeneous cytosols, commonly of relatively high MT content, are usually quantitated using an SC(0). See the section Selected Problems in Utilization of Radioimmunoassay, below, for further discussion on assay of serum and urine.

plied by DuPont/Sorvall (Wilmington, DE) and used with rotor and carriers that accommodate 10 × 75 mm tubes, i.e., Sorvall HS-4 rotor with four 24-place carriers (96-tube total rotor capacity)

Vacuum aspirator system: This is an improvised system consisting of two 500-ml side-arm flasks, each stabilized by attachment to a ring stand and each assembled with a single-hole rubber stopper containing a short piece of glass tubing. One of the flasks serves as a trap to prevent escape of radioactive solution from the system; it has a short piece of Tygon tubing attached at one end to the glass tubing in the stopper, and at the other end to the outlet of the vacuum source. The other flask, into which the aspirated solution is collected for discard as radioactive waste, has a length of Tygon tubing connected at one end to the piece of glass tubing in the stopper, and contains a Pasteur pipette at the other end as the intake for the supernatant aspirated from the assay tubes in step 5 of the following procedure

Procedure. This assay is described for an SC(0) quantitation. The entire assay from step 1 until the supernatants are aspirated in step 5 is performed with solutions at 4–8° and with assay tubes maintained at 4–8°. All solutions are prepared as specified in the section Reagents and Equipment and they are added in the following sequence:

Step 1. a. All the assay tubes are marked with a permanent identifying mark, in the sequence of eight assay tubes each for max and NSB determination for the standard curve, four assay tubes for each of the varying known concentrations of the competitor MT for the standard curve, and six assay tubes for each of the unknown concentrations to be quantitated (two of the six tubes are designated for use with the NSB reagent in step 3).

b. To the octuplets for determination of max and NSB for the standard curve is added 400 μl TGA. To all other assay tubes is added 300 μl TGA.

c. To all tubes is added 50 μl primary antibody, with the exception of tubes designated for NSB determinations; to these latter tubes is added 50 μl NSB reagent.

d. Into tubes for competitor MT is added 100 μl/assay tube of the competitor, either the known MT used for determination of the standard curve, or the MT to be quantitated. In the former case, the 100 μl will contain the varying concentrations of the known competitor MT required to develop the standard curve; in the latter case, the 100 μl will contain that volume of physiological fluid or cytosol homogenate that will produce a response in the central region of the standard curve (see Fig. 2). At this point in the assay procedure, the volume of the reaction solution in every tube is 450 μl.

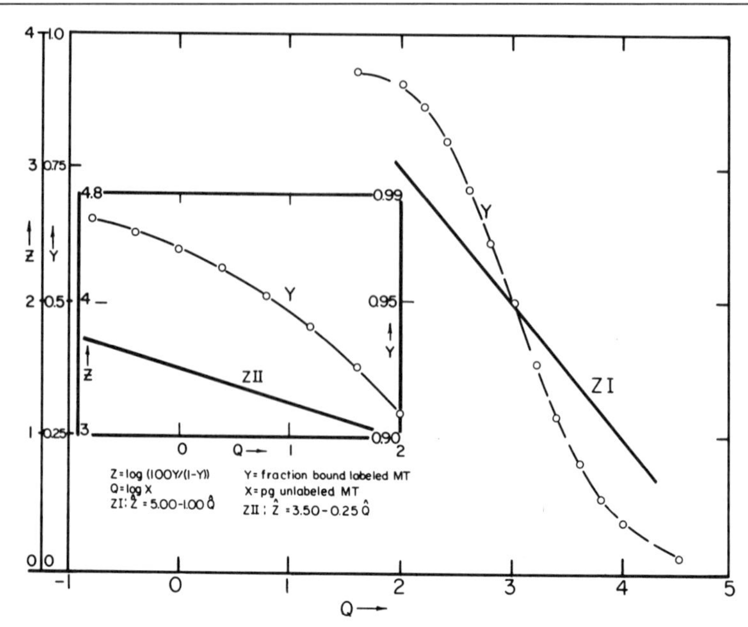

FIG. 2. Typical standard curve for quantitation of MT by radioimmunoassay. The sigmoid response (Y vs Q) is linearized when transformed to logit-log form (Z vs Q), the latter regression being inverse-variance weighted. Accuracy in quantitation of unknowns is usually ±5% over the central region. The characteristic correlation coefficient (in excess of 0.99) of the regression decreases to 0.80–0.85 for the extension (insert developed over the concentration range 100–10 pg MT and to 0.5 over the range 10–1 pg MT. Y, Fraction bound labeled MT; $Z = \log[100Y/(1 - Y)]$; X, picograms competitor MT; $Q = \log X$; ZI, $\hat{Z} = 5.00 - 1.00\hat{Q}$; ZII, $\hat{Z} = 3.50 - 0.25\hat{Q}$. [Reproduced from J. S. Garvey, *in* "Biological Markers of Environmental Contaminants" (L. R. Shugart and J. McCarthy, eds.), Chap 16, p. 255. Lewis Publishers, CRC Press, Boca Raton, Florida, 1990, with publisher's permission.]

e. All tubes are mixed, both by vortexing at the completion of steps (a)–(d) for each rack of assay tubes, and by a brief centrifugation when steps (a)–(d) are completed for all racks of assay tubes. The centrifuge is operated at 4–8° for a brief period of time to attain a speed of about 3000 rpm, then turned off. The purpose here is to dislodge solution deposited along the walls of the tubes and to transfer these deposits to the bottom of the tubes. The tubes are then covered and the racks are stored at 4–8° for 16–18 hr.

Step 2. After the incubation that permits the binding of primary antibody and antigen, 50 μl ^{125}I-labeled MT is added to every tube and mixing is performed as in Step 1(e). At this point in the assay, the total assay volume is 500 μl, the primary antibody added in step 1 is diluted by a factor of 20,000, and competition between labeled and unlabeled MT is initiated and allowed to continue for 6–8 hr at 4–8°.

Step 3. After the above competition step, 100 μl second antibody reagent, anti-rabbit IgG, is added to every tube at the predetermined concentration needed to precipitate all the rabbit IgG in the assay. The tubes are mixed as in previous steps 1 and 2 and stored undisturbed for a period of 44–48 hr.

Step 4. TGA, 2 ml, is added to every assay tube. The purpose of this addition is to resuspend the precipitate and to increase the volume as an aid to uniform decantation of supernatant from the pellet of precipitate that is collected by centrifugation at 7000 rpm (about 9400 *g*) for 40 min. As the rotor decelerates from about 1000 rpm, the brake is turned off as a precautionary measure to maintain the precipitates as a uniform pellet in the bottom of the tubes.

Step 5. The aspirator system described in the section Reagents and Equipment is assembled and connected with a vacuum source at one end, and tubing with a Pasteur pipette at the other end of the system. To avoid warming when centrifugation is completed in step 4, the rotor carriers are placed immediately into a tub of ice. The tubes are removed individually from the carriers for rapid but thorough removal of the supernatant without loss of any precipitate. A technique of supernatant removal is developed that is effective, e.g., the tube, held in one hand, is tilted at a slight angle so that the liquid flows into the Pasteur pipette held in the other hand. The pipette tip is held so that it touches the wall of the tube at the level of liquid surface. A continuous, swift but even flow of supernatant is drained into the pipette by a smooth operation of advancing the position of the pipette tip and changing the angle at which the tube is held until all the liquid is aspirated from the tube. Although this step is simple in execution, it must be faultless since precipitates are easily dislodged during the handling of the tubes. And, if there is warming of the tubes, dispersion and disassociation of the precipitate may occur, which leads to establishment of a reequilibrium of the reaction and thus to aberrant results. After the supernatants are aspirated from a rotor-load of assay tubes, the assay tubes are examined for drops of solution along the walls of the tube. A cotton-tipped applicator stick is used for removal of solution, with obvious care to avoid any cross-contamination with the fluid and any removal of the precipitate.

Step 6. The ^{125}I radioactivity in the precipitate is determined by γ counting of the tubes for no less than 50,000 cpm/assay tube, and the counting time should range from 2 to 3 min. The elapsed time for the counting of the tubes for the standard curve and for the unknowns is noted to allow proper accounting for the decay of ^{125}I. A standard response curve is shown in Fig. 2 and the statistical analysis of the data is discussed in the following step 7.

Step 7. As described in the previous steps 1–6, the usual standard curve for quantitation of unknowns in the RIA is developed with varying amounts of a known MT competing with a constant amount of ^{125}I-labeled MT for the specific MT antibody. This curve is sigmoid when the fraction (Y) of labeled MT bound is plotted as a function of the competitor MT concentration [$Q = \log(\text{pg MT})$]. The central linear segment usually terminates at 100 pg competitor MT (Y about 0.9) and 10,000 pg competitor MT (Y about 0.2). Characteristically, data are transformed to the logit-log form: $Z = \log[100Y/(1 - Y)]$; $Q = \log(\text{pg MT})$; 100 is a factor of convenience. The regression developed (the standard curve for quantitation) is inverse-variance weighted.[14-16] The variance of Z (var Z) is expressed as $(a + bY)/Y^2(1 - Y)^2$, where a and b are functions of the variance of NSB and of max, respectively. The minimum of var Z, as a function of Y, usually ranges from $Y = 0.35$ to $Y = 0.45$. Since the weights are reciprocals of the variance, the range of Y values that will be the principal determinants of the weighted regression is usually 0.2 to 0.7. Accuracy over this range is usually $\pm 5\%$. In most cases, short extensions of the principal segment can be developed with reduced slope and reduced accuracy in quantitation of unknowns (Fig. 2).

When quantitating cytosols prepared from liver or kidney tissue, the content of tissue in the sample to be assayed typically is about 1 mg (wet weight). With such dilute cytosols, the binding by non-MT constituents in the sample is sufficiently minor that it is adequate to quantitate using an SC(0); the minor influence of nonspecific binding should be first confirmed.

Selected Problems in Utilization of Radioimmunoassay

Detection of Normal Physiological Levels of Metallothionein. Metallothionein concentration in typical human physiological fluids is generally less than 2 pg/μl in serum and less than 5 pg/μl in urine.[16] Thus, quantitation of MT in either serum or urine requires use of relatively large volumes of these physiological fluids. The standard curve for quantitation of serum is usually developed with the addition of 100 μl of a control serum of known MT content to each tube of competitor MT. For quantitation of urine, the usual addition of a control urine of known MT content is 50 μl.

[14] D. Rodbard and J. E. Lewald, *Acta Endocrinol. Suppl.* **64**(147), 79 (1970).

[15] D. Rodbard, *in* "Competitive Protein Binding Assays" (W. D. Odell and W. H. Daughaday, eds.), p. 204. Lippincott, Philadelphia, Pennsylvania, 1971.

[16] J. S. Garvey, *Environ. Health Perspect.* **54,** 117 (1984).

In developing the SC(100) or SC(50), respectively, the known control MT concentration is added to that of the competitor MT. The control fluids are selected from a panel of candidate fluids, and those of lowest response are then pooled for use in the assay. The precise MT content of the control fluid is determined by use of a standard curve developed over the range 0–20 pg MT; this curve is best expressed as a linear–linear function (fraction bound labeled MT *vs* pg competitor MT), the regression going through the origin (1, 0). Control fluid candidates are selected from appropriate subjects (young, healthy nonsmokers). The final pooled fluid for use in the assay usually has an MT content in the range 0–1 pg MT/μl (serum) and 1–2 pg MT/μl (urine). To enhance accuracy, multiple responses are determined over the 0 to 20-pg range for the standard curve and multiple dilutions of the control fluid are assayed over that range of responses. Although this procedure produces control fluid MT concentrations with about 10% accuracy, it is too time consuming to use for the unknown samples.

Identification of Epitopes of Metallothionein. The competitive binding of RIA is used to obtain information about the antigenic determinants (epitopes) of MT, the regions of the antigen that are the principal sites for binding with antibody. An example of the procedure using the polyclonal antibody is an experiment performed in collaboration with Dr. Dennis Winge (University of Utah).[17] Winge prepared and quantitated by amino acid compositional analysis certain rat MTs, native or chemically modified, and made them available for assay in the author's laboratory. These proteins, described in Table I, included native MT isoforms, the apoproteins, the β domain (residues 1–29), the α domain (residues 30–61), a peptide from the β domain (residues 1–25), and isoforms with chemically modified amino terminus or with lysines blocked. The proteins were used as competitors at several 10-fold dilutions in the competitive binding RIA. As the results show (Table I), the competing antigens fall into two classes, characterized by the relatively low (class 1) or high (class 2) concentration required to reduce binding of the labeled antigen to the 50% level. Class 1 antigens included native Cd,Zn-MT-1 and -2, the apoproteins following oxidation, Cd-MT-1 following removal of zinc from Cd,Zn-MT-1, and a peptide (OT-1) comprising the amino-terminal residues 1–25. Class 2 antigens included the α domain of MT-1 and MT-2, the oxidized α domain of MT-1, an oxidized MT-2 with the terminal methionine cleaved by CNBr treatment, a carboxymethylated MT-2 with terminal methionine cleaved by CNBr, a citraconylated and oxidized MT-1 with lysines blocked, and various peptide mixtures.

[17] D. R. Winge and J. S. Garvey, *Proc. Natl. Acad. Sci. U.S.A.* **80,** 2472 (1983).

TABLE I
RESPONSES OF METALLOTHIONEINS AND METALLOTHIONEIN-DERIVED PEPTIDES IN COMPETITIVE BINDING RADIOIMMUNOASSAY[a]

Class	Competing antigen	Fraction (y) of ^{125}I-labeled MT bound as a function of competing antigen concentration (pmol)					Picomoles at $Y = 0.5$
		0.1	1	10	100	1000	
1	1	0.81	0.50	0.16			1.0
	2	0.81	0.50	0.22			1.0
	3	0.92	0.60	0.19			1.6
	4	0.94	0.68	0.20			2.4
	5	0.90	0.55	0.14			1.3
	6	0.84	0.63	0.23			2.1
	11	0.84	0.50	0.16			1.0
2	7			0.88	0.61	0.25	190
	8			0.78	0.42	0.13	61
	9		0.96	0.80	0.41	0.10	59
	10		0.96	0.78	0.33		43
	12			0.92	0.62		100
	13			0.59	0.15	0.02	15
	14		0.93	0.75	0.42		57
	15		0.98	0.86	0.47	0.12	87

[a] Values were determined by inverse-variance weighted logit-log regression[14-16] based on RIA data, each antigen acting as competitor with ^{125}I-labeled rat metallothionein (MT) for the polyclonal antibody produced in rabbits.[2] The relationship developed in each case is $Z = a + bQ$, derived from the data X (picomoles competing antigen), Y (fraction of labeled antigen bound), and the transformations $Q = \log(100x)$, and experimental $Z = \log[100Y/(1 - Y)]$. Expressed in terms of Y values, the standard errors at 1 pmol for the class 1 antigens ranged from 0.02 to 0.06, and for the class 2 antigens from 0.01 to 0.09. All antigens had a measured response at 10 pmol, and differences in Y values (derived from the regression) may be evaluated here. All members of class 1 differ significantly at 10 pmol from all members of class 2 ($p < 0.01$ to $p < 0.001$), except for antigen 13 ($p < 0.05$). A similar distinction between the two classes is found in comparison of concentrations needed to attain the 50% bound level. The antigen types were as follows: 1, native MT-1; 2, native MT-2; 3, oxidized MT-1; 4, oxidized MT-2; 5, Cd-MT-1 (Zn^{2+} removed from the native molecule); 6, a Cd-MT + inactivated subtilisin; 7, α domain of MT-1; 8, α domain of MT-2; 9, oxidized α domain of MT-1; 10, a Cd-MT following subtilisin/PMSF treatment and consisting of a mixture of five to six peptides from the β domain, plus the α domain; 11, tryptic peptide OT-1 (residues 1–25); 12, CNBr-cleaved oxidized MT-2; 13, CNBr-cleaved carboxymethylated MT-2; 14, citraconylated oxidized MT-1; 15, an equimolar mixture of antigens 12 and 14, representing complementation of two altered molecules, one with the amino-terminal methionine cleaved and the other with the amino groups of lysines blocked by citraconic anhydride. (Modified from Winge and Gárvey[17] with publisher's permission.)

The results of the experiment indicated that the principal antigenic determinants of mammalian MTs are in the β domain, since binding was as strong with the β domain as competing antigen as it was with the complete molecule (apothionein or metal chelated). The specificities recognized by the antibody were restricted to the amino-terminal sequence of residues 1–25, since the peptide OT-1 bound the antibody as well as did the complete molecule. The α domain was minimally reactive. Contributing to the binding reaction were both the amino-terminal sequence of residues (1–5) and the region of residues 20–25, since CNBr treatment (cleavage of methionine) and citraconylation (blockage of the amino groups of lysine) significantly diminished the binding compared to that of the native molecule (See Table I). Also, the presence or absence of metals (alteration of conformation) was not a factor of influence in the antibody–antigen reaction.

Quantitative precipitation, which is another form of antigen–antibody binding, demonstrated that the ratio by weight in moles of Ag:Ab in precipitates as 1.5 : 1; this finding is interpreted as resulting from a two-epitope recognition by the antibody, one epitope being of dominant reactivity.

Theoretical analysis of hydrophilicity was performed to determine surface-exposed regions of MT and thus likely sites of binding with the antibody, as was analysis of primary structure to predict secondary structure. The Hopp–Woods hydrophilicity scale and the procedure of sequential hexapeptide analysis[18] were used in the former study, and the complete Chou–Fasman protocols[19,20] were used for the latter study. Both the invariant residues comprising the amino terminus (residues 1–7) and the region of residues 20–25 are prominently hydrophilic; the latter region exhibits the maximal hydrophilicity and is presumably the immunodominant epitope (see Fig. 3A).[16] The Chou–Fasman protocols lead to a predicted structure that favors reverse turns for both the mentioned principal determinants of mammalian MTs (Fig. 3B), and indicates that reverse turns and random coils dominate the secondary structure, with the few candidate β turns and helices concluded to be unstable. As purified and sequence MTs and candidate MTs have been produced by various investigators and made available to the author for study, they have been analyzed by RIA and ELISA (described below). In the case of mammalian MTs (human, rat, horse), their degree of immunological reactivity has proven well correlated with preservation of the invariance of the amino-terminal residues

[18] T. P. Hopp and K. R. Woods, *Proc. Natl. Acad. Sci. U.S.A.* **78,** 3824 (1981).

[19] P. Y. Chou and G. D. Fasman, *J. Mol. Biol.* **116,** 135 (1977).

[20] P. Y. Chou and G. D. Fasman, *Adv. Enzymol.* **47,** 45 (1978).

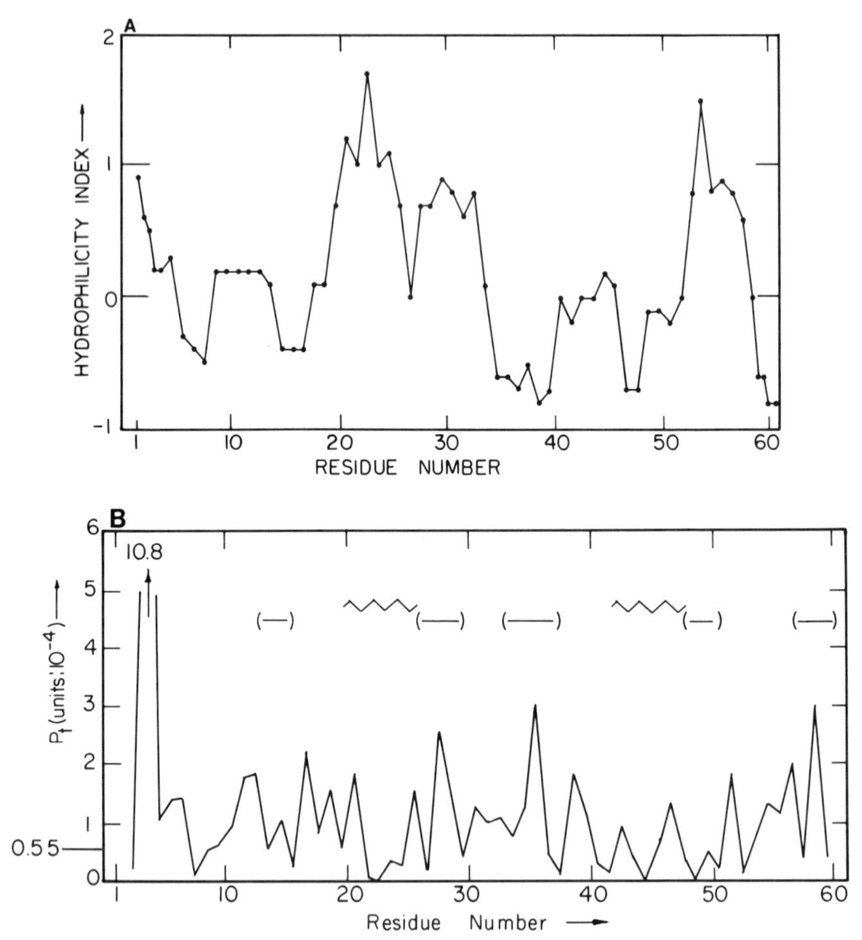

FIG. 3. (A) Hydrophilicity profile of human MT-2. Hydrophilicity indices (kcal/mol free energy change as solvent changes from water to an organic solvent) are indicated for sequential hexapeptides according to the protocols of Hopp and Woods.[18] Terminal regions are completed by values for penta-, tetra-, tri-, and dipeptides. The Hopp–Woods indices for the amino acids commonly found in mammalian MTs are as follow: 3 (D, E, K, R); 0.3 (S); 0.2 (N, Q); 0 (G, P); −0.4 (T); −0.5 (A); −1 (C); −1.3 (M); −1.5 (V); −1.8 (I, L). Other indices include those for H, Y, F, and W (−0.5, −2.3, −2.5, and −3.4, respectively). (Modified from Garvey[16] with publisher's permission.) (B) Reverse turn profile of human MT-2 and candidate helices and β chains. Probabilities (p_t) for sequential tetrapeptides to form a reverse turn according to the protocols of Chou and Fasman[19,20] are plotted with values indicated in the center of each tetrapeptide. Analysis of 29 proteins leads to assignment of a frequency of a given amino acid to appear in a specific position (1, 2, 3, or 4) of a reverse turn (similar frequencies are developed for appearance in a specific position of a helix or β chain). The continued product of site frequencies for a set of four residues represents the probability that the set will form a reverse turn; the average value is 0.55×10^{-4}, and a value in excess of 10^{-4}

(MDPNCSC)[20a] and preservation of the residues in positions 20–25 (KCKECK in human MT-2); some substitutions in this region, such as in rat and equine MTs, may occur with minimal influence on regional antigenicity. Thus, the complete lack of reactivity of chicken MT[21] with the polyclonal antibody (Fig. 4A[22]) is not unexpected from the variation of residues of chicken MT with respect to mammalian MTs in the regions of the principal epitopes of mammalian MTs (Fig. 4B). Two substitutions and one insertion occur in the amino-terminal region and four substitutions occur in the region of residues 23–30. Some mammalian MTs (rat, equine) exhibit substitutions of less severity in this region without significant modification of antigen–antibody reactivity; presumably, the changes in the case of chicken MT influence tertiary structure sufficiently to seriously lower avidity of the antibody in the mentioned antigenic sites. These results complement numerous similar studies of mammalian MTs, nonmammalian MTs, and various non-MT metal-binding proteins that demonstrate the sensitivity and utility of the antibody as a probe of protein structure.

Development of Enzyme-Linked Immunosorbent Assay

Since the RIA is a labor-intensive method, requiring a time span of 3–4 days for completion, and is quite complex when all aspects are properly performed, it is not a routine type of assay to be popularized for widespread use. With these practical considerations in mind, an ELISA was

[20a] Single-letter abbreviations used for amino acids: A, alanine; R, arginine; N, asparagine; D, aspartic acid; C, cysteine; Q, glutamine; E, glutamic acid; G, glycine; H, histidine; I, isoleucine; L, leucine; K, lysine; M, methionine; F, phenylalanine; P, proline; S, serine; T, threonine; W, tryptophan; Y, tyrosine; V, valine.

[21] C. C. McCormick, C. S. Fullmer, and J. S. Garvey, *Proc. Natl. Acad. Sci. U.S.A.* **85,** 309 (1988).

[22] J. H. R. Kägi and Y. Kojima, *Experientia, Suppl.* **52,** 25 (1987).

indicates a significant potential for forming a reverse turn. Also indicated are the segments that satisfy the primary conditions for formation of a helix (〰) or β chain (—). The measure here is the conformational parameter of the sequence that is the average value of the conformational parameters of the member residues. Conformational parameters are derived in the Chou–Fasman analysis from the observed frequency of appearance of an amino acid in a specific conformation. To be predicted as a stable conformation, the indicated helices and chains must satisfy certain boundary conditions regarding appearance of various dipeptides and tripeptides in the conformation, and the relative magnitude of the reverse turn potential involving the same residues. Since these conditions compromise the stable appearance of both indicated candidate helices and at least three of the indicated candidate β chains, their appearance is unlikely. The two-domain structure of MT compromises the appearance of the other two β chain candidates. (Modified from Garvey[16] with publisher's permission.)

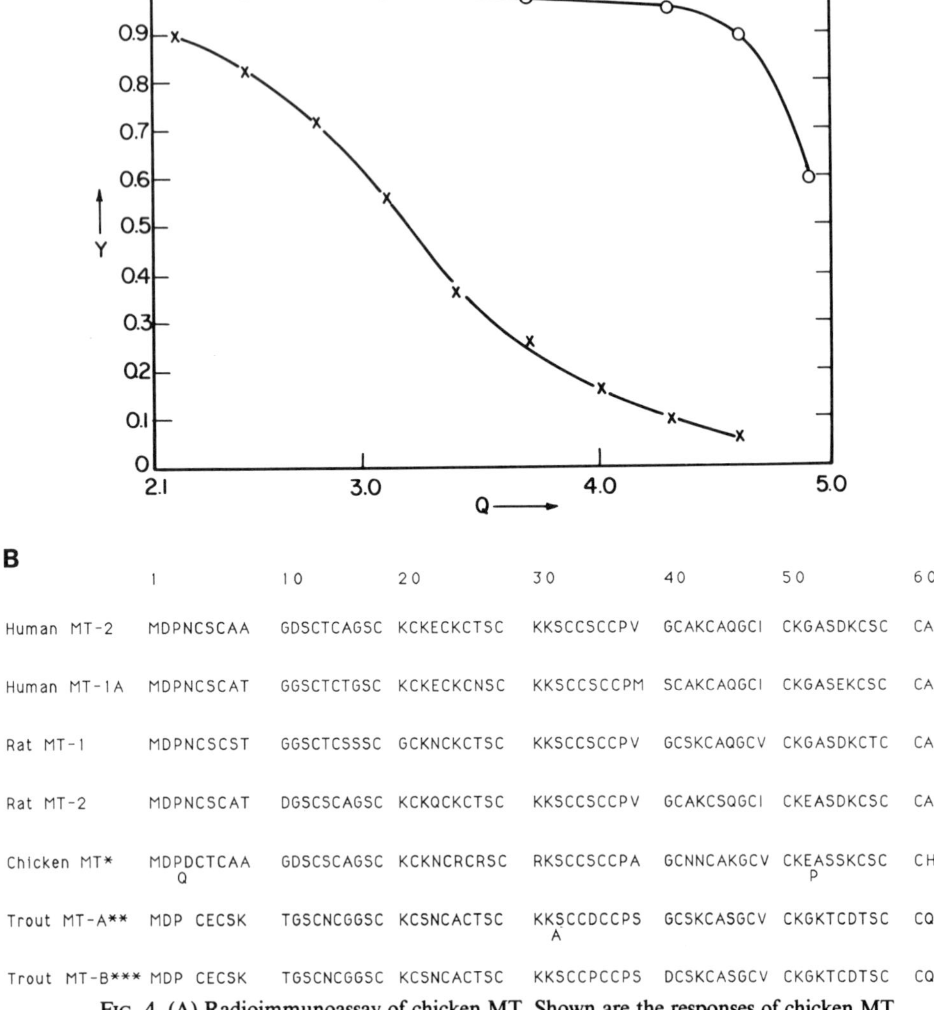

FIG. 4. (A) Radioimmunoassay of chicken MT. Shown are the responses of chicken MT (O) and rat liver MT (X) isoforms 1 and 2 (a 1 : 1 mix of the latter isoforms) in the competitive binding double-antibody RIA. *Y* is the fraction of ^{125}I-labeled antigen (rat MT-1) bound; *Q* is log(pg MT). The displacement (two orders of magnitude in competitor antigen concentration) of the response of chicken MT to the polyclonal antibody raised against rat MT, compared to the response of the rat MT, indicates a major change in the region of the two principal antigenic determinants of mammalian MTs (residues 1–5 and 20–25). (Modified from McCormick *et al.*[21] with publisher's permission.) (B) Amino acid sequences of human MT-2, human MT-1A, rat MT-1, rat MT-2, chicken MT, and trout MTs A and B. (*), Insertions Q and P make the sequences MDPQD- and CKEPA-; 63 residues. (**), Insertion A makes the sequence KKASC-; 61 residues; (***), 60 residues. Sources: Human and rat isoforms (Ref. 22, pp. 34 and 35); chicken MT (Ref. 21, p. 311); trout MTs [K. Bonham, M. Zafarullah, and L. Gedamu, *DNA* **6,** 519, 524 (1987)].

developed for detection and quantitation of MT.[23] The previously developed RIA was used as a reference assay to characterize its performance. In the protocol for the ELISA, the same immunological reagents are used as in the RIA, but fluorescence replaces radioactivity for detection of the reference MT in the specific binding reaction. Affinity-purified goat anti-rabbit IgG is used to prepare the second antibody as an alkaline phosphatase conjugate.[24] Initial optimizations are performed for concentrations to be used of the conjugate, coating antigen, primary antibody, and substrate; from a protocol developed with these reagents at optimized concentrations, additional optimizations for each step of the protocol are made, e.g., in selection of buffer concentrations and pH values and in selection of incubation times.

Protocol for ELISA. The described basic protocol is an abbreviated description from Ref. 23. Wells replace the assay tubes of the RIA, and the format for these reproduces that of the RIA, with the exception that there are positional effects that must be considered for precision in results. These effects are considered in Ref. 23, but not here.

Step 1. The wells of a 96-well polystyrene microtiter plate (Dynatech, Chantilly, VA; black, Cat. No. 011-101-7800) are coated with the optimized MT concentration, i.e., 100 ng MT/ml of coating buffer (approximately 100 m*M* sodium carbonate/bicarbonate, pH 9.6) added at 150 μl/well. The plate is tightly sealed and stored overnight at 4°. Complete decantation of the MT solution is achieved and the wells are washed 3 times with PBS[24a] that contains 0.05% (v/v) Triton X-405, 0.01% (w/v) sodium azide, and 0.01% (w/v) bovine serum albumin (BSA). The modified PBS is hereafter referred to as buffer and it is used for dilutions of all reagents and for washing wells; if an alternative buffer is used, it is specified. The first two additions of buffer are allowed to remain in the wells for 5 min and the third remains for 15 min; between washes, the plate is covered to protect it from dust contamination.

Step 2. After removal of the third addition of buffer (step 1), 50 μl of a competitor solution is added to each well. Competitor solutions are known concentrations of MT for development of an SC(0) following the RIA protocol for concentration range, and they are also unknown concentrations of MT to be quantitated. Primary antibody reagent, 50 μl, e.g., No. 4 or No. 7 polyclonal antiserum (see Fig. 1), optimized at a 1 : 3000 dilution in buffer, is added to all the wells except those of the SC(0) determination and of the unknowns that are used for determination of NSB. To the NSB

[23] D. G. Thomas, H. L. Linton, and J. S. Garvey, *J. Immunol. Methods* **89,** 239 (1986).

[24] E. Engvall, *Scand. J. Immunol. Suppl.* **8,**(7) 25 (1978).

[24a] Phosphate-buffered saline (PBS) is prepared as 0.386 g $NaH_2PO_4 \cdot H_2O$, 8.5 g NaCl, and 1.926 g $Na_2HPO_4 \cdot 7H_2O$ in 1 liter doubly distilled H_2O, and is adjusted to pH 7.2.

wells is added 50 μl buffer. After the additions of primary antibody and buffer to the previously added 50 μl competitor, each well contains 100 μl of solution. The plate is sealed with a cover that adheres tightly to prevent loss of solution by evaporation and is incubated at 37° for 3–4 hr. The washings are then repeated as described in step 1.

Step 3. Second antibody–enzyme conjugate, diluted with PBS as determined from a previous optimization, e.g., 1:3000 in buffer, is added as 100 μl to each well. The plate is reincubated for an additional 3–4 hr at 37°, after which the three washings are repeated as described in step 1.

Step 4. Substrate solution, prepared as 2×10^{-4} *M* 4-methylumbelliferyl phosphate in 10 m*M* sodium carbonate/bicarbonate buffer, pH 10, containing 1 m*M* $MgCl_2$, is added to each well in a volume of 150 μl. Storage of the plate thereafter is either at ambient temperature or at 37°, depending on the more suitable kinetics.

Step 5. Fluorescence of the cleaved substrate (4-methylumbelliferone) is measured in the automatic fluorometric plate reader in which optimization of the reagents was previously performed (e.g., Dynatech Microfluor reader, excitation 362 nm, emission 440 nm). The recorded responses of this reader are in relative fluorometric units (f.u.) to a maximum of 4094 f.u. Readings are usually made at two or three intervals of time within a 2-hr incubation time.

Typical Standard Curve and Quantitation of Metallothionein

An SC(0) is obtained by using the same reagents, primary antibody, and competitor MT (rat liver MT) as in the RIA. The logit-log regression that is typical of the RIA is similarly obtained in the ELISA and with a similar range for the competitor antigen, i.e., 20,000 to 100 pg. Cytosols prepared from tissues, e.g., liver and kidney, with a significant content of MT permit quantitation of MT using an SC(0), a condition paralleling that discussed for the RIA. Accordingly, when such cytosols were prepared and assayed by both RIA and ELISA, the results obtained for MT concentrations in the two assays demonstrated no statistical significance in differences. Quantitation of samples of unknowns in the central region of the respective standard curves is characterized by standard errors of the mean values of quadruplicates of 5–10% (RIA) and 10–15% (ELISA).[23]

As discussed for the RIA measurement of physiological fluids, serum, or urine, that contain very low concentrations of MT, a standard curve is developed with each competitor MT well also containing a control fluid that has been previously assayed for MT content. The procedure developed for the RIA for quantitation of unknown concentrations of MT (see protocol outlined in the section Selected Problems in Utilization of Radioimmunoassay) is applicable for the ELISA for cytosols and urine, but additional treatment of serum is required.[23]

Monoclonal Antibodies

Rationale for Obtaining Monoclonal Antibodies of Specificity for Metallothionein

The impetus for development of MAbs with specificity for MT was the potential for an "unlimited supply." It was anticipated that, contrary to polyclonal techniques, such MAbs would recognize antigenic determinants other than those on the most immunodominant β domain. According to the hydrophilicity profile (Fig. 3A), other exposed regions of the MT secondary structure than the amino-terminal residues and the region of residues 20–25 are potential epitopes, although of less affinity for the antibody. In theory, the MAb technique permits cells of rare occurrence that comprise the complete repertoire of cells stimulated by antigen to be utilized for hybridoma production. Whereas the previous part discussed methodology developed for polyclonal antibodies, the following monoclonal antibody production utilized the rat–rat hybridoma methodology well known from published procedures by Professor H. Bazin, University of Louvain. Moreover, Bazin has been generous in providing cell lines to investigators, including the present author.

Thus, in this part, attention will be given only to methodology of production and characterization unique to MT, since the basic procedures for hybridoma production[25,26] and MAb purification[27] were used as detailed in the noted references. If the reader should lack the presumed familiarity with the basic principles and practices of monoclonal antibody production, Ref. 28 will be found useful.

Preparation of Antigen and Immunization

Human liver tissue obtained post mortem from local hospitals was processed into MT isoforms 1 and 2 by the same isolation protocols used for rat MTs[12]; these protocols were supplemented in recent years by chromatography on two HPLC columns.[23] Purity of the MTs was assessed by

[25] L. De Clercq, F. Cormont, and H. Bazin, this series, Vol. 121, p. 234.

[26] H. Bazin, *in* "Hybridoma Formation" (A. H. Bartel and Y. Hirshaut, eds.), p. 337. Humana Press, Clifton, New Jersey, 1987.

[27] F. Cormont, P. Manouvriez, L. De Clercq, and H. Bazin, this series, Vol. 121, p. 622.

[28] J. W. Goding, "Monoclonal Antibodies: Principles and Practice," 2nd Ed., Chaps. 2 and 3, p. 5. Academic Press, New York, 1986.

nondenaturing polyacrylamide gels stained with either Coomassie Blue, silver stain, or a combination of these two stains.[29] Quantitation was obtained by amino acid compositional analysis, absorbance at 220 nm[30] and protein estimation.[13]

Two panels of MAbs, A and B, were developed as a consequence of a different rationale regarding both the preparation of the immunogen and the immunization protocol. Female LOU/MN rats (age 3–6 months) were used for all immunizations. These rats were bred in an isolation unit in the animal care facility of the Department of Biology, Syracuse University, Syracuse, New York, from stock obtained in late 1984 from Dr. Carl A. Hansen (Veterinary Resources Branch, Division of Research Services, National Institutes of Health, Bethesda, MD). The LOU rat is available from commercial suppliers and the origin of the lymphoid tumors in the strain is well described.[31] For panel A, MTs were conjugated to keyhole limpet hemocyanin (KLH) by the bifunctional reagent *m*-maleimidobenzoyl-*N*-hydroxysuccinimide ester[32] and the injection of 50 μl MT was intrasplenic.[33] Spleen cell isolation and fusion with IR983F cells were performed 4 days later. For panel B, the MTs were emulsified with an equal volume of the following adjuvant mixture: MPL (monophosphate lipid A) + TDM (trehalose dimycolate) + CWS (cell wall skeleton) in 2% oil and Tween 20. The mixture was purchased from RIBI Immunochem Research (Hamilton, MT), and reconstituted according to the instructions provided with the product. The emulsion consisting of approximately 100 μg MT was injected into multiple sites (subcutaneous, intramuscular, intraperitoneal) as previously described for rabbits.[12] After a time interval of 3–4 months, 50 μg MT, unconjugated and without adjuvant, was injected into the spleen. The spleen cell isolation and fusion with IR983F cells were performed at 3 days after the intrasplenic injection. The aim of the intrasplenic route of injection is to provide an extremely effective use of immunogen and satisfy the need to use judiciously the isolated pure human MTs. Additionally, use of the intrasplenic route of injection is supported by the definitive finding made in rats that a population of splenic lymphocytes is noncirculating.[34]

[29] L.-Y. Lin and C. C. McCormick, *Comp. Biochem. Physiol. C: Comp. Pharmacol. Toxicol.* **85C,** 75 (1986).

[30] M. Vašák, J. H. R. Kägi, and H. A. O. Hill, *Biochemistry* **20,** 2852 (1981).

[31] H. Bazin, W. S. Pear, and J. Sumegi, *Adv. Cancer Res.* **50,** 279 (1988).

[32] F.-T. Liu, M. Zinnecker, T. Hamaoka, and D. H. Katz, *Biochem. J.* **18,** 690 (1979).

[33] M. Spitz, this series, Vol. 121, p. 33.

[34] D. Gray, I. C. M. MacLennan, B. Platteau, H. Bazin, J. Lortan, and G. D. Johnson, *Exp. Med. Biol.* **186,** 437 (1985).

Fusion, Culture, and Cloning

Panel A MAbs were obtained by procedures aimed at a precise duplication of the published procedures for rat–rat hybridomas,[25,26] but the screening method was the fluorometric ELISA[23] described in the section Polyclonal Antibodies. This solid-phase assay for detection of antibody specific for MT had been developed in advance of the MAb project, both in anticipation of the need for a rapid and sensitive screening method for MAbs and as an alternative methodology to the RIA, using the polyclonal antibody. For hybridoma screening, the format of the ELISA[23] was modified by omission of competing antigen. Since the MAb is from the rat species, the second antibody was anti-rat IgG diluted 1 : 1000 in PBS. It was purchased from Hyclone (Logan, UT; Cat. No. EA-1036-X), specified to be >99% in monospecificity for rat IgG. It was a conjugate of β-galactosidase rather than of alkaline phosphatase, this chosen to avoid the possibility of a high background from endogenous alkaline phosphatase when ascites is the source of the MAbs. The substrate was 4-methylumbelliferyl-β-D-galactoside (purchased from Sigma, St. Louis, MO) and used as 1 mg/10 ml PBS, adjusted to pH 7.8, and containing 1 mM $MgCl_2$.

Panel B MAbs were obtained by procedures used for panel A MAbs, with the exception of the variation in immunization mentioned above and with some alterations that follow.

Growth Stimulants

Three growth stimulants were investigated for their influence on the outgrowth of hybridomas and for their ability to increase the number of specific hybridomas producing MAbs of defined specificity in binding to MT. Postfusion, a portion of the cell suspension was cultured in the usual complete medium with HAT[25,26] and, in addition, some cell suspensions were prepared in each of the three media that differed by addition of one of the following: (1) endothelial cell growth supplement (CR-ECGS, purchased from Collaborative Research, Bedford, MA), used at a concentration of about 430 μg/ml cell suspension; (2) crystalline porcine insulin, "for research purposes," glucagon free, a gift from Dr. William Bromer, Eli Lilly and Company (Indianapolis, IN), and used at a concentration of 0.1 M; and (3) STM, an extract of *Salmonella typhimurium* mitogen, purchased from RIBI Immunochem Research (Hamilton, MT), and used at a concentration of 10 μl/ml cell suspension. Each of the three supplements has been used to promote the growth of murine MAb-producing hybrid-

TABLE II
EFFECT OF GROWTH STIMULANT ON GROWTH AND SPECIFIC ANTIBODY PRODUCTION[a]

	Without feeder layer		With feeder layer	
Stimulant[b]	Growth (%)	Positive in screen with MT (%)	Growth (%)	Positive in screen with MT (%)
ECGS	None	—	60	76
Insulin	14	38	75	33
STM	3	67	58	43
Control	None	—	46	41

[a] By hybridoma cultures with and without feeder layer.
[b] ECGS, Endothelial cell growth supplement; insulin, crystalline porcine insulin; STM, *Salmonella typhimurium* mitogen.

omas, i.e., lipopolysaccharide and endothelial cell growth factor[35] and insulin.[36] Each cell suspension was plated in two ways: the first at a volume of 400 μl/well in 48-well plates that had been prepared 48 hr earlier with LOU/MN peritoneal cells (1×10^5), cultured to provide a "feeder layer," and the second at a volume of 100 μl/well in 96-well plates lacking the "feeder layer." The results shown in Table II at 10 days of postfusion culture confirm the need for a feeder layer, regardless of the presence or absence of one of the added growth stimulants. Thus, in a comparison of positive clones on a feeder layer, the results for growth show the following: insulin superior to ECGS, STM superior to control. The yield of specific clones shows the following: ECGS superior to STM, control superior to insulin. A factor that contributed significantly to the tabulated results for specificity is the use of conservative scoring of specific clones.[37] Supernatant secretion of specific antibody *vs* MT was scored positive if the ELISA result was three standard deviations above the mean for the negative control of myeloma cell (IR983F) supernatant. This strategy was probably important in early initial screening (10 days) with human MT as antigen and the later assessment for clones that showed binding to domains and peptides used as antigens.

[35] R. J. Westerwoudt, this series, Vol. 121, p. 3.
[36] A. H. Bartel, C. Feit, and Y. Hirshaut, *in* "Monoclonal Antibodies: Standardization of Their Characterization and Use," (M. Barme and W. Hennessen, eds.), Develop. Biol. Standard. Vol. 57. Karger, Basel and New York, 1984.
[37] J. Y. Douillard and T. Hoffman, this series, Vol. 92, p. 168.

Steady Culture Conditions and Cloning

The hybridomas were never weaned from HT,[38] and a 6.5% CO_2 atmosphere for incubation was maintained constant, thus reducing changes in culture conditions as much as possible. Each positive well was expanded gradually to a flask-type culture, from which a portion of the hybridoma culture suspension was cryopreserved as a precaution against inadvertent loss. Cloning by limiting dilution involved a series of dilutions prepared for each MT-specific culture, so that by calculation of cell number, 0.5 cells/well was cultured in a 96-well plate. Additionally, use was made of a modified procedure[39] that involved microscopic observing of the dilute cell suspension, and was aimed at visualization before culture to ensure the presence of a single hybridoma cell in each well of the 96-well plate, in which 2–3 drops of complete medium had been previously placed. For both methods of cloning, a microscopic examination was used to observe that a single hybridoma cell yielded a single clone. The assurance of monoclonality is obtained by two subclonings. This work-intensive process of cloning can be reduced considerably in time and labor if the cell-sorting capability of a flow cytometer can be employed for cloning of individual antigen-binding cells; however, the author has no experience from which to recommend this method of cloning in the case of MT-binding hybridoma cells.

Bulk In Vitro Culture and Ascites Production

Both methods were employed with each panel of MAbs and with the inherent advantages/disadvantages of each method, e.g., the less concentrated antibody secretion being an obvious disadvantage in necessitating large-scale *in vitro* culture, but an advantage in that it avoided animal usage. In both methods, there are other serum proteins present with the antibody (since these hybridomas have not been adapted to serum-free *in vitro* culture). Although MT is present as an endogenous constituent of ascites, it is absent from *in vitro* cultures of hybridomas. The volume of ascites obtained was variable, sometimes being less than 5 ml and of low antibody concentration, due to the solid tumor growth that developed more rapidly than ascites fluid and necessitated euthanasia. It was found that the injection volume in which the suspension of hybridoma cells was administered to the rat was an important factor in obtaining ascites fluid; the volume must be 10–15 ml for the optimal yield of 20–30 ml ascites

[38] H. Zola and D. Brooks, *in* "Monoclonal Hybridoma Antibodies: Techniques and Applications" (J. G. R. Hurrel, ed.), p. 1. CRC Press, Boca Raton, Florida, 1982.

[39] A. Boeyé, this series, Vol. 121, p. 332.

fluid/rat at a usual concentration of 1–2 mg specific antibody/ml. Adjuvant was not used in the ascites production with the MAb-secreting hybridomas of the present study. Use of adjuvant, considered routine as a priming agent with murine hybridomas, had not been proven for rats.[26] This has changed with a recent report recommending a 2-ml injection of a mixture containing equal volumes of pristane and Freund's incomplete adjuvant, administered intraperitoneally at the time of transfer of the hybridoma cells[40]; this procedure is reported to improve the yield of ascites fluid in both volume and concentration of specific antibody. At this writing, the reported study needs confirmation for hybridomas produced with MT as immunogen.

Purification of Monoclonal Antibody

There are numerous reports in the literature of MAb purification by chromatography, using soft gel or HPLC as one or two column procedures, with or without prior salt precipitation to provide an enrichment of the class or subclass of Ig known to be that of the MAb. As noted in the comments on the ELISA, for screening of the MAbs at various stages of their production, constant use was made of an enzyme-conjugated anti-rat IgG as the second antibody. An Ouchterlony assay was performed, with the culture supernatant fluid in an agar diffusion/precipitation reaction with class-specific antiserum to confirm the positive screening of IgG class by ELISA, and with subclass-specific antiserum to determine subclass (predominantly IgG_{2a} for both panel A and panel B). A number of purification strategies were tried in the author's laboratory, but despite strict adherence to published procedures, the experience led to disappointing results in a yield of MAb that was low (indicating a loss of MAb during purification) and, moreover, contamination with other proteins than those of the class/subclass of MAb (especially albumin) was often high. This background of experience that yielded completely unsatisfactory results was followed by the use of Mark 1 and Mark 3 proteins as affinity columns, and these were confirmed as useful.[27] Briefly, Mark 1 and Mark 3 are myelomas of BALB/c mice that secrete MAb with specificity for binding to the two alleleic forms of κ chain of rat Ig. Since the light-chain class of rat Igs is about 95% κ, the two mentioned MAbs are extremely useful as an affinity column to bind Ig.

Both CNBr activation of Sepharose as described for protein conjugation in the published procedures,[26,27] as well as Reacti-Gel (purchased from Pierce Chemical, Rockford, IL, and requiring no activation) have been

[40] J.-P. Kints, P. Manouvriez, and H. Bazin, *J. Immunol. Methods* **119,** 241 (1989).

used successfully for the two required columns, (1) the rat Ig affinity matrix to purify Mark proteins from *in vitro* culture of the myeloma cell lines, Mark 1 and Mark 3, and (2) the Mark affinity matrix to bind MT-specific MAb. As mentioned in the previous discussion of bulk production of MAb as ascites fluid, normal rat protein that is present will copurify with the MAb. The preparation of an affinity matrix with MT, or with peptides containing the epitopes for binding of the MAb, seems an obvious solution in a purification protocol, but limited success has been obtained to date. Although the current application of the ascites-produced MAb has been mostly for immunohistochemical determination of MT (see [15] in this volume), and this has been achieved without purification of ascites fluid, it is likely that the future expanded production of MT-specific MAbs will be by large-scale *in vitro* culture of the hybridomas.

Storage of Monoclonal Antibody

Supernatants have been stored while maintained sterile at 4–8, −20, and −70°, without loss of binding activity in the ELISA when examined after a period of 1 year. Ascites fluid has been collected in plastic tubes to which both a protease inhibitor (phenylmethylsulfonyl fluoride, PMSF) and a bacteriostatic agent (ethylmercurithiosalicylic acid, sodium salt) have been routinely added. Routine storage of ascites fluid is at −70° unless lyophilized for ambient shipment and subsequent storage at −20°. No decrease has been found in the binding activity by ELISA after 3–4 years of continuous storage. After the ambient storage occurred by overseas shipment and storage at −20° for 2 years, MAbs in ascites have been reconstituted and used successfully for immunohistochemical determination of MT (see [15] in this volume).

Characterization of Monoclonal Antibody

The format of the ELISA for the following results was fluorometric. The MAb reacts with a previously coated antigen (without competing antigen). The complex subsequently reacts with the enzyme–antibody conjugate that hydrolyzes the substrate to a fluorescent product whose fluorescence is read. The format is often termed "antigen capture."

Panel A. Nine clones, as described above, showed approximately the same specificity profile of binding in the ELISA as did the polyclonal antibodies, e.g., complete cross-reactivity with all the mammalian species of MT investigated [human, rat, equine, rabbit, and bovine (calf), including two isoforms of each species, excepting equine MT]. They also showed an equal binding with the β domain, as with the complete molecule, in the case of rat MT-1. All clones failed to bind the α domain of rat MT-1 and rat MT-2, and also the cleavage product of the latter two molecules, when treated with CNBr. Trout MT-2, *Neurospora crassa* MT, and a synthetic peptide (provided by Dr. Ben Dunn, University of Florida, Gainesville, FL) of the same sequence of residues (16–28; AGSCKCKECKCTS) as human MT-2 also lacked binding. Supernatants from the *in vitro* culture of the hybridomas, diluted 1 : 100, and ascites fluid, diluted 1 : 3000, were about equal in binding in the ELISA to a 1 : 5000 dilution of the polyclonal antibody. Negative controls for the MAbs were IR983F myeloma supernatant and trout MT-2 hybridoma ascites fluid (available from a fusion experiment performed separately from the human MT being described here); preimmune serum was the negative control for the polyclonal antibody. Precipitation was lacking in the Ouchterlony assay with any of the cross-reacting MTs and hybridoma supernatants. If the reactions had been positive, evidence would have been provided for recognition of more than one epitope by the MAb panel members. Nor was it possible to use the hybridomas in the form of either supernatant or ascites fluid in the RIA. Since there was no evidence of additivity in binding[41] in an ELISA in which mixtures of the separate MAbs were used, and likewise no evidence of an increased response in the RIA with mixed MAbs, the conclusion is that the epitope recognized by all the nine MAbs comprising the panel is the same. The location of the epitope is with certainty the invariant amino terminus of mammalian MTs, beginning with acetylated methionine, but otherwise not precisely defined as to length.

Panel B. The changes that were made in the production of panel B MAbs compared to panel A MAbs collectively were aimed at obtaining early hybridomas. On the basis of specificity for human MT in the ELISA, more than 50 hybridomas were saved for cloning. At the stages of cloning and subcloning, screening was directed toward obtaining hybridomas with binding specificities other than for the epitope at the amino-terminus invariant region of mammalian MT recognized by the panel A MAbs. Screening was most intensive with pure α domain, but rat MT with the methionine cleaved by CNBr was usually included also. Both the α and β domains prepared from rat MT-1 by Dr. Dennis Winge (University of Utah, Salt Lake City, UT) were provided to the author as the apoprotein, as well as copper-chelated and cadmium-chelated forms; the preparations were then used for MAb screening. Dr. Dean Wilcox (Dartmouth College, Hanover, NH) provided synthetic α and β domains, these based on the amino acid sequence of human liver MT-2. An affinity ranking of MAbs positive for each domain was obtained by using serial dilutions of the

[41] B. Friguet, L. Djavadi-Ohaniance, J. Pages, A. Bussard, and M. Goldberg, *J. Immunol. Methods* **60**, 351 (1983).

supernatant fluid of hybridomas cultured to confluency.[42] Hybridomas to be compared were screened similarly with respect to each other from the time of their scoring for positive growth, initial antigen-specific binding to MT, and all aspects of cloning and subcloning. As shown in Figs. 5A and B, the binding to the α and β domains of the apoprotein was about the same for the supernatant of a clone. Moreover, the relative affinity of individual clones plotted as a series resulted in the same ranking of the members of the series for the metal-chelated domains prepared from the native molecule, and for the synthetic α and β domains. Additionally, there was a binding that was similar for trout (see Fig. 4B for the primary sequence). Since the binding of the MAb to the trout MT was not performed in the same assays as were the α and β domains, the degree of binding cannot be used for direct comparison with that observed in the cases of α and β domains of rat and human MTs. The tentative conclusion from the findings is that the interdomain region of the MT molecule, i.e, residues 29–33 (see the hydrophilicity profile, Fig. 3A) is recognized by the individual members of panel B. Additionally, a formatting of an ELISA with a pair of MAbs, one from panel A and one from panel B, gives neither evidence of competitive inhibition, nor of additivity. Yet, because MT-2 was the immunogen for both immunizations and was used in the initial screening of both panels, the latter assay for extending the characterization of panel B seemed appropriate, and the more so since the epitope is defined for panel A as being at the amino terminus.

Summary and Perspective

The experience with both polyclonal and monoclonal antibody production, as well as characterization, has been sufficiently extensive to suggest the following as applying to MT. Despite the advantages/disadvantages of each for many soluble proteins, if by necessity one approach must be utilized for MT, it should be polyclonal. A polyclonal antibody reagent is usually regarded as being adaptable to various types of assay because of high avidity and because different classes/subclasses of Ig may be represented. On the contrary, a monoclonal antibody reagent requires that the assay used for screening reproduce as closely as possible the format of the assay that is to be used, in order to detect and nurture throughout the screening process the single specificity that is the goal of the work-intensive process. Both points have been well demonstrated, particularly the versatility of the polyclonal antibody for use in various optimized assays such as the RIA and ELISA, in immunohistochemistry, and in immunodiffusion,

[42] V. van Heyningen, this series, Vol. 121, p. 472.

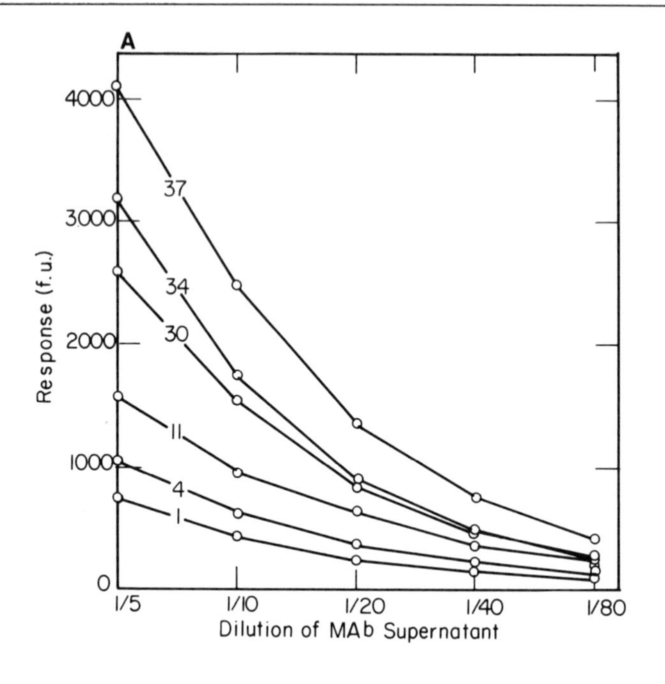
A
4000
3000
2000
1000
0
Response (f.u.)
37
34
30
11
4
1
1/5
1/10
1/20
1/40
1/80
Dilution of MAb Supernatant

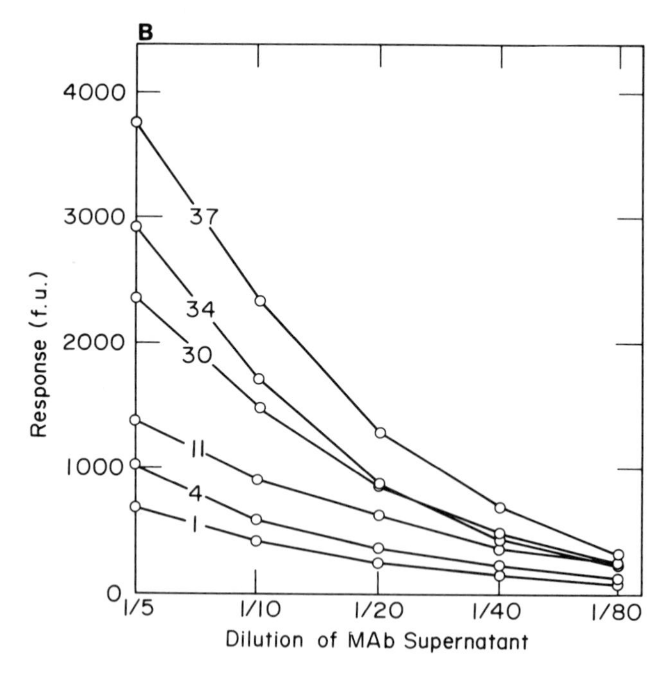
B
4000
3000
2000
1000
0
Response (f.u.)
37
34
30
11
4
1
1/5
1/10
1/20
1/40
1/80
Dilution of MAb Supernatant

immunoblotting, and immunoprecipitation. The monoclonal antibodies are quite limited at this time in their usefulness and this condition may continue. Regarding the results of epitope discrimination, the polyclonal and monoclonal results are supportive of each other. It is the β domain of MT that is recognized as the more immunodominant of the two domains and this is a definitive result in view of the fact that the panel B of monoclonal antibodies binds to some degree with both α and β domains. This binding is weak with respect to that observed with panel A, in which the region of MT that is recognized for binding is the invariant amino terminus, complementing the fact that both principal determinants for binding by the polyclonal antibody are located on the β domain.

All the immunological findings, experimental to determine the epitopes and theoretical to determine the secondary structural features in the regions of those epitopes, are supported by investigations that define mammalian MT structure. An example is X-ray crystallography,[43] that indicates that the two domains have limited contact and that the charged residues in positions 20, 22, and 25 (lysines in the case of human MT) may be in proximity to the amino terminus in solution. Thus, the two epitopes recognized by the polyclonal antibody of IgG class may be more easily accommodated by the antibody because of their flexible, rather than rigid, topography. And the fact that the entire MT secondary structure is one of reverse turns and random coils is compatible with mobility, an aspect emphasized by others in defining antigenicity of proteins.[44,45] Spectro-

[43] W. F. Furey, A. H. Robbins, L. L. Clancy, D. R. Winge, B. C. Wang, and C. D. Stout, *Science* **231,** 704 (1986).

[44] J. A. Tainer, E. D. Getzoff, H. Alexander, R. A. Houghten, A. J. Olson, R. E. Lerner, and W. A. Hendrickson, *Nature (London)* **312,** 127 (1984).

[45] E. Westhoff, D. Altschuh, D. Moras, A. C. Bloomer, A. Mondragon, A. Klug, and M. H. V. Van Regenmortel, *Nature (London)* **311,** 123 (1984).

FIG. 5. (A) Relative binding affinity of MAbs to the α fragment of MT. A *Panel B* of rat–rat hybridomas selected in an ELISA for binding to human MT-2 (the immunogen) were subcloned twice and expanded by *in vitro* culture. Supernatants from confluent cultures were tested for binding to the α fragment of rat apo-MT-1 (provided by Dr. Dennis Winge, University of Utah Medical School). The ELISA procedure was essentially as previously described.[24] The response in fluorometric units (f.u.), corrected for the response lacking MAb (20 f.u.), is measured after a 45-min ambient temperature incubation with substrate. Not shown are the responses of MAbs 32 (similar to MAb 34), 2 (similar to MAb 11), 22 and 33 (intermediate to responses of MAbs 11 and 4), and 3 and 18 (similar to MAb 1). (B) Relative binding affinity of MAbs to the β fragment of MT. The same *Panel B* of rat–rat hybridomas, whose response in the ELISA to the α fragment of MT is illustrated in (A), was similarly analyzed for response to the β fragment of MT. The relative responses of the 12 MAbs parallel that described in (A).

scopic studies have characterized the β domain as having higher kinetic properties of bound metals, i.e., orders of magnitude higher, than does the α domain,[46] and the folding as a structural feature of the MT molecule is little affected by metal binding.[47] Both these findings are suggestive that the apo-MT and the native MT have like immunoreactivity, which was true of all the immunochemical findings using either the polyclonal or the monoclonal antibodies. In addition to the unique specificity of the immunological approach to the quantitation of MT, i.e., distinguishing mammalian MT from nonnmammalian MTs and other metal-binding proteins, importantly it is the only approach that provides sufficient sensitivity to quantitate the very small amounts of MT found in normal human serum and urine.[48,49]

Acknowledgments

The author acknowledges previous support from the NIEHS in the form of Grant R01-1629 and previous students and staff: R. J. Vander Mallie, C. C. Chang, H. J. Linton, D. G. Thomas, J. Klimczak, A. Zelazowski, L. Giddings, A. D. Dingman, and D. Campanile. Dr. W. E. Vannier is thanked for his critical review of the manuscript and for valuable comments. Karen Gregorich is thanked for preparation of the final manuscript.

[46] J. D. Otvos, H. R. Engseth, D. G. Nettlesheim, and C. R. Hilt, *Experientia, Suppl.* **52,** 171 (1987).

[47] M. Vašák, *in* "Progress in Inorganic Biochemistry and Biophysics, Volume 1, Zinc Enzymes" (I. Bertini, C. Luchinat, W. Maret, and M. Zeppezauer, eds.), p. 595. Birkhäuser Verlag, Basel, 1986.

[48] M. P. Waalkes, J. S. Garvey, and C. D. Klaassen, *Toxicol. Appl. Pharmacol.* **79,** 524 (1985).

[49] H. H. Dieter, L. Muller, J. Abel, and K. H. Summer, *Experientia, Suppl.* **52,** 351 (1987).

[21] Epitope Mapping of Metallothionein Antibodies

By KATSUYUKI NAKAJIMA, KEIJI SUZUKI, NORIKO OTAKI, and MASAMI KIMURA

Introduction

Metallothioneins (MTs) have no easily detectable function or property other than their metal-binding activity. Therefore, the methods of determining MT levels in biological tissues and fluids are limited to immunoassay or quantitation of metals in MTs.

The accuracy and sensitivity of immunoassay primarily depend on specificity and affinity of the antibodies used. It is well known that because MTs are small peptides and their interspecies differences are slight, they are very weak immunogens. Therefore, aggregated or conjugated forms of MTs have been employed as immunogen for antibody production, which make MTs denatured and different from native conformational features.

Immunological studies using rabbit polyclonal antisera and murine monoclonal antibodies raised against rat MT showed that they were cross-reactive with various kinds of mammalian MTs (Table I[1-9]). With the isolated domains and peptide fragments of rat MT available, the immuno-reactive properties of the domains and the fragments are compared with those of the native protein.

As the sensitive competitive radioimmunoassay (RIA) for MT using both polyclonal and monoclonal antibodies[1-10] was developed, complete cross-reactivity has been observed for MT isolated from the livers of rat, mouse, rabbit, monkey, horse, and human, and for cadmium, zinc, copper, thionein, and apothionein. The synthetic fragments of the antigenic sites of MT molecule cross-reactive with MT antibodies were prepared and the amino acid sequence homology was compared with various kinds of MTs for the antigenic determination of MT antibodies.

Prediction of Metallothionein Antigenic Determinants from Amino Acid Sequence

The elucidation of MT antigenic structures is still a difficult, uncertain task. To delineate antigenic determinants precisely, it is necessary to prepare a large number of well-characterized chemical derivatives and peptide fragments from native MT and then to test these derivatives for immunological activity.

In recent years a number of systems have been developed to predict

[1] C. Toyama and Z. A. Shaikh, *Biochem. Biophys. Res. Commun.* **84,** 907 (1978).
[2] R. J. Vander Mallie and J. S. Garvey, *Immunochemistry* **15,** 857 (1978).
[3] C. C. Chang, R. L. Vander Mallie, and J. S. Garvey, *Toxicol. Appl. Pharmacol.* **55,** 94 (1980).
[4] C. Toyama and Z. A. Shaikh, *Fundam. Appl. Toxicol.* **1,** 1 (1981).
[5] S. Ohi, G. Cardenosa, R. Pine, and P. C. Huang, *J. Biol. Chem.* **256,** 2180 (1981).
[6] R. K. Mehra and I. Bremner, *Biochem. J.* **213,** 459 (1983).
[7] T. Masui, T. Utakoji, and M. Kimura, *Experientia* **39,** 182 (1983).
[8] N. Ikei, T. Kodaira, F. Shimizu, K. Nakajima, C. Toyama, and H. Saito, *Nippon Eiseigaku Zasshi* **39,** 476 (1984).
[9] Y. Kikuchi, N. Wada, M. Irie, H. Ikebuchi, J. Sawada, T. Terano, S. Nakayama, S. Iguchi, and Y. Okada, *Mol. Immunol.* **25,** 1033 (1988).
[10] T. Umeyama, K. Saruki, K. Imai, H. Yamanaka, K. Suzuki, N. Ikei, T. Kodaira, K. Nakajima, H. Saito, and M. Kimura, *Prostate (N.Y.)* **10,** 257 (1987).

TABLE I
CHARACTERISTICS OF METALLOTHIONEIN ANTIBODIES

Isolation of MT from[a]	Preparation of MT for immunization	Immunized to	Antibody cross-reactive with	Ref
Rat liver MT-2 ($CdCl_2$ treated)	Polyacrylamide gel slice for MT carrier	Rabbit	MT-1 and MT-2 from rat, rabbit, human liver	1
Rat liver Cd-BP-1 ($CdCl_2$ treated)	Glutaraldehyde polymerized	Rabbit	1. Cd-BP-1 from rat, hamster, equine, human liver 2. Apothionein of rat liver Cd-BP-1	2
Rat liver Cd-BP-1 and Cd-BP-2 ($CdCl_2$ treated)	Glutaraldehyde polymerized	Rabbit	Rat and human CdBP	3
Rat liver MT-2 ($CdCl_2$ treated)	MT–rabbit serum albumin conjugate	Rabbit	MT-1 and MT-2 from rat, rabbit, human liver, but not with crab MT-1	4
Rat liver MT-A and MT-B ($CdCl_2$ treated)	Glutaraldehyde polymerized	Rabbit	MT in epithelial cells of rat organs	5
Rat liver MT-1 ($CdCl_2$ treated)	MT–rabbit IgG conjugate	Sheep	Zn,Cu,Cd-MT-1, but not enough with Zn,Cd-MT-2	6
Chinese hamster lung fibroblast	MT alone	Mouse (monoclonal)	ACM-1 (clone name): MT-1 and MT-2 from ch, mouse, rat, pig, bovine, monkey liver, but not yeast MT	7
Rat liver MT-2 ($CdCl_2$ treated)	MT–*Ascaris* protein conjugate	Rabbit	MT-1 and MT-2 from mouse, rat, rabbit, monkey, pig, human, dog liver	8
Rat liver MT-2 ($ZnCl_2$ treated)	MT–KLH conjugate	Mouse, rabbit	Cd,Zn-MT-1 and Cd,Zn-MT-2 from rat, mouse, rabbit, human liver	9

[a] BP, Binding protein; KLH, keyhole limpet hemocyanin.

protein conformational features from amino acid sequences, but none of them was specifically oriented to the prediction of antigenic determinants. Hopp and Woods[11] established a method that was not based on predictions of particular structural features, but rather sought a simple correlation between surface location of stretches of peptide chain and the likelihood of

[11] T. P. Hopp and K. R. Woods, *Proc. Natl. Acad. Sci. U.S.A.* **78**, 3824 (1981).

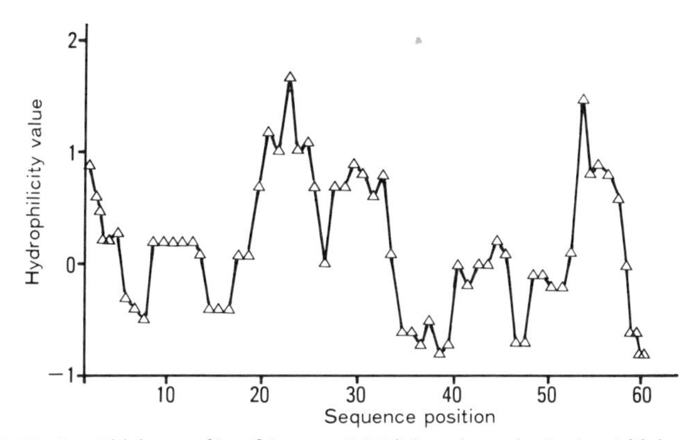

FIG. 1. Hydrophilicity profile of human MT-2 based on the hydrophilicity values [free energy change (kcal/mol; 1 cal = 4.18 J) for a residue, as the solvent is changed from water to an organic solvent].

antibody binding. The charged, hydrophilic amino acid side chains are common features of antigenic determinants. Antigenic determinants can be predicted by analyzing amino acid sequences to find the point of greatest local hydrophilicity. This is accomplished by assigning each amino acid a numerical value (hydrophilicity value) based on solvent parameters and then repetitively averaging these values along hexapeptides in the polypeptide. The region of highest average hydrophilicity is invariably located in or adjacent to an antigenic determinant. A plot of the hydrophilicity profile of the residues in human MT-2 is shown in Fig. 1.[12] Values for rat, rabbit, and human MT isoforms were similar and showed a maximum in the region of residues 20–25. An antigenic determinant would be predicted near residues 22 and 23 from hydrophilicity data. This agrees with the results of the modification studies, which suggest immunoreactivity in the region of residues 20–25. Three other possible regions exist, one near the amino terminus, a second at residues 27–32, and a third at residues 52–58. These results suggest that the polypeptide segments encompassing residues 1–5 and 20–25 are exposed to solvent and are thus prominent surface features on the MT molecules.These calculations do not take into account the contribution from the metal ions, which may affect the hydrophilicity of the folded polypeptide chain. The aggregated and conjugated forms of MTs that were prepared for immunization might have different hydrophilicity from the native molecule, especially in the region of residues 20–25, because the following studies of epitope mapping did not show any cross-reactivity with residues 20–25.

[12] D. R. Winge and J. S. Garvey, *Proc. Natl. Acad. Sci. U.S.A.* **80,** 2472 (1983).

Preparation of Antigen and Immunization Procedure for Rabbits

The antigen used for MT polyclonal antibody production is mostly rat MT-1 or MT-2. Rat liver MTs are either polymerized or covalently coupled to a larger carrier protein, using a bifunctional reagent such as glutaraldehyde. An example of the method of MT polyclonal antibody production in rabbit (OAL-JI) described by Ikei and co-workers[8,10] is as follows.

Rat MT-1 is isolated from rat liver administered $CdCl_2$ according to the method of Toyama and Shaikh.[1] Twenty milligrams of highly purified rat liver MT-1 and 15 mg of the *Ascaris* extract are dissolved in 5 ml of 0.1 *M* phosphate buffer (pH 7.0). To this mixture, 175 μl of 14% glutaraldehyde solution in distilled water is added and allowed to stand at room temperature for 5 hr. The reaction mixture is dialyzed against distilled water four times in 4 days at 4° and after lyophilization. The conjugated MT-1 solution is emulsified with an equal volume of Freund's complete adjuvant (Difco Laboratories, Detroit, MI). Male New Zealand rabbits are immunized at multiple sites subcutaneously with 100 μg of the conjugate/rabbit once every 2 weeks for 12 weeks. After this immunization the conjugate immunogen that is not emulsified with adjuvant is injected in ear veins four times. Ten days after the final booster, all blood is removed.

Preparation of Murine Monoclonal Antibodies[9]

Rat MT-2 conjugated to keyhole limpet hemocyanin is used as an immunogen. BALB/c mice are immunized by four subcutaneous injections of 50 mg of antigen in Freund's complete adjuvant (FCA) at 3-week intervals. Three days after the final injection (intraperitoneal) of the antigen in saline, the spleen cells are fused with the myeloma cell line P3/NS/1-Ag4-1 with polyethylene glycol.

Hybridoma cells from ELISA-positive cells are cloned twice by limiting dilution. Cloned hybridoma cells (2.4×10^6) are implanted (ip) into pristane-treated BALB/c mice to obtain the ascites. The monoclonal antibody is purified by using ammonium sulfate precipitation and gel filtration on Sephacryl S-400 superfine.

Radioimmunoassay

Reactivities of the antibodies with native MTs, fragment peptides, and synthetic peptides are determined by competitive RIA.[1-4,6,8,9] The double-antibody method for RIA developed by Ikei *et al.*[8] is described as follows.

Anti-rat MT-1 rabbit serum is titrated in order to determine the concentration of immunoglobulin to be used for the radioimmunoassay. MT radioimmunoassay is carried out with a 10 × 75 mm glass tube, to which is added the following: 200 μl of standard diluent, standard or MT fragment peptide, and ^{125}I-labeled Tyr-MT [about 10,000 counts per min (cpm)]. The mixture is incubated at 4° for 48 hr, and after this first incubation, anti-rabbit IgG goat serum in 100 μl of standard diluent, 100 μl of normal rabbit sera, each adequately diluted, and 200 μl of 12.5% polyethylene glycol are added to the tube. After a 30-min incubation at room temperature, the reaction mixture is centrifuged at 3000 rpm at 4° for 30 min. The supernatant is decanted and the radioactivity of this liquid and the precipitate is counted in a γ counter.

The ratio of the radioactivity of the bound radiolabeled antigen in the presence of inhibitor (MTs and peptide fragments) to that in the absence of inhibitor (B/B_0) is plotted against the concentration of inhibitor added, and the concentration of the inhibitor giving 50% inhibition (IC_{50}) is determined. When inhibition is incomplete, the IC_{50} value is defined as the concentration of inhibitor giving half-maximal inhibition.

Preparation of Native Metallothioneins and Fragment Peptide[12]

BALB/c mouse Cd,Zn-MT-1 and -2, rabbit Cd,Zn-MT-1 and -2, Wistar rat Cd,Zn-MT-1 and -2, and Wistar rat Zn-MTs are isolated and purified from the livers injected subcutaneously with $CdCl_2$. The α fragments (residues 30–61) are prepared from both isoforms of MT, digested with subtilisin, designated Cd,Zn-MT-1 and Cd,Zn-MT-2.

The tryptic peptide (residues 1–25) is prepared by proteolysis of performic acid-oxidized thionein 2 with trypsin at a protease/thionein ratio (w/w) of 1:50 in 0.1 *M* pyridine acetate (pH 5.0) at 37° for 48 hr. The tryptic peptides are resolved by gel filtration on a Sephadex G-50 column (1.5 × 160 cm) equilibrated with 1 *M* acetic acid. The peptide eluting first is shown to be the tryptic peptide.

Preparation of Synthetic MT Peptides[13,14]

Synthetic peptides, Ac-(hMT-2, residues 1–29)-OH, Ac-(rMT-1, 1–7)-OH, H-(rMT-1, 12–20)-OH, H-(rMT-1, 20–27)-OH, Ac-(rMT-1, 33–36)-H, H-(rMT-1, 40–49), and H-(rMT-1, 56–61)-OH are prepared by the conventional solution method (Table II). Furthermore, within the region

[13] Y. Okada, S. Iguchi, S. Nakayama, Y. Kikuchi, M. Irie, J. Sqawada, H. Ikebuchi, and T. Terao, *Chem. Pharm. Bull.* **36,** 3614 (1988).

[14] Y. Kikuchi, M. Irie, H. Ikebushi, J. Sawada, T. Terao, S. Nakayama, S. Iguchi, and Y. Okada, *J. Biochem.* **107,** 650 (1990).

TABLE II
PEPTIDE FRAGMENTS OF RAT MT-1 USED FOR COMPETITIVE RADIOIMMUNOASSAYS

Sequence	Residue
1. NH_2-terminal domain residues	1–25
2. COOH-terminal domain residues	30–61
3. Met-Asp-Pro-Asn-Cys-Ser-Cys	1–7
4. NH_2-Ser-Cys-Thr-Cys-Ser-Ser-Ser-Cys-Gly-COOH	12–20
5. NH_2-Gly-Cys-Lys-Asn-Cys-Lys-Cys-Thr-COOH	20–27
6. Ac-Cys-Cys-Ser-Cys-NH_2	33–36
7. NH_2-Gly-Cys-Ser-Lys-Cys-Ala-Gln-Gly-Cys-Val-COOH	40–49
8. NH_2-Lys-Cys-Thr-Cys-Cys-Ala-COOH	56–61

of Ac-(MT-1, 1–7)-OH, Ac-(MT-1)-OH, Ac-(MT-1, 1–2)-OH, Ac-(MT-1, 1–3)-OH, Ac-(MT-1, 1–4)-OH, Ac-(MT-1, 1–5)-OH, Ac-(MT-1, 1–6)-OH, and deacetylated peptides H-(MT-1, 1–7)-OH and H-(MT, 2–7)-OH are prepared using the same peptide synthetic method (Table III).

Specificity of Rabbit Polyclonal Anti-Rat Metallothionein Antibodies

Most of the MT antibodies reported in Table I show that MT-1 and MT-2 from rat liver MT exhibit identical cross-reactivities with the rabbit antisera prepared against rat liver MT-1. Competition is apparent not only with the native isoforms of Cd-, Zn-, and Cu-thionein, but also with isoforms that have been denatured by performic acid oxidation. The α fragment of MT represents the COOH terminal; half of the molecule (residues 30–61) does not compete efficiently with ^{125}I-labeled intact thionein.[12] The concentration of α fragment from both MT-1 and MT-2 needed to reduce ^{125}I-labeled MT to 50% bound is at least 50 times that needed in the case of native or denatured intact thionein. Denaturation of the α fragment by performic acid oxidation does not alter the ability of the α polypeptide to compete with the labeled protein.

The NH_2-terminal peptide (residues 1–25)[12] and Ac-(MT-1, 1–7)[14,15] compete with ^{125}I-labeled thionein as effectively as native MT. Removal of the NH_2 terminal-acetylated methionine in thionein 2 by digestion with CNBr greatly diminishes the immunoreactivity of the resulting polypeptide. Other peptide fragments (4–8 shown in Table II) do not show cross-reactivity. Therefore immunological reactivity of MT clearly resides in the NH_2-terminal domain.

Vander Mallie and Garvey[2] have found that purified MT from human,

[15] D. R. Winge, W. R. Gray, A. Zelazowski, and J. S. Garvey, *Arch. Biochem. Biophys.* **245,** 254 (1986).

TABLE III
IC_{50} VALUES OF SYNTHETIC PEPTIDES AND NATIVE METALLOTHIONEINS IN COMPETITIVE RADIOIMMUNOASSAY

Inhibitor	IC_{50} (nM)[a]		
	MT 189-14-7	RS-30	RS-5
Ac-(MT-1)-OH	>140,000	>140,000	>64,000
Ac-(MT, 1–2)-OH	>140,000	>140,000	>41,000
Ac-(MT, 1–3)-OH	>140,000	38,000	>31,000
Ac-(MT, 1–4)-OH	25,000	9.7	150
Ac-(MT, 1–5)-OH	590	3.8	6.3
Ac-(MT, 1–6)-OH	410	3.8	6.3
Ac-(MT, 1–7)-OH	180	1.6	3.2
H-(MT, 1–7)-OH	>27,000	5,500	3,200
H-(MT, 2–7)-OH	>27,000	27,000	>58,000
Mouse Cd,Zn-MT-1	110	3.4	8.2
Mouse Cd,Zn-MT-2	110	8.2	11
Rabbit Cd,Zn-MT-1	190	6.9	7.7
Rabbit Cd,Zn-MT-2a	190	86	16
Rat Cd,Zn-MT-1	190	1.8	4.8
Rat Cd,Zn-MT-2	260	1.6	11

[a] The concentration required to give half-maximal inhibition. Data cited from Kikuchi *et al.*[14]

equine, and rat liver MT-1 and MT-2 shows complete cross-reactivity, which implies that the antigenic determinants are preserved in the primary structure of these molecules. The amino acid sequences of NH_2-terminal segment (residues 1–7) are identical with most vertebrate MT,[16] predicting the antigenic site and its cross-reactivity.

Specificity of Murine Monoclonal Anti-Rat Metallothionein Antibody

Epitope mapping for monoclonal antibody, MT 189-14-7, raised by Kikuchi *et al.*,[14] shows that the epitope of MT to a murine monoclonal anti-MT antibody (MT 189-14-7) is located within the NH_2 terminal-acetylated heptapeptide Ac-(MT, 1–7)-OH, common to various animal MTs. Subsequently, the epitope structure was determined more precisely by using smaller synthetic peptides of various lengths (Table III). Inhibitory activity of the NH_2 terminal-acetylated pentapeptide Ac-(MT, 1–5)-OH or hexapeptide Ac-(MT, 1–6)-OH is almost the same as that of Ac-(MT, 1–7)-OH. The IC_{50} value of Ac-(MT, 1–5)-OH is comparable to that of

[16] J. H. R. Kägi and Y. Kojima, *Experientia, Suppl.* **52**, 25 (1987).

Ac-(MT, 1–4)-OH: 140-fold greater than that of Ac-(MT, 1–7)-OH, showing that the NH_2 terminal-acetylated tetrapeptide Ac-(MT, 1–4)-OH has much lower reactivity with the antibody. Other smaller peptides, Ac-(MT, 1–3)-OH, Ac-(MT, 1–2)-OH, and Ac-(MT-1)-OH, exhibit little reactivity with the antibody (Table III).

The importance of the acetyl residue at the NH_2 terminus is examined by using deacetylated peptides. The binding of labeled antigen to the MT 189-14-7 antibody is not inhibited by the deacetylated heptapeptide H-(MT, 1–7)-OH and another deacetylated peptide, H-(MT, 2–7)-OH (Table III).

These results suggest that the major epitope structure of MT to the MT 189-14-7 antibody is located within the NH_2 terminal-acetylated pentapeptide, AcMDPNC.[16a]

Comparison of Antigenic Determinant Specificities between Polyclonal Rabbit Antisera and Murine Monoclonal Antibody

To compare the specificities of polyclonal rabbit antisera with that of the murine monoclonal antibody, MT 189-14-7, the reactivities of two rabbit anti-MT antisera (RS-30 and RS-5) that had been prepared by immunization with the same antigen used in the mouse were investigated.[14] These rabbit antisera gave higher apparent association constants than the murine monoclonal antibody.

Various animal MTs and synthetic peptides (Table III) were tested for their ability to compete with rat ^{125}I-labeled Cd,Zn-MT-2 in RIA. The IC_{50} values for the various MTs differed slightly from each other, but the differences were not remarkable except for the rabbit Cd,Zn-MT-2. The properties of these two rabbit anti-MT antisera were similar to that of the MT 189-14-7 antibody. The binding of labeled antigen to RS-30 was reduced to a level of almost 10% B/B_0 (but not to 0%) by the addition of Ac-(MT, 1–4)-OH, Ac-(MT, 1–5)-OH, or Ac-(MT, 1–6)-OH. The IC_{50} values for Ac-(MT, 1–5)-OH, Ac-(MT, 1–6)-OH, and rat MT-2 were 9.7, 3.8, 3.8, and 1.6 nM, respectively, whereas the IC_{50} of the acetylated tripeptide Ac-(MT, 1–3)-OH was much greater than those of the acetylated tetra-, penta-, and hexapeptides. These results suggest that the amino-terminal tetrapeptide is the major epitope structure of MTs to the RS-30 and RS-5 antisera in the RIA system. These polyclonal anti-MT sera showed only marginal binding activity to H-(MT, 1–7)-OH. The IC_{50} values of H-(MT, 1–7)-OH were about 3400-fold greater than that of Ac-(MT, 1–7)-OH (Table III), suggesting again that the presence of an

[16a] M, Methionine; D, aspartic acid; P, proline; N, asparagine; C, cysteine.

acetyl residue at the NH_2 terminus is essential for epitope recognition by these polyclonal antibodies.

These data demonstrate that the amino-terminal region of rat MT can be an epitope to antibodies produced in the rabbit antibody response, as previously demonstrated in the mouse.

On the other hand, RS-5 antiserum showed a different reactivity pattern for various MTs and synthetic peptides. Although RS-5 was inhibited by the acetylated hepta- and pentapeptides, and by autologous rabbit MT-1 and MT-2, the inhibition was partial (60%). Therefore, the data shown in Table III indicate that RS-5 contained another population of antibodies that are not reactive with the NH_2-terminal oligopeptides, and that the rabbit produced antibodies reactive with another epitope that was present on rat MTs but absent on rabbit MTs. As shown in Table III, the IC_{50} value of Ac-(MT, 1–4)-OH was much greater than that of Ac-(MT, 1–7)-OH. Thus, it seems that RS-5 antiserum contains antibodies specific for the NH_2 terminal-acetylated peptides, but their specificity is slightly different from that of RS-30.

Cross-Reactivity with Soybean Metallothionein Using Polyclonal Anti-Rat Metallothionein Antibody (OAL-JI)[17]

Preparation of Tissues for Immunohistochemistry

Fresh germinating seeds and 2-mm length root tips are fixed in 10% (v/v) formalin for 24 hr, washed in running tap water for 4 hr, and then dehydrated with a graded series of 70% (v/v) ethanol to anhydrous ethanol, subsequently with xylene, paraffinized, and embedded in paraffin. Paraffinized tissues are sectioned at a thickness of 3 μm and mounted on glass slides. Finally, the tissue sections are dried at 37° overnight.

Indirect Immunoperoxidase Technique for Metallothionein Stain

Three-micrometer sections are deparaffinized, dehydrated with xylene and ethanol, respectively, and the sections incubated with the following reagents in sequence, subsequent to a phosphate-buffered saline (PBS) (pH 7.2) washing step (three times for 5 min each prior to each reagent): 3% H_2O_2 in methanol (to quench endogenous peroxidase activity in the tissues) for 30 min; normal goat serum (diluted 50-fold by PBS) (to block nonspecific binding sites) for 20 min; polyclonal rabbit anti-rat MT conjugated to *Ascaris* (diluted 400-fold by PBS) as primary antibody for 1 hr; biotinylated anti-rabbit IgG from goat (diluted 200-fold by PBS) as second-

[17] P. Chongpraditnum, K. Suzuki, U. Kawaharada, K. Nakajima, and M. Chino, in preparation.

ary antibody for 1 hr; avidin biotinylated horseradish peroxidase complex (ABC reagent prepared 30 min prior to use) as the peroxidase substrate for 1 hr; and 0.05% 3,3′-diaminobenzidine tetrachloride (DAB, prepared freshly in the dark container) as a chromagen in 0.05 *M* Tris buffer, pH 7.2, containing 0.01% H_2O_2, for 2–3 min. Finally, the immunostained sections are rinsed with distilled water for 5 min, subsequently counterstained with Mayer's hematoxylin for 3 sec, and washed again with distilled water.

Immunohistochemical Control

The specificity of the rabbit anti-rat MT conjugated to *Ascaris* is ascertained by three kinds of examinations. First, the antibody is replaced with either the normal rabbit serum or PBS, which results in negligible staining of MT in both seeds and roots. Second, 48-hr preabsorbed primary antibody with rat MT is used in place of the primary antibody, resulting in a negative immunostain of MT. Third, the preabsorbed rabbit anti-rat MT conjugated to *Ascaris* with 400-fold diluted *Ascaris* is used as the primary antibody, resulting in a positive stain of MT in 6-day-old germinating seeds.

Immunoperoxidase Staining of Metallothionein in Soybean Roots and Soybean Seeds

Metallothionein is present in ground meristem, protoderm, and root cap, especially the columella of the root tip, but not in the procambium (Fig. 2). This result corresponds with MT localization in embryos showing the positively immunostained portions of root tips developed from a common initial embryo, which is also cross-reactive to anti-rat MT. Observation of immunostaining for MT in root tips of soybean nontreated and treated with 400 parts per billion (ppb) copper for 1 month suggests that MT localizes only at the columella of nontreated root tip, while it scatters all over the root cap at high intensity in the treated one.[18] Seeds (Toyosuzu variety) water-immersed for 2 hr either with or without copper or zinc treatments show that MT localizes mostly in the embryo and partially in the surface of cotyledon, epidermis, and palisade parenchyma cells, as well as some cells next to the veins, as shown in Fig. 2.

FIG. 2. Immunoperoxidase staining of metallothionein in soybean roots (A) and soybean seeds (B).

FIG. 3. Immunoperoxidase staining of metallothionein in rat testis, with (A) or without (B) pretreatment by 0.1% trypsin after deparaffinization.

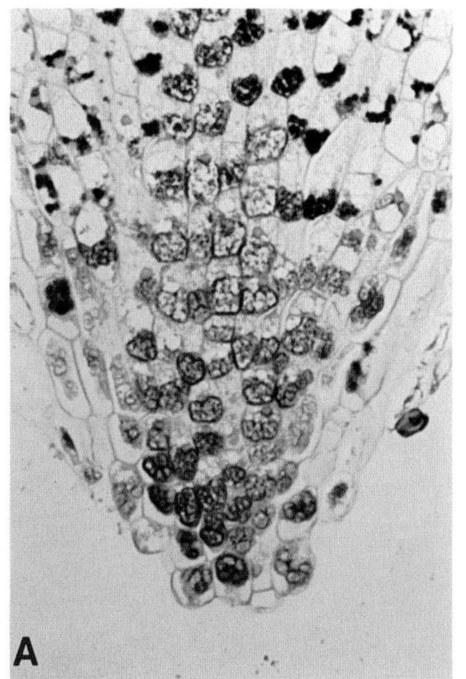

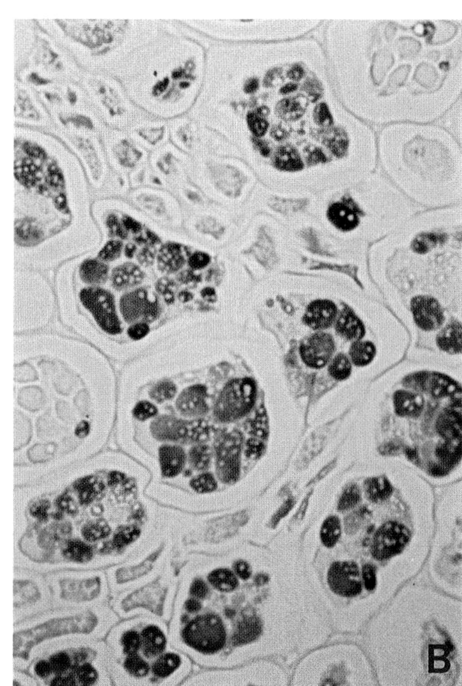

FIG. 2

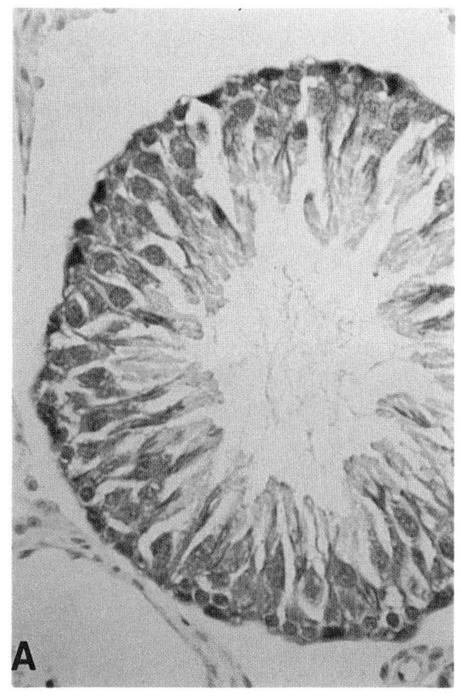

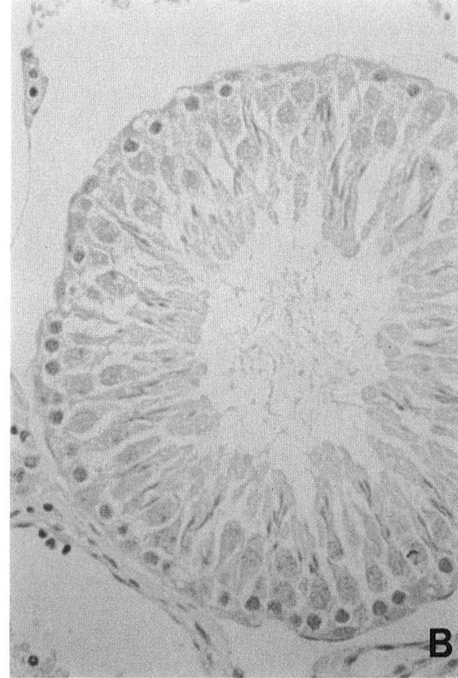

FIG. 3

Isolation of Metallothionein Gene from Soybean Roots[19]

The oligonucleotide of 21-mer NH_2-terminal residue (1–7) of mammalian MT is synthesized and hybridized with soybean DNA (λgt11 cDNA). The base pairs of clone 21-1-A (0.7K) are hybridized with 21-mer probe, showed at open reading frame 237 base and clarified 79-amino acid sequence (Table IV), including 14 cysteine residues (18%) of soybean MT. The cysteine residues are localized to the NH_2 terminal and COOH terminal, predominantly, and show an MT-like metal-binding capacity.

Using this cDNA probe, 0.8-kilobase (kb) mRNA is detected by Northern hybridization at soybean roots.

Masked Metallothionein in Rat Testis

Several reports[20,21] have indicated the presence of MT in rodent testis. But the major metal-binding proteins (MBP) in rat testis do not appear to be MT, but rather are reported as TMBP-2 and TMBP-3.[22] Furthermore, the protein previously clarified as testicular MT may be a breakdown product of a higher molecular weight zinc-binding protein and apparent absence of MT has been insisted on by several authors.

The MT contents in rat testis extracts assayed by RIA showed an extremely high amount of MT, over 200 μg/g wet weight, which revealed the highest MT content among rat organs. The gel-filtration pattern of zinc and MT in the testis extract was very similar to that of human kidney, suggesting the presence of MT.

Shiraishi *et al.* reported[23,24] an extremely high MT mRNA content in rat testis, which was very similar to the results obtained by RIA using antibody OAL-JI. Although RIA showed extremely high MT content, the immunohistochemical studies showed the MT levels to be only slightly positive in rat testis Sertoli cells. Pretreatment with a protease, such as 0.05% trypsin (GIBCO, Grand Island, NY) in 0.05 *M* Tris-HCl buffer incubated for 30 min at room temperature, or 0.05% pronase (Sigma, St. Louis, MO, type XXV) in 0.05 *M* Tri-HCl buffer incubated for 10 min at

[18] J. L. Casterline and N. M. Barnett, *Plant Physiol.* **69,** 1004 (1982).

[19] I. Kawashima, Y. Iguchi, M. Chino, M. Kimura, and N. Shimizu, *Proc. Annu. Meet. Plant Physiol. Tokyo 30th Symp. Jpn. Soc.* **1,** Cp02 (1990).

[20] K. G. Danielson, S. Ohi, and P. C. Huang, *Proc. Natl. Acad. Sci. U.S.A.* **79,** 2301 (1982).

[21] H. Nishimura, N. Nishimura, and C. Toyama, *J. Histochem. Cytochem.* **38,** 927 (1990).

[22] C. D. Klaassen and M. P. Waalkes, *Experientia, Suppl.* **52,** 273 (1987).

[23] N. Shiraishi, H. Hayashi, Y. Hiraki, K. Aono, Y. Itano, F. Kosaka, S. Noji, and S. Taniguchi, *Toxicol. Appl. Pharmacol.* **98,** 501 (1989).

[24] M. R. Bhave, M. J. Wilson, and M. P. Waalkers, *Toxicology,* 231 (1988).

TABLE IV
AMINO ACID SEQUENCES OF METALLOTHIONEINS[a]

Rat MT-1 (61)

1 10 20
Ac-Met-Asp-Pro-Asn-Cys-Ser-Cys-Ser-Thr-Gly-Gly-Ser-Cys-Thr-Cys-Ser-Ser-Ser-Cys-Gly
21 30 40
Cys-Lys-Asn-Cys-Lys-Cys-Thr-Ser-Cys-Lys-Lys-Ser-Cys-Cys-Ser-Cys-Cys-Pro-Val-Gly
41 50 60
Cys-Ser-Lys-Cys-Ala-Gln-Gly-Cys-Val-Cys-Lys-Gly-Ala-Ser-Asp-Lys-Cys-Thr-Cys-Cys
61
Ala

Equine MT-1 (61)

1 10 20
Ac-Met-Asp-Pro-Asn-Cys-Ser-Cys-Pro-Thr-Gly-Gly-Ser-Cys-Thr-Cys-Ala-Gly-Ser-Cys-Lys
21 30 40
Cys-Lys-Glu-Cys-Arg-Cys-Thr-Ser-Cys-Lys-Lys-Ser-Cys-Cys-Ser-Cys-Cys-Pro-Gly-Gly
41 50 60
Cys-Ala-Arg-Cys-Ala-Gln-Gly-Cys-Val-Cys-Lys-Gly-Ala-Ser-Asp-Lys-Cys-Ser-Cys-Cys
61
Ala

Human MT-1 (61)

1 10 20
Ac-Met-Asp-Pro-Asp-Cys-Ser-Cys-Ala-Thr-Gly-Gly-Ser-Cys-Thr-Cys-Ala-Gly-Ser-Cys-Lys
21 30 40
Cys-Lys-Glu-Cys-Lys-Cys-Thr-Ser-Cys-Lys-Lys-Ser-Cys-Cys-Ser-Cys-Cys-Pro-Val-Gly
41 50 60
Cys-Ala-Cys-Lys-Ala-Gln-Gly-Cys-Ile-Cys-Lys-Gly-Ala-Ser-Glu-Lys-Cys-Ser-Cys-Cys
61
Ala

Chicken MT (63)

1 10 20
Met-Asp-Pro-Gln-Asp-Cys-Thr-Cys-Ala-Ala-Gly-Asp-Ser-Cys-Ser-Cys-Ala-Gly-Ser-Cys
21 30 40
Lys-Cys-Lys-Asn-Cys-Arg-Cys-Arg-Ser-Cys-Arg-Lys-Ser-Cys-Cys-Ser-Cys-Cys-Pro-Ala
41 50 60
Gly-Cys-Asn-Asn-Cys-Ala-Lys-Gly-Cys-Val-Cys-Lys-Glu-Pro-Ala-Ser-Ser-Lys-Cys-Ser
61
Cys-Cys-His

(continued)

TABLE IV *(continued)*

oybean MT (79)

1 10 20
let-Ser -Cys-Cys-Gly-Gly-Asn-Cys-Gly-Cys-Gly-Ser -Ser -Cys-Lys-Cys-Gly-Asn-Gly-Cys
21 30 40
ly-Gly-Cys-Lys-Met-Tyr-Pro-Asp-Leu-Ser -Tyr-Thr-Glu-Ser -Thr-Thr-Thr-Glu-Thr-Leu
41 50 60
al -Met-Gly-Val-Ala-Pro-Val-Lys-Ala-Gln-Phe-Glu-Gly-Ala-Glu-Met-Gly-Val-Pro-Ala
61 70
lu-Asn-Asp-Gly-Cys-Lys-Cys-Gly-Pro-Asn-Cys-Ser -Cys-Asn-Pro-Cys-Thr-Cys-Lys

ea (*Pisum sativum* L.) MT (75)

1 10 20
let-Ser -Gly-Cys-Gly-Cys-Gly-Ser -Ser -Cys-Asn-Cys-Gly-Asp-Ser -Cys-Lys-Cys-Asn-Lys
21 30 40
rg-Ser -Ser -Gly-Leu-Ser -Tyr-Ser -Glu-Met-Glu-Thr-Thr-Glu-Thr-Val-Ile -Leu-Gly-Val
41 50 60
ily-Pro-Ala-Lys-Ile -Gln-Phe-Glu-Gly-Ala-Glu-Met-Ser -Ala-Ala-Ser -Glu-Asp-Gly-Gly
61 70
ys-Lys-Gys-Gly-Asp-Asn-Cys-Thr-Cys-Asp-Pro-Cys-Asn-Cys-Lys

Vheat germ Ec (59)

1 10 20
ily-Cys-Asn-Asp-Lys-Cys-Gly-Cys-Ala-Val-Pro-Cys-Pro-Gly-Gly-Thr-Gly-Cys-Arg-Cys
21 30 40
hr-Ser -Ala-Arg-Ser -Gly-Ala-Ala-Ala-Gly-Glu-His-Thr-Thr-Cys-Gly-Cys-Gly-Glu-His
41 50
ys-Gly-Gly-Asn-Pro-Cys-Ala-Cys-Gly-Gly-Glu-Gly-Thr-Pro-Ser -Gly-Cys-Ala-Asn

'east MT (53)

1 10 20
iln-Asn-Glu-Gly-X -Glu-Cys-Gln-Cys-Gln-Cys-Gly-Ser -Cys-Lys-Asn-Asn-Glu-Gln-Cys
21 30 40
iln-Lys-Ser -Cys-Ser -Cys-Pro-Thr-Gly-Cys-Asn-Ser -Asp-Asp-Lys-Cys-Pro-Cys-Gly-Asn
41 50
ys-Ser -Glu-Glu-Thr-Lys-Lys-Ser -Cys-Cys-Ser -Gly-Lys

[a] Homologous sequences are underlined.

room temperature, or 0.4% pepsin (Sigma) in 0.01 *N* HCl incubated for 30 min at 37°, before staining (after deparaffinization), showed extremely positive MT staining in rat testis (Fig. 3), but not much change in other rat organs.

This indicated that the surface area of the MT molecule, especially the NH_2-terminal region of the MT molecule, was masked by some intracellular proteins that caused the poor immunostaining. In the case of rat testis extracted by 5 m*M* Tris-HCl buffer (pH 8.6), some of the protease might change the MT molecule, making possible the MT antibody–antigen reaction under homogenization condition and showing extremely high levels of MT in the tissue.

Amino Acid Sequence and Metallothionein Antigenicity

A comparison of the known sequences of MTs,[16] isoforms, and species differences indicated that the peptide regions implicated in this study to be involved in antibody reactivity show marked homology.

The cross-reactivity was unaffected whether the protein was native or denatured. The antigenicity is thus independent of the degree of folding of the protein and, although all antigenic sites depend to some degree on conformation or topography, this favors a sequential rather than a discontinuous nature of the determinants. The region including the NH_2-terminal residues 1–7 and the acetylated methionine was shown to be the most important in the interaction of the molecule with the antisera. The amino acid sequence of this region is homologous in the various vertebrate MTs known to cross-react with rabbit anti-rat MT antiserum (Table IV).

The sequential homology of Pro-Asn-Cys-Ser-Cys, residues 3–7, which did not involve acetylated methionine, but were located within the NH_2-terminal residues 1–7 found in soybean roots and seeds, showed very high reactivity with rat MT antibody (OAL-JI). This result indicated the possibility that the different MT antigenic regions exist inside the NH_2-terminal residues 1–7. Although the synthetic peptide Pro-Asn-Cys-Ser-Cys showed negative cross-reactivity with MT polyclonal and monoclonal antibodies, some conformational features of soybean MT might provide the positive cross-reactivity with NH_2-terminal residues 3–7 of vertebrate MT. The plant MTs, especially wheat germ Ec protein,[25] did not show any cross-reactivity and amino acid sequences were not homologous.

Chicken MT[26] and pea MT[27] showed similar but less reactive cross-

[25] T. Hofman, D. I. C. Kells, and B. G. Lane, *Can. J. Biochem. Cell Biol.* **62,** 908 (1984).

[26] C. C. McCormick, C. S. Fullmer, and J. S. Garvey, *Proc. Natl. Acad. Sci. U.S.A.* **85,** 309 (1988).

reactivity with polyclonal antibody (OAL-JI), with a similar sequence—Asp(Asn)-Cys-Thr-Cys. Yeast MT,[16] other plant MT,[28] crab MT-1,[29] and other kinds of MTs with no amino acid sequential homology showed no cross-reactivity with rat MT antibodies (OAL-JI).

Conclusion

The antigenic determinants of vertebrate MT have been determined by a competitive radioimmunoassay and immunohistochemical studies using anti-rat MT polyclonal and monoclonal antibodies. The results indicate that the antigenic reactivity of rat MT with rabbit antisera resides predominantly in the NH_2-terminal domain of the molecule, with the COOH terminal showing only a minimal interaction with the antisera.

Using the monoclonal antibody raised against rat MT showed that the epitope to the monoclonal antibody was located within the NH_2-terminal heptapeptide, including acetylated methionine.

Two possibilities for the antigenic determinants include the NH_2-terminal heptapeptide having MT (1–5) residues with acetylated methionine and MT(3–7) residues without acetylated methionine.

Most of the antibodies reported in Table I seemed to cross-react with the NH_2-terminal region residues 1–7, including acetylated methionine, but the antibody (OAL-JI) raised by our laboratory showed positive cross-reactivity with soybean Pro-Asn-Cys-Ser-Cys, which is the only homologous sequence with rat MT and is equivalent to vertebrate MT NH_2-terminal residues 3–7.

The well-characterized rabbit antisera and murine monoclonal anti-MT antibody, both antibodies specific for the NH_2-terminal region and having similar reactivities, may be useful tools in studying the biology of MT.

[27] M. Evans, L. N. Gatehouse, J. A. Gatehouse, N. J. Robinson, and R. R. D. Croy, *FEBS Lett.* **262,** 29 (1990).

[28] J. R. de Miranda, M. A. Thomas, D. A. Thurman, and A. B. Tomsett, *FEBS Lett.* **260,** 277 (1990).

[29] K. Lerch and D. Ammer, *J. Biol. Chem.* **257,** 2420 (1982).

[22] Separation and Quantification of Isometallothioneins by High-Performance Liquid Chromatography–Atomic Absorption Spectrometry

By CURTIS D. KLAASSEN and LOIS D. LEHMAN-MCKEEMAN

Introduction

Over the last 30 years, a variety of methods have been developed to quantitate metallothionein (MT) in biological tissues and fluids. Many of these procedures have relied on the metal-binding properties of MT, using mercury,[1] cadmium,[2] or silver[3] to displace constituent metal and saturate MT metal-binding sites. Several radioimmunoassays have been developed to quantitate MT,[4,5] and because of the high sulfhydryl content of MT, electrochemical procedures have also been utilized.[6]

It is well established that there are two major isoforms of MT in all mammals. These proteins, referred to as MT-1 and MT-2, are similar with respect to metal-binding capacities and sulfhydryl content.[7] Furthermore the antigenic determinants are highly conserved between MT-1 and MT-2 and across many species.[8] Whereas the metal-saturation assays, radioimmunoassays, and electrochemical procedures are useful for quantitating total MT, these methods fail to distinguish the MT isoforms.

MT-1 and MT-2 differ slightly in amino acid content and can be separated electrophoretically or chromatographically based on differences in their isoelectric points. Generally, MT-2 is more acidic than MT-1, as the isoelectric points for rat MT-1 and MT-2 are 4.7 and 4.2, respectively.[9] The chromatographic differences between the MT isoproteins can be exploited to separate MT-1 and MT-2, after which the proteins can be quantified with the more classical procedures described above. The methods presented herein outline an assay in which MT isoforms are quantitated as cadmium-saturated proteins following separation by anion-exchange high-performance liquid chromatography (HPLC). An atomic

[1] J. Piotrowski, W. Bolanowska, and A. Sapata, *Acta Biochim. Pol.* **20,** 207 (1973).
[2] D. L. Eaton and B. F. Toal, *Toxicol. Appl. Pharmacol.* **66,** 134 (1982).
[3] A. M. Scheuhammer and M. G. Cherian, *Toxicol. Appl. Pharmacol.* **82,** 417 (1986).
[4] F. O. Brady and R. L. Kafka, *Anal. Biochem.* **98,** 89 (1979).
[5] J. S. Garvey, R. J. Vander Mallie, and C. C. Chang, this series, Vol. 84, p. 121.
[6] R. W. Olafson and R. G. Sim, *Anal. Biochem.* **100,** 343 (1979).
[7] D. R. Winge, K. B. Nielson, R. D. Zeikus, and W. R. Gray, *J. Biol. Chem.* **259,** 11419 (1984).
[8] D. R. Winge and J. S. Garvey, *Proc. Natl. Acad. Sci. U.S.A.* **80,** 2472 (1983).
[9] M. G. Cherian, *Biochem. Biophy. Res. Commun.* **61,** 920 (1974).

absorption (AA) spectrophotometer used to detect cadmium is connected directly to the HPLC instrument, allowing for online quantitation of cadmium-saturated MTs.

Isolation and Purification of Metallothionein Standards

Rat liver MT-1 and MT-2, used as standards for the HPLC-atomic absorption spectrometry (AAS) method, are isolated from cadmium-treated rats by standard chromatographic procedures,[10] as outlined below.

1. To induce hepatic MT, dose rats daily for 3 days with $CdCl_2$ (20 μmol Cd/kg, subcutaneously) and, on the fourth day, inject rats with ^{109}Cd (1 μCi/kg, intravenously).
2. Four hours after injecting with ^{109}Cd, remove livers and homogenize in 1 vol of ice-cold Tris–acetate buffer (10 m*M*, pH 7.4, at 4°). All remaining procedures should be conducted at 4°.
3. Centrifuge the liver homogenates at 10,000 *g* for 15 min. Heat the supernatant from this centrifugation step in a water bath maintained at 80° for 5 min, and then cool on ice for 10 min. Centrifuge the heat-denatured samples at 20,000 *g* for 20 min.
4. The supernatant resulting from centrifugation of the heat-denatured samples is applied directly to a Sephadex G-75 gel-filtration column (60 × 2.6 cm) previously equilibrated with 10 m*M* Tris–acetate buffer [containing 0.02% (w/v) sodium azide, pH 7.4], and eluted at a flow rate of 30 ml/hr. Fractions are collected in 10-min intervals (5-ml fractions) and the elution of ^{109}Cd determined with an autogamma scintillation counter.
5. Fractions eluting with a retention coefficient of 1.8–2.2 are considered to represent total MT. These fractions are pooled and applied to a DEAE-Sephadex A-25 column. MT-1 and MT-2 are eluted with a linear gradient of Tris–acetate buffer (10–240 m*M*, pH 7.4, 0.02% sodium azide) at a flow rate of 30 ml/hr. Fractions are collected at 10-min intervals, after which the elution of ^{109}Cd in MT-1 and MT-2 is determined.
6. Fractions containing MT-1 or MT-2 are pooled, desalted on Sephadex G-25 columns, and then lyophilized. It is essential that all Tris salt is removed from the purified MTs, because Tris buffer can interfere with the quantitation of MT-1 and MT-2 standards with the Kjeldahl method for determining nitrogen.
7. The desalted, lyophilized proteins are stored desiccated at −20°. Under these conditions, the MTs are stable for many months.

[10] M. Vašák, this volume [6].

Quantitation of Metallothionein Standards

To set up standard curves for the HPLC–AAS assay, it is essential to determine the concentration of the MTs in solution accurately. Because the proteins lack aromatic amino acids, the Lowry protein assay is not appropriate. Likewise, Coomassie Blue in the concentrations used for the Bradford protein assay does not readily bind to MT, thereby rendering this common protein assay also inappropriate. The concentration of the MT standards can be accurately measured by the Kjeldahl method for the determination of nitrogen. A microassay described by Lang[11] is particularly useful, as less sample is required for analysis.

HPLC Apparatus

The HPLC method is a gradient one, requiring a system with at least two HPLC pumps and a gradient controller.The atomic absorption detector is used for MT quantitation. However, it is useful to equip the HPLC system with a UV detector (214 nm), as this will detect proteins, and serves to monitor the HPLC system for problems. To determine simultaneously the atomic absorption of metals, the outlet of the HPLC system (analytical column or UV detector) is connected directly to the nebulizer uptake capillary of the AA spectrophotometer. This interface is achieved with a stainless steel-to-Teflon flange coupling the adaptor between the stainless steel tubing of the HPLC system and the Teflon tubing attached to the nebulizer uptake capillary of the AA spectrophotometer. These adaptors can be made to order by many chromatographic supply companies. The length of the Teflon tubing connecting the HPLC to the AA should be the shortest distance possible to optimize cadmium peak shapes. When operating independently, the optimum uptake rate of the nebulizer capillary is at least 7 ml/min. To accommodate the flow rate of the mobile phase in the HPLC system, the uptake rate should be adjusted to 1 ml/min.

It should be noted that the chromatographic baseline obtained from the AA spectrophotometer output is never stable and flat, as is typical of other HPLC detectors. To reduce the fluctuation in the baseline, the burner and nebulizer should be kept clean at all times, and the temperature of the flame should not be allowed to change. Under these conditions, the baseline fluctuation is kept to a minimum and the peak areas for MT-1 and MT-2 can be accurately integrated, even when present in levels near the detection limit.

[11] C. A. Lang, *Anal. Chem.* **30,** 1691 (1958).

HPLC Method

Chromatography is performed on an anion-exchange HPLC column. This method was originally developed using a DEAE-5PW column (7.5 cm × 7.5 mm; Waters Associates, Milford, MA).[12] The mobile phases (A, 10 m*M* Tris-HCl, pH 7.4 at room temperature; B, 200 m*M* Tris-HCl, pH 7.4 at room temperature) are filtered and decreased daily by vacuum filtration through a 0.45-μm pore membrane filter (Metricel, Fisher Scientific, St. Louis, MO). MT-1 and MT-2 are eluted with a linear gradient from 0 to 40% B in 12 min at a flow rate of 1 ml/min. After completion of the gradient, the column is cleaned by pumping 100% mobile phase B for 3 min, and then the mobile phase is switched back to 100% A to reequilibrate the column in preparation for the next injection.

Saturation of Metallothionein Standards and Tissue Cytosols with Cadmium

The conditions required to fully saturate MT standards with cadmium were originally determined empirically. As shown in Fig. 1, the purified cadmium-induced MT standards contained both cadmium and zinc. The zinc present in the MTs can be displaced with increasing concentrations of cadmium, and with a final concentration of 50 ppm cadmium, all zinc is removed from the standards. Addition of higher concentrations of cadmium (75, 100 ppm cadmium) did not increase the amount of cadmium bound to MTs, but did alter peak shape and retention times of MT-1 and MT-2. From this experiment, 50 ppm cadmium was used to saturate both MT standards and tissue cytosols for HPLC-AAS quantitation. Under these conditions, approximately 7 g-atoms cadmium was bound/mol MT, a finding that is consistent with the metal-binding properties of MT. Typical standard curves range from 1 to 40 μg MT-1 and MT-2.

Sample Preparation for Quantitation of MT-1 and MT-2 by HPLC-AAS

A scheme outlining the preparation of tissue samples for HPLC-AAS analysis of MT-1 and MT-2 is shown in Fig. 2. This scheme also applies to preparation of cadmium-saturated MT standards (beginning with addition of 50 μl cadmium). Briefly, tissue cytosols are prepared by homogenization in Tris-HCl buffer followed by two centrifugation steps. The supernatant

[12] L. D. Lehman and C. D. Klaassen, *Anal. Biochem.* **153,** 305 (1986).

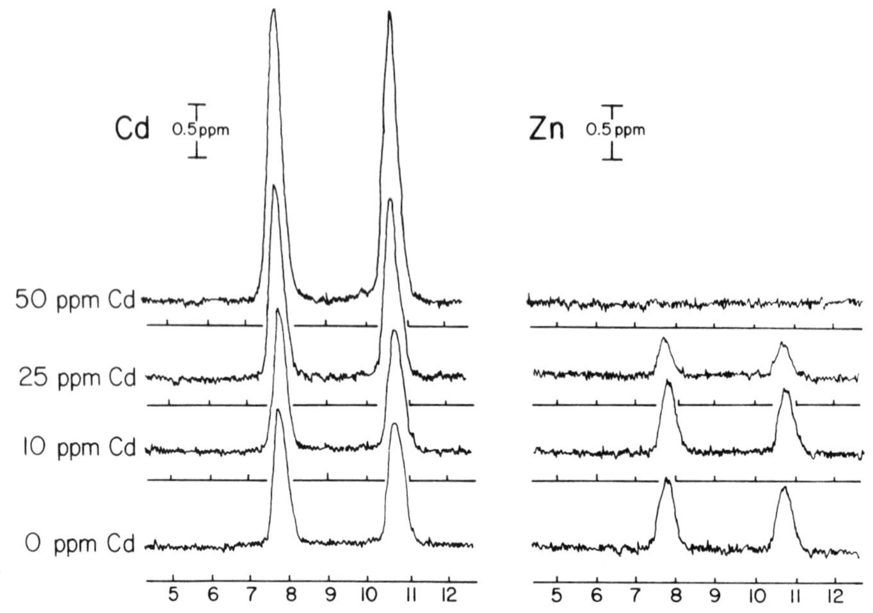

FIG. 1. Saturation of purified rat liver MT-1 and MT-2 with cadmium. Cadmium was added to solutions of MT standards in final concentrations of 0, 10, 25, and 50 ppm and allowed to incubate at room temperature for 15 min. Standards were heat denatured prior to injection. A 100-μl aliquot (containing 33.8 and 33.4 μg MT-1 and MT-2, respectively) was injected onto the anion-exchange column and either cadmium or zinc was detected with the AA spectrophotometer. Increasing concentrations of cadmium displaced zinc in a stepwise manner until the standards were fully saturated with cadmium at 50 ppm. (From Ref. 12, with permission from the publisher.)

resulting from centrifugation at 100,000 g can be stored frozen for several months at $-20°$. It is best to complete both centrifugation steps prior to freezing the samples. The samples are placed in 1.5-ml microcentrifuge tubes for the cadmium saturation and heat-denaturation steps. As described above, tissue cytosols and standards are saturated with cadmium by the addition of 50 ppm cadmium (final concentration). Samples are allowed to incubate at room temperature for 15 min, after which they are placed in a boiling water bath for 1 min. The heat-denatured samples are centrifuged in a microcentrifuge (10,000 g for 3 min), and the resulting supernatant is injected onto the HPLC column. Injection volumes can range from 25 to 500 μl.

A representative cadmium chromatogram is shown in Fig. 3. The sample shown here is of hepatic MT-1 and MT-2. However, the method is applicable for quantitating MTs in other tissues, and has been used to

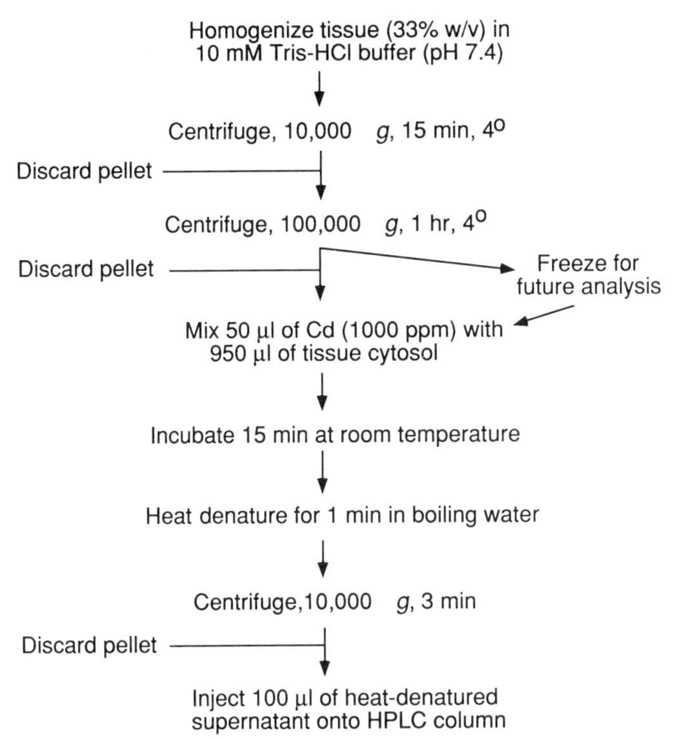

FIG. 2. Schematic representation of the preparation of tissue cytosols followed by the cadmium saturation procedure used for the HPLC-AAS assay. Standards are treated identically to tissue samples beginning with the addition of cadmium.

analyze MTs in kidney and pancreas.[13] For all tissues, a small peak that elutes just before MT-2 has been observed. The identity of this peak has not been confirmed; however, a third metal-containing peak eluting between MT-1 and MT-2 has been observed with both conventional anion-exchange chromatography[14] and other HPLC techniques.[15] This peak is not included in the quantitation of MT-1 and MT-2 because it does not coelute with either authentic standard.

Analysis of HPLC-AAS Data

Because MT standards are saturated with cadmium prior to analysis and detected as cadmium-containing peaks, the peak areas reported from

[13] L. D. Lehman-McKeeman and C. D. Klaassen, *Toxicol. Appl. Pharmacol.* **88,** 195 (1987).

[14] K.-L. Wong and C. D. Klaassen, *J. Biol. Chem.* **254,** 12399 (1979).

[15] K. T. Suzuki, H. Uehara, H. Sunaga, and N. Shimoto, *Toxicol. Lett.* **24,** 15 (1985).

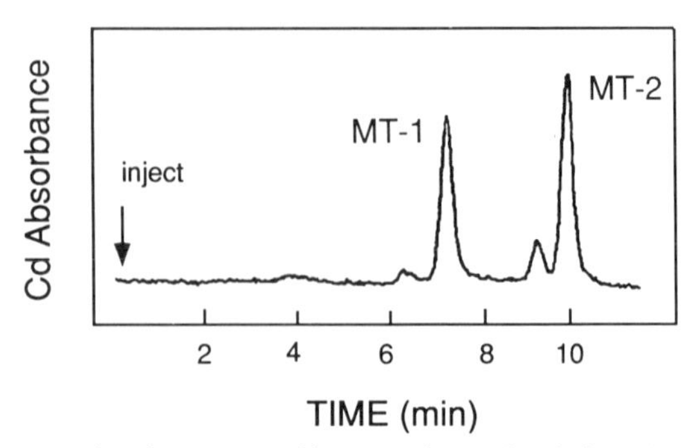

FIG. 3. Representative chromatographic separation and cadmium detection of MT-1 and MT-2. The sample analyzed was hepatic cytosol from an untreated 10-day-old rat. MT-1 and MT-2 elute at approximately 7 and 10 min, respectively. A cadmium-containing peak that eluted immediately before MT-2 was typically seen in all liver cytosols of immature, cadmium- or zinc-treated animals. This peak, however, was not included in the analysis because it does not coelute with either MT standard. The baseline obtained from the AA spectrophotometer is never a flat line, as is typical of other HPLC detection methods (see also Fig. 1).

the AA spectrophotometer can be equated to micrograms MT, for establishing standard curves. For sample analysis, tissue dilution resulting from the homogenization procedure, as well as HPLC injection volume, must be considered. To quantitate MT-1 and MT-2 in tissues where a 33% tissue homogenate and 100 μl HPLC injection are made, the following equation is used:

$$\mu\text{g MT/g tissue} = \frac{(\mu\text{g MT} \times 3\text{ ml/g tissue})}{0.1\text{ ml}} \times 1.05$$

where micrograms MT-1 or MT-2 is determined from the standard curve, 3 ml/g tissue represents the initial dilution of the tissue homogenization, 0.1 ml represents the HPLC injection volume, and 1.05 is the sample dilution that occurs during the cadmium saturation step (950 μl/ml).

With this method, 0.50 μg MT has been detected above the noise level of the AA spectrophotometer. Given that samples are prepared as 33% homogenates and that the maximum injection volume is 500 μl, a detection limit of 3 μg MT/g tissue was established for the HPLC–AAS method. Therefore, the method readily detects MT-1 and MT-2 in untreated and induced animals.

Validation of the HPLC-AAS method has indicated that the method is an extremely reproducible procedure. Whereas the procedure has been used extensively for quantitating hepatic MTs, it can also be used to

TABLE I
CONCENTRATIONS OF METALLOTHIONEIN DETERMINED BY HPLC-AAS AND CADMIUM–HEMOGLOBIN RADIOASSAY[a]

Rat	MT-1	MT-2	Total	Cd/Heme
Control[b]	ND[c]	13.6 ± 1.9	13.6 ± 1.9	10.1 ± 1.3
Cd treated[d]	138 ± 25	150 ± 25	289 ± 44	297 ± 30
Zn treated[e]	73.3 ± 20	125 ± 14	198 ± 26	225 ± 22

[a] Results expressed as micrograms MT per gram liver and represent the mean ± SE of five rats. (From Ref. 12 with permission from the publisher.)
[b] Rats were injected with saline (2 ml/kg) and livers were removed 24 hr later for MT analysis.
[c] Not detected.
[d] Rats were injected with cadmium as $CdCl_2$ (30 μmol/kg; 2 ml/kg) and livers were removed 24 hr later for MT analysis.
[e] Rats were injected with zinc as $ZnCl_2$ (100 μmol/kg; 2 ml/kg) and livers were removed 24 hr later for MT analysis.

quantitate MT-1 and MT-2 in other tissues. In kidneys, where constituent levels of MTs are higher than in liver, both MT-1 and MT-2 can be quantitated in untreated animals.[13] In other tissues, where control levels are much lower than in liver or kidney, basal levels of MT-1 and MT-2 cannot be determined directly in cytosol, but concentration (via lyophilization) can facilitate detection of the proteins. To date, the HPLC–AAS method has not been widely used to evaluate MT-1 and MT-2 in cells grown in culture.

The HPLC-AAS method has also been validated by comparing to other cadmium saturation techniques. A comparison of data obtained from the HPLC–AAS method and the cadmium-hemoglobin radioassay[16] for hepatic MT levels in cadmium- and zinc-treated rats is shown in Table I. These two cadmium saturation techniques are in good agreement, with the HPLC–AAS method offering the advantage of separating MT-1 and MT-2.

Lastly, the anion-exchange HPLC–AAS method is a nondenaturing assay that can also be utilized as a rapid, simple method for separation of MT-1 and MT-2. It is quite obvious that the 15-min run time is significantly shorter than the time required for conventional column chromatography procedures. Moreover, the proteins can be separated directly on the anion-exchange column, requiring no gel-filtration step. The short analysis time is a distinct advantage for conducting experiments concerning the

[16] D. L. Eaton and M. G. Cherian, this volume [13].

relative rates of synthesis and degradation of MT-1 and MT-2.[17] The metal composition of unsaturated MTs can also be evaluated with this method by simply changing the lamp on the AA spectrophotometer to the metal of interest. This approach is particularly useful for evaluating the metal composition of MTs induced by an organic compound. However, HPLC–AAS is more cumbersome than HPLC–ICPES (inductively coupled plasma emission spectrometry) methods for evaluating MT-1 and MT-2, as many metals can be detected simultaneously with the ICP.[18]

In summary, the cadmium-saturated HPLC–AAS method offers a rapid, simple procedure for quantitating MT-1 and MT-2 induced in many tissues by a variety of compounds.

Acknowledgment

Supported by USPH Grant ES-01142.

[17] L. D. Lehman-McKeeman, G. K. Andrews, and C. D. Klaassen, *Toxicol. Appl. Pharmacol.* **92,** 1 (1988).

[18] K. T. Suzuki, this volume [23].

[23] Detection of Metallothioneins by High-Performance Liquid Chromatography–Inductively Coupled Plasma Emission Spectrometry

By KAZUO T. SUZUKI

Metallothionein (MT) can be separated on a high-performance liquid chromatography (HPLC) column using various separating principles such as gel filtration or ion exchange. Then the metallothionein can be detected using specific and/or nonspecific detectors in either fractionated eluate (offline detection) or eluate directly, without fractionation (online detection).

Ultraviolet absorption at 254 and/or 280 nm is often monitored using a flow cell as a nonspecific online detection. An atomic emission spectrometer (AES) can be utilized as an element-specific detector by introducing the eluate of an HPLC column directly into the nebulizer of the spectrometer.[1-4]

[1] D. M. Fraley, D. Yates, and S. E. Manahan, *Anal. Biochem.* **51,** 2225 (1979).

In this chapter, an analytical method that can be applied to characterize MTs from various biological sources is presented as the HPLC/ICPES (inductively coupled plasma emission spectrometry) method, which consists of HPLC as a separating tool and ICPES as an element-specific online detector.[5,6]

ICPES as Element-Specific Detector

Atomic emission spectrometry with inductively coupled argon plasma excitation is abbreviated as ICPES, ICPAES, or ICP. This new analytical method for metals and other elements is now widely applied to detect and quantify multielements not only in inorganic materials but also in biological samples. The details of its principle and instrumentation can be found in many books and review articles.[7]

The advantages of ICPES over other analytical methods for elements are as follows:

1. Many elements can be detected at the parts per billion (ppb, 10^{-9}) level.
2. Dynamic ranges for quantification of many elements are wide and a linear relationship is obtainable at six orders of magnitude.
3. Stable plasma allows accurate measurement for a prolonged analytical period.
4. Simultaneous or rapid detection of multielements is possible.
5. Interferences such as chemical and ionization interactions are low compared with other methods.

These advantages, especially over those of atomic absorption spectrometry (AAS), made it attractive to use ICPES as an element-specific detector after HPLC instead of AAS.[8] Two types of ICPES are available: sequential and simultaneous detection of multielements. Multielements can be detected by using sweeping wavelengths in sequential detection type or they can be detected simultaneously by using a polychromator in the simulta-

---

[2] M. Morita, T. Uehiro, and K. Fuwa, *Anal. Chem.* **52,** 349 (1980).

[3] R. E. Majors, H. G. Barth, and C. H. Hochmüller, *Anal. Chem.* **56,** 300R (1984).

[4] H. G. Barth, W. E. Barber, C. H. Hochmüller, R. E. Majors, and F. E. Regnier, *Anal. Chem.* **58,** 211R (1986).

[5] K. T. Suzuki, H. Sunaga, E. Kobayashi, and N. Sugihira, *J. Chromatogr.* **400,** 233 (1987).

[6] H. Sunaga, E. Kobayashi, N. Shimojo, and K. T. Suzuki, *Anal. Biochem.* **160,** 160 (1987).

[7] A. Montaser and D. W. Golightly, eds., "Inductivity Coupled Plasmas in Analytical Atomic Spectrometry." VCH Publishers, New York, 1987.

[8] C. D. Klaassen and L. D. Lehman-McKeeman, this volume [22].

neous detection type. Digital data for multielements are usually obtained using ICPES.

The ICPES data for multielements in a continuous eluate from an HPLC column are delivered as digital data and can be stored on a floppy disk. Acquisition and storage of digital data for multielements are determined by the type of computer used. Software for graphic demonstration of data has been developed only for simultaneous detection types of ICPES instruments. Because the sequential detection type of ICPES instrument acquires data for each element with a time delay, each element can be detected with time delay for the eluate of an HPLC. Therefore, the sequential detection type of ICPES instrument cannot be used as a multielement detector of an HPLC.

Because ICPES detects emission lines spectrometrically, emission lines shorter than 210 nm cannot be detected due to the presence of interfering oxygen absorption. Emission lines of light elements exist mostly in the short-wavelength region and a vacuum ultraviolet ICPES must be used to detect light elements such as sulfur at 180.7 nm and carbon at 193.1 nm.

Connection between HPLC and ICPES

When two independent instruments must be connected, an appropriate interface is required. In fact, an interface is used to adjust the difference between the flow rate from an HPLC column (<1.0 ml/min for an analytical column) and the uptake rate of an atomic absorption spectrophotometer ($\sim$4–5 ml/min).[9,10] However, because the uptake rate for ICPES is comparable to the flow rate from HPLC, the two machines can be connected without any interfaces while maintaining the best conditions for the respective instruments. This is an additional advantage of ICPES over AAS as an element-specific detector for use with HPLC.

Due to the slow uptake rate in ICPES, the peak intensity is affected by the nature of sample solution, e.g., viscosity and acidity. To overcome this disadvantage, special torches for use with high salt solutions or organic solvents have been invented and are available commercially.

A torch appropriate for the composition of elution buffer should be selected for HPLC/ICPES. Two limitations should be mentioned when using different buffer compositions. (1) Buffers should not contain the elements to be determined. For example, if sulfur and phosphorus distri-

[9] N. Yoza and S. Ohashi, *Anal. Lett.* **6,** 595 (1973).

[10] K. T. Suzuki, T. Motomura, Y. Tsuchiya, and M. Yamamura, *Anal. Biochem.* **107,** 75 (1980).

butions are to be determined, reducing agents such as 2-mercaptoethanol and phosphate buffer must be avoided. (2) Buffers should not contain powerful ligands, because redistribution of metals during separation procedures is always a matter of concern.

Data other than element concentrations can be obtained by connecting the corresponding flow cells between the HPLC column and the nebulizer of the ICPES instrument. Absorption intensity at 254 nm is often monitored to detect the mercaptide bond in MT, which causes absorption due to charge transfer between metal and sulfur bonding. Although this absorption can be detected in the case of the Cd–S bond, it cannot be observed in the case of the Zn–S bond. Absorption intensity is monitored more often at 280 nm than at 254 nm in order to detect proteins and other biocomponents by utilizing absorption bands of aromatic amino acids. As MT is known to contain no aromatic amino acids, the presence of absorption at 254 nm and the absence of absorption at 280 nm can be used as convenient evidence for cadmium-containing MTs. The refractive index (RI) can also be monitored using a flow cell, and conductivity data through a flow cell are useful for ion-exchange chromatography. As the sample solution is supplied at a constant flow rate to an ICPES instrument through an HPLC column, a flat baseline is obtained, thereby improving the signal-to-noise *(S/N)* ratio also observed by HPLC/AAS.

Detection of Metallothioneins

Metallothionein has been characterized by its unique structure and can be detected based on the following structural properties: (1) it is a low-molecular-weight protein; (2) it contains cysteinyl residues in high quantity; (3) it binds heavy metals in high quantity; and (4) all metals are bound through sulfhydryl groups and all sulfhydryl groups in MT coordinate with metals. In addition to the structural properties, the following properties are often used to characterize or to detect MTs: MT is stable to heat treatment, i.e., resistant to heat denaturation (70° for 10 min) and metals bound to MT are easily replaced by metals with a higher affinity for MT.

Molecular sizes of MTs are estimated from their elution volumes on a gel-filtration column by HPLC, and quantities of sulfur and metals bound to MT are obtained simultaneously by detection using ICPES. Therefore, the single experimental procedure, namely, analysis of a biological sample solution by HPLC/ICPES, can give useful information to characterize MT: molecular size, number of metal species, quantities or at least relative ratios of each metal and sulfur.

Experimental Procedure to Detect Metallothionein

Figure 1 illustrates a typical distribution profile of MT-containing liver supernatant on a gel-filtration column by HPLC/ICPES.[6] Rats were injected with cadmium at a dose of 1.0 mg Cd/kg body weight daily for 10 days and the livers were homogenized in 3 vol of the extraction buffer (0.1 *M* Tris-HCl buffer, pH 7.4, containing 0.25 *M* glucose) using a glass–Teflon homogenizer under nitrogen in an ice-water bath. The homogenate was centrifuged at 170,000 *g* for 60 min at 2° and the resulting supernatant was analyzed by HPLC/ICPES.

Analytical conditions are as follows. An aliquot (0.1 ml) of the supernatant is applied on an SW column (TSK gel G3000SW, 600 × 7.5 mm i.d. with a guard column of 75 × 7.5 mm i.d.; Tosoh, Tokyo, Japan). The column is eluted with 10 m*M* Tris-HCl buffer [pH 8.0, containing 0.1% (w/v) sodium azide] at a flow rate of 1.0 ml/min with an HPLC series instrument 340 (Beckman, Berkeley, CA).

Sodium azide in the buffer works not only to preserve the column from microorganisms but also to maintain the salt concentration of the elution buffer. The buffer concentration must be increased to 50 m*M* when sodium azide is not added in the buffer, otherwise MT would be eluted faster, resulting in a poor separation from high-molecular-weight components.[11]

The buffer solution is connected online and passed through a degasser (Shodex Degas, Showa Denko K.K., Tokyo, Japan). Absorbances at 254 and 280 nm are determined with a dual-wavelength ultraviolet detector (model 152; Altex, Berkeley, CA).

The plasma spectrometer is JY48 (Seiko Instruments & Electronics, Ltd., Tokyo, Japan), which contains an inductively coupled plasma [Plasmatherm; HFP-2500F/APCS-3 radio frequency (rf) generator, with ICP 2500/AMN-PS-1 plasma and matching system), a concentric glass pneumatic nebulizer, and a JY48PVH polychromator (Paschen-Runge, Rowland circle of 1 m, 2550 grooves/mm holographic master grating]. The operational parameters for the plasma are as follows: rf power forward, 1.3 kW; reflected power, <5 W; nebulizer gas pressure, 31 psi; sample uptake, 2.1 ml/min; coolant gas flow, 18 liters/min; middle gas flow, 0 ml/min; observation height, 16 ± 5 mm above coil top.

The outlet tube of an HPLC column is connected directly to the nebulizer of an ICPES machine. Acquisition of ICPES data is performed every 2 sec for 30 min (in total 900 data points for each element/30 min) and the data are recorded on a floppy disk. Conversions of the stored data to distribution profiles of respective elements are carried out by an in-

[11] K. T. Suzuki, H. Sunaga, and T. Yajima, *J. Chromatogr.* **303,** 131 (1984).

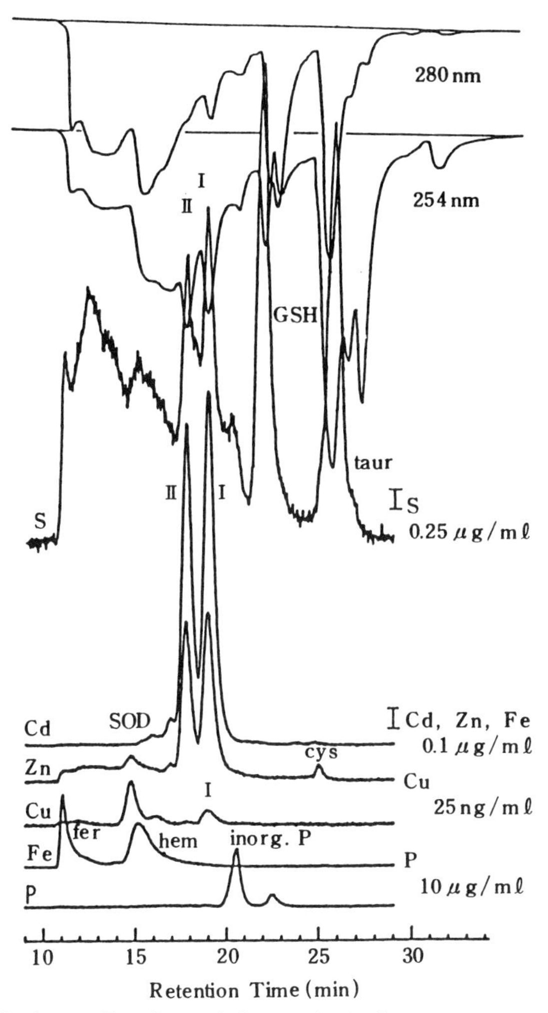

FIG. 1. Distribution profiles of several elements in the liver supernatants of cadmium-injected rats on an SW column by HPLC/ICPES.[6] An aliquot (0.1 ml) of the liver supernatant prepared from cadmium-injected rats was applied to an SW column (TSK gel G3000SW, 600 × 7.5 mm i.d. with a guard column of 75 × 7.5 mm i.d.; Tosoh) and the column was eluted with 10 m*M* Tris-HCl buffer, pH 8.0, containing 0.1% NaN_3, at a flow rate of 1.0 ml/min. Peaks are assigned as follows: I and II, MT-1 and MT-2; GSH, glutathione; taur, taurine; SOD, superoxide dismutase; fer, ferritin; hem, hemoglobin; cys, zinc bound to cysteine; inorg. P, inorganic phosphate.

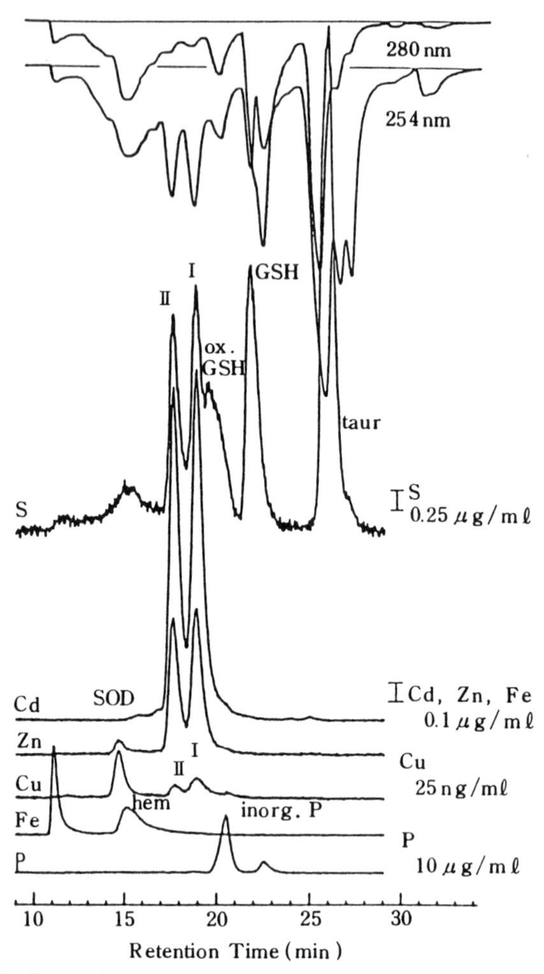

FIG. 2. Distribution profiles of several elements in the heat-treated liver supernatants of cadmium-injected rats on an SW column by HPLC/ICPES.[6] The liver supernatant, corresponding to that in Fig. 1, was heated at 70° for 10 min in an atmosphere of nitrogen, cooled in an ice-water bath, and then centrifuged at 2300 *g* for 10 min. An aliquot (0.1 ml) of the supernatant was applied on an SW column as described in the legend to Fig. 1. ox. GSH, Oxidized glutathione; other abbreviations as in Fig. 1.

house-developed program using a personal computer (PC-9801; NEC Corp., Tokyo, Japan) and an XY plotter (FP5301R; Graphtec Corp., Tokyo, Japan). The program for distribution profiles is available at present only for simultaneous detection-type instruments from Nippon Jarrell-Ash (Kyoto, Japan) for their ICPES machines (ICAP-750, Nippon Jarrell-Ash) and with some modification for ICAP-61 (Thermo Jarrell-Ash Corp., Franklin, CA).

The gel-filtration column used for separation of MT in Fig. 1 has a cation-exchange property in addition to a gel-filtration property when the column is eluted with alkaline buffer solution.[6] Mammalian MTs consist of two major isoforms that can be separated on an ion-exchange column and are named MT-1 and MT-2 according to the order of elution from an anion-exchange column. Metallothionein is separated into the two isoforms and eluted from the column in the order of MT-2 and MT-1 in Fig. 1.

The two isoforms of MT are separated from most biocomponents and the corresponding peaks of the metals (cadmium, zinc, and copper) and sulfur and the absorbance at 254 nm are detected at the same retention times of 17.9 (MT-2) and 18.9 min (MT-1). The characteristic properties of MT are visualized by a single analytical procedure as shown in Fig. 1, namely a low-molecular-weight protein with high cysteinyl residues without aromatic amino acids and with high metal content through mercaptide bonding.

Heat treatment of supernatants is often carried out to characterize MT or to remove heat-unstable proteins. Figure 2 shows the distribution profile of the heat-treated supernatant. The liver supernatant used for Fig. 1 is heated at 70° for 10 min and then centrifuged at 2300 *g* for 10 min. An aliquot (0.1 ml) of the heat-treated supernatant is analyzed by HPLC/ICPES in the same way. Heat-denatured proteins are removed and the two isoforms are more clearly observed as shown in Fig. 2. However, several unintended changes are shown to occur by this procedure; copper is taken up by the isoform MT-2 and glutathione (21.9 min) is partly oxidized and eluted at 20.1 min.

[24] Electrochemical Detection of Metallothionein

By ROBERT W. OLAFSON and PER-ERIC OLSSON

Introduction

Due to a lack of a measurable biological activity the quantitation of metallothionein (MT) has proven to be problematical and investigators have been forced to consider unique structural aspects of the molecule in order to provide analytical procedures. Thus efforts to detect MT have relied heavily on estimates of metal binding and sulfhydryl content, or on immunological assays. Each of the different approaches has strengths and weaknesses that should be taken into account prior to implementation. For example, the immunological approach uses a radioimmunoassay (RIA)[1] or enzyme-linked immunosorbent assay (ELISA),[2] which combine the exquisite specificity inherent in antibodies with the high sensitivity of either isotope counting or enzyme-catalyzed reactions but may suffer from a lack of interspecies cross-reactivity. On the other hand, assay methods utilizing the metal-binding characteristic of MTs rely on saturation of metal-binding ligands with group IB and IIB heavy metals.[3-5] When employed with radionuclides, this analytical approach can be very sensitive and unaffected by species differences. However, difficulties in achieving conditions where the radionuclide binds specifically to MT requires a substantial allocation of time, either to verify the species being measured in advance or alternatively to employ a routine chromatographic step. Electrochemical quantitation of MT by differential pulse polarography (DPP) is a technique that takes advantage of the high cysteine content of MT and overcomes many of the shortcomings of alternate methods of detection. Capable of picomole-level sensitivity, the technique is very rapid, highly reproducible, and species independent, providing a useful procedure for following MT concentrations during isolation from tissues[6,7] in environmental samples[8-11] or during physiological experiments.[12-16] First described by Brdicka in 1933[17] and further investigated by Palecek and Pechan,[18] the method was originally adopted for MT analysis by Kehr[19] in 1973. The technique was

[1] R. J. Vander Mallie and J. S. Garvey, *Immunochemistry* **15,** 857 (1978).

[2] R. J. Cousins, this volume [19].

[3] R. W. Chen and H. E. Ganther, *Environ. Physiol. Biochem.* **5,** 378 (1975).

[4] J. K. Piotrowski, W. Balanowska, and A. Sapota, *Acta Biochim. Pol.* **20,** 207 (1973).

[5] A. M. Scheuhammer and M. G. Cherian, this volume [12].

[6] R. W. Olafson, R. G. Sim, and K. G. Boto, *Comp. Biochem. Physiol. B: Comp. Biochem.* **22B,** 407 (1979).

[7] R. W. Olafson, *Int. J. Pept. Protein Res.* **24,** 303 (1984).

[8] P.-E. Olsson and C. Haux, *Aquat. Toxicol.* **9,** 231 (1986).

[9] C. Hogstrand, G. Lithner, and C. Haux, *Mar. Environ. Res.* **28,** 179 (1989).

[10] P.-E. Olsson and C. Hogstrand, *J. Chromatogr.* **402,** 293 (1987).

[11] M. Roch, J. A. McCarter, A. T. Matheson, M. J. R. Clark, and R. W. Olafson, *Can. J. Fish. Aquat. Sci.* **39,** 1596 (1982).

[12] R. W. Olafson, *J. Biol. Chem.* **256,** 1263 (1981).

[13] R. W. Olafson, S. Loya, and R. G. Sim, *Biochem. Biophys. Res. Commun.* **95,** 1495 (1980).

[14] R. W. Olafson, *J. Nutr.* **113,** 268 (1983).

[15] R. W. Olafson, *Can. J. Biochem. Cell Biol.* **63,** 91 (1985).

[16] S. J. Hyllner, T. Anderson, C. Haux, and P.-E. Olsson, *J. Cell. Physiol.* **139,** 24 (1989).

[17] R. Brdicka, *Collect. Czech. Chem. Commun.* **5,** 112 (1933).

[18] E. Palecek and Z. Pechan, *Anal. Biochem.* **42,** 59 (1971).

[19] P. F. Kehr, Thesis, Purdue University, West Lafayette, Indiana (1973).

subsequently refined by Olafson,[12,20] Thompson and Cosson,[21] and Olsson and Haux.[8] The technique discussed in this chapter describes the DPP procedure as it has been developed for both the Princeton Applied Research Corporation (PARC) model 174/70 drop timer and the newer model 303A static mercury drop electrode (SMDE) and should thus be applicable to alternative polarographic systems.

Principle

Voltametry comprises a group of electrochemical procedures based on the potential current behavior of a polarizable electrode in an analyzed supporting electrolyte. Theoretically any species can be analyzed if it undergoes oxidation or reduction within the working potential range of the electrode system employed. The redox reaction taking place at the electrode is therefore controlled by variation in the applied electrode potential. The DPP procedure is a special case utilizing a dropping mercury electrode where the difference between two current measurements taken at two times is repetitively made at a renewed mercury droplet while the instrument scans a potential range. The basis of the Brdicka DPP assay for protein is the generation of catalytic double waves from SS/SH-containing proteins in a hexamminecobalt chloride supporting electrolyte. Although the complex electrochemical reactions at the electrode have not been explained in detail, it is generally accepted that cobalt complexes with ionized sulfhydryl groups catalyzing the evolution of protons.[22] Bridicka[23] has reported that on decomposition of metallic cobalt from the complex into cobalt amalgam that the ionized sulfhydryl groups take up protons from the supporting electrolyte. It is proposed that the coordination bond to the sulfur anion remains preserved on deposition of cobalt, facilitating proton reduction. The wave height is therefore a function of the number of thiol groups adsorbed on the electrode surface and is controlled by the recombination rate of ionized sulfhydryl groups and protons (Fig. 1). It is significant to note that the electrochemical reactions occur regardless of whether the sulfur-containing side chains are thiols, disulfides, or complexed with metals.

Since DPP quantitation of MT is not a specific procedure, as in the case of RIA, it is critical that the contribution of other cysteine-containing proteins be determined and eliminated before making determinations

[20] R. W. Olafson and R. G. Sim, *Anal. Biochem.* **100,** 343 (1979).
[21] J. A. J. Thompson and R. P. Cosson, *Mar. Environ. Res.* **11,** 137 (1984).
[22] M. Kuik and K. Krassowski, *Bioelectrochem. Bioenerg.* **9,** 419 (1982).
[23] R. Brdicka, M. Brezina, and V. Kalous, *Talanta* **12,** 1149 (1965).

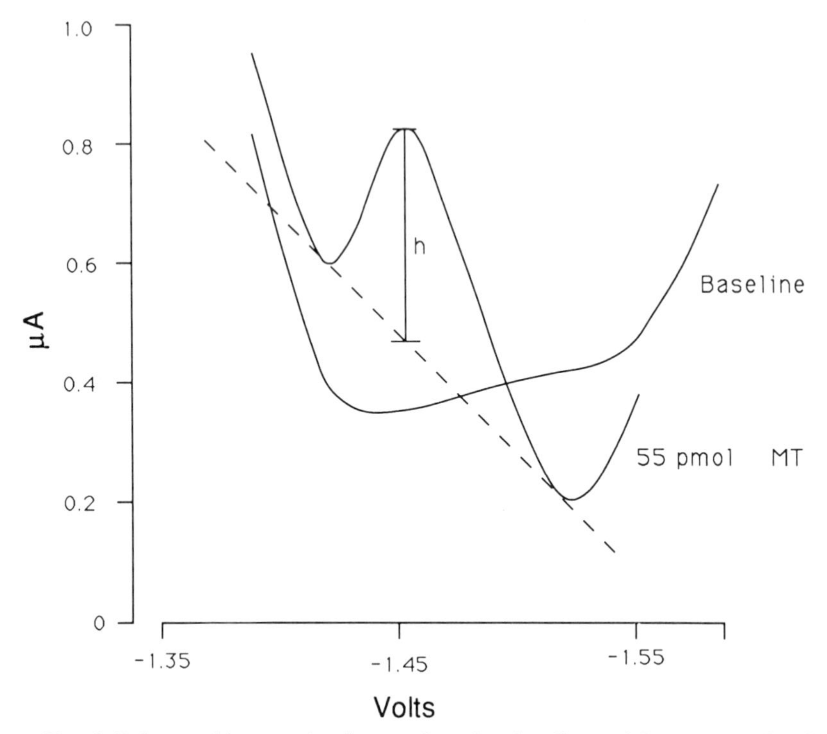

FIG. 1. Polarographic scan showing an electrolyte baseline and B wave resulting from 55 pmol of purified rainbow trout MT-1. The amount of MT was calculated by measuring the height (*h*) from the tangent of the two minima and dividing it by a constant (6420 μA/μmol MT) derived from a standard curve. [Reprinted, with permission, from P.-E. Olsson (1987).[24]]

from complex samples. In most cases, this is achieved by heat denaturation, leaving the heat-stable MT in solution (Fig. 2). Although MT is not the only heat-stable protein remaining in solution, it is usually the only species left having significant polarographic activity. In the case of well-characterized eukaryotic MTs this is readily established by chromatography of the heat-denatured supernatant on a Sephadex G-75 column where polarographic activity should be found in a symmetrical peak with an apparent molecular weight of 10,000.[12] As with all nonspecific methods, further verification of the identity of uncharacterized MTs may require ion-exchange chromatography and amino acid analysis. For example, the specificity of differential pulse polarographic analysis of perch MT was tested by monitoring polarographic activity during chromatographic purification and subsequent comparison of the DPP method to RIA. Ion-ex-

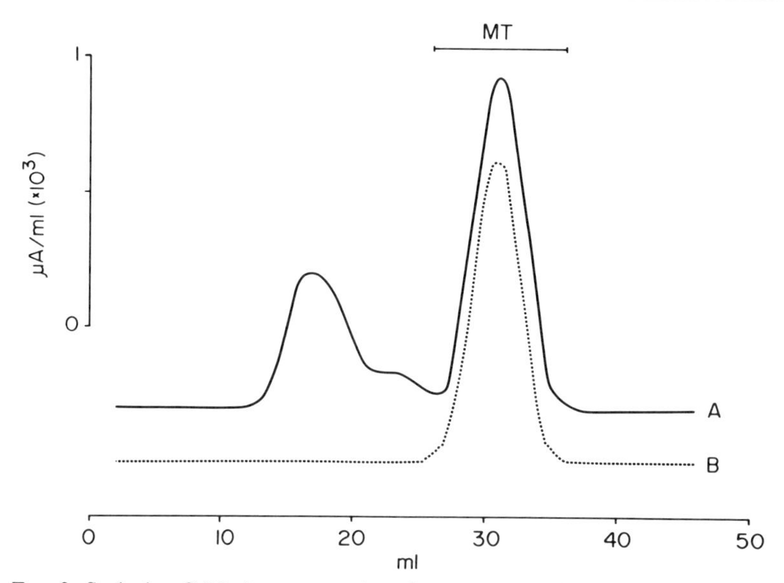

FIG. 2. Sephadex G-75 chromatography of supernatant before (A) and after (B) heat denaturation at 95° for 5 min. The heat-denatured material was removed by centrifugation at 10,000 *g* at 4° for 20 min. Two identical samples were applied to the column and polarographic activity was measured in the eluted fractions. [Reprinted, with permission, from P.-E. Olsson (1987).[24]]

change fast protein liquid chromatography (FPLC) separation of a polarographically active peak obtained from gel-permeation chromatography of heated-denatured cytosol confirmed that MT constituted essentially the only polarographically active cytosolic species following heat denaturation.[24] Specificity of the DPP technique was further evaluated by comparing it to RIA performed on the isolated perch MT.[25] The MT determination using heat denaturation and DPP was shown to be strongly correlated to the MT values found by RIA assay of perch liver homogenates ($r = 0.92$, $P < 0.001$). The presence of low-molecular-weight sulfhydryl-containing species, such glutathione and free cysteine in tissue samples, does not interfere with the DPP assay, as concentrations of the former compound an order of magnitude above that found in tissues had no affect on the polarographic response to standard MT.[12] This appears to be due to an unusually low response to low-molecular-weight compounds (Table I).

[24] P.-E. Olsson, Ph.D. Thesis, Göteborgs Universitet, Göteborg, Sweden (1987).

[25] C. Hogstrand and C. Haux, *Mar. Environ. Res.* **28,** 187 (1989).

TABLE I
WAVE HEIGHTS AND POTENTIALS FOR SELECTED SULFHYDRYL COMPOUNDS

	Wave potential (V vs $Ag^+/AgCl$)	Wave height ($\mu A/\mu mol$)
Crab[a] Cd-MT-1	−1.47	555
Bovine serum albumin	−1.44	975
Synthetic peptide[b]	−1.43	7.86
Glutathione	−1.41	0.17
Insulin	−1.52	15.31
2-Mercaptoethanol	−1.46	0.01
Cysteine	−1.41	1.13
Cystine	−1.39	1.92

[a] *Scylla serrata.*
[b] Ser-Cys-Thr-Cys-Thr-Ser-Ser-Cys-Ala.

Materials

Chemicals. Standard rabbit MT-1 and Triton X-100 were obtained from Sigma Chemical Company (St. Louis, MO), hexamminecobalt chloride was obtained from BDH Chemicals (Toronto, Ontario, Canada). Vacuum-distilled mercury was purified as described by Whitnack and Sasseli.[26] All other chemicals were of analytical grade.

Biological Material. Metallothionein was prepared and characterized from mouse and crab as described by Olafson *et al.*[6,12] and from perch as described by Olsson and Hogstrand.[10]

Equipment. Differential pulse polarography using the Brdicka procedure was performed using a PARC model 174A polarographic analyzer equipped with either a model 174/70 drop timer or a model 303A SMDE. Both instruments were equipped with an Ag/AgCl reference electrode and platinum counter electrode. Current profiles were plotted on a Houston Instruments model 2000 X-Y recorder. The SMDE electrode system consisted of a beveled capillary working electrode, a platinum counterelectrode, and an Ag/AgCl reference electrode, as supplied by Princeton Applied Research Company (Princeton, NJ). All polarography was performed under the following scan conditions: range of scan, −1.35 to −1.60 V vs Ag/AgCl; pulse amplitude, 50 mV; scan rate, 5 mV/sec; drop time, 0.5 sec. For the model 174/70 drop timer the dropping electrode had a flow rate (M) of 0.827 mg mercury/sec at a constant height (h) of 42 cm. For the model 303A SMDE the drop size was set to medium and mode setting

[26] G. C. Whitnack and R. Sasseli, *Anal. Chim. Acta* **47,** 267 (1969).

DME. Other settings for the polarographic analyzer were as follows: display direction, positive; output offset, off; low-pass filter, off. All work was performed at 25°. The SMDE capillary electrodes had to be washed at regular intervals. The electrode was disassembled according to the manufacturer's protocol and all mercury trapped in the capillary was removed by flushing with air. The capillary was then filled to two-thirds with a 10% hydrofluoric acid solution, leaving the beveled tip immersed in the HF solution. After 15–20 sec the capillary was washed with deionized water and the procedure repeated several times, followed by drying at 60° for 1 hr. The capillary was filled to two-thirds with 5% dichlorodimethylsilane [DDS, $(CH_3)_2 SiCl_2$] and left with the beveled tip in the solution for 1 hr. The DDS solution was then removed by flushing with air and the capillary was allowed to dry at 60° for 1 hr prior to reinstallation. Reconditioning of dropping electrodes for the model 174/70 was found to be troublesome and fresh capillaries were made when the instrument showed signs of instability.

Supporting Electrolyte. Different supporting electrolyte solutions have been used depending on the electrode system employed. When using the model 174/70 drop timer, the Brdicka supporting electrolyte was used as described by Palecek and Pechan[18] but without surface-active agent. The original supporting electrolyte, used with the model 174/70 drop timer, contains 1 *M* NH_4Cl, 1 *M* NH_4OH, and 0.6 m*M* $[Co(NH_3)_6]Cl_3$. On the other hand, when the model 303A SMDE is used it is recommended that the Brdicka supporting electrolyte be modified by doubling the cobalt concentration to 1.2 m*M* $[Co(NH_3)_6]Cl_3$ while keeping the ammonium ion concentration at 1 *M*. In addition, when using the SMDE electrode system, a 100-μl aliquot of Triton X-100 (working solution of 1.25×10^{-2} %, v/v) is added to 10 ml of electrolyte to suppress a maximum that appears close to the cobalt peak. The electrolyte should be made fresh weekly and stored in the dark at 4°. Since this is a temperature-dependent reaction the electrolyte must be reequilibrated to 25° for analysis.[20]

Procedures and Discussion

Sample Preparation. It should be noted that samples from chromatographic or electrophoretic isolation procedures can be introduced directly into the polarographic cell without prior treatment. Very few interferences from buffers or reagents have been observed, but alterations in baselines are occasionally seen, requiring inclusion of appropriate buffers in standard solutions. Tissue sample preparation involves homogenization of approximately 200 mg of tissue in 10 ml saline (0.9% NaCl) followed by heat denaturation at 90–95° for 5 min. Precautions should be taken not to heat denature large tissue samples in a small volume of saline, because this leads to loss of MT by coprecipitation. The latter problem can be avoided by including a centrifugation step (10,000 *g*, 4°, 20 min) prior to heat denaturation. The denatured homogenates are cooled on ice and filtered through a 0.45-μm membrane filter (Millipore, Bedford, MA) to remove flocculent material. The filtrate is then ready for analysis and can be injected directly into the polarographic cell. An alternative protocol may be employed, replacing the filtration step by a centrifugation step. In this instance the heat-denatured samples were centrifuged at 10,000 *g* at 4° for 20 min to remove the heat-denatured proteins. Since only microliter quantities are needed for polarographic analysis it was possible to process a large number of samples by centrifugation in 1.5-ml tubes.

Polarographic Analysis. For use with the SMDE, electrolyte (10 ml) was dispensed into the polarographic cell and 100 μl of Triton X-100 was added. Nitrogen was passed through the solution (2–4 min) to remove oxygen and mix the sample. Subsequent purges after addition of standard or sample were generally 15–30 sec. A typical baseline and response to addition of 55 pmol of standard MT is shown in Fig. 1. Determination of the linear working range and detection limit was performed every time the capillary was changed or washed, as it was found that these manipulations affected the electrode sensitivity. Standard MT, used to calibrate the instrument, was either purified in the laboratory or was commercially obtained rabbit MT-1.

The basis for the quantitation of thiolic compounds by the polarographic method lies in the linear relationship between the second of the two wave maxima, designated A and B. The A wave lies at about −1.35 V vs Ag/AgCl; the B wave is found at −1.47 V. The linearity and shape of the A wave is poor, whereas the B wave gives a distinct maximum and is more reliable for quantitation of MT. It should be noted that crude homogenates that contain uncharacterized low-molecular-weight components will increase the current at the starting potential. Although these contaminants do not elicit a response at the MT peak potential in the presence of protein, they induce a shift in the catalytic wave to higher current values. For this reason the wave height is measured from the tangent of the two minima, as shown in Fig. 3. Analysis of heat-denatured tissue homogenates using the model 174/70 drop timer has demonstrated linearity of the polarographic response to at least 250 nmol MT.[12] Standard addition, as shown in Fig. 3C, is recommended for quantitation of MT by wave height measurement using the DPP technique.

The linear working range using the model 174/70 drop timer has been determined to be from about 0.01 to 25 μM MT (Fig. 3). The sensitivity of the model 174/70 drop timer is approximately 788 μA/μmol MT. The

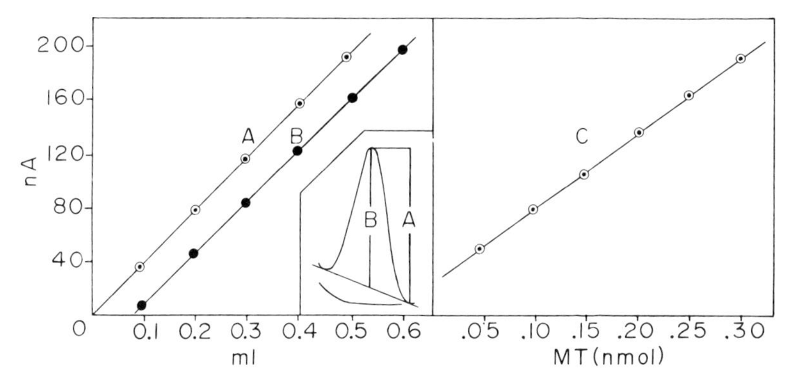

FIG. 3. Polarographic response to repeated addition of tissue homogenate with and without addition of standard MT. Increasing volumes of hepatic tissue homogenate (16 mg/ml) were assayed in separate supporting electrolytes using the model 174/70 drop timer. The amount of MT was calculated by measuring the wave height from the baseline (A) and the tangent of the two minima (B). Duodenal homogenate (16 mg/ml) containing increasing amounts of standard protein was assayed using the tangent procedure (C). The *y* intercept in (C) indicates the duodenal MT response. [Reprinted, with permission, from R. W. Olafson, *J. Biol. Chem.* **256,** 1263 (1981).]

linear working range for the model 303A SMDE is from about 1 to 20 n*M* MT and a sensitivity of about 6420 μA/μmol MT.[8] Thus, with the model 174/70 and 100 pmol MT can be readily detected while the newer model 303A SMDE allows detection to 10–15 pmol MT. Approximately 10 times higher sensitivity can thus be obtained from the newer electrode within a narrower working range.

Conclusion

A pulse polarographic method for determination of MT has been developed that is suitable for picomole-level detection of MT under a variety of conditions. Using a hexaamminecobalt chloride supporting electrolyte, thiol- or disulfide-containing proteins have been found to produce a differential pulse polarographic wave that is proportional in height to the protein concentration. The method is specific for MT after removal of contaminating proteins in tissue homogenates by heat denaturation. This electrochemical approach to quantitation of MT has been used in a large number of studies of MT regulation.[16–20,27] The species independence of the technique renders it particularly attractive in field studies of MT responses to environmental heavy-metal pollution.

[27] P.-E. Olsson, M. Zafarullah, and L. Gedamu, *Biochem. J.* **257,** 555 (1989).

Section IV

Isolation and Purification of Metallothioneins

[25] Purification and Quantification of Metallothioneins by Reversed-Phase High-Performance Liquid Chromatography

By MARK P. RICHARDS

Introduction

Metallothionein (MT) continues to generate a great deal of interest among investigators because of the many unique characteristics exhibited by this family of metal-binding proteins and the genes that encode them and because of the putative role of MT in the intracellular metabolism of heavy metals. Consequently, there is an on-going need for new analytical techniques applicable to the isolation and quantification of MT that are both rapid and sensitive.

Since its initial isolation and characterization from equine renal cortex tissue, a number of techniques have been developed for the purification and quantification of MT from a variety of tissues. A two-step, low-pressure, column chromatographic procedure generally involving a combination of gel-permeation and anion-exchange modes has been widely used to isolate MT from tissue extracts (cytosol) and continues to be the method of choice for the purification of MT on a preparative scale. However, this approach is time consuming and requires relatively large amounts of sample. Moreover, it is not readily adaptable to the quantification of MT in physiological fluids or tissue and cell extracts, especially under noninduced conditions.

Several techniques have been specifically developed to quantify MT. These include (1) competitive displacement of MT-bound metal (i.e., cadmium, mercury, and silver saturation assays),[1-3] (2) electrochemical (polarographic) quantification of MT via its sulfhydryl groups,[4] (3) polyacryalmide gel electrophoresis (PAGE; both nondenaturing and SDS-PAGE),[5,6] and (4) immunological assays [radioimmunoassay (RIA) and enzyme-linked immunosorbent assay (ELISA)].[7-9] Of these, the immunoassay tech-

[1] A. J. Zelazowski and J. K. Piotrowski, *Acta Biochim. Pol.* **24,** 97 (1977).

[2] D. L. Eaton and B. F. Toal, *Toxicol. Appl. Pharmacol.* **66,** 134 (1982).

[3] A. M. Scheuhammer and M. G. Cherian, *Toxicol. Appl. Pharmacol.* **82,** 417 (1986).

[4] R. W. Olafson and R. G. Sim, *Anal. Biochem.* **100,** 343 (1979).

[5] L. Y. Lin and C. C. McCormick, *Comp. Biochem. Physiol. C: Comp. Pharmacol. Toxicol.* **85C,** 75 (1986).

[6] Y. Aoki, M. Kunimoto, Y. Shibata, and K. T. Suzuki, *Anal. Biochem.* **157,** 117 (1986).

[7] J. S. Garvey, R. J. Vander Mallie, and C. C. Chang, this series, Vol. 84, p. 121.

[8] R. K. Mehra and I. Bremner, *Biochem. J.* **213,** 459 (1983).

METHODS IN ENZYMOLOGY, VOL. 205

niques offer the highest degree of sensitivity (ca. 10^{-12} g). However, it is not always possible to quantify separately all of the individual MT isoforms with these assays nor do these assays provide any information about the metal composition of the individual MT isoforms.

High-performance liquid chromatography (HPLC) has been applied to the separation of MT isoforms from different tissues for a variety of eukaryotic species. Several modes of HPLC have been employed, including gel permeation,[10] ion exchange,[11] reversed phase,[11-13] and a combination of gel permeation and ion exchange via column switching.[14] Of these, reversed-phase HPLC (RP-HPLC) is the most powerful technique because it can effectively separate peptides that differ in their composition by a single amino acid. The major advantages of HPLC techniques are that they can, in one step, isolate MT isoforms from complex mixtures such as tissue extracts (cytosol). Only small amounts of sample are required and, depending on the type of detection method used, a relatively high degree of sensitivity can be achieved for the quantification of MT (ca. 10^{-6} g). Moreover, Suzuki *et al.*[10,11,14] and others[15-17] have extended the utility of HPLC by feeding the effluent from an HPLC column directly into the nebulizer of an atomic absorption spectrophotometer (AAS) in order to characterize the metal composition of the separated MT isoforms. The following sections describe the application of RP-HPLC procedures to isolate, characterize, and quantify MT isoforms and their associated metals.

Reagents

Water

The use of ultrapure water (<18 MΩ resistivity) is essential in the preparation of all buffers and solutions employed in RP-HPLC. Such precautions can substantially eliminate sample and buffer contamination leading to spurious peaks and elevated baseline ultraviolet (UV) absorbance during the chromatographic run.

[9] A. Grider, K. J. Kao, P. A. Klein, and R. J. Cousins, *J. Lab. Clin. Med.* **113,** 221 (1989).
[10] K. T. Suzuki, *Anal. Biochem.* **102,** 31 (1980).
[11] K. T. Suzuki, H. Sunaga, Y. Aoki, and M. Yamamura, *J. Chromatogr.* **281,** 159 (1983).
[12] S. Klauser, J. H. R. Kägi, and K. J. Wilson, *Biochem. J.* **209,** 71 (1983).
[13] M. P. Richards and N. C. Steele, *J. Chromatogr.* **402,** 243 (1987).
[14] K. T. Suzuki, H. Sunaga, and T. Yajima, *J. Chromatogr.* **303,** 131 (1984).
[15] L. D. Lehman and C. D. Klaassen, *Anal. Biochem.* **153,** 305 (1986).
[16] H. Van Beek and A. J. Baars, *J. Chromatogr.* **442,** 345 (1988).
[17] M. P. Richards, *J. Chromatogr.* **482,** 87 (1989).

Solvents and Chemicals

All organic solvents used to modify the mobile phase or in sample preparation should be HPLC grade or better and all chemicals used in preparation of buffers should be certified American Chemical Society (ACS) or ultrapure grade to minimize sample contamination and their contribution to baseline UV absorbance.

Metallothionein Standards

Currently, there are two types of purified MT available commercially (Sigma Chemical Co., St Louis, MO). These include horse kidney and rabbit liver MTs as well as the MT-1 and MT-2 isoforms further purified by anion-exchange chromatography of rabbit liver MT. Because it has been reported that lot-to-lot variation can be significant,[16] these sources should be checked for purity and fully characterized prior to their use as standards.

Instrumentation

High-Performance Liquid Chromatography System

A basic HPLC system is required with the following capabilities: (1) either dual pumps and a control system or a single pump equipped with a proportioning valve for gradient elution, (2) a sample injector, (3) a fixed or variable-wavelength detector for monitoring UV absorbance at 200–220 nm and (4) a data-handling system capable of collecting and integrating data from the detector(s). We have used a Waters (Waters/Millipore, Milford, MA) liquid chromatograph equipped with dual M6000A pumps, a WISP 710 autosampler, a model 720 system controller and a model 730 data module (plotter/integrator) to perform the isolation and quantification of MT isoforms. Ultraviolet absorbance (214 nm) is monitored with a model 441 fixed-wavelength detector. A computer-based pump control and data collection/integration system (Baseline 810; Waters/Millipore) is also used to simultaneously monitor and integrate UV and AAS absorbance signals. A fraction collector (FRAC-100; Pharmacia LKB, Piscataway, NJ) is used to collect the effluent from the column when further analysis of individual peaks is required.

Atomic Absorption Spectrophotometer (AAS)

To determine the metal content of separated MT isoforms, the effluent from the HPLC column is fed via Teflon tubing (0.009 in., i.d.) directly into the nebulizer of an AAS (model 5000; Perkin Elmer, Norwalk, CT) set to accept a flow rate of 3.0 ml/min. The tubing is flanged at one end so as to fit over the stainless steel capillary of the nebulizer. A single-slotted burner head with an air/acetylene oxidizing (lean blue) flame is used and the spectrophotometer set to 213.9, 228.8, and 324.8 nm for the determination of zinc, cadmium, and copper, respectively. A 1-V output signal from the AAS is used both for detection and peak area integration.

Sample Preparation

Tissue and Cell Extracts

Soluble extracts of tissues and cultured cells are prepared by homogenizing tissue samples in an appropriate buffer such as 10 m*M* Tris-HCl, pH 8.6, or 10 m*M* NaH_2PO_4, pH 7.0. The proportions of tissue to buffer will depend on the concentration of MT in the sample. That is, for samples containing low levels of MT (i.e., a noninduced tissue) a concentrated homogenate is recommended. Homogenate concentrations ranging from 1 : 1 to 1 : 10 (grams tissue : volume of buffer) are appropriate for RP-HPLC. The inclusion of 2-mercaptoethanol in the homogenizing buffer should be considered in light of the fact that this reducing agent has been reported to prevent oxidation of MT.[18] Furthermore, the inclusion of 5 m*M* 2-mercaptoethanol in the buffers used to prepare soluble extracts of tissue samples for MT analysis has been shown to be compatible with subsequent RP-HPLC.[16]

A preliminary fractionation step such as heat treatment or solvent extraction is recommended prior to RP-HPLC, especially if concentrated (i.e., 1 : 1) homogenates are prepared. To reduce the amount of extraneous protein in the samples to be analyzed by RP-HPLC, tissue or cell homogenates can be heat treated (i.e., 60° for 10 min), following which they are subjected to centrifugation to remove the heat-coagulated protein from the homogenate. The resulting supernatant (heat-treated cytosol) should be filtered through a 0.22-μm pore membrane (i.e., Millex-GV filter; Millipore Corp., Bedford, MA) prior to RP-HPLC analysis. An alternative to heat treatment of the homogenate is solvent extraction such as the chloroform–ethanol extraction method of Kimura *et al.*[19] This extraction procedure has been used in conjunction with RP-HPLC to isolate and characterize MT isoforms from the tissues of a variety of eukaryotic species.[12] The use of heat treatment or solvent extraction of the sample prior

[18] D. T. Minkel, K. Poulsen, S. Wielgus, C. F. Shaw III, and D. H. Petering, *Biochem. J.* **191,** 475 (1980).

[19] M. Kimura, N. Otaki, and M. Imano, *Experientia, Suppl.* **34,** 163 (1979).

to chromatography will not only prolong the useful lifetime of the analytical column used in RP-HPLC but may also improve the quality of the RP-HPLC separation of MT isoforms. However, it is also possible to inject untreated cytosol samples directly onto the reversed-phase column if they are sufficiently dilute (i.e., 1 : 5).[16]

Purification of Metallothionein and Metallothionein Isoforms Prior to Reversed-Phase High-Performance Liquid Chromatography

Purification of MT and each isoform (MT-1 and MT-2) prior to RP-HPLC can be achieved using a two-step column chromatographic fractionation of tissue cytosol not subjected to heat treatment. A typical purification procedure is as follows:

1. Gel-permeation chromatography is performed on a Sephadex G-75 column (5.0 cm i.d. × 60 cm), eluted with 10 m*M* Tris-HCl, pH 8.6. Fractions (10 ml) are collected and analyzed for zinc (or an appropriate metal) using AAS to detect the MT peak.
2. Those fractions comprising the MT peak from the gel-permeation separation are pooled and applied to a DEAE-Sephadex A-25 or DEAE-Sephacel column (2.6 cm i.d. × 20 cm) and eluted with a 500-ml linear gradient of 10–300 m*M* Tris-HCl, pH 8.6. Fractions (5 ml) are collected and analyzed for zinc (or an appropriate metal) using AAS to detect the peaks corresponding to each of the MT isoforms (MT-1 and MT-2).

Fractions containing the MT isoforms are pooled and stored at −20° prior to RP-HPLC. Ultrafiltration (YM2 membrane, M_r 1000 cutoff; Amicon, Danvers, MA) is useful not only to concentrate the MT isoforms but also to exchange the buffer for one compatible with the buffer system to be used for RP-HPLC (see below) or for deionized (DI) H_2O for subsequent lyophilization of the purified isoforms.

Reversed-Phase High-Performance Liquid Chromatography Procedures

Analytical Column

The type of column that we use routinely to separate the MT isoforms is a μBondapak C_{18} polyethylene cartridge (8 mm i.d. × 100 mm, 10-μm particle size) housed in a Z-module or RCM 8 × 10 radial compression device (Waters/Millipore).[13,17] In order to prolong the lifetime of the analytical column, a guard column consisting of the same material as is packed in the analytical column (i.e., μBondapak C_{18}) is recommended. The

TABLE I
CHROMATOGRAPHIC CONDITIONS APPLIED TO RP-HPLC SEPARATION OF METALLOTHIONEIN ISOFORMS

Column characteristics	Mobile phase	Elution profile	Flow (ml/min)	Refs.
		Acidic buffer system		
Aquapore RP-300 (RP-8) C_8 (Brownlee Laboratories, Santa Clara, CA), 10-μm particle size, 4.6 × 250 mm	A: 0.1% (w/v) TFA B: 60:40 (v/v) acetonitrile:0.1% TFA	Two-step linear gradient: 0–10 min: 0–30% B; 10–70 min: 30–45% B	1.0	*a*
		Neutral buffer system		
Aquapore RP-300 (RP-8) C_8, 10-μm particle size, 4.6 × 250 mm	A: 50 m*M* Tris-HCl, pH 7.5 B: 60:40 (v/v), acetonitrile:50 m*M* Tris-HCl, pH 7.5	One-step linear gradient: 0–40 min: 5–20% B	1.0	*a*
Lichrosorb RP-300 (RP-18) C_{18} (Bischoff Analysentechnik, Leonberg, Germany), 10-μm particle size, 2.1 × 250 mm	A: 25 m*M* Tris-HCl, pH 7.5 B: 60:40 (v/v) acetonitrile:25 m*M* Tris-HCl, pH 7.5	One-step linear gradient: 0–100 min: 0–40% B	2.0	*b*
Ultrapore RPSC C_3 (Beckman, Berkeley, CA), 4.6 × 75 mm	A: 50 m*M* Tris-HCl, pH 7.2 B: 80:20 (v/v) *n*-propanol:50 m*M* Tris-HCl, pH 7.2	One-step linear gradient: 0–100 min: 0–40% B	1.0	*c*

μBondapak C_{18} (Waters/Millipore, Milford, MA), 10-μm particle size, 8 × 100 mm (radial compression)	A: 10 mM NaH_2PO_4, pH 7.0 B: 60:40 (v/v) acetonitrile:10 mM NaH_2PO_4, pH 7.0	Two-step linear gradient: 0–5 min: 0–10% B; 5–20 min: 10–25% B	3.0	*d,e*
Lichrospher 100 (RP-8) C_8 (Merek, Rahway, NJ), 5-μm particle size, 4 × 125 mm	A: 50 mM Tris-HCl, pH 7.0 B: Methanol	One-step linear gradient: 0–20 min: 0.5–30% B	1.5	*f*
Bio-Rad RP-318 (Bio-Rad, Richmond, CA) C_{18}, 5-μm particle size, 10 × 250 mm	A: 10 mM NaH_2PO_4, pH 7.0 B: 60:40 (v/v) acetonitrile:10 mM NaH_2PO_4, pH 7.0	Two-step linear gradient: 0–10 min: 0–10% B; 10–20 min: 10–20% B	2.0	*g*
Pecosphere 3C C_{18} (Perkin-Elmer, Norwalk, CT), 3-μm particle size, 4.6 × 83 mm	A: 10 mM NaH_2PO_4, pH 7.0 B: 60:40 (v/v) acetonitrile:10 mM NaH_2PO_4, pH 7.0	Two-step linear gradient: 0–5 min: 0–10% B; 5–20 min: 10–25% B	1.0	*h*

[a] S. Klauser, J. H. R. Kägi, and K. J. Wilson, *Biochem. J.* **209,** 71 (1983).
[b] P. E. Hunziker and J. H. R. Kägi, *Biochem. J.* **231,** 375 (1985).
[c] K. Suzuki, H. Sunaga, Y. Aoki, and M. Yamamura, *J. Chromatogr.* **281,** 159 (1983).
[d] M. P. Richards and N. C. Steele, *J. Chromatogr.* **402,** 243 (1987).
[e] M. P. Richards, *J. Chromatogr.* **482,** 87 (1989).
[f] H. Van Beek and A. J. Baars, *J. Chromatogr.* **442,** 345 (1988).
[g] J. Hildago, A. Dingman, and J. S. Garvey, *Rev. Esp. Fisiol.* **45,** 255 (1989).
[h] D. E. Laurin, K. C. Klasing, and R. L. Baldwin, *BioChromatography* **4,** 254 (1989).

column is maintained at ambient temperature and eluted at a flow rate of 3.0 ml/min. Steel columns containing different types of silica-based reversed-phase packings including C_{18}, C_8, and C_3, with particle sizes ranging from 3 to 10 μm, have also been used to separate MT isoforms (Table I). However, there are several advantages to using radially compressed cartridge columns to effect the separation of MT isoforms. The flow through a polyethylene cartridge column can be maintained at a relatively high rate (3.0 ml/min) without sacrificing column efficiency. Therefore, the time required to separate the MT isoforms is reduced compared to steel columns, which are typically run at lower flow rates (i.e., 1.0–2.0 ml/min). Moreover, the higher flow rate is better suited to detection methodology involving AAS (see Detection by Atomic Absorbance, below). Also, cartridge columns are generally more durable and cost less than steel columns.

Gradient Elution

The mobile phase used to elute MT isoforms from the reversed-phase column consists of an equilibration buffer (buffer A) such as 10 m*M* NaH_2PO_4, pH 7.0, and an elution buffer (buffer B) that contains an organic modifier such as 60% acetonitrile dissolved in buffer A. Metallothioneins are eluted with a gradient formed by mixing an increasing amount of buffer B with buffer A over a specified period of time. For example, MT isoforms can be eluted from the μBondapak C_{18} radially compressed cartridge column with a two-step, linear gradient consisting of 0–10% B from 0 to 5 min followed by 10–25% B from 5 to 20 min at a flow rate of 3.0 ml/min.[13,17] Following gradient elution of MT, the column is purged with 100% B for 5 min and equilibrated with 100% A for 10 min prior to injection of the next sample. The initial step of this two-part gradient is particularly useful when isolating MT isoforms from crude mixtures such as tissue extracts because it aids in removing extraneous protein prior to the elution of the MT isoforms. By varying the steepness (increase in percentage B over time) and duration of the gradient, it is possible to effect significant changes in the resolution of individual MT isoforms. Factors such as the complexity of the mixture containing MT and the elution characteristics of the individual MT isoforms will dictate the type of gradient to be employed.

In addition to acetonitrile, other organic modifiers such as methanol and *n*-propanol have been included in the mobile phase used to elute MT isoforms from reversed-phase columns (Table I). One reported drawback to the use of methanol as an organic modifier is that it has been implicated in the incomplete resolution of rat liver MT isoforms.[16] The use of *n*-propanol apparently gives comparable resolution of MT isoforms to that achieved with acetonitrile,[11] although it has been suggested to reduce the useful lifetime of the reversed-phase column.[16]

Both the equilibration and elution buffers should be filtered and degassed by vacuum filtration through a 0.45-μm pore membrane (i.e., type HA membrane; Millipore Corp.) prior to use in RP-HPLC. Also, sparging both buffers with argon or helium to remove dissolved oxygen may prove to be a useful precaution to attain a reducing environment during RP-HPLC. This is an important consideration in light of the observed susceptibility of MTs to oxidation during isolation.[18]

The choice of buffer used to elute the MT isoforms should be considered carefully. Both acidic and neutral buffer systems have been employed to separate MT isoforms by RP-HPLC (Table I). The acidic buffer system most commonly used consists of 0.1% (w/v) trifluoroacetic acid (TFA) with acetonitrile as the organic modifier. Because of the low pH (ca. 2.0) of the TFA buffer, metals will dissociate from the MT peptide chain and apothionein will be detected.[12] There are advantages to using the TFA buffer system: (1) It is a volatile buffer, so that collected fractions containing purified MT isoforms can be lyophilized in a salt-free form. This is particularly useful if subsequent amino acid composition or sequence analyses are to be performed on the individual MT isoforms. (2) Any differences in chromatographic properties can be directly attributed to differences in the peptide chain and not to differences in metal binding. There are also disadvantages to using the TFA buffer system: (1) Because of the low pH of the TFA buffer system, there is a substantial decrease in the absorbance (at 200–220 nm) per microgram of protein due to a disruption of the metal–thiolate complexes characteristic of MT that contribute substantially to the absorbance in this range.[12] This results in a reduction in the sensitivity of detection methods based on UV absorbance. (2) Because the metals dissociate from the MT protein, it is not possible to derive any information about the metal composition of the individual MT isoforms. (3) The recovery of MT protein from reversed-phase columns eluted with an acidic buffer system is apparently lower than that achieved under neutral conditions.[12] (4) Since TFA can contribute substantially to UV absorbance (at 200–220 nm), highly purified TFA (sequanal grade) must be used so as to minimize its contribution to baseline UV absorbance.

Alternatively, neutral buffer systems can be applied to the separation of MT isoforms (Table I). Because silica-based reversed-phase columns are unstable above pH 8.0, the pH of the buffers used should, in practice, not exceed 7.5. Several investigators have employed 50 m*M* Tris, pH 7.0–7.5, with acetonitrile, methanol, or *n*-propanol to effect the separation of MT isoforms (Table I). The advantage of Tris buffer is that it is compatible with the most commonly used methods for preparation of tissue cytosol that employ Tris-buffered solutions. However, Tris does contribute significantly to the absorbance in the 200 to 220-nm range.[15] Therefore, sensitivity will be reduced somewhat because of elevated baseline UV absorbance. Sodium phosphate, on the other hand, is an excellent buffer because it exhibits negligible absorbance in the 200 to 220-nm range. Neutral buffer systems, in general, offer the user several advantages over acidic ones: (1) Because the metal complement of MT remains largely intact at neutral pH, the native peptide conformation as well as the original metal composition of the MT molecule would be expected to be preserved. (2) Any studies involving the metal composition of MT isoforms require the use of a neutral buffer. Therefore, the neutral buffer can be used when either UV or atomic absorbance modes of detection are employed (see below). (3) Because the metal–thiolate complexes remain intact at neutral pH, the UV absorbance per microgram of MT protein is substantially higher than that observed when acidic buffers are employed. As a result, neutral buffer systems offer a higher level of sensitivity for the UV absorbance detection of MT isoforms. (4) At neutral pH, less organic solvent is required to elute the MT isoforms than under acidic conditions. (5) Recoveries are generally quite high (>90%) when a neutral buffer is employed.

Thus, the choice of a RP-HPLC column, an acidic versus a neutral buffer system, the type of organic modifier for the mobile phase, and the type of gradient are all to be considered when formulating a strategy to isolate MT isoforms from a particular sample. Since no single configuration will be applicable to all sample types, modifications in these parameters will have to be made routinely.

Detection by Ultraviolet Absorbance

Absorbance in the UV range (200–220 nm) can be used as both a qualitative as well as a quantitative means of detection of MT isoforms. By monitoring absorbance at 214 nm it is possible to demonstrate differences in the isoform pattern for MTs purified from the livers of different eukaryotic species (Fig. 1). Another commonly used wavelength for monitoring MT isoforms separated by RP-HPLC is 220 nm.[12] The choice of wavelength will depend on the type of detector available, the nature of the mobile phase, and the specific peak of UV absorbance for each MT isoform species.

It is important to point out that the order of elution of the two major MT isoforms (MT-1 and MT-2) is reversed at neutral pH for RP-HPLC. That is, the MT-2 isoform elutes prior to the MT-1 isoform. This could lead to some confusion with existing nomenclature, which is based on the order of elution from an anion-exchange column at neutral pH or above.

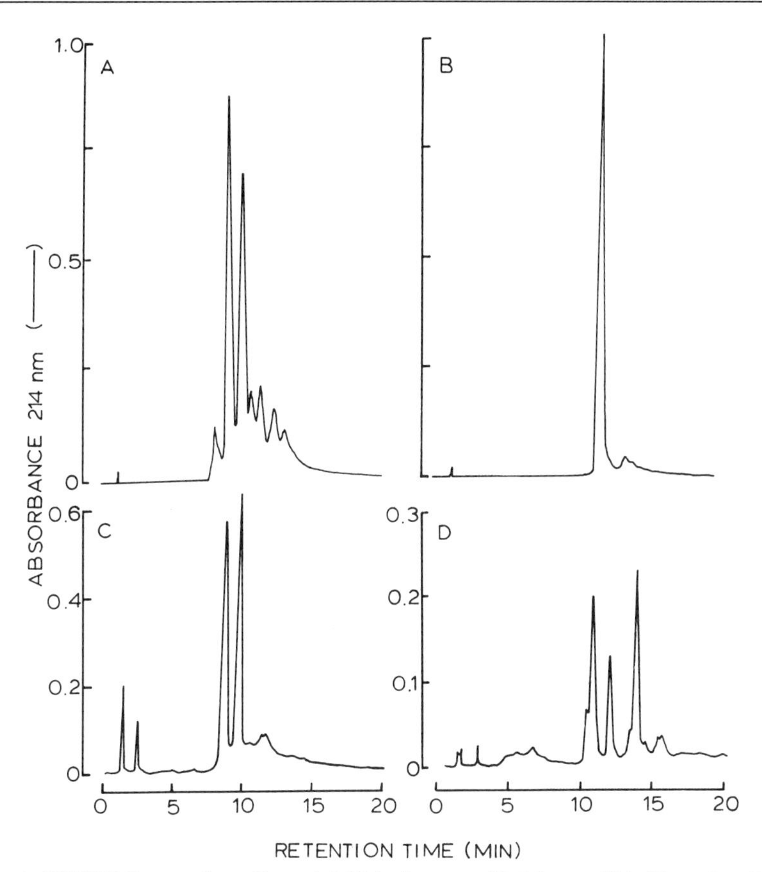

FIG. 1. RP-HPLC separation of hepatic MT isoforms purified from rabbit (A), turkey (B), rat (C), and pig (D) liver. The separations were performed with a μBondapak C_{18} (10-μm particle size) radially compressed cartridge column using a two-step linear gradient (0–6% from 0 to 5 min and 6–15% from 5 to 20 min) of acetonitrile in 10 mM NaH_2PO_4, pH 7.0, at ambient temperature and at a flow rate of 3.0 ml/min. The column effluent was monitored for UV absorbance (214 nm). Injections were 100 μl containing 50–100 μg of purified MT in 10 mM NaH_2PO_4, pH 7.0.[13]

Detection by Atomic Absorbance

As for UV absorbance, detection of MT-associated metals by AAS can be both qualitative as well as quantitative. An AAS can be used to survey the metal composition of individual MT isoforms separated by RP-HPLC (Fig. 2). This is particularly useful for determining metal distribution among individual MT isoforms as well as selective accumulation of metal by a particular isoform species. The use of a neutral buffer is essential, not

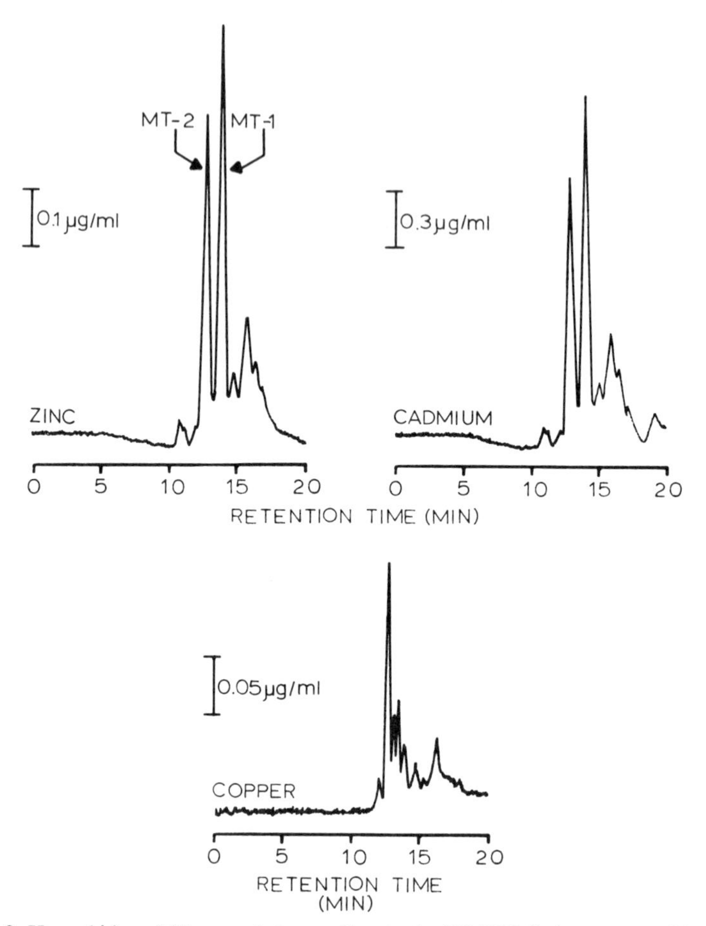

FIG. 2. Horse kidney MT separated according to the RP-HPLC chromatographic conditions described in the legend to Fig. 1, except that the effluent from the column was fed directly into the nebulizer of an atomic absorption spectrophotometer set to detect the atomic absorbances for zinc (213.9 nm), cadmium (228.8 nm), and copper (324.8 nm). A separate injection of 100 μg of MT in 100 μl of 10 mM NaH_2PO_4, pH 7.0, was made for the determination of each metal. The vertical bars represent the level of atomic absorbance for the indicated concentration of an aqueous metal standard solution. The peaks corresponding to the two major MT isoform species (MT-1 and MT-2) are indicated.[17]

only in the preparation of the sample to be analyzed, but also in the elution of MT from the RP-HPLC column for subsequent detection by AAS.

Quantification of Metallothionein Isoforms

Because the integrated peak area for UV absorbance as well as that for atomic absorbance is a linear function of the quantity of MT (determined gravimetrically) or the quantity of MT-bound metal (determined by AAS) injected onto the reversed-phase column, it is possible to quantify MT isoforms and their associated metals using previously established standard curves (Fig. 3). To establish a standard curve, it is recommended that a source of MT composed of a singular isoform species be used. The use of standards composed of multiple species is not recommended because questions concerning differential recovery and detection of each species can complicate the quantification procedure. Since the commercially available sources of MT (including the prepurified MT isoforms from rabbit liver) are composed of multiple species,[13] it is necessary to isolate a singular isoform for this purpose. This can be easily done using RP-HPLC itself. We routinely use an avian MT source (turkey liver MT) because it is composed of a single predominant isoform species (MT-2) that is easily purified.[13,17] Moreover, as shown in Fig. 3, the same standard (MT-2) can be used to quantify MT (based on absorbance at 214 nm) and MT-bound zinc (based on atomic absorbance). Rat MT isoforms (MT-1 or MT-2) are also suitable as standards since both isoforms are composed of a singular species.[12] It is recommended that metal-containing MT standards be used for quantification of MT-bound metal as opposed to aqueous metal standards. The use of an MT-bound metal standard versus an aqueous one takes into account any effects of protein-binding on the detection and subsequent quantification of the metal following RP-HPLC. The specific metal content of a prospective MT standard can be accurately determined by AAS prior to its use for RP-HPLC. It is not yet known if separate homologous standard curves are necessary to accurately quantify each MT isoform and its associated metals. Although it is certainly feasible to establish a separate standard curve for each isoform, additional work is required to evaluate the necessity of this approach.

It may be possible to use the same MT standard containing more than one metal to quantify more than one metal bound to MT in a particular sample. Ideally, an "enriched" MT standard should be prepared for each metal to be assayed. This can be done *in vivo* by injection of the metal into a particular animal species with subsequent purification of the metal-containing MT isoforms. Alternatively, it may also be possible to produce "customized" MT standards *in vitro* by first removing all metals from MT by lowering the pH followed by reconstitution at neutral pH with the appropriate metal(s). The former *(in vivo)* approach has been successfully employed to generate turkey liver MT-2 standard,[17] whereas the feasibility of the latter *(in vitro)* approach remains to be determined. Moreover, the utility of "multipurpose" MT standards to quantify more than one MT-bound metal also remains to be established.

Using a neutral buffer (NaH_2PO_4, pH 7.0) and a purified MT standard (turkey liver MT-2), we have estimated the lower limit of detection by UV

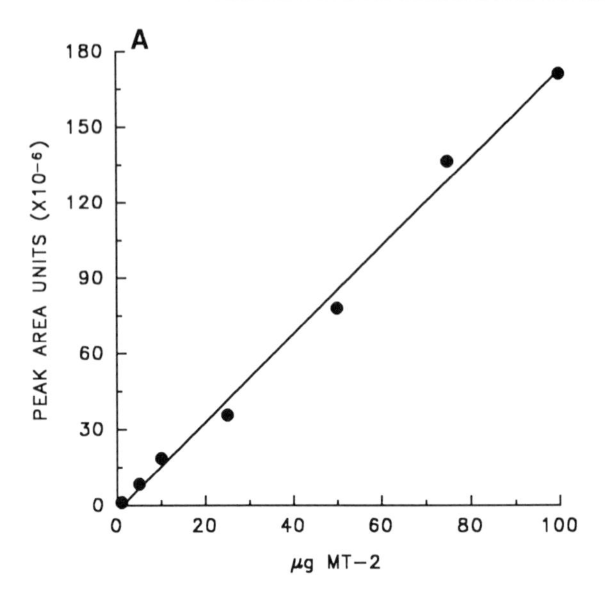

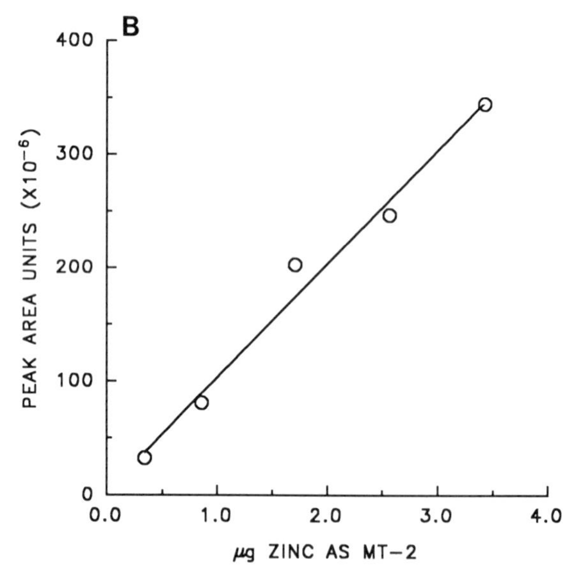

FIG. 3. Standard curves showing (A) the linear relationship between the quantity of MT-2 injected onto the column and the integrated peak area for UV absorbance at 214 nm ($r^2 = 0.997$) and (B) the linear relationship between the quantity of zinc (bound to MT-2) injected onto the column and the integrated peak area for the atomic absorbance (at 213.9 nm) for zinc ($r^2 = 0.991$). The RP-HPLC chromatographic conditions used were the same as those specified in the legend to Fig. 1. Purified turkey liver MT-2 served as the standard. Samples of MT were dissolved in 100 μl of 10 mM NaH_2PO_4, pH 7.0.[13,17]

TABLE II
RECOVERY ESTIMATES OF METALLOTHIONEIN FROM RP-HPLC WITH NEUTRAL BUFFER SYSTEM[a,b]

Method and Isoform	Species	Recovery (%)
[^{35}S]Cysteine-labeled MT		
MT-1	Rat	92 ± 7
MT-2	Rat	98 ± 5
MT-bound zinc		
MT-1	Turkey	91 ± 2
MT-2	Turkey	97 ± 7
A_{214} peak area units		
MT-1	Rat	96 ± 3
MT-2	Rat	97 ± 2
Overall estimates[c]		
MT-1	Rat/turkey	93 ± 1
MT-2	Rat/turkey	97 ± 0.3
MT[d]	Rat/turkey	95 ± 1

[a] The following RP-HPLC conditions were used: (1) a μBondapak C_{18} cartridge column housed in a Z-module radial compression device (Waters/Millipore, Milford, MA); (2) buffer A: 10 m*M* NaH_2PO_4, pH 7.0; buffer B: 60% acetonitrile in buffer A; (3) a two-step linear gradient: 0–10% B from 0 to 5 min and 10–25% B from 5 to 20 min; (4) flow rate: 3.0 ml/min; (5) ambient temperature; (6) injection volume: 100 μl.

[b] Values represent the mean ± SEM of three observations.

[c] Values represent the mean ± SEM for three methods of recovery estimation.

[d] Values represent the mean ± SEM for both isoforms combined.

absorbance (214 nm) to be approximately 1 μg of MT injected onto the reversed-phase column. Similar results have been reported by others.[20] A lower limit of detection by AAS for MT-bound zinc was estimated to be 0.1 μg. This is consistent with the lower detection limit determined for UV absorbance (1 μg), assuming that 1 μg of MT fully saturated with zinc would bind approximately 0.07 μg. These limits of detection will depend on such factors as the age and condition of the analytical column, the type and sophistication of the instrumentation used, the nature of the sample containing MT, and the type of buffer used to elute MT from the reversed-phase column.

Recovery of Metallothionein Isoforms

The recovery of MT from the reversed-phase column is an important consideration in the validation of RP-HPLC for the quantification of MT. An example of three different methods used to assess recovery of MT from RP-HPLC is summarized in Table II.

[20] D. E. Laurin, K. C. Klasing, and R. L. Baldwin, *BioChromatography* **4,** 254 (1989).

1. Purified rat liver MT-1 and MT-2, labeled *in vivo* with [^{35}S]cysteine, were individually injected onto the column (μBondapak C_{18} radially compressed cartridge), eluted with a neutral buffer (NaH_2PO_4, pH 7.0) as described above (Gradient Elution), and 1-ml fractions were collected and analyzed for ^{35}S using liquid scintillation counting. The recovery (%) is determined as follows:

$$\text{Percentage} = \left(\frac{\text{counts recovered in MT isoform peak fractions}}{\text{counts injected onto column}}\right) \times 100$$

2. Purified turkey liver MT-1 and MT-2 isoforms were analyzed for their zinc contents using AAS. A known amount of zinc (either as MT-1 or MT-2) was injected onto the column, eluted as described above, and fractions (1 ml) collected and analyzed for zinc using AAS. The recovery (%) is determined as follows:

$$\text{Percentage} = \left(\frac{\mu\text{g zinc in MT isoform peak fractions}}{\mu\text{g zinc injected onto column}}\right) \times 100$$

3. Samples of rat liver MT-1 and MT-2 were injected onto the column and eluted as described above. Absorbance at 214 nm was monitored and the MT peak area determined by integration. A second injection was made of an identical quantity to the first injection containing a known amount of peak area units. The recovery (%) is determined as follows:

$$\text{Percentage} = \left(\frac{\text{integrated area of MT isoform peak}}{\text{area units injected onto column}}\right) \times 100$$

An overall estimate of recovery obtained by an average of all three methods is 95% (Table II), which agrees well with recoveries reported by others using neutral buffer systems with RP-HPLC.[11,16,20] Klauser *et al.*[12] reported a considerably lower recovery (65%) for MT isoproteins from RP-HPLC using an acidic buffer system. This discrepancy indicates that the type of column and buffer system used for RP-HPLC, as well as the nature of the sample matrix, may affect the recovery of MT. Clearly, such considerations need to be taken into account when estimating MT recovery from RP-HPLC.

Applications of Reversed-Phase High-Performance Liquid Chromatography

Reversed-phase HPLC can be applied to the analysis of a wide range of sample types for their content of MT isoforms and MT-associated metals. It is a useful technique to assess the complexity of the isoform pattern for

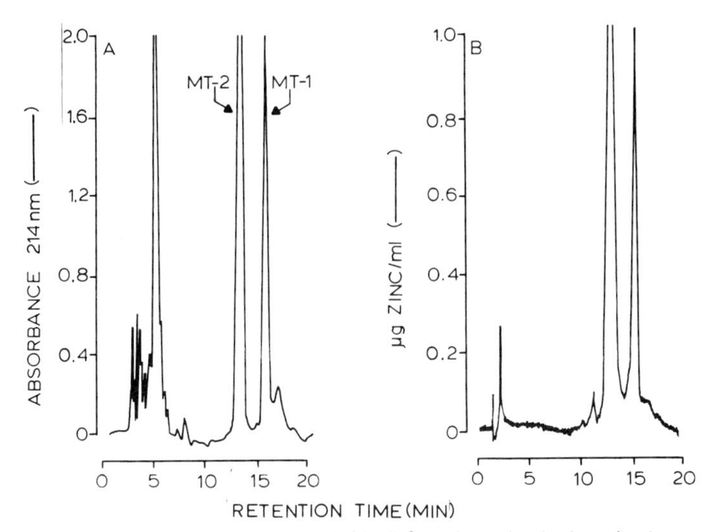

FIG. 4. Turkey liver heat-treated cytosol (100 μl) from hens that had received two consecutive intraperitoneal injections of zinc (10 mg/kg body weight at 24-hr intervals) subjected to RP-HPLC with on-line detection for UV absorbance at 214 nm (A) and atomic absorbance for zinc at 213.9 nm (B). The RP-HPLC chromatographic conditions were the same as those described in the legend for Fig. 1. The two major MT isoform species are designated as MT-1 (eluting at 14.83 min) and MT-2 (eluting at 12.93 min).[17]

MTs purified from the livers of a variety of animal species (Fig. 1). Furthermore, it is an excellent method to assess quality of as well as to purify MT to be used as a standard reference material. The use of a neutral buffer (Tris) followed by rechromatography with an acidic buffer (TFA) has proven to be a useful two-step RP-HPLC procedure for the characterization of individual isoforms of MT isolated from human liver.[21,22] More complex samples such as tissue extracts (cytosol) can be directly analyzed by RP-HPLC both for MT isoforms as well as for MT-bound metal (Fig. 4). It has been estimated that as little as 1 g of organ tissue is required to simultaneously quantitate MT isoforms and MT-bound metals using RP-HPLC,[16] whereas Hunziker and Kägi[21] found that 0.1 g of human liver is sufficient to characterize the pattern of MT isoforms. Moreover, extracts of cultured cells can be assayed for their content of MT isoforms using RP-HPLC.[17,20] Figure 5 presents data for turkey embryo hepatocytes cultured in the presence of 0–50 μM added zinc for 24 hr and then assayed

[21] P. E. Hunziker and J. H. R. Kägi, *Biochem. J.* **231,** 375 (1985).
[22] P. E. Hunziker and J. H. R. Kägi, *Experientia, Suppl.* **52,** 257 (1987).

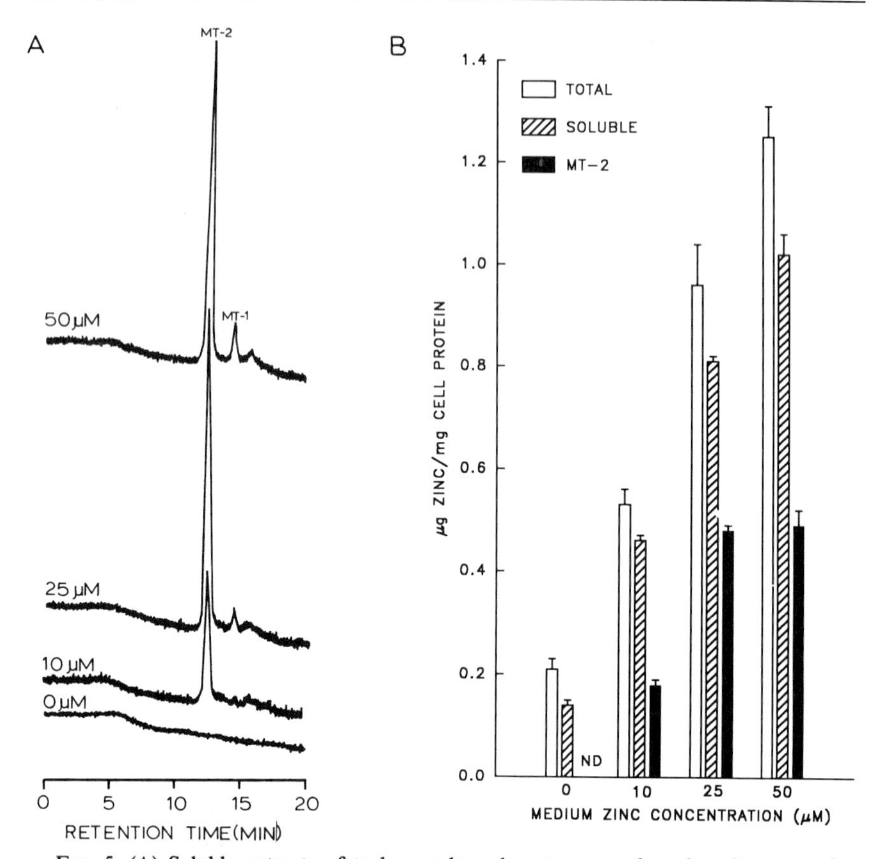

FIG. 5. (A) Soluble extracts of turkey embryo hepatocytes cultured under serum-free conditions subjected to RP-HPLC with atomic absorbance detection of MT-bound zinc at 213.9 nm. The RP-HPLC chromatographic conditions were the same as those described in the legend to Fig. 1. Hepatocytes were cultured in the presence of 0, 10, 25, and 50 μM supplemental zinc in the medium for 24 hr. (B) The cellular distribution of zinc in the intact hepatocyte (open bars), the soluble extract (striped bars) as determined by AAS, and the amount of zinc bound specifically to MT-2 (black bars) as determined by RP-HPLC with on-line AAS detection. Each value, expressed as micrograms zinc per milligram total cellular protein, represents the mean ± SEM of triplicate plates of cells. ND, A value below the limits of AAS detection for the RP-HPLC technique.[17]

for total cellular and cytosolic (soluble) zinc by AAS and for MT-2-bound zinc content using RP-HPLC with AAS detection. Thus, RP-HPLC is a technique with a great deal of versatility for the analysis of MT isoforms and their associated metals in a variety of sample matrices.

Advantages and Limitations of Reversed-Phase High-Performance Liquid Chromatography

There are several advantages of using RP-HPLC to assay a sample for MT isoforms. The amount of sample required for the RP-HPLC analysis is considerably less than that required for classical chromatographic techniques. It is possible to achieve a reasonably high level of detection sensitivity, both for MT (ca. 10^{-6} g) and for MT-bound metal (ca. 10^{-7} g zinc) with this technique. Moreover, because of its superior resolving capability, RP-HPLC can, in a single step, be used to rapidly and reproducibly isolate and quantify individual MT isoforms from a variety of sample matrices, including tissue extracts.

There are, however, several limitations of RP-HPLC that should be considered. One important limitation of RP-HPLC that has been encountered concerns the detection of copper-induced or copper-enriched MTs. It has been reported that MT isoforms containing substantial amounts of copper cannot be adequately separated and detected using RP-HPLC at neutral pH.[17,22] Considering that copper-enriched MTs are inherently unstable, are prone to oxidation, and require anaerobic conditions for their isolation, this finding is not really surprising. It may be possible to quantitate copper-induced MT isoforms by RP-HPLC using an acidic (TFA) buffer system; however, this approach would preclude any determination by AAS detection of the metal content and distribution among different isoforms. When copper is not the predominant metal bound to MT, RP-HPLC at neutral pH may be a useful method to detect and quantitate MT as well as MT-bound copper. The fact that copper could be detected in the sample of horse kidney MT analyzed (Fig. 2) may indicate that MTs stabilized by the presence of zinc and/or cadmium can be analyzed for copper as well. Glennas *et al.*[23] have also found that RP-HPLC was not appropriate for the isolation and detection of gold-containing MT from human epithelial cells exposed to aurofin. Thus the metal composition of the MTs to be isolated and quantified will determine whether or not RP-HPLC is an appropriate analytical method.

Because of the superior resolution achieved with RP-HPLC, it is possible to obtain MT species that may, in fact, not be true isoforms. Species such as those resulting from oxidation (i.e., dimers and polymerized products),[24] different conformations, or differing metal composition (metalloforms) are possible. Therefore, careful characterization of each peak must be performed to determine the exact nature of the species isolated. The

[23] A. Glennas, P. E. Hunziker, J. S. Garvey, J. H. R. Kägi, and H. E. Rugstad, *Biochem. Pharmacol.* **35,** 2033 (1986).

[24] K. T. Suzuki and M. Yamamura, *Biochem. Pharmacol.* **29,** 689 (1980).

characterization of such species by RP-HPLC in both acidic and neutral buffer systems is a useful way to help establish their identity as isoforms.[12,21,22] However, it is possible for individual MT isoforms to coelute during RP-HPLC with an acidic buffer (TFA).[25]

Finally, despite the relatively high level of sensitivity of detection, the use of RP-HPLC to isolate and quantify MT isoforms is, in practice, only applicable to samples containing significant amounts of MT. Moreover, this technique in its current form is not applicable to the quantitation of MT isoforms in samples such as extracellular fluids (i.e., plasma, bile, urine, etc.) or extracts from tissues in which MT levels are quite low (i.e., $< 10\ \mu g/g$ fresh tissue).

Future Directions

Although RP-HPLC has been used for a number of years to isolate and quantify MT isoforms, new modifications are required to extend the utility of RP-HPLC. The following are some suggested modifications intended to increase the sensitivity of RP-HPLC for the detection and quantification of MT isoforms.

Sample Preparation

Acetone or ammonium sulfate fractionation and heat treatment can be used prior to RP-HPLC to reduce the amount of extraneous protein injected onto the column. Fractionation techniques such as these may also be used to enrich a sample for a particular component such as MT by selective precipitation of proteins. Moreover, reversed-phase column enrichment techniques have been employed to detect peptides present in trace amounts by loading larger volumes onto the column. Peptides such as MT would be concentrated at the top of the column prior to elution with an appropriate gradient. This procedure could also be done prior to RP-HPLC by using a solid-phase extraction device such as a Sep-Pak cartridge (Waters/Millipore). Such techniques may be useful for the RP-HPLC determination of MT isoforms in tissue extracts or fluids in which MT levels are low.

Column Technology

Improvements in column packing materials continue to add new matrices to the family of reversed-phase packings. Resins that are more resistant to basic pH (>8.0) offer new opportunities to modify the RP-HPLC

[25] L. Y. Lin, W. C. Lin, and P. C. Huang, *Biochim. Biophys. Acta* **1037,** 248 (1990).

procedure and perhaps improve the resolution of MT isoforms. The use of microbore column technology may prove to be a valuable addition for the quantification of MT isoforms present in low amounts. This technology offers the advantages of high recovery, less dilution of sample, and higher sensitivity compared to conventional RP-HPLC.

Detection Methodology

New methods are needed to extend the lower limit of detection for MT isolated by RP-HPLC, e.g., fluorescence detection following derivatization (either precolumn or postcolumn) of MT with an appropriate fluorescent label. A number of compounds for labeling specific groups on proteins are available. Compounds such as fluorescamine and *o*-phthalaldehyde are used to generally label amino groups on peptides, whereas a number of sulfhydryl-specific fluorescent compounds are also available. If the appropriate techniques can be worked out, the use of these compounds would greatly increase the sensitivity of detection of MT isoforms. Another possibility for increasing the detection sensitivity involves the application of on-line electrochemical detection methodology to the RP-HPLC separation of MTs. If feasible, such an approach would not only offer increased sensitivity but might also offer information concerning the redox state of the sulfhydryl groups contained in the MT molecule. Finally, the application of flow-through radioactivity detectors for the detection of radiolabeled (either the peptide or metal) MTs may prove to be a useful alternative to UV and AAS detection methods.

Standardization

Improvements in standardization will increase the accuracy of quantitative estimates of MT isoforms resolved by RP-HPLC. Because recoveries from different sample matrices can vary, the need for internal standards is evident. It may be possible to purify heterologous MT isoform species using RP-HPLC that could be used as internal standards, assuming that their retention times under the specific conditions employed for RP-HPLC are sufficiently different from those MT isoforms to be quantified. The ability to add an internal MT standard previously characterized by RP-HPLC would permit correction for losses encountered during sample preparation and RP-HPLC. Clearly, more work needs to be done to assess the feasibility of such an approach.

Combination of Quantification Techniques

The possibilities of combining other quantification techniques with RP-HPLC should be explored. For example, the combination of the cadmium saturation assay coupled with anion-exchange HPLC separation of MT isoforms has proven to be a viable means of quantification of individual MT isoforms in the rat.[15] A similar combination with RP-HPLC has yet to be explored.

Concluding Remarks

Reversed-phase HPLC is clearly a technique that has a great deal of promise for studying aspects of MT, such as its synthesis, degradation, metal composition, structure, and function. It is a rapid, highly reproducible, and sensitive technique well suited to the purification, characterization, and quantification of MT isoforms and their associated metals. The amount of sample required is considerably less than that required for classical chromatographic techniques used to isolate MT. With appropriate improvements it may be possible to further extend the utility of the analytical RP-HPLC technique to cover a range of applications from the purification of MT isoforms on a preparative scale to the isolation and quantification of MT isoforms present at low levels in tissues and physiological fluids.

Acknowledgment

Mention of a trade name, proprietary product, or specific equipment does not constitute a guarantee or warranty by the U.S. Department of Agriculture and does not imply its approval to the exclusion of other suitable products.

[26] Purification of Metallothionein by Fast Protein Liquid Chromatography

By PER-ERIC OLSSON

Introduction

Isolation of metallothionein (MT) and its isoforms usually involves a three-step chromatographic procedure involving a size-exclusion step, an ion-exchange step, and a buffer-desalting step using a volatile buffer prior to lyophilization and amino acid analysis.[1] Characteristics used to isolate MTs by this procedure include a molecular weight of approximately

[1] P.-E. Olsson and C. Haux, *Inorg. Chim. Acta, Bioinorg. Chem.* **107,** 67 (1985).

6000[2,3] and an acid isoelectric point resulting in binding to anion-exchange resins.[4] Metallothioneins are thus routinely isolated from Sephadex G-75 gel-permeation fractions containing protein with a molecular weight of approximately 10,000. Pooled fractions are further separated using anion-exchange chromatography from which different MT isoforms can be eluted with a salt gradient.[5] Finally the separated isoforms of MT are subjected to a desalting step on Sephadex G-50 equilibrated with a volatile mobile phase such as an ammonium buffer to facilitate lyophilization prior to amino acid analysis.[1]

Conventional procedures for the isolation of MTs, using the above methodologies, have in some instances failed to separate iso-MTs[6,7] or have been so time consuming as to be unsuitable for analytical purposes. Two alternative chromatographic systems have been developed to reduce isolation time and enhance the resolution of different iso-MTs.[8,9] One system utilizes reversed-phase columns coupled to high-performance liquid chromatography (HPLC)[8,10] and the other system involves ion-exchange chromatography using fast-protein liquid chromatography (FPLC)[9] or HPLC.[8] Both systems have been used to study the biochemical properties of MTs, achieve enhanced separation, and for antibody production.[11-16] In the present chapter the separation of teleost MT by FPLC using the Mono Q anion-exchange media is described and compared to conventional anion-exchange chromatography using a DEAE-Sephadex A-25 resin.

[2] J. H. R. Kägi, S. R. Himmelhoch, P. D. Whanger, J. L. Betune, and B. L. Vallee, *J. Biol. Chem.* **249,** 3537 (1974).

[3] W. F. Furey, A. H. Robbins, L. L. Clancy, D. R. Winge, B. C. Wang, and C. D. Stout, *Science* **231,** 704 (1986).

[4] M. Webb, ed., *in* "The Chemistry, Biochemistry and Biology of Cadmium." p. 195. Elsevier/North-Holland, 1979.

[5] F. E. Reginer, this series, Vol. 104, p. 170.

[6] P.-E. Olsson and C. Haux, *Aquat. Toxicol.* **9,** 231 (1986).

[7] J. Overnell and T. L. Coombs, *Biochem. J.* **183,** 277 (1979).

[8] K. T. Suzuki, H. Sunaga, Y. Aoki, and M. Yamamura, *J. Chromatogr.* **281,** 159 (1983).

[9] P.-E. Olsson and C. Hogstrand, *J. Chromatogr.* **402,** 293 (1987).

[10] R. W. Olafson, *Int. J. Pept. Protein Res.* **24,** 303 (1984).

[11] H. Neuberger and M. L. Moniek, *in* "Trace Element Analytical Chemistry in Medicine and Biology Volume 5" (P. Brätter and P. Schramel, eds.), p. 174. de Gruyer, Berlin and New York, 1988.

[12] M. W. Brown, D. Shurben, J. F. de L. G. Solbe, A. Cryer, and J. Kay, *Comp. Biochem. Physiol. C: Comp. Pharmacol. Toxicol.* **87C,** 65 (1987).

[13] S. Klauser, J. H. R. Kägi, and K. J. Wilson, *Biochem. J.* **209,** 71 (1983).

[14] C. G. Norey, W. E. Lees, B. M. Darke, J. M. Stark, T. S. Baker, A. Cryer, and J. Kay, *Comp. Biochem. Physiol. B: Comp. Biochem.* **95B,** 597 (1990).

[15] K. T. Suzuki, *Anal. Biochem.* **102,** 31 (1980).

[16] R. W. Olafson, W. D. McCubbin, and C. M. Kay, *Biochem. J.* **251,** 691 (1988).

Rationale for Column Selection

The most important question confronting the investigator contemplating chromatography is that of the nature of the column to be used. By including the FPLC system the choice is broadened beyond conventional Sephadex and DEAE columns to include the Superose gel-exclusion columns as well as Mono Q anion-exchange columns. Since Sephadex G-75 has a working range of about 1×10^3 to 5×10^4 Da and Superose 12 has a working range from 1×10^3 to 3×10^5 Da, chromatography using Sephadex usually gives better separation of MTs from both high-molecular-weight (HMW) material and low-molecular-weight (LMW) material subjected to matrix inclusion. On the other hand, due to the even particle size (30 μm) of the Superose it achieves shorter elution times than the Sephadex G-75 fine media (particle size 20–50 μm).

Protein separation by ion-exchange chromatography depends on the differential charge between different proteins at a given pH. Although the total charges of the molecules to be separated are important, ion-exchange surfaces can readily recognize the differences between two proteins that are identical in every respect except charge distribution.[5] Protein retention on an ion-exchange column is the result of electrostatic interactions in which the retention increases in proportion to the charge density on both the ion-exchange matrix and the protein. In contrast to a weak ion-exchange matrix (such as DEAE) a strong ion-exchange matrix (such as Mono Q) remains fully ionized over the entire pH range from 2 to 12. The strong ion-exchange resins are also thought to exhibit slightly better selectivity than the weak ones.[5] The Mono Q resin is based on Monobeads that bind negatively charged components through quaternary amine groups.[17] The very narrow particle size distribution of the Monobeads results in low back pressure, allowing short elution times.

Materials

Chemicals. The Sephadex G-75, DEAE-Sephadex A-25, Superose 12, and Mono Q columns were obtained from Pharmacia LKB Biotechnology (Uppsala, Sweden). YM2 membranes were obtained from Amicon (Danvers, MA). Tris was obtained from Merck (Darmstadt, Germany) and ammonium bicarbonate from Sigma (St. Louis, MO). All other chemicals were analytical grade.

Apparatus. Anion-exchange chromatography was carried out on an FPLC system equipped with two high-precision P-500 pumps, a liquid

[17] "FPLC Ion Exchange and Chromatofocusing Principles and Methods." Pharmacia Fine Chemicals AB, Uppsala, Sweden.

chromatography controller LCC-500 PLUS, a dual-path monitor UV-2, a two-channel recorder REC-482, a valve V-7, and a fraction collector FRAC-100. The Mono Q column was an HR 5/5 (5 × 50 mm) and the dimensions of the DEAE-Sephadex column were 20 × 2.6 cm.

Procedures and Discussion

Sample Preparation. Metallothionein is prepared by intraperitoneal injections of juvenile perch *(Perca fluviatilis)* with cadmium chloride, repeatedly over a 2-week period, to yield a total dose of 3 mg Cd/kg body weight. One week after the last injection the fish are killed and the livers excised and washed in 10 m*M* Tris-HCl buffer (pH 7.6). The livers are homogenized in the same buffer (20%, w/v), using a glass–Teflon homogenizer. A glass–Teflon homogenizer is preferred to a Polytron homogenizer since it results in reduced air oxidation of the proteins.

The homogenates are centrifuged at 10,000 *g* for 20 min at 4° and the supernatant is recentrifuged at 105,000 *g* for 120 min at 4°. The supernatant is loaded onto a Sephadex G-75 column (60 × 5.0 cm) and eluted with 10 m*M* Tris-HCl (pH 7.6). Absorbance is measured at 254 and 280 nm and fractions of 10 ml are collected. The 254-nm peak eluting at an approximate molecular weight of 10,000 is pooled and used for anion-exchange experiments using either a DEAE-Sephadex A-25 column (Fig. 1) or a Mono Q column (Fig. 2). Because MT has been found to undergo chemical alteration on repeated freeze-thawing and will be proteolytically degraded if left at 4°, it is recommended that fresh material be used for all analytical purposes, or that aliquots be stored at −20°.

Choice of Buffer. It is vital that all buffers used in the present procedure are passed through a 0.22-μm sterile filter and are degassed before use. The Sephadex G-75-isolated MT samples in the present experiment are passed through a 0.22-μm sterile filter before being applied to the anion-exchange column. The buffers are supplemented with 0.01% NaN_3 except in the case of the ammonium bicarbonate buffer used in the Sephadex G-50 column to elute protein for amino acid analysis.

The most widely used technique for eluting ion-exchange columns is by gradient at a fixed pH. Although pH gradients have been successful, they are more difficult to reproduce and generally give poorer resolution than salt gradients as both ionic strength and pH are varied. Two different gradient systems have been employed in this laboratory, one using NaCl to generate a gradient and keeping the Tris-HCl buffer concentration at a fixed level (data not shown), the other increasing the concentration of the Tris buffer to elute the different MT isoforms (Fig. 2).[9]

Chromatography. The Mono Q column is stored in 24% (v/v) ethanol

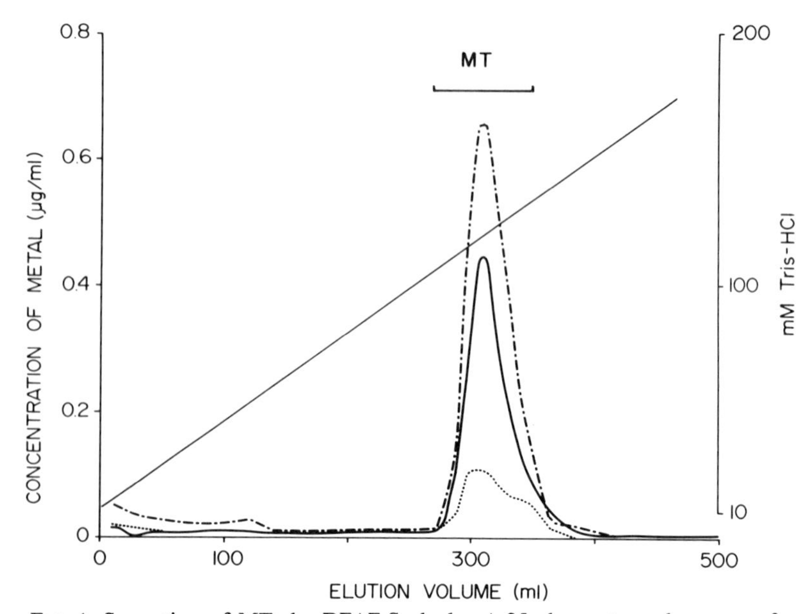

FIG. 1. Separation of MTs by DEAE-Sephadex A-25 chromatography was performed using a linear Tris-HCl gradient from 10 m*M* Tris-HCl (pH 8.1) to 200 m*M* Tris-HCl (pH 8.1). The absorbance at 254 and 280 nm was measured and 5-ml fractions collected. Cadmium (—), copper (· · · · ·), and zinc (– · – · –) were measured in each fraction. The column was washed with 100 ml starting buffer after application of the sample and prior to starting the gradient. [Reprinted, with permission, from P.-E. Olsson and C. Haux, *Aquat. Toxicol.* **9,** 231, (1986).]

and must be equilibrated before use. Equilibration is done by pumping 5 ml of 100% starting buffer (buffer A) through the column followed by 10 ml of 100% elution buffer (buffer B), and finally reequilibration with 5 ml 100% starting buffer.

Chromatography is performed under the following conditions: Flow rate, 1.0 ml/min; buffer A, 10 m*M* Tris-HCl (pH 7.6); buffer B, 400 m*M* Tris-HCl (pH 7.6). The LCC-500 controller unit is programmed to give a stepwise Tris-HCl gradient (see Fig. 2). Five milliliters (approximately 4 mg of MT) of Sephadex G-75-purified perch MT is loaded at time zero and the absorbance monitored at 254 and 280 nm. One-milliliter fractions are collected and the cadmium content of each fraction is determined using air–acetylene flame atomic absorption spectrophotometry (Fig. 2).

In most species at least two isoforms of MT have been identified.[1,4] However, when conventional anion-exchange chromatography is utilized to separate perch MT, only one MT-containing peak is observed eluting from the DEAE-Sephadex A-25 (Fig. 1). Chromatography of an aliquot of

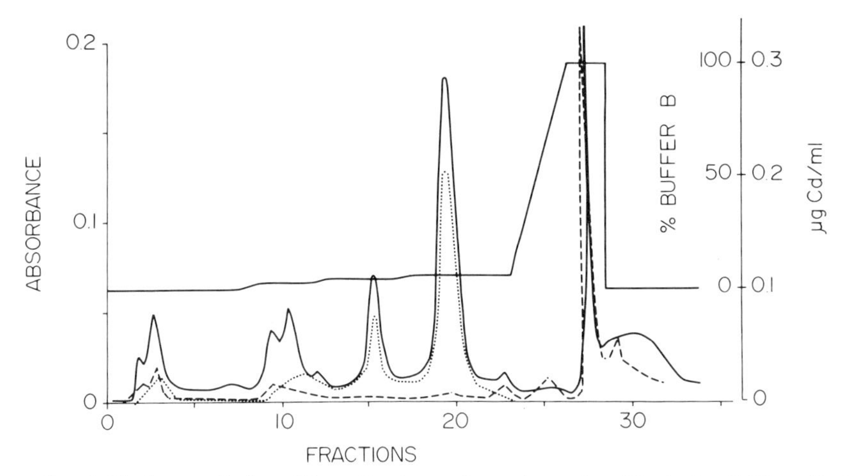

FIG. 2. Elution profile from Mono Q column chromatography of Sephadex G-75-purified perch MT. The absorbance of 254 nm (—) and 280 nm (- - -) and the cadmium content (· · · · ·) are indicated. The LCC-500 controller unit was programmed to give the following gradient after application of the sample: 0% B for 7 min, 0 to 2% B in 1 min, 2% B for 3 min, 2 to 4% B in 1 min, 4% B for 4 min, 4 to 5.5% B in 1 min, 5.5% B for 6 min, 5.5 to 100% B in 3 min, 100% B for 2 min, 0% B for 7 min. The solid line denotes the theoretical gradient. [Reprinted, with permission, from P.-E. Olsson and C. Hogstrand, *J. Chromatogr.* **402,** 293, (1987).]

the same sample, using a Mono Q column coupled to an FPLC, results in the appearance of two distinct MT peaks (Fig. 2).

The major cadmium-containing peaks from the Mono Q column are concentrated by ultrafiltration over a YM2 membrane and applied to a Superose 12 column equilibrated with 10 m*M* ammonium bicarbonate (pH 8.1). After final purification the proteins are lyophilized and characterized as MTs by amino acid analysis.[9]

There are certain variables that will affect the elution pattern of MT. Changing the pH of the Tris-HCl buffer results in a shift in the elution position of both MT-1 and MT-2. Both isoforms elute at a higher ion concentration at a higher pH. Thus, no change in the relative elution positions is observed when isolating perch MT at different pH. However, by altering the shape of the gradient it is possible to change the relative positioning of the two MT isoforms. The first isoform elutes at about 26 m*M* Tris-HCl (pH 7.6), while the second elutes at about 32 m*M* Tris-HCl (pH 7.6). A linear salt gradient from 10 to 400 m*M* Tris-HCl results in poor resolution of the two isoproteins. However, by utilizing a step gradient and flattening the ramp as soon as the first protein elutes it is possible to achieve total separation of the two proteins.

Loading Capacity. The loading capacity of the Mono Q HR 5/5 column is approximately 25 mg protein.[17] However, when loading samples close to the loading capacity of the column proteins often begin breaking through with the nonretained material. On the other hand, when loading small amounts of sample recovery may decline because the surface area-to-solute mass ratio in the column becomes very large, and the column is underloaded. In an underloaded column the small number of imperfections in a support that irreversibly adsorbs or denatures protein can dominate the separation.[5]

Column Maintenance. The Mono Q column should be stored in 24% ethanol when it is not used. When there is a pressure buildup in the column it is advisable to clean the column by washing with 1–4 ml of 70% acetic acid followed by rinsing with 10 ml doubly distilled water. If further washing is required, the column is washed with 1–5 ml of 2 *M* NaOH followed by 1 *M* HCl and 2 *M* NaCl with a 10-ml doubly distilled water rinse between each step. Following this the column can be reequilibrated with starting buffer or stored in 24% ethanol.

Comments. An FPLC chromatographic method using the Mono Q HR 5/5 column has been developed that is suitable for analytical purposes as well as preparative isolation of MTs.[9] The Mono Q column has been used to isolate MT for antibody production.[14] Furthermore, it can readily achieve separation of MT isoforms that will elute in the same position after DEAE-Sephadex A-25 chromatography.

Acknowledgments

I wish to thank Dr. R. W. Olafson (Department of Biochemistry and Microbiology, University of Victoria, Victoria, Canada) for a critical review of the manuscript.

[27] Purification of Human Isometallothioneins

By PETER E. HUNZIKER

Introduction

As all mammalian metallothioneins (MTs), human MT can be isolated as two charge-separable isoforms designated as MT-1 and MT-2. The purification protocol as described by Bühler and Kägi contains two precipitation steps followed by gel-filtration and ion-exchange chromatography.[1]

[1] R. H. O. Bühler and J. H. R. Kägi, *FEBS Lett.* **39**, 229 (1974).

While the MT-2 fraction is usually homogeneous, the MT-1 fraction represents a mixture of several isoforms that cannot be separated either by their molecular weight or by their charge. By using reversed-phase high-performance liquid chromatography (HPLC) at neutral pH all isoforms are separated in one single chromatographic step. This procedure allows the analysis of nanogram quantities and is easily scaled up for the purification of milligram quantities of protein.

Preparation of Human Metallothioneins

Depending on the available amount of tissue, crude MT is prepared in slightly different ways:

1. For small amounts of tissue of less than 30 g wet weight, the tissue sample is cut into small pieces and homogenized in 2 vol (w/v) of buffer (25 m*M* Tris/HCl, 250 m*M* sucrose, pH 7.5) in a Waring blender for 1 min. The homogenate is centrifuged at 4° for 90 min using a Ti50 rotor at 40,000 rpm. The resulting supernatant is loaded on a Sephadex G-50 column (3.5 × 90 cm) equilibrated with 25 m*M* Tris/HCl, pH 7.5, and eluted at a flow rate of 43.3 ml/hr. Fractions are monitored by atomic absorption flame spectroscopy for metal content, and MT-containing fractions are pooled and used for the separation of iso-MTs by preparative HPLC.

2. For amounts of tissue up to 2 kg the homogenate is precipitated with ethanol/chloroform according to Bühler and Kägi.[1] The resulting pellet is resuspended in 300 ml 25 m*M* Tris/HCl, pH 7.5, centrifuged at 4° for 10 min at 35,000 *g*, and loaded on a Sephadex G-75 column (6 × 150 cm) equilibrated with 25 m*M* Tris/HCl, pH 7.5, and eluted at a flow rate of 73 ml/hr. Metallothionein-containing fractions identified by their metal content are pooled and used for the separation of iso-MTs by preparative HPLC.

3. For the isolation of MT from cultured cells, procedure 1 was used except that the cells were suspended in 5 ml 25 m*M* Tris/HCl, pH 7.5, without considering wet weight or cell number. The cell suspension was harvested either by ultrasonification or by repeated thawing and freezing in liquid nitrogen. Gel filtration was carried out on a Sephadex G-50 column (1.5 × 90 cm) at a flow rate of 13.6 ml/hr.

Separation of Human Isometallothioneins

Without any concentrating step the pooled MT-containing G-50/G-75 fractions are loaded at a flow rate of 2 ml/min on a preparative Vydac C4

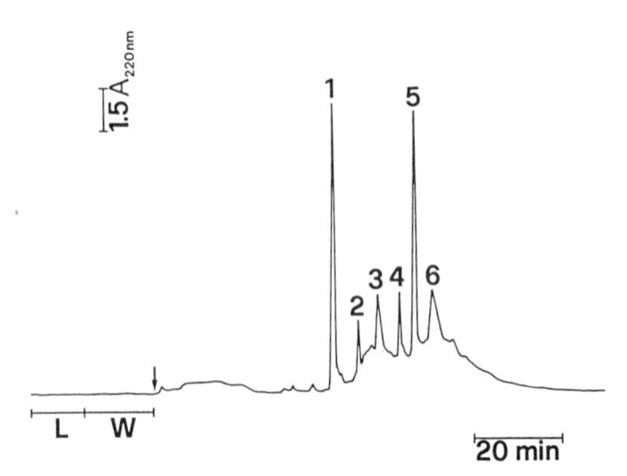

FIG. 1. Preparative HPLC at neutral pH of human hepatic MT obtained by gel filtration. L, 25 ml containing 1.4 mg MT was loaded as described in the text; W, the column was washed with 50 ml 25 m*M* Tris/HCl, pH 7.5, at a flow rate of 3 ml/min. The arrow indicates the start of the gradient. Chromatography was performed as described in the text. The numbering of the iso-MT fractions is as follows: 1, MT-2; 2, MT-1e; 3, MT-1h; 4, MT-1i; 5, MT-1k; 6, MT-1l. [P. E. Hunziker and J. H. R. Kägi, *in* "Current Topics in Nutrition and Disease," Vol. 18, p. 349. Alan R. Liss, New York, 1988.]

column (2.1 × 25 cm; The Separations Group, Hesperia, CA) equilibrated with 25 m*M* Tris/HCl, pH 7.5, either by using the pump for buffer A of the HPLC equipment or by an additional HPLC pump connected between the column and the manual injector. After the sample is loaded the column is washed with 50 ml of 25 m*M* Tris/HCl, pH 7.5, and elution is performed by a gradient formed between buffer A (25 m*M* Tris, pH 7.5) and buffer B (same as A, containing 60% acetonitrile) at a flow rate of 3 ml/min (Fig. 1). The gradient used was as follows: 8% of buffer B for 10 min, 8 to 33% of buffer B within 76 min. Collected iso-MT fractions are analyzed for homogeneity by chromatography as apo-MTs at low pH on an Aquapore RP-300 column (Applied Biosystems, San Jose, CA).[2] Nonhomogeneous fractions are further purified by chromatography at pH 7.5 under the same conditions used for preparative HPLC following a 1 : 1 dilution of the sample with buffer A. Depending on the amount of MT recovered from gel filtration or preparative HPLC, smaller columns with an inner diameter of 10.0, 4.6, or 2.1 mm may be used for the separation of isoforms. Sample loading and elution are carried out under the same conditions as for preparative HPLC except that the flow rates are adjusted to the size of the columns, i.e., 1.5, 0.6, and 0.2 ml/min, respectively.

[2] P. E. Hunziker, this volume [53].

Discussion

A typical preparative HPLC elution pattern of human hepatic MT is shown in Fig. 1. Iso-MTs are typically eluted between 40 and 70 min on the column used. However, other column matrices, such as RP-18 or RP-8, have also been used.[3] All of them showed the same good resolution of isoforms as with the C4 column, indicating that the separation of human iso-MTs is not limited to only one particular type of reversed-phase HPLC column.

One disadvantage of the described method is its failure to resolve Cu-MT. However, with an indirect method including the removal of copper and the reconstitution of the apo-MT with either zinc or cadmium, Cu-MT is converted to a sample that is suitable for the separation of the isoforms by HPLC.[2,4,5]

[3] P. E. Hunziker and J. H. R. Kägi, *Biochem. J.* **231**, 375 (1985).
[4] M. Vǎsák, this volume [54].
[5] P. E. Hunziker and I. Sternlieb, *Eur. J. Clinical Invest.*, in press.

[28] Isoelectric Focusing of Mammalian Metallothioneins

By MONICA NORDBERG

Introduction

This chapter deals with methods for analyzing mammalian metallothioneins (MTs). Potential pitfalls and necessary precautions in isolation and characterization of the MTs by isoelectric focusing are discussed. Isoelectric focusing of metallothioneins is used for physical identification of the different forms of metallothionein, separation of these forms from each other, and isolation of MTs from other proteins. On isolation of MTs from mammalian organs and tissues the MTs are accompanied by other proteins that conceal the presence of the MTs, for example, when ultraviolet (UV) monitoring is used. Isoelectric focusing has only been used for preparation of smaller (microgram) quantities of the protein when a high grade of purity and determined MT were necessary for further analyses.

Identified MTs have been used for further studies of MT metabolism and kinetics in the mammalian system. The biological function of MTs has been widely discussed. The simplest means of determining MT function is to study the metabolism and kinetics of the protein or proteins as has been done in metal toxicology[1-4] and molecular biology.[5] Purified MTs have also been used in developing radioimmunological methods for tracing MT in biological systems (see [20]).

Equipment and Method

Equipment for isoelectric focusing has been described in various publications. Nordberg *et al.*[6] first described the technique with special reference to isolation and characterization of mammalian metallothionein, followed by others,[7-9] and in relation to nonmammalian MTs.[10]

Pharmacia LKB (Uppsala, Sweden) provided columns (mostly 110-ml volume) equipped with a high-voltage power supply. If possible the column should be connected to a cooling unit or, if such a unit is not available, to ordinary tap water. Because the temperature affects the isoelectric point (p*I*) and is directly dependent on the correlation to p*I*, it is necessary to record the temperature. After harvesting, the pH of the column for each fraction is determined at the same temperature as that for focusing. The general technique for isoelectric focusing is described by Vesterberg and Svensson[11] and Haglund.[12]

The method is based on two gradients. One gradient, 0–50% (w/v) sucrose, constitutes the base of the pH gradient. The pH gradient is built up by mixing ampholytes (also obtained from LKB). The ampholytes are chemically identical to proteins but have no peptide bond. The ampholytes have an NH_2 group as the N terminus and a COOH group equal to the C terminus.

By mixing ampholytes of different composition and different pH values, it is possible to construct a pH gradient. A description of the method is supplied by the manufacturer. The pH gradient can be broad,

[1] G. F. Nordberg, *Environ. Physiol. Biochem.* **2,** 7 (1972).
[2] M. G. Cherian, *Biochem. Biophys. Res. Commun.* **61,** 920 (1974).
[3] M. Nordberg and G. F. Nordberg, *Environ. Health Perspect.* **12,** 103 (1975).
[4] T. Jin, G. F. Nordberg, and M. Nordberg, *Pharmacol. Toxicol.* **61,** 89 (1987).
[5] S. Bohm, A. Berghard, C. Peresmetoff-Morath, and R. Toftgård, *Cancer Res.* **50,** 1626 (1990).
[6] G. F. Nordberg, M. Nordberg, M. Piscator, and O. Vesterberg, *Biochem. J.* **126,** 491 (1972).
[7] M. Nordberg, G. F. Nordberg, and M. Piscator, *Environ. Physiol. Biochem.* **5,** 396 (1975).
[8] T. Syversen, *Arch. Environ. Health* **30,** 158 (1975).
[9] M. Wikner, G. F. Nordberg, M. Nordberg, and J. S. Garvey, *Arct. Med. J.* **47,** 179 (1988).
[10] M. Nordberg, J. Nuottaniemi, M. G. Cherian, G. F. Nordberg, T. Kjellström, and J. S. Garvey, *Environ. Health Perspect.* **65,** 57 (1986).
[11] O. Vesterberg and J. Svensson, *Acta Chem. Scand.* **20,** 820 (1966).
[12] H. Haglund, *Methods Biochem. Anal.* **19,** 1 (1970).

TABLE I
SETUP FOR ISOELECTRIC FOCUSING OF MAMMALIAN METALLOTHIONEINS

Column: 110 ml
Sucrose density gradient: 0–50% (w/v)
Carrier ampholytes: Ampholine, 40% (w/v)
On the bottom and in the central tube (anode): 20 ml of 50% (w/v) sucrose containing 0.7 ml 1 M H_2SO_4
On the top, 8 ml of deionized water with 0.05 ml ethylenediamine added, 500 V, 24 hr; 600–700 V (2–3 mA), 24 hr maximum 1 W
Dense solution: 2 ml, pH 3–5
Less dense solution: 0.2 ml, pH 3–5; 0.2 ml, pH 6–8

e.g., 2–10 for estimation of unknown p*I* or narrower and differing by 0.5 pH units for closer characterization. Proteins with a difference of 0.02 pH units can successfully be separated. Determination of pH should be performed with a precision of ±0.01 pH unit.

Table I shows a typical setup for pH gradients and Table II[13,14] presents different mixtures. The pH gradient is built up, together with the sucrose gradient, by mixing in a gradient mixer especially designed for isoelectric focusing equipment.

Pitfalls

It is imperative to avoid the MT coming in contact with or passing the acidic part of the gradient. If this occurs the metals are stripped off the protein and no true protein bonds will show up in the separation. Caution should also be exercised to avoid overheating. Overheating causes a visible white ring in the gradient and concomitant destruction of protein.

For the successful separation of MTs, the protein should not be added directly to the gradient mixer. The gradient should be built up and preestablished over a period of 24 hr before the MT sample is added.

First, the gradient is built up and high voltage is applied. After establishing the pH gradient, the voltage is increased, usually to 600–700 V. When the current no longer increases, the separation is complete. The current is switched off and as a precaution the electrodes are removed one by one. The next phase is the most critical for the method. At that point on the gradient where a pH close to the p*I* for the MT occurs (about one-third

[13] M. Nordberg, B. Trojanowska, and G. F. Nordberg, *Environ. Physiol. Biochem.* **4,** 149 (1974).

[14] M. Nordberg, *Environ. Health Perspect.* **54,** 13 (1984).

TABLE II
DIFFERENT pH MIXTURES FOR DENSE AND LESS DENSE SOLUTIONS INCLUDING MAIN METAL FOR PROTEIN STUDIED

Solution			
Dense	Less dense	Metal	Ref.
2 ml, pH 3–5	0.2 ml, pH 3–5 0.2 ml, pH 6–8	Cd/Zn	*a*
0.4 ml, pH 4–6 1.6 ml, pH 3.5–5	0.1 ml, pH 3.5–5 0.4 ml, ph 4–6 0.4 ml, pH 6–8	Cd/Zn	*b*
1 ml, pH 3–10 0.5 ml, pH 9–11 Add 0.5 ml, pH 9–11 on top of gradient	0.2 ml, pH 4–6 0.2 ml, pH 6–8 0.2 ml, pH 9–11	Hg	*c*

[a] Nordberg *et al.* (1972).[6]
[b] Nordberg, 1984.[14]
[c] Nordberg *et al.* (1974).[13]

from the top of the column), a suitable pH for MT is available. Calculate what the sucrose concentration is at this point and make an equivalent solution of sucrose. Take the MT solution to be separated, usually obtained after Sephadex gel chromatography with very low ionic strength and low in buffer content, and add this to the sucrose solution using a syringe and a long plastic tube. Add the solution to the gradient in the column. Insert the plastic tube into the gradient and slowly pour the test solution into it. Remove the tube. The procedure at this point is critical. If possible, no turbulence and no air bubbles should occur. Remove the syringe and tube before proceeding. Apply the electrodes one by one and then apply a high voltage (about 500 V). The temperature in the column should be kept as low as possible. By using a high-voltage aggregate with a constant energy supply, it is possible to avoid overheating and precipitation of the protein. An energy supply of not more than 1 W is recommended. A temporary drop in voltage may occur as current increases within 1 hr. Separation takes about 24 hr and should be performed at a current of up to at least 600 V. When no further increase in current is observed, even though the voltage is increased, the separation is complete.

Due to the high voltage used it is necessary to take precautionary steps to avoid accidents, e.g., signs stating that the equipment is connected to a high-voltage power source.

Table I gives a general set-up for separating MT with unknown p*I*. For

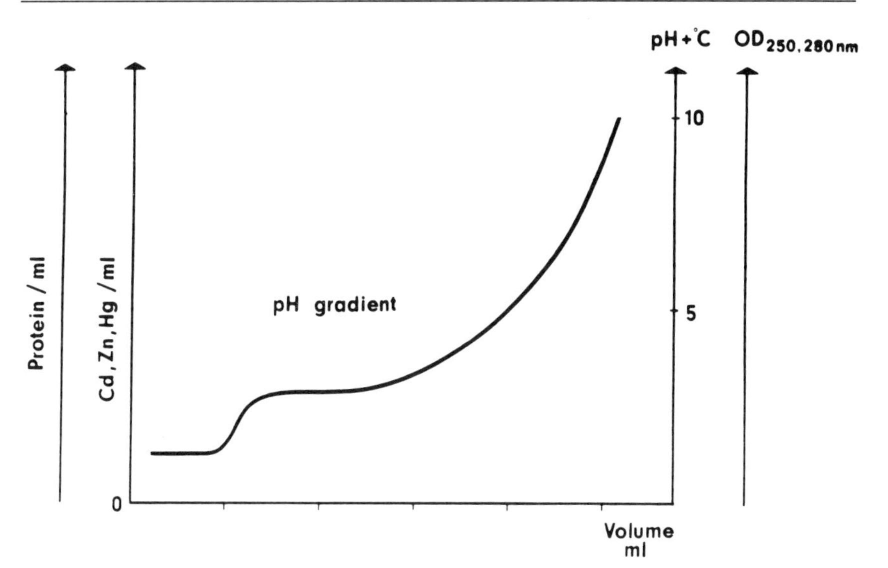

FIG. 1. Typical presentation of results after isoelectric focusing of MT.

this the dense solution should consist of ampholytes with pH 3–10. Table II shows the gradients used in relation to the metal content in MT.

When the column is ready for harvesting the following procedure takes place: First, the central tube is closed and sulfuric acid is removed using a syringe attached to a plastic tube. Avoid any unnecessary movement of the column in order to eliminate the risk of turbulence. Open the bottom tube cautiously and collect fractions of 2–2.5 ml at the rate of 1 ml/min. Keep the fractions at the same temperature as that in the separation procedure and measure the pH of every fraction to three decimal places (thousandths) after short intervals. The pH does not stabilize immediately, which makes it necessary to establish a constant time interval for pH readings of each fraction. Every fraction taken should be monitored for the detection of the different forms of MT. Analyses of the metals cadmium and zinc have been used for monitoring procedures. Determination of protein concentration should be performed for each of the different fractions. A typical example of the data presentation is given in Fig. 1. To remove the ampholytes from the protein use Sephadex chromatography.

Results and Discussion

Reported p*I* values for MT, or likely MT forms, at 4 and 8° vary for different organs. In liver, the p*I* for MT-1, which is rich in cadmium, is around 3.9; MT-2 p*I* is 4.5, with cadmium and zinc in equal proportions, and zinc-MT is probably about 6. For kidney metallothionein, the corresponding figures are higher and are reported at 8° to be 4.1, 4.4, and 6, respectively. For a protein high in inorganic mercury and even if not fully characterized by amino acid composition, the p*I* is reported to be 10.[13,15]

Due to the somewhat complicated procedure and the fact that these procedures are not usually available in instructions for isoelectric separation, the method for isoelectric focusing of mammalian MTs discussed here is not widely used. However, this method is recommended as a viable technique for obtaining purified forms of MTs and for collecting more information for the identification of MT.

[15] J. H. R. Kägi and M. Nordberg, *Experientia, Suppl.* **34,** (1979).

[29] Purification of Vertebrate Metallothioneins

By KAZUO T. SUZUKI

The animal kingdom can be subdivided into about 30 groups, with the phylum Vertebrata, animals with backbones, being the most complex. About 20,000 animal species within the phylum Vertebrata are divided into the two superclasses Pisces and Tetrapoda, and the superclasses are divided into three and four classes, respectively.

The fishes (the superclass Pisces) are divided into three classes, the class Agnatha (jawless fishes), the class Chondrichthyes (cartilaginous fishes), and the class Osteichthyes (bony fishes). The lamprey and the hagfish in the class Agnatha are the most primitive of the fishes and therefore the most primitive within the phylum Vertebrata. The sharks, skates, rays, and ratfishes belong to the class Chondrichthyes and these fishes are characterized by an absence of bone. The bony fishes (Osteichthyes) are the most common and most fish MTs have been studied in these fishes.

The class Amphibia and the class Reptilia include approximately 3000 and 7000 species, respectively. In animal species of these two classes metallothioneins (MTs) have been detected among several representative animals that can be observed in everyday life. However, in these two classes the details of MTs, such as amino acid sequences, are not yet known.

The birds (the class Aves) include approximately 8000 species and are arranged into 31 orders. Some species of birds are commonly used as livestock or are found in our natural habitats. However, avian MTs have been studied only in some species considered useful as indicator animals for contaminated wildlife and as livestock or experimental animals.

Animals of the class Mammalia include more than 4000 species. Although MT studies on animal species are limited mostly to the rodents, horses, and humans, general concepts regarding MT are derived from mammalian MTs.

This chapter deals with purification methods for MTs found in vertebrate animals. Metallothioneins of nonmammalian vertebrates are reviewed first to determine whether purification methods can be generalized for all vertebrate MTs. Vertebrate MTs are biosynthesized by direct translation of MT mRNA and not by enzymatic reactions, indicating that all vertebrate MTs belong to class I MT,[1] and have some common structural properties that should be considered in separation and purification strategies.

Overview of Vertebrate Metallothioneins

Fish Metallothioneins

Limited data are available for jawless and cartilaginous fishes except for those of dogfish.[2] Dogfish MT isolated from normal livers contains copper and zinc, the former metal being predominant as was also observed in our preliminary study on lamprey eel. High contents of copper seem to be common in lower vertebrate MTs, as shown later, and also in the livers of fetal and neonatal mammals.

Most of available data for fish MTs are those of bony fishes. Rainbow trout is most commonly used for MT studies of bony fishes, and its MT gene structure has been revealed[3] and cell lines have been established.[4] Fish MTs in the hepatopancreas are isolated or detected under conditions similar to those employed for mammalian MTs. Catfish MT is purified by Sephadex G-75 gel filtration and DEAE-Sephadex A-25 column chromatography and preparative polyacrylamide gel electrophoresis[5] and gibel MT is detected at an isometallothionein level by applying high-performance liquid chromatography/inductively coupled plasma emission spec-

[1] B. A. Fowler, C. E. Hildebrand, Y. Kojima, and M. Webb, *Experientia, Suppl.* **52,** 19 (1987).

[2] J. Hidalgo and R. Flos, *Comp. Biochem. Physiol. C: Comp. Pharmacol. Toxicol.* **83C,** 99 (1986).

[3] M. Zafarullah, K. Bonham, and L. Gedamu, *Mol. Cell. Biol.* **8,** 4469 (1988).

[4] J. Price-Haughey, K. Bonham, and L. Gedamu, *Environ. Health Perspect.* **65,** 141 (1986).

[5] A. Chatterjee and I. B. Maiti, *Mol. Cell. Biochem.* **78,** 55 (1987).

trometry (HPLC/ICPES) under column-switching conditions.[6] However, MTs in the gill must be detected and separated with care, as described later for the lung MT (see the section Protection from Oxidation, below).

Amphibian and Reptilian Metallothioneins

Metallothioneins in animal species of these two classes have not been studied extensively. However, amphibian MTs were studied to determine phylogenetic differences in metal compositions and numbers of isoforms. Metallothionein was shown to be present in the livers of adult amphibians as a single isoform of Cu,Zn-, Cu-, or Zn-binding protein by HPLC/atomic absorption spectrometry (AAS).[7] Although amphibian MTs in normal adults of 10 species consisted of only a single isoform, one more isoform was shown to be induced by exposure to cadmium in bullfrogs.[8] Analytical procedures employed for amphibian MTs were similar to those for mammals.[8]

Reptilian MTs were reported only for three animal species. Metallothionein in the tortoise is a mixture of three isoforms and contains zinc and copper.[9] Injection of cadmium or copper did not induce MT synthesis in this animal and the metals were incorporated into preexisting MT by replacing zinc. Injection of cadmium into the alligator gave MT rich in cadmium, MT-1 being the major isoform.[10] Reptilian MTs are also isolated according to procedures similar to those of mammalian MTs, although reptilian livers are sometimes hard to homogenize.

Avian Metallothioneins

Although MT is detected mostly by gel-filtration chromatography, in several species of birds, including wild ones, intensive studies are restricted to those of chicken[11,12] and to a lesser extent turkeys[13] and quail.[14] Avian

[6] K. T. Suzuki, H. Sunaga, E. Kobayashi, and S. Hatakeyama, *Comp. Biochem. Physiol. C: Comp. Pharmacol. Toxicol.* **87C,** 87 (1987).

[7] K. T. Suzuki, *Experientia, Suppl.* **52,** 265 (1987).

[8] K. T. Suzuki and H. Akitomi, *Comp. Biochem. Physiol. C: Comp. Pharmacol. Toxicol.* **75C,** 211 (1983).

[9] M. Yamamura and K. T. Suzuki, *Comp. Biochem. Physiol. C: Comp. Pharmacol. Toxicol.* **79C,** 63 (1984).

[10] J. U. Bell and J. M. Lopez, *Comp. Biochem. Physiol. C: Comp. Pharmacol. Toxicol.* **82C,** 123 (1985).

[11] D. Wei and G. A. Andrews, *Nucleic Acids Res.* **16,** 537 (1988).

[12] C. C. McCormick, C. S. Fullmer, and G. S. Garvey, *Proc. Natl. Acad. Sci. U.S.A.* **85,** 309 (1988).

[13] M. P. Richards and N. C. Steele, *J. Chromatogr.* **402,** 243 (1987).

[14] M. Yamamura and K. T. Suzuki, *Comp. Biochem. Physiol. B: Comp. Biochem.* **77B,** 101 (1984).

MTs in the normal liver consist of a single or predominantly a single isoform, and one more isoform is inducible by exposure to metal ions.[14] The major chicken MT consists of 63 amino acid residues[11,12] including 1 histidine residue,[11,14] as is the case in reptilian MTs.[9] These properties are different from those of mammalian MTs. However, the amino acid sequence of avian and mammalian MTs is conserved and avian MTs can be purified by the same procedures used to purify those of mammals.

Mammalian Metallothioneins

Although livers are most frequently used in the study of mammalian MT, as in the case in other vertebrates, kidneys have also been studied because of their toxicological importance,[15] followed by the gastrointestinal tract, pancreas, spleen, lung, and brain. Metallothioneins in body fluids are low in concentration but have attracted attention because of their toxicological and physiological significance.[15,16] Metallothioneins from different tissues within the same animal species are believed to have a similar primary structure, but they differ in metal composition and may be different in relative ratios of isoforms. Purification procedures for MTs from different tissues within the same mammalian species seem to be more complicated than those for MTs in the same tissue from different vertebrate species other than mammals.

Metallothionein in the liver of the fetus and newborn is known to be high in concentration and copper is the major metal bound to the prenatal and neonatal MTs, along with zinc. Content of copper in liver MTs of developing mammals is high, as in the case of lower vertebrates and these MTs must be handled with care to avoid copper-induced oxidation. Specific care should be paid to respective MTs in different tissues from the same mammalian species and these guidelines are applicable to the purification of MT in all vertebrate animal species.

General Precautions during Purification of Vertebrate Metallothioneins

Protection from Oxidation

Metallothioneins in animal species of the phylum Vertebrata contain 20 cysteinyl residues, and all sulfhydryl groups are coordinated with heavy metals. Although sulfhydryl groups are stabilized by coordination with heavy metals, this functional group is easily oxidized to form disulfide bonds intra- and intermolecularly. Abundant sulfhydryl groups in MTs

[15] M. Webb, *Experientia, Suppl.* **52,** 109 (1987).
[16] I. Bremner, *Experientia, Suppl.* **52,** 81 (1987).

make this protein susceptible to oxidation reactions, especially in the presence of oxygen during isolation and storage procedures.

It is essential to remove oxygen from buffer solutions for homogenization, isolation, and storage. Tissue samples should be stored free from oxygen, if possible, in an inert gas atmosphere. The samples should be homogenized under an inert gas atmosphere after removing dissolved oxygen by bubbling nitrogen through the solution for several minutes. Biological samples, such as lungs and gills, that contain air should be degassed under reduced pressure before homogenization; lungs float on a buffer solution when the tissues contain air, but sink into a buffer solution after air is removed under reduced pressure. A Polytron-type homogenizer (Kinematica, Switzerland) must be used with care, especially for samples susceptible to air oxidation and a glass–Teflon (Potter–Elvehjem) homogenizer is recommended in that case.

A reducing agent such as 2-mercaptoethanol can be used to protect samples from air oxidation at a concentration of 0.5% (v/v). However, reducing agents may work as chelating agents and the metals with low affinities, such as zinc, that are bound to MTs may be removed by a reducing agent with strong chelating capacity, such as dithiothreitol. Addition of reducing agents should be the second choice, with removal of or protection from oxygen as the first choice.

2-Mercaptoethanol can be used to reduce not only intermolecular but also intramolecular disulfide bonds in MTs because this agent is small in size and can reach the folded disulfide bonds to which other reducing agents, such as glutathione and cysteine, do not have access.[17,18] Further, 2-mercaptoethanol does not remove metals bound to MTs, in addition to its capacity to reduce disulfide bonds in MTs.

Supernatants can be stored in a sealed tube with an inert gas atmosphere at a temperature lower than $-20°$. Repeated freezing and thawing of stock sample solutions results in oxidation of MT; therefore it is preferable to store samples as individual aliquots.

Displacement of Metals in Metallothioneins

Metals bound to MT are easily displaced by metals with higher affinities. Zinc in zinc-containing MT is replaced with cadmium at the same molar ratio by adding cadmium into the MT solution. Cadmium and zinc bound to MT are replaced by copper.[18]

Cupric ions are reduced to cuprous ions at the expense of sulfhydryl groups, indicating that copper-containing MT produced by replacement of

[17] K. T. Suzuki and M. Yamamura, *Biochem. Pharmacol.* **29,** 689 (1980).
[18] K. T. Suzuki and T. Maitani, *Biochem. J.* **199,** 289 (1981).

cadmium and/or zinc with copper is partially oxidized and contains disulfide bonds. As copper is also coordinated with sulfhydryl groups, copper-containing MT prepared by using cupric ions has fewer available sulfhydryl ligands than that prepared by using cuprous ions.[18] Unintended replacement of cadmium and zinc in MT often occurs during isolation procedures due to the presence of contaminating cupric ions in buffer solutions and columns and also to the presence of endogenous copper in biological samples. In fact, copper is incorporated into MT to an appreciable extent by the simple heat treatment of liver supernatant; the isoform MT-2 is more susceptible to the replacement with copper than MT-1.[19]

Metallothionein binds heavy metals through the mercaptide bond (metal–sulfur bond) and heavy metals bound to MT are fully coordinated with sulfhydryl groups. The disulfide bond does not serve as a ligand for heavy metals in MTs; therefore, formation of disulfide bonds means that the capacity of MT to bind heavy metals is decreased by oxidation. When, for example, cadmium- and zinc-containing MT is partially oxidized, zinc is always liberated from the oxidized MT as a result of its lower affinity for MT than cadmium, thus giving an MT with a higher cadmium-to-zinc ratio than the original one. The relative ratio of metals, such as the cadmium-to-zinc ratio, seems to be related to the synthesis of MT in the organ,[20] and may have other biological significance not yet understood.

Multiplicity of Isoforms

Metallothionein in typical mammals consists of two isoforms, MT-1 and MT-2, named in the order of their elution from an anion-exchange column. The two isoforms contain the same number of cysteinyl residues occupying the same positions along the polypeptide chain. Therefore, the differences in electric charge and hydrophobicity of two isoforms are caused by the difference in amino acid composition other than cysteinyl residues. All vertebrate MTs contain 20 cysteinyl residues out of 61 amino acids, except for avian MTs. Hence, the difference in chemical properties of MTs among isoforms and among animal species is attributable to the difference in amino acid compositions other than cysteinyl residues.

Although the difference in biological roles of each isoform has not yet been explained, the number of isoforms, including their relative ratios and differences in metal composition, needs to be determined in the study of MT. The most popular separation procedure for isoforms of MT is anion-exchange chromatography for the MT fraction after the fraction is sepa-

[19] K. T. Suzuki, this volume [23].

[20] K. T. Suzuki, *in* "Biological Roles of Metallothionein" (E. C. Foulkes, ed.), p. 215. Elsevier/North-Holland, New York, 1982.

rated by gel-filtration chromatography. Metallothionein is well separated from high- and low-molecular-weight substances on a Sephadex G-75 column and elutes as a single peak. However, the MT in this fraction is not eluted with the same isoform ratio throughout. The faster eluting MT fraction contains a higher ratio of the MT-2 isoform than the MT-1 isoform, indicating that the column has not only gel-filtration capacity but also cation-exchange capacity under alkaline buffer conditions.[21] This implies that a fractionated tube is not representative of the whole fraction from the viewpoint of isoform ratio. This relationship is shown in Fig. 1.

The minor constituents of isoforms, oxidized forms, and isoforms of different metal compositions must be separated with care due to this ion-exchange property of gel-filtration columns. Metallothionein may be expressed in different ratios of isoforms depending on the variety of organs used. Within the same organ the ratio depends on physiological, biological, and other conditions. An MT sample not representative of the whole MT isoform composition may yield misleading results regarding the roles and meanings of isoforms.

Multiplicity of Metal Compositions

Metal compositions in the same MT isoform differ depending on the biological, physiological, and other conditions even if the isoform is induced in the same organ. Metallothionein in the kidneys of rats increases in concentration after birth to a high plateau level by 25 weeks of age, copper being the major metal bound to the MT, followed by zinc. Injections of cadmium into rats induce MT synthesis in the kidney and the MT found in the kidney is rich in cadmium and copper at the beginning of repeated injections. However, copper content in the kidney MT decreases suddenly when cadmium is accumulated at a concentration of more than 80–100 μg/g kidney and the kidney MT becomes rich only in cadmium.[20] These changes in metal composition are toxicologically meaningful.

Metallothioneins binding only zinc and/or copper are less stable than cadmium-containing MT. Even if cadmium is bound to MT, the presence of copper destabilizes MT owing to the easy oxidation through the redox reaction of copper. Metallothioneins in prenatal and neonatal livers of mammals contain a high content of copper. Likewise, livers of lower vertebrates are rich in MTs having a high copper content. These MTs should be protected from oxidation by air. At present it is not possible to separate MTs of different metal compositions having the same amino acid sequences.

[21] K. T. Suzuki, T. Motomura, Y. Tsuchiya, and M. Yamamura, *Anal. Biochem.* **107,** 75 (1980).

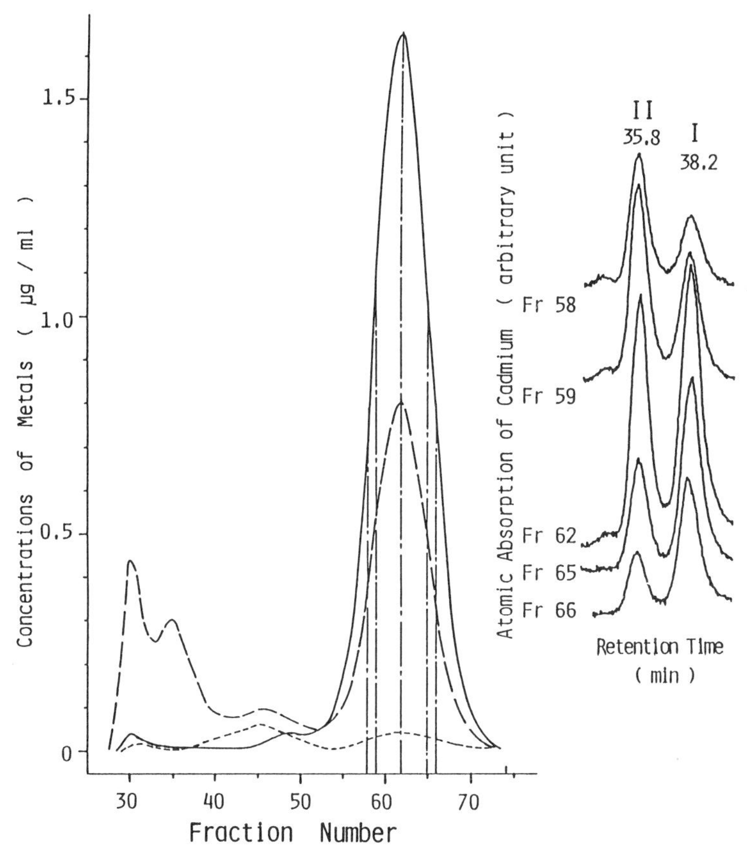

FIG. 1. Separation of MT into the two isoforms on a Sephadex G-75 column.[21] An aliquot (4 ml) of the liver supernatant prepared from cadmium-injected rats was applied to a Sephadex G-75 column (90 × 2.6 cm i.d.) and the column was eluted with 50 m*M* Tris/HCl buffer solution, pH 8.6, at a flow rate of 62 ml/hr. Five-milliliter fractions were collected (cadmium, —; zinc, — —; copper, - - -) and the MT fraction centered at tube 62 was analyzed by HPLC/AAS.[22] An aliquot (1 ml) of tubes 58, 59, 62, 65, and 66 on the Sephadex G-75 column was applied to a TSK gel G-3000-SW column (600 × 21.5 mm i.d. with a guard column of 100 × 21.5 mm i.d.; Tosoh, Tokyo, Japan) and the column was eluted with 50 m*M* Tris/HCl buffer, pH 8.6, containing 0.1% NaN_3 at a flow rate of 3.7 ml/min. Concentration of cadmium was monitored by AAS. I and II, The two isoforms MT-1 and MT-2, respectively.

Purification Procedure of Vertebrate Metallothioneins

Purification procedures depend on the subsequent scientific questions under investigation, sources of biological materials, and other factors. A typical flow diagram for the purification of MT is shown in Fig. 2 and specific comments are given below.

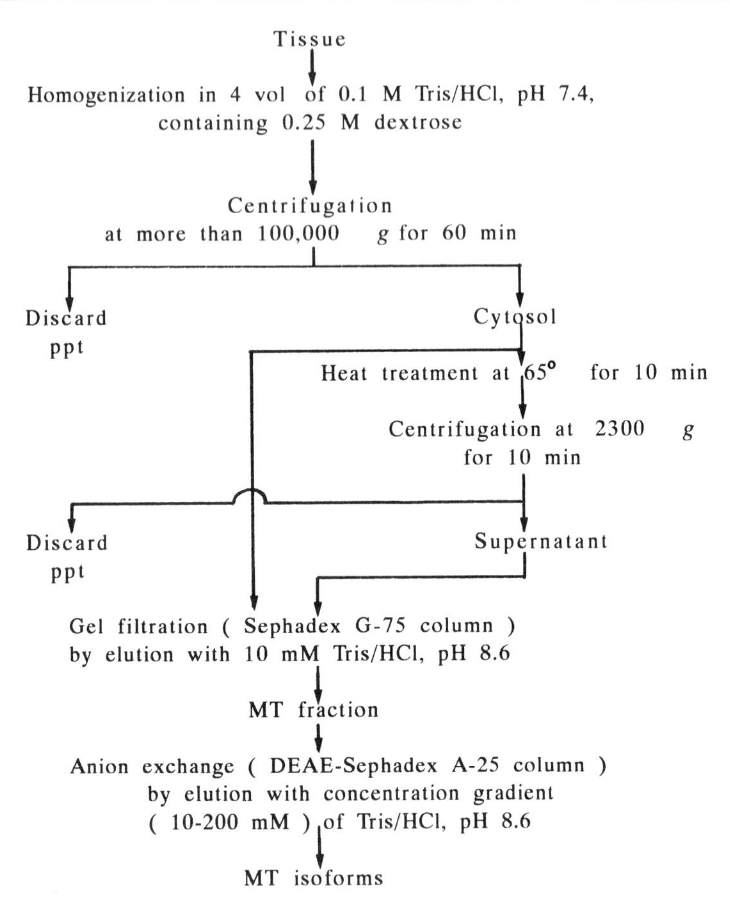

FIG. 2. A typical flow diagram for purification of vertebrate MTs.

Homogenization Procedures

Metallothioneins are most susceptible to air oxidation during homogenization procedures. Oxygen should be removed from tissues, such as lungs, and also from buffer solutions, as described above. Homogenization should be carried out in an atmosphere of nitrogen. Tissues can be homogenized using a Polytron homogenizer in a long-necked vessel such as a test tube or a centrifuge tube by introducing nitrogen onto the surface of the buffer solution during homogenization. When tissues containing less glutathione than liver are homogenized, a glass–Teflon homogenizer is recommended instead of a Polytron homogenizer.

Addition of reducing agents such as 2-mercaptoethanol and dithiothreitol helps to prevent oxidation, especially in case of copper-containing MTs. However, the use of these agents may alter metal compositions of MTs from their native states.

Buffers of various compositions have been used for homogenization. The buffer shown in Fig. 2 (0.1 *M* Tris-HCl buffer, pH 7.4, containing 0.25 *M* dextrose) has been used successfully as an extraction buffer in our laboratory. Relative ratios of buffer volumes to sample weights depend on the organs and contents of MT; 2 to 9 volumes (v/w) of buffer are used in most cases. The Tris-HCl buffer is chosen because of its low contamination with metals and coordination capacity, whereas dextrose is used because of its reducing capacity and low interference in metal detection procedures.

As MTs in homogenates are easily degraded enzymatically, homogenization must be carried out using ice-water cooling. The recovery of MT is reduced if low volumes of buffer solution are used for homogenization.

Centrifugation Procedures

To remove particulate matters the homogenate is centrifuged either at $>100{,}000\ g$ for 60 min or in two steps, $20{,}000\ g$ for 20 min and then $100{,}000\ g$ for 60 min. Metallothionein is recovered more efficiently with less oxidation using direct ultracentrifugation.

Heat Treatment of Supernatant

The supernatant fraction obtained by centrifugation of the homogenate can be treated at 65° for 10 min to remove heat-unstable high-molecular-weight proteins, thereby leaving the heat-stable MT intact in the soluble fraction. This heat treatment reduces dramatically the total amount of protein in the supernatant fraction.

The endogenous reducing agent, glutathione, remains in the heat-stable soluble fraction and MT does not seem to be exposed to oxygen; but the heat treatment should be carried out in an atmosphere of nitrogen.

Metal compositions may be affected by the heat treatment. In fact, copper is incorporated into one of the isoforms, MT-2, during the heat treatment of liver supernatant; the metal is assumed to originate from heat-denatured proteins.[19]

When heat treatment is incomplete, partially denatured proteins may appear at a lower temperature or at low buffer concentrations during column chromatography. Lipids and fluffy material floating on the surface are removed, together with heat-unstable proteins, making it easier to use the heat-treated supernatant as a source for further purification. The change in elution profiles of chromatography should be compared before and after heat treatment and heat treatment should be omitted when the change is critical for purification purposes.

Gel-Filtration Chromatography

A Sephadex G-75 column is most frequently used for separating MT from other components. Selecting the proper column size and flow rate yields a better separation of MT. Overloading and faster flow rate generally result in poorer separation. The gel should be packed after a complete degassing procedure: the gel, swollen in an elution buffer, is degassed completely in a round flask under reduced pressure, using an aspirator at 80°. The elution buffer (10 m*M* Tris-HCl, pH 8.6) should be degassed to keep the column material free from air and to protect MT from oxygen. Better separation of MT is assured for samples of lower protein content, such as a heat-treated supernatant.

Anion-Exchange Chromatography

Metallothionein can be separated into isoforms by anion-exchange chromatography on a DEAE-Sephadex A-25 column or its equivalent, including anion-exchange columns for HPLC. Hard gels such as a DEAE-Toyopearl 650M column (Tosoh) for liquid chromatography or an ES-502N column (Asahi Chem. Ind., Kawasaki, Japan) for HPLC can be used repeatedly and are more convenient than soft gels. Most vertebrate MTs are eluted with a linear concentration gradient of Tris-HCl buffer (pH 8.6) from 10 to 200 m*M*. A concentration gradient of Tris-HCl buffer rather than NaCl is better for detecting metals in the eluate without interference using atomic absorption spectrophotometry. The MT fraction eluted from a gel-filtration column can be applied directly on an anion-exchange column. The MT can be concentrated on the anion-exchange column when the gel-filtration column is eluted with 10 m*M* Tris-HCl buffer, pH 8.6.

Other Columns for Chromatography of Metallothioneins

A Sephadex G-75 column separates MTs from other high- and low-molecular-weight components and can be used to separate MT from a large volume of crude supernatant. For gel filtration using liquid chromatography, hard gels such as a Toyopearl HW-55 column (Tosoh) can be used as an alternative to soft gels (Sephadex G-75 column). Gel-filtration columns and reversed-phase columns for HPLC are another choice for the final purification of MT rather than repeating gel-filtration and/or anion-exchange chromatography. Larger amounts of column materials relative to sample volumes and elution at slower flow rates or slower gradient slopes guarantee better separation of MT from other biocomponents.

Criteria of Purity

The purity of MTs isolated at the isometallothionein level can be assessed by a number of methods used to determine the purity of proteins.[22,23]

[22] C. C. McCormick and L.-Y. Lin, this volume [11].
[23] Y. Aoki and K. T. Suzuki, this volume [16].

[30] Purification of Invertebrate Metallothioneins

By G. ROESIJADI and B. A. FOWLER

Introduction

The ubiquitous nature of metallothionein (MT) or MT-like metal-binding proteins in almost all major animal groups studied thus far is now widely recognized (see reviews on earlier work in Roesijadi,[1] Kagi and Kojima,[2] and Engel and Brouwer[3]). First reported to occur in invertebrates in a study on the American oyster *Crassostrea virginica* in 1975,[4] these proteins have been identified in most major invertebrate groups and include nematodes, annelids, mollusks, arthropods, pogonophora, and echinoderms (Table I). Most of these proteins in invertebrate species remain uncharacterized. With the exception of a relatively early study by Olafson *et al.*,[5] which resulted in the purification of cadmium-induced crab MTs and subsequent characterization of their primary and secondary structures,[6,7] procedures that have suitably purified invertebrate MTs for biochemical characterization have only recently been reported.[8-12] Thus,

[1] G. Roesijadi, *Mar. Environ. Res.* **4,** 167 (1981).
[2] J. H. R.Kägi and Y. Kojima, *Experientia, Suppl.* **52,** 25 (1987).
[3] D. W. Engel and M. Brouwer, *Ad. Comp. Environ. Physiol.* **5,** 54 (1989).
[4] J. L. Casterline and G. Yip, *Arch. Environ. Contam. Toxicol.* **3,** 319 (1975).
[5] R. W. Olafson, R. G. Sim, K. G. Boto, *Comp. Biochem. Physiol. C: Comp. Pharmacol.* **62B,** 407 (1979).
[6] K. Lerch, D. Ammer, and R. W. Olafson, *J. Biol. Chem.,* **257,** 2420 (1982).
[7] J. D. Otvos, R. W. Olafson, and I. M. Armitage, *J. Biol. Chem.,* **257,** 2427 (1982).
[8] J. Overnell, *Environ. Health Perspect.* **65,** 101 (1986).
[9] G. Roesijadi, S. L. Kielland, and P. Klerks, *Arch. Biochem. Biophys.* **273,** 403 (1989).
[10] M. Brouwer, D. R. Winge, and W. R. Gray, *J. Inorg. Biochem.* **35,** 289 (1989).
[11] L. W. Slice, J. H. Freedman, and C. S. Rubin, *J. Biol. Chem.* **265,** 256 (1990).
[12] M. Imagawa, T. Onozawa, K. Okumura, S. Osada, T. Nishimura, and M. Kondo, *Biochem. J.* **268,** 237 (1990).

TABLE I
INVERTEBRATE SPECIES FOR WHICH METALLOTHIONEIN OR SIMILAR PROTEINS HAVE BEEN REPORTED[a]

Phylum	Class	Species and common name
Echinoderms	Echinoids	*Strongylocentrotus purpuratus* (sea urchin embryo)[b,c]
		Anthocidaris crassispina (sea urchin egg)[d]
	Asteroids	*Asterias rubens* (starfish)[e]
Pogonophora		*Riftia pachyptila* (hydrothermal vent worm)[f]
Arthropods	Insects	*Drosophila melanogaster*[g,h]
		Sarcophaga peregrina (fleshfly)[i]
		Baetis thermicus (mayfly)[j]
	Crustaceans	*Scylla serrata* (crab)[k,l]
		Cancer magister (crab)[k]
		Cancer pagurus (crab)[m]
		Carcinus maenus (crab)[n]
		Callinectes sapidus (crab)[o]
		Rithropanopeus harrissii (crab larvae)[p]
		Homarus americanus (American lobster)[q]
		Austropotamobius pallipes (crayfish)[r]
		Palaemon elegans (shrimp)[s]
		Acetes sibogae (shrimp)[k]
Mollusks	Bivalves	*Crassostrea virginica* (American oyster)[t]
		Crassostrea gigas (Japanese oyster)[u]
		Ostrea edulis (European oyster)[u]
		Ostrea lutaria (New Zealand oyster)[v]
		Mytilus edulis (blue mussel)[w]
		Mytilus galloprovincialis (Mediterranean mussel)[x]
		Macoma balthica (clam)[y]
		Protothaca staminea (clam)[z]
		Mercenaria mercenaria (clam)[aa]
		Calyptogena magnifica (clam)[ab]
		Placopecten magellanicus (scallop)[ac]
		Anodonta cygnea (freshwater clam)[ad]
		Unio elongatus (freshwater clam)[ad]
	Gastropods	*Crepidula fornicata* (limpet)[ae]
		Patella vulgata (limpet)[af]
		Patella intermedia (limpet)[af]
		Littorina littorea (marine snail)[ag]
		Helix pomatia (land slug)[ah]
		Cepaea hortensis (land slug)[ah]
		Arianta abustorum (land slug)[ah]
		Arion lusticanus (land slug)[ai]
	Polyplacophora	*Cryptochiton stelleri* (chitron)[k]
Annelids	Oligochaetes	*Eisenia foedita* (earthworm)[aj]
		Monopylephorus cuticulatus (marine worm)[ak]
		Limnodrilus hoffmeisteri (freshwater worm)[al]

(continued)

TABLE I (continued)

Phylum	Class	Species and common name
	Polychaetes	*Alvinella pompejana* (hydrothermal vent worm)[f]
		Neanthes arenaceodentata (bristle worm)[am]
		Eudystylia vancouveri (fanworm)[an]
Nematodes		*Caenorhabditis elegans* (soil nematode)[ao,ap]

[a] Of the 46 invertebrate species listed, amino acid sequences have thus far been determined for MTs of the sea urchin, *S. purpuratus,* fruit fly *D. melanogaster,* crab *S. serrata,* lobster *H. americanus,* oyster *C. virginica,* and nematode *C. elegans* by peptide or nucleotide sequencing techniques.

[b] P. Harlow, E. Watkins, R. D. Thornton, and M. Nemer, *Mol. Cell. Biol.* **9,** 5445 (1989).

[c] M. Nemer, D. G. Wilkinson, E. C. Travaglini, E. J. Sternberg, and T. R. Butt, *Proc. Natl. Acad. Sci. U.S.A.* **82,** 4992 (1985).

[d] H. Ohtake, T. Suyemitsu, and M. Koga, *Biochem. J.* **211,** 109 (1983).

[e] P. J. den Besten, H. J. Herwig, D. I. Zandee, and P. A. Voogt, *Mar. Environ. Res.* **28,** 163 (1989).

[f] M.-A. Cosson-Mannevy, R. Cosson, and F. Gaill, *C. R. Acad. Sci. Ser. 3* **302,** 347 (1986).

[g] A. Debec, R. Mokdad, and M. Wegnez, *Biochem. Biophys. Res. Commun.* **127,** 143 (1985).

[h] D. Lastowski-Perry, E. Otto, and G. Maroni, *J. Biol. Chem.* **260,** 1527 (1985).

[i] Y. Aoki, K. T. Suzuki, and K. Kubota, *Comp. Biochem. Physiol. C: Comp. Pharmacol. Toxicol.* **77C,** 279 (1984).

[j] Y. Aoki, S. Hatakeyama, N. Kobayashi, Y. Sumi, T. Suzuki, and K. T. Suzuki, *Comp. Biochem. Physiol. C: Comp. Pharmacol. Toxicol.* **93C,** 345 (1989).

[k] R. W. Olafson, R. G. Sim, and K. G. Boto, *Comp. Biochem. Physiol. B: Comp. Biochem.* **62B,** 407 (1979).

[l] K. Lerch, D. Ammer, and R. W. Olafson, *J. Biol. Chem.* **257,** 2427 (1982).

[m] J. Overnell and E. Trewhella, *Comp. Biochem. Physiol. C: Comp. Pharmacol.* **64C,** 69 (1979).

[n] J. R. Jennings, P. S. Rainbow, and A. G. Scott, *Mar. Biol.* **50,** 141 (1979).

[o] M. A. Wiedow, T. J. Kneip, and S. J. Garte, *Environ. Res.* **28,** 164 (1982).

[p] B. M. Sanders, K. D. Jenkins, W. G. Sunda, and J. D. Costlow, *Science* **222,** 53 (1983).

[q] M. Brouwer, D. R. Winge, and W. R. Gray, *J. Inorg. Biochem.* **35,** 289 (1989).

[r] R. Lyon, M. Taylor, and K. Simkiss, *Comp. Biochem. Physiol. C: Comp. Pharmacol. Toxicol.* **74C,** 51 (1983).

[s] S. L. White and P. S. Rainbow, *Comp. Biochem. Physiol. C: Comp. Pharmacol. Toxicol.* **83C,** 111 (1986).

[t] G. Roesijadi, S. L. Kielland, and P. Klerks, *Arch. Biochem. Biophys.* **273,** 403 (1989).

[u] J. M. Frazier and S. G. George, *Mar. Biol.* **76,** 55 (1983).

[v] R. P. Sharma, *Bull. Environ. Contam. Toxicol.* **30,** 428 (1983).

[w] F. Noel-Lambot, *Experientia* **32,** 324 (1976).

[x] A. Viarengo, M. Pertica, G. Mancinelli, G. Zanicchi, J. M. Bouguegneau, and M. Orunesu, *Mol. Physiol.* **5,** 41 (1984).

[y] C. Johansson, D. J. Cain, and S. N. Luoma, *Mar. Ecol. Prog. Ser.* **28,** 87 (1986).

[z] G. Roesijadi, *Biol. Bull.* **158,** 233 (1980).

[aa] W. E. Robinson, M. P. Morse, B. A. Penney, J. P. Karareka, and E. U. Meyhöfer, *in* "Marine Pollution and Physiology: Recent Advances" (F. J. Vernberg, F. P. Thurberg, A. Calabrese, and W. B. Vernberg, eds.), p. 83. University of South Carolina Press, Columbia, South Carolina, 1985.

(continued)

only a few invertebrate metal-binding proteins have been purified to the extent that their characterization as MTs (criteria of Fowler *et al.*[13]) has been possible. As a result, the progress on the biochemistry and structure–function relationships for invertebrate MTs has occurred at a slower pace than that seen with higher animals. It appears that most previous studies on purification and characterization of invertebrate MTs have been on preparations of MT that included copurifying substances. These substances have been especially recalcitrant to removal by the conventional size-exclusion and ion-exchange chromatographic procedures routinely used to isolate the major isoforms of mammalian MTs. Thus, previous analyses of amino acid composition for invertebrate MTs have often included aromatic amino acid residues and reduced values for cysteine, due to the presence of these coeluting substances. Studies that have incorporated reversed-phase high-performance liquid chromatography (HPLC) in purification schemes appear to have successfully addressed this problem.[9-12] These have resulted in preparations of individual isoforms of sufficient purity to allow sequencing and other forms of biochemical analysis. Unlike studies on mammals, in which it appears that reversed-phase HPLC is most useful for resolution of heterogeneity among MT isoforms, among invertebrates it appears that this isolation procedure may be necessary for

[13] B. A. Fowler, C. E. Hildebrand, Y. Kojima, and M. Webb, *Experientia, Suppl.* **52,** 19 (1987).

[ab] G. Roesijadi, J. S. Young, E. A. Crecelius, and L. E. Thomas, *Biol. Soc. Wash. Bull.* **6,** 311 (1985).
[ac] B. A. Fowler and D. Gould, *Mar. Biol.* **97,** 207 (1988).
[ad] L. Tallandini, A. Cassini, N. Favero, and V. Albergoni, *in* "Heavy Metals in the Environment." CEP Consultants, Edinburgh, 1983.
[ae] F. L. Harrison, K. Watness, D. A. Nelson, J. E. Miller, and A. Calabrese, *Estuaries* **10,** 78 (1987).
[af] A. G. Howard and G. Nickless, *Chem.-Biol. Interact.* **16,** 107 (1977).
[ag] W. J. Langston and M. Zhou, *Mar. Biol.* **92,** 505 (1986).
[ah] R. Dallinger, B. Berger, and A. Bauer-Hilty, *Mol. Cell. Biochem.* **85,** 135 (1989).
[ai] R. Dallinger, H. H. Janssen, A. Bauer-Hilty, and B. Berger, *Comp. Biochem. Physiol. C: Comp. Pharmacol. Toxicol.* **92C,** 335 (1989).
[aj] K. T. Suzuki, M. Yamamura, and T. Mori, *Arch. Environ. Contam. Toxicol.* **9,** 415 (1980).
[ak] K. A. Thompson, D. A. Brown, P. M. Chapman, and R. O. Brinkhurst, *Trans. Am. Microsc. Soc.* **101,** 10 (1982).
[al] P. L. Klerks and J. S. Levinton, *in* "Ecotoxicology: Problems and Approaches" (S. A. Levin, M. A. H., J. R. Kelly, and K. D. Kimball, eds.). Springer-Verlag, New York, 1989.
[am] K. D. Jenkins and A. Z. Mason, *Aquat. Toxicol.* **12,** 229 (1988).
[an] J. S. Young and G. Roesijadi, *Mar. Pollut. Bull.* **14,** 30 (1983).
[ao] L. W. Slice, J. H. Freedman, and C. S. Rubin, *J. Biol. Chem.* **265,** 256 (1990).
[ap] M. Imagawa, T. Onozawa, K. Okumura, S. Osada, T. Nishimura, and M. Kondo, *Biochem. J.* **268,** 237 (1990).

purification of MT from other substances. In recent studies, data obtained on these purified proteins have been coupled with molecular techniques for construction of probes for analyses of expression and further characterization of both protein structure and molecular regulation.[11,12,14] These and the earlier studies indicate considerable diversity in the structure of MTs of invertebrate species.

Purification Procedures

Development of the now conventional technique[15] of using Sephadex gel-based size-exclusion and DEAE ion-exchange chromatography of cytosol for isolation of major MT isoforms was accompanied by rapid progress in understanding the properties and behavior of MTs in higher animals. Invertebrate MTs, as with vertebrate MTs, are enriched from cytosol by size-exclusion chromatography. This procedure has been routinely applied in the numerous studies whose goal was to examine the dynamics of metal-binding to the MT "pool." However, with the exception of the study of Olafson *et al.*,[5] application of the ion-exchange procedures to invertebrate metal-binding proteins resulted in only partial success in achieving purity of MT preparations. The oyster *C. virginica* was reported originally to have characteristics similar to those of the "copper-chelatin" later shown to be copper-MT.[16,17] Similarly, in mussels,[18-22] metal-binding proteins with reduced cysteine and low amounts of aromatic amino acid residues were reported following purification by size-exclusion and anion-exchange chromatography. Levels of the aromatic residues negatively correlated with the cysteine content, suggesting that removal of components containing aromatics is related to increased cysteine content and the extent of purification of preparations containing putative MTs.[20,22,23] It has been suggested

[14] M. E. Unger, T. T. Chen, C. C. Fenselau, C. M. Murphy, M. M. Vestling, and G. Roesijadi, *Biophys. Biochim. Acta* (in press).

[15] Z. A. Shaikh and O. J. Lucis, *Experientia* **27**, 1024 (1971).

[16] R. Premakumar, D. R. Winge, R. D. Wiley, and K. V. Rajagopalan, *Arch. Biochem. Biophys.* **170**, 278 (1975).

[17] D. R. Winge, B. L. Geller, and J. Garvey, *Arch. Biochem. Biophys.* **208**, 160 (1981).

[18] S. G. George, E. Carpene, T. L. Coombs, J. Overnell, and A. Youngson, *Biochim. Biophys. Acta* **580**, 225 (1979).

[19] G. Roesijadi and R. E. Hall, *Comp. Biochem. Physiol. C: Comp. Pharmacol.* **70C**, 59 (1981).

[20] A. Viarengo, M. Pertica, G. Mancinelli, G. Zanicchi, J. M. Bouquegneau, and M. Orunesu, *Mol. Physiol.* **5**, 41 (1984).

[21] J. Frazier, S. G. George, J. Overnell, T. L. Coombs, and J. H. R. Kägi, *Comp. Biochem. Physiol. C. Comp. Pharmacol. Toxicol.* **80C**, 257 (1985).

[22] G. Roesijadi, *Environ. Health Perspect.* **65**, 45 (1986).

[23] B. A. Fowler, D. W. Engel, and M. Brouwer, *Environ. Health Perspect.* **65**, 63 (1986).

that these copurifying substances may have functionally relevant interactions with MT.[23]

A discussion of basic techniques and problems associated with purification of nonmammalian MTs[24] was organized as part of the Conference on High-Affinity Metal-binding Proteins in Nonmammalian Species (Research Triangle Park, 1984, collected papers published in Environmental Health Perspectives, 1986, Vol. 65). Recent studies on invertebrate MTs that have resulted in MTs of sufficient purity to allow sequencing[9-12] have at the final purification step utilized reversed-phase HPLC based on modifications of the procedure originally described by Klauser et al.[25] Purification procedures have included acetone precipitation of the cytosol[9,10] and size-exclusion and ion-exchange chromatography[9-12] prior to the final purification step. Ion-exchange HPLC[9] and fast protein liquid chromatography (FPLC)[1] have also been used in these preliminary chromatographic steps. Methods recently developed for purification of molluskan MTs are described in detail below.[9]

For semipreparative-scale isolations, 30 g oyster soft tissues is homogenized on ice in 2 vol of ice-cold buffer of 20 m*M* Tris-HCl, 275 m*M* NaCl, 5 m*M* dithiothreitol (DTT) at pH 8.6, 0.1 m*M* phenylmethylsulfonyl fluoride (PHSF). Following centrifugation at 80,000 *g* for 90 min at 4°, the supernatant containing the cytosolic fraction is then subjected to sequential acetone precipitation using 45 and 80% final acetone concentrations (volume to volume, assuming additive volume) at −20°. The supernatant from the 45% acetone precipitation is brought up to 80% acetone, and the 80% acetone pellet is redissolved in 20 m*M* Tris-HCl, 2 m*M* DTT, pH 7.5. Metallothioneins in this redissolved acetone pellet are purified by gel chromatography on Sephadex G-75 in the same buffer. The protein peak (~10 kDa) containing MTs is further purified by anion-exchange HPLC (TSK DEAE-5PW column: Tosoh, Tokyo, Japan). Buffer A is 20 m*M* Tris-HCl, pH 7.5; buffer B is 400 m*M* Tris-HCl, pH 7.5. Metallothioneins are eluted with a linear gradient of 0 to 50% buffer B over 30 min. This step separates the MTs into two peaks previously designated as Cd-BP1 and Cd-BP2 (cadmium-binding proteins) (Fig. 1[9]). For final purification, individual peaks from the previous step are precipitated in 80% acetone, redissolved in 0.1% trifluoroacetic acid, and chromatographed on a C_4 reversed-phase column (Synchropak RP-4 column: Synchrom, Inc., Linden, IA). Buffer A is 0.1% trifluoroacetic acid; buffer B is 50% 0.1% trifluoroacetic acid plus 50% acetonitrile (volume to volume, assuming additive volume). Metallothioneins are eluted with a linear gradient of 0 to 70% buffer B over 30 min. This results in the purification of the two forms

[24] D. R. Winge and M. Brouwer, *Environ. Health Perspect.* **65,** 211 (1986).

[25] S. Klauser, J. H. R. Kägi, and K. J. Wilson, *Biochem. J.* **209,** 71 (1983).

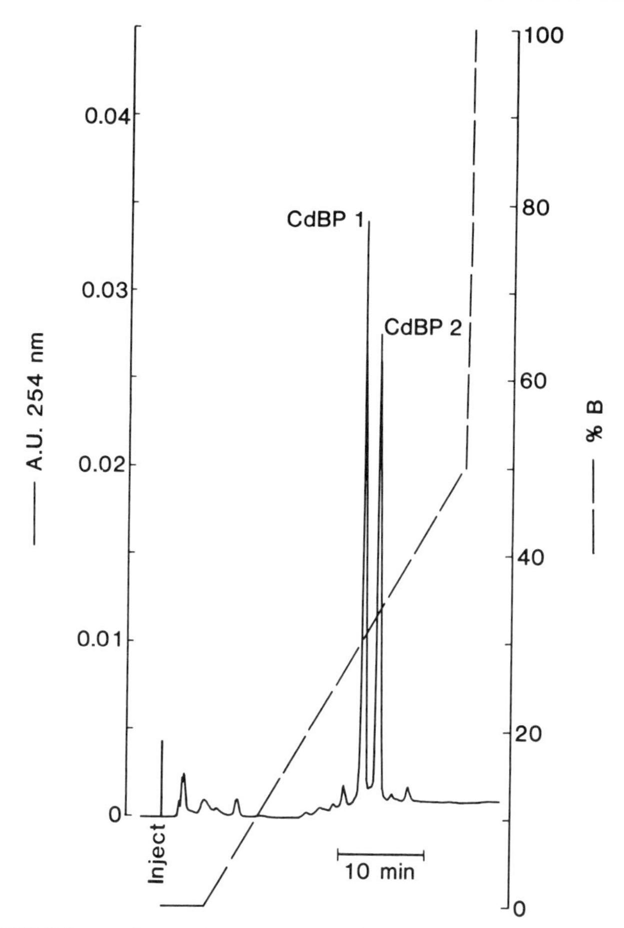

FIG. 1. HPLC ion-exchange chromatogram (TSK DEAE-5PW column; Tosoh, Tokyo, Japan) of oyster MTs previously enriched by acetone precipitation and Sephadex G-75 size-exclusion chromatography. (Reproduced with permission from Roesijadi *et al.*[9])

of oyster MT (Fig. 2[9]). Proteins obtained at this stage are sufficiently pure for amino acid sequencing and other methods of characterization.

Amino acid analysis of the two proteins following the ion-exchange HPLC step results in the detection of low amounts of the aromatic residues tyrosine and phenylalanine and is consistent with previous reports of copurifying substances through this step. These are removed by reversed-phase HPLC.

These methods have been used in purifying MTs for analysis of the primary structure of the two oyster MT isoforms by N-terminal amino acid

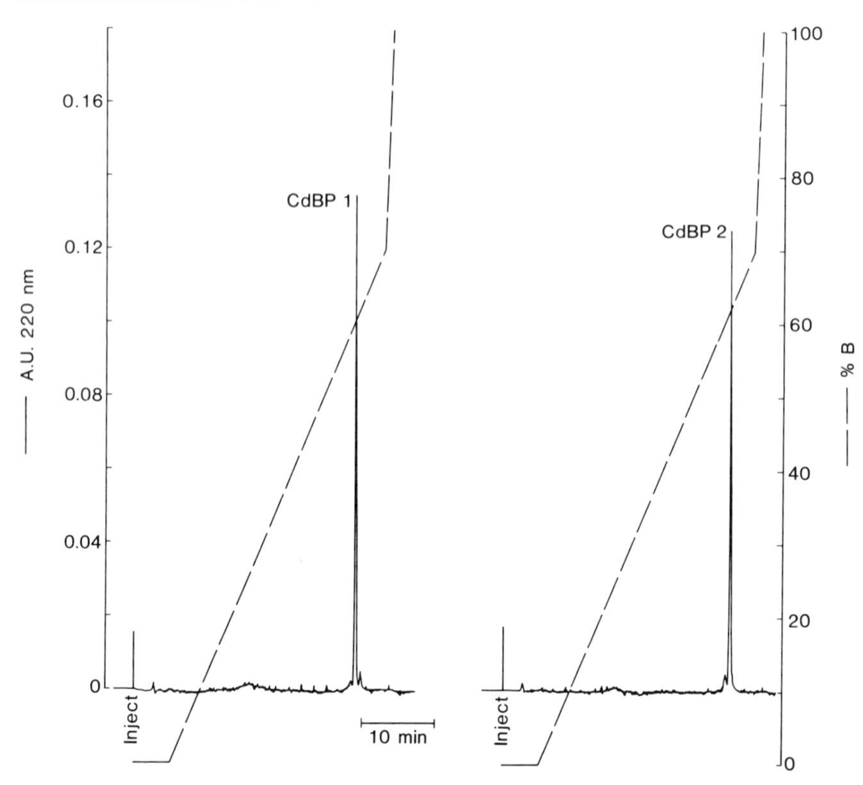

FIG. 2. Reversed-phase HPLC (Synchropak RP-4 column; Synchrom, Inc., Linden, IA) of oyster MTs previously isolated by ion-exchange HPLC (see Fig. 1). (Reproduced with permission from Roesijadi *et al.*[9])

sequencing and tandem mass spectrometry.[9,14,26] The two isoforms possess the same amino acid sequence and differ only in the presence of an *N*-acetyl group on Cd-BP2. Thus, the ion-exchange separation described here is sufficiently powerful to resolve two proteins that differ only in a single positive charge (Fig. 2), and the reversed-phase separation is able to discriminate on the basis of the increased hydrophobicity imparted by the acetyl group (Fig. 3).

Although developed using oyster MTs, the procedures described above should be adaptable to other species with minor modifications of elution gradients to accommodate the differential behavior of various MTs. The procedures described here are basically similar to those described for lob-

[26] G. Roesijadi, M. M. Vestling, C. M. Murphy, P. L. Klerks, and C. C. Fenselau, *Biochim. Biophys. Acta* (in press).

sters[10] and nematodes[11,12] and were modified from those developed for mammalian MTs.[25]

For use in the isolation and analysis of small amounts of MTs in physiological experiments, the procedures described above have been modified for analytical-scale purifications of the two forms of oyster MT. This procedure allows processing of samples as small as 0.1 g and utilizes HPLC for all chromatographic steps. The key modification is the substitution of gel-permeation HPLC (TSK SW-3000 or SW-2000 columns; Tosohaas) for the Sephadex gel chromatography. The overall recovery of oyster MTs in this procedure is 73%, which reflects 90% recovery at each of the individual steps (i.e., acetone precipitation, size-exclusion HPLC, and ion-exchange HPLC). The final reversed-phase purification step is not required when determination of the metal composition of individual MT forms is desired. The individual proteins are clearly resolved and the integrity of metal composition is maintained after the anion-exchange step (see Fig. 1).[9] The acidic conditions of the reversed-phase step result in dissociation of metals from the MTs and is contraindicated when the determination of metals bound to MTs in the objective. Reversed-phase HPLC in a neutral buffer system (20 m*M* Tris-HCl, pH 7.5), which would maintain the metal composition, did not resolve the two forms observed in the acidic conditions. Thus, with these proteins, the degree of purity desired must be evaluated in the context of the information expected to be derived as a result of the purification.

Diversity of Invertebrate Metallothioneins

Unlike the vertebrate MTs, which show considerable sequence homology, invertebrate MTs differ significantly from those of vertebrates and each other.[27] To date, MTs considered to belong to both class I and class II MTs have been identified among the invertebrates. However, phylogenetic comparisons based solely on the amino acid sequences should be regarded with caution because apparently nonhomologous MTs may have a high degree of similarity in the metal-responsive elements in noncoding sequences in the genome; e.g., comparisons with vertebrate class I MTs and sea urchin class II MTs.[28]

A neglected feature to date is the fact that invertebrate MTs characterized to date possess free N terminals that lack the *N*-acetylmethionine considered to be present in all vertebrate MTs. Whether this difference represents any functional significance remains to be determined. However, N-terminal amino acid sequences, especially the N-terminal residue,

[27] M. Nemer, D. G. Wilkinson, E. C. Travaglini, E. J. Sternberg, and T. R. Butt, *Proc. Natl. Acad. Sci. U.S.A.* **82,** 4992 (1985).

[28] P. Harlow, E. Watkins, R. D. Thornton, and M. Nemer, *Mol. Cell. Biol.* **9,** 5445 (1989).

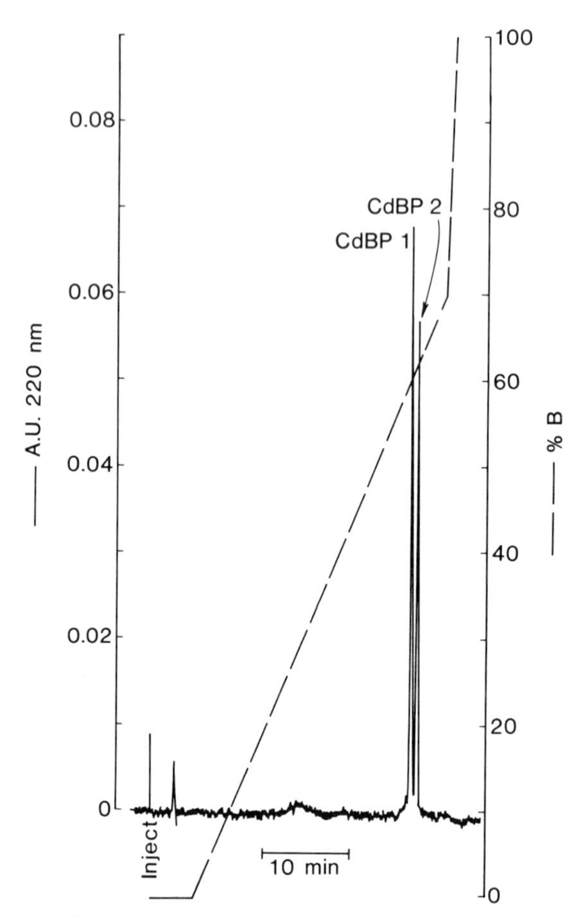

FIG. 3. Reversed-phase HPLC (Synchropak RP-4 column; Synchrom) of oyster MTs previously isolated by ion-exchange HPLC (see Fig. 1), then recombined to show the degree of resolution effected in this system. The proteins differ only by the presence of an *N*-acetyl group on Cd-BP2. (Reproduced with permission from Roesijadi *et al.*[9])

among the MTs of invertebrates are sufficiently different that any functional significance may not be explainable by the same general rule. Proline, valine, and serine have been identified at the N-terminal position, thus far. In the nematode *Caenorhabditis elegans*[11,12] and oyster *Crassostrea virginica*,[9] the N-terminal residues are the penultimate residues in the nucleotide sequence, the initiator methionine apparently having been processed off the nascent polypeptide during cotranslational processing.[29,30] The oyster MT is unique in that it possesses an MT that comes in both acetylated and nonacetylated forms.[26]

Apart from the N-acetylation described above, invertebrate MTs have also been reported to occur as isoforms. The crab MTs originally described by Olafson *et al.*[5] exist as two isoforms. Metallothioneins for a different crustacean, a lobster, have been shown to exist as three isoforms,[10] two that are inducible and involved in metal detoxification and one that is not induced and can function as a copper donor for hemocyanin, the oxygen-carrier protein in these animals. This is one of the few examples of direct evidence for MT involvement in a physiologically relevant function.

In mollusks, the existence of multiple forms of metal-binding proteins has been recognized for some time.[18] These MTs can occur as molecular weight variants that exhibit internal heterogeneity. Thus, in oysters,[23] mussels,[18,21] and a gastropod,[31] major forms of both 20K and 10K have been observed. Each of these variants can consist of up to four individual proteins.[21] The 20K forms most likely represent different dimeric structures, although this possibility has yet to be confirmed. To date, little work has been conducted on the functional significance of the occurrence of these forms. An intriguing observation is the fact that in the oyster *Crassostrea virginica* different populations that exhibit only the apparent 10K form or both 10K and 20K forms exist. The basis for this intraspecific diversity in MTs in different populations of oysters is not currently known.

Summary

Purification procedures that appear to be generally applicable to invertebrate MTs have only recently been developed and are described here. Thus far, few invertebrate MTs have been purified and characterized, although proteins that exhibit similarities to MT have been identified in numerous invertebrate species. A greater understanding of MTs of this group, which comprises most of the animals in the animal kingdom, will provide the basis for increased understanding of the evolution of this ubiquitous protein. Additionally, specific invertebrate species may provide useful models for probing MT function and contribute to a greater understanding of the biological role of MT.[3]

Acknowledgments

We thank M. E. Unger for reviewing an earlier version of this manuscript.

---

[29] S. Huang, R. C. Elliot, P.-S. Liu, R. K. Koduri, J. L. Weickmann, J.-H. Lee, L. C. Blair, P. Ghosh-Dastidar, R. A. Bradshaw, K. M. Bryan, B. Einarson, R. L. Kendall, K. H. Kolacz, and K. Saito, *Biochemistry* **26**, 8242 (1987).

[30] S. M. Arfin and R. A. Bradshaw, *Biochemistry* **27**, 7980 (1988).

[31] W. J. Langston and M. Zhou, *J. Mar. Biol. Assoc. (U.K.)* **67**, 585 (1987).

[31] Purification of Yeast Copper-Metallothionein

By ULRICH WESER and HANS-JÜRGEN HARTMANN

Introduction

The formation of Cu-thionein in copper-supplemented yeast cells *Saccharomyces cerevisiae)* was discovered in 1975.[1-3] Due to the multiplicity of the *CUP1* gene these cells are able to enrich copper from the culture medium very efficiently, up to millimolar concentrations.[4-6] Unlike Class I metallothioneins, in which zinc, cadmium, and copper are simultaneously bound, copper is exclusively coordinated in the yeast protein *in vivo*. The isolated protein was found to lack the first eight amino acids predicted by the nucleotide sequence of the *CUP1* locus. It is characterized as a 53-residue polypeptide (M_r 5655) containing 12 cysteines and 8 copper atoms/molecule.[4] The copper is coordinated in the form of Cu–thiolate clusters as deduced from electronic absorption, circular dichroism, fluorescence, electron paramagnetic resonance (EPR), X-ray photoelectron, and X-ray absorption measurements.[7-10]

Fermentation

Bakers' yeast or the copper-resistant strain *Saccharomyces cerevisiae* X-2180-1Aa is cultivated in a medium composed of 50 g yeast extract, 100 g gelatin hydrolysate (GIBCO, Paisley, Scotland), 350 g glucose, 3 g NaCl, 3 g KH_2PO_4, and 1–1.5 m*M* $CuSO_4$ in 6 liters of water. After 48 hr

[1] R. Prinz and U. Weser, *FEBS Lett.* **54,** 224 (1975).
[2] R. Prinz and U. Weser, *Z. Physiol. Chem.* **356,** 767 (1975).
[3] R. Premakumar, D. R. Winge, R. D. Wiley, and K. V. Rajagopalan, *Arch. Biochem. Biophys.* **170,** 278 (1975).
[4] D. R. Winge, K. B. Nielson, W. R. Gray, and D. H. Hamer, *J. Biol. Chem.* **260,** 14464 (1985).
[5] M. Karin, R. Najarian, A. Haslinger, P. Valenzuela, J. Welch, and S. Fogel, *Proc. Natl. Acad. Sci. U.S.A.* **81,** 337 (1984).
[6] T. R. Butt, E. J. Sternberg, J. A. Gorman, P. Clark, D. Hamer, M. Rosenberg, and S. T. Crooke, *Proc. Natl. Acad. Sci. U.S.A.* **81,** 3332 (1984).
[7] U. Weser, H.-J. Hartmann, A. Fretzdorff, and G.-J. Strobel, *Biochim. Biophys. Acta* **493,** 465 (1977).
[8] H. Rupp, R. Cammack, H.-J. Hartmann, and U. Weser, *Biochim. Biophys. Acta* **578,** 462 (1979).
[9] U. Weser and H.-J. Hartmann, *in* "Copper Proteins and Copper Enzymes" (R. Lontie, ed.), p. 151. CRC Press, Boca Raton, Florida, 1984.
[10] A. Richter and U. Weser, *Inorg. Chim. Acta* **151,** 145 (1988).

the cells are harvested by centrifugation at 3000 *g* for 15 min and washed three to four times, each in 10 vol of deionized water. The total copper concentration in the yeast was found to reach up to 3 mg Cu/g wet cells.

Isolation

Copper-loaded washed yeast cells are suspended in an equal volume of 20 m*M* Tris/HCl buffer, pH 7, and ruptured either in a BIOX X-press (frozen state, −20°) or in a Manton-Gaulin 15 M high-pressure homogenizer at 700 bar (suspended state, 4–10°). The homogenate is centrifuged at 10,000 *g* for 1 hr. In earlier work the supernatant obtained from homogenized bakers' yeast was subjected to ammonium sulfate precipitation.[1,2] The precipitate formed in 50–80% ammonium sulfate was dissolved and dialyzed. Ion-exchange chromatography on DEAE-23 cellulose revealed four Cu-thionein fractions of different specific copper content. Based on an apparent molecular mass of 10 kDa the main fraction contained 10 g-atoms of copper/molecule. Because of the ellipsoid shape of copper-thionein, it elutes during gel filtration like a protein of M_r 10,000. Therefore a molecular mass of 10,000 Da was used at that time as a convenient calculation unit.

Similar results are obtained with a modified method of preparation. After heating the homogenate to 60° for 3 min the supernatant is subjected to ion-exchange chromatography and gel filtration. Homogeneous Cu-thionein containing from 8 to 13 g-atoms of copper/10,000 g of protein are obtained.[7,8]

The isolation procedure currently employed is based on the exclusive application of gel chromatography.[4,11] Cell paste (25 ml) derived from 6 liters of cultured yeast is ruptured in an X-press at −20°, diluted two-fold with 10 m*M* Tris/HCl buffer, pH 7, and centrifuged for 1 hr at 48,000 *g*. For large scale preparations the suspended yeast cells are homogenized using a Manton–Gaulin high-pressure homogenizer at 700 bar. The supernatant is applied to a Sephadex G-75 column (4 × 80 cm) equilibrated with 10 m*M* Tris/HCl, pH 7, in the presence of 0.1% (v/v) 2-mercaptoethanol to avoid uncontrolled oxidation of thiolate sulfur. The lyophilized Cu-thionein fraction from the M_r 10K region is chromatographed on Sephadex G-50 (2.5 × 80 cm) equilibrated with N_2-saturated 10 m*M* Tris/HCl, pH 7. Depending on the efficacy of the homogenization, approximately 4 ± 2 mg of purified Cu-thionein is recovered from 1 g of wet cells. The detection and localization of Cu-thionein during the preparation is controlled by copper analyses in the respective chromatographic fractions. Alternatively, physicochemical methods, including luminescence emis-

[11] U. Weser and H.-J. Hartmann, *Biochim. Biophys. Acta* **953**, 1 (1988).

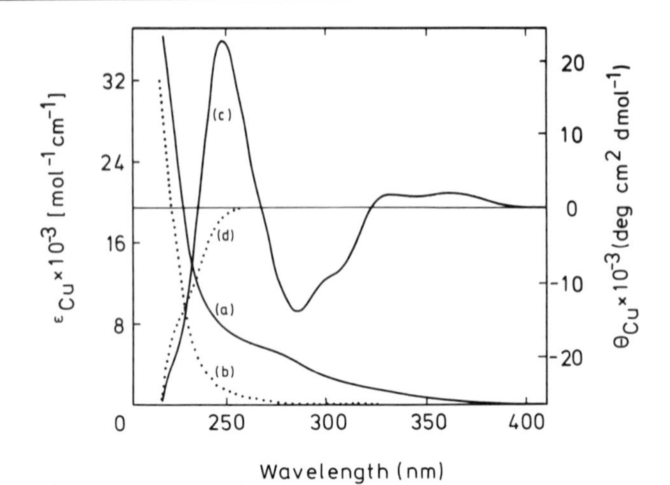

FIG. 1. Electronic spectra (a, b) and circular dichroic spectra (c, d) of yeast Cu_8-thionein. The chiroptical and UV electronic properties between 400 and 240 nm are exclusively assigned to the Cu(I)–thiolate bonding in the protein. Cotton extremes are seen at 359, 328, 283 and 245 nm. Adjustment to pH 0.2 leads to the essential loss of all spectroscopic signals (b, d).

sion, circular dichroism, and UV electronic absorption, are useful tools to identify and characterize the protein in the course of the isolation process (Figs. 1 and 2). The characteristic orange-red luminescence near 600 nm of the intact Cu(I)-thiolate chromophores of Cu-thionein is detectable after irradiation with UV light in the 300 to 360-nm region.[10,12] Owing to the fact that 80–90% of the intracellular copper is found coordinated in Cu-thionein, extended X-ray absorption fine structure (EXAFS) measurements were applied for the detection of copper–sulfur coordination in intact lyophilized yeast cells.[13] As deduced from earlier spectroscopic techniques on the isolated protein the EXAFS spectra of intact yeast indicate that the accumulated copper is univalent and is exclusively coordinated to sulfur atoms at a distance of 2.19 Å. However, there are discrepancies in the coordination number of copper and sulfur, which was estimated to be 2 in the intact cells and 3 in the isolated protein.

The phenomenon of a Cu-thionein release from washed, copper-loaded yeast cells into the culture medium or even water may also be considered a most elegant isolation technique for this protein.[14] The Cu-thionein excre-

[12] M. Beltramini and K. Lerch, *FEBS Lett.* **127,** 201 (1981).
[13] A. Desideri, H.-J. Hartmann, S. Morante, and U. Weser, *Biol. Met.* **3,** 45(1990).
[14] K. Felix, H.-J. Hartmann, and U. Weser, *Biol. Met.* **2,** 50 (1989).

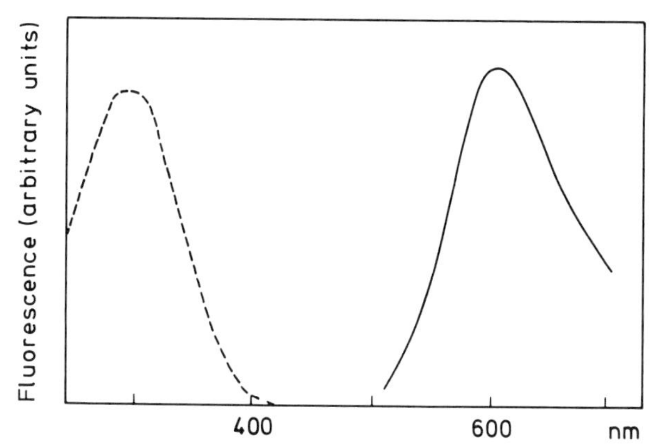

FIG. 2. Excitation and emission spectra of yeast Cu-thionein.

tion is observed for more than 96 hr when the cells are repeatedly suspended and maintained under agitation at 24° (Fig. 3). Although only about 10% of the initial intracellular copper is released at this time $1-1.5 \times 10^{-4}$ M of thionein-copper (0.08–0.12 mg Cu-thionein/ml) is present extracellularly under these conditions. A five- to six-fold increase of Cu-thionein excretion is caused by the addition of 1 mM glutathione to the

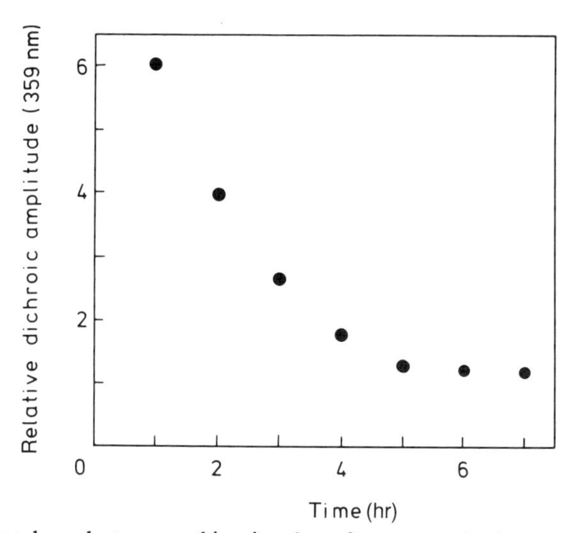

FIG. 3. Time-dependent copper-thionein release from copper-loaded yeast. Dichroic amplitude at 359 nm of excreted yeast Cu-thionein after repeated incubation of 1 g wet cells in 2 ml phosphate-balanced saline (140 mM NaCl, 5 mM Na_2HPO_4, 2 mM KH_2PO_4, pH 7.4) at 24°. The supernatants of the centrifuged suspensions were directly used for the measurements.

suspension buffer. The excreted protein is directly detectable in the cell-free supernatant by both its luminescent and dichroic properties. Due to the specific ellipticities of Cu-thionein expressed for copper (i.e., $\theta_{Cu}^{359} = 1700 \pm 100$ deg cm^2 dmol^{-1}) the latter technique can be used for quantification. After chromatography of the cell-free supernatant on Sephadex G-75 and G-50 the same molecular mass, stoichiometry of 8 copper atoms/mol protein, and amino acid composition were found compared to that of the intracellular yeast Cu-thionein. Moreover, no detectable differences were seen in the physicochemical properties, including luminescence emission, UV electronic absorption, and chiroptical data.

Both the structure and the stoichiometry of copper to sulfur are the subject of debate. When ion-exchange chromatography was performed during the purification procedure sometimes a ratio of copper to sulfur of 1 : 2 was found (i.e., 6 copper/mol protein), indicating the loss of 2 metal atoms/mol. Consequently, two types of copper must be present in the Cu_8-thionein, which is prepared exclusively by gel filtration. Differently bound copper in Cu_8-thionein is confirmed by the rapid reactivity of two out of eight coppers with the specific Cu(I) chelators, bathocuproine sulfonate and cuproine.[11] The two reactive coppers do not contribute to the chiroptical properties of the protein. The exact coordination of this circular dichroic silent copper is unknown. Considerable efforts are needed to determine the exact structure and/or possible interconversion reactions of these Cu-thiolate centers in order to explain the diversities in the reported copper-to-sulfur ratios.

Acknowledgment

The continued support of the Cu-thionein work by the Deutsche Forschungsgemeinschaft is gratefully acknowledged.

[32] Purification of *Neurospora crassa* Copper-Metallothionein

By KONRAD LERCH

Metallothioneins are a class of low-molecular-weight cysteine-rich proteins that bind large amounts of metal ions such as Zn^{2+}, Cu^{2+}, Cd^{2+}, or Hg^{2+}.[1,2] The biosynthesis of these proteins is transcriptionally regulated by

[1] J. H. R. Kägi and Y. Kojima, eds., *Experientia, Suppl.* **52,** (1987).

[2] D. H. Hamer, *Annu. Rev. Biochem.* **55,** 913 (1986).

heavy metal ions, glucocorticoids, and a number of other factors.[3] In contrast to MTs from mammalian sources, fungal MTs contain exclusively copper,[4] and Cu-MTs have been isolated from the yeast *Saccharomyces cerevisiae,*[5] the ascomycete *Neurospora crassa,*[6] and the basidiomycete *Agaricus bisporus.*[7] The Cu-MTs isolated from *N. crassa* and *A. bisporus* consist of only 25 amino acids and show a striking homology in primary structure to the amino-terminal parts of mammalian MTs. They both contain six atoms of copper per molecule, which are bound to seven cysteinyl residues in the form of a single Cu(I)-thiolate cluster.[8,9] The *Cu-MT* gene from *N. crassa* has been cloned and characterized.[10]

Methods of Analysis

High-Performance Liquid Chromatography Analysis. Neurospora crassa Cu-MT can be easily detected by reversed-phase high-performance liquid chromatography (HPLC) (Lichrosorb RP-18, Merck, Darmstadt). A suitable buffer system includes buffer A [0.01 *M* sodium perchlorate in 0.1% (w/v) phosphoric acid, pH 2.1] and buffer B [as in buffer A, except in 60% (v/v) acetonitrile]. Chromatography is carried out with a linear gradient of buffer B (0.4%/min) at room temperature with a flow rate of 1.0 ml/min.

Metal Analysis. Metal analyses can be performed either by atomic absorption spectrometry or by a colorimetric test[11] using Chelex 100-treated buffers and acid-washed glassware.

Amino Acid Analysis. Samples are hydrolyzed for 22 hr at 110° *in vacuo* in 6 *N* HCl. For the determination of cysteine, samples are oxidized with performic acid[12] before acid hydrolysis.

Growth of *Neurospora crassa*

Neurospora crassa wild-type strain Sing is grown in liquid cultures (800 ml in 2-liter Erlenmeyer flasks with intrusions) of Vogel's medium

[3] M. Karin, *Cell (Cambridge, Mass.)* **41,** 9 (1985).
[4] K. Lerch, *Met. Ions Biol. Syst.* **13,** 299 (1981).
[5] D. R. Winge, K. B. Nielson, W. R. Gray, and D. H. Hamer, *J. Biol. Chem.* **260,** 14464 (1985).
[6] K. Lerch, *Nature (London)* **284,** 368 (1980).
[7] K. Münger and K. Lerch, *Biochemistry* **24,** 6751 (1985).
[8] M. Beltramini and K. Lerch, *Biochemistry* **22,** 2043 (1983).
[9] T. A. Smith, K. Lerch, and K. O. Hodgson, *Inorg. Chem.* **25,** 4677 (1986).
[10] K. Münger, U. A. Germann, and K. Lerch, *EMBO J.* **4,** 2665 (1985).
[11] W. N. Poillon and C. R. Dawson, *Biochim. Biophys. Acta* **77,** 27 (1963).
[12] C. H. W. Hirs, this series, Vol. 11, p. 59.
[13] H. J. Vogel, *Microbiol. Genet. Bull.* **13,** 42 (1952).

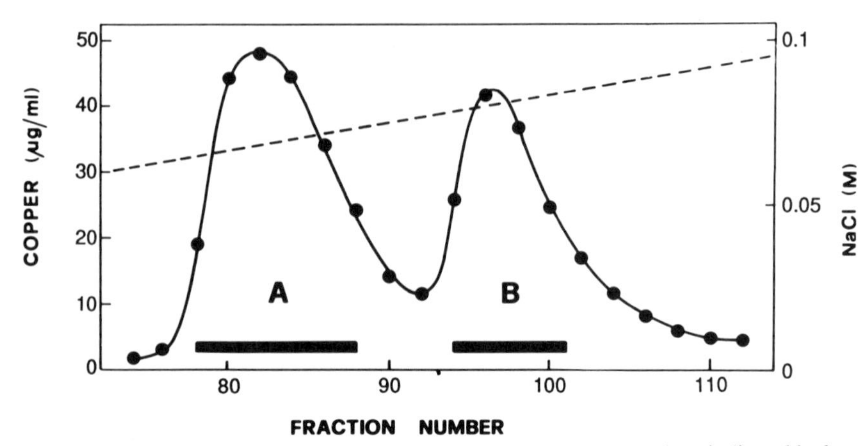

FIG. 1. DEAE-cellulose chromatography of *N. crassa* Cu-MT. Fractions indicated by bars (designated as A and B) were pooled.

N[13] supplemented with 0.5 m*M* $CuSO_4$. After 3 days of cultivation in the dark at 25°, the mycelia are collected by filtration through cheesecloth, washed extensively with distilled water, and lyophilized. Freeze-dried mycelia can be stored at −20° for more than a year.

Purification Procedure

Step 1. Extraction. All steps described in the purification procedure are carried out at 2–4°. To avoid air oxidation of Cu-MT, all buffers are kept under a constant stream of nitrogen or argon. Lyophilized mycelia (30 g) are ground to a fine powder in a Waring blender and suspended in 280 ml of 0.1 *M* sodium phosphate, pH 7.2. After stirring for 10 min, the cell debris is removed by centrifugation for 20 min at 14,000 *g*. The supernatant is passed through glass wool in a funnel to remove lipids.

Step 2. Heat Treatment. The clear supernatant from step 1 is gently stirred for 10 min at 60° under a blanket of argon. The resulting precipitate is then removed by centrifugation for 20 min at 20,000 *g*.

Step 3. Gel Filtration. The slightly yellowish supernatant from step 2 is applied to a 5.0 × 160 cm column of Sephadex G-50 fine equilibrated in 10 m*M* Tris-HCl, pH 8.0. The column is run at 200 ml/hr and fractions of 20 ml are collected. The front peak containing high-molecular weight proteins is discarded.

Step 4. DEAE-Cellulose Chromatography. Neurospora crassa Cu-MT is characterized by an unusual luminescence on UV excitation.[14] This

TABLE I
PURIFICATION OF *Neurospora crassa* COPPER-METALLOTHIONEIN

Step	Volume (ml)	Total protein (mg)	Total copper (μg)	Copper/protein ratio ($\times 10^3$)
Crude extract	265	4370	23600	5.4
Heat-treated extract	245	1320	10800	8.2
Sephadex G-50	546	147	4100	27.9
DEAE-cellulose				
Buffer A	80	11.4	1950	171
Buffer B	60	4.3	740	172

unique property is very useful for following the elution of the protein in steps 3 and 4. The Cu-MT-containing fractions from step 3 are immediately applied to a 2.5 × 6 cm column of DEAE-cellulose (DE-32; Whatman, Clifton, NJ) equilibrated in 10 m*M* Tris-HCl, pH 8.0. Cu-MT is eluted with a linear gradient from 10 m*M* Tris-HCl, pH 8.0, to 10 m*M* Tris-HCl, 0.1 *M* NaCl, pH 8.0 (250 ml of buffer each). A typical elution profile is shown in Fig. 1. For spectroscopic studies, the protein is rechromatographed on a Sephadex G-25 column in 10 m*M* sodium phosphate, pH 7.5.

The purification is summarized in Table I. The overall procedure takes approximately 3 days.

Properties

Purity. The purity of *N. crassa* Cu-MT is best assessed by HPLC and amino acid analysis. The elution profile displayed in Fig. 1 shows a major (A) and a minor (B) copper-containing peak. The first fraction is homogeneous by the criteria of HPLC (Fig. 2a) and amino acid analysis.[15] In contrast, the second fraction is heterogeneous as judged by HPLC analysis (Fig. 2b) and represents a mixture of four different Cu-MT species. They all have the same copper content; however, peaks 2–4 (Fig. 2b) contain an additional methionine residue at the amino-terminal end. Moreover, the Cu-MT species eluting as peaks 3 and 4 are characterized by a blocked amino terminus.[16]

[14] M. Beltramini and K. Lerch, *FEBS Lett.* **127,** 201 (1981).
[15] K. Lerch, *Chem. Scr.* **21,** 109 (1983).
[16] K. Lerch, unpublished data (1981).

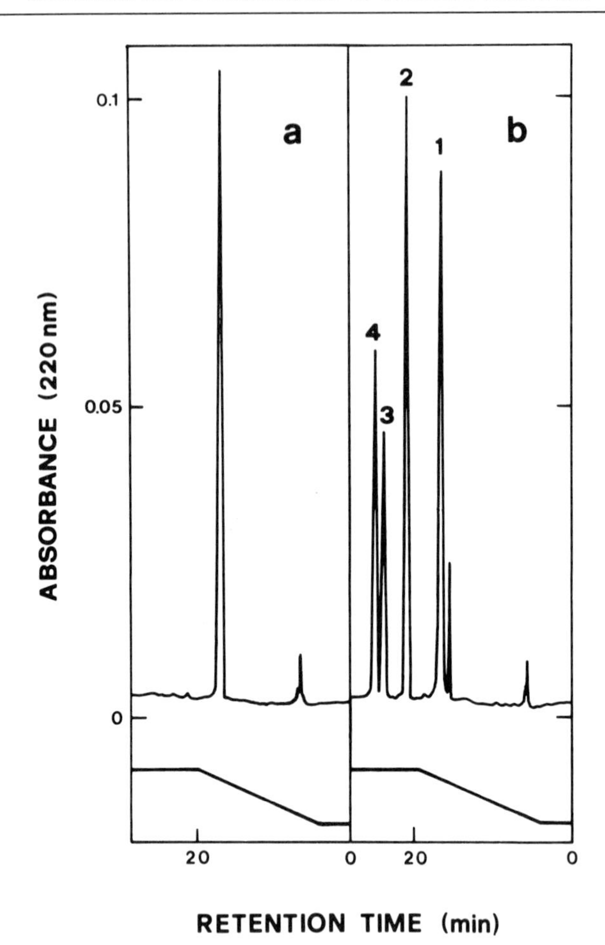

FIG. 2. HPLC profiles of MT-containing fractions after DEAE-cellulose chromatography (Fig. 1). Samples containing (a) 2 μg and (b) 5 μg of copper were applied, respectively.

Stability. *Neurospora crassa* Cu-MT can be stored in 0.1 *M* sodium phosphate, pH 7.5, at −80° or as a lyophilized powder at −20°. Under these conditions, MT preparations are stable for periods of over 1 year.

Physical and Chemical Properties. The molecular weight of *N. crassa* Cu-MT is 2600, including 6-g-atoms of copper/mol. The protein is devoid of free sulfhydryl and disulfide groups, indicating that all seven cysteinyl residues are involved in copper binding.[15] Like the zinc- and cadmium-binding mammalian MTs,[1] *N. crassa* Cu-MT has an asymmetric shape with a frictional ratio (f/f_{min}) of 1.4.[15]

Spectroscopic Properties. The absorption spectrum of freshly isolated *N. crassa* Cu-MT shows a broad band with a characteristic shoulder at 250 nm. In contrast, the circular dichroism profile is rather complex, exhibiting at least five Cotton maxima between 230 and 400 nm.[8] On UV excitation, *N. crassa* Cu-MT shows a unique luminescence characterized by a large Stokes shift with a maximum at 565 nm.[14] Lifetimes of 10.3 and 3.4 μ were measured for the protein in the absence and in the presence of molecular oxygen, respectively.[17] Results from X-ray absorption edge and extended X-ray absorption fine structure (EXAFS) spectroscopy of *N. crassa* Cu-MT are consistent with an average copper environment of three to four sulfur atoms at 2.20 Å, and one to two copper atoms at 2.71 Å, indicating a very compact and rigid Cu(I)–thiolate cluster for the protein.[9] Proton nuclear magnetic resonance (NMR) measurements of *N. crassa* Cu-MT show a marked absence of regular secondary structures, again supporting a tertiary structure determined primarily by metal ligation.[18] Although the isolated MT from *N. crassa* contains copper exclusively, the apoprotein binds group IIB metal ions such as Zn^{2+}, Cd^{2+}, Hg^{2+}, Ni^{2+}, and Co^{2+} with an overall metal-to-protein stoichiometry of 3. The spectroscopic properties of these derivatives are remarkably similar to those reported for the mammalian MTs.[19]

Acknowledgment

This work was supported by the Swiss National Foundation Grants 3.420-0.78, 3.709-1.80, and 3.285-0.82.

[17] M. Beltramini, G. M. Giacometti, B. Salvato, G. Giacometti, K. Münger, and K. Lerch, *Biochem. J.* **260,** 189 (1989).
[18] J. A. Malikayil, K. Lerch, and I. A. Armitage, *Biochemistry* **28,** 2991 (1989).
[19] M. Beltramini, K. Lerch, and M. Vašák, *Biochemistry* **23,** 3422 (1984).

[33] Purification of Prokaryotic Metallothioneins

By ROBERT W. OLAFSON

General Considerations

Metallothioneins (MTs) have been reported from a number of prokaryotic sources, including several species of *Synechococcus*[1,2] and *Pseudomonas putida,*[3] and appear to be found in all cyanobacteria investigated to date.[1,2,4] The best characterized prokaryotic MT has been isolated from the former *Synechococcus* sp.[4,5] The general approach to isolation of cyanobacterial metallothionein differs somewhat from the usual methodology for isolation of eukaryotic MTs in that ion-exchange chromatography has provided only broadly resolving peaks. The most useful technique for purification of MT-containing fractions from gel-permeation chromatography was found to be reversed-phase chromatography, which elicited a large array of highly purified MT species. None of the latter however, showed the markedly different amino acid compositions consistent with the designation isoprotein, as seen in eukaryotic species. With those proteins isolated by reversed phase, the variation in elution position was attributed to a combination of metal speciation and disulfide aggregation products.[3] Thus this methodology is capable of resolving metalloproteins that are virtually identical in amino acid composition.

Procedures

The Biological Sources. Synechococcus strains TX-20 (Stevens and Myers, 1976)[6] and UTEX-625 (Starr, 1964)[7] have been used extensively as a source of MT,[2,4] but as indicated above, MTs may be expressed in all cyanobacteria. Some strains of bacteria are capable of developing considerable resistance to cadmium and zinc as a function of MT induction[3,8] which can be very useful in enhancement of yield. Evidence also exists that these resistant strains develop as a function of episomal gene amplification and must be maintained in the presence of metal to maintain the capability of high-level resistance (R. W. Olafson, unpublished results).

The *Synechococcus* species were grown in axenic cultures in BG-11 medium[9] at 27° under fluorescent light (50 W/m^2) with slow addition of filtered air. Cells were harvested by centrifugation and broken in a French pressure cell at 0° in the presence of 0.5 *M* Tris-HCl, pH 8.6. This buffer was found necessary as cyanobacterial cell lysates can have a very high pH in the absence of buffer. Protease inhibitors have not been used routinely

[1] R. W. Olafson, K. Abel, and R. G. Sim, *Biochem. Biophys. Res. Commun.* **89,** 36 (1979).
[2] R. W. Olafson, *Int. J. Pept. Protein Res.* **24,** 303 (1984).
[3] D. P. Higham and P. J. Sadler, *Science* **225,** 1043 (1984).
[4] R. W. Olafson, W. D. McCubbin, and C. M. Kay, *Biochem. J.* **251,** 691 (1988).
[5] R. W. Olafson, *J. Electroanal. Chem.* **253,** 111 (1988).
[6] C. L. R. Stevens and J. Myers, *J. Physicians* **7,** 350 (1976).
[7] R. C. Starr, *Am. J. Bot.* **51,** 1034 (1964).
[8] R. W. Olafson, S. Loya, and R. G. Sim, *Biochem. Biophys. Res. Commun.* **95,** 1495 (1980).
[9] R. Rippka, J. Devuelles, J. B. Waterbury, M. Herdman, and R. Y. Stanier, *J. Gen. Microbiol.* **111,** 1 (1979).

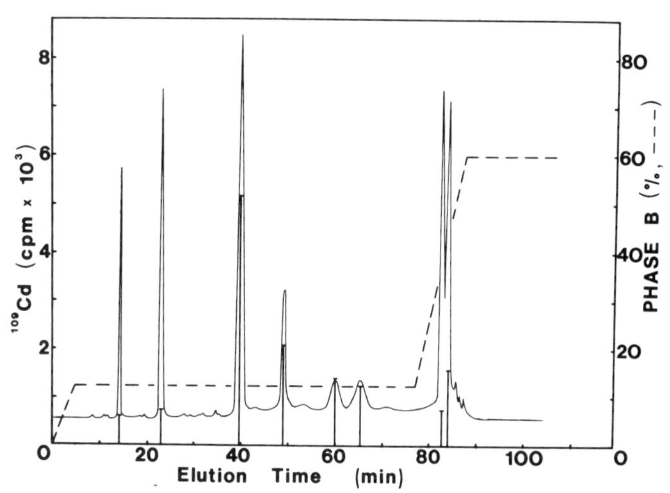

FIG. 1. Reversed-phase HPLC of the M_r 10,000 fraction obtained from gel-permeation chromatography of a *Synechococcus* cell lysate. Polarographic response to sulfhydryl groups was measurable in all but the first two peaks. ^{109}Cd counts are indicated by the histogram bars. [Reprinted with permission from R. W. Olafson, *Int. J. Pept. Protein Res.* **24,** 303 (1984).]

in this procedure, relying instead on rapid implementation of the following molecular sizing procedure employing gel-permeation chromatography in the cold.

Isolation Procedures. Cell lysates are clarified by centrifugation at 40,000 *g* for 15 min at 4° and applied to a Sephadex G-75 column equilibrated with 10 m*M* NH_4HCO_3 containing 5 m*M* 2-mercaptoethanol. The column is developed at the fastest flow rate allowable without column packing. Cytosol from 5 to 10 g of wet cells can be chromatographed on a 5 × 100 cm column. Detection of the eluted material was conveniently carried out by ultraviolet spectrophotometry at 250 nm and verification of the presence of MT made by differential pulse polarography[10] or be detection of the ^{109}Cd radionuclide, added at the stage of cell lysis. The pooled protein is then lyophilized directly in the presence of the volatile buffer.

The final purification involves reversed-phase high-performance liquid chromatography (HPLC), which is carried out on a 300-Å pore size propylsilane column (2.1 × 100 mm C_3) (RP300; Brownlee Aquapore, Santa Clara, CA) using a gradient HPLC system.[2] The column is developed at ambient temperature using 20 m*M* KH_2PO_4, pH 7.0, as solvent A and acetonitrile as organic modifier in solvent B. The effluent is monitored at

[10] R. W. Olafson and P.-E. Olsson, this volume [24].

250 nm, as in the preceding method. Up to 3 mg of the above lyophilized powder, dissolved in a minimum volume of buffer A, can be applied to the reversed-phase column. New columns have been observed to bind MT more strongly than used columns, such that the organic modifier concentration necessary for elution may decrease with successive applications of sample. Once the new column equilibrates the prokaryotic MTs can be resolved using a very shallow acetonitrile gradient. As can be seen in Fig. 1, a rapid ramp applied over the first 5 min to 10% acetonitrile results in the elution of several nonmetallothionein contaminants while MTs are eluted almost isocratically in the subsequent 70-min ramp to 12% acetonitrile. Depending on metal speciation, upward of seven MT isoforms have been isolated from cadmium-exposed *Synechococcus* species during this portion of the ramp.[4] An additional complex that has not been well investigated is eluted in a final ramp to 60% organic modifier and includes both MTs and pigmented contaminants (Fig. 1). Once the conditions for the shallow ramp have been attained the elution times are highly reproducible, using material isolated by gel-permeation chromatography. Fractions were routinely desalted by Sephadex G-25 gel-permeation chromatography and stored as lyophilized powder. No evidence of denaturation was found using the reversed-phase protocol, as assessed by spectropolarometry or cyclic voltametry.[4]

[34] Purification of Canine Hepatic Lysosomal Copper-Metallothionein

By RICHARD J. STOCKERT, ANATOL G. MORELL, and IRMIN STERNLIEB

Introduction

The hepatic copper concentration of the normal dog (91 – 358 μg/g dry tissue) is 10 times greater than that of normal human subjects.[1] Of this total 75% is in the cytosol and the rest is distributed among particulate fractions. The proportion is reversed in Bedlington terriers affected by hepatic copper toxicosis, a metabolic disorder that is transmitted in autosomal recessive fashion.[1,2] In these mutant dogs copper accumulates progressively to toxic concentrations in their livers, causing pathological

[1] D. C. Twedt, I. Sternlieb, and S. R. Gilbertson, *J. Am. Vet. Med. Assoc.* **175,** 269 (1979).
[2] J. Ludwig, C. A. Owen, S. S. Baram, J. T. McCall, and R. M. Hardy, *Lab. Invest.* **40,** 82 (1980).

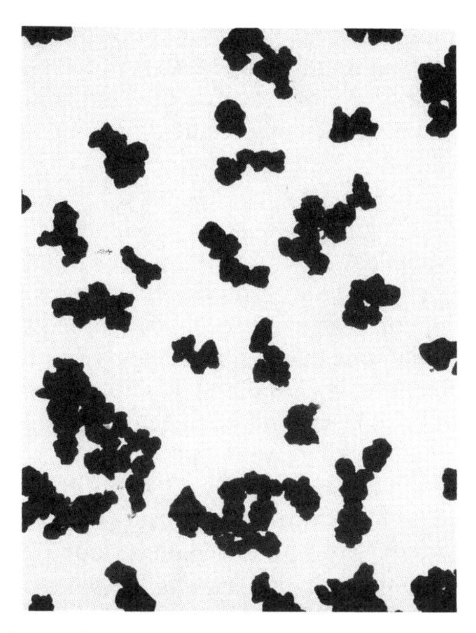

FIG. 1. Unfixed electron-dense granules from the pellet obtained below a 59% sucrose gradient after centrifugation at 65,000 *g* for 1 hr of the low-speed fraction (350 *g*) of liver homogenate from an affected female Bedlington terrier. The granules were placed on a Formvar-coated grid and examined in a Siemens Elmiskop 1a electron microscope (Siemens A. G., Berlin, Germany) without staining. Magnification: ×1350. [From G. F. Johnson, A. G. Morell, R. J. Stockert, and I. Sternlieb, *Hepatology* **1,** 245 (1989) with permission.]

changes in hepatic function and structure. The final result is cirrhosis, hepatic insufficiency, hemolytic anemia, and premature death.[1,2] Other breeds may also exhibit varying degrees of copper overload (West Highland White terriers,[3] Skye terriers,[4] Doberman pinschers[5]), but their biochemical defects have not been elucidated so far.

Excess copper in hepatocellular lysosomes can be demonstrated histochemically using rhodanine or rubeanic acid (for copper), or orcein or Victoria blue (for copper-binding protein).[6] In affected Bedlington terriers these organelles are encountered as prominent, unusually dark granules in histologic sections.[1] Their high density facilitates their isolation in pure

[3] L. P. Thornburg, G. Rottinghaus, M. McGowan, K. Kupka, S. Crawford, and S. Forbes, *Vet. Pathol.* **27,** 81 (1990).

[4] S. Haywood, H. C. Rutgers, and M. R. Christian, *Vet. Pathol.* **25,** 408 (1988).

[5] G. F. Johnson, D. A. Zawie, S. R. Gilbertson, and I. Sternlieb, *J. Am. Vet. Med. Assoc.* **180,** 1438 (1982).

[6] G. F. Johnson, S. R. Gilbertson, S. Goldfischer, P. S. Grushoff, and I. Sternlieb, *Vet. Pathol.* **21,** 57 (1984).

fractions from pellets of Bedlington liver homogenates by gradient centrifugation (Fig. 1). The lysosomal copper-binding liver protein (Ly-CuLP) isolated from this fraction was identified as metallothionein (MT) by amino acid analysis and sequencing studies.[7,8]

Preparation of LyCuLP

Liver specimens are obtained at surgery or necropsy from Bedlington terriers with hepatic copper concentrations that ranged from 3865 to 5550 μg/g dry tissue.[7] Copper is assayed spectrophotometrically following wet ashing with dicyclohexanone oxalyldihydrazone.[9] Soluble fractions are analyzed for copper and zinc by atomic absorption spectrophotometry. Results of the two methods are within 5% of each other for copper determinations.

Samples of liver are homogenized in 3 vol of 0.005 *M* Tris-HCl, pH 8.6, containing 0.15 *M* NaCl using a Polytron (Beckman, Fullerton, CA) for 30 sec at speed 4 at 4°. The homogenate is centrifuged at 350 *g* (at 4°) for 10 min and the pellet, which accounts for the major portion of the hepatic copper content (Table I), is used as the source for the purification of LyCuLP. The pellet is suspended in 3 ml homogenization buffer/g wet liver and recentrifuged at 350 *g* (at 4°) for 10 min. The washed pellet is resuspended in 5 ml/g liver in 0.1 *M* Na_2CO_3, pH 10.5, containing 0.15 *M* NaCl at 4°. The polymerized LyCuLP is solubilized by reduction with 2-mercaptoethanol (2-ME) to a final concentration of 1%, and stirring for 30 min at 4°. The extract is clarified by centrifugation at 105,000 *g* (at 4°) for 30 min. After addition of an equal volume of ice-cold acetone to the supernatant and centrifugation at 13,000 *g* (at 4°) for 20 min to remove the precipitate, which contains only traces of copper, more ice-cold acetone is added to a final concentration of 75% and LyCuLP is recovered by centrifugation at 13,000 *g* (at 4°) for 10 min. This crude preparation of LyCuLP is dissolved in water containing 1% (v/v) 2-mercaptoethanol and clarified by centrifugation at 20,000 *g* (at 4°) for 20 min. Soluble protein is resolved on a Sephadex G-50 column (1.5 × 90 cm) equilibrated with 0.005 *M* Tris-HCl, pH 7.6, 0.015 *M* NaCl containing 0.01% 2-mercaptoethanol. Low-molecular-mass (10 kDa) fractions with a ratio A_{250}/A_{280} greater than 1.8, indicating a high copper content, are pooled and lyophi-

[7] G. F. Johnson, A. G. Morell, R. J. Stockert, and I. Sternlieb, *Hepatology* **1,** 243 (1981).

[8] K. Lerch, G. F. Johnson, P. S. Grushoff, and I. Sternlieb, *Arch. Biochem. Biophys.* **243,** 108 (1985).

[9] A. G. Morell, J. Windsor, I. Sternlieb, and I. H. Scheinberg, *in* "Laboratory Diagnosis of Liver Diseases" (F. W. Sunderman and F. W. Sunderman, Jr., eds.), p. 196. Warren H. Green, St. Louis, Missouri, 1968.

TABLE I
RECOVERY OF COPPER AND ZINC IN A PREPARATION OF LYCULP[a]

Fraction	Copper (μg/g wet liver)	Zinc (μg/g wet liver)	Copper recovered (%)
Homogenate	1800	31.0	
350 *g* pellet	1340	4.6	100
2-ME soluble	1240	3.5	92
75% acetone insoluble	880	—	65
Sephadex G-50 eluate	800	—	59

[a] From G. F. Johnson, A. G. Morell, R. J. Stockert, and I. Sternlieb, *Hepatology* **1,** 245 (1989), with permission.

lized. Purity of the final preparation is indicated by a single band following SDS-PAGE on a 12% (w/v) gel stained for protein with Coomassie Brilliant Blue, and for copper by incubation for 30 min in a 0.2% aqueous solution of diethyl dithiocarbamate (DTC).

Characteristics

A molecular mass of 10 kDa was indicated by sodium dodecyl sulfate-polyacrylamide gel electrophoresis (SDS-PAGE).[7] This value was compared to that obtained by gel filtration of the linearized polypeptide. For the preparation of an alkylated derivative of LyCuLP, copper was displaced with *p*-mercuribenzoate (PMB). The PMB derivative was alkylated with ^{3}H-labeled iodoacetic acid. Apo-LyCuLP containing 1.5 μmol of PMB/mg was dissolved in 1.0 ml of 6 *M* guanidine hydrochloride in 0.5 *M* Tris-HCl, pH 7.8, and 0.1 *M* 2-mercaptoethanol and incubated under nitrogen at 22° overnight. Following the addition of 45 μmol of ^{3}H-labeled iodoacetic acid, the alkylation reaction was allowed to continue for 90 min at 22°. The resulting alkylated LyCuLP was isolated by Sephadex G-25 gel filtration with the recovery of 1.1 μmol of [^{3}H]acetate/mg protein. Gel filtration in 6 *M* guanidine hydrochloride indicated that the polypeptide contained 54 amino acid residues.[7] The values obtained by SDS-PAGE and by gel filtration are in reasonable agreement with those obtained for other MTs.[10]

Amino acid analysis of solubilized LyCuLP shows a 27% cysteine content, and almost complete absence of methionine, histidine, and aromatic amino acids.[7] Sequence analysis of the tryptic peptides of LyCuLP

[10] J. H. R. Kägi and Y. Kojima, *Experientia, Suppl.* **52,** 25 (1987).

showed all the cysteine residues matching those of other mammalian MTs.[8] A high degree of sequence identity corresponded to the flanking regions with 94.6% sequence homology between LyCuLP and human MT-2. However, the common *N*-acetylmethionine structure of mammalian MTs is absent from LyCuLP and the amino terminus is heterogeneous and blocked. The data suggest the existence of two isoforms of LyCuLP.[8]

The 7 or 8 g-atoms of copper, all electron paramagnetic resonance silent, cannot be removed by dialysis or by passage through a Chelex 100 ion-exchange column. This copper does not react directly with DTC, indicating a tight copper–protein bond. In contrast to MTs isolated from other species, there is virtually no zinc present in purified LyCuLP. The ultraviolet spectrum of purified LyCuLP shows higher absorption at 250 than at 280 nm, characteristic of the copper–sulfur chromophore (A_{250} of 6.080/mol copper). Acidification to pH 1.0 abolishes the chromophore, as evidenced by a change in the UV spectrum. X-Ray edge and extended X-ray absorption fine-structure spectroscopic studies (EXAFS) suggest that the univalent copper in LyCuLP is arranged in an adamantane-like cluster.[11]

The copper-free apoprotein is unstable, as indicated by a broad Sephadex G-50 elution profile. To prepare a stable PMB-derivatized apoprotein capable of binding copper, as indicated by the restoration of the characteristic chromophore in the ultraviolet region of the optical spectrum, a 1% solution of lyophilized LyCuLP, containing 1.5 μmol of copper/mg protein, was prepared in water containing 8 μmol of PMB and 3 μmol of ethylenediaminetetraacetic acid (EDTA)/mg protein. The solution was incubated at 4° for 30 min, after which the PMB–LyCuLP derivative was separated from the unreacted PMB, EDTA, and EDTA–copper chelate by gel filtration on Sephadex G-25 equilibrated with 0.1 *M* Na_2CO_3, pH 10.5, containing 0.15 *M* NaCl. Residual copper was removed by precipitation of the PMB–LyCuLP by adjusting the pH to 4.25 with 0.2 *N* acetic acid and washing the precipitate with 0.1 *N* sodium acetate, pH 4.25, containing 0.01 *M* EDTA. To displace the PMB, the copper-free precipitate was dissolved in carbonate buffer containing 1% 2-mercaptoethanol. The apoprotein was recovered in the void fraction of a Sephadex G-25 column. The holoprotein was reconstituted by adding 1.5 μmol of copper sulfate, dissolved in carbonate–mercaptoethanol buffer, per milligram apoprotein. After a 10-min incubation at 22° the holoprotein was recovered by gel filtration on Sephadex G-25 equilibrated with carbonate buffer. The similarity of the kinetics of the reaction with DTC of the reconstituted holoprotein with those of native LyCuLP suggests that copper is strongly

[11] J. H. Freedman, L. Powers, and J. Peisach, *Biochemistry* **25,** 2342 (1986).

chelated by both forms.[7] Increased binding of metal to reconstituted holoprotein can be observed in the presence of greater than stoichiometric concentrations of copper. This excess copper, however, reacts promptly with DTC, indicating binding to low-affinity sites.

It is still uncertain whether the accumulation of polymerized Cu-MT in the hepatocellular lysosomes of certain Bedlington terriers is the result of unregulated synthesis of a normal protein that is sequestered by lysosomes secondarily, or whether the accumulation of hepatic copper induces synthesis and polymerization of cytosolic MT that overwhelms the catabolic and excretory capacities of the lysosomes.

[35] Metallothioneins of Monocytes and Lymphocytes

By MASAMI KIMURA

Many experimental studies on the synthesis of metallothionein (MT) in animals and cells have been reported, but the induction of this protein in peripheral blood lymphocytes and monocytes has not been well researched. In 1979, Phillips[1] reported the zinc-induced synthesis of a low-molecular-weight zinc-binding protein by human lymphocytes. He showed that the incubation of lymphocytes with zinc-transferrin induced the synthesis of this protein, and that it might be characterized as metallothionein by electrophoretic as well as immunological methods, and by such properties as absorbance maximum at 220 nm, zinc content of 5.8 mol/6600 Da, and sulfhydryl content of 20.2 mol/6600 Da. Detectable synthesis occurred 5 hr after initiation of lymphocyte culture and reached a maximum after 24–30 hr. However, the abnormal lymphocytes from donors with chronic lymphocytic leukemia or lymphoblastic leukemia did not synthesize metallothionein (presented as peak-3 protein in Sephadex G-75 gel filtration) in response to increased zinc in the incubation medium to the same extent as normal lymphocytes. It was pointed out that lymphocyte metallothionein might serve in the homeostatic regulation of lymphocyte zinc metabolism.

In the same year, Hildebrand and Cram[2] reported the distribution of cadmium in human blood cultured in the presence of low levels of $CdCl_2$. In erythrocytes, cadmium was bound to several macromolecular species, none of which corresponded to specific, inducible, cadmium-binding me-

[1] J. L. Phillips, *Biol. Trace Elem. Res.* **1,** 359 (1979).

[2] C. E. Hildebrand and L. S. Cram, *Proc. Soc. Exp. Biol. Med.* **161,** 438 (1979).

tallothionein. In contrast, in lymphocytes most of the cadmium was found associated with a low-molecular-weight metallothionein. They suggested that cadmium exposure might alter normal immunocompetence associated with lymphocytes on the basis of these studies, which showed the accumulation of cadmium by lymphocytes and their sequestering of cadmium in metallothionein.

Differences in the inducibility of metallothionein between human peripheral blood lymphocytes and other cell lines such as human hepatocytes (Chang liver) and cervical carcinoma cells (HeLa) were studied by Kobayashi and Kimura.[3] From the results obtained using Sephadex G-75 gel filtration of the cells, it appeared that human lymphocytes have lower or slower inducibility of metallothionein synthesis due to cadmium exposure than other cells.

The cadmium-induced metallothionein synthesis in cultured peripheral human blood cells such as lymphocytes, monocytes, and polymorphonuclear cells (granulocytes) was further studied by Enger *et al.*[4] The response of monocytes and lymphocytes to cadmium cytotoxicity was similar: Monocytes were somewhat more sensitive than lymphocytes. Polymorphonuclear cells were found to be more resistant to cadmium than were mononuclear cells (25% monocytes, 75% lymphocytes). Most of the cadmium incorporated by blood cells was taken up by nucleated cells. Polymorphonuclear cells incorporated more cadmium at high doses than did mononuclear cells, but cadmium was not bound to metallothionein in polymorphonuclear cells following exposure to cadmium. In mononuclear cells, however, metallothionein synthesis was induced rapidly following exposure to cadmium.

Metal cation influences on the immune response have been reported in a wide variety of experimental systems. These influences can either result in the augmentation or suppression of immunological activities. Most recently it was shown that a metal cation-induced metallothionein may play a biological role in immune systems. Oberbarnscheidt *et al.*[5] demonstrated that the supernatant obtained from human peripheral mononuclear cells stimulated by lipo-polysaccharide (LPS) or concanavalin (Con A) increased the MT concentration in a human immunoglobulin M (IgM)-secreting B cell line (RPMI 1788). However, LPS and Con A failed directly to increase MT concentration in RPMI 1788, whereas they increased MT concentration in human peripheral mononuclear cells. Lynes *et al.*[6] noted

[3] S. Kobayashi and M. Kimura, *Toxicol. Lett.* **5,** 357 (1980).

[4] M. D. Enger, C. E. Hildebrand, and C. C. Stewart, *Toxicol. Appl. Pharmacol.* **69,** 214 (1983).

[5] J. Oberbarnscheidt, P. Kind, J. Abel, and E. Gleichmann, *Res. Commun. Chem. Pathol. Pharmacol.* **60,** 211 (1988).

[6] M. A. Lynes, J. S. Garvey, and D. A. Lawrence, *Mol. Immunol.* **27,** 211 (1990).

that thionein, either as apothionein or when complexed as Cd,Zn-MT, Zn-MT, or Cd-MT, is capable of inducing lymphocyte proliferation. Some metal influences on lymphocytes might be through a thionein intermediary.

As described above, metallothionein is probably not only important in biological functions for metal homeostasis and protection against metal toxicity but also for the immune response.

Separation of Human Lymphocytes and Monocytes[7]

Human peripheral blood is taken from healthy adults for routine preparation of lymphocytes and monocytes. The heparinized blood is diluted two-fold with phosphate-buffered saline (PBS), and centrifuged at 400 g_{max} for 35 min on Ficoll-Paque (specific gravity, 1.077). The mononuclear cells are collected from the top of the Ficoll layer. After washing with RPMI 1640 medium, the cells are incubated at 37° for 1 hr in plastic dishes with the same medium supplemented with 10% fetal calf serum (FCS). Nonadherent cells (lymphocytes) are removed and the cells adhering to the plastic substratum (monocytes) are collected with a rubber policeman.

Immunofluorescent staining with mouse anti-human Leu M1 (for macrophages), anti-human Leu 1 (for T cells), or anti-human HLA-DR IgG (for B cells and macrophages) as the first antibody and fluorescein isothiocyanate-conjugated rabbit anti-mouse IgG serum as the second antibody, indicates that the purities of the lymphocyte and monocyte fractions obtained by this method are typically 75 and 80%, respectively.

Separation of Human T and B Cells by Fluorescence-Activated Cell Sorter[8]

Peripheral blood mononuclear cells are isolated from the blood of normal healthy adults by centrifugation through a Ficoll-Paque step gradient. Adherent monocytes are then separated from lymphocytes by incubation on a plastic substratum at 37° for 1 hr in a CO_2 incubator.

The cells are incubated for 30 min on ice with a saturating concentration of Fl-anti-Leu 1 (fluorescein-conjugated mouse monoclonal antibodies to human T cells) or anti-Leu 12 (mouse monoclonal antibodies to human B cells), then washed twice in RPMI 1640 containing 10% heat-inactivated FCS. FCS is heat inactivated by incubating it at 56° for 30 min. For indirect immunofluorescent staining lymphocytes treated with anti-

[7] T. Sone, S. Koizumi, and M. Kimura, *Chem.-Biol. Interact.* **66,** 61 (1988).

[8] H. Yamada, S. Minoshima, S. Koizumi, M. Kimura, and N. Shimizu, *Chem.-Biol. Interact.* **70,** 117 (1989).

Leu 12 are incubated with Fl-anti-mIg [$F(ab')_2$ fragments of sheep antibodies raised against mouse immunoglobulin] for another 30 min on ice. Cells are washed as above. Flow cytometry and sorting are performed on a fluorescence-activated cell sorter (FACS 440; Beckton Dickinson, Mountain View, CA). An argon ion laser is used for excitation of fluorescein at a wavelength of 488 nm and a constant power output of 300–500 mW. A band-pass filter BP530/30 is used to prevent incident light from interfering with emitted light. Calcium- and magnesium-free phosphate-buffered saline is used as a sheath fluid. A plastic nozzle with a 50-μm pore is used. Cells are sorted into glass Petri dishes filled with 1 ml of RPMI 1640 medium containing 10% heat-inactivated FCS. The cells are collected by centrifugation and washed twice with cysteine-free RPMI 1640 supplemented with 10% heat-inactivated FCS. Viability of the sorted cells is 96% by Trypan Blue exclusion. When the cell population is sorted again after the first cell sorting to examine their purity, more than 96% of the populations is recovered.

Anti-Leu 1 is used for purification of human T cells because it specifically binds T cells in normal peripheral blood, but does not stimulate their proliferation. Anti-Leu 4, another T cell-specific monoclonal antibody, is reported to be a mitogenic stimulant of the target cells. On the other hand, anti-Leu 12 is used to separate B cells from non-B cells. The fluorescently labeled cells are separated from nonfluorescent cells via FACS. Approximately 80% of the cells are fluorescently labeled with anti-Leu 1 and 10% with anti-Leu 12.

Separation of Human Granulocytes[4]

The red cell pellets obtained from human peripheral blood using Ficoll-Hypaque are diluted gradually with 20 vol of 0.83% (w/v) NH_4Cl, and incubated 20 min at 37° to promote red cell lysis. Unlysed cells are collected by centrifugation at 150 *g* for 10 min at 20–22°. These cells are washed by suspension and centrifugation two times with 50 ml of αMOPS [Gibco αMEM (minimal essential medium; GIBCO, Grand Island, NY) buffered with 4 g/liter MOPS (morpholinopropanesulfonic acid) and containing 100 units/ml of penicillin and 100 μg/ml of streptomycin] and one time with RPMI 1640 and suspended in RPMI 1640 for culture.

Synthesis of Metallothionein in Human Blood Cells and Lymphoid Cells[5,7,8]

Lymphocytes or monocytes (8.8×10^6 cells/2 ml) are incubated with 8.9 μM $CdCl_2$ plus $^{109}CdCl_2$ (5 μCi) at 37° under a 5% (v/v) CO_2 atmosphere for 18 hr in RPMI 1640 medium supplemented with 10% (v/v)

FCS, 100 units/ml of penicillin, and 100 μg/ml of streptomycin. Alternatively the cells are incubated with 8.9 μM $CdCl_2$ plus [^{35}S]cysteine (10 μM) or [^{35}S]cysteine alone in cysteine-free medium.

For one assay, T cells or B cells (5×10^5 cells) are labeled with 10 μCi of [^{35}S]cysteine in the presence of 10 μM $CdCl_2$ in cysteine-free RPMI 1640 containing 10% FCS at 37° for 18 hr (0.5×10^6 cells/0.2 ml of medium/cm^2) in 5% CO_2 and 100% humidity.

Lymphoid cells are cultured in the same medium under the above conditions. The cells (10^6 cells/ml) are labeled with 10 μCi of [^{35}S]cysteine for 18 hr in the presence or absence of 8.9 μM $CdCl_2$ in cysteine-free RPMI 1640 medium with 10% fetal calf serum, 100 units/ml of penicillin, and 100 μg/ml of streptomycin.

The human peripheral mononuclear cells (2.5×10^6 cells in 10 ml) are incubated for 48 hr with LPS (10–40 μg/ml dissolved in 0.15 M NaCl) or Con A (2.5 or 5 μg/ml dissolved in balanced salt solution)[8a] in RPMI 1640 medium supplemented with 10% FCS, glutamine (200 mM), penicillin (2×10^4 IU/100 ml), and gentamycin sulfate (10 mg/100 ml). At the end of the incubation period, cell suspensions are transferred to Falcon (Oxnard, CA) tubes and centrifuged (10 min, 250 g, 4°). After centrifugation, the supernatant is removed, sterilized by filtration, and used immediately in further experiments. The cells are stored at −70° until MT determination.

Identification and Determination of Metallothioneins Synthesized in Human Blood Cells[7]

The labeled cells are lysed by freezing and thawing twice in 20 mM ammonium carbonate or 10 mM Tris-HCl (pH 7.4)/0.15 M NaCl/0.02% NaN_3 and centrifuged at 15,000 g_{max} for 2 min. The supernatant is chromatographed on a Sephadex G-75 column. Radioactivity of ^{35}S is counted by a liquid scintillation spectrometer and ^{109}Cd is measured by an autogamma scintillation spectrometer.

The fraction obtained from the cadmium-exposed and [^{35}S]cysteine-labeled cells is lyophilized, carboxymethylated, and analyzed by SDS-polyacrylamide gel electrophoresis as described elsewhere in this volume [17].

A portion of the extract (40 μl) is mixed with 10 μl of normal rabbit serum or anti-MT serum and incubated at 4° for 1 hr. Goat anti-rabbit IgG serum (7.5 μl) is then added and the mixtures are further incubated for 6 hr. Precipitates are collected by centrifugation, washed twice with the same buffer, and finally dissolved in the same buffer for electrophoresis.

The amounts of metallothioneins induced in cultured cells and tissues

[8a] R. Dulbeco and P. Vogt, *J. Exp. Med.* **99**, 167 (1954).

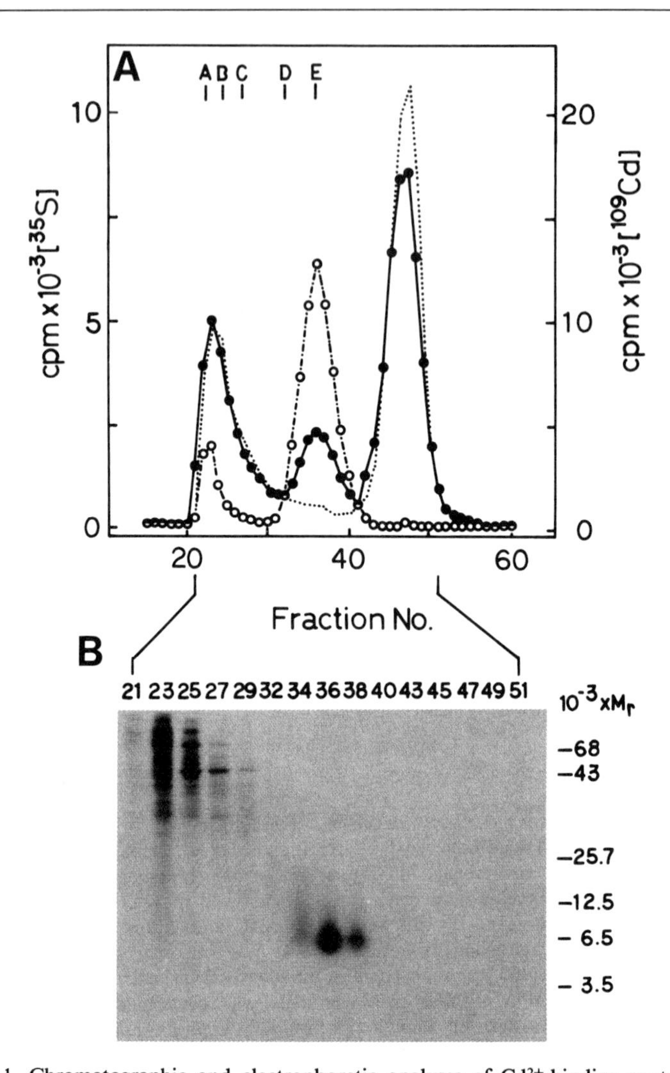

FIG. 1. Chromatographic and electrophoretic analyses of Cd^{2+}-binding proteins from human peripheral blood lymphocytes. (A) Extracts of the lymphocytes incubated with 8.9 μM $^{109}CdCl_2$ (○), 8.9 μM $CdCl_2$ and [^{35}S]cysteine (●), or [^{35}S]cysteine (···) for 18 hr were chromatographed on a Sephadex G-75 column. The letters A–E indicate the positions of molecular weight markers: A, Blue dextran (2000K); B, bovine serum albumin (68K); C, ovalbumin (43K); D, chymotrypsinogen A (25.7K); E, ribonuclease A (13.7K). (B) The fractions obtained from the extract of the Cd^{2+}-exposed and [^{35}S]cysteine-labeled cells (A) were analyzed by SDS–15% polyacrylamide gel electrophoresis. The fluorogram prepared from the gel is shown. The positions of molecular weight marker proteins are indicated on the right. The markers used were bovine serum albumin (68K), ovalbumin (43K), chymotrypsinogen A (25.7K), cytochrome *c* (12.5K), trypsin inhibitor (6.5K), and insulin B chain (3.5K).

by a variety of substances, including several metals, hormones, and immunostimulatory substances, such as lipopolysaccharide and interferon, can also be determined by radioimmunoassay as well as the cadmium saturation method as described elsewhere in this volume [17].

Results and Discussion

In our laboratory, nonadherent lymphocytes separated from human peripheral blood were cultured with 8.9 μM ^{109}Cd (1 μg/ml of Cd^{2+}) for 18 hr. The extract prepared from the cells was chromatographed on a Sephadex G-75 column (Fig. 1A). The majority of ^{109}Cd (79% of total ^{109}Cd in the extract) was found to be associated with a low-molecular-weight peak (fractions 31–42 in Fig. 1A). Its molecular weight was estimated to be 13,000 relative to the marker proteins. Another culture was incubated with the same concentration of cold Cd^{2+} and [^{35}S]cysteine, and its extract was analyzed in the same manner. [^{35}S]Cysteine was also incorporated into the low-molecular-weight peak (Fig. 1A). In contrast, no corresponding peak was observed when [^{35}S]cysteine-labeled cells not exposed to Cd^{2+} were analyzed (Fig. 1A). These results suggest that the Cd^{2+}-binding peak represents the proteins that were newly synthesized after Cd^{2+} exposure. The chromatographic fractions from the ^{35}S-labeled, Cd^{2+}-exposed cells were electrophoresed on a sodium dodecyl sulfate (SDS)-polyacrylamide gel (Fig. 1B). The major components of the Cd^{2+}-binding peak had a mobility quite similar to thioneins (apoproteins of MTs) of HeLa cells[9] (see Fig. 2A.).

To confirm the inducible nature of the MT-like proteins from human lymphocytes, extracts prepared from Cd^{2+}-exposed and unexposed lymphocytes were analyzed by gel electrophoresis. The proteins were detected in Cd^{2+}-exposed lymphocytes (Fig. 2A, lane 3), but not in unexposed control lymphocytes (Fig. 2A, lane 2). Monocytes separated from human peripheral blood were also found to synthesize the same proteins in response to Cd^{2+} (Fig. 2A, lane 5; see control in lane 4). The induced protein bands consist of at least two components, which differ slightly from HeLa thioneins (Fig. 2A, lane 1) in mobility. Antigenicities of these proteins were determined with anti-mouse MT-2 serum, which cross-reacts with human MTs.[10] The proteins from both lymphocytes and monocytes were specifically immunoprecipitated by this serum (Fig. 2B, lanes 2 and 5), but not by normal rabbit serum (Fig. 2B, lanes 1 and 4). We thus conclude that both

[9] S. Koizumi, N. Otaki, and M. Kimura, *J. Biol. Chem.* **260,** 3672 (1985).

[10] S. Koizumi, T. Sone, N. Otaki, and M. Kimura, *Biochem. J.* **227,** 879 (1985).

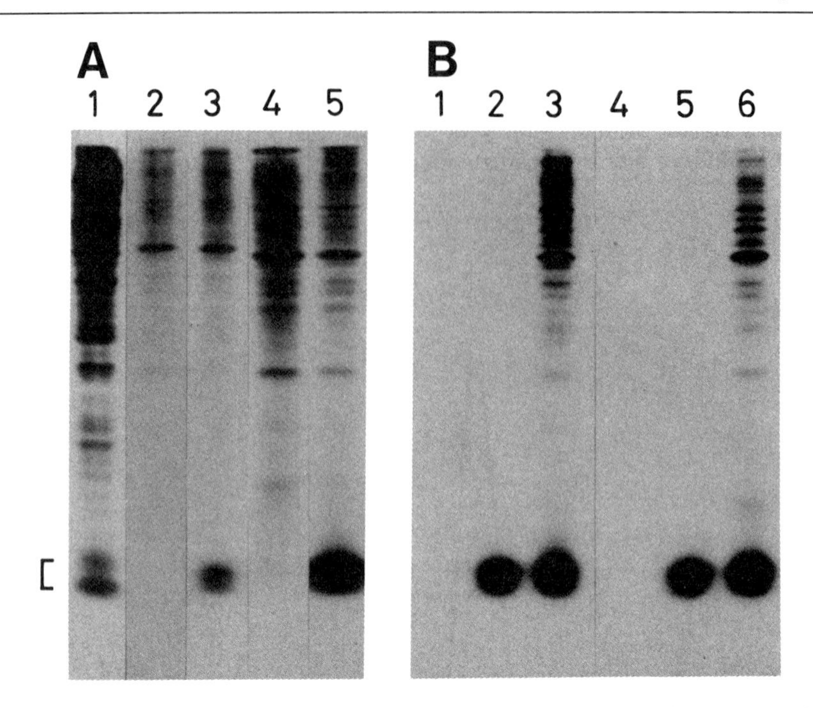

FIG. 2. Cd^{2+}-induced synthesis of MTs in human lymphocytes and monocytes. (A) SDS-15% polyacrylamide gel electrophoresis of [^{35}S]cysteine-labeled proteins from lymphocytes and monocytes cultured with or without 8.9 μM $CdCl_2$ for 18 hr. The fluorogram prepared from the gel is shown. Lane 1, Cd^{2+}-exposed HeLa cells; lane 2, unexposed lymphocytes; lane 3, Cd^{2+}-exposed lymphocytes; lane 4, unexposed monocytes; lane 5, Cd^{2+}-exposed monocytes. The bracket indicates the position of HeLa cell thioneins in lane 1. Molecular weights of the two thionein bands are 6100 (upper band) and 4900 (lower band), respectively. (B) SDS-15% polyacrylamide gel electrophoresis of immunoprecipitated MTs. Immunoprecipitates from [^{35}S]cysteine-labeled extracts of Cd^{2+}-exposed lymphocytes and monocytes were electrophoresed and the gel was subjected to fluorography. Lanes 1-3, lymphocytes; lanes 4-6, monocytes. Lanes 1 and 4, precipitates by normal rabbit serum; lanes 2 and 5, precipitates by rabbit anti-mouse liver MT-2 serum; lanes 3 and 6, untreated extract.

human peripheral blood lymphocytes and monocytes synthesize MTs in response to Cd^{2+}.

Incorporation of Cd^{2+} and relative metallothionein synthesis (thionein/nonthionein $\times$ 100) in lymphocytes and monocytes from six different cell donors (adult males) are shown in Table I. The mean values for relative metallothionein synthesis are 20.2 ± 2.3 and 34.4 ± 5.4 in lymphocytes and monocytes, respectively.

The amounts of metallothionein induced in both cell types with a 12-hr exposure to 10 μM $CdCl_2$ were determined by the ^{109}Cd saturation

TABLE I
INCORPORATION OF Cd^{2+} AND RELATIVE METALLOTHIONEIN SYNTHESIS IN MONOCYTES AND LYMPHOCYTES[a]

Donor	Cells	Incorporation of Cd^{2+} (10^{-12} mol/10^6 cells)	Relative MT synthesis (%)
A	Monocytes	109.8 ± 9.3	16.2 ± 2.0
	Lymphocytes	31.0 ± 0.9	19.9 ± 2.3
B	Monocytes	119.7 ± 26.8	51.3 ± 12.5
	Lymphocytes	41.2 ± 3.9	19.2 ± 1.1
C	Monocytes	79.0 ± 6.5	10.5 ± 0.4
	Lymphocytes	30.1 ± 1.2	14.5 ± 1.4
D	Monocytes	ND[b]	19.2 ± 6.1
	Lymphocytes	26.0 ± 2.5	13.2 ± 0.9
E	Monocytes	178.2 ± 51.0	74.0 ± 7.6
	Lymphocytes	51.2 ± 4.6	34.0 ± 5.8
F	Monocytes	108.2 ± 8.1	19.0 ± 4.0
	Lymphocytes	31.7 ± 2.1	20.3 ± 2.5

[a] Cultures were run in triplicate for monocytes and lymphocytes. Values were expressed as the mean ± SD.
[b] ND, Not determined.

method.[11] Metallothionein was induced in both lymphocytes and monocytes about 10-fold: 57×10^{-5} ($+Cd^{2+}$) versus 6×10^{-5} ($-Cd^{2+}$) molecules/cell for monocytes; 18×10^{-5} ($+Cd^{2+}$) versus 2×10^{-5} ($-Cd^{2+}$) for lymphocytes. Polymorphonuclear cells expressed relatively little MT (0.6×10^{-5} molecules/cell). Variation in MT levels in peripheral blood mononuclear leukocytes (85% lymphocytes and 15% monocytes) before and after cadmium exposure (10 μM) for 12 hr in culture for seven normal adult donors was measured. The mean values of the basal and induced levels were 2.7 ± 0.9 and $22.1 \pm 7.4 \times 10^{-5}$ MT (molecules/cell), 13-fold, which reflected an approximately 2-fold variation in both basal and induced levels of MT. Many factors, such as genetic differences in metal uptake or MT expression and differences in copper, zinc, or glucocorticoid levels, probably account for the observed variation between donors in both basal and induced levels of MT. Such variation could also reflect experimental error and/or the specific immune and endocrine status of the individual at the time of blood collection.

T and B cells were fractionated from human peripheral blood lymphocytes with a fluorescence-activated cell sorter (FACS).[8] Each cell population was cultured in a medium containing [^{35}S]cysteine. The proteins

[11] C. B. Harley, C. R. Menon, R. A. Rachubinski, and E. Nieber, *Biochem. J.* **262**, 873 (1989).

synthesized in the presence or absence of Cd^{2+} were analyzed by SDS-polyacrylamide gel electrophoresis. The purified T cells and unfractionated lymphocytes produced low-molecular-weight proteins in response to 10 μM Cd^{2+}. These protein bands were identified as MTs by immunoprecipitation. Also, the purified B cells synthesized MTs in the presence of 10 μM Cd^{2+}. These results demonstrate that both T and B cells have the ability to synthesize MTs in response to Cd^{2+}. It is of interest to determine whether T and B cells have different capacities to synthesize MTs. However, no apparent difference in the amounts of synthesized MTs was found between T and B cells in the present study.

Mayo and Palmiter[12] have reported that a mouse thymoma cell line and a mouse lymphoma cell line produce almost no MT-1 mRNA in response to Cd^{2+} and dexamethasone. In order to know whether suppression of MT production such as this can also occur in human lymphoid cell lines, we examined MT induction by Cd^{2+} in 3 human T cell lines and 10 human B cell lines.[7] In response to 8.9 μM Cd^{2+}, all of the cell lines tested synthesized thioneins, as shown in Table II. Both of the two major isoforms are expressed in all of the cell lines tested, as in peripheral blood lymphocytes. The level of relative thionein synthesis [(thionein/nonthionein proteins $\times$ 100] differs among the cell lines. Some of the B cell lines, including RPM1 8226 (myeloma), Raji, P3HR-1 (derived from Burkitt's lymphoma), Wa, and KCCL (Epstein–Barr virus transformed), showed relatively little expression of MTs. From the experimental results, in which all of human lymphoid cell lines have the capacity to synthesize MTs, they can also induce MT mRNA in response to Cd^{2+} and thus differ from mouse myeloma and lymphoma cell lines.

Harley *et al.*[11] have reported levels of MT mRNA determined in parallel with measurements of MT protein in human peripheral blood leukocytes by semiquantitative dot-blot analysis by using synthetic oligonucleotides complementary to MT mRNA. Metallothionein mRNA levels of cultured mononuclear cells reached a peak after 6–12 hr when incubated with 10 mM $CdCl_2$. Continued accumulation of MT protein after the point at which MT mRNA reaches a plateau presumably reflects the long half-life of Cd-MT protein ($>$ 1 day) compared with MT mRNA ($<$ 1 day). Metallothionein mRNA levels corresponded qualitatively to expression of protein in these cells. Moreover, individual variation in MT mRNA levels in peripheral blood mononuclear leukocytes before and after exposure to 10 μM Cd^{2+} for 6 hr was determined for 39 normal adult donors aged 20–40 years. Both basal and induced MT mRNA levels varied markedly from individual to individual, giving rise to a large range in induction ratio

[12] K. E. Mayo and R. Palmiter, *J. Biol. Chem.* **256,** 2621 (1981).

TABLE II
METALLOTHIONEIN SYNTHESIS IN HUMAN LYMPHOID CELL LINES[a]

Cell	Type	Origin	Relative MT synthesis (%)	
			Experiment 1	Experiment 2
CEM-5	T	Leukemia	30.0	22.3
M-8	T	Leukemia	27.8	24.7
Molt-4	T	Leukemia	27.6	23.7
RPM1-8226	B	Myeloma	5.9	8.8
Raji	B	Burkitt lymphoma	6.0	5.7
P3HR-1	B	Burkitt lymphoma	3.5	1.7
EB-SAS	B	EBV transformed[b]	27.6	36.2
EB-GOT	B	EBV transformed	22.7	17.6
Wa	B	EBV transformed	10.6	6.8
EB-YN	B	EBV transformed	18.6	ND[c]
EB-HU	B	EBV transformed	21.8	ND
Cold-47	B	EBV transformed	15.0	ND
KCCL	B	EBV transformed	7.7	ND

[a] [^{35}S]Cysteine-labeled extracts from the lymphoid cells cultured with or without 8.9 μM $CdCl_2$ for 18 hr were analyzed on SDS–15% polyacrylamide gels. The thionein bands in the fluorogram were densitometrically quantified. Relative thionein synthesis represents (thionein/nonthionein proteins) × 100.

[b] EBV, Epstein–Barr virus.

[c] ND, Not determined.

[4.3- to 38-fold; mean = 17 ± 8 (SD)]. They explained that such variation in both basal and induced MT mRNA levels reflected environmental or experimental (intraindividual) and possibly genetic (interindividual) differences.

Metallothionein is inducible by heavy metals, glucocorticoid hormones, interferon, and interleukin 1 *in vitro.* As described above, human peripheral blood leukocytes and monocytes are able to induce the synthesis of MT by exposure to heavy metals such as Zn^{2+} and Cd^{2+}. T and B cells can also synthesize MT after exposure to these metals. Metallothionein could play a role in maintaining the homeostasis of essential metals in cells and in preventing toxicity caused by metals incorporated into cells. However, it has been reported that thioneins complexed with certain metal cations are detrimental to the normal cellular activity of lymphocytes.[6]

On the other hand, products obtained from peripheral mononuclear cells and murine spleen cells stimulated by LPS or Con A are able to induce MT concentrations in the human IgG-secreting B cell line RPM1

1788.[5] But LPA or Con A themselves do not change the MT concentration in RPM1 1788. Also, others have reported that LPS or endotoxin added to human hepatic Chang liver cells fails to enhance MT concentration but the supernatants from LPS- or endotoxin-stimulated rat macrophages do.[13,14] These results strongly suggest that MT has an important function in the immune system during immunostimulation. Such mechanisms of MT function in the immune system require further investigation.

[13] Y. Iijima, T. Takahashi, T. Fukushima, S. Abe, Y. Itano, and F. Kosaka, *Toxicol. Appl. Pharmacol.* **89,** 135 (1987).
[14] Y. Iijima, T. Fukushima, and F. Kosaka, *Biochem. Biophys. Res. Commun.* **164,** 114 (1989).

[36] Role of Metallothionein in Essential, Toxic, and Therapeutic Metal Metabolism in Ehrlich Cells

By SUSAN K. KREZOSKI, C. FRANK SHAW III, and DAVID H. PETERING

Introduction

Ehrlich ascites tumor cells grown in animals or culture have been used to study the cellular roles of metallothionein (MT) in zinc metabolism and its interactions with copper, platinum, and cadmium.[1-4] Because the basal level of Zn-MT in these cells is easily measured by flame atomic absorption spectrophotometry after Sephadex G-75 chromatography, it is possible to examine the physiological properties of native metallothionein in the absence of inducers. Such characteristics are thought to relate to some of its normal functions in cellular metal metabolism. In addition, one can investigate the reactions of metallothionein with metallic species in relationship to the toxic effects of such reagents. The methods described below illustrate

[1] A. J. Kraker, G. Krakower, C. F. Shaw III, D. H. Petering, and J. S. Garvey, *Cancer Res.* **48,** 3381 (1988).
[2] A. Kraker, S. Krezoski, J. Schneider, D. Minkel, and D. H. Petering, *J. Biol. Chem.* **260,** 13710 (1985).
[3] A. Kraker, J. Schmidt, S. Krezoski, and D. H. Petering, *Biochem. Biophys. Res. Commun.* **130,** 786 (1985).
[4] J. Koch, S. Wielgus, B. Shankara, L. A. Saryan, C. F. Shaw III, and D. H. Petering, *Biochem. J.* **189,** 95 (1980).

types of experiments that can be done with cultured Ehrlich cells to address some of the cellular properties of metallothionein.

Growth of Ehrlich Cells

Ehrlich cells can readily be grown in mice and in culture. Therefore, it is possible to compare results of similar experiments done under these two conditions.

Mice

Outbred female Swiss mice are inoculated intraperitoneally with 1×10^7 log-phase cells in 0.1 ml phosphate-buffered saline (PBS) (0.15 *M* NaCl and 10 m*M* sodium phosphate, pH 7.4), which has been taken either from another mouse or from cells growing in culture. The cells remain and proliferate in the intraperitoneal cavity as an ascites tumor. After 8 to 11 days, the animals are killed by cervical dislocation and the fluid tumor contents in their abdomen removed by sterile syringe into a 50-ml centrifuge tube containing 25 ml cold PBS to prevent coagulation. Alternatively, a drop of sterile sodium heparin solution in the tube can be used to inhibit coagulation.

After centrifugation at 1000 *g* at 4° for 6 min to remove ascites fluid, the cells are washed twice in cold sterile PBS prior to use. All further work is performed on ice. Typically, 10 ml of 10^8 cells/ml is obtained per mouse as determined either by manual or automatic cell-counting procedures. Relative numbers are viable and defective cells are assessed by microscopic inspection of cells placed on a hemocytometer and counting the cells that exclude and accumulate 1 drop Trypan Blue dye (0.04%, v/v)/ml cells, respectively.

Cell Culture

Ehrlich cells grow easily in suspension as well as in monolayer culture. Eagle's minimum essential medium with Earle's salts obtained from GIBCO, Inc. (Grand Island, NY) supplemented with 18 m*M* 4-morpholinopropanesulfonic acid (MOPS) (3.76 g/liter), 50 mg/liter streptomycin sulfate, and 50 mg/liter penicillin G (MEM) and 2.5% (v/v) fetal calf serum are used for both cultures. Both cultures are maintained at 37°; the monolayer culture is propagated in a 6% CO_2 atmosphere. In both monolayer and suspension cultures the medium is changed daily. Under these conditions, cells typically double in 24 hr. The concentration of cells in suspension is kept between 3×10^5 and 2×10^6 cells/ml.

Isolation of Metallothionein

Cell Lysis and Chromatography

Freshly washed cells are suspended in ice-cold distilled water (about two v/v), supplemented with 5 m*M* 2-mercaptoethanol (2-ME) just before mixing, to a concentration of $1-2 \times 10^8$/ml, and then lysed gently by mechanical homogenization with a hand-held tissue homogenizer. The presence of 2-ME ensures that the sulfhydryl groups of metallothionein are not oxidized during initial isolation procedures.[5] All solutions used in the purification before the ion-exchange step contain 2-ME. Nuclei and other organelles remain intact during homogenization. One can subsequently study metal distribution and localization in cytosol and particulate fractions. Alternatively, sonication of cells for 1 min with a Branson sonifier with a 40% pulse instead of homogenization produces a cell supernatant in which compartmentalization has been lost. After centrifugation at 20,000 *g* at 4° for 20 min in a refrigerated centrifuge, the resultant cytosol or supernatant can be frozen at this point.

A number of contributions to this volume describe methods to isolate and purify metallothionein. Briefly, to chromatograph 1.0 ml of supernatant or cytosol from 1×10^8 cells, a 1×40 cm Sephadex G-75 column is equilibrated at 4° in either 5 or 20 m*M* Tris, pH 7.8, and 1 m*M* 2-ME.[5] The supernatant with known total concentrations of metals of interest is applied and 1.0-ml fractions collected. Metal analysis reveals the presence of two major bands, one following the excluded volume representing the proteins of greater than 40,000 Da not well separated by this column, and a second at 10,000 Da, the metallothionein peak. Metallothionein containing zinc, copper, or cadmium runs as a 10,000-Da band. If other metals are bound to the protein, such as Au(I) thiomalate, the cluster structure of the protein is altered and the protein migrates more rapidly over the column.[6] Integration of the metal contents of the various parts of this chromatographic profile is compared with the initial amounts of metals placed on the column to determine if complete recovery of metals is obtained in this procedure. The metallothionein band from 10^8 cells can be further purified using a DEAE ion-exchange column equilibrated with 5 m*M* Tris, pH 7.8, and a Tris gradient from 5 to 250 m*M*, pH 7.8, with no 2-ME. A BioGel (Bio-Rad, Richmond, CA) TSK DEAE-5W high-perfor-

[5] D. T. Minkel, K. Poulsen, S. Wielgus, C. F. Shaw III, and D. H. Petering, *Biochem. J.* **191,** 475 (1980).

[6] J. E. Laib, C. F. Shaw III, D. H. Petering, M. K. Eidness, R. C. Elder, and J. S. Garvey, *Biochemistry* **24,** 1977 (1985).

mance liquid chromatography (HPLC) ion-exchange column, 7.5 × 75 mm, run at 1.0 ml/min is employed for this step. Alternatively, a liquid chromatography column of DEAE Sephadex A-25 ion-exchange beads, 1.5 × 10 cm, can be used. Following this procedure and collecting 1.0-ml fractions, two metal-bearing peaks, eluting at approximately 4 and 6 mS conductivity, are separated corresponding to the MT-1 and MT-2 isoproteins as determined by atomic absorption spectrophotometry.

Metal Content of Metallothionein, Total Metal-Binding Sites, and Apometallothionein

Metal Content

Different methods have been developed to measure metallothionein efficiently in multiple samples of a previously characterized source of the protein. Some detect only protein (e.g., immunochemical techniques).[7] Others determine only total metal-binding capacity of metallothioneins bearing metals that are displacable by Cd^{2+} or some other high-affinity metal ion (cadmium-heme method).[8] To obtain information on the particular metal content and degree of saturation of metallothioneins and other sites of metal binding, Sephadex G-75 chromatography of solubilized cell fractions has been done, followed by flame atomic absorption spectrophotometry, as detailed in Section II. Starting with 1×10^8 cells, one can readily measure by flame atomic absorption spectrophotometry the chromatographic profiles of zinc, copper, and cadmium from 1.0 ml cytosol passed over a 1 × 40 cm column equilibrated with 10 m*M* Tris-HCl, pH 7.4, at room temperature. Should it be necessary to measure more than these three metals or to detect metals less sensitive in flame atomic absorption analysis, such as iron or platinum, either more cells are needed or flameless atomic absorption spectrophotometric techniques may be employed. Typically, one finds about 0.7 μg zinc in MT/10^8 untreated cells.[2]

Total Binding Sites and Apoprotein

To assess the degree of metal saturation of native Ehrlich cell metallothionein, which has only zinc bound to it, samples of the cell homogenate are divided in two. One-half is incubated with $CdCl_2$ in order to displace all of the zinc from metallothionein and to occupy any previously vacant sites in the protein. The other serves as an untreated control sample

R. J. Vander Mallie and J. S. Garvey, *J. Biol. Chem.* **254,** 8416 (1979).

[8] D. L. Eaton and B. F. Toal, *Toxicol. Appl. Pharmacol.* **66,** 134 (1982).

of cells. Typically, 2.0 ml of Ehrlich cells homogenate, prepared as described in the section Isolation of Metallothionein is divided into two parts (approximately 10^8 cells/ml). One part receives 0.1 μmol $CdCl_2/10^8$ cells and is incubated for 20 min at 25°; the other is simply placed on ice. At the end of this time period, both samples are centrifuged at 20,000 *g* for 20 min. The isolation of metallothionein is completed as described above. After chromatography, a comparison is made of the total equivalents of metal bound to metallothionein in each sample. Extra cadmium in the MT fraction relative to the amount of zinc in the untreated sample indicates the presence of apoprotein. Apoprotein is common in tumor cell samples, although not in control Ehrlich cells.[9]

Distribution of Toxic Metals in Cells

Ehrlich cells have been used to examine the kinetics of intracellular distribution of metals in relationship to toxicity following exposure to toxic metallic species. There are two general approaches that can be taken in this type of study. The first is to expose cells in culture or in animals intermittantly or continually to the metal ion or complex of interest.[4] Assuming the metallic reagent does not kill the cells, its speciation and cellular effects can then be followed over time. In this way one can approximate low-level, chronic exposure to the metal. However, the sequence of events involved in the intracellular movement of metals, their localization in sites such as metallothionein, and in the toxicity that they cause is blurred because continual uptake of metal into the cells is superimposed on the behavior of previously accumulated metal.

An alternative is to expose cells to the metallic species for a short time (<1 hr), during which some of it is taken up by the cells. Then, if the cells are in culture, they are washed, placed in fresh growth medium without the metal, and the movement, speciation, and cellular effects followed as a linear sequence of events starting at time zero when the cells were rinsed free of exogenous metal. In this type of experiment, one can kinetically distinguish features of metal distribution that may relate to toxicity from those that occur too slowly to be part of the mechanism. The drawback of this approach is that a larger concentration of toxic metal may be needed for the short-term exposure to elicit toxicity than is required during continual exposure.

[9] A. Pattanaik, D. H. Petering, J. S. Garvey, and A. J. Kraker, to be submitted for publication.

Time-Dependent Distribution of Metals

To follow the kinetics of metal distribution in Ehrlich cell cytosol of cells exposed continually to a toxic metal over the course of the experiment, multiple spinner bottles, each containing 500 ml of 4×10^6 cells/ml, are set up at 37° and inoculated with the toxic metal at time zero. At prescribed times all cells are removed from a single bottle, washed, lysed, and chromatographically analyzed for metal distribution according to the procedures as described above in Section II.

The pulse exposure method begins with incubation of all of the cells with the toxic metal for a short period beginning at time zero.[2] During this time cells are suspended in MEM without calf serum at a concentration of 10^7 cells/ml. After they have incubated with periodic mixing for 30 min at 25°, the cells are quickly pelleted by centrifugation at 1000 *g* at 4° for 6 min, resuspended in cold saline, and centrifuged again. Finally, cells are suspended in fresh, complete medium (MEM plus 2.5% fetal calf serum), which does not contain the toxic metal, at a concentration of 1×10^6 cells/ml. The suspended cells are divided equally among a set of suspension (spinner) bottles in volumes of 500 ml, each, and incubated at 37°. Cellular metal profiles are measured over time as during the continuous exposure method described above, with the first one done immediately on allocation of cells to the spinner bottles.

At each time point in either version of the experiment, the total number of cells in the suspension and their fractional viability is determined and used both to normalize metal distribution data (micrograms of metal in particular chromatographic band per certain number of cells) and as an indicator of the amount of proliferation and viability of the culture.

Turnover of Metal and Biodegradation of Protein

The kinetic lability of Zn-MT in chemical ligand (L) substitution reactions has focused attention on the possible similar reactivity of this protein in cells.[10-12] Reaction (1) is an example of this type of reaction, shown for one of the bound metal ions in metallothionein,

$$\text{Zn-MT} + \text{L} \rightarrow \text{Zn-L} + \text{MT} \qquad (1)$$

However, this net reaction might proceed through the initial degradation

[10] T.-Y. Li, A. J. Kraker, C. F. Shaw III, and D. H. Petering, *Proc. Natl. Acad. Sci. U.S.A.* **77**, 6334 (1980).

[11] J. D. Otvos, D. H. Petering, and C. F. Shaw III, *Comments Inorg. Chem.* **9**, 1 (1989).

[12] S. Krezoski, J. Villalobos, C. F. Shaw III, and D. H. Petering, *Biochem. J.* **255**, 483 (1988).

of the protein followed by the formation of Zn-L [reactions (2) and (3)]:

$$\text{Zn-MT} \rightarrow \text{Zn} + \text{amino acids} \quad (2)$$

$$\text{Zn} + \text{L} \rightarrow \text{Zn-L} \quad (3)$$

Biodegradation of MT protein is also a basic consideration in the understanding of the steady state dynamics of binding of toxic metal ions to metallothionein in tissues of whole organisms.[13]

Turnover of Zinc in Metallothionein

To follow the kinetics of loss of zinc from Ehrlich cell Zn-MT or to produce cellular apometallothionein, cells are incubated in zinc-deficient media.[12] In the face of an extracellular zinc deficit, there is a specific loss of zinc from the cytosolic metallothionein pool that occurs before proliferation slows or other effects are noted, which produces apo-MT.

The zinc-deficient medium is prepared in one of the two ways. First, Chelex 100 (Bio-Rad, Richmond, CA), a resin bearing metal-chelating iminodiacetate groups, can be added directly to the cell suspension (2×10^6 cells/ml, 500 ml) at a concentration of 10 g/100 ml of standard culture medium (MEM + 2.5% fetal calf serum). The cells are allowed to incubate in the deficient medium for varying time periods and then are prepared for column chromatography as described above. Cells incubated in the presence of Chelex 100 beads are easily separated from the beads because the beads settle much more rapidly than the cells. To do this, stirring is stopped and the cells and beads are allowed to sediment for 3 min. Then the cells are carefully decanted into 50 ml centrifuge tubes and any beads transferred with the cells allowed to settle for another 3 min. After a second, careful decantation the preparation is completed as described above. Alternatively, the culture medium can be preincubated with Chelex 100 (10 g/100 ml) for more than 2 hr at 37° and then used after the resin is allowed to settle and the medium is carefully decanted. Analyses show that Chelex 100 lowers zinc in the final growth medium by 80% and reduces calcium but does not alter magnesium, copper, or iron content.

In order to minimize the removal of other cations such as Mg^{2+} from the culture medium by Chelex 100, the resin is preincubated for a minimum of 2 hr in 10- or 20-g aliquots with 1 or 2 liters of water, respectively, containing the salts and buffer used in the MEM solution. After this time the beads are settled and concentrated into 20 or 40 ml of that solution in a 125-ml bottle and autoclaved for 20 min.

The methods described above are used to follow the time-dependent

[13] R. W. Chen, P. D. Whanger, and P. H. Weswig, *Biochem. Med.* **12,** 95 (1975).

distribution of zinc after cells are placed in the zinc-depleted medium at 37°. The rate of zinc loss can be calculated and compared with the rate of basal biosynthesis of MT, measured according to procedures described below.

Turnover of Copper and Cadmium in Metallothionein

Addition of Cu(II) 3-ethoxy-2-oxobutyraldehyde bis(thiosemicarbazone) (CuKTS) or Cd^{2+} to Ehrlich cells rapidly leads to a one-for-one exchange of these metals for zinc in Zn-MT.[2,14] If excess copper or cadmium is present, it localizes first in the high-molecular-weight fraction of metal separated by Sephadex G-75 chromatography and then moves into the metallothionein fraction as more MT protein is formed. At longer times, both copper and cadmium are lost from metallothionein and from these cells. By comparing rates of acquisition and loss of copper or cadmium over time with the rates of synthesis and degradation of MT protein, the underlying basis of these changes in metal content can be addressed.

Biodegradation and Biosynthesis of Metallothionein Protein

The biodegradation rate constant for Zn-MT is calculated after monitoring the time-dependent loss of [^{35}S]cysteine from the metallothionein pool of Ehrlich cells at 37°.[12] Cultured cells (1×10^6 cells/ml, 500 ml) are labeled with [^{35}S]cysteine (10 μCi/100 ml) for 24 hr, pelleted, and washed three times with MEM and suspended (8×10^7 cells/100 ml) in fresh, complete medium at time zero in the presence or absence of 2–4 nmol/ml unlabeled cystine, which is added to inhibit reincorporation of ^{35}S from degraded protein into newly synthesized metallothionein.

To obtain kinetic data, 20-ml aliquots of cells are removed at increasing time intervals (0, 2, 4, 16, and 24 hr), their numbers counted, and the aggregate of cells collected at each time point by centrifugation. They are resuspended in 0.5 ml PBS, frozen in liquid N_2 in cryotubes, thawed, and finally acidified to pH 2.2 by addition of cold trichloroacetic acid (final concentration, 2.2%, w/v). Among cellular proteins, metallothionein alone remains soluble. The pH of 0.09 ml the supernatant is raised to neutrality and 0.01 ml dithiothreitol (final concentration, 10 m*M*) added before chromatography at room temperature over a small Sephadex G-10 column (0.5×10 cm) equilibrated with 10 m*M* Tris, pH 7.4. Collecting 0.045-ml fractions, this procedure rapidly separates MT protein from any amino

[14] D. H. Petering, S. Krezoski, J. Villalobos, C. F. Shaw III, and J. D. Otvos, *Experientia, Suppl.* **52,** 573 (1987).

acid or peptide forms of [^{35}S]cysteine.[12,15] The amount of label in the faster eluting protein band is determined by liquid scintillation counting. This is normalized to the total number of cells used at each time point and the resulting concentrations of label, representative of MT protein in the cells, plotted logarithmically as a function of time. The assumption of first-order decay of labeled metallothionein is correct, so that a rate constant (k_d) can be directly calculated.

In the steady state of protein turnover, zero-order MT protein synthesis (k_s) equals the rate of biodegradation (k_d):

$$k_s = k_d[\text{MT protein}] \tag{4}$$

Thus, knowing k_d and the concentration of metallothionein in some units, the value of k_s is immediately defined.[12]

Transport of Zinc and Uptake into Apometallothionein

Apometallothionein can be generated in Ehrlich cells placed in the presence of Chelex 100 or in media pretreated with this resin.[12] Because its biodegradation rate is much slower than its rate of appearance under zinc-deficient conditions and because there is also a constitutive rate of synthesis of MT protein, it is possible to examine the uptake of zinc added back to these cells into apometallothionein. Apometallothionein appears to be a preferred site of binding of zinc entering cells. Furthermore, the rate of formation of Zn-MT from Zn^{2+} or its metal complexes is fast. Hence, the observed slow rate of reconstitution of Zn-MT in Ehrlich cells under these conditions is thought to represent rate-limiting transport or subsequent facets of zinc metabolism.[16] The advantage of this type of transport experiment is that it is realistically based on cellular demand for the metal ion that arises because of the presence of a thermodynamically favorable binding site for Zn^{2+} in the cells.

To carry out this experiment, 8×10^8 cells are treated for 3 hr with Chelex-treated media (as indicated above in Turnover of Metal and Biodegradation of Protein) to remove zinc specifically from metallothionein. Afterward they are rapidly centrifuged from the zinc-depleted medium and resuspended in the complete culture medium at a concentration of 2×10^6 cells/ml and maintained at 37°. Then, cytosolic distributions of zinc are measured at 0, 1, 2, 3, and 4 hr to follow the kinetics of reappearance of zinc in the metallothionein fraction.

[15] S. R. Patierno, N. R. Pellis, R. M. Evans, and M. Costa, *Life Sci.* **32,** 1629 (1983).

[16] D. H. Petering, A. Pattanaik, P. Chen, S. Krezoski, and C. F. Shaw III, *in* "Metal Binding in Sulfur-Containing Proteins" (M. Stillman, K. Suzuki, and C. F. Shaw III, ed.), VCH Publishers, New York, to be published.

Assays of Toxicity of Metals

To understand the role of metallothionein in the protection of cells from deleterious metal exposure, it is important to correlate metal distribution with some indices of toxicity. Specifically, one can determine to what extent a metal is sequestered in metallothionein in relation to the degree of toxicity that the cells experience. Only some generally used assays of toxicity are described here.

Cell Death and Inhibition of Proliferation

The simplest and most general assay of toxicity is the measurement of effect of a toxic metal on cell population growth and viability. Because Ehrlich cells grown in suspension, it is convenient to follow the change in cell number and resistance to Trypan Blue uptake over time as indications of toxicity (see the section, Growth of Ehrlich Cells, above).

Inhibition of DNA Synthesis

Another widely used test for cytotoxicity assesses the ability of the cells to carry out DNA synthesis.[17] In a short assay, 7 ml of 1×10^7 cells/ml suspended in PBS is incubated with 1 μCi (0.5 μmol) [^{3}H]thymidine at 37°. At 5, 10, and 20 min, 2 ml of cells is removed and rapidly cooled in ice. They are centrifuged at 1000 *g*, washed in PBS, and recentrifuged. The cells are lysed in 3 ml H_2O and the macromolecules precipitated by making the suspension 0.6 *N* in HCO_4. After another centrifugation, the acid-soluble supernatant is decanted and the pellet washed in 1.0 ml 0.2 *N* $HC1O_4$. A final centrifugation provides the precipitate including ^{3}H-labeled DNA. The two supernatants from the precipitation and wash steps are combined and contain intracellular labeled thymidine. After solubilization of the DNA in 0.075 *N* KOH, both fractions are counted for radioactivity. On normalization of each to the number of cells used, the sum of the concentrations of label from both supernatant and precipitate fractions is a measure of transport of the precursor into cells during the incubation time period. The amount of label in the DNA polymer (precipitate) is proportional to the amount of synthesis of DNA over the measured time period. The results are given as counts (disintegrations)/minute/unit number of cells at each sampling time.

Acknowledgments

The authors gratefully acknowledge the support from NIH Grant ES-04026 for this work.

[17] L. A. Saryan, D. T. Minkel, P. J. Dolhun, B. L. Calhoun. S. Wielgus, M. Schaller, and D. H. Petering, *Can. Res.* **39**, 2457 (1989).

[37] Purification of Mammalian Metallothionein from Recombinant Systems

By MARY J. CISMOWSKI and P. C. HUANG

Protocols for the purification of large quantities of native metallothioneins (MTs) from various higher eukaryotes have been well documented.[1] The ability to purify milligram quantities of MT from mammalian livers has allowed for detailed structural analysis of MT by both nuclear magnetic resonance (NMR)[2,3] and X-ray diffraction[4] techniques. These structural studies raise some intriguing questions about the role each individual amino acid plays in the metal-binding function of MT.

The genetic manipulation of metallothionein coding sequences by site-specific mutagenesis is a powerful method for analyzing these structure–function relationships. For example, the systematic replacement of each of the 20 conserved cysteine residues in MT should allow for the analysis of the relative contribution individual cysteines make toward both *in vivo* and *in vitro* functions of MT. Detailed structural analysis of site-specific MT mutants requires the efficient expression and purification of genetically manipulated MT sequences from heterologous systems. In this chapter, we describe two such systems for analysis of recombinant MTs, one using *Escherichia coli* as host and one using *Saccharomyces cerevisiae* as host. The expression and partial purification of a mammalian MT fusion protein from *E. coli* is described, and the expression and purification of both wild-type mammalian MT and MTs with amino acid replacements from *S. cerevisiae* are described. Potential applications for each system are discussed.

Expression and Partial Purification of Metallothionein Fusion Protein in *Escherichia coli*

The stable expression of native mammalian metallothionein in *E. coli* has met with little success.[5] As an alternative, MT expression was accomplished by fusing Chinese hamster MT coding sequences to a gene native to

[1] M. Vašák, this volume [5].
[2] J. D. Otvos and I. M. Armitage, *Proc. Natl. Acad. Sci. U.S.A.* **77,** 7094 (1980).
[3] K. Wuthrich, this volume [59].
[4] A. H. Robbins and C. D. Stout, this volume [58].
[5] Y.-M. Hou, R. Kim, and S.-H. Kim, *Biochim. Biophys. Acta* **951,** 230 (1988).

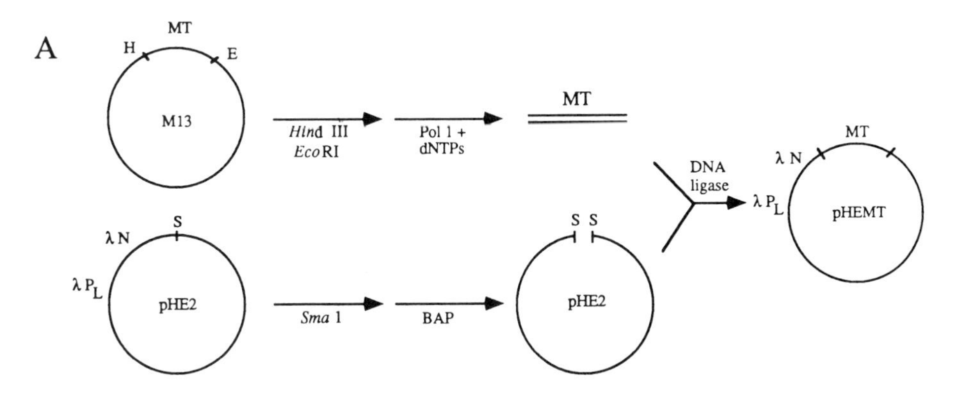

B

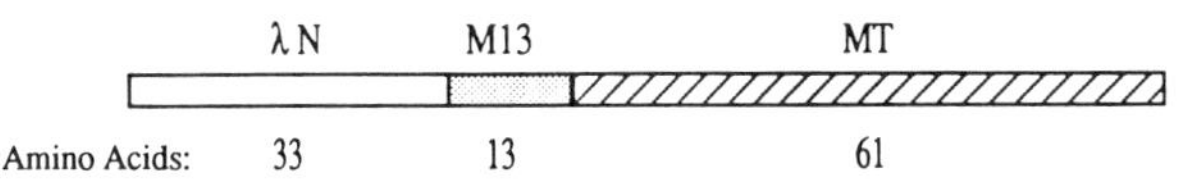

FIG. 1. (A) Construction of *E. coli* expression plasmid pHEMT. H, *Hin*dIII; E, *Eco*RI; S, *Sma*I; Pol 1, *E. coli* DNA polymerase I large fragment; BAP, bacterial alkaline phosphatase; λ P_L, leftward promoter from phage λ; λ N, N-terminal sequence of the λ N gene. (B) Schematic of λ N-MT fusion construct.

E. coli cultures. We describe here a system for the stable expression and partial purification of a λ N-MT fusion protein from *E. coli*.

Plasmid Construction

A λ N-MT *E. coli* expression plasmid is constructed as shown in Fig. 1A. A Chinese hamster MT-2 cDNA sequence is excised from an M13 vector[6] by digestion with *Eco*RI and *Hind*III restriction enzymes, which cleave in the multilinker sequence of M13. The 235-bp fragment containing MT is made blunt ended by incubation with *E. coli* DNA polymerase I large fragment and purified from a 6% (w/v) polyacrylamide gel.[7] The isolated fragment is then inserted into the *Sma*I site of pHE2[8] in frame with the λ N protein and downstream of the λ P_L promoter. The introduction of MT cDNA sequences at the *Sma*I site creates a hybrid MT sequence with 33 codons from the λ N protein, 13 codons from the M13 multilinker, and 61 codons from MT (Fig. 1B).

[6] R. Pine, M. Cismowski, S. W. Liu, and P. C. Huang, *DNA* **4,** 115 (1985).
[7] H. O. Smith, this series, Vol. 65, p. 371.
[8] S. W. Liu and G. Milman, *J. Biol. Chem.* **258,** 7469 (1983).

Partial Purification of Metallothionein Fusion Protein

Growth of Cells. The pHE2 plasmid contains sequences encoding the λcI_{857} temperature-sensitive repressor; incubation of *E. coli* containing pHEMT at 42° leads to rapid induction of the fusion protein. The following protocol is used for the partial purification of the MT fusion protein from a 100-ml culture of *E. coli;* the protocol can be scaled up or down as required. The 100-ml culture is grown to mid log phase at 30° and 250 rpm in 10 g/liter Bacto-tryptone, 5 g/liter Bacto-yeast extract, 10 g/liter NaCl, 0.5% (w/v) glucose, pH 7.5. $ZnCl_2$ and [^{35}S]cysteine are added to the culture at final concentrations of 600 μM and 1 μCi/ml, respectively, and the culture is rapidly heated to 42° and grown for 2 hr. Cells are collected by centrifugation at 4300 *g* at 4° and resuspended in 6 ml 5 m*M* Tris-HCl, pH 8.0.

Lysis and Removal of Nucleic Acids. All subsequent procedures are carried out at 0–4°. Cells are lysed by sonication and centrifuged at 4300 *g* to pellet debris. The cleared lysate is brought to 1.5% (v/v) streptomycin sulfate by addition from a 40% (v/v) stock solution and mixed slowly by inversion for 60 min. The sample is centrifuged at 12,000 *g* and the supernatant is brought to 50 m*M* NaCl.

P11 Phosphocellulose Chromatography. The supernatant (approximately 6 ml) is loaded onto a 3-ml P11 phosphocellulose column equilibrated in 5m*M* Tris-HCl, pH 8.0, 50 m*M* NaCl. The column is washed stepwise with buffers containing 5 m*M* Tris-HCl, pH 8.0, plus 100 m*M* NaCl (50 ml), 200 m*M* NaCl (9 ml), 300 m*M* NaCl (9 ml), and 400 m*M* NaCl (9 ml). The hybrid MT elutes in the last two high-salt washes, presumably due to the basic properties of the λ N protein, and is detected by sodium dodecyl sulfate-polyacrylamide gel electrophoresis and autoradiography. The hybrid MT fractions are pooled, concentrated to approximately 3 ml by lyophilization, and dialyzed extensively against 5 m*M* Tris-HCl, pH 8.0. The sample is again concentrated by lyophilization to approximately 0.5 ml.

DEAE-Sephadex A-25 Chromatography. The P11 purified fraction is loaded onto a 3-ml DEAE-Sephadex A-25 column equilibrated with 10 m*M* Tris-HCl, pH 8.0. The column flow-through, containing the MT fusion protein, is collected and pooled with a 20-ml column wash of 10 m*M* Tris-HCl, pH 8.0. This sample is concentrated by lyophilization to approximately 0.3 ml. Fractions from the purification steps are analyzed by electrophoresis on a 16% (w/v) SDS-polyacrylamide gel followed by Coomassie Blue R250 staining and autoradiography. The MT fusion protein in the DEAE fraction is approximately 65% pure by Coomassie Blue staining, and is the only [^{35}S]cysteine-labeled protein detected by autoradiography.

Potential Uses of Metallothionein Fusion Proteins

Although we were unable to verify stoichiometric metal binding in the partially purified MT fusion protein, the protein was strongly immunoreactive with antisera raised against rat liver MT,[9] both by enzyme-linked immunosorbent assay (ELISA) and immunoblotting analysis (unpublished results, 1991). This is consistent with the extensive cross-reactivity seen between most MTs,[10] and indicates that the fusion protein shares some of the structural properties of native MTs. It is likely, therefore, that this fusion protein would be useful as an antigenic agent for the production of anti-MT antibodies. As native MT contains only one methionine at the N terminus, it is possible that a "native" MT can be recovered from the partially purified fusion protein by cyanogen bromide cleavage. Finally, as this fusion construct is stable in *E. coli,* it is possible that other fusion constructs may also be stable; this would be useful in the construction of fusions for intracellular targeting assays.

Expression and Purification of Metallothionein from *Saccharomyces cerevisiae*

Because of the disadvantages inherent in working with a fusion protein, an alternative system was established for the expression and purification of MT in its native form using the building yeast *S. cerevisiae* as host. This expression system has proven useful in the *in vivo* analysis of wild-type and mutant yeast copper-thioneins.[11,12] Our laboratory has used this system to express and purify mammalian MT and MT mutants for structural and functional characterization.[13-16]

Plasmid Construction

High-copy-number yeast expression plasmids containing wild-type and mutant Chinese hamster MT-2 cDNAs are constructed as shown in Fig. 2A. Plasmids pYSK102 and pYSK54 were generously provided by Dr.

[9] S. Ohi, G. Cardenosa, R. Pine, and P. C. Huang, *J. Biol. Chem.* **256,** 2180 (1981).

[10] J. S. Garvey, this volume [20].

[11] D. J. Ecker, T. R. Butt, E. J. Sternberg, M. P. Neeper, C. Debouck, J. A. Gorman, and S. T. Crooke, *J. Biol. Chem.* **261,** 16895 (1986).

[12] A. R. Thrower, J. Byrd, E. B. Tarbet, R. K. Mehra, D. H. Hamer, and D. R. Winge, *J. Biol. Chem.* **263,** 7037 (1988).

[13] I. K. Rhee, K. S. Lee, and P. C. Huang, *Protein Eng.* **3,** 205 (1990).

[14] M. J. Cismowski and P. C. Huang, *Biochemistry* (in press).

[15] M. Chernaik and P. C. Huang, *Proc. Natl. Acad. Sci. U.S.A.* **88,** 3024 (1991).

[16] M. J. Cismowski, S. S. Narula, M. Chernaik, I. M. Armitage, and P. C. Huang, *J. Biol. Chem.* (in press).

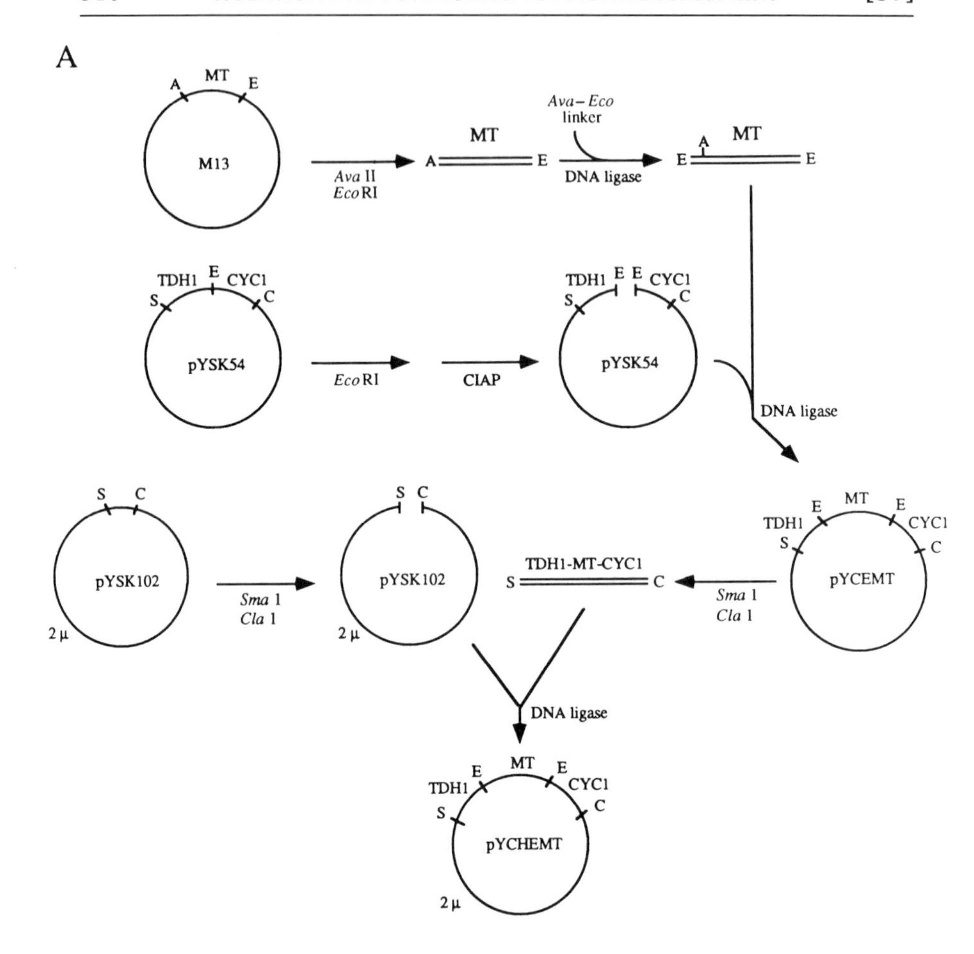

B

5' A A T T C G G A C G T G C C A C C A T G 3'
3' G C C T G C A C G G T G G T A C C T G 5'

FIG. 2. (A) Construction of *S. cerevisiae* expression plasmid pYCHEMT. A, *Ava*II; E, *Eco*RI; S, *Sma*I; C, *Cla*I; CIAP, calf intestine alkaline phosphatase; *TDH1*, yeast glyceraldehyde-3-phosphate dehydrogenase promoter sequences; *CYC1*, yeast cytochrome *c* oxidase terminator sequences; 2μ, yeast 2-μm β sequences. (B) Sequence of *Ava–Eco* linker used to reconstruct the 5′ ends of MT fragments.

David Ecker of Smith, Kline and French laboratories. MT cDNA sequences are excised fom M13 using *Ava*II and *Eco*RI. MT fragments are isolated from low-melting-point agarose. The *Ava*II restriction site is immediately downstream of the ATG initiation codon for MT. This initiation codon and a consensus ribosome-binding site sequence[17] are added by ligation of a synthetic oligonucleotide linker (Fig. 2B). The restored MT sequences, with *Eco*RI cohesive ends, are then ligated to *Eco*RI-digested pYSK54 to form pYCE(MT). Transcription cassettes containing the *TDH1* constituitive promoter, the MT-coding sequences, and the *CYC1* terminator are then excised from pYCE(MT) using *Sma*I and *Cla*I. These cassettes are purified from agarose and ligated to a gel-purified *Sma*I–*Cla*I vector fragment of pYSK102 to form pYCHE(MT).

Purification of Chinese Hamster Metallothionein from Saccharomyces cerevisiae

The pYCHE(MT) expression plasmids contain MT sequences under the control of a constituitive promoter; 2μ sequences on the plasmid lead to an intracellular plasmid copy number of approximately 50.[11] When introduced into a yeast strain auxotrophic for tryptophan and lacking its own copper-thionein, AB-DE1 (courtesy of Dr. David Ecker), the MT-containing plasmids direct the expression and accumulation of intracellular MT to approximately 1–2% total protein, as judged by Coomassie Blue staining. Cells grown in the presence of low levels of zinc or cadmium accumulate MT, with the added metal being the predominant chelated species.

Growth of Cells. Saccharomyces cerevisiae strain AB-DE1 carrying a pYCHE(MT) plasmid is grown while shaking at 30° and 150 rpm to late log phase in 6 liters of trp$^-$ medium supplemented with either 10 μM $ZnCl_2$ or 10 μM $CdCl_2$. trp$^-$ medium consists of 1.7 g/liter yeast nitrogen base, 5 g/liter ammonium sulfate, 5 g/liter casamino acids, 20 g/liter glucose, 30 mg/liter adenine, 20 mg/liter histidine, 30 mg/liter arginine, 40 mg/liter leucine, and 40 mg/liter lysine. The first three ingredients are autoclaved together for 30 min to heat inactivate tryptophan prior to the addition of the remaining ingredients. Cells are harvested by centrifugation at 5000 *g* and washed once with H_2O. The washed cell pellet is stored in 10-g aliquots at −80° until use. Approximately 40–50 g of wet cells is obtained under these growth conditions.

Cell Lysis. All subsequent operations are carried out at 0–4° unless otherwise indicated. All buffers used are degassed and/or N_2 bubbled immediately prior to use.

One batch of cells is thawed and resuspended in an equal volume of 10 m*M* Tris-HCl, pH 8.0, 10 m*M* dithiothreitol (DTT), 2 m*M* phenylmethylsulfonyl fluoride (PMSF), 50 ppm Antifoam A (Sigma, St. Louis, MO). Approximately 8 ml 0.5- mm glass beads is added to the cell suspension

[17] M. Kozak, *Cell (Cambridge, Mass.)* **44,** 283 (1986).

and cells are lysed by vortexing continuously on the maximum setting for 20 min. The lysate is centrifuged at 27,000 *g* for 60 min and the supernatant is brought to 0.2 *M* sucrose.

Sephadex G-75 Chromatography. Approximately 10 ml of the lysate supernatant is loaded onto a Sephadex G-75 column (65 × 1.5 cm) and eluted in 10 m*M* Tris-HCl, pH 7.5, 0.1% 2-mercaptoethanol at approximately 0.5 ml/min. Fractions (5 ml) are collected and MT is detected by zinc or cadmium atomic absorption.[18] Metallothionein typically elutes from this column at approximately 1.8 × void volume (V_0).

DEAE-Sephadex A-25 Chromatography. Pooled fractions containing MT are loaded onto a DEAE-Sephadex A-25 column (6 × 0.5 cm) equilibrated in 10 m*M* Tris-HCl, pH 7.5. Metallothionein is eluted by a linear gradient (total volume 250 ml) of 10 to 200 m*M* Tris-HCl, pH 7.5, in the absence of 2-mercaptoethanol at 1 ml/min. Fractions (4 ml) are collected and monitored by atomic absorption.

Reversed-Phase High-Performance Liquid Chromatography (HPLC). Fractions containing MT from the DEAE column are pooled, concentrated by ultrafiltration (Diaflo YM2 membrane; Amicon, Danvers, MA) under N_2 pressure (15–40 psi), and chromatographed on a C_{18} reversed-phase HPLC column (250 × 4.6mm) run at ambient temperature. The column is equilibrated in 10 m*M* NaH_2PO_4, pH 7.5, and MT is eluted by a linear gradient of 0–60% (v/v) CH_3CN in NaH_2PO_4, pH 7.5, at 1 ml/min.[19] Metallothionein elution is monitored spectroscopically at 214 nm for cells grown in zinc and at both 214 and 250 nm for cells grown in cadmium. Monitoring Cd-MT at 250 nm, a wavelength at which cadmium–thiolate ligands absorb, is useful as there is little or no background absorption from other protein species. Wild-type MT typically elutes from this column at approximately 20% CH_3CN.

Sephadex G-75 Chromatography. Metallothionein recovered from the HPLC column is >98% pure as judged by both Coomassie Blue and silver staining. An additional Sephadex G-75 chromatography step is performed, however, both to exchange buffers and to remove any minor amounts of MT dimer formed during purification.[20] The HPLC-purified sample is lyophilized briefly to remove CH_3CN, brought to 0.1 *M* sucrose, and loaded onto a Sephadex G-75 column (85 × 0.75 cm) run at ambient temperature in 10 m*M* Tris-HCl, pH 7.5. Fractions of 1.5 ml are collected at <0.5 ml/min and monitored by atomic absorption spectrophotometry as described above. Those fractions containing MT are pooled and concentrated by ultrafiltration under N_2 pressure. Purified MT is stored at

[18] C. D. Klaassen and L. D. Lehman-McKeeman, this volume [22].

[19] M. P. Richards, this volume [25].

[20] D. G. Nettesheim, H. R. Engeseth, and J. D. Otvos, *Biochemistry* **24,** 6744 (1985).

$-20°$ under N_2. Purity is confirmed by sodium dodecylsulfate-polyacrylamide gel electrophoresis after iodoacetamide treatment[21] followed by silver staining.[22] The yield of purified MT, as determined by quantitation at 220 nm in 20 m*M* HCl,[14,23] is approximately 1.5 mg/liter of log-phase culture.

Uses for Recombinant Metallothionein from Saccharomyces cerevisiae

Using the above procedure, we were able to purify both wild-type Chinese hamster MT and mutant MTs with various amino acid substitutions, including the purification of milligram quantities of MTs with cysteine-to-tyrosine changes. These purified MTs have yielded information about the relative contribution individual residues make toward formation of stable metal-bound MT.[14-16] By manipulating the metal content in the culture media, MTs in differing metal bound states can be isolated. Growth of yeast in $^{113}CdCl_2$, for example, has produced MTs uniformly labeled with ^{113}Cd for NMR analysis, without the usual requirement for ^{113}Cd enrichment by denaturation and isotope exchange. This is important, especially in our study of MT mutants, as protein refolding after denaturation may be affected by the introduced mutation. Finally, this expression and purification system should prove useful in the structural analysis of MTs for which tissue is difficult to obtain, such as human MTs.

Acknowledgments

This work was supported by a grant from the National Institutes of Health (RO1GM32606) to P.C.H. and a predoctoral fellowship from the National Science Foundation to M.J.C.

[21] T. Sone, K. Yamaoka, Y. Minami, and H. Tsunoo, *J. Biol. Chem.* **262,** 5878 (1987).
[22] F. Otsuka, S. Koizumi, M. Kimura, and M. Ohsawa, *Anal. Biochem.* **168,** 184 (1988).
[23] R. H. O. Buhler and J. H. R. Kägi, *Experientia, Suppl.* **34,** 211 (1979).

[38] Cadmium-Binding Peptides from Plants

By WILFRIED E. RAUSER

One reaction of plants to excess cadmium is the formation of cadmium-binding complexes. These complexes generally contain predominantly cadmium, when this metal is given to the plant, a range of cysteine-rich polypeptides, and acid-labile sulfide. The physiologically relevant aspect of metal binding is only evident with complexes. Study of the polypeptides requires dissociation of the complexes to liberate the constituent metal-free polypeptides. To date, the primary structures of these polypeptides show a series of molecules of poly(γ-glutamylcysteinyl)glycine or $(\gamma\text{-EC})_n$G,[1] where n is 2 to 11 depending on the source.[1a-4] These molecules are related to glutathione (γ-ECG). In certain species of the family Fabaceae (Leguminosae) glutathione is partly or totally replaced by homoglutathione where β-alanine substitutes for glycine so that the polypeptides of the complexes are poly(γ-glutamylcysteinyl)-β-alanines.[5] All these cysteine-rich polypetides belong to the superfamily of proteins called metallothioneins (MT).[6] Because they differ markedly from equine renal MT, they are classified as class III MTs.

Agreement on a practical name for the class III MTs is lacking. The chemical name [e.g., $(\gamma\text{-EC})_n$G] applies when the primary structure is established. In situations where the polypeptides are only partly characterized by amino acid composition or even less stringently by their chromatography as thiols, naming them as if they had a known primary structure is difficult to justify. One solution in these cases (as in this chapter) is to use the trivial names phytochelatin (PC)[2] and homophytochelatin (hPC),[5] corresponding to poly(γ-glutamylcysteinyl)glycine and poly(γ-glutamylcysteinyl)-β-alanine, respectively. The prefix phyto is, however, not exactly accurate for the microbes *Schizosaccharomyces pombe* and *Candida glabrata* because these organisms, which produce $(\gamma\text{-EC})_n$Gs, are usually not classified along with plants (Kingdom Plantae: algae through flowering species).

Various aspects of the metal-binding polypeptides and their complexes from plants and certain fungi are considered in this volume[7-9] and in recent reviews.[10,11] Isolation of cadmium-binding complexes and quantitative analyses of thiols, including class III MTs, are described below.

[1] E, Glutamic acid; C, cysteine; G, glycine.

[1a] N. Kondo, K. Isobe, K. Imai, T. Goto, A. Murasugi, C. Wada-Nakagawa, and Y. Hayashi, *Tetrahedron Lett.* **25,** 3869 (1984).

[2] E. Grill, E.-L. Winnacker, and M. H. Zenk, *Science* **230,** 674 (1985).

[3] J. C. Steffens, D. F. Hunt, and B. G. Williams, *J. Biol. Chem.* **261,** 13879 (1986).

[4] P. J. Jackson, C. J. Unkefer, J. A. Doolen, K. Watt, and N. J. Robinson, *Proc. Natl. Acad. Sci. U.S.A.* **84,** 6619 (1987).

[5] E. Grill, W. Gekeler, E.-L. Winnacker, and M. H. Zenk, *FEBS Lett.* **205,** 47 (1986).

[6] Y. Kojima, this volume [2].

[7] E. Grill, E.-L. Winnacker, and M. H. Zenk, this volume [39].

[8] N. Mutoh and Y. Hayashi, this volume [40].

[9] Y. Hayashi, M. Isobe, N. Mutoh, C. W. Nakagawa, and M. Kawabata, this volume [41].

[10] W. E. Rauser, *Annu. Rev. Biochem.* **59,** 61 (1990).

[11] J. C. Steffens, *Annu. Rev. Plant Physiol. Plant Mol. Biol.* **41,** 553 (1990).

Isolation of Cadmium-Binding Complexes

Several protocols are available for isolating cadmium-binding complexes (Cd-BC).[10] Some are patterned after the isolation of MT. The crude extract is first chromatographed on Sephadex G-75 or G-50 and the Cd-BC further purified by anion-exchange chromatography. The gel filtration requires a small volume of extract, which can be met by lyophilizing the tissue prior to extraction, or concentrating the extract by lyophilization, ultrafiltration, or precipitation with $(NH_4)_2SO_4$. Solubility can become limiting in highly concentrated extracts, and elevated viscosity can affect loading of the gel-filtration column and distort resolution.

An alternative protocol is to first chromatograph the crude extract on an anion-exchange column and then to obtain the Cd-BC by gel filtration. By this means larger volumes of dilute extract can be applied rapidly to the anion exchanger to effectively concentrate the Cd-BC. Extracts of roots exposed to cadmium also contain abundant brown materials, perhaps products of polyphenol oxidases, which are retained more effectively by the strong anion exchanger QAE than by DEAE, and are thus largely removed prior to gel filtration. The extraction and isolation of Cd-BC from roots of maize by such an alternative protocol are described below.

Extraction

Reagent

Extraction buffer: 100 m*M* Tris-HC1, pH 8.6, 100 m*M* KCL, 5m*M* ascorbic acid, 5 m*M* thiourea, 5 m*M* 2-mercaptoethanol, 1 m*M* KCN

This buffer is prepared by dissolving the required quantities of Tris and KCl in water, adjusting the pH, and bringing to final volume. The solution is cooled to 0–4° overnight. About 30 min prior to extraction, the solution is purged with N_2 for 10 min. The remaining compounds are added with stirring while a stream of N_2 is passed over the solution surface. The container is sealed.

Procedure

The following is adapted from an earlier procedure.[12]

1. Cadmium-exposed roots are washed in ice-cold deionized water, blotted, weighed, frozen in liquid N_2 in a stainless steel beaker, and pulverized in liquid N_2 with a pestle. Such pulverized roots can be used directly or stored frozen in a sealed container.

[12] W. R. Bernhard and J. H. R. Kägi, *Experientia, Suppl.* **52,** 309 (1987).

2. About 100 to 150 g of pulverized roots are placed in a stainless steel Waring blender and homogenized to a fine powder in liquid N_2.

3. The homogenate is transferred to a 1-liter heavy-duty Pyrex bottle. The flask is capped loosely to allow N_2 to dissipate at room temperature.

4. When most of the liquid N_2 has dissipated, a volume of cold extraction buffer equivalent to the fresh weight of roots used is added slowly to the root powder. There is effervescence as the buffer freezes and further N_2 is dissipated. Cap the flask loosely.

5. When vapors have dissipated, the cold flask is transferred to water at room temperature to cause melting. Pressure increases are released occasionally by loosening the cap. The contents of the flask are shaken until a cold slurry results.

6. The slurry is strained through four layers of cheesecloth previously moistened with buffer and the expressed fluid is collected in a flask. A stream of N_2 is directed over the cheesecloth. The tissue residue is moistened with 50 ml extraction buffer and the fluid expressed.

7. The extract is transferred to centrifuge tubes under a stream of N_2 and centrifuged at 48,000 *g* for 6 min at 0–2°.

8. The supernatants are collected under a stream of N_2 in a graduated cylinder on ice. The volume can be measured and a 2-ml sample removed for cadmium analysis. The supernatant is then rapidly pumped onto an anion-exchange column. Steps 6 through 8 should be done as quickly as possible.

Chromatography

Reagents

Ion-exchange buffers

Buffer A: 100 m*M* Tris-HCl, pH 8.6, 100 m*M* KCl, 5 m*M* ascorbic acid, 5 m*M* thiourea
Buffer B: the same as buffer A except for 2 *M* KCl

Both buffers are prepared in the same manner as the extraction buffer
Gel-filtration buffer: 10m*M* Tris-HCl, pH 8.6, 1 *M* KCl. Purge with N_2 prior to use
Desalting buffer: 0.5 m*M* Tris, unadjusted pH near 8

Procedure

The following is similar to an earlier procedure.[13]

1. Prepare a 10 × 1.6 cm column of QAE-Sephadex A-25 in water and

[13] W. E. Rauser, *Plant Physiol.* **74,** 1025 (1984).

equilibrate with buffer A overnight in the cold at a flow rate of 20 ml/hr. Keep a head space of N_2 over buffer A.

2. Pump the centrifuged extract onto the column at a flow rate of 85–90 ml/hr; maintain a slow stream of N_2 over the extract.

3. When the extract has been applied, wash the column with buffer A at a flow rate of 85–90 ml/hr for 2 hr, then at a flow rate of 35–40 ml/hr for 2 hr. The column effluent from steps 2 and 3 is collected in bulk.

4. Apply a 500-ml linear gradient from 0.1 to 2 *M* KCl using buffers A and B at a flow rate of 30 ml/hr. Bubble a gentle stream of N_2 into the mixing chamber. Collect 130 drop fractions (8–8.8 ml) in an atmosphere of N_2. The fraction collector can be placed in a sealable box flushed with N_2.

5. Determine the cadmium content of the first 20 to 30 fractions. The anion-exchange column can be allowed to run dry during the night without affecting the regeneration of the exchanger.

6. Combine the cadmium-rich fractions and reduce the volume to 3–5 ml by ultrafiltration under N_2. A large-diameter (76 mm) stirred cell containing a YC05 ultrafilter (Amicon Corp., Danvers, MA) effectively retains Cd-BC.

7. The concentrate is transferred quantitatively to a 96 × 2.5 cm column of Sephadex G-50 fine. The column is equilibrated and developed with cold N_2-purged 1 *M* KCl buffer at a flow rate of 20 ml/hr. Collect 100-drop fractions (6.1–6.9 ml) in an atmosphere of N_2.

8. Determine the absorbance at 254 nm and the cadmium content of the fractions. Keep tubes 40 through 80 under N_2 and covered as much as possible. They can reach room temperature.

9. The selected cadmium-rich fractions are combined, desalted on a BioGel P-2 column (Bio-Rad, Richmond, CA) equilibrated with desalting buffer, and lyophilized. The volumes of the column bed, sample, elution buffer, and wash are variable according to the manufacturer's suggestions.

Results and Comments

The compositions of extraction buffers for optimum maintenance of complexes have not been investigated systematically. Adequacy has been assumed when a large percentage of the cadmium in the extract behaves as an anionic species. The extraction conditions used here (N_2, ascorbic acid, thiourea, KCN) were designed to minimize oxidative changes that could lead to metal loss and intra- and intermolecular disulfide bridging. Since the Cd-BCs from roots of maize, tomato, and the grasses *Agrostis gigantea* and *Deschampsia caespitosa* are strongly anionic, use of 100 m*M* KCl in the extraction buffer and ion-exchange buffer A was warranted to reduce retention of materials not binding cadmium. Whether this applies to extracts from other plant sources and for metal complexes other than cadmium needs to be investigated.

Anion-exchange chromatography of a crude extract is shown in Fig. 1A. Total recovery of cadmium was 95%. Six percent of the cadmium eluted during sample application (fractions 1–31), 1% during washing (fractions 32–60), and 93% with the KCl gradient (fractions 61–80). These recoveries were corrected for the cadmium in the KCl. The thiourea and 2-mercaptoethanol did not permit measurement of the UV profile of the chromatogram. The top 2 cm of the QAE-Sephadex A-25 column became intensely dark blue with application of the extract. This part of the exchanger was discarded because the pigment was not removed during regeneration. The column effluent was yellowish; purple material of increasing intensity eluted starting with fraction 71.

The cadmium-rich fractions 65 to 73 (fraction Q, Fig. 1A) were concentrated by ultrafiltration, giving a wine-red fluid that entered the gel bed

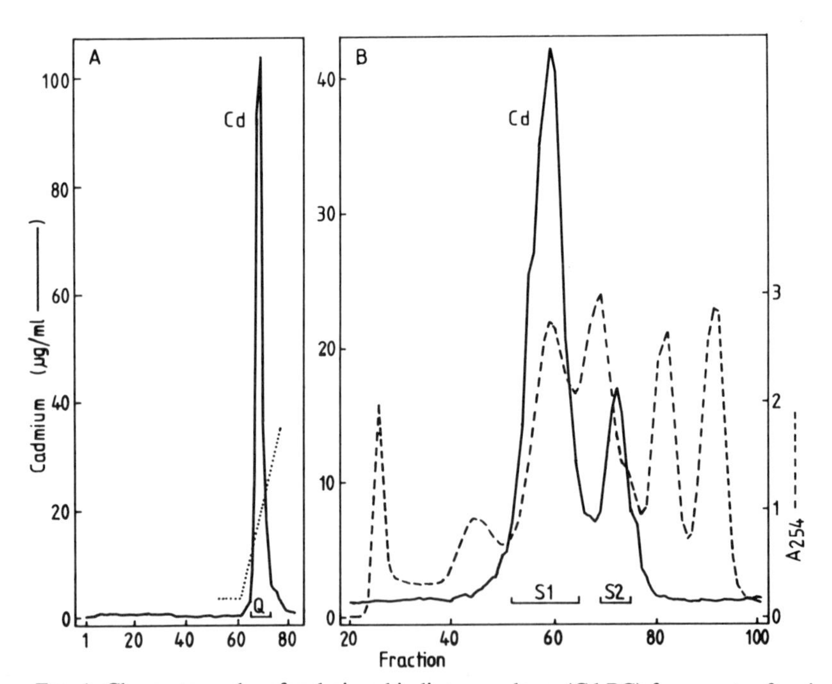

FIG. 1. Chromatography of cadmium-binding complexes (Cd-BC) from roots of maize. (A) Crude extract (272 ml) from 150 g of roots exposed to 3 μM cadmium for 7 days was applied to a column of QAE-Sephadex A-25. The column was washed and Cd-BC eluted with a linear gradient in KCl. The beginning part of the gradient is indicated by the dotted line, where the conductivity changed from 1.5 to 12 dS/m. Fraction Q was concentrated by ultrafiltration and chromatographed on a column of Sephadex G-50 fine (B). Fraction S1 was the source of prepurified Cd-BC. The void volume of the column was 179 ml.

uniformly and readily. Essentially all of the cadmium applied was recovered during gel filtration (Fig. 1B). Two cadmium-containing peaks were resolved. The first Cd-BC, fraction S1, eluted at a relative elution volume (V_e/V_0) of 2.3 and contained 69% of the cadmium. The second Cd-BC, fraction S2, contained 18% of the cadmium. Desalting and lyophilization of these fractions resulted in 19 mg of light-gray fluffy material from fraction S1 and 14 mg of dark-brown, somewhat sticky, crystalline material from fraction S2. The material from fraction S1 was designated prepurified Cd-BC for two reasons. First, a major portion of the cadmium was found in this well-resolved fraction. Second, the UV spectrum showed an inflection at 254 nm that disappeared by 60% on acidification and reappeared to 90% of the initial on back titration to pH 8.6, features typical of the cadmium–mercaptide chromophore. The electronic absorption of the material in fraction S2 did not change with pH, as if cadmium–thiolate coordination was absent.

Only the material in fraction S1 has been analyzed by high-performance liquid chromatography (see below). The nature of the Cd-BC in fraction S2 requires investigation. The prevalent view is that poly(γ-glutamylcysteinyl)glycines bind cadmium to form Cd-BC. Little attention has been given to the possibility that glutathione forms a Cd-BC. Preliminary evidence indicates that glutathione occurs in fraction S2. Freedman *et al.*[14] found that 60% of the cytoplasmic copper in hepatoma cells occurred as copper–glutathione. Demonstration of this complex required use of nonchelating buffer, anaerobiosis, and the absence of thiol-reducing agents. These are valuable insights into any changes that might be made to extraction and purification protocols for metal-binding complexes from plants.

The resolution of fraction S1 from UV-absorbing materials (Fig. 1B) depended on the inclusion of 1 *M* KCl in the gel-filtration buffer. Separation was also enhanced by removal of materials through the prior anion-exchange chromatography. The larger scale separation described can be used for smaller sample sizes. The major limitation is dilution of the metal, thus the sizes of the gel-filtration column and fractions would have to be reduced accordingly. The radioisotope ^{109}Cd was used to locate ^{109}Cd-BCs during analytical gel filtration of cell extracts.[15]

Analysis of Thiols and Phytochelatins

Class I and II MTs can be studied as separate metal-binding molecules. The metal serves to identify the MTs. In plants, however, a mixture of class III MT molecules forms metal-binding complexes. The constituent poly-

[14] J. H. Freedman, M. R. Ciriolo, and J. Peisack, *J. Biol. Chem.* **264,** 5598 (1989).

[15] E. Delhaize, P. J. Jackson, L. D. Lujan, and N. J. Robinson, *Plant Physiol.* **89,** 700 (1989).

peptides can only be studied following dissociation of the complexes, thereby causing loss of the metal and a major tag of the molecules. To identify apopolypeptides as phytochelatins (PC) requires, at a minimum, demonstration that glutamic acid, cysteine, and glycine predominate, and that glutamic acid and cysteine are equimolar and present at 2–11 times the level of glycine. For homophytochelatins (hPC) the same criteria apply after substituting β-alanine for glycine.

The thiol groups in PCs and hPCs provide the basis for assaying the molecules. The sulfhydryl groups can be derivatized with monobromobimane[16] to give highly sensitive fluorescent adducts that are then separated by high-performance liquid chromatography (HPLC). An alternative is to react the thiol groups with 5,5′-dithiobis(2-nitrobenzoic acid) (DTNB, Ellman's reagent) after the molecules have been separated by HPLC. The product of the reaction is 5-thio-2-nitrobenzoic acid, which absorbs broadly at 412 nm.[17]

Characteristics of the DTNB postcolumn reaction for thiols are described below for prepurified Cd-BC from maize and for fresh crude extracts of maize roots. A method similar to that described here has been used extensively in analyses of PCs and hPCs in many different plants[18] and the enzyme involved in PC biosynthesis from glutathione.[19]

High-Performance Liquid Chromatography

Reagents

Solvent A: 0.1% (v/v) trifluoroacetic acid in water

Solvent B: 70% (v/v) acetonitrile and 0.1% (v/v) trifluoroacetic acid in water

Solvent C: 20% (v/v) acetonitrile and 0.1% (v/v) trifluoroacetic acid in water. This solvent is used in instruments with ternary solvent systems

DTNB solution: 1.8 m*M* in 300 m*M* potassium phosphate, pH 7.8–8.0, and 15 m*M* K_2 EDTA. Stock phosphate–EDTA buffer can be stored at 0–4°. The DTNB solution is prepared fresh daily, passed through a 0.45-μm filter, and kept on ice in darkness

Standard solutions: A few milliliters of 1 m*M* glutathione (GSH) in water are prepared fresh daily and a portion diluted to 0.1 m*M* for injection

[16] G. L. Newton, R. Dorian, and R. C. Fahey, *Anal. Biochem.* **114,** 383 (1981).

[17] G. L. Ellman, *Arch. Biochem. Biophys.* **82,** 70 (1959).

[18] W. Gekeler, E. Grill, E.-L. Winnacker, and M. H. Zenk, *Z. Naturforsch. C: Biosci.* **44,** 361 (1989).

[19] E. Grill, S. Löffler, E.-L. Winnacker, and M. H. Zenk, *Proc. Natl. Acad. Sci. U.S.A.* **86,** 6838 (1989).

Equipment

A model 5000 Varian liquid chromatograph (Varian Instruments, Inc., Sunnyvale, CA), a manual Rheodyne injector valve (Rheodyne, Cotati, CA) with a 500-μl sample loop, a model 2050 Varian UV detector, and a strip chart recorder are used to separate and analyze the polypeptides. Only an instrument generating reproducible linear gradients with solvents A and B is required. The Varian instrument allowed use of solvents A, C, and B to generate the gradients.

The postcolumn reaction requires a model 350 metering pump (Scientific Systems, Inc., State College, PA), a 10-μl mixing chamber (The Lee Co., Westbrook, CT), a 0.8-mm i.d. × 3.05 m knitted Teflon capillary tube (Supelco, Inc., Bellefonte, PA) followed by a 1.95-m length of 0.3-mm i.d. Teflon tubing, a model 1740 visible detector with a 405-nm filter (Bio-Rad Laboratories, Richmond, CA), and a Varian 4290 integrator. For the solvent system described, a 10-μm C_{18} (4.6 × 250 mm) Nucleosil column (Macherey-Nagel, Dueren, Germany) is used and protected by a 2 × 20 mm precolumn containing 30-to 40-μm pellicular Perisorb RP-18 (Merck, Darmstadt, Germany). The Nucleosil column and Teflon capillary tubes are held in a column heater at 37°. Solvents A, B, and C are sparged with helium. In this system the column effluent is first monitored at 220 nm and flows at 1 ml/min into the mixing chamber, where the DTNB solution is added at 0.5 ml/min. Color development occurs in the 1.8-ml reactor for 1.2 min before the A_{405} of derivatized material is measured and the signal integrated.

Procedure

The column is equilibrated for about 15 min with solvent A at a flow of 1 ml/min while DTNB solution is pumped into the mixing chamber at a flow of 0.5 ml/min. The A_{405} is set to 0.03. Glutathione standard or sample is injected and a linear gradient from 0 to 20% acetonitrile in 40 min is applied. The acetonitrile is then increased to 70% in 2 min, the column is washed with solvent B for 5 min, and then equilibrated with solvent A.

Results and Comments

The absorbance of the DTNB solution rose slowly with time; on ice the change was much less than at varying room temperature. Color development in the reactor at 37° for 1.2 min was nearly complete. Addition of a second knitted Teflon capillary tube gave only a 5% increase in integrated signals with an associated peak broadening that was disadvantageous.

Chromatography of pre-purified Cd-BC (fraction S1, Fig. 1B) and its reaction with DTNB are shown in Fig. 2A. The Cd-BC was resolved into

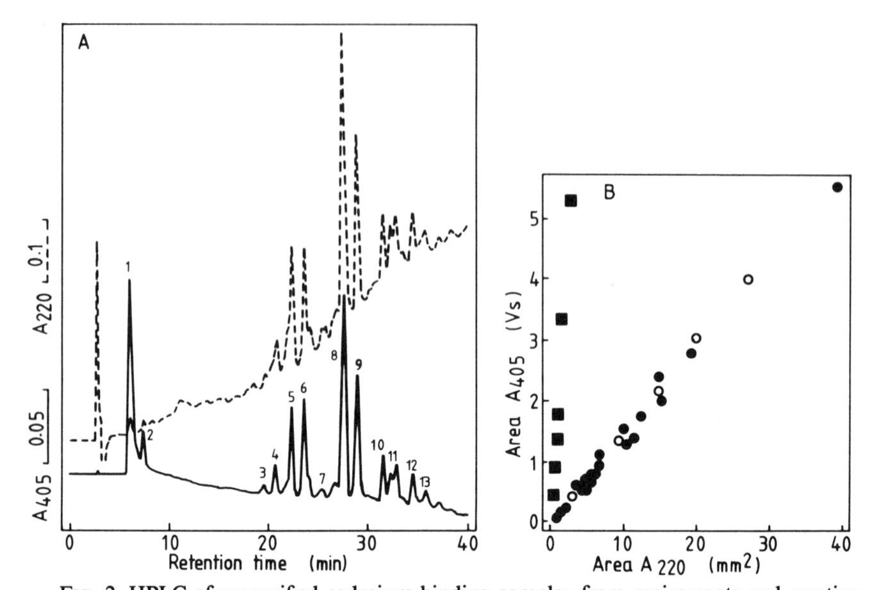

FIG. 2. HPLC of prepurified cadmium-binding complex from maize roots and reaction with DTNB. (A) The elution profile of 100 μg prepurified Cd-BC is given by A_{220} and after derivatization of thiols with DTNB by A_{405}. The entire thiol profile is shown shifted to the left by 1.2 min (the residence time in the postcolumn reactor) to remove the systematic delay in detecting a polypeptide via the DTNB reaction. The relationship between the area of UV peaks and the counterpart thiols is shown in (B). Samples of 80, 160, and 240 μg prepurified Cd-BC were chromatographed as in (A). The solid squares are the responses for peaks 1 and 2, the solid circles for peaks 3 through 13. Responses for glutathione (5, 15, 25, 35, and 45 nmol) are given by the open circles. Vs, volt·seconds.

various components that reacted with DTNB. The thiol profile had a declining baseline from A_{405} of 0.03 at the start to 0 or below by 40 min, as if increasing concentrations of acetonitrile reduced the DTNB reaction. If this were so, then glutathione would not be a suitable quantitative calibrating compound because it eluted at 6.7 min with minimal acetonitrile, whereas most of the polypeptides eluted with 10–15% acetonitrile.

The amount of UV-absorbing material in peaks 1 through 13 (Fig. 2A) was determined by measuring the areas on the A_{220} trace and standardizing to two absorbance units full scale, comparable to the integrated signals for thiols. The relationship between the area in the UV and the counterpart thiol reaction is shown in Fig. 2B for three increasing quantities of Cd-BC. For the 30 pairs of absorbances of peaks 3 through 13, the absorbances of the DTNB reaction increased linearly ($r = 0.9957$) with the amount of UV-absorbing material. Concentrations of acetonitrile up to 20% did not affect the degree of reaction between DTNB and thiols. Glutathione reacted quantitatively in the same way as did the thiols in peaks 3 through 13. Glutathione was a suitable calibrating substance for quantitating the various thiols. The linear response for thiols reached a maximum at 55 nmol/peak; the peak detection threshold was 0.2 nmol.

Peaks 1 and 2 gave a 14.7-fold greater yield of A_{405} per unit of A_{220} than did peaks 3 through 13 (Fig. 2B). Their inordinately low A_{220} values suggested that these thiols were not peptides. Sulfide and 2-mercaptoethanol were suspected. Cadmium-binding complexes can contain acid-labile sulfide and 2-mercaptoethanol was used in the initial stages of Cd-BC isolation. When a solution containing 100 μg Cd-BC (as in Fig. 2A) was acidified with trifluoroacetic acid to 0.1% (v/v) and N_2 gently blown over it prior to chromatography, most of thiol peak 1 was eliminated, yet thiol peak 2 remained unchanged. Addition of $(NH_4)_2S$ to Cd-BC enhanced thiol peak 1, addition of 2-mercaptoethanol enlarged thiol peak 2.

The following elution times, in minutes and corrected for delay in the postcolumn reactor, were found for various thiols: cysteine, 3.2; sulfide, 6.0; glutathione, 6.7; 2-mercaptoethanol, 7.6; dithioerythritol (DTE), 10.8; dithiothreitol (DTT), 13.9. The elution times for the first four thiols remained unchanged whether a gradient in acetonitrile or isocratic flow of solvent A was used. Isocratic flow of solvent A increased the elution time of DTE to 11.7 min and of DTT to 18.2 min. Glutathione disulfide (GSSG) eluted in the gradient at 12.5 min without reacting with DTNB.

The chromatogram of Cd-BC shown in Fig. 2A showed a prominent thiol peak 1. Because sulfide and glutathione elute close to each other they are poorly resolved when one is abundant. Glutathione must consequently be measured independently; the enzymatic recycling assay was used.[20] No glutathione was found in the preparation of Cd-BC. Since the Cd-BC was purified in alkaline buffer, any acid-labile sulfide was released only in a closed system on mixing of the injected complex with acidic mobile phase during HPLC. The preparation shown in Fig. 2A contained 21% of its thiol as acid-labile sulfide.

Thiols and Phytochelatins in Crude Extracts

Reagent

Extraction solution: 100 mM HCl stored at 0–4°

Procedure

1. The roots of 5-day-old maize seedlings are exposed to 3 μM cadmium for 24 hr. The root systems of 10 seedlings are rinsed briefly in ice-cold 0.5 mM $CaSO_4$, blotted, the apical 10 cm excised, and weighed.
2. The roots are homogenized with a volume of extraction solution

[20] M. E. Anderson, this series, Vol. 113, p. 548.

equal to twice the fresh weight of the sample. A small mortar and pestle packed in ice can be used.

3. The homogenate is transferred to a tube and centrifuged in a microcentrifuge at 6000 rpm for 5–10 min at 4°.

4. The supernatant is passed through a low-retention 0.45-μm Millex HV filter (Millipore Corp., Bedford, MA) and stored on ice until used for HPLC and glutathione determination. Storage beyond 2 hr is not recommended.

5. Analyze 250 μl of the filtered supernatant by HPLC and postcolumn derivatization with DTNB. Ten to 25 μl of the same supernatant suffices for glutathione determinations by the enzymatic recycling assay.

6. To standardize the thiol response, inject 5, 10, or 20 nmol glutathione and develop the gradient for the initial 8 min prior to equilibrating the column with solvent A.

Results and Comments

The HCl extraction solution precipitated more extraneous proteins than did 5% (w/v) 5-sulfosalicylic acid; thiol profiles were the same. Measurement of glutathione in 100 m*M* HCl by the enzymatic recycling assay[20] requires substitution of 5-sulfosalicylic acid by the HCl. Extraction of roots with 1 *M* NaOH-containing $NaBH_4$ and subsequent acidification[21] was, in our hands, cumbersome and not reproduced easily.

Postcolumn derivatization of thiols in a crude extract of maize is shown in Fig. 3. Some of the major thiol peaks had only small UV peaks, others were shoulders to larger peaks or parts of large UV peaks. Retention times and standard additions identified cysteine and sulfide plus glutathione. Identification of the remaining thiols was achieved in part.[22] Phytochelatin standards were not available and cadmium could no longer serve to identify PCs because it was not bound by peptides in the acidic mobile phase. A guide to a provisional identification of PCs came from the thiol profile of the apical 1 cm of maize roots exposed to cadmium (Fig. 4A). In addition to cysteine and sulfide plus glutathione, the apices contained only peaks 1, 2, and 3, whereas additional small and large thiol peaks occurred in the mature region of the same roots (Fig. 4B). For roots not exposed to cadmium, both root regions had thiol profiles containing cysteine, sulfide + glutathione, and a small amount of peak 1. Exposure to cadmium for as little as 15 min caused an increase in peak 1 and appearance of peaks 2 and 3. The other thiols in the mature root region were detected only after 2 hr of cadmium treatment. Thiol peaks 1, 2, and 3 were

[21] E. Grill, E.-L. Winnacker, and M. H. Zenk, *Proc. Natl. Acad. Sci. U.S.A.* **84,** 439 (1987).
[22] A. Tukendorf and W. E. Rauser, *Plant Sci.* **70,** 155 (1990).

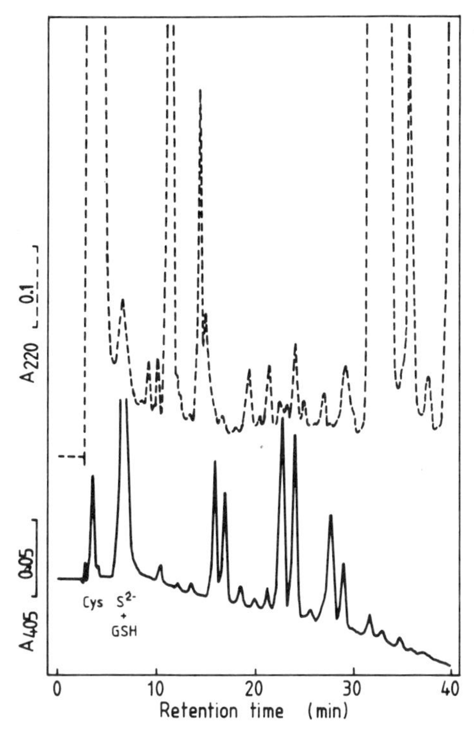

FIG. 3. HPLC of a crude extract of maize roots and the reaction with DTNB. A crude extract was prepared with the apical 10 cm from 10 roots of maize seedlings exposed to 3 μM cadmium for 24 hr. After centrifugation and filtration, 250 μl of extract was injected onto the column. The position of the thiol profile (A_{405}) relative to the UV profile (A_{220}) was corrected for the delay required for derivatization with DTNB.

designated putative PCs. The UV portions equivalent to thiol peaks 1, 2, and 3 were isolated from a crude extract without addition of DTNB, lyophilized, and rechromatographed. The amino acid analyses (Table I) showed that only cysteine, glutamic acid, and glycine were present. Cysteine and glutamic acid were equimolar and occurred at two, three, and four times the level of glycine. Thiol peaks 1, 2, and 3 (Fig. 4) were designated PC_2, PC_3, and PC_4, respectively. The identity of the other thiols remains unknown. Attempts to rechromatograph the materials in the major thiol peaks eluting after the PCs (Fig. 4B) resulted in considerable losses, as if these thiols were particularly sensitive to degradation.

The thiol profiles of fresh extracts of samples containing mature root regions (Figs. 3 and 4B) were similar to that of prepurified Cd-BC (Fig. 2A). However, the Cd-BC lacked PC_2. Peaks 5 and 8 of Cd-BC had elution

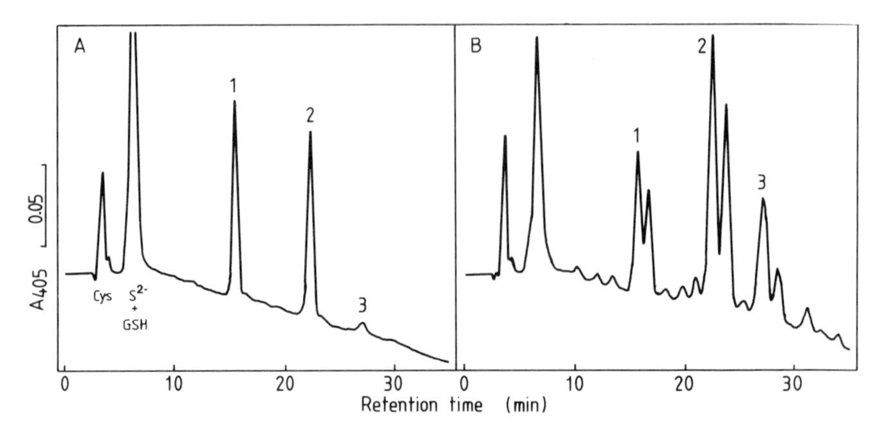

FIG. 4. Thiol profiles of crude extracts of maize roots following HPLC. Twenty seedlings exposed to 3 μM cadmium for 24 hr were used. (A) The root apices (0-to1-cm region) with a fresh weight of 0.116 g were homogenized with 0.928 ml 100 mM HCl and 250 μl was analyzed. (B) The 1-to10-cm regions (1.244 g fresh weight) from the same roots were homogenized with 2.488 ml extraction solution and 250 μl was analyzed. The contents (nmol thiol/g fresh weight) of the various components were (A) cysteine, 149; glutathione, 607; peak 1, 261; peak 2, 265; peak 3, 18. The equivalents in (B) were 49, 112, 76, 155, and 85, respectively. (These data are excerpted from Tukendorf and Rauser.[22])

times equivalent to PC_3 and PC_4, respectively. Perhaps PC_2 was absent from roots of maize exposed to cadmium for 7 days or the smallest PC was lost during desalting of fraction S1. Direct analyses by HPLC and DTNB postcolumn derivatization are warranted for fraction S1 prior to desalting and for fresh roots exposed to cadmium for 7 days.

The PCs were quantitated by relating the integrated thiol response to that of standard glutathione. Quantitation of PCs in this way did not, however, indicate to what degree they participated in metal binding *in vivo.* Quantitation of Cd-BC could only assess the extent of metal binding. Sulfide was not quantitated in chromatograms of crude extracts because of unknown losses of acid-labile sulfide during homogenization of the tissue in 100 mM HC1.

The simplicity of the thiol profile in the apical 1 cm of maize roots and the relative abundance of PCs (Fig. 4) made this tissue attractive for experimentation. One caution is that maize root tips may be covered with copious amounts of mucigel, depending on the culture method. Difficulty in filtering the extracts may be encountered as well as clogging of the inlet frit of the HPLC precolumn.

Most of our experience with thiol analyses has been with freshly prepared extracts stored on ice for no more than 2 hr. Extracts stored overnight at 0–4° lost about 15% of their glutathione and some degradation of

TABLE I
AMINO ACID COMPOSITION OF PUTATIVE PHYTOCHELATINS FROM MAIZE ROOTS[a,b]

Amino acid	Percentage of residue		
	Peak 1	Peak 2	Peak 3
Cysteine[c]	38.1 (1.8)	43.5 (3.1)	43.5 (4.0)
Glutamic acid	41.1 (2.0)	42.4 (3.0)	45.7 (4.2)
Glycine	20.8 (1)	14.1 (1)	10.8 (1)

[a] Excerpted from Tukendorf and Rauser.[22]

[b] The UV fractions corresponding to peaks 1, 2, and 3 (Fig. 4) from 5.2 g fresh weight of maize roots exposed to 3 μM cadmium for 24 hr were collected without addition of DTNB. The eluates were lyophilized, rechromatographed, and peaks recollected for amino acid analysis. The data are expressed as percentage residues with calculated residues per mole in parentheses.

[c] Determined as cysteic acid after performic acid oxidation.

other thiols was evident. Freeze-drying small root samples prior to extraction caused losses of thiols. Rapid freezing of extracts and maize tissues in liquid N_2 and storage therein or at $-80°$ is acceptable.

Acknowledgments

This research was supported by Operating Grant OPG 0004921 from the Natural Sciences and Engineering Research Council of Canada. I thank Dr. Anna Tukendorf, Maria Curie-Sklodowska University, Lublin, Poland, for her assistance with the development and evaluation of the postcolumn derivatization of thiols.

[39] Phytochelatins

By ERWIN GRILL, ERNST-LUDWIG WINNACKER, and MEINHART H. ZENK

Heavy metal-inducible and metal-complexing systems are ubiquitous in eukaryotes. The low-molecular-weight proteins metallothioneins[1] in animals and the phytochelatin (PC) peptides[2,3] in plants are induced by

[1] J. H. R. Kägi and M. Nordberg, eds., *Experientia, Suppl.* **34** (1979).

[2] E. Grill, E.-L. Winnacker, and M. H. Zenk, *Science* **230,** 674 (1985).

[3] W. Gekeler, E. Grill, E.-L. Winnacker, and M. H. Zenk, *Z. Naturforsch. C: Biosci.* **44,** 361 (1989).

heavy metal ions such as Cd^{2+}, Zn^{2+}, Pb^{2+}, and Cu^{2+}, and complex these ions by thiolate coordination. Both systems seem to play a role in the detoxification and homeostasis of heavy metal ions.[4,5] In contrast to metallothioneins, however, PCs are not primary gene products.

Phytochelatins represent a class of peptides consisting of linear chains of γ-glutamylcysteinyl units with a carboxyl-terminal glycine in PC or β-alanine in homo-PC. The general formula is $(\gamma\text{-Glu-Cys})_n$-Gly ($n = 2-11$) and $(\gamma\text{-Glu-Cys})_n$-$\beta$-Ala ($n = 2-6$)[2,6-8]. A non-ribosome-dependent biosynthesis of the molecules was indicated by the isopeptide linkages of the γ-glutamyl residues in the primary structures. PCs are formed by the action of the constitutive enzyme phytochelatin synthase.[9] The enzyme catalyzes the transpeptidation of glutamylcysteinyl moieties of glutathione (GSH) onto another GSH molecule or the growing PC peptide, thus generating the set of homologous molecules.

Assay of Phytochelatins

Principle

Phytochelatins can be assayed by virtue of their high level of reduced sulfur groups, their metal-binding properties, and the chromatographic behavior of the set of peptides.

Analysis of Sulfhydryl Groups

Assays of reduced thiols have been recently reviewed in this series.[10] Sulfhydryl groups are detected with 2 m*M* Ellman's reagent[11] in 0.1 *M* degassed potassium phosphate buffer, pH 7.3, and 10 m*M* EDTA. After 10 min of reaction the absorption is recorded at 410 nm against a blank sample.

[4] D. H. Hamer, Annu. *Rev. Biochem.* **55,** 913 (1986).
[5] E. Grill, E.-L. Winnacker, and M. H. Zenk, *Proc. Natl. Acad. Sci. U.S.A.* **84,** 439 (1987).
[6] N. Kondo, K. Imai, M. Isobe, T. Goto, A. Murasugi, C. Wada-Nakagawa, and Y. Hayashi, *Tetrahedron Lett.* **25,** 3869 (1984).
[7] J. C. Steffens, D. F. Hunt, and B. G. Williams, *J. Biol. Chem.* **261,** 13879 (1986).
[8] E. Grill, W. Gekeler, E.-L. Winnacker, and M. H. Zenk, *FEBS Lett.* **205,** 47 (1986).
[9] E. Grill, S. Löffler, E.-L. Winnacker, and M. H. Zenk, *Proc. Natl. Acad. Sci. U.S.A.* **86,** 6838 (1989).
[10] P. C. Jocelyn, this series, Vol. 143, p. 44.
[11] G. L. Ellman, *Arch. Biochem. Biophys.* **82,** 70 (1959).

Analysis of Heavy Metals

Heavy metal ions such as ions of cadmium, zinc, and copper are quantified by atomic absorption spectroscopy[12] (PE-1100B, operated in flame mode; Perkin-Elmer, Norwalk, CT) of solutions acidified by 0.3% (v/v) H_2SO_4. Precipitated material is removed prior to the measurement by centrifugation at 11,000 *g* and 4° for 10 min.

These two assays are simple and fast and are routinely employed during purification of the PC–metal complex from plant tissue. A more specific analysis of the individual PC peptide is based on high-performance liquid chromatography (HPLC) under PC complex-dissociating conditions, combined with postcolumn derivatization of sulfhydryl groups.

High-Performance Liquid Chromatography Analysis

Reagents

Alkaline solution: 1 *N* NaOH containing 1 mg/ml $NaBH_4$ (freshly prepared solution)
HCl (3.6 *N*)
Solvent A: 0.05% (v/v) phosphoric acid in water
Solvent B: 20% (v/v) acetonitrile in solvent A
Reagent solution: 75 μM Ellman's reagent in 0.05 *M* potassium phosphate buffer, pH 8.0 (freshly prepared)

Procedure. The following steps are carried out in microfuge tubes (Eppendorf, Hamburg, Germany). The sample (0.25 ml) is mixed with an equal volume of alkaline solution. After 5 min of incubation at room temperature the solution is acidified by adding 0.10 ml of 3.6 *N* HCl and left on ice for 15 min. Precipitated protein is removed by centrifugation (11,000 g, 4°, 5 min) and aliquots of the supernatant (10–100 μl) are analyzed by HPLC. Phytochelatin assays of plant tissue are performed by freezing the material in liquid nitrogen and grinding it with a glass rod or a mortar and pestle. The alkaline $NaBH_4$-containing solution (0.4 ml) is added to 0.40 g of the tissue powder and the suspension is sonicated (sonicator equipped with a microtip; Branson, Danburg, CT) by three pulses of 5 sec each at maximal energy setting. The cell debris is subsequently sedimented by centrifugation at 11,000 g at 4° for 5 min. The supernatant (0.5 ml) is acidified with 0.10 ml 3.6 N HCl solution, incubated on ice and further treated as described above. Phytochelatin peptides are separated on a reversed-phase column (Nucleosil C_{18}, 10 μm; 4 × 250

[12] W. Slavin, this series, Vol. 158, p. 117.

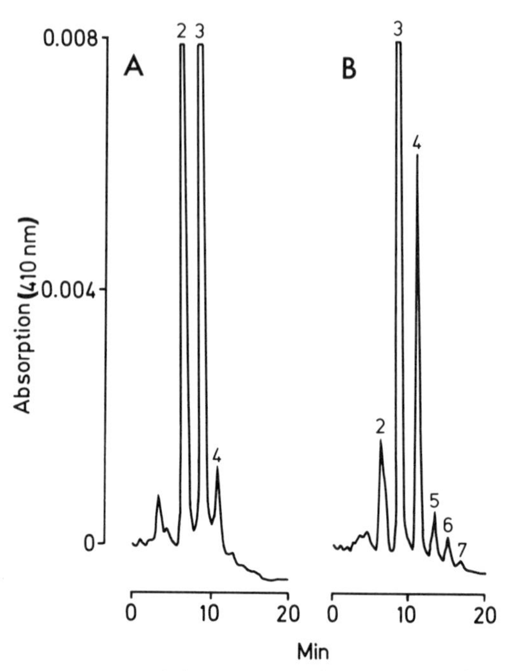

FIG. 1. HPLC chromatogram of zinc-phytochelatin (A) and cadmium-phytochelatin complex (B) under complex-dissociating conditions. Sulfhydryl groups were derivatized with Ellman's reagent[11] and detected at 410 nm. The numbers (2–7) indicate n of the individual PC peptide (γ-Glu-Cys)$_n$-Gly.

mm; Macherey and Nagel, Düren, Germany) with a linear gradient of solvent A to solvent B at a flow rate of 2 ml/min. All solutions used for HPLC are degassed and continuously purged with argon gas.

Detection of the purified PCs is performed at 220 nm. Postcolumn derivatization of the sulfhydryl groups is performed for the PC assay of crude plant extracts. The derivatization is accomplished by mixing the eluate with the reagent solution (2 ml/min). The mixture is passed through a reaction loop (5-ml capacity) at room temperature and the absorption is then recorded at 410 nm. Signals are calibrated with GSH and purified individual PC peptides. A typical chromatogram is shown in Fig. 1.

Isolation of Phytochelatin–Metal Complex

All steps are carried out at 4° unless otherwise stated and all solutions are degassed prior to use. The plant material, preferably cell suspension cultures, is exposed to heavy metal ions for 2–5 days. Maximal PC induction is observed with cadmium ion concentrations of 0.1–0.3 mM. The following procedure is recommended for Cd-PC complexes, but Zn-PC and Cu-PC complexes have also been isolated with this protocol.[13]

Cultivation of Plant Cells

Cell suspension cultures of *Rauwolfia serpentina* are cultivated in Linsmaier and Skoog medium.[14] Cell suspensions are diluted 10-fold with fresh medium after cultivation on a gyratory shaker (100 rpm) for 1 week at 23°. The cultures are exposed to heavy metal ions during the log phase of growth, on the third or fourth day of the culture period, by administration of 1 *M* $Cd(NO_3)_2$ solution to a final concentration of 0.2 m*M*. After 2 to 5 days of cadmium exposure the cells are harvested by suction filtration, washed with water, and frozen in liquid nitrogen. The suspension culture yields about 0.3–0.4 kg fresh weight of cells/liter of culture, corresponding to 10–15 g dried material.

Step I: Extract. Frozen plant tissue (0.3 kg) is thawed in 0.15 liters of 20 m*M* Tris-HCl buffer, pH 8.6, containing 10 m*M* 2-mercaptoethanol. The homogenate is pressed through four layers of cheesecloth and the extract is cleared by centrifugation (10,000 *g*, 30 min). The extract is readjusted to pH 8.6 with 1 *N* NaOH and fourfold diluted with water to a conductivity below 1 kS.

Step 2: Ion-Exchange Chromatography. This step serves primarily to concentrate the proteins in the extract. The extract is applied to a DEAE column (1.5 × 25 cm; DEAE-BioGel A, Bio-Rad, Richmond, CA) at a flow rate of 0.2 liter/hr. The column has been equilibrated with 10 m*M* tris-HCl buffer, pH 8.6, and 1 m*M* 2-mercaptoethanol. This buffer (50 ml) is also used to wash the column after sample application and bound protein is subsequently eluted in the same buffer supplemented with 0.5 *M* NaCl. The elution is easily monitored by the brownish color of the protein solution that is collected for the next purification step.

Step 3: Salt Fractionation. Solid ammonium sulfate (ground to a fine powder) is slowly added to the eluted protein solution (approximately 30 ml) with stirring, to give a final value of 80% saturation [0.6 g $(NH_4)_2SO_4$/ml eluate]. This step is carried out at room temperature. Precipitated proteinaceous material is removed by centrifugation (9000 *g*, 10 min). The supernatant is diluted twofold with 10 m*M* Tris-HCl buffer, pH 8.0, and concentrated by ultrafiltration (YM2 membrane; Amicon, Danvers, MA). A considerable level of pentapeptide PC is lost during ultrafiltration due to insufficient retention of small PC complexes. Dilution and concentration of the preparation is repeated until the ammonium sulfate is reduced to

[13] E. Grill, *Mol. Cell. Biol.* **98** (Suppl.), 283 (1989).
[14] E. M. Linsmaier and F. Skoog, *Physiol. Plant,* **18,** 100 (1965).

TABLE I
ISOLATION OF PHYTOCHELATIN[a]

Purification step	Volume (ml)	Protein (mg)	Cadmium (mg)	Yield (% Cd)
Crude extract	330	225	18.6	100
DEAE eluate	18	91.2	17.6	95
Salt fractionation	21.8	1.7	16.4	88
Ultrafiltration	6.1	1.1	14.0	75
Gel filtration	52.4	0.02[b]	11.1	60

[a] Cell suspension cultures of *Rauwolfia serpentina* were exposed to 0.2 m*M* $Cd(NO_3)_2$ for 5 days. Cd-PC was isolated from 0.28 kg of fresh cell material corresponding to 10.5 g dry weight.

[b] Standard protein assays are not able to quantify PC properly. In this case protein was determined according to Bradford.[15] The lyophilized fraction yielded 59.1 mg pure Cd-PC as determined by amino acid analyses.

about one-tenth of its initial level and the sample is then concentrated to 5–7 ml.

Step 4: Gel Filtration. The final concentrate is chromatographed on a gel-filtration column (Sephadex G-50, 2.5 × 52 cm) at a flow rate of 90 ml/hr. Ammonium acetate buffer, 10 m*M*, pH 7, is used for chromatography. The absorption of the eluate is monitored at 254 nm and fractions of 6 ml are collected. Fractions containing metal–PC complexes are identified by atomic absorption spectrometry and by assay for sulfhydryl groups using Ellman's reagent.[11] The sulfhydryl- and cadmium-containing fractions correspond to the major UV-absorbing fraction. They are pooled and lyophilized, and yield about 50–80 mg of Cd-PC complex. A typical purification is shown in Table I.

Isolation of Metal-Free Phytochelatin Peptides

Principle

Metal-free PC peptides can be isolated by using conditions in which the metal–peptide complex dissociates and by subsequent separation of the components. This can be achieved either by using efficient metal chelators [e.g., ethylenediaminetetraacetic acid (EDTA) or diethyl dithiocarbamate] or by acidification, and then removal of the metal ions as a sulfide precipitate or by chromatographic procedures. In our hands, acidification of the

[15] M. M. Bradford, *Anal. Biochem.* **72,** 248 (1976).

PC solution after reduction of the complex, followed by HPLC separation, gave the best results and is outlined below.

Procedure

Cd-PC (50 mg) is dissolved in 0.9 ml of 10 m*M* Tris-HCl, pH 8.0. The pH is checked and, if necessary, adjusted to pH 8 by addition of 1 *N* NaOH. $NaBH_4$ (1 mg) is added to the solution and the sample is incubated for 15 min at 40°. The PC solution is put on ice and 6 *N* HCl is added to destroy excess $NaBH_4$ and to acidify the solution to pH 1–2. The solution is spun at 11,000 g for 15 min and the supernatant is injected onto a semipreparative HPLC column (Nucleosil C_{18}, 25 × 250 mm; Macherey and Nagel). Individual PC peptides are eluted by a linear gradient (40 min) of 0–20% acetonitrile (v/v) and 0.1% trifluoroacetic acid (v/v) in water at a flow rate of 10 ml/min. The elution is recorded at 220 nm and peptidic fractions are collected. In order to isolate the mixture of metal-free peptides present in the PC complex, a step gradient is used, consisting of the same solvents as mentioned before. After application of the sample the column is washed for 12 min with solvent lacking acetonitrile (flow rate, 10 ml/min) and the PC peptides are eluted in solvent containing 20% acetonitrile. The collected fractions are lyophilized and yield 39 mg of metal-free peptides as a mixture of the individual PCs from 50 mg Cd-PC complex.

Fingerprinting of Peptides

The metal-free forms of the isolated PC peptides can be characterized by partial hydrolysis. The sulfhydryl groups of individual peptides are oxidized with performic acid[16] and the amino terminus of the PC molecule is dinitrophenylated.[17] Samples corresponding to 20–50 μg of peptide are hydrolyzed in 100 μl 1 *N* HCl for 18 min at 95°. The acid is then completely removed in a stream of nitrogen.

The hydrolysates thus obtained are subjected to two-dimensional separation on silica plates (No. 5748; Merck, Darmstadt, Germany). In the first dimension, separation is achieved by electrophoresis in acidic buffer (2.8% formic acid, 7% acetic acid in water) for 1.5 hr at 25 V/cm. In the second dimension, the peptide fragments are resolved by chromatography in 1-propanol:ammonia:water (4:1.5:1). Fragments with an intact amino terminus are visible as yellow spots and are eluted with 50% aqueous ethanol in water. Equimolar amounts of peptides [quantified by the absorption of the

[16] C. W. M. Hirs, *J. Biol. Chem.* **219,** 611 (1956).
[17] F. Sanger, *Biochem. J.* **45,** 563 (1949).

dinitrophenyl (DNP) moiety at 360 nm] are analyzed for their amino acid content after hydrolysis. By this technique the unhydrolyzed PC molecule and the peptide lacking the carboxyl-terminal glycine are not separated.[18] Fast atomic bombardment has been used as an elegant alternative method for fingerprinting and sequencing of PC peptides.[7]

Reconstitution of Phytochelatin–Metal Complexes

Principle

The metal-free PC peptides can be reassociated with heavy metal ions simply by mixing metal ions with the reduced peptides. Extra care should be taken to avoid oxidation of the sulfhydryl groups during reconstitution. We performed the reconstitution in an anaerobic chamber under a reducing atmosphere [5% hydrogen and 95% nitrogen (v/v) over palladium catalyst]. For most purposes, working under a stream of nitrogen with nitrogen-purged solutions is sufficient, particularly with high concentrations of peptide material.

Procedure

The metal-free peptides (10 mg) are dissolved in 10 m*M* Tris-HCl, pH 8, the pH is readjusted, and the sample is reduced with $NaBH_4$ as outlined under Isolation of Metal-Free Phytochelatin Peptides, above. A second aliquot of $NaBH_4$ (1 mg) is added and the incubation continued for another 15 min. The excess reducing agent is destroyed by acidification with 6 *N* HCl to pH 1–2 and 25 μmol of Cd^{2+} is added, corresponding to a ratio of metal to sulfhydryl group of approximately 1:1.5. Tris base (2 *M*) is added to the solution to a final concentration of 0.2 *M*. The pH of the PC sample is adjusted to pH 8 using 6 *N* NaOH. The PC-Cd complex precipitates during the process of neutralizing the acidic solution but redissolves easily under slightly alkaline conditions. The reconstituted complex is incubated for another hour and then removed from the anaerobic chamber. The Cd-PC is chromatographed on a gel-filtration column (Sephadex G-50, 1.5×50 cm) in 10 m*M* Tris-HCl, pH 8, and 0.1 *M* KCl (flow rate, 20 ml/hr and 4 ml/fraction). The fractions strongly absorbing UV at 254 nm are desalted using a Sephadex G-25 column (1.5×25 cm) and lyophilized. Starting from 10 mg of metal-free peptide we finally obtained 9.2 mg of Cd-PC complex.

[18] W. Gekeler, E. Grill, E.-L. Winnacker, and M. H. Zenk, *Arch. Microbiol.* **150,** 197 (1988).

Properties of Phytochelatin–Metal Complexes

Phytochelatin peptides form complexes with Cd^{2+} of M_r 2.5K and 3.6K.[2] In fission yeast a Cd-PC complex of M_r 1.8K was observed.[19] These complexes appear to consist of a mixture of the PC peptides of various lengths. The metal-to-sulfur ratio is 1:2 for complexes with ions of cadmium,[20] zinc, and lead,[2,13] whereas a 1:1 ratio was determined for copper complexes.[13,21] In addition, acid-labile sulfide was identified in the complexes isolated from yeast[21,22] and plants[7,18,23] capable of forming Cd–S crystalline structures.[24] The level of sulfide varied widely, from below 1 to 30% of total sulfur in the complexes isolated from plants.[18] We also observed small amounts of cysteine and GSH in metal complexes. These findings could reflect different phases of PC complex catabolism in which PC-Cd molecules are processed similarly to GSH–S conjugates.

Acknowledgments

This research was supported by the Bundesministerium für Forschung und Technologie and the Fonds der Chemischen Industrie. We thank Dr. Michael Kertesz for reading the manuscript.

[19] A. Murasugi, C. Wada, and Y. Hayashi, *J. Biochem. (Tokyo)* **90,** 1561 (1981).
[20] H. Lue-Kim and W. E. Rauser, *Plant Physiol.* **81,** 896 (1986).
[21] D. R. Winge, R. N. Reese, R. K. Mehra, E. B. Tarbet, A. K. Hughes, and C. T. Dameron, *Mol. Cell. Biol.* **98** (Suppl.), 301 (1989).
[22] A. Murasugi, C. Wada, and Y. Hayashi, *J. Biochem. (Tokyo)* **93,** 661 (1983).
[23] D. N. Weber, C. F. Shaw, and D. H. Petering, *J. Biol. Chem.* **262,** 6962 (1987).
[24] C. T. Dameron, R. N. Reese, R. K. Mehra, A. R. Kortan, P. J. Carroll, M. L. Steigerwald, L. E. Brus, and D. R. Winge, *Nature (London)* **338,** 596 (1989).

[40] Sulfur-Containing Cadystin–Cadmium Complexes

By NORIHIRO MUTOH and YUKIMASA HAYASHI

Cadystins[1] are heavy metal-chelating peptides whose structural formula is (γ-Glu-Cys)$_n$-Gly ($n = 2,3,4, \ldots$).[2,3] They are synthesized in some fungi[1,4] and in plants[3,5,6] on exposure to heavy metal salts. They form

[1] A. Murasugi, C. Wada, and Y. Hayashi, *J. Biochem. (Tokyo)* **90,** 1561 (1981).
[2] N. Kondo, K. Imai, M. Isobe, T. Goto, A. Murasugi, C. W. Nakagawa, and Y. Hayashi, *Tetrahedron Lett.* **25,** 3869 (1984).
[3] E. Grill, E.-L. Winnacker, and H. M. Zenk, *Science* **230,** 674 (1985).

complexes with the heavy metal ions by sequestering them from the intracellular environment. The fission yeast mutants that cannot synthesize cadystins are hypersensitive to heavy metal salts.[7] Among the heavy metal ions, cadmium ion is the strongest inducer for cadystin synthesis.[8-10] Cadystin–cadmium complexes formed in the fission yeast *Schizosaccharomyces pombe,* in which cadystins with *n* equal to 2 or 3 are dominant, are called Cd-BP1 and Cd-BP2.[1] Cd-BP1 contains inorganic sulfur in addition to cadystins and cadmium.[11,12] Cd-BP2 is composed of cadystins and cadmium and no other components are found in the complex. By definition, Cd-BP1 means the cadystin–cadmium complexes containing inorganic sulfur. Because Cd-BP1's have greater Stokes radii than Cd-BP2's, they can be separated from each other by Sephadex G-50 column chromatography.[1] Cd-BPs can also be separated from cellular proteins or other low-molecular-weight components by gel-filtration chromatography. Cd-BP1 is composed of several species of complexes and each complex has a specific UV spectral property.[13,14] They are nanometer-scale quantum semiconductor particles. Electron transitions seen in ultraviolet (UV) spectra depend on the particle size of the Cd–S core,[15] although the exact structure of the complexes is uncertain. The larger Cd-BP1 has a higher molar ratio of cadmium to cadystin and of labile sulfur to cadmium. The cadmium-to-cadystin ratio of Cd-BP2 is lower than that of any species of Cd-BP1. Inorganic sulfur stabilizes the complexes and the complex containing more inorganic sulfur is more stable than the complex containing less inorganic sulfur.[14]

[4] R. K. Mehra, E. B. Tarbet, W. R. Gray, and D. R. Winge, *Proc. Natl. Acad. Sci. U.S.A.* **85,** 8815 (1988).

[5] J. C. Steffens, D. G. Hunt, and B. G. Williams, *J. Biol. Chem.* **261,** 1389 (1986).

[6] W. Gekeler, E. Grill, E.-L. Winnacker, and H. M. Zenk, *Z. Naturforsch. C: Biosci.* **44,** 361 (1989).

[7] N. Mutoh and Y. Hayashi, *Biochem. Biophys. Res. Commun.* **151,** 32 (1988).

[8] Y. Hayashi, C. W. Nakagawa, and A. Murasugi, *Environ. Health Perspect.* **65,** 13 (1986).

[9] E. Grill, E.-L. Winnacker, and M. H. Zenk, *FEBS Lett.* **197,** 115 (1986).

[10] N. Mutoh, C. W. Nakagawa, M. Kawabata, and Y. Hayashi, unpublished observation.

[11] A. Murasugi, C. Wada, and Y. Hayashi, *J. Biochem. (Tokyo)* **93,** 661 (1983).

[12] Y. Hayashi, C. W. Nakagawa, D. Uyakul, K. Imai, M. Isobe, and T. Goto, *Biochem. Cell Biol.* **66,** 288 (1988).

[13] R. N. Reese, R. K. Mehra, E. B. Tarbet, and D. R. Winge, *J. Biol. Chem.* **263,** 4186 (1988).

[14] R. N. Reese and D. R. Winge, *J. Biol. Chem.* **263,** 12832 (1988).

[15] C. T. Dameron, R. N. Reese, R. K. Mehra, A. R. Kortan, P. J. Carroll, M. L. Steigerwald, L. E. Brus, and D. R. Winge, *Nature (London)* **338,** 596 (1989).

Purification of Cd-BP1

Strains and Growth Condition

Schizosaccharomyces pombe L972h$^-$ or its derivatives are used. Virtually any growth media for yeast culture may be used. A typical rich medium is YPD medium [2% (w/v) polypeptone, 1% (w/v) yeast extract, 2% (w/v) glucose]. Cells are grown at 30° with vigorous shaking. *Schizosaccharomyces pombe* is more sensitive to cadmium ion at higher temperature (37°); however, cadystins are induced at this temperature. Cadystins are induced by adding cadmium salt, usually $CdCl_2$, to a logarithmically growing culture to a final concentration of 0.5 m*M*. Although a high concentration of cadmium inhibits the growth of *S. pombe*, cadystins can be induced at up to 4 m*M* $CdCl_2$ (we have not tried concentrations of cadmium ion higher than 4 m*M*). The lowest cadmium concentration we have tested was 50 μM; cadystins are induced at this concentration of cadmium, although the amount of cadystins synthesized was reduced. To obtain a large amount of Cd-BP1, incubation with cadmium ion is continued for 17 hr because this longer incubation leads to greater accumulation of Cd-BP1 in *S. pombe*.[11] Cells are harvested by centrifugation at 3000 *g* at 4° for 5 min and washed twice with buffer A (50 m*M* Tris-HCl, pH 7.6, 100 m*M* KCl, 1 μM $CdCl_2$). Cell pellets can be stored at −70° for at least 1 year without any change in cadystin–cadmium complexes.

Preparation of Cell Extract

Cells are disrupted by grinding with quartz sand. Three grams (wet weight) of cell pellet is ground with 6 g of acid-washed quartz sand for 10 min with a mortar and a pestle. Three milliliters of buffer A is added to extract water-soluble components. The extract is centrifuged at 5000 *g* at 4° for 5 min to remove cell debris and quartz sand. The precipitate is reextracted with 3 ml of buffer A, which has been used to wash the quartz sand in the mortar. The cell extract can be further clarified by centrifugation at 100,000 *g* at 4° for 60 min. The cell pellet used to prepare the cell extract can be scaled up (the largest pellet we have used had a 50 g wet weight).

Sephadex G-50 Column Chromatography

In order to purify Cd-BP1 and Cd-BP2, cell extract is applied to a Sephadex G-50 (superfine) column. A 2 × 60 cm column is used. Three milliliters of cell extract can be applied to this column without any interfer-

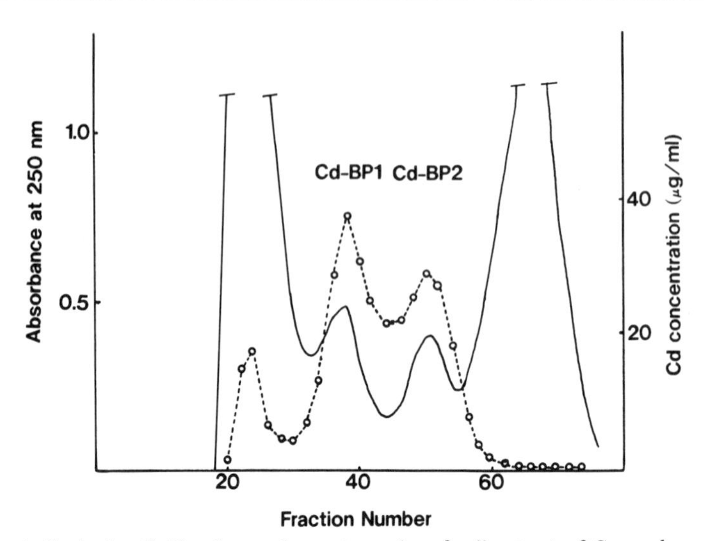

FIG. 1. Sephadex G-50 column chromatography of cell extract of *S. pombe* grown in YPD medium containing 0.5 m*M* $CdCl_2$. —, Absorbance at 250 nm; O–O, Cd^{2+} concentration.

ence in the separation of Cd-BP1 and Cd-BP2. Each 2-ml fraction is collected and the absorbance at 250 nm is recorded. Cd-BP1 is eluted around fraction 40 and Cd-BP2 is eluted at fractions 48–52 (Fig. 1). Most proteins are eluted in the void volume and low-molecular-weight components are eluted after fraction 55. Cadmium concentration in each fraction can be determined by atomic absorption spectrophotometry. Almost all of the cadmium ions are eluted with Cd-BP1 and Cd-BP2, but small fractions of cadmium ions are eluted in the void volume and are probably bound nonspecifically to cellular proteins. Cadmium ions not complexed with cadystins or proteins are undetectable in the induced cells. A significant cadmium peak is observed in the glutathione region (at about fractions 55–60) when cadmium ions are mixed with the extract of uninduced cells and chromatographed. Cd-BP1 and Cd-BP2 are pure enough for most experiments at this stage.

DEAE-Toyopearl 650M Column Chromatography

Cd-BP1 and Cd-BP2 can be purified further by DEAE-Toyopearl 650M (purchased from Tosoh, Tokyo, Japan) ion-exchange chromatography. The ion-exchange resin is packed in a 1 × 2 cm Sepacol Mini-pp disposable column (purchased from Seikagaku Kogyo, Tokyo, Japan).

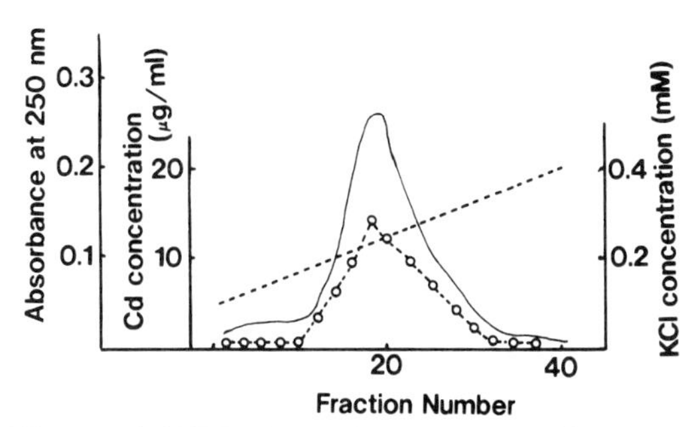

FIG. 2. DEAE-Toyopearl 650M column chromatography of Cd-BP1 obtained from Sephadex G-50 column chromatography. —, Absorbance at 250 nm; O–O, Cd^{2+} concentration.

Cd-BP1 and Cd-BP2 fractions obtained from Sephadex G-50 column chromatography are diluted with equal amounts of 1 m*M* Tris-HCl, pH 7.6, 1 μ*M* $CdCl_2$ and applied to the ion-exchange column. Cd-BPs are eluted by a KCl gradient of 100 to 400 m*M* in 25 m*M* Tris-HCl, pH 7.6, 1 μ*M* $CdCl_2$. Cadmium salt is added to the buffer to stabilize the complexes. In some cases, crude cell extract is directly applied to the ion-exchange chromatography column to purify the cadystin–cadmium complexes.[13,16] Cd-BP1 is eluted at a KCl concentration of about 230 m*M* (Fig. 2) and Cd-BP2 is eluted at a concentration of about 180 and 240 m*M*.[12] The broad peak of Cd-BP1 is derived from the heterogeneity of Cd-BP1 (described below). The largest Cd-BP1 obtained from a cadmium-hypersensitive mutant complemented with a recombinant plasmid, which has a 1.5-kilobase (kb) AT-rich sequence of *S. pombe* genomic DNA with unknown function, is eluted at 280 m*M* KCl.[10]

Characterization of Cd-BP1

Optical Properties of Cd-BP1

Cd-BP1 is composed of heterogeneous complexes and they are nanometer-scale semiconductor particles.[15] Each complex has a unique cadmium-to-inorganic sulfur ratio and shows a specific UV absorbance spectrum in the range of 260–320 nm.[13,14] Complexes with higher Stokes radii have optical transitions at longer wavelengths. The UV spectrum can be measured from the sample obtained by either Sephadex G-50 column chromatography or DEAE-Toyopearl 650M column chromatography. Salt

[16] A. Murasugi, C. Wada, and Y. Hayashi, *Biochem. Biophys. Res. Commun.* **103,** 1021 (1981).

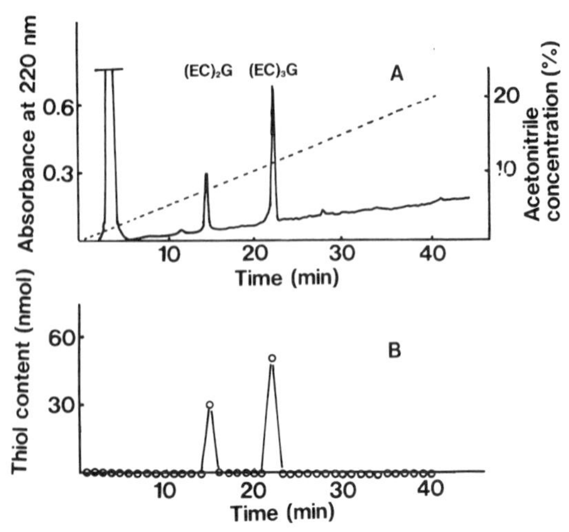

FIG. 3. HPLC profile of cadystins in Cd-BP1 obtained from ODS80 TM HPLC column. (A) Absorbance at 220 nm. (B) Thiol content.

has no effect on the spectral properties of Cd-BP1 in the range used by these preparations.[17] The UV spectrum is changed by lowering the pH because of the dissociation of the complexes,[12,13,16] and the original shoulder at 265 nm in the UV spectrum is not obtained after reneutralization in the presence of excess cadmium ion.

Assays for Components of Cd-BP1

Reversed-phase HPLC is used to determine the content of each homolog of cadystins in Cd-BPs. Trifluoroacetic acid (TFA) (5%, v/v) is added to a Cd-BP solution. After centrifugation and filtration to remove insoluble materials, the sample is applied to an ODS80 TM HPLC column (Tosoh) and eluted by a linear gradient of 0–20% (v/v) acetonitrile in 0.05% TFA at a flow rate of 1 ml/min. Absorbance at 220 nm is recorded. (γ-Glu-Cys)$_2$-Gly and (γ-Glu-Cys)$_3$-Gly are eluted at about 8 and 13% acetonitrile, respectively (Fig. 3). For the determination of thiol content, a 1-ml fraction is collected; 50 μl 1 M K_2HPO_4 and 7 μl of thiol reagent, 5,5′-dithiobis(2-nitrobenzoic acid)[18] (4 mg/ml in 45 mM phosphate buffer, pH 7.0), are added. Absorbance at 412 nm is recorded and calibrated against a standard curve for the reduced form of glutathione.

When Cd-BP1 is acidified to release cadmium from the cadystin–cadmium complexes, an odor resembling that of H_2S is usually noticed.

[17] Y. Hayashi and M. Kawabata, unpublished observation.

[18] G. L. Ellman, *Arch. Biochem. Biophys.* **82,** 70 (1959).

This acid-labile sulfur in Cd-BP1 can be determined according to King and Morris,[19] using an iodometrically standardized Na_2S solution. The labile sulfur in purified ferredoxin can be determined by this method. In the determination of the inorganic sulfur content in Cd-BP1, it is necessary to compensate for the inhibitory effect of cadmium ion on the color development of the reaction. With a cadmium concentration of 10–20 μg/ml, color development is reduced to 80% of that obtained using a sample without cadmium.

Cadmium content is determined by atomic absorption spectrophotometry.

Reconstitution of Cd-BP1

Cadystin–cadmium complexes are dissociated by lowering the pH of the solution, i.e., below pH 5. The exact pH required to dissociate a complex depends on the size of complex.[12,16] Cd-BP1 can be reconstituted from cadystin, cadmium, and inorganic sulfur (Na_2S) by incubating them together. When the complex is reconstituted with an excess amount of Na_2S at room temperature, a complex is formed having a large Stokes radius. A complex with a small Stokes radius is reconstituted by incubating with a stoichiometric amount of Na_2S at low temperature (4°) for 16 hr.[14] Complexes with large radii can be formed by incubating complexes with small radii with an excess amount of Na_2S.[11,14] *In vivo* formation of Cd-BP1 depends on the concentration of cellular cadmium and sulfide production per cell also increases with the increasing concentration of cadmium ions in the culture medium.[20] This indicates that the reconstitution condition mimicks the formation of Cd-BP1 *in vivo.* Cd-BP2 can be reconstituted easily from cadystin and cadmium by mixing in acidic solution and neutralization by alkaline solution.

[19] T. E. King and R. O. Morris, this series, Vol. 10, p. 634.

[20] A. Murasugi, C. W. Nakagawa, and Y. Hayashi, *J. Biochem. (Tokyo)* **96,** 1375 (1984).

[41] Cadystins: Small Metal-Binding Peptides

By Y. HAYASHI, M. ISOBE, N. MUTOH, C. W. NAKAGAWA, and M. KAWABATA

Cadmium-binding peptides (Cd-BP1 and Cd-BP2) were induced in fission yeast on exposure to cadmium ion.[1] Metal-chelating peptides, cadystins, were first observed in these Cd-BPs as the peptide component having the general structure (γ-Glu-Cys)$_n$Gly ($n = 2,3,4, \ldots$).[2,3] These peptides were also detected in plants, other yeast, and protozoa as the peptide component of metal complexes, although the trivial names given were different from each other: phytochelatin,[4] γ-Glu peptide,[5] poly(γ-Glu-Cys)Gly,[6] and others.[7] In this chapter we describe the induction and isolation of metal–cadystin complexes, properties of the metal complexes, and the chemical structure of cadystins.

Induction by Administration of Cadmium Ions

Preparation of Cadmium–Peptide Complexes

The fission yeast *Schizosaccharomyces pombe* L972 h$^-$ is grown to the early stationary phase of culture by shaking at 30° in YPD medium [1% (w/v) yeast extract, 2% (w/v) polypeptone and 2% (w/v) dextrose (Wako Pure Chem., Osaka, Japan)], and this culture is used as the preculture. The main culture of 500 ml is started in a 3-liter flask by addition of preculture to 0.3 or 1.2×10^7 cells/ml; $CdCl_2$ is added at various concentrations (0.1–1 m*M*). After an appropriate incubation time (2–24 hr), cells are harvested by centrifugation, washed once with 0.14 *M* NaCl and twice with distilled water, and stored at −30° until use. Frozen cells are ground in a chilled mortar and pestle with acid-washed quartz sand (three times the weight of the cells), and extracted with 5 vol of 50 m*M* Tris-HCl (pH 7.5)–0.1 *M* KCl. The extract is centrifuged at 17,000 *g* for 20 min at 4°,

[1] A. Murasugi, C. Wada, and Y. Hayashi, *J. Biochem. (Tokyo)* **90,** 1561 (1981).
[2] N. Kondo, M. Isobe, K. Imai, T. Goto, A. Murasugi, and Y. Hayashi, *Tetrahedron Lett.* **24,** 925 (1983).
[3] N. Kondo, K. Imai, M. Isobe, T. Goto, A. Murasugi, C. W. Nakagawa, and Y. Hayashi, *Tetrahedron Lett.* **25,** 3869 (1984).
[4] E. Grill, E.-L. Winnacker, and M. H. Zenk, *Science* **230,** 674 (1985).
[5] R. N. Reese, R. K. Mehra, E. B. Tarbet, and D. R. Winge, *J. Biol. Chem.* **263,** 4186 (1988).
[6] P. J. Jackson, C. J. Unkefer, J. A. Doolen, K. Watt, and N. J. Robinson, *Proc. Natl. Acad. Sci. U.S.A.* **84,** 6619 (1987).
[7] D. W. Weber, C. F. Shaw III, and D. H. Petering, *J. Biol. Chem.* **262,** 6962 (1987).

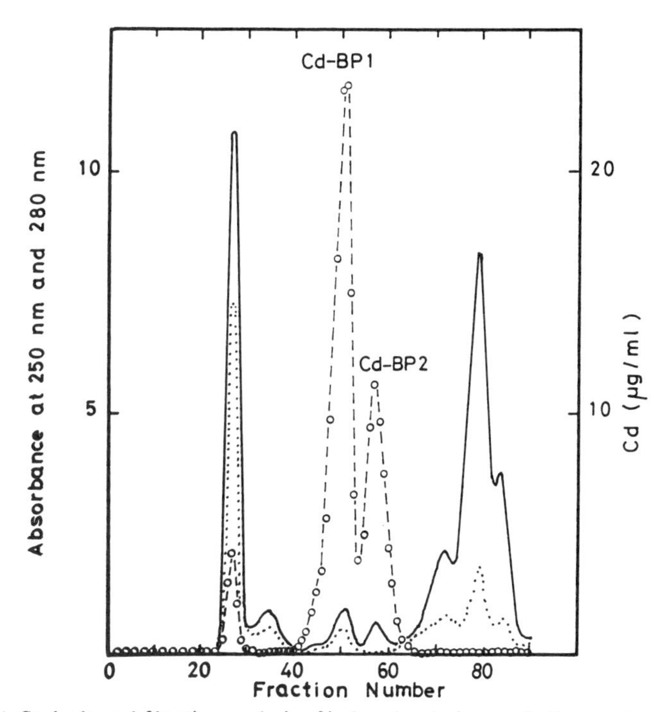

FIG. 1. Sephadex gel-filtration analysis of induced cadmium-cadystin complexes (Cd-BP1 and Cd-BP2). The overnight culture of the fission yeast is diluted with the fresh YPD medium containing 0.5 mM $CdCl_2$ (1.2×10^7 cells/ml) and incubated for 17 hr with vigorous shaking at 30°. Cell extract is prepared as described in the text and about one-third volume of the extract is applied to the Sephadex G-50 SF column. Fractions (1.5 ml) are collected and absorbances at 250 nm (—) and 280 nm (····) and cadmium content (O—O) in each fraction are determined.

the supernatant is applied to Sephadex G-50 Superfine (Pharmacia, Piscataway, NJ) column (1.6×65 cm), and eluted with the extraction buffer. Two peaks of cadmium-binding peptides are detected in the low-molecular-weight region (Fig. 1). The time course study of Cd-BP formation indicates that the amount of Cd-BP per cell reaches saturation after 10 hr of 1 mM $CdCl_2$ administration,[1] although cell growth continues and the total amount of Cd-BP per culture still increases (Fig. 2). This is consistent with the time needed for maximum cadmium uptake per cell, implying that cellular cadmium uptake occurs as long as Cd-BP synthesis continues.

Subspecies of Cadmium-Binding Peptides

Cd-BP1 and Cd-BP2 from 0.1 mM $CdCl_2$-induced cells are rechromatographed in a Sephadex G-50 SF column, and the purified Cd-BP1 or

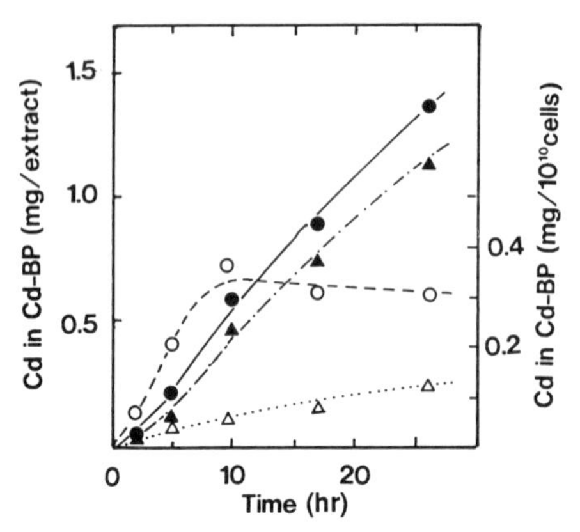

FIG. 2. Increases of Cd-BPs with time in extracts from 500 ml of cultured cells. Amounts of Cd-BPs are expressed as their cadmium amounts. (●), Cd-BPs in the total cell extract; (○), Cd-BPs/10^{10} cells; (▲), Cd-BP1 in the total cell extract; (△), Cd-BP2 in the total cell extract.

Cd-BP2 is diluted and applied to a DEAE-Toyopearl (Tosoh, Tokyo, Japan) 650S column (0.9 × 10 cm) equilibrated with 25 m*M* Tris-HCl (pH 7.5)–50 m*M* KCl–10 μM $CdCl_2$. The subspecies of Cd-BP1 and Cd-BP2 are isolated by KCl gradient elution (Fig. 3). Each subspecies is rechromatographed on the DEAE-Toyopearl column and some properties of the cadmium–cadystin complex are analyzed.

Determination of Heavy Metals, Sulfhydryl Groups, and Acid-Labile Sulfides

The content of heavy metals, e.g., cadmium, zinc, and copper, in a solution containing metal–peptide complexes is determined with a Perkin-Elmer (Norwalk, CT) atomic absorption spectrophotometer or Hitachi (Tokyo, Japan) Z-7000 polarized Zeeman atomic absorption spectrophotometer.

Sulfhydryl group is determined by Ellman's method, using 5,5′-dithiobis(2-nitrobenzoic acid) (DTNB).[8]

The acid-labile sulfide content is determined according to King and Morris,[9] except for the addition of an appropriate amount of $CdCl_2$ (5–50 mg/ml) to the standard Na_2S reaction mixture. This modification is necessary because the coupling reaction is affected by heavy metal ions,[9] and cadmium–peptide complexes contain cadmium ions. With the use of an Na_2S solution standardized iodometrically, the effect of cadmium ions is

[8] G. L. Ellman, *Arch. Biochem. Biophys.* **82,** 70 (1959).

[9] T. King and O. Morris, this series, Vol. 10, p. 634.

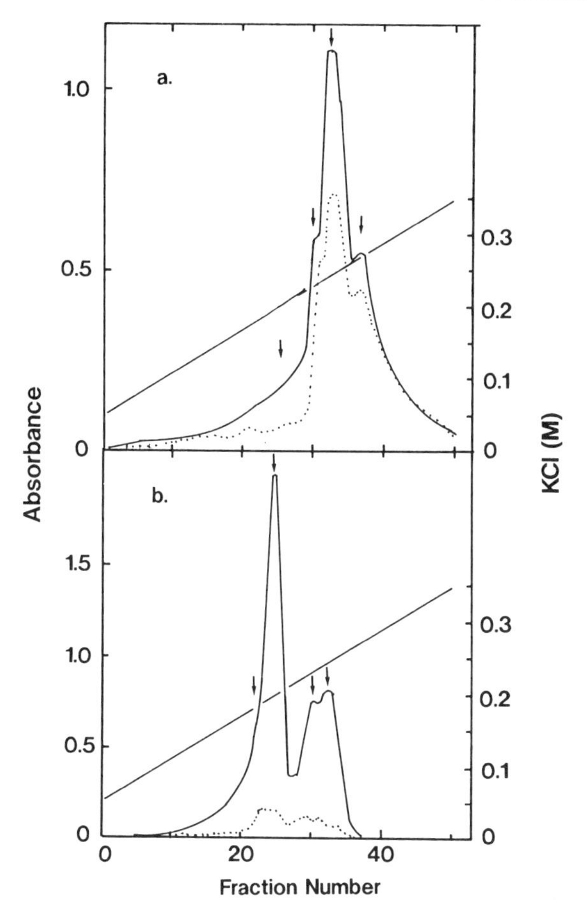

FIG. 3. Separation of cadmium–cadystin complex subspecies by ion-exchange chromatography. The Cd-BP1 or Cd-BP2 fraction from the Sephadex G-50 SF column is chromatographed on a DEAE-Toyopearl 650S column (Tosoh). Absorbances are read at 250 nm (—) and at 280 nm (····). (a) Cd-BP1 subspecies separation. The arrows indicate 0.19, 0.22, 0.25, and 0.27 *M* subspecies, respectively. (b) Cd-BP2 subspecies separation. The arrows indicate 0.18, 0.20, 0.24, and 0.26 *M* subspecies.

checked; the presence of 20 μg/ml Cd^{2+} in the sample decreased the efficiency of the coupling reaction to 81%.[10]

Peptide content in metal–peptide complexes can be determined by one of the following methods: (1) nitrogen content determined by the Kjeldahl–Nessler method, (2) quantitative amino acid analysis, or (3)

[10] A. Murasugi, C. Wada, and Y. Hayashi, *J. Biochem. (Tokyo)* **93,** 661 (1983).

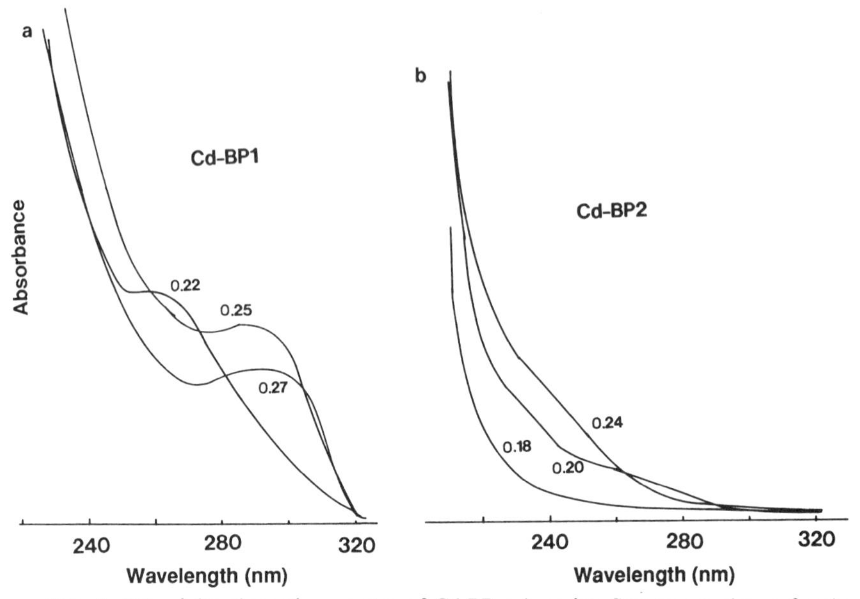

FIG. 4. Ultraviolet absorption spectra of Cd-BP subspecies. Spectra are shown for the eluted fractions from the DEAE-Toyopearl column. The numbers by the curves indicate the salt concentration of the elution fractions normalized to contain cadmium (2 μg/ml) and 25 mM Tris-HCl (pH 7.5)–50 mM KCl. (a) Cd-BP1 subspecies. (b) Cd-BP2 subspecies.

sulfhydryl groups or the absorption units of peptides fractionated by high-performance liquid chromatography (HPLC).

Ultraviolet Spectra of Cadmium–Cadystin Complexes

Cd-BPs purified by rechromatography are dialyzed against 25 mM Tris-HCl (pH 7.5)–50 mM KCl–1 μM $CdCl_2$, diluted to the same Cd^{2+} concentration (2 μg Cd/ml) with the same buffer, and ultraviolet (UV) spectra are determined.

Among Cd-BP1 subspecies, the higher the eluting salt concentration from the anion-exchange column, the longer the wavelength of the optical transition characteristic for cadmium–sulfide in Cd-BP1 (Fig. 4a). The physical basis of the optical properties relate to the quantum nature of Cd–S crystallite that is formed in these Cd-BP1 species.[11]

The UV spectra of Cd-BP2 subspecies reveal that these are cadmium complexes with characteristic cadmium–thiolate charge-transfer transi-

[11] C. T. Dameron, R. N. Reese, R. K. Mehra, A. R. Kortan, P. J. Carroll, M. L. Steigerwald, L. E. Brus, and D. R. Winge, *Nature (London)* **338,** 596 (1989).

tions (Fig. 4b). The Cd-BP2 subspecies are devoid of the labile sulfide and do not show absorbances between 260 and 320 nm due to CdS.

The effects of salt concentration on the UV spectra of these cadmium complex subspecies are determined by increasing the KCl concentration in a sample having the same cadmium–cadystin concentration. With Cd-BP1 subspecies, very little absorbance shift is observed over the entire UV

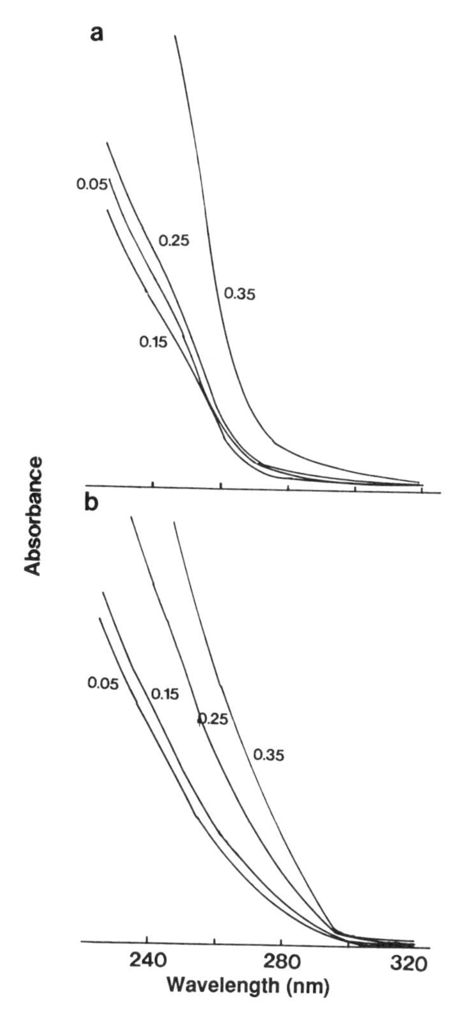

FIG. 5. Effect of the salt concentration on UV spectra of the cadmium–cadystin complexes. Absorbance shifts are observed in (a) 0.20 *M* and (b) 0.24 *M* subspecies of Cd-BP2 (cadmium, 1 μg/ml) with increasing concentration of KCl in 25 m*M* Tris-HCl (pH 7.5) as indicated by the number by the respective curve.

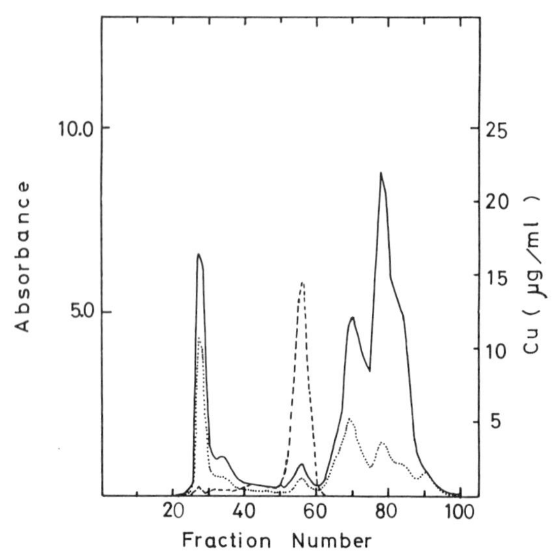

FIG. 6. Induction of the copper–cadystin complex. $CuCl_2$ is added to the YPD medium at 2.5 m*M* and culture continued for 17 hr. The cell extract from a 500-ml culture was applied to a Sephadex G-50- SF column as described in the text. Absorbances at 250 nm (—) and at 280 nm (· · · ·) and copper concentration (---) in each fraction are determined.

range. On the other hand, some Cd-BP2 subspecies show significant absorbance shifts with increasing salt concentration (Fig. 5).

Cadystin Induction by Other Heavy Metals and Oligosaccharides

Induction

Cadystin synthesis can be induced by other heavy metals in the fission yeast, although the synthesized amounts of cadystin are less than 10% of that by Cd^{2+}. With 2.5 m*M* $CuCl_2$ in YPD medium, copper–cadystin complex in the cell extract is observed (Fig. 6) by analysis of Sephadex gel filtration at the region around M_r 3000, which is an intermediate molecular weight between the values observed for the two cadmium–cadystin complexes (Cd-BP1 and Cd-BP2).[1]

With 2 m*M* $ZnCl_2$ in the medium, only a small amount of cadystin synthesis is detected after 10 hr of culture. However, with the simultaneous addition of oligosaccharides, such as hydrolysates of pectin, mannan, chitin, and chitosan, at 50 μg/ml the induction of cadystin synthesis is significant (about 20–40% of that induced by 0.1 m*M* cadmium) and a zinc–cadystin complex peak is observed at around M_r 1200.[12] The induction by

[12] Y. Hayashi, S. Morikawa, N. Mutoh, M. Kawabata, and Y. Hotta, submitted.

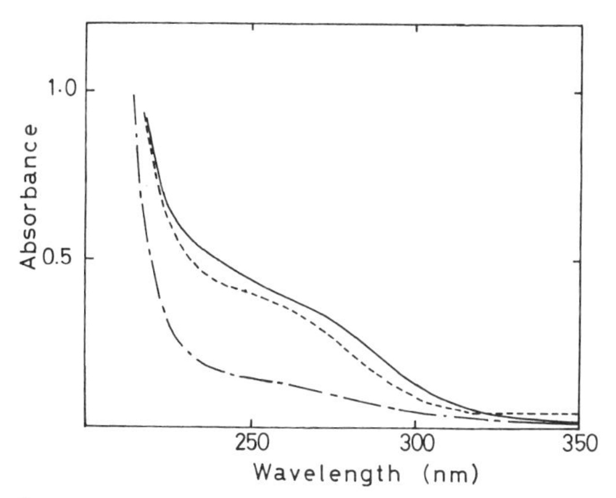

FIG. 7. Ultraviolet absorption spectra of copper–cadystin complexes. UV spectra are recorded for the native copper–cadystin complex (---), Cu(I)-replaced complex (—), and Cu(II)-replaced complex (- · - ·). Copper-replaced complexes are prepared by incubation of Cd-BP1 with an excess amount of Cu(I) or Cu(II) ion, and the replaced complexes are isolated by the use of a BioGel P-2 column or Sephadex G-10 column. Cadystin amounts in complexes are about 100 μg/ml in 5 m*M* Tris-HCl (pH 7.5)–10 m*M* KCl.

oligosaccharides is observed even without addition of $ZnCl_2$, and the zinc–cadystin complex is also detected at the same region in Sephadex G-50 SF column fractions. Oligosaccharides of fungal and plant origin have been identified that can act as chemical signals to activate a broad spectrum of genes for the defense mechanism in plants.[13] The induction of cadystin synthesis by oligosaccharides could be one of the defense reactions of the fission yeast. However, because the induction of cadystin synthesis in the fission yeast by oligosaccharides is accompanied by the appearance of a significant peak of zinc–cadystin complex even without special addition of zinc to YPD medium, the defense signal of oligosaccharides not only induces the cadystin synthesis but may increase the influx of Zn^{2+} from the usual YPD medium (Zn^{2+} concentration is 0.3–0.4 m*M*).

Ultraviolet Absorption Spectra of Copper–Cadystin and Zinc–Cadystin Complexes

When the UV absorption spectrum of the native copper–cadystin complex is compared with the spectra of the reconstituted Cu(I)–cadystin and Cu(II)–cadystin complex, the spectrum of the native copper–cadystin corresponds to that of Cu(I)–cadystin complex (Fig. 7). This indicates that

[13] C. A. Ryan, *Biochemistry* **27**, 8879 (1988).

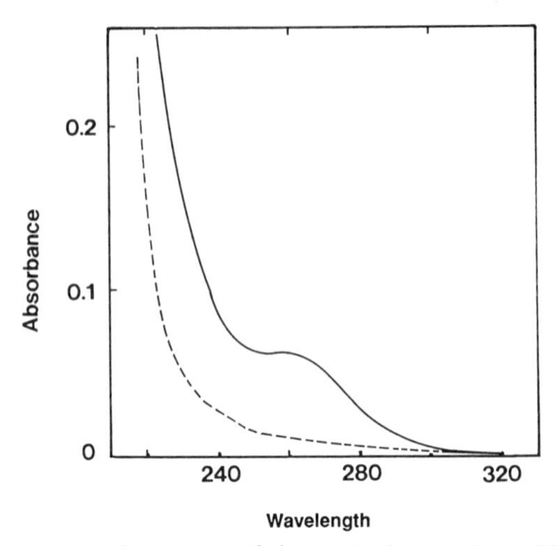

FIG. 8. Ultraviolet absorption spectra of zinc–cadystin complexes. The UV spectra are recorded for the native zinc–cadystin complex (—) and reconstituted zinc–cadystin complex (---) from free cadystin and excess zinc ions followed by BioGel P-2 isolation. The amount of complex is normalized as zinc (1 μg/ml) in 25 mM Tris-HCl (pH 7.5)–50 mM KCl.

the ionic state of copper in the copper–cadystin complex in the cell is Cu(I), like copper–thionein in *Neurospora* and in mammals. In the spectrum of the native copper–cadystin complex, the optical transition at 265 nm is also observed as in Cd-BP1 species (Fig. 7). The UV spectrum of the native Zn^{2+}–cadystin complex induced by the oligosaccharides and zinc also shows the optical transition at 265 nm, and this transition cannot be observed in the reconstituted zinc–cadystin complex (Fig. 8).

Chemical Structure of Cadystins

Purification of Cadystins

The metal–cadystin complexes from the Sephadex G-50 SF column fraction are purified by rechromatography on the same gel-filtration column and by an anion-exchange column [DE-52 (Whatman, Clifton, NJ) or DEAE-Toyopearl 650M]. The purified metal–cadystin complex is concentrated by ultrafiltration using SPECTRA/POR membrane (Spectrum, Los Angeles, CA) (molecular weight cutoff, 1000). The metal complex is acidified by the addition of trifluoroacetic acid (TFA) to 5%, centrifuged at 15,000 *g*, filtered through Chromatodisc 13A (Kurabou), and applied to an HPLC column [TSK- ODS80 (Tosoh, Tokyo, Japan), 0.46 × 25 cm or Beckman Ultrasphere ODS (San Ramon, CA), 1.0 × 25 cm with TSK-

ODS80 precolumn, 0.46 × 1.5 cm]. Cadystin species are eluted from the column by a linear gradient of 0–20% acetonitrile in 0.05% TFA in 40 min (flow rate: 1 ml/min for 0.46-cm diameter column or 2 ml/min for 1.0-cm diameter column). Depending on the column size 1- or 2-ml fractions are collected, and the contents of SH groups are determined in an aliquot of each fraction to ascertain the position of cadystin species and to determine the amounts of cadystins synthesized.

Amino Acid Analyses of Cadystins

Cadystin species (HPLC purified) are lyophilized and oxidized by performic acid, and digested in 5.7 *N* HCl–0.02% (v/v) phenol for 24 hr at 120°. Samples are lyophilized again to remove HCl completely, and coupled with phenyl isothiocyanate.[14] Phenylthiocarbamyl-amino acids are analyzed in HPLC (TSK-ODS80, 0.46 × 25 cm with a precolumn) by a linear gradient of A solution [0.14 *M* sodium acetate (pH 6.3)–0.05% triethylamine/acetonitrile (940:60)] and B solution [acetonitrile/H_2O (60:40)] in 50 min from 100% A to 100% B. The results of quantitative amino acid analyses give the equimolar amounts of glutamic acid and cysteinic acid, and the respective amount of glycine according to cadystin species.

Chemical Structure of Cadystins

Cadmium–cadystin complex is desulfurized with Raney nickel W-2 at 50° for 12 hr in 50 m*M* Tris-HCl, pH 7.6.[2] The solvent is evaporated *in vacuo* and the catalyst is dissolved in a small volume of 6 *N* HCl below 30° and then the mixture is lyophilized. The residue is desalted in a BioGel P-2 column with 3 *M* acetic acid. Dethiocadystin thus obtained is separated by HPLC (ODS column) as previously described.[2,3] Amino acid analysis of dethiocadystins shows equimolar amounts of glutamate and alanine, and the respective amount of glycine, depending on the cadystin species. Then the chemical structures of cadystins are determined by carboxypeptidase P digestion of dethiocadystins and chemical analyses of the resulting dipeptide as follows.[15] The smallest dethiocadystin is digested with carboxypeptidase P for 96 hr at 35° in 0.1 *M* sodium acetate (pH 5.2). It yields glycine and two molar equivalents of dipeptides containing glutamate and alanine in 1:1 ratio. The dried dipeptide is esterified in 10 μl of a mixture of methanol, trimethyl orthophosphate, and thionyl chloride (80:20:5 w/w) for 4 hr at 40°, and dried *in vacuo*. The residue is reduced in 10 μl of 0.3 *M* $LiBH_4$ in tetrahydrofolic acid (THF) for 12 hr at 80° in a sealed tube. After

[14] R. L. Heinrikson and S. C. Meredith, *Anal. Biochem.* **136,** 65 (1984).

[15] N. Kondo, M. Isobe, K. Imai, and T. Goto, *Agric. Biol. Chem.* **49,** 71 (1985).

evaporation to dryness, the residue is hydrolyzed with 50 μl of 6 N HCl for 12 hr at 105°. The hydrolysate gives 4-amino-5 hydroxyvaleric acid and 2-aminopropanol in nearly quantitative yield without formation of alanine or 2-amino-5-hydroxyvaleric acid. Thus, the sequence of the dipeptide should be γ-Glu-Ala. The deduced general structures for cadystins are (γ-Glu-Cys)$_n$Gly ($n = 2,3,4, \ldots$).[3] These structures for cadystins are confirmed by comparing the behavior in HPLC (ODS column), CD spectra, and ^{1}HNMR spectra of natural cadystins vs chemically synthesized cadystins.[3,15]

Possible Functions of Cadystins in Cells

Cadystins induced on exposure to heavy metals work in detoxification of excess heavy metals. However, cadystin induction caused by the administration of oligosaccharides may be different from heavy metal detoxification, as follows. (1) The influx of Zn^{2+} increases. The chemical signals of the oligosaccharides presumably increase the influx of Zn^{2+} to the cell. The excess Zn^{2+} resulting from the increased influx is stored as the complex form with cadystin and can be used for the activation of zinc enzymes, which may be newly synthesized as the result of the defense reaction. (2) Active radicals are detoxified. Like the biological function of metallothionein, cadystins may have a role in the elimination of harmful oxygen radicals or hydroxy radicals because of their strong reduction activities. The reducing activities of cadystins stronger than glutathione indicate that the role of cadystins in the damaged cell could be mainly detoxification of reactive radicals.

[42] Isolation of Metallothionein from Ovine and Bovine Tissues

By P. D. WHANGER

Although metallothionein (MT) was first isolated and characterized from the kidney of a large animal, the equine species,[1] most of the work on its metabolism and involvement in metal metabolism has been performed with small animals, mainly the rat.[2] There are some major differences in the number of MT species between various animals.[2] Thus, a study of MT

[1] J. H. R. Kägi and B. L. Vallee, *J. Biol. Chem.* **235,** 3460 (1960).
[2] J. H. R. Kägi and Y. Kojima, eds., *Experientia, Suppl.* **52** (1987).

metabolism in large animals was necessary to obtain information applicable to these animals.

Supplies and Reagents

A Potter–Elvehjem homogenizer, centrifuge, chromatography columns, freeze drier, fraction collector, an amino acid analyzer, equipment for electrophoresis on acrylamide gels, an absorption spectrophotometer, and an atomic absorption spectrophotometer are the necessary equipment for the isolation and characterization of MT. Sephadex G-75, DEAE-Sephacel, BioGel P-10, Tris-HCl, isotonic saline, and dialysis tubing are the necessary reagents for these procedures.

Procedures

Methods should be employed to minimize metal contamination, as was discussed in a previous volume of this series.[3] Since dietary zinc was shown to accumulate in tissue MT,[4,5] animals were fed elevated amounts of this metal in the diet to increase these levels in tissues. Dietary levels of zinc to 2000 mg/kg feed can be fed for 2 months without any affect on the performances of sheep and cattle. Excess zinc was shown to accumulate in MT in liver, kidney, pancreas, and small intestinal epithelia, but not in heart, testis, or epithelia of abomasum and rumen of cattle and sheep.[5]

About 50 g of tissue is homogenized in 2 vol of cold isotonic saline with a Potter–Elvehjem homogenizer. Homogenates and successive supernatant fractions are centrifuged (4°) at 37,000 *g* for 30 min and at 160,000 *g* for 90 min to obtain the tissue cytosols. The tissue cytosols are eluted through a Sephadex G-75 column (5.0 × 100 cm) with 0.05 *M* Tris-HCl buffer, pH 8.4. Fractions of 15 ml are collected at a rate of 120 ml/hr. The eluted fractions are monitored for zinc by atomic absorption spectrophotometry. The MT fractions are then combined, dialyzed against 0.003 *M* Tris-HCl, pH 8.4, for 12 hr. This is then chromatographed on a DEAE-Sephacel (Sigma Chemical Co., St. Louis, MO) column (2.0 × 30 cm) with a gradient of 0.003 *M* (350 ml) to 0.3 *M* (350 ml) Tris-HCl buffer, pH 8.4, at a flow rate of 20 ml/hr. About 4 ml/fraction is collected and the zinc monitored by atomic absorption spectrophotometry. A typical elution pattern is shown in Fig. 1. This pattern is very similar for both ovine and

[3] J. F. Riordan and B. L. Vallee, eds., this series, Vol. 158.

[4] R. L. Kincaid, W. J. Miller, P. R. Fowler, R. P. Gentry, D. L. Hampton, and M. W. Neathery, *J. Dairy Sci.* **59,** 1580 (1976).

[5] P. D. Whanger, S.-H. Oh, and J. T. Deagen, *J. Nutr.* **111,** 1196 (1981).

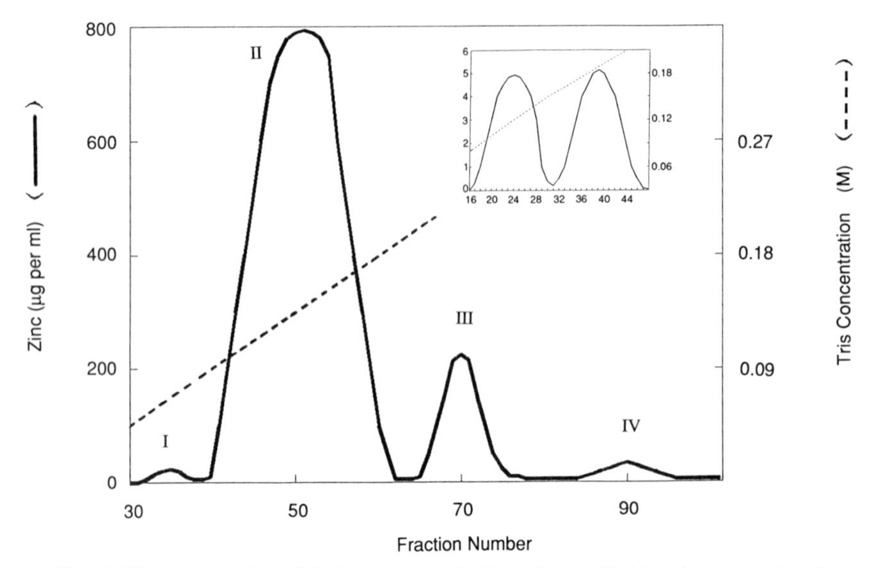

FIG. 1. Chromatography of the bovine or ovine hepatic metallothionein preparation from the gel-filtration step on DEAE-Sephacel. The inset is the pattern from a rat preparation.

bovine preparations. There is one predominant MT species, with up to three minor ones. For comparison, the pattern for rat liver is shown in the inset (Fig. 1), indicating a major difference between these two species of animals. The number of MT species in tissues of various animals has been discussed.[2,6]

The individual MT fractions are combined, dialyzed, and rechromatographed on DEAE-Sephacel columns (1.5 × 20 cm) with a salt gradient of 0.003 *M* (160 ml) to 0.3 *M* (160 ml) Tris-HCl buffer, pH 8.4, at a flow rate of 26 ml/hr, collecting 5 ml/fraction. The individual fractions are combined, dialyzed against distilled water, and freeze dried. The freeze-dried samples are dissolved in 0.05 *M* Tris-HCl, pH 8.4, and then chromatographed in this same buffer on BioGel P-10 (Bio-Rad Laboratories, Richmond, CA) columns (2.0 × 60 cm) at a flow rate of 11 ml/hr with collection of 3.5-ml fractions. All these chromatographic procedures are conducted at 4°. The resulting purified MT preparations can then be used for metal and amino acid analyses. In addition to zinc, the absorbance at 240 nm can also be monitored to follow the purification of MT.

Each of these preparations yields one band on disc gel electrophoresis.

[6] P. D. Whanger, S.-H. Oh, and J. T. Deagen, *J. Nutr.* **111,** 1207 (1981).

The amino acid analyses on the hydrolyzed MT preparations with an amino acid analyzer give the amino acid content expected for MT.[6] The cysteine content ranges from 20.4 to 35.7% for the various MTs in bovine tissues and from 27.3 to 35.6% for the various MTs in ovine tissues. The lower cysteine content is found in the minor MT species (I, III, and IV; Fig. 1). The minor MT species appear to have a low content of aromatic amino acids.

A procedure was developed in the author's laboratory using hydroxylapatite after the first DEAE-Sephacel step to obtain pure MT from rat tissues.[7] However, this method could not be used for the minor MT species (I, II, and IV; Fig. 1) from bovine or ovine tissues, and it was suggested that their amino acid content may account for this behavior.[6] This hydroxylapatite step, however, could be used for the major MT species (II; Fig. 1) from both ovine and bovine tissues.

The gram atoms of zinc per mole MT ranged from 0.5 to 6.4 for the various preparations from bovine tissues and from 2.0 to 6.3 for the various preparations from ovine tissues. Again, the lower zinc content was present in the minor MT species (I, II, and IV; Fig. 1).

Metals in Metallothioneins

When animals are fed zinc in the diet, this is essentially the only metal present in MT. There are traces of copper but never on a molar basis to MT.

When cadmium is included in the diet, this metal is also incorporated into MT, but not at the exclusion of zinc. The accumulation of cadmium in MT will also cause zinc to accumulate in MT.[8] Thus, it is impossible to biologically incorporate only cadmium into tissue MT. Of course, this can be done easily by *in vitro* methods.[3]

It is impossible to incorporate only copper biologically into MT unless there is an inducer for MT. This is apparently because copper is a poor inducer of MT, at least in the ovine and bovine species. One way to incorporate copper into MT has been to elevate the zinc status of cattle,[9] as demonstrated in Fig. 2. When only elevated copper was included in the diet, this metal accumulated with high-molecular-weight proteins (void volume of Sephadex G-75) in tissues. However, when elevated zinc was also included in the diet, copper instead accumulated with MT. This pattern was also demonstrated for the kidneys and pancreas. The molar

[7] R. W. Chen and P. D. Whanger, *Biochem. Med.* **24,** 71 (1980).
[8] J. T. Deagen, S.-H. Oh, and P. D. Whanger, *Biol. Trace Elem. Res.* **2,** 65 (1980).
[9] P. D. Whanger and J. T. Deagen, *Biol. Trace Elem. Res.* **28,** 69 (1990).

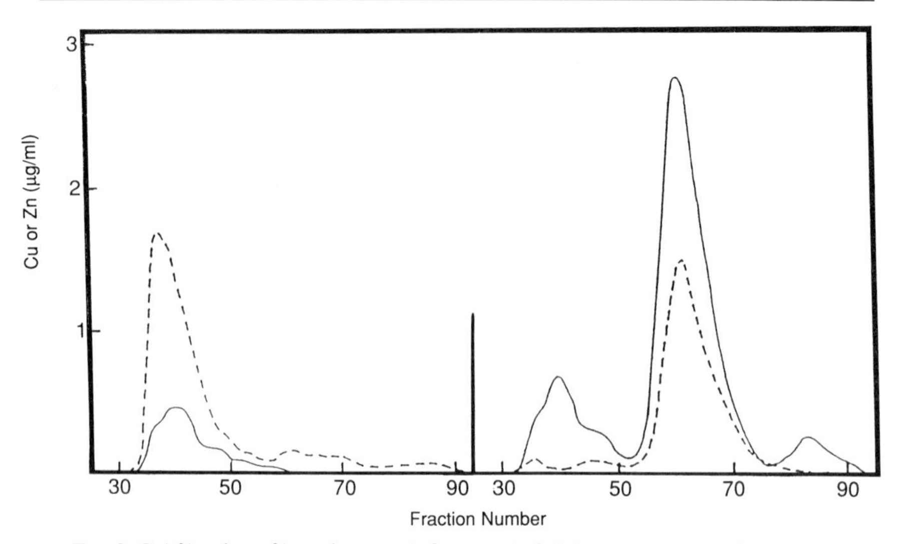

FIG. 2. Gel filtration of hepatic cytosols from cattle fed elevated copper (left) in the diet or copper plus zinc (right). (—), Zinc; (---), copper.

ratios of copper and zinc in MT vary with the relative amounts of these metals in the diet. However, it is impossible to biologically obtain MT with only copper because zinc will also be present. In order to obtain MT with only copper, this has to be done *in vitro* by metal substitution similar to that used for cadmium.

Extra precaution must be taken in the purification of MTs containing copper because the cysteine in copper-MT is susceptible to oxidation.[10] Measures must be taken to avoid this problem. Also, the gels should be treated to reduce the interaction with copper during purification of the copper-MT. Treatment with sodium borohydride appears to be the simplest method that has been used.[11] Although this was not done in the above procedures for purification of MT from animals fed elevated zinc in the diet, it is probably prudent to use borohydride-treated gels for this isolation.

Half-Life of Metallothionein

The degree of saturation of MT with metals influences its turnover.[12] The more metal-saturated MTs have a longer half-life than the less satu-

[10] U. Weser, H.-J. Hartmann, A. Fretzdorff, and G.-J. Strobel, *Biochim. Biophys. Acta* **493,** 464 (1977).

[11] B. Lonnerdal and T. Laas, *Anal. Biochem.* **72,** 527 (1976).

rated ones. Also, the type of metal influences this half-life. Cadmium-MT has a longer half-life than when zinc is the predominant metal bound to it.[12] The half-life of MT in bovine and ovine tissues is markedly longer than that in tissues of rats or chicks.[13] The half-life of MT in ovine and bovine livers was found to be 22 to 24 days, whereas in rats and chickens it ranged from 10 to 65 hr, depending on the zinc status of the animals.[13] The half-life was determined in sheep and cattle MT from the zinc turnover, and is assumed to be valid because the turnover of zinc parallels that of the protein moiety in adults animals.[14] Adult sheep and cattle were used in these studies.

[12] P. D. Whanger, J. W. Ridlington, and C. L. Holcomb, *Ann. N.Y. Acad. Sci.* **355,** 336 (1980).
[13] P. D. Whanger and J. W. Ridlington, *in* "Biological Roles of Metallothionein" (E. C. Foulkes, ed.), p. 263. Elsevier/North-Holland, New York, 1982.
[14] S.-H. Oh and P. D. Whanger, *Am. J. Physiol.* **237,** E18 (1979).

[43] Metallothioneins and Other Zinc-Binding Proteins in Brain

By M. Ebadi

Introduction

With the exception of calcium and magnesium, zinc is the most abundant cation in the brain. In addition, the distribution of zinc in the brain is nonuniform and its concentration is highest in the cerebellum, hippocampus, retina, and the pineal gland. Furthermore, the mammalian hippocampi not only contain high concentrations of zinc, but also exhibit subregional variation in this essential element, with concentration being highest in the hilar region and lowest in the fimbria.

In addition to protein-bound zinc designated as "structural pool of zinc" and to enzyme-bound zinc designated as "metabolic pool of zinc," the brain contains a unique synaptosomal pool of zinc designated as "vesicular pool of zinc." This latter and restricted pool of zinc is found in zinc-containing neurons of the limbic, cerebrocortical, and corticifugal systems.[1-3]

[1] C. J. Frederickson and G. Danscher, *in* "Nutritional Modulation of Brain Function" (J. E. Morley, M. B. Sterman, and J. H. Walsh, eds.), p. 289. Academic Press, San Diego, 1988.
[2] J. Perez-Clausell, *J. Comp. Neurol.* **267,** 153 (1988).
[3] C. J. Frederickson, *Int. Rev. Neurobiol.* **31,** 145 (1989).

The concentration of zinc in the gray matter is 0.15–0.2 m*M*, which is significantly higher than the concentration of acetylcholine, a classical transmitter, or than the concentrations of enkephalin or cholecystokinin, two peptide transmitters. Therefore, in addition to its established roles with zinc metalloenzymes,[4] it has been proposed that zinc participates in neurosecretion, neuromodulation, and neurotransmission. For example, zinc participates in the synthesis of γ-aminobutyric acid and in the action of glutamate receptor, two widely utilized inhibitory and excitatory transmitters in the brain.[5]

All living organisms have developed efficient mechanisms to utilize essential trace elements such as zinc and copper in expressing their biological functions, and to minimize the cytotoxicity of nonessential posttransition metal ions such as cadmium and barium. The ability to bind excess metals and to limit concentration of "free" ions is achieved mostly by cysteine-rich polypeptides such as γ-glutamyl peptide (phytochelatin) found in plants, and metallothioneins found in animals.[6] The brain regions contain metallothioneins and other low-molecular-weight zinc-binding proteins of unknown nature. The biochemical properties of neuronal metallothioneins seem to be similar to those of metallothioneins found in nonneuronal tissues. However, the regulation of the synthesis and most probably the functions of neuronal metallothioneins appear to be different from those found in the peripheral tissues. In view of the fact that neither essential elements such as zinc and copper nor nonessential elements such as cadmium and barium traverse the brain readily and rapidly, it is very doubtful that the brain metallothioneins play major roles in acute metal detoxification—a task that is carried out efficiently by the hepatic and the renal metallothioneins. On the other hand, a complex structure such as brain, possessing an estimated 30,000 mRNAs and having one-third of the mammalian genome exclusively dedicated to its function,[7] may have developed unique processes for transporting, compartmenting, releasing, and utilizing zinc, copper, calcium, and other essential elements, commensurate with the diversified and vital functions endowed and invested in its various regions.

[4] B. L. Vallee and A. Galdes, *Adv. Enzymol.* **56,** 284 (1984).

[5] M. Ebadi, *in* "Selected Topics from Neurochemistry" (N. N. Osborne, ed.), p. 341. Pergamon, London, 1985.

[6] J. H. R. Kägi and Y. Kojima, *Experientia, Suppl.* **52,** 25 (1987).

[7] J. G. Sutcliffe, *Annu. Rev. Neurosci.* **11,** 157 (1988).

Classification of Zinc-Binding Proteins in Brain

Zinc-binding proteins in mammalian brains may be divided into three major categories of (1) zinc-metallothioneins, (2) zinc-metalloenzymes, and (3) zinc-containing structural proteins other than metalloenzymes or metallothioneins.

Neuronal Metallothioneins

Metallothioneins have been identified first in rat brain,[8-10] and later characterized further in monkey brain,[11] in bovine pineal gland,[12] retina,[13] cerebellum,[14] and hippocampus.[14,15] In addition, hippocampal neurons in primary culture,[16] retinoblastoma Y79 cell line,[17] and neuroblastoma IMR-32 cell line,[18] are all able to synthesize and express metallothionein isoforms.

Events Leading to Discovery of Metallothionein-Like Protein in Rat Brain

The discovery of a low-molecular-weight zinc-binding and zinc-inducible protein in the rat brain in our laboratory[8] and its designation as metallothionein-like protein[19] was a fortuitous observation that took place while attempting to comprehend the mechanism of convulsive seizures associated with Pick's disease, a rare neurological disorder, which allegedly results in higher than normal accumulation of zinc in the hippocampus, causing inhibition of γ-aminobutyric acid (GABA) transmission. In a study, Itoh and Ebadi[20] showed that the intracerebroventricular (icv) administration of zinc sulfate (0.3 μmol/10 μl) caused seizures that were

[8] M. Itoh, M. Ebadi, and S. Swanson, *J. Neurochem.* **44,** 823 (1983).
[9] M. Ebadi, *Biol. Trace Elem. Res.* **11,** 101 (1986).
[10] M. Ebadi, V. K. Paliwal, T. Takahashi, and P. L. Iversen, *in* "Metal Ion Homeostasis: Molecular Biology and Chemistry" (D. H. Hamer and D. R. Winge, eds.), p. 257. Alan R. Liss, New York, 1989.
[11] S. Gulati, V. K. Paliwal, M. Sharma, K. D. Gill, and R. Nath, *Toxicology* **45,** 53 (1987).
[12] A. Awad, P. Govitrapong, Y. Hama, M. Hegazy, and M. Ebadi, *J. Neural Transm.* **76,** 129 (1989).
[13] T. Takahashi, V. K. Paliwal, and M. Ebadi, *Neurochem. Int.* **13,** 525 (1988).
[14] S. M. Sato, J. M. Frazier, and A. M. Goldberg, *J. Neurosci.* **4,** 1662 and 1671 (1984).
[15] V. K. Paliwal and M. Ebadi, *Exp. Brain Res.* **75,** 477 (1989).
[16] P. Thakran, M. P. Leuschen, and M. Ebadi, *In Vivo* **3,** 191 (1989).
[17] C.-Z. Ou, T. E. Donnelly, and M. Ebadi, *Pharmacologist* **30,** 141 (1988).
[18] M. Ebadi, T. Takahashi, and P. Timmins, *Biol. Trace Elem. Res.* **22,** 233 (1989).
[19] M. Ebadi, *Fed. Proc.* **43,** 3317 (1984).
[20] M. Itoh and M. Ebadi, *Neurochem. Res.* **7,** 1287 (1982).

prevented by GABA, but not by other putative transmitters. During the course of our investigation some interesting observations were made. The administration of substantially larger doses of zinc sulfate (up to 100 mg/kg), either intravenously or intraperitoneally, did not cause convulsive seizures in rats. Since the convulsive dose of 0.3 μmol/10 μl of zinc sulfate was considerably lower than the concentrations of zinc found in 14 regions of rat brain,[21] we postulated that most of the zinc in the brain was bound and did not exist in "free" form. Based on these observations and conclusions, we searched for and identified a metallothionein-like protein in the rat brain.[8,10,19]

Induction of Rat Brain Metallothionein by Using Surgically Implanted Minipump

The intraperitoneal administration of zinc sulfate (7.5 mg/kg) given either in a single dose or daily for 10 days does not stimulate the synthesis of brain metallothionein.[9] On the other hand, the intracerebroventricular administration of zinc sulfate (0.20 μmol/μl/hr for 48 hr) not only induces the synthesis of brain metallothionein, but also induces the synthesis of hepatic metallothionein. The results of this study add further credence to a hypothesis that the steady state concentration of zinc in the brain is firmly regulated and is not disturbed readily. Indeed, unbound and "free" zinc in higher than physiological concentration is an extremely neurotoxic substance[22] with an inherent ability to inhibit an extensive number of sulfhydryl-containing enzymes (e.g., glutamic acid decarboxylase) and of sulfhydryl-containing receptor sites (e.g., glutamate and aspartate receptors).[23-25] Therefore, it is believed that the excess and unbound zinc most probably flows out from the brain via cerebrospinal fluid and reenters into plasma by transport through the choroid plexus, which lies outside of the blood–brain barrier.

Since the intraperitoneal administration of zinc sulfate does not stimulate the synthesis of brain metallothionein, this protein is induced by administering zinc sulfate intracerebroventricularly using a surgically implanted minipump (Alzet model 2M14; Alza Corp., Palo Alto, CA). In short, rats are anesthetized with sodium pentobarbital (35 mg/kg body

[21] M. Ebadi, M. Itoh, J. Bifano, K. Wendt, and A. Earle, *Int. J. Biochem.* **13,** 1107 (1981).
[22] M. Yokoyama, H. Koh, and D. W. Choi, *Neurosci. Lett.* **71,** 351 (1986).
[23] M. Ebadi, R. J. White, and S. Swanson, *Neurol. Neurobiol.* **11A,** 39 (1984).
[24] M. Ebadi and R. Pfeiffer, *Neurol. Neurobiol.* **11B,** 307 (1984).
[25] M. Ebadi and Y. Hama, *Adv. Exp. Med. Biol.* **203,** 557 (1986).

weight). The dorsal surface of the head is shaved, and a midline incision is made through the skin and underlying periosteum of the skull. The skin and periosteum are reflected and held in place with hemostats. A small hole is drilled with a dental drill approximately 2 mm posterior to the coronal suture and 3 mm lateral to the midline. The hole should be slightly larger than a 21-gauge hypodermic needle. A second hole of the same size is drilled into the dorsal surface of the skull 3 to 4 mm posterior to the first hole and 3 mm lateral to the midline. A small stainless steel machine screw is threaded into the second hole and screwed into the skull so that the head of the screw is 1.5 mm above the surface of the skull.

A 21-gauge hypodermic needle is gently squeezed in a pair of wire cutters to prepare a 15-mm length of hypodermic tubing. The tip of the needle is bent, allowing it to break off without crimping the needle. A second area of the needle is scored in a similar manner 15 mm below the broken end so that the length of the hypodermic tubing is 15 mm. This hypodermic tubing is then held in the flame of a gas burner, and the temper of the steel is changed so that the softened metal can be bent without cracking. The tubing is placed in a vise and a right angle bend made in it. One end of the angle will be 2.5 mm in length while the other will be 12.5 mm.

A short piece of plastic tubing is fitted over the long end of the hypodermic tubing, and the plastic tubing and hypodermic tubing are filled with saline (control group) or zinc sulfate (test group) by means of a syringe. The 2.5-mm end of the hypodermic tubing is placed in the remaining hole in the dorsal surface of the skull of the rat. As mentioned, this hole is 2 mm posterior to the coronal suture and 2 mm lateral to the midline. If properly placed, the end of the hypodermic tubing should rest in the lateral ventricle.[26] The long end of the hypodermic tubing is placed underneath the projecting head of the machine screw and secured in place with Durelon carboxylate cement (Espe GmbH, Seefeld/Oberbayern, Germany). The dental cement is allowed to dry thoroughly, and a subcutaneous tunnel is made between the scapulae of the rat with a blunt forceps. The minipump is then inserted into the subcutaneous pocket that had been made between the scapulae, and the wound is closed with silk sutures.

In addition to implanting a single pump for delivery of zinc sulfate (test) or saline (control), a dual pump system may be implanted, wherein one pump may deliver zinc sulfate and the second pump may deliver [^{35}S]cysteine; or one pump may deliver zinc sulfate and the second pump may deliver a protein synthesis inhibitor such as actinomycin D.

[26] E. P. Nobel, R. J. Wurtman, and J. Axelrod, *Life Sci.* **6**, 281 (1967).

Synthesis of Metallothionein in Hippocampal Neurons in Primary Culture

The metallothionein synthesis has been studied in newborn (7–8 days old) rat hippocampal neurons in primary culture.[16]

Tissue Preparation and Cell Culture Conditions. The animals are killed by decapitation and the brains quickly and aseptically removed into sterile Petri dishes containing Ca^{2+}- and Mg^{2+}-free Hanks' balanced salt solution (GIBCO, Grand Island, NY) with 100 U/ml penicillin and 100 μg/ml streptomycin. The hippocampi are dissected free of meningeal vessels and cut into explants of 1-mm diameter. Hippocampal explants from rat pups are plated in 60-mm Permanox Lux dishes (NUNC, Inc., Naperville, IL) that have been coated previously with a solution containing carbodiimide (130 μg/ml) and collagen (500 μg/ml) (Vitrogen-Flow) according to the protocol of Macklis *et al.*[27] Explants from adult rats are plated in similar dishes treated with poly (L-lysine) (Sigma, St. Louis, MO) according to Pettman *et al.*[28] Cultures are fed twice a day during the first week with 9 drops (330 μl) of Iscove's modification of Dulbecco's MEM (IMDM), containing 40.3 ml/liter $NaHCO_3$, 30 m*M* D-glucose, 293.2 mg/L-glutamine, 24.5 m*M* KCl, 100 mU/liter insulin, 7 μM *p*-aminobenzoic acid, 100 μg/ml transferrin, 10 mg/ml bovine serum albumin (BSA), 10^{-12} *M* β-estradiol, 100 g/ml gentamicin, and 3 μg/ml Fungizone, completely removing the spent medium from the dishes before feeding.[29] Thereafter, cultures are fed every other day with 770 μl of fresh medium, following the same procedure. Cultures are equilibrated with 90% O_2–10% CO_2 (v/v) in a humid incubator at 37° and monitored daily for survival and growth using a Leitz phase-contrast microscope (American Optical Corp., Buffalo, NY).

Staining Procedure. Neurons and their processes are identified by direct histochemical staining for cholinesterase according to the method of Karnovsky and Roots.[30] Briefly, the presence of cholinesterase is detected by its ability to hydrolyze the thiocholine ester used as substrate. The thiocholine thus liberated reduces ferricyanide to ferrocyanide. The latter compound combines with Cu^{2+} to form the insoluble copper ferrocyanide (Hatchett's Brown). On the tenth day *in vitro,* cultures derived from rat hippocampi originally plated on poly(L-lysine) are exposed to various concentrations of zinc and incubated with [^{35}S]cysteine in order to ascertain its incorporation into metallothionein. Spent medium is aspirated off

[27] J. D. Macklis, R. L. Sidman, and H. D. Shine, *In Vitro* **21,** 189 (1985).

[28] B. Pettman, T. C. Lewis, and M. Sensenbrenner, *Nature (London)* **281,** 378 (1979).

[29] V. Silani, G. Pezzoli, E. Motti, A. Falini, A. Pizzuti, C. Ferrante, A. Zecchinelli, F. Marossero, and G. Scarlato, *Appl. Neurophysiol.* **51,** 10 (1988).

[30] M. J. Karnovsky and L. Roots, *J. Histochem. Cytochem.* **12,** 219 (1964).

and fresh, unsupplemented medium containing a different concentration of zinc plus 10 μCi [^{35}S]cysteine/ml is added to the dishes. Control dishes do not receive zinc. In preliminary experiments, incubation was carried out for 72 hr, but subsequently the medium was aspirated off after 48 hr and the cultures were rinsed three times with 1 ml of Ca^{2+}- and Mg^{2+}-free phosphate-buffered saline (PBS). The cells are harvested with a rubber policeman and washed twice with PBS at 4°. Cell extracts are prepared by freezing and thawing followed by sonication. An aliquot is taken for protein determination and the remaining suspension is centrifuged at 13,600 *g* for 3 min in a Fisher (Pittsburgh, PA) microcentrifuge (model 235B).

Aliquots of the resulting supernatant are chromatographed on a Sephadex G-75 column (56 × 0.9 cm) using 10 m*M* Tris-HCl, pH 7.5, containing 1 m*M* dithiothreitol as the eluent.[9] The radioactivity of each fraction is determined in a Packard (Downers Grove, IL) liquid scintillation counter (model Tri-Carb 4530). In order to verify the authenticity of the metallothionein peaks isolated from the hippocampal primary cultures by gel-filtration chromatography, zinc-induced hepatic metallothionein is used as a marker protein according to earlier studies conducted in our laboratory.[8,9]

Zinc sulfate at a concentration of 1×10^{-9} to 1×10^{-6} *M* stimulated the incorporation of [^{35}S]cysteine and the synthesis of metallothionein in a time-dependent fashion and the maximum effects were seen 48 hr after incubation with zinc.[16]

Synthesis of Metallothionein in Neuroblastoma IMR-32 Cell Line in Culture: Lack of Induction by Dexamethasone

Although neuroblastoma cells are of tumor origin, they possess several features in common with neurons,[31-33] thus making them a useful model to study the effects of substances on the nervous system. The Chang liver cells may be used as control.

Human neuroblastoma IMR-32 cells are cultured in Eagle's minimum essential medium with nonessential amino acids containing 10% heat-inactivated fetal calf serum, 100 U/ml penicillin, and 100 μg/ml streptomycin at 37° in a humid atmosphere of 5% (v/v) CO_2 in air. Semiconfluent plates (2×10^6 cells/4 ml of culture medium/60 mm in diameter) of neuroblastoma or Chang cells are used in all experiments.

After aspiration of the culture media of the semiconfluent plates, fresh culture medium plus 5 μCi [^{35}S]cysteine/ml, 1 μM zinc or cadmium plus 5 μCi [^{35}S]cysteine/ml, or 2.5–100 μM dexamethasone plus 5 μCi [^{35}S]cys-

[31] G. Augusti-Tocco and G. Sato, *Proc. Natl. Acad. Sci. U.S.A.* **64,** 311 (1969).
[32] K. N. Prasad, *Biol. Rev.* **50,** 129 (1975).
[33] S. C. Haffke and N. W. Seeds, *Life Sci.* **16,** 1649 (1976).

teine/ml, is added to each semiconfluent plate. The control plates do not receive cadmium, zinc, or dexamethasone, but otherwise receive identical treatments. After 24 hr of incubation, the medium is aspirated and the cells are rinsed with 1 ml phosphate buffer solution without Ca^{2+} and Mg^{2+} [PBS(−)], followed by treatment with 1 ml 0.05% (w/v) trypsin and 0.02% ethylenediaminetetraacetic acid (EDTA, v/v) solution. After addition of

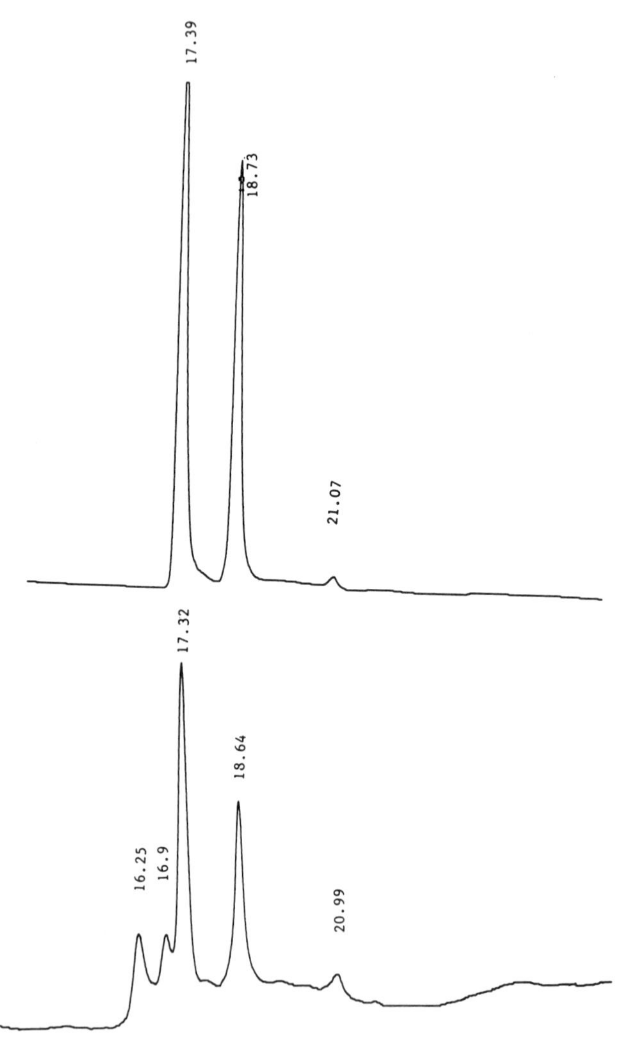

FIG. 1. Comparative high-performance liquid chromatography (HPLC) profiles of hepatic metallothionein (top) and brain metallothionein-like protein (bottom). One group of rats was administered intraperitoneally with 7.5 mg/kg of $ZnSO_4$ and killed 18 hr thereafter. The

1 ml of ice-cold PBS(−), the cells are harvested by pipetting, and the collected cells are washed with ice-cold PBS(−) twice and resuspended in 1 ml of 10 m*M* Tris-HCl, pH 7.5, containing 1 m*M* dithiothreitol. The cell extracts are prepared by freezing and thawing, followed by sonication. The extracts are centrifuged at 13,600 *g* for 3 min in a Fisher microcentrifuge (model 235B).

Aliquots of the resulting supernatant are chromatographed on a Sephadex G-75 column (0.9 × 56 cm) using 10 m*M* Tris-HCl, pH 7.5, containing 1 m*M* dithiothreitol as the eluant, as described by Ebadi.[9] The radioactivity of each fraction is determined in a Packard liquid scintillation counter (model TRI-CARB 4530).

The control Chang cells or the human neuroblastoma IMR-32 cells, untreated with either metals or glucocorticoid hormone, are able to incorporate [^{35}S]cysteine slightly (250–900 cpm $\times 10^{-3}$). The incubation of Chang liver cells with 100 μM zinc for 24 hr stimulated 6.7-fold over the control value the incorporation of [^{35}S]cysteine into metallothionein over the control group, whereas the incubation of neuroblastoma cells with an identical amount of zinc, under identical conditions, stimulated the synthesis of metallothionein only 2.3-fold over the control value. The incubation of Chang liver cells with 1 μM cadmium for 24 hr stimulated 5.0-fold over the control value the incorporation of [^{35}S]cysteine into metallothionein; whereas the incubation of neuroblastoma cells with an identical amount of cadmium, under identical conditions, stimulated the synthesis of metallothionein only 2.25-fold over the control value. As expected, 10 μM dexamethasone induced the synthesis of metallothionein in the Chang cells, whereas dexamethasone in concentrations ranging from 2.5 to 100 μM had no significant effects on the same parameter in the neuroblastoma cells. The minimum tested concentration of cadmium stimulating the synthesis of metallothionein in the Chang liver cells was 1 μM, and 7.5 μM caused cell death. The minimum tested concentration of zinc stimulating the synthesis of metallothionein was 100 μM in the neuroblas-

second group of rats was administered $ZnSO_4$ intracerebroventricularly with 0.22 μmol/μl/hr/48 hr by Alzet minipumps and then decapitated. The zinc-binding proteins were determined by HPLC using the technique of Klauser *et al.*,[34] with only one modification of using 10 m*M* Tris-HCl buffer instead of 50 m*M* Tris, as recommended by Suzuki *et al.*[35] These studies have shown the adult rat brain contains three zinc-binding proteins. The intracerebroventricular (icv) administration of zinc stimulates the synthesis of only one of these proteins, producing metallothionein-like protein isoforms 1 and 2, with retention times of 17.32 and 18.64 min, respectively. The synthesis of the other two zinc-binding proteins, with retention times of approximately 16.25 and 16.9 min, are not stimulated by intracerebroventricular administration of zinc. (Data from Ebadi.[9])

toma IMR-32, and 150 μM caused cell death. The neuroblastoma IMR-32 exhibits a lesser degree of tolerance to the toxic effects of either cadmium or zinc and demonstrated susceptibility to morphological damage or lethal effects with substantially smaller amounts of the metals.[18]

Amino Acid Composition of Zinc-Induced Metallothionein Isoforms in Rat Brain

Purification of the zinc-inducible rat brain metallothionein by ion-exchange chromatography on DEAE-Sephadex A-25 columns produces two isoforms, eluting, respectively, at 68 and 130 nM of Tris–acetate buffer (pH 7.5). Furthermore, the zinc-induced metallothionein also produces two distinct isoforms on reversed-phase high-performance liquid chromatography that exhibit retention times of 17.32 and 18.64 min, respectively (Fig. 1[9]). Brain metallothionein was characterized further by studies showing that the zinc-induced metallothionein incorporated a large quantity of [^{35}S]cysteine and that isoforms MT-1 and MT-2 contain 17 and 18 cysteine residues, respectively, while being devoid of any arginine, histidine, leucine, phenylalanine, or tyrosine (Table I).[36]

Regulation of Zinc Metallothionein Synthesis in Rat Brain

Although the metallothionein-like proteins isolated from the rat brain share certain biochemical properties with those isolated and characterized in the peripheral organs, there is no reason to believe that these proteins are regulated by identical mechanisms. The evidence gathered thus far indicates that the brain metallothionein has unique properties, summarized in Table II.[8-10,18,37,38]

Cadmium stimulates the synthesis of metallothionein in neuroblastoma IMR-32 cells. However, the lack of induction of brain metallothionein by cadmium most probably is related to its unique pharmacokinetic properties. Since cadmium is extremely neurotoxic in nature, it does not accumulate in brain in sufficient amounts to stimulate the synthesis of metallothionein *in vivo*. Indeed, Arvidson and Tjalve,[39] who studied the distribution of intravenously administered ^{109}Cd, have shown that within the peripheral sensory and autonomic ganglia (e.g., trigeminal and superior

[34] S. Klauser, J. H. R. Kägi, and K. J. Wilson, *Biochem. J.* **209,** 71 (1983).
[35] K. Suzuki, H. Sunaga, and I. Yajima, *J. Chromatogr.* **303,** 131 (1984).
[36] M. Ebadi and D. Babin, *Neurochem. Res.* **14,** 69 (1989).
[37] M. Ebadi, *Biol. Trace Elem. Res.* **11,** 117 (1986).
[38] V. K. Paliwal, P. Iversen, and M. Ebadi, *Neurochem. Int.* **17,** 441 (1990).
[39] B. Arvidson and H. Tjalve, *Acta Neuropathol.* **69,** 111 (1986).

TABLE I
AMINO ACID COMPOSITION OF ZINC-INDUCED BRAIN METALLOTHIONEIN-LIKE ISOFORMS[a]

	Isoform 1		Isoform 2	
Amino acids	Residues	Total (%)	Residues	Total (%)
Aspartic acid	5	8.4	4	6.5
Threonine	3	5.0	3	4.9
Serine	8	13.3	9	14.8
Glutamic acid	3	5.0	4	6.6
Proline	2	3.3	2	3.3
Glycine	7	11.7	5	8.2
Alanine	4	6.7	5	8.2
Valine	2	3.3	1	1.6
Cysteine[b]	17	28.3	18	29.5
Methionine[c]	1	1.7	1	1.6
Isoleucine	—	—	—	—
Lysine	8	13.3	9	14.8
Histidine	—	—	—	—
Arginine	—	—	—	—
Leucine	—	—	—	—
Tyrosine	—	—	—	—
Phenylalanine	—	—	—	—
Total:	60	100	61	100

[a] From Ref. 36. Minimum molecular weight (assuming no tryptophan): MT-1, 6900.83; MT-2, 7076.10.
[b] Determined as cysteic acid.
[c] Determined as methionine sulfone.

cervical ganglia) the uptake of cadmium was high, whereas the uptake of cadmium in the brain and spinal cord did occur, but was low.

Regional Distribution of Metallothionein in Bovine Brain

Bovine Pineal Zinc Metallothionein. The pineal gland may be involved in regulating zinc homeostasis, and zinc in turn may participate in modulating events associated with pineal functions. A study by Markowitz *et al.*[40] using human subjects reported that the peak serum level of zinc occurred at 9:30 AM, and a midtrough was evident at 8:00 PM, with a peak–trough difference of 19 μg/dl. In the same study, these authors showed that not only the serum calcium exhibited a circadian rhythm but

[40] M. E. Markowitz, J. F. Rosen, and M. Mizruchi, *Am. J. Clin. Nutr.* **41,** 689 (1985).

TABLE II
REGULATION OF ZINC METALLOTHIONEIN SYNTHESIS IN RAT BRAIN

Parameter	Rat hepatic metallothionein and Chang liver cells	Rat brain metallothionein-like protein and neuroblastoma IMR-32 cells
Synthesis stimulated following administration of Zn, given either on acute or chronic basis	Yes	No
Synthesis stimulated following intracerebroventricular administration of zinc	Yes	Yes
Synthesis stimulated by Cd^{2+} in *in vivo* system	Yes	No
Basal level of metallothionein	Very low but is stimulated dramatically	Relatively high but is stimulated further modestly
Basal metallothionein-2 mRNA and its degradation	Liver lower and faster than the brain	Brain higher and slower than the liver
Synthesis stimulated following ethanol	Yes	No
Synthesis in neuroblastoma IMR-32 cells stimulated by Zn or Cd^{2+}	Yes	Yes
Synthesis in neuroblastoma IMR-32 cells stimulated by dexamethasone	Yes	No
Neonatal level of metallothionein	High in liver and declines	Low in brain and increases
Known enzymatic activity	No	No
Isoforms	Two in rat liver	Two in rat brain
Amino acid residues	60, 60	60, 61
Cysteine residues	20, 20	17, 18
Immunoreactivity	Bovine hippocampal metallothionein cross-reacts with antibody formed against sheep and rat hepatic metallothioneins	

also that this rhythm was identical in peak–trough concentrations of calcium to that exhibited by zinc. Consequently, the authors suggested that a common mechanism may be responsible for the regulation of zinc and calcium rhythms. Although the precise factor (be it central or peripheral) responsible for exhibiting the specific circadian pattern of zinc has not been delineated, it may involve the same oscillator regulating temperature and/or cortisol in human. Indeed, in human a reversal of the sleep/wake cycle that disengages these oscillators[41] has been shown to shift the diurnal zinc pattern.[42]

The presence of a high concentration of zinc in the bovine pineal gland[12,43] prompted us to investigate the existence of a zinc-binding protein in this organ. Studies involving gel filtration using a Sephadex G-75 column and a 105,000 *g* supernatant fraction revealed two zinc-binding protein peaks that did bind 1.7 and 3.7 μg Zn^{2+}/mg protein, respectively. Furthermore, purification of the protein peak with an elution volume (V_e/V_0) of 2.06 on anion-exchange chromatography (DEAE-Sephadex A-25 columns) yielded a single protein peak that did bind 10 μg/mg protein. The comparative high-performance liquid chromatographic (HPLC) profiles of the zinc-induced hepatic metallothionein isoform 1 (retention time, 17.39 min) and of the bovine pineal metallothionein-like protein isoform 1 (retention time, 17.49 min) are similar (Fig. 2).

Bovine Retinal Zinc-Metallothionein. In all vertebrates, the retina is derived embryonically from an outpocketing of the central nervous system, and the mature retina and brain have several features in common.[44] Zinc, which occurs abundantly in the brain, also exists in an extremely high concentration in the retina. For example, the highest concentration of zinc found in any normal living tissue occurs in the choroid layer of the retina. Nocturnal animals with keen vision for hunting, like the fox, have a zinc concentration of 138,000 ppm or 13.8% by weight in the iridescent layer tapetum lucidum of the choroid.[45]

Although the function(s) of zinc in such high concentration in eye tissue is not known for certain, abundant evidence gathered thus far indicates that an alteration in the steady state concentration of zinc could lead to visual impairment. For example, maternal zinc deficiency during pregnancy in rats affects the morphogenesis of the optic vesicle and closure of the choroid fissure, leading to anophthalmia or colobomatous microphthalmia.[46] In addition, zinc deficiency causes cataracts in rainbow trout.[47] Cataract is a complication of acrodermatitis enteropathica,[48] a genetic

[41] M. E. Moore-Ede, C. A. Czeisler, and G. S. Richardson, *N. Engl. J. Med.* **309,** 469 (1983).
[42] D. M. Lanuza and S. F. Marotta, *Aerosp. Med.* **45,** 864 (1974).
[43] P. Y. Wong and K. Fritze, *J. Neurochem.* **16,** 1231 (1969).
[44] C. J. Barnstable, *Mol. Neurobiol.* **1,** 9 (1987).
[45] E. J. Underwood, "Trace Elements in Human and Animal Nutrition" (M. Mertz, ed.), 4th Ed., p. 196. Academic Press, New York, 1977.
[46] J. M. Rogers and L. S. Hurley, *Development* **99,** 231 (1987).
[47] H. G. Ketola, *J. Nutr.* **109,** 965 (1979).
[48] P. Racz, B. Kovacs, L. Varga, E. Ujlaki, E. Zombai, and S. Karbuczky, *J. Pediat. Ophthalmol. Strab.* **16,** 180 (1979).

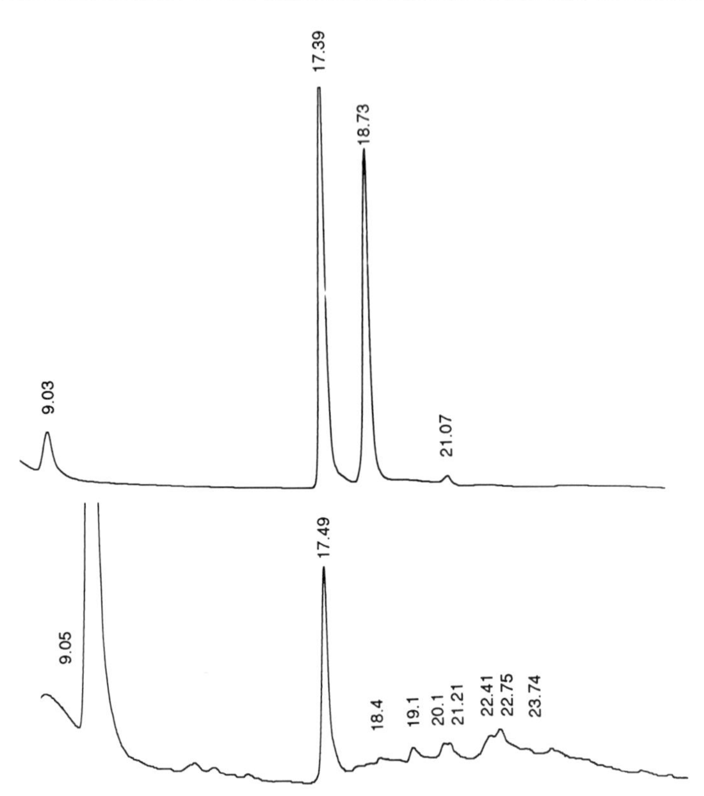

FIG. 2. Comparative HPLC profiles of the native bovine metallothionein-like isoform 2 (bottom) and of the zinc-induced hepatic metallothionein isoform 2 (top). The zinc-binding proteins were determined by HPLC using the technique of Klauser *et al.*,[34] with only one modification of using 10 m*M* Tris-HCl buffer instead of 50 m*M* Tris, as recommended by Suzuki *et al.*[35] All absorption trials were recorded at 220 nm. The zinc-induced hepatic metallothionein produces two isoforms with retention times of 17.39 and 18.73 min, respectively. The zinc-unstimulated pineal gland produces only one isoform with a retention time of 17.49 min, corresponding to the hepatic metallothionein isoform 2, which eluted at 130 n*M* Tris-acetate buffer and to the pineal metallothionein-like isoform 2, which eluted at 125 m*M* Tris-acetate buffer. However, their respective positions are reversed by using reversed-phase high-performance chromatography. These illustrations represent typical HPLC profiles, which have been repeated in identical fashion in five separate experiments. The minor protein peaks with retention times of 19.1, 20.1, 21.21, 22.45, and 23.74 min, respectively, are zinc-binding ligands of unknown nature. (Data from Awad *et al.*[12])

disorder with abnormal absorption of zinc.[49] In addition to cataracts, acrodermatitis enteropathica is accompanied by photophobia, blepharitis, conjunctivitis, and linear subepithelial corneal opacities.[51-53] These visual abnormalities are reversible by administration of zinc.[54-57]

Alcoholic liver disease causes zinc deficiency[58] and simultaneously produces abnormal dark adaptation.[59] In the synthesis of visual pigment (rhodopsin) by the rod, retinaldehyde must be constantly available. Vitamin A alcohol (retinol) is oxidized to photochemically active vitamin A aldehyde (retinaldehyde) by alcohol dehydrogenase, a zinc-requiring enzyme,[4,60] which apparently becomes relatively inactive in the alcohol-induced zinc-deficient state.[61]

Inherited retinal dystrophy, which results in spontaneous degeneration of retina,[62,63] is accompanied by an elevated concentration of zinc, but not copper.[64] Furthermore, the concentration of serum zinc is elevated in patients with retinitis pigmentosa.[65]

It is evident that the steady state concentration of zinc in the eye must be maintained, because zinc deficiency causes visual impairment and conditions such as retinal dystrophy and retinitis pigmentosa are associated with elevated levels of zinc. It may be argued that in order to maintain such a delicate balance of zinc in this tissue, it must exist in some reversibly found form, because the free zinc concentration is not regulated and is also known to be neurotoxic.[22-25]

Consequently, the extremely high concentration of zinc in the retina goaded us to search for a low-molecular-weight metal-binding protein, using the well-characterized hepatic metallothionein as a standard.[13]

We have measured the subcellular distribution of zinc in the bovine retina and have found the distribution to be nonuniform, with the outer

[49] P. M. Barnes and E. J. Moynahan, *Proc. R. Soc. Med.* **66,** 327 (1973).
[50] E. J. Moynahan, *Lancet* **2,** 399 (1974).
[51] L. Wirsching, Jr., *Acta Ophthalmol.* **40,** 567 (1962).
[52] R. S. Warshawsky, C. W. Hill, D. J. Doughman, and J. E. Harris, *Arch. Ophthalmol.* **93,** 194 (1975).
[53] C. S. Matta, G. V. Falker, and C. H. Ide, *Arch. Ophthalmol.* **93,** 140 (1975).
[54] B. Portnoy and M. Molokhia, *Br. J. Dermatol.* **91,** 701 (1974).
[55] G. Michaelsson, *Acta Derm. Venereol.* **54,** 377 (1974).
[56] N. Thyresson, *Acta Derm. Venereol.* **54,** 383 (1974).
[57] K. H. Neldner and K. M. Hambidge, *N. Engl. J. Med.* **292,** 879 (1975).
[58] B. L. Vallee, W. E. C. Wacker, A. F. Bartholomay, and E. D. Robin, *N. Engl. J. Med.* **255,** 403 (1956).
[59] S. A. Morrison, R. M. Russell, E. A. Carney, and E. V. Oaks, *Am. J. Clin. Nutr.* **31,** 276 (1978).
[60] B. L. Vallee, *in* "Zinc Enzymes, Volume 5, Metal Ions in Biology Series" (T. G. Spiro, ed.), p. 3. Wiley, New York, 1983.
[61] N. W. Solomons and R. M. Russell, *Am. J. Clin. Nutr.* **33,** 2031 (1980).
[62] M. C. Bourne, D. A. Campell, and K. Tansley, *Br. J. Ophthalmol.* **22,** 613 (1938).
[63] M. C. Bourne and H. Gruneberg, *J. Hered.* **30,** 130 (1939).
[64] C. D. Eckhert, *J. Hered.* **72,** 130 (1981).
[65] I. B. Mozha, *Vestn. Oftal'mol.* **5,** 59 (1974).

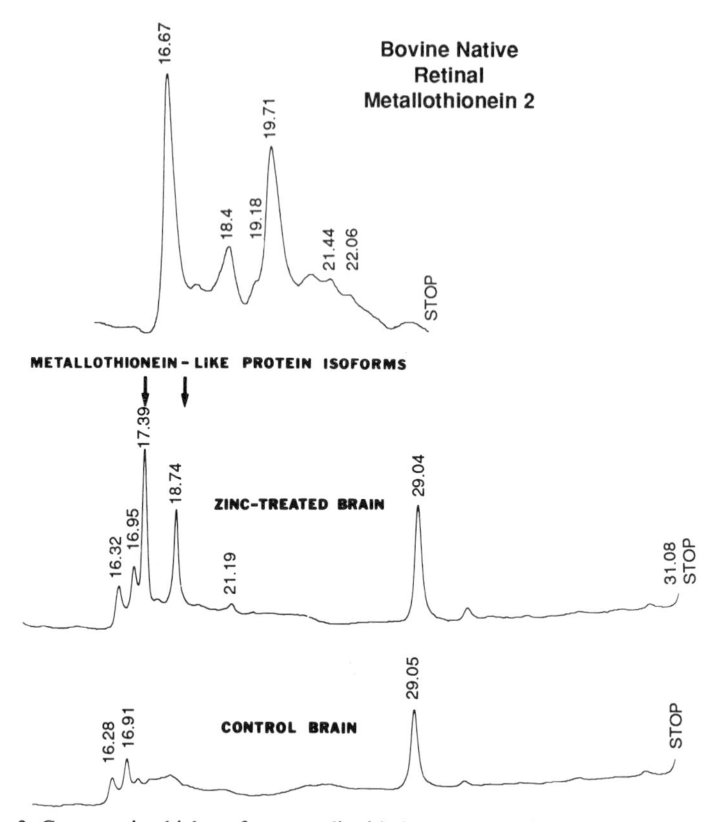

FIG. 3. Comparative high-performance liquid chromatographic (HPLC) profiles of zinc-induced rat hepatic metallothionein isoform 2 and the native bovine metallothionein-like isoform 2. The metallothionein isoform was determined by HPLC according to the procedures of Klauser *et al.*[34] with only one modification of using 10 m*M* Tris-HCl buffer as recommended by Suzuki *et al.*[35] The metallothionein-like isoforms and other zinc-binding proteins were analyzed for zinc using atomic absorption spectrometry and protein. The retinal protein peak with retention time of 16.67 min and the rat protein peak with retention time of 16.47 min are identical metallothioneins. On the other hand, the retina contains at least two other major zinc-binding proteins, with retention times of 18.4 and 19.71 min, that appear on chromatographs when absorption is recorded at 220 nm. The biochemical nature of these proteins is not known.[13]

rod segments and the pellet 1 fraction, known to be highly enriched in photoreceptor cell synaptosomes, containing the highest amounts of 0.230 ± 0.040 and 0.119 ± 0.04 μg Zn/mg protein, respectively. In addition, the bovine retina contains a low-molecular-weight metallothionein- like protein that exhibits an elution volume (V_e/V_0) of 1.9 on gel-permeation chromatography and which produces only one isoform on reversed- phase high-performance liquid chromatography with a retention time of 16.67 min (Fig. 3).

Bovine Hippocampal Zinc Metallothionein. The "hippocampal formation" or "hippocampal region" consists of the hippocampus (sometimes called Ammon's horn), the dentate gyrus, and subiculum. On the basis of architectonic differences, the hippocampus may be subdivided into different fields (CA_1, CA_2, CA_3, and CA_4) running along its length. Therefore, in outline, this design appears in the sequential connections given in Scheme I.

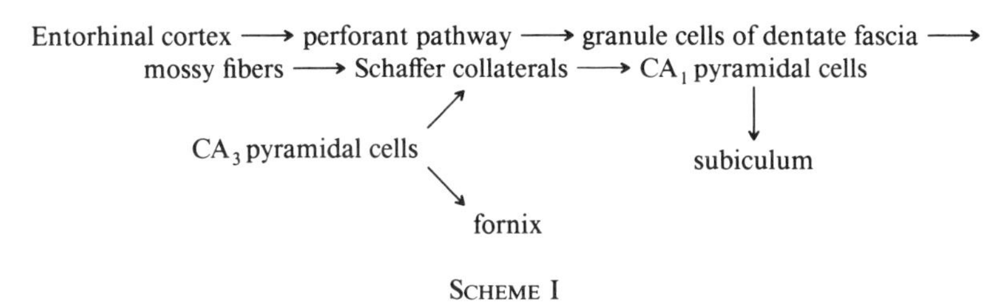

SCHEME I

At each of these synaptic relays, there is divergence and convergence so that each operates integrationally.[66-68]

An original observation of Maske[69] that the hippocampus contained a high concentration of dithizone-chelatable zinc gained greater significance when it was recognized that the major portion of this zinc was associated with the hippocampal mossy fibers.[70,71] Furthermore, studies involving Timm sulfide silver staining[72] and stable isotope dilution technique[73,74] revealed that the concentration of zinc in the hippocampus exhibited regional variation, being highest in the hilar region and lowest in the fimbria. For example, the amount of zinc associated directly with mossy fiber axons was estimated to be approximately 8% of the total zinc in the hippocampus, and the concentration of mossy fiber zinc was estimated to be 300–350 μM.[74]

[66] D. G. Jones and B. J. Smith, *Prog. Neurobiol.* **15,** 19 (1980).
[67] L. W. Swanson, T. J. Teylor, and R. F. Thompson, *Neurosci. Res. Prog. Bull.* **20,** 612 (1981).
[68] B. H. Bland, *Prog. Neurobiol.* **26,** 1 (1986).
[69] H. Maske, *Naturwissenschaften* **42,** 424 (1955).
[70] T. McLardy, *Confin. Neurol.* **20,** 1 (1960).
[71] F.-M. S. Haug, *Histochemie* **8,** 355 (1967).
[72] G. Danscher, E. J. Fjerdingstad, E. Fjerdingstad, and K. Fredens, *Brain Res.* **112,** 442 (1976).
[73] C. J. Frederickson, G. A. Howell, and M. H. Frederickson, *Exp. Neurol.* **73,** 812 (1981).
[74] C. J. Frederickson, M. A. Klitenick, W. I. Manton, and J. B. Kirkpatrick, *Brain Res.* **273,** 335 (1983).

A suggestion by Crawford and Connor[75] that zinc may be involved in the maturation and function of the mossy fiber pathway was highlighted further by Frederickson *et al.*,[73] who reported that the concentration of zinc in the mossy fiber region increased with age. Moreover, the dynamic involvement of zinc in the synaptic activities of the hippocampus became substantiated by studies reporting that the uptake of exogenous zinc and the release of the previously bound ^{65}Zn from the hippocampal mossy fiber neuropil were facilitated by electrical stimulation.[76] Furthermore, the cholinergic deafferentation of the hippocampus[77] or the selective kainic acid lesion of either granule cells or CA_3 pyramidal cells[78] altered the concentration of zinc in the hippocampal mossy fibers. In addition, a study by Baimbridge *et al.*[79] has shown that the kindling of a number of sites, including the commissure, amygdala, septum, and perforant path, results in a reduction of a calcium-binding protein in the hippocampus, suggesting that this cytosolic metal-binding protein does indeed respond to alteration in the level of excitability and synaptic activity of neurons.

Although zinc-deficiency states result in a reduction in the activity of many zinc-dependent enzymes, the bound zinc in these enzymes does not react freely with histochemical reagents (neo-Timm, selenium, dithizone, or quinoline), and appears to be hidden to histochemical visualization.[1,3]

The "chelatable" pool of zinc, which is also called "storage granule zinc," "neuronal zinc," or "vesicular zinc," and which has been proposed to be involved in hippocampal neurotransmission, (1) has a restricted distribution in brain by being highest in the hippocampus, (2) is completely restricted to regions of neuropils, (3) is present with synaptic vesicles in the hippocampal mossy fiber axon terminals, (4) is released by a Ca^{2+}-dependent mechanism, and (5) is found in the synaptic clefts, presumably by exocytosis of the zinc-laden vesicles[1-3] (Fig. 4).

We hypothesized that low-molecular-weight zinc-binding proteins may exist in the hippocampus in order to regulate the steady state concentration of zinc. In an attempt to investigate this hypothesis and the dynamic metabolism of zinc, we have searched for and have identified a metallothionein-like protein in bovine hippocampus that exhibits an elution volume (V_e/V_0) of 2.0 on gel-filtration chromatography and which produces four isoforms, of which two, on a reversed-phase high-performance liquid chromatography, show retention times of 15.70 and 16.37 min, respec-

[75] I. L. Crawford and J. D. Connor, *J. Neurochem.* **19,** 1451 (1972).
[76] G. A. Howell, M. G. Welch, and C. J. Frederickson, *Nature (London)* **308,** 736 (1984).
[77] G. R. Stewart, C. J. Frederickson, G. A. Howell, and F. H. Gage, *Brain Res.* **290,** 43 (1984).
[78] J. P. Kesslak, C. J. Frederickson, and F. H. Gage, *Exp. Brain Res.* **67,** 77 (1987).
[79] K. G. Baimbridge, I. Mody, and J. J. Miller, *Epilepsia* **26,** 460 (1985).

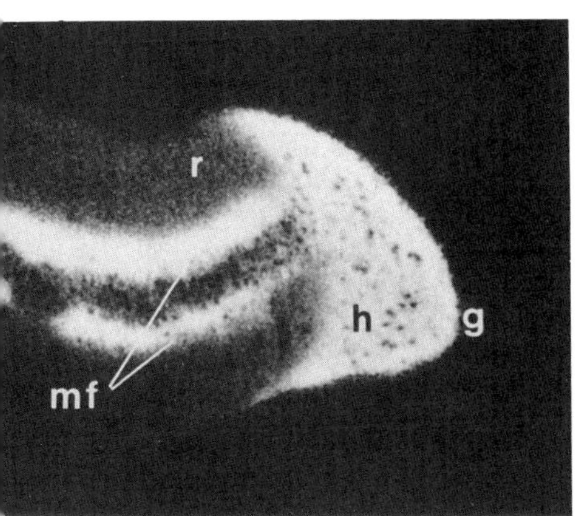

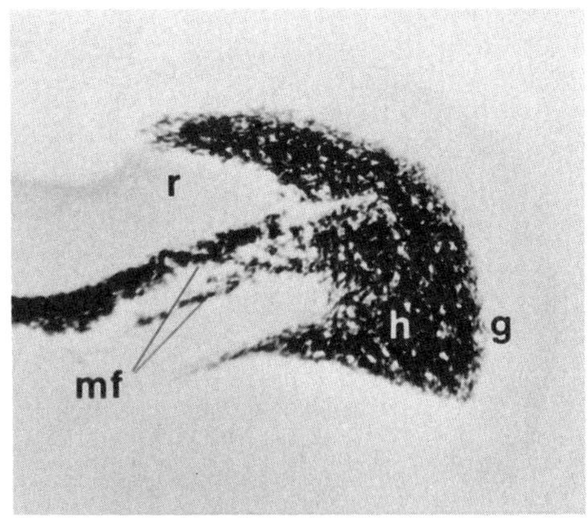

FIG. 4. Zinc in axonal boutons is shown by 6-methoxy-8-*p*-toluenesulfonamide quinoline (TSQ) fluorescence (left) and neo-Timm's (right) in the dentate gyrus and hippocampus of a rat. Neural perikarya in the granule cell stratum and within the hilus (arrows) are unstained. g, Granule stratum; h, hilus; mf, mossy fiber neuropil; r, stratum radiatum. (Slide courtesy of Christopher J. Frederickson and Ref. 1.)

tively (Fig. 5). The hippocampal metallothionein isoform 2 contains a cysteine- to-zinc ratio of 2.8 to 1.0, has an apparent molecular weight of 9500, and, as judged by studies involving ultraviolet (UV) spectral analysis, lacks aromatic amino acids, but possesses metallomercaptide bonds. Furthermore, antibodies formed against both the sheep and rat hepatic metallothioneins cross-reacted completely with the hippocampal metallothionein, suggesting that the immunologically dominant regions of the NH_2-terminal domain (residues 1–29) of neuronal metallothioneins are conserved.[10]

Zinc Metallothionein-Modulated Enzymes in Brain

Zinc, which is the fourth most abundant ion in the brain, is involved in an impressive array of physiological states, including maintaining the functions of enzymes and nucleic acids. The importance of these metabolic processes in central nervous maturation suggests that zinc deficiency may affect neural development.

The roles of zinc in more than 235 metalloenzymes can be divided into 4 major categories.[4,60]

1. Catalytic role: Zinc participates directly and is involved in catalysis by enzymes such as carbonate dehydratase, carboxypeptidase, and aldolase.

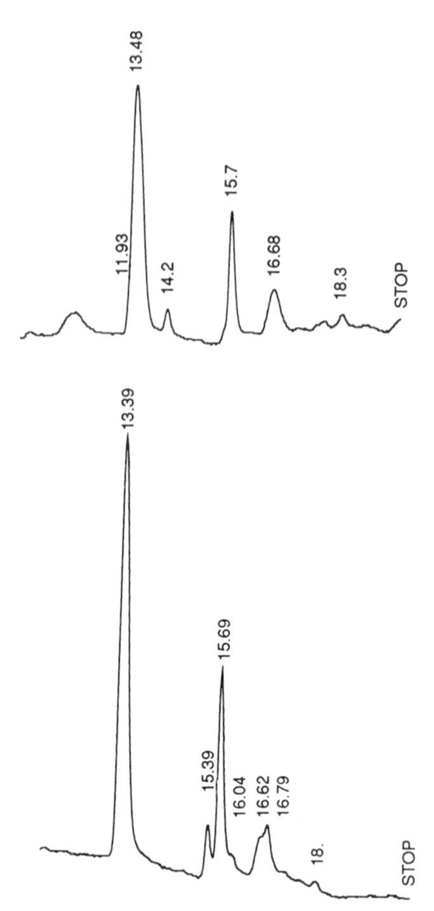

FIG. 5. Comparative HPLC profiles of zinc-induced hepatic metallothionein isoform 1 (bottom) and native hippocampal metallothionein isoform 1 (top) with retention times of 15.69 and 15.70 min, respectively, are shown. The zinc-binding proteins were determined using the technique of Klauser *et al.*[34] and Suzuki *et al.*[35]

2. Structural role: Zinc participates mostly, but not exclusively, in maintaining the quaternary structure of oligomeric holoenzymes such as aspartate carbamyoltransferase.

3. Regulatory (modulatory) role: Zinc regulates enzymatic activity by either activating enzymes (e.g., bovine lens leucine aminopeptidase) or inhibiting them (e.g., porcine kidney aminopeptidase).

4. Noncatalytic and nonstructural role: Zinc is involved in an obscure fashion in certain metalloenzymes such as human alcohol dehydrogenase

or *Escherichia coli* alkaline phosphatase, but the involvement is of neither a catalytic nor structural nature.

In the brain, zinc participates in the synthesis of pyridoxal phosphate, which in turn is involved in the synthesis of dopamine, norepinephrine, serotonin, taurine, histamine, and γ-aminobutyric acid.[5] The crucial role that pyridoxal phosphate plays in the synthesis of neurotransmitters may be judged by the fact that 30–40% of all neurons in the brain utilize γ-aminobutyric acid.

γ-Aminobutyric acid (GABA), a major inhibitory neurotransmitter in the vertebrate system, is synthesized under the catalytic activity of a rate-limiting enzyme, glutamate decarboxylase (GAD). A positive correlation exists between the activity of glutamate decarboxylase and the concentration of GABA. Furthermore, glutamate decarboxylase has an absolute requirement for pyridoxal phosphate (PLP), as shown in studies carried out both *in vitro* and *in vivo*.[80–82] Furthermore, studies have shown that two forms of GAD are found in the rat brain. One form (GAD A) does not require PLP for maximal activity, whereas another form (GAD B) does. Furthermore, the ratio between GAD A and GAD B is nonuniform throughout brain areas, and the hippocampus contains twice as much GAD B (the PLP-requiring GAD) as GAD A.[82] A reduction in the concentration of pyridoxal phosphate resulting from dietary deficiency of pyridoxine causes an inhibition in the activity of glutamate decarboxylase and a reduction in the concentration of GABA. Indeed, glutamate decarboxylase is more susceptible to vitamin B_6 deficiency in comparison to other vitamin B_6-requiring enzymes, such as aromatic L-amino-acid decarboxylase. Altogether, these observations suggest that in the brain, pyridoxal phosphate controls the activity of glutamate decarboxylase, which in turn influences the central nervous system (CNS) excitability through the synthesis of GABA, an inhibitory transmitter.

Synthesis of Pyridoxal Phosphate

Pyridoxal phosphate is not supplied by diet and must be synthesized in the body. The naturally occurring vitamin B_6 derivatives, namely pyridoxine, pyridoxal, and pyridoxamine are absorbed efficiently and rapidly from the gastrointestinal tract. On the other hand, the phosphorylated derivatives, pyridoxine phosphate, pyridoxal phosphate, and pyridoxamine

[80] J. Y. Wu and E. Roberts, *J. Neurochem.* **23,** 759 (1974).
[81] J. Y. Wu, E. Wong, K. Saito, E. Roberts, and A. Schousboe, *J. Neurochem.* **27,** 653 (1976).
[82] L. A. Denner and J. Y. Wu, *J. Neurochem.* **44,** 957 (1985).

phosphate, are not transported, to any great extent, across most cellular membranes. In the body, the absorbed pyridoxine, pyridoxal, and pyridoxamine are converted, respectively, to pyridoxine phosphate, pyridoxal phosphate, an pyridoxamine phosphate by pyridoxal kinase.[83]

Among divalent cations tested (Co^{2+}, Cu^{2+}, Fe^{2+}, Mg^{2+}, Mn^{2+}, Ni^{2+}), Zn^{2+} was the most effective activator of hepatic pyridoxal kinase.[83] Furthermore, the human hepatic pyridoxal kinase exhibits a complete dependence on zinc.[84] Moreover, zinc activates pyridoxal kinase in the rat brain and the ED_{50} values for stimulation in the cerebellum, superior colliculus, and the hippocampus are 5.5×10^{-6}, 8.5×10^{-6}, and 17.0×10^{-6} *M*, respectively.[21] Finally, Churchich *et al.*[85] have shown that metallothionein-activated pyridoxal kinase may be isolated from sheep and pig brains. On the other hand, apoprotein of metallothionein inhibited the activity of pyridoxal kinase. This fact supports the contention that zinc activates pyridoxal kinase according to the following mechanisms.

$$\text{Zn-metallothionein} \rightleftharpoons Zn^{2+} + \text{metallothionein}$$

$$Zn^{2+} + \text{ATP} \rightleftharpoons \text{ZnATP}$$

$$\text{Zn-ATP} + \text{pyridoxal kinase} \rightleftharpoons \text{Zn-ATP-pyridoxal kinase}$$

Zinc as Endogenous Modulator of Glutamate Receptors

The mammalian CNS contains large amounts of chelatable Zn^{2+}. In the hippocampus, much of the Zn^{2+} is localized in synaptic vesicles in excitatory boutons and is likely released into the synaptic cleft during excitation, especially at higher rates of neuronal firing. The extracellular accumulation of synaptically released Zn^{2+} may be substantial. For example, Assaf and Chung[86] found that exposure to high concentration of K^+ induced the release of 18% of the Zn^{2+} contained in rat hippocampal slices, an amount of Zn^{2+} that if it were distributed evenly in tissue extracellular space would produce a concentration of 300 μM.

Because Zn^{2+} selectively attenuates the neuroexcitatory and neurotoxic effects of *N*-methyl-D-aspartate (NMDA) receptor agonists on cortical neurons, Peters *et al.*[87] and Westbrook and Mayer[88] proposed that zinc might be an important endogenous modulator of glutamate receptors. Further-

[83] D. B. McCormick, M. E. Gregory, and E. E. Snell, *J. Biol. Chem.* **236,** 2076 (1961).
[84] A. H. Merrill, J. M. Henderson, E. Wang, B. W. McDonald, and W. J. Millikan, *J. Nutr.* **144,** 1664 (1984).
[85] J. E. Churchich, G. Scholz, and F. Kwok, *Biochem. Int.* **17,** 395 (1988).
[86] S. Y. Assaf and S. H. Chung, *Nature (London)* **308,** 734 (1984).
[87] S. Peters, J. Koh, and D. W. Choi, *Science* **236,** 589 (1987).
[88] G. L. Westbrook and M. L. Mayer, *Nature (London)* **328,** 640 (1987).

more, Forsythe *et al.*[89] have shown that Zn^{2+} is able to block NMDA receptor-mediated excitatory postsynaptic potentials in the hippocampal neurons. Moreover, Reynolds and Miller,[90] by studying the effects of divalent cations on ^{3}H-labeled MK801 binding to the NMDA receptor, and Christine and Choi,[91] by studying the effects of zinc on NMDA receptor-mediated channel currents in cortical neurons, concluded that zinc may reduce NMDA receptor-activated channel currents on cortical neurons by acting at two different sites, one outside of the membrane field and affecting opening frequency, and the other inside the channel and interfering directly with the passage of ions (Fig. 6[92]). In this manner, Zn^{2+} coreleased with glutamate from presynaptic terminals may dynamically modulate NMDA receptor-activated currents.

It is postulated that zinc metallothionein, by regulating the steady state concentration of zinc, influences the synthesis of GABA and the action of NMDA receptors in the hippocampus according to Scheme II.

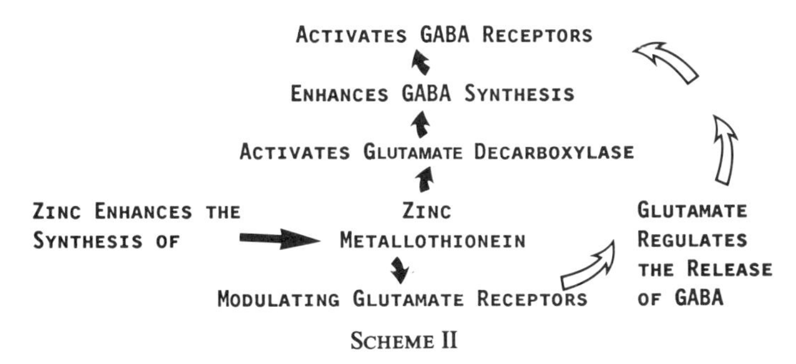

SCHEME II

Indeed, Harris and Miller,[93] by using hippocampal neurons in primary culture, have shown that [^{3}H]GABA release was stimulated by the excitatory amino acid neurotransmitter glutamate, as well as by *N*-methyl-D-aspartate, and that [^{3}H]GABA release was blocked by Zn^{2+}.

Metalloprotein in Brain Unrelated to Metalloenzymes or Metallothionein

Studies by Baudier *et al.*[94] and Baudier and Gerard,[95] using brain, have shown that the S100 protein fraction, which is a mixture of S100a ($\alpha\beta$) and

[89] I. D. Forsythe, G. L. Westbrook, and M. L. Mayer, *J. Neurosci.* **8,** 3733 (1988).
[90] I. J. Reynolds and R. J. Miller, *Mol. Pharmacol.* **33,** 581 (1988).
[91] C.-W. Christine and D. W. Choi, *J. Neurosci.* **10,** 108 (1990).
[92] M. L. Mayer, L. Vyklicky, Jr., and E. Sernagor, *Drug Div. Res.* **17,** 263 (1989).
[93] K. M. Harris and R. J. Miller, *Brain Res.* **482,** 23 (1989).
[94] J. Baudier, K. Haglid, J. Haeck, and D. Gerard, *Biochem. Biophys. Res. Commun.* **114,** 1138 (1983).

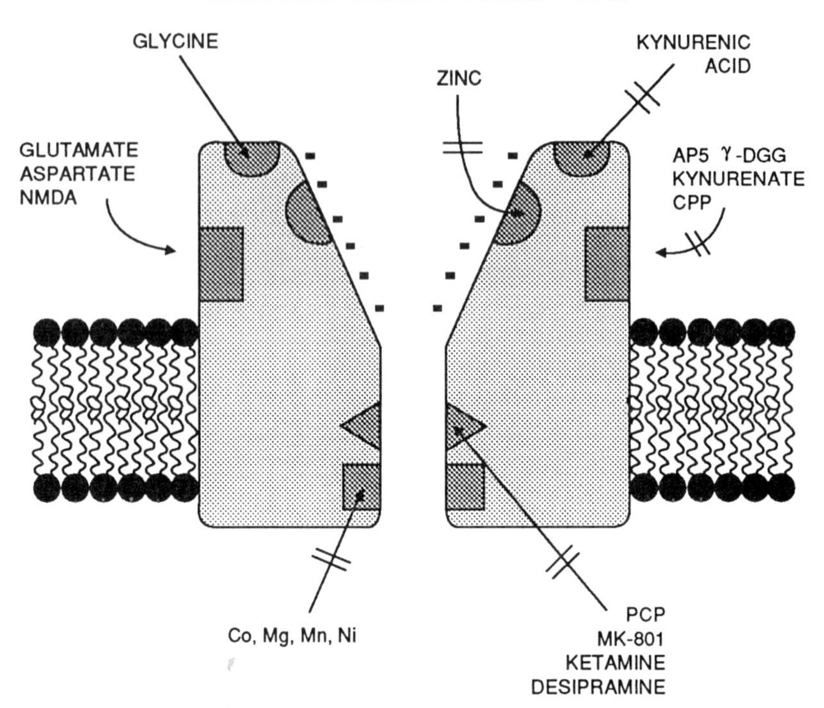

FIG. 6. Model for Zn^{2+} action on the NMDA receptor–channel complex. Speculative model shows two sites of action, one influencing the coupling of the NMDA receptor to its channel and the other within the channel itself, perhaps similar to the Mg^{2+} site. Some of the Zn^{2+} released by the presynaptic terminal might permeate the open NMDA channel, perhaps triggering changes in the postsynaptic cell. The PCP site is illustrated as a separate site for convenience; it could be the same as the Zn^{2+} or the Mg^{2+} site. (Slide courtesy of M. L. Mayer, L. Vyklicky, Jr., and E. Sernagor, Ref. 92.)

S100b (*ββ*) components, should no longer be considered only as a "calcium-binding protein" but also as a zinc-binding protein. The S100b protein exhibits two sets of zinc-binding sites with K_D values of 10^{-7} to 10^{-6} *M*. Studies by Hesketh[96] have shown that at least one zinc-binding site exists on tubulin. Furthermore, a zinc-dependent interaction of bovine S100b protein with the tubulin–microtubule system has been reported. Finally, Baudier *et al.*[97] have shown that zinc binding on rat S100b protein

[95] J. Baudier and D. Gerard, *Biochemistry* **22,** 3360 (1983).

[96] J. E. Hesketh, *Int. J. Biochem.* **15,** 743 (1983).

[97] J. Baudier, G. Labourdette, and D. Gerard, *J. Neurochem.* **44,** 76 (1985).

regulates calcium binding by increasing the calcium affinity of the protein and reducing the antagonistic effects of K^+ on calcium binding. The implication of these findings in the neurobiology of zinc, which may be far reaching, awaits clarification.

In conclusion, the concentration of zinc has been shown to be altered in an extensive number of disorders of the central nervous system, including alcoholism, Alzheimer-type dementia, Down's syndrome, epilepsy, Friedreich's ataxia, Guillain–Barré syndrome, hepatic encephalopathy, multiple sclerosis, Pick's disease, retinitis pigmentosa, retinal dystrophy, schizophrenia, and Wernicke–Korsakoff syndrome.[23-25] The status of zinc metallothioneins and other low-molecular-weight zinc-binding proteins in these conditions, diseases, disorders, or syndromes is not known.

Acknowledgments

The excellent secretarial assistance of Mrs. Dorothy Panowicz is gratefully acknowledged. Studies cited were supported in part by grants from USPHS NS 08932 and ES 03949.

[44] Detection of Metallothionein in Brain

By KATSUYUKI NAKAJIMA, KEIJI SUZUKI, NORIKO OTAKI, and MASAMI KIMURA

Introduction

The concentration of zinc in the brain is known to be high and also shows regional variations.[1] Both an excess and a deficiency of zinc have been reported to be associated with several kinds of neurological disorders.[2] These studies of zinc deficiency and excess on neurological disorders indicate that the steady state concentration of zinc must be firmly regulated, and the zinc-binding proteins may play a crucial role in maintaining zinc homeostasis in the brain.

The existence of metallothionein (MT) in brain was studied by the cadmium–heme method[3] and radioimmunoassay (RIA) by several researchers.[4-10] Ebadi *et al.*[11-13] have isolated MT-like proteins that had

[1] M. Ebadi, M. Itoh, J. Bifano, K. Wendt, and A. Earle, *Int. J. Biochem.* **13,** 1107 (1981).
[2] M. Ebadi and R. F. Pfeiffer, *in* "The Neurobiology of Zinc" (C. J. Frederickson, G. A. Howell, and E. J. Kasarskis, eds.), Part B, p. 307. Alan R. Liss, New York, 1984.
[3] S. Onosaka and G. Cherian, *Toxicology* **22,** 91 (1981).
[4] F. O. Brady and R. L. Kafka, *Anal. Biochem.* **98,** 89 (1979).
[5] R. J. Vander Mallie and J. S. Garvey, *J. Biol. Chem.* **254,** 8416 (1979).

amino acid compositions different from hepatic MT, although they also described some similarities. Kano *et al.*[14] have isolated silver-containing MT-like proteins from rat brains after treatment with silver sulfadiazine.

There have been no reports concerning the immunocytochemical studies of brain MT localization, even though the existence of MT was predicted by immunochemical and biochemical methods. Therefore, we have applied the RIA and immunoperoxidase method, using the same specific anti-MT antibody, to measure the amounts of MT and to detect the localization of MT in human and rat brains.

Preparation of Metallothionein-1 as Immunogen

Rat MT-1 is isolated from rat liver after administration of $CdCl_2$ according to the method of Toyama and Shaikh.[15] Twenty milligrams of highly purified rat liver MT-1 and 15 mg of the extract of *Ascaris* are dissolved in 5 ml of 0.1 *M* phosphate buffer (pH 7.0). To this mixture 175 μl of 14% (v/v) glutaraldehyde in distilled water is added and allowed to stand at room temperature for 5 hr. The reaction mixture is dialyzed against distilled water changed four times for 4 days at 4° and after lyophilization. The conjugated MT-1 solution is emulsified with an equal volume of Freund's complete adjuvant (Difco Laboratories, Detroit, MI). Male New Zealand white rabbits are immunized at multiple sites subcutaneously with 100 μg of the conjugate/rabbit once every 2 weeks for 12 weeks. After this immunization the conjugate immunogen that was not emulsified with adjuvant is injected into ear veins four times. Ten days after the final booster, all blood is drawn out.

Preparation of Tyrosine-Metallothionein[16]

The *m*-maleimidobenzoyl-*N*-hydroxysuccinimide ester (MBS) used for this study is purchased from Pierce (Rockford, IL). The tyrosine MT-1 is

[6] C. C. Chang, R. J. Vander Mallie, and J. S. Garvey, *Toxicol. Appl. Pharmacol.* **55,** 94 (1980).

[7] C. Toyama and Z. A. Shaikh, *Fundam. Appl. Toxicol.* **1,** 1 (1981).

[8] J. S. Garvey and C. C. Chang, *Science* **214,** 805 (1981).

[9] D. R. Winge and J. S. Garvey, *Proc. Natl. Acad. Sci. U.S.A.* **80,** 2472 (1983).

[10] R. K. Mehra and I. Bremner, *Biochem. J.* **213,** 459 (1983).

[11] M. Itoh, M. Ebadi, and S. Swansan, *J. Neurochem.* **41,** 823 (1983).

[12] M. Ebadi, *Trans. Soc. Neurosci.* **14,** 1062 (1984).

[13] M. Ebadi and J. C. Wallwork, *Biol. Trace Elem. Res.* **7,** 129 (1985).

[14] K. Kano, K. Itoh, Y. Nishino, J. Kodama, Y. Itokawa, and Y. Kojima, *SEIKAQ* **60,** 785 (1988).

[15] C. Toyama and Z. A. Shaikh, *Biochem. Biophys. Res. Commun.* **84,** 907 (1978).

[16] T. Ikei, F. Shimizu, T. Kodaira, K. Nakajima, C. Toyama, H. Saito, N. Otaki, and M. Kimura, "Second International Meeting on Metallothionein and Other Low Molecular Weight Metal Binding Proteins" August, Univ. of Zurich, Zurich, 21–24, 1985.

prepared by reacting MT-1 with MB-tyrosine. In this reaction, the ratios of molar doses of MB-tyrosine to MT-1 are 2.5, 5, 20, 100, and 200. The experiment described below is with ratios of MB-Tyr to MT-1 of 200.

L-Tyrosine (10.35 mg) is dissolved in 9.5 ml of 0.1 *M* phosphate buffer, pH 8.0. MBS (89.78 mg) dissolved in dimethylformamide (DMF) (1 ml) is added dropwise, and the mixture stirred at room temperature for 20 min. Then the reaction mixture is washed with 3 ml dichloromethane to remove unreacted MBS. The aqueous layer is transferred to another tube and the dichloromethane layer is washed with a small amount of 0.1 *M* phosphate buffer. Both washings are combined. The absence of MBS in this solution is confirmed by thin-layer chromatography.

The above reaction mixture of MB-Tyr and MT-1 (2 mg) is combined and deoxygenated by a stream of nitrogen gas. After 2 hr, the reaction mixture is desalted on Sephadex G-50 equilibrated with 20 m*M* ammonium carbonate. The desired fractions, which are monitored by absorbance at 250 and 280 nm, are combined and lyophilized. The number of tyrosines incorporated into MT-1 is determined by amino acid analysis.

Radioiodination of Tyr–MT-1 by Chloramine-T Method

Chloramine-T (20 μg) in 20 μl of 0.05 *M* phosphate buffer, pH 7.4, is added to a solution of Tyr–MT-1 (5 μg) and $Na^{125}I$ (1 mCi) in 0.05 *M* phosphate buffer (20 μl). The mixture is stirred vigorously for 25 sec at room temperature. The reaction is terminated by adding 100 μg of tyrosine in 100 μl of 0.05 *M* phosphate buffer. The mixture is loaded on a Sephadex G-50 column that was previously equilibrated with 0.05 *M* phosphate buffer, pH 7.4, containing 0.1% (w/v) gelatin, 0.15 *M* NaCl, and 0.02% (w/v) NaN_3. One-milliliter fractions are collected.

Radioimmunoassay of Metallothionein[17]

Anti-rat MT-1 rabbit serum is titrated in order to determine the adequate concentration of immunoglobulin to use for the radioimmunoassay. Metallothionein radioimmunoassay is carried out with a 10 × 75 mm glass tube, to which is added the following: 200 μl of standard diluent, 50 m*M* phosphate buffer containing 0.25% bovine serum albumin (BSA), 10 m*M* EDTA-2Na, and 0.01% NaN_3, pH 7.4, 100 μl of anti-MT rabbit serum (initial dilution factors at 20,000) diluted in standard diluent, standard or unknown sample, and ^{125}I-labeled-Tyr-MT (about 10,000 cpm). The mixture is incubated at 4° for 48 hr, and after this first incubation anti-rabbit

[17] N. Ikei, T. Kodaira, F. Shimizu, K. Nakajima, D. C. Tohyama, H. Saitoh, M. Kimura, and N. Otaki, *Rinsho Kensa* **33,** 215 (1989).

IgG goat serum in 100 μl of standard diluent, 100 μl of normal rabbit serum, each adequately diluted, and 200 μl of 12.5% polyethylene glycol are added to the tube. After a 30-min incubation at room temperature, the reaction mixture is centrifuged at 3000 rpm at 4° for 30 min. The supernatant is decanted and the radioactivity of this liquid and the precipitate are counted in a γ counter.

Tissue Extraction and Sample Dilution

Rat and human brains are homogenized with 5- to 10-fold 50 m*M* Tris-HCl (pH 8.5) and centrifuged at 40,000 *g* for 30 min at 4°. The supernatant is collected and heat treated in boiling water for 3 min, and chilled to 4°. The extract is recentrifuged at 20,000 *g* for 20 min at 4°, the supernatant is separated, and it is the starting material for the tissue dilution experiment.

Tissue-extracted samples are diluted adequately with standard diluent for radioimmunoassay.

Standard Metallothionein Purified from Rat, Mouse, and Other Species

Rat MT-1, rat MT-2, and mouse liver MT-2 are purified according to the method previously described for each.[15] Each standard solution of MT is stocked at 1 μg/ml concentration in standard diluent at −80°.

Preparation of Brain for Immunoperoxidase Staining

Human and rat brain specimens are fixed in 10% buffered formalin or in 0.1 *M* phosphate buffer (pH 7.4) containing 0.3% (v/v) picric acid and 4% (v/v) paraformaldehyde for 4–24 hr, respectively. After fixation, specimens are rinsed three times with ice-cold phosphate-buffered saline (pH 7.2) over a period of 3 days, dehydrated with a graded series of 30% ethanol to anhydrous ethanol, transferred to xylene, and embedded in paraffin. Tissue sections are cut at a thickness of 6 μm for immunoperoxidase staining.

Preparation of Antiserum for Immunoperoxidase Staining

Five milliliters of antiserum against rat MT-1 already described is diluted 400 times by PBS and incubated with 25 mg of *Ascaris* proteins for an overnight period at 4° to deplete the cross-reactivity of the MT-conjugated *Ascaris* protein. The supernatant is used for staining after centrifugation (3000 rpm, 10 min, 4°).

Preparation of Immunocytochemical Control Serum

The specificity of the rabbit anti-rat MT-1 antiserum is ascertained in two ways. First, the antibody is replaced with preimmune, normal rabbit serum, which results in negligible staining of peroxidase both in liver and kidney. Second, antiserum is preabsorbed with highly purified rabbit MT-1. An aliquot (2.0 ml) of 400-fold diluted rabbit anti-rat MT-1 antiserum is mixed with three different amounts (10, 100, and 1000 μg) of rabbit MT-1, followed by incubation at 4° for 48 hr. The serum is centrifuged at 10,000 *g* for 10 min using a Hitachi (Tokyo, Japan) centrifuge at 4° and the supernatant is used for immunohistochemical staining. The serum absorbed with 1 mg of rabbit liver MT-1 shows significant loss of specific staining in the MT-positive tissues.

Indirect Immunoperoxidase Procedure Specific for Metallothionein[18,19]

After deparaffinization, human and rat brain tissue sections are immersed in 0.1% (v/v) H_2O_2–methanol solution for 10 min to block the activity of endogenous peroxidase and are then washed in running water for 3 min, followed by a brief wash with PBS. The sections are incubated with normal goat serum for 30 min rinsed, and allowed to react with either rabbit anti-rat MT-1 antiserum (400-fold diluted) or normal rabbit serum in a moisturized box at 37° for 1 hr. After rinsing the sections with five changes of PBS for 3 min each, they are allowed to react with biotinylated labeled goat anti-rabbit IgG solution at 37° for 30 min. After rinsing with five changes of PBS (3 min each), the sections are reacted with avidin–biotin peroxidase (Vector Laboratories, Burlingame, CA) at 37° for 30 min. After rinsing with five changes of PBS (3 min each), a 3,3′-diaminobenzidine solution, which consists of 25 mg of 3,3′-diaminobenzidine, 65 mg of sodium azide, and 1 μl of hydrogen peroxide in 100 ml of 0.05 *M* Tris-HCl buffer (pH 7.6), is applied to the section. After the reaction, the sections are washed in running water and hematoxylin–eosin staining is used for nuclear staining of the sections.

Sensitivity of Immunoperoxidase Staining

The sensitivity of MT detection by immunoperoxidase staining regarding the amount of MT in specific tissues or cells is difficult to determine, because localization of MT within the tissues and cells is not uniform.

[18] C. Toyama, H. Nishimura, and N. Nishimura, *Acta Histochem. Cytochem.* **21,** 91 (1988).

[19] T. Umeyama, K. Saruki, K. Imai, H. Yamanaka, K. Suzuki, N. Ikei, T. Kodaira, K. Nakajima, H. Saitoh, and M. Kimura, *Prostate (N.Y.)* **10,** 257 (1987).

TABLE I
METALLOTHIONEIN CONCENTRATION IN TISSUES BY RADIOIMMUNOASSAY AND IMMUNOHISTOCHEMICAL STUDIES[a]

Species	Method	Blood vessel	Brain	Heart	Liver	Kidney	Prostate	Testis	Intestine
Rat	RIA (μg/g wet wt)	1.9 ± 0.4 (n = 10)	3.6 ± 0.4 (n = 10)	0.8 ± 0.2 (n = 10)	12.0 ± 4.6 (n = 15)	62.7 ± 10.8 (n = 15)	20.5 ± 5.2 (n = 10)	220 ± 40 (n = 9)	12.6 ± 4.8 (n = 18)
	Immuno-histochemical	—	—	—	+	++	++	—	+
Human	RIA (μg/g wet wt)	12.4 ± 5.6 (n = 10)	33.6 ± 10.2 (n = 6)		756 ± 460 (n = 4)	228 ± 59 (n = 16)	33.6 ± 7.9 (n = 15)		
	Immuno-histochemical	±	++	±	+++	+++	++	—	+

[a] Male Wistar rats (body wt 250–300 g) and randomly sampled human tissues were used for the studies.

However, it is necessary to determine if MT is actually not present or if the lack of sensitivity of this method yields negative results.

The tissue MT content was assayed by RIA as described above, using the supernatants of the extracted tissues in 0.05 m*M* Tris-HCl buffer, pH 8.6.

Immunohistochemical studies were done by the method described above.

These results (Table I) indicated when the MT staining of various tissues and cells appeared negative even though there existed a few micrograms of MT per gram of wet weight, except in the case of testis. Metallothionein in testis tissue cannot be stained by this method, but pretreatment of 0.1% trypsin in 0.05 *M* Tris-HCl buffer revealed MT staining to be strongly positive.

These results may be related to the differences in sensitivity of peroxidase activity detected by visible light and γ-ray counting. Tissues containing MT of over 10 μg/g wet weight were MT positive with immunoperoxidase staining.

Gel Filtration of Extract from Human Brain

One gram of tissue is homogenized with 9 ml of ice-cold 5 m*M* Tris-HCl buffer, pH 8.6, and centrifuged at 100,000 *g* for 30 min. The supernatant is used for Sephadex G-75 gel filtration (column 2.5 × 90 cm). A two-milliliter aliquot from the gel filtration is assayed for zinc content by the atomic absorption spectrophotometer and for MT by RIA. The elution patterns of zinc and MT from the human brain extract seem to be very similar to those from the human kidney extract (Fig. 1). These results suggest the presence of MT in brain.

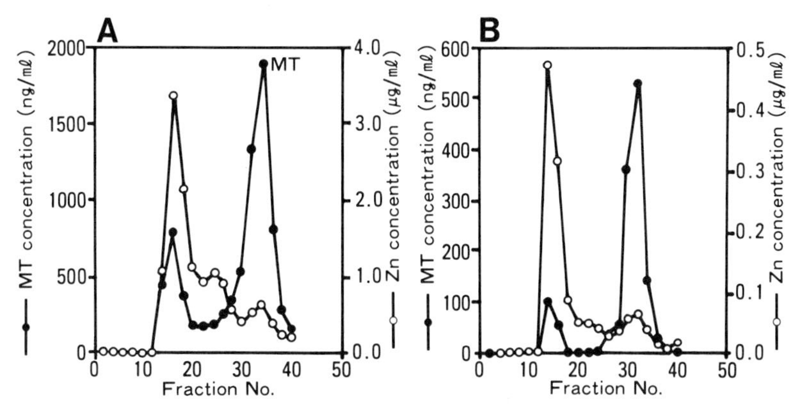

FIG. 1. Gel-filtration chromatogratography (Sephadex G-75) of metallothionein and zinc extracted from human kidney (A) and brain (B). Elution buffer, 5 m*M* Tris-HCl, pH 8.6.

Localization of Metallothionein in Brain

Zinc-binding proteins have been detected in rat brain,[11,20] bovine hippocampus,[21,22] and bovine cerebellum.[14] Ebadi *et al.*[11-13] detected three separate zinc-binding proteins in rat brains. The low-molecular-weight (6200) zinc-binding protein resembled some but not all aspects of hepatic MT, consisting of 60 amino acid residues with 12 cysteines and no histidine, arginine, leucine, tyrosine, or phenylalanine.

Using the same MT antibody, MT immunoreactivity in vertebrate brain was studied by RIA and immunoperoxidase staining. The MT content in rat brain was 3–5 μg/g wet weight and ependyma, choroid plexus, and blood vessels of rats showed MT-positive staining, but glial cells are negative using immunostaining. On the other hand, the cells of very strongly stained MT were observed in whole human brain, which represented an MT content of over 20 μg/g wet weight (Fig. 2). The positively stained cells were identified as brain glial cells (astrocytes). The increase in glial cell number also showed intense MT-positive staining, especially in protoplasmic and fibrillar astrocytes. MT-positive staining was observed in the nucleus, cytoplasm, and processes of these astrocytes. In five cases of Alzheimer's disease glial cell proliferation was observed. The increase in glial cell number was observed and showed intense MT-positive staining in brains of Alzheimer's patients.

The MT-positive staining of the human brain glial cell (astrocyte) was

[20] R. W. Chen and H. E. Ganther, *Environ. Physiol. Biochem.* **5,** 235 (1975).
[21] M. Ebadi, *Fed. Proc.* **43,** 3317 (1984).
[22] M. Karin and H. R. Herschman, *Eur. J. Biochem.* **107,** 395 (1980).

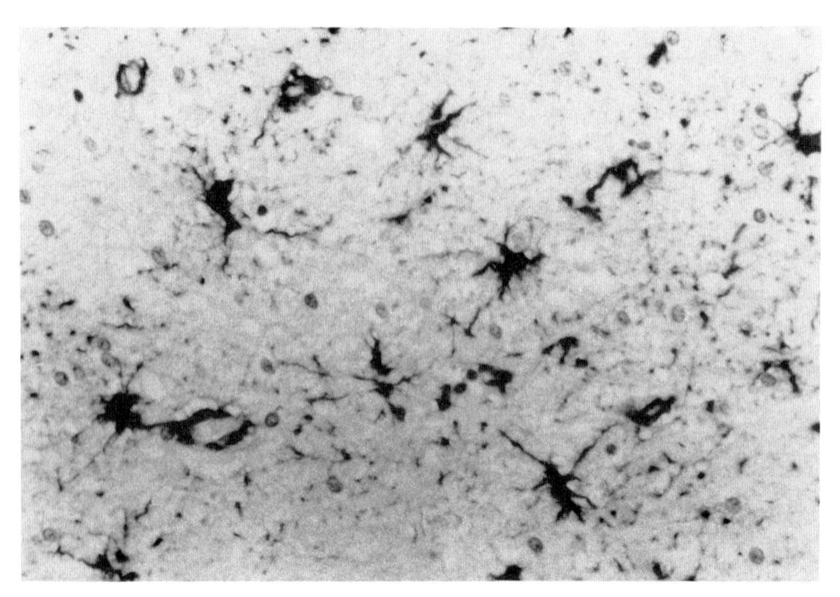

FIG. 2. Astrocytes of human brain show strongly positive immunostaining for MT. (ABC method.)

completely blocked by rabbit MT-1 purchased from Sigma (St. Louis, MO) by using the antiserum absorbed by MT-1 described as an immunocytochemical control serum. Metallothionein was also detected in vascular adventitia cells in human brain.

In cases of brain injury, such as Alzheimer's disease and apoplexy, glial cell proliferation was observed at the injured region by MT-positive staining of glial cells. The same glial cells of human brain specimens were stained positive with the antibody against glial fibrillar acidic protein, which is known as a typical glial cell protein. The results confirmed that the cells stained by MT antibody were indeed glial cells (astrocytes).

As there are many reports concerning zinc storage in the hippocampus,[23-25] the human hippocampus was stained with this MT antibody; it was found that only glial cells in the hippocampus showed MT positive as in other regions of human brains. In the case of rat brain, no MT-positive glial cells were observed with the immunoperoxidase method. The reason for this might be related to the sensitivity of the immunoperoxidase method.

[23] K. Fleischhauer and E. Horstmann, *Z. Zellforsch. Mikrosk. Anat.* **46,** 598 (1957).

[24] H. Maske, *Naturwissenschaften* **42,** 424 (1955).

[25] N. Otsuka and M. Kawamoto, *Histochemie* **6,** 267 (1967).

Conclusion

The amount of MT in human and rat tissues was measured by RIA and by using the same tissues simultaneously immunostained to detect the localization of MT. The tissues containing over 10 μg MT/g wet weight showed MT positive in immunoperoxidase staining.

The localization of MT in brain was first reported by Nakajima *et al.*,[26] using immunohistochemical studies of human and monkey brains. Metallothionein was localized at the glial cells (astrocytes) predominantly in the brain. The role of MT in the glial cells is unknown, but MT may play some important role for zinc metabolism and neurological homeostasis.

[26] K. Nakajima, M. Adachi, M. Kimura, K. Kobayashi, and K. Suzuki, *J. Trace Elem. Exp. Med.* **2,** 104 (1989).

Section V

Chemical Characterization of Metallothioneins

[45] Cysteine Modification of Metallothionein

By PETER E. HUNZIKER

Introduction

For amino acid analysis, peptide mapping, and sequence analysis a variety of methods for the modification of the cysteine residues of metallothionein (MT) have been used.[1-4] Because of the high number of cysteine residues in the protein each of the different methods is attended by a different number of problems. Performic acid oxidation, which quantitatively converts cysteine and cystine to cysteic acid, results in a highly negatively charged product that is unsuitable for peptide mapping by reversed-phase high-performance liquid chromatography (HPLC). Automated sequence analysis of this derivative is not possible because of the inability of many systems to identify phenylthiohydantoin (PTH)-cysteic acid. The oxidized MT is, however, the derivative of choice for amino acid analysis because of its simple production and the well-known characteristics of cysteic acid using ion-exchange chromatography.[5] The carboxymethyl derivative is suitable for amino acid analysis but problems arise during Edman degradation because of the tendency of N-terminally located carboxymethyl cysteine to undergo cyclization under the acidic conditions employed in the cleavage reaction.[6,7] The same difficulties in sequence analysis occur with carboxyamidated MT. In contrast to the carboxymethylated protein this derivative is suitable for peptide mapping by HPLC.[3,4] However, the occasionally observed partial desamidation of the carboxyamido cysteine makes the comparison of different isoforms difficult. Pyridylethylation of MT results in a derivative that has been used for primary structure determinations.[1,8] Because of its insolubility in aqueous buffers comparative studies between different isoforms by peptide mapping are difficult. In contrast, S-methylation of MT results in a product that is freely water soluble. This derivative of MT has been found to be most suitable for primary structure analysis of this unusual protein.

[1] M. M. Kissling and J. H. R. Kägi, FEBS *Lett.* **82,** 247 (1977).
[2] D. R. Winge, K. B. Nielson, R. D. Zeikus, and W. R. Gray, *J. Biol. Chem.* **259,** 11419 (1984).
[3] W. R. Bernhard, M. Vašák, and J. H. R. Kägi, *Biochemistry* **25,** 1975 (1986).
[4] K. Münger and K. Lerch, *Inorg. Chim. Acta* **151,** 11 (1988).
[5] Y. Kojima and P. E. Hunziker, this volume [48].
[6] D. G. Smyth and S. Utsumi, *Nature (London)* **216,** 332 (1967).
[7] A. F. Bradbury and D. G. Smyth, *Biochem. J.* **131,** 637 (1973).
[8] Y. Kojima, C. Berger, B. L. Vallee, and J. H. R. Kägi, *Proc. Natl. Acad. Sci. U.S.A.* **73,** 3413 (1976).

S-Methylation

The preparation of apo-MT is described elsewhere.[9] S-Methylation is carried out using a modified procedure of Heinrikson.[10] Dried apo-MT is dissolved in 500 μl of argon-purged buffer [6 *M* guanidine (Pierce, Rockford, IL; "sequenal grade"), 0.5 *M* Tris/HCl, 5 m*M* EDTA, pH 8.6, adjusted with HCl] containing 10 m*M* dithiothreitol (1.5 mg/ml). For complete reduction and S-methylation the protein concentration should not exceed 140 μ*M*. After incubation at room temperature under argon for 3–4 hr, 250 μl of 272 m*M* methyl-4-nitrobenzenesulfonate (Aldrich, Milwaukee, WI) in acetonitrile (59 mg/ml) is added. The sample tube is again purged with argon and incubated at 37° for 1–2 hr. The reaction is stopped by adding 10 μl of 2-mercaptoethanol. S-Methylated MT is recovered by desalting on a Sephadex G-25 column equilibrated with 0.1% (v/v) trifluoroacetic acid.

Discussion

S-Methylation is only rarely used in cysteine modification, because the methylated products tend to be insoluble in water.[10] This difficulty is not encountered with S-methylated apo-MT, which is freely soluble.

Under the conditions given the yield of modification is typically 98 ± 2%. In certain cases some precipitation of reagent is observed during the first 30 min of incubation at 37° due to the evaporation of acetonitrile. The precipitate is redissolved by adding one or two drops of acetonitrile. However, care should be taken not to use higher concentrations of the organic solvent, because the reaction mixture tends to be two-phasic when concentrations of more than 33% (v/v) of acetonitrile are used.

Since none of the problems encountered with other derivatives of this protein have been observed, S-methylation is a valuable tool for comparing different isoforms by peptide mapping as well as for the chemical characterization of MT.

[9] P. E. Hunziker, this volume [53].

[10] R. L. Heinrikson, *J. Biol. Chem.* **246**, 4090 (1971).

[46] Ligand Substitution and Sulfhydryl Reactivity of Metallothionein

By BY C. FRANK SHAW III, M. MERAL SAVAS, and DAVID H. PETERING

Introduction

Metallothionein (MT) is classified as a metal-binding protein. It also has a remarkable density of sulfhydryl groups, which may participate in its cellular chemistry through nucleophilic and redox reactions with electrophilic species.[1] The detailed examination of its reactions with metal ions, with reagents that compete for these metal ions, and with electrophilic reagents is needed to define the range of chemical reactivity that metallothionein may be able to display in cells.[1]

Ligand Substitution Reactions

Survey of Reactions

Metallothionein undergoes ligand substitution reactions as shown in reaction (1).[2-4] Competing ligands include apoproteins and polydentate ligands such as ethylenediaminetetraacetic acid (EDTA). In order for reactions to occur, the competing chelating agent must have a sufficiently large apparent stability constant under the conditions of reaction so that ligand exchange may take place. The log apparent stability constant per Zn for Zn_7-MT is 11.2 at pH 7.0 and 25°. Those for Cd_7-MT and the Cu(I) protein are orders of magnitude larger. Table I[5-7] lists a number of chelating agents that do react with MT-bound metals under particular conditions.

$$Zn_7\text{-MT} + 7L \rightleftharpoons 7ZnL + MT \quad (1)$$

[1] J. D. Otvos, D. H. Petering, and C. F. Shaw III, *Comments Inorg. Chem.* **9**, 1 (1989).

[2] T.-Y. Li, A. J. Kraker, C. F. Shaw III, and D. H. Petering, *Proc. Natl. Acad. Sci. U.S.A.* **77**, 6334 (1980).

[3] A. Udom and F. O. Brady, *Biochem. J.* **187**, 387 (1980).

[4] D. H. Petering, A. Pattanaik, P. Chen, S. Krezoski, and C. F. Shaw III, *in* "Metal Binding in Sulfur-Containing Proteins" (M. Stillman, K. T. Suzuki, and C. F. Shaw III, eds.). VCH Publishers, New York, in press.

[5] W. A. Nelson, M.S. Dissertation, University of Wisconsin—Milwaukee, Wisconsin (1986).

[6] G. Bachowski, M.S. Dissertation, University of Wisconsin—Milwaukee, Wisconsin (1984).

[7] M. M. Savas, D. H. Petering, and C. F. Shaw III, *Inorg. Chem.* (accepted for publication).

METHODS IN ENZYMOLOGY, VOL. 205

TABLE I
CHELATING AGENTS FOR LIGAND SUBSTITUTION REACTIONS WITH METALLOTHIONEINS

MT-bound metal	Ligand	Method of observation[a]	Number of kinetic steps	Ref.
Zn	Nitrilotriacetate	A	2	4
	N-(2-Hydroxyethyl)ethylenediamine triacetate	A	1	5
	Ethylenediaminetetraacetate	A	1	5
	trans-1,2-Diaminocyclohexane *N,N,N′N′*-tetraacetate	A	1	5
	Ethylene glycol-bis(β-aminoethyl ether) *N,N,N′,N′*- tetraacetate	A	1	5
	Diethylenetriaminepentaacetate	A	1	5
	N,N,N-2,2′,2″-Terpyridyl-3-ethoxy-2-oxobutyraldehyde bis(N^4-dimethylthiosemicarbazone) [$\epsilon = 12{,}800\ M^{-1}\ cm^{-1}$ (445 nm)]	B	2	6
	Pyridylazoresorcinol [$\epsilon = 43{,}300\ M^{-1}\ cm^{-1}$ (484 nm)]	B	2	7
Cd	Nitrilotriacetate	A	2	
	EDTA	A	1	2, 6
	3-Ethoxy-2-oxobutyraldehyde bis(N^4-dimethylthiosemicarbazone) [$\epsilon = 15{,}399\ M^{-1}\ cm^{-1}$ (434 nm)]	B	1	4, 6
Cu	2,9-Dimethyl-4,7-phenylsulfonate-1,10-phenanthroline [$\epsilon = 13{,}500\ M^{-1}\ cm^{-1}$ (480 nm)]	B	—	4

[a] Method A: change in intensity of metal–thiolate charge transfer band of metallothionein; Method B: formation of product metal complex with distinctive spectrum.

Methods of Observation of Ligand Substitution in Homogeneous Metal Metallothioneins

Reactions of metallothioneins with organic chelating agents take minutes to hours to complete, depending on the particular metal bound to metallothionein, the nature of the competing ligand, and the conditions of the reaction. Since the sulfhydryl groups in Cu-MT and apometallothionein are sensitive to oxygen over this time period, it is important to compare results of reactions carried out aerobically with ones done in the absence of oxygen. Similar data will demonstrate that the observed ligand substitution process is not being driven by prior oxidation of the thiol groups. In general, even Cu-MT is reasonably stable in air at 25° during the course of reaction.

Physical Separation

In some reactions the formation of product metal–ligand complex cannot be directly proven by the method of analysis of the reaction. In such cases, it is important to demonstrate that ligand exchange has occurred by gel-exclusion chromatography, which separates protein from metal complex.[2]

As an example of this procedure, 60 μM zinc as Zn_7-MT is mixed at time zero with a pseudo first-order excess of EDTA (1 mM) in 10 mM Tris-Cl and 100 M KCl at pH 7.5 and 25°. Over a 2-hr period, 0.5-ml aliquots are removed at intervals and immediately placed onto a 1×25 cm Sephadex G-25 column, equilibrated at room temperature with the same buffer. At a flow rate of 30 ml/hr and collecting 0.5-ml fractions, the column separation is complete within 15 min. Analysis of the metal content of the fractions collected will verify that the metals have, indeed, been extracted from metallothionein and transferred into a lower molecular weight complex with the competing ligand.

Charge-Transfer Bands of Metallothionein

Homogeneous Zn-, Cd-, and Cu-MT display metal–thiolate charge-transfer bands at 220, 254, and 284 nm, respectively. Apoprotein absorbs below 240 nm and has a molar absorptivity less than that for Zn-MT ($\Delta\epsilon = 12{,}300\ M^{-1}\text{cm}^{-1}$ at 220 nm). Loss in absorbance of these chromophores can, therefore, be used as an indicator of ligand substitution as long as the ultraviolet absorbance of the competing ligand or product complex does not obscure this change. Generally, the ligands studied by this method (designated A in Table I) do not absorb above 250 nm. To utilize them, spectrophotometric measurements on the reaction mixtures are made against blanks containing the same concentrations of all reactants, excluding the protein. In addition, it may be useful to monitor some of the reactions of Zn-MT at wavelengths above 220 nm, where the absorbance of the ligand and the protein are smaller, thereby increasing the signal-to-noise ratio of the spectral measurements. In some cases the absorbance of the competing ligand decreases more rapidly than Zn-MT, affording some gain in sensitivity for observations of changes in metal–protein ligation.

TABLE II
MOLAR DIFFERENCE ABSORPTIVITY OF Zn_7-MT VS APOMETALLOTHIONEIN[a]

Wavelength (nm)	ϵ (M^{-1} cm^{-1})[b]
220	12,000
224	11,400
228	10,000
232	8,000
236	5,900
240	4,100

[a] pH 7.0, 50 mM potassium phosphate buffer.
[b] Values are per zinc.

Depending on the ligand, the concentrations of reactants, and other conditions of reaction, particular reactions may approach equilibrium and not go to completion. To determine if this occurs, one can use the equilibrium absorbance at 254 or 284 nm to determine whether some Cd- or Cu-MT remains in the reaction mixture. However, with Zn-MT it is necessary to use a difference molar absorptivity for Zn_7-MT and apo-MT at the wavelength of interest to calculate the extent of reaction (Table II). An alternative is to add diethylenetriaminepentaacetic acid (DTPA) to the equilibrium mixture (Fig. 1). It reacts irreversibly with the remainder of the MT-bound zinc. The loss in absorbance in the presence of the competing ligand before addition of DTPA represents the metal that is lost in the

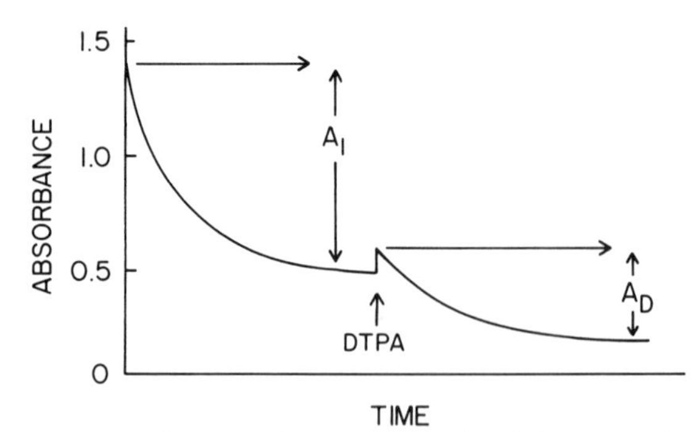

FIG. 1. Use of DTPA (diethylenetriaminepentaacetic acid) to determine the equilibrium saturation of Zn_7-MT with zinc. A_I is the absorbance change induced when a ligand, L (e.g., EGTA), partially removes MT-bound Zn ions in an equilibrium process; A_D is the absorbance change after excess DTPA completely removes the residual Zn ions from MT.

establishment of equilibrium (ΔA_L). The change in absorbance at 220 nm (ΔA_D) after addition of DTPA is directly proportional to the zinc present in MT at equilibrium. Then, the fraction of the initial zinc in Zn-MT remaining in the protein at equilibrium is equal to $\Delta A_D/(\Delta A_D + \Delta A_L)$.

A typical experiment is the following: 15 mM EGTA is added to Zn_7-MT, which is 23 μM in zinc, in 0.1 M KCl and 50 mM phosphate buffer at pH 7.1 and 22°. After equilibrium is attained as measured by the lack of change in absorbance at 220 nm, 0.2 mM DTPA is added in a negligible volume and further changes in the absorbance are observed as the removal of zinc from the protein is completed.

Competing Ligands that Form Chromophoric Metal Complexes

Some ligands react with Zn(II), Cd(II), or Cu(I) to form spectrophotometrically distinct metal complexes that absorb outside of the spectral range of the corresponding metallothioneins. When these chelating agents are used, the metal transfer can be observed directly and the extent of ligand substitution determined by use of Beer's law and the molar absorptivity constants listed in Table I. This method is designated B in Table I.

Activation of Enzyme Activity

For metal ion transfer from metallothionein to apometalloproteins it is important both to measure actual exchange of the metal and to monitor the reconstitution of holoprotein activity in order to specify whether exchange does lead to regain in function.[2,8] An illustration of the latter has 0.1 μM apocarbonate dehydratase reacting with 0.1 μM zinc in Zn-MT in 0.2 M phosphate buffer at 25° and pH 7.1.[2] Over the course of 30 min 0.1-ml aliquots of the reaction mixture are periodically taken and added to 2.9 ml of an assay solution for Zn–carbonate dehydratase. The dilution effectively stops the bimolecular ligand substitution process. Then the linear rate of hydrolysis of p-nitrophenylacetate is compared to a standard curve of rates for known concentrations of carbonate dehydratase (ϵ = 57,000 $M^{-1}cm^{-1}$ at 280 nm) to define the concentration of active enzyme at each time point.

The assay mixture for carbonate dehydratase is made by dissolving 0.04 g of p-nitrophenylacetate in 3 ml of reagent grade acetone and adding this solution to 97 ml of 0.20 M phosphate buffer, pH 7.1. The assay measures hydrolysis of the ester at 25° by following the rate of formation of p-nitrophenylate anion at 412 nm under saturation conditions in which

[8] A. J. Kraker, G. Krakower, C. F. Shaw III, D. H. Petering, and J. S. Garvey, *Cancer Res.* **48**, 3381 (1988).

TABLE III
SELECTIVITY OF LIGANDS IN LIGAND SUBSTITUTION REACTIONS

Ligand	Selectivity	
	Thermodynamic	Kinetic
Nitrilotriacetate	Zn $\gg$ Cd, Cu(I)	Zn $\approx$ Cd
EDTA	Zn $\approx$ Cd	Zn $\gg$ Cd
2,9-Dimethyl-4,7-phenylsulfonyl-1,10-phenanthroline (BCS)	Cu(I) > Zn, Cd	Cu(I) > Zn, Cd

the rate of reaction is proportional to the concentration of active enzyme present.

Ligand Substitution with Mixed Metal Metallothioneins

In principle, one can use the above methods to examine the ligand substitution reactions of metallothioneins containing more than one type of metal ion. Since metal-binding ligands are not absolutely specific for one metal ion, it is necessary to determine whether the agent in use reacts with one or more of the metals bound to MT in order to interpret the ligand-exchange results.

Major thermodynamic and kinetic differences do exist in the reactions of ligands with the various metal ions that bind to metallothionein. These distinctions can be used to achieve selective reactions (Table III). For example, in the reaction of Zn_4,Cu_6-MT from bovine calf with competing ligands, nitrilotriacetate reacts exclusively with zinc.[4]

Kinetic Analysis of Reaction Rates

To assess rates of reactions of metallothionein described in this chapter, it is possible to use any of the methods outlined above because the reactions are sufficiently slow, occurring over periods of minutes to hours. Although there are multiple metal ions bound to metallothionein via 20 sulfhydryl groups, reactions of its bound metals or thiols generally occur with simple kinetics. For example, under pseudo first-order conditions for the competing reagents, nitrilotriacetate, 3-ethoxybutyraldehyde bis(N^4-dimethylthiosemicarbazone), or 5,5′-dithiobis(2-nitrobenzoic acid) react with Zn_7-MT with two first-order rate constants; EDTA reacts with one pseudo first-order rate constant (Table I). To explain this kinetic behavior, one can argue that the reactive sites are independent of one another but react at one or two rates characteristic of all of the sites. Alternatively, because of the grouping of metals and thiolate ligands in metallothionein into two structurally independent domains, it is possible that the simple kinetics are due to the cooperative reaction of the two metal clusters with competing ligands or sulfhydryl reagents. In this model the first site of attack in a cluster is rate limiting, opening the rest of the structure to rapid reaction.

Cluster Reactivity

To test whether biphasic kinetics are due to the differential reactivity of the two clusters in metallothionein, the reactions of isolated α domains, Cd_4-α or Zn_4-α, can be examined and compared with the properties of similar reactions of the holometallothionein structure.[7] The isolated cadmium domain can be prepared as follows, based on the methods of Winge and Miklossy and Stillman and Zelazowski.[9,10] Zn_7-MT-2 (1 – 1.5 m*M* in zinc) from rabbit liver is reacted under argon with 4.5 Eq of $CdCl_2$, added in three successive aliquots. After 30 min of incubation at room temperature, a 10-fold excess of EDTA is added and the mixture is incubated for another hour. The chelating agent specifically removes zinc from the mixed metal metallothionein. Then, subtilisin (20 : 1, protein : enzyme) is used to hydrolyze apoprotein for 18 hr at room temperature under argon. To isolate the α cluster resulting from this treatment from smaller fragments of protein and enzyme, the mixture is chromatographed over Sephadex G-50 (1.5×120 cm) equilibrated and eluted with 10 m*M* NH_4HCO_3 buffer.

Sulfhydryl Reactions

Survey of Reactions

The sulfhydryl reagents that react with the sulfhydryl groups of MT, leading to the displacement of the bound metals, may be broadly classified into two groups, organic electrophiles and metals. Table IV lists the reagents whose reactions have been examined kinetically. It does not list reagents studied in cell culture or qualitatively *in vitro*. Although the four reagents examined in detail each show two kinetically resolvable steps, the mechanistic details vary considerably.

[9] D. R. Winge and K.A. Miklossy, *J. Biol. Chem.* **257,** 3471 (1982).
[10] M. J. Stillman and A. J. Zelazowski, *Biochem. J.* **262,** 181 (1989).

TABLE IV
SULFHYDRYL REAGENTS FOR *in Vitro* STUDIES WITH METALLOTHIONEIN

Reagent	Method of observation[a]	Kinetic steps[b]
5,5′-Dithiobis(2-nitrobenzoic acid)	B	2
N-Ethylmaleimide	B, C	2
Gold sodium thiomalate	C	2
cis-Platin [$(NH_3)_2PtCl_2$]	A, C	2

[a] Method A: change in intensity of metal–thiolate charge transfer band of MT; method B: formation of a product or loss or reactant with distinctive spectrum; method C: use of a metallochromic reagent to monitor displaced metal ions A.

[b] Biphasic reactions may result from two independent and parallel kinetic steps or from sequential steps such as an induction phase followed by a second reaction step. The slope of the resulting $\ln(A_\infty - A_t)$ vs time plot distinguishes the two cases.

Ellman's Reagent (ESSE)

Ellman's reagent (5,5′-dithiobis(2-nitrobenzoic acid) is widely used to measure the reactivity of protein sulfhydryl groups.[11] The aromatic disulfide is readily cleaved by thiols (RSH) to yield the chromophore 5-thio-2-nitrobenzoate (ES^-), which has an intense yellow-orange color with an absorbance maximum at 412 nm:

$$\mathrm{ESSE + RSH \rightarrow RSSE + ES^- + H^+} \tag{2}$$

which can be monitored directly without interference from any metallothionein species. The reaction has been applied to the analysis of the reactivity of metal-bound thiolate groups of MT. It was possible to subject this reaction to careful kinetic study, since it is slower than that with free sulfhydryl groups and unhindered (e.g., surface) cysteine residues of proteins.[12] In addition, it serves as a model for the reactions of medicinal alkylating reagents and reactive oxygen species with MT *in vivo.*

The reagent itself is subject to base-catalyzed cleavage, yielding the sulfenic acid ESOH and ES^-; this places limitations on its use at pH values above approximately pH 8.[13] While a kinetic control at the same pH has been used for studying the reactions of albumin with ESSE,[13] the effect of Cd^{2+} released from MT is to accelerate (catalyze) the base hydrolysis.

[11] G. L. Ellman, *Arch. Biochem. Biophys.* **82,** 70 (1959).

[12] T.-Y. Li, D. T. Minkel, C. F. Shaw III, and D. H. Petering, *Biochem. J.* **193,** 441 (1981).

[13] J. M. Wilson, D. Wu, R. Motiu-DeGrood, and D. J. Hupe, *J. Am. Chem. Soc.* **102,** 359 (1980).

Thus, the rate of the MT reaction at high pH is the difference between two large rate measurements.[14] ESSE can be used successfully in a variety of buffers, including Tris, phosphate, and 4-morpholinopropanesulfonic acid (MOPS).

The following stock solutions have been typically used to study reaction 2: buffer, 5 mM Tris-HCl, pH 7.4, containing 100 mM KCl; ESSE, a 10 mM stock solution in Tris (this is best prepared by making a concentrated solution in 0.10 M phosphate buffer, pH 7.0, which is then diluted with Tris to the desired concentration); Zn_7-MT, 9×10^{-4} M in zinc, in the Tris buffer; Cd_7-MT, 4×10^{-5} M in cadmium, in the Tris buffer. By adding appropriate aliquots of the three stock solutions, the following concentrations of reactants are obtained in a 0.5-ml cuvette: MT, 20 μM cadmium or zinc = 57 μM thiolate; ESSE, 0.5 to 6.0 mM. A blank containing ESSE at the same concentration in buffer is prepared in advance.

Immediately after adding the protein to the reaction mixture, the cuvette is placed into a spectrophotometer set to record the absorbance change at 412 nm as a function of time. The absorbance rises to a maximum over a period of 15–45 min, depending on the ESSE concentration. If the reaction is monitored for longer periods, a slow loss of the 412 absorbance may be noted due to the aerobic reoxidation of ES^- to ESSE. The resulting plot of A_{412} versus time can be further analyzed by plotting it as $\ln(A_{inf} - A_t)$ vs time, which reveals the biphasic kinetics of the reactions. Two rate constants, k_f and k_s, can be calculated as described in the section, Kinetic Analysis of Reaction Rates.

N-Ethylmaleimide

N-Ethylmaleimide (NEM) is another commonly used thiol reagent that can be used to examine the reactivity of the metal-bound thiolates of MT.[15-17] The structure of the reactive center in this case is an activated double bond to which a thiol or thiolate can add.

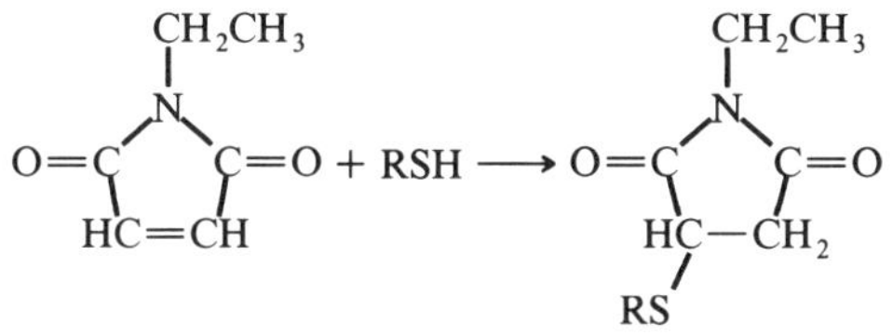

$$\text{NEM: } O{=}C\,(N{-}CH_2CH_3)\,C{=}O \text{ ring with } HC{=}CH \; + \; RSH \longrightarrow O{=}C\,(N{-}CH_2CH_3)\,C{=}O \text{ ring with } (RS)HC{-}CH_2 \qquad (3)$$

[14] M. M. Savas, C. F. Shaw III, and D. H. Petering, unpublished observations (1990).
[15] D. G. Smyth, A. Nagamatsu, and J. S. Fruton, *J. Am. Chem. Soc.* **82,** 4600 (1960).
[16] J. F. Riordan and B. F. Vallee, this series, Vol. 25, p. 449.
[17] P. C. Jocelyn, this series, Vol. 143, p. 44.

In the case of MT the reaction leads to the displacement of metals from protein because the NEM–thiol adduct is a thioether, which has a much lower metal-binding affinity than the original thiolate. The complete alkylation can be described as

$$M_7\text{-MT} + 20\text{NEM} \rightarrow 7M^{2+} + (\text{NEM-S})_{20}\text{-MT} \quad (4)$$

Reactions can be followed by the loss of the 302-nm absorbance band ($\epsilon_{302} = 593\ cm^{-1}M^{-1}$) of NEM, or as described below, by use of a metallochromic dye that chelates the displaced metal ions.

A surprising feature of the reaction is that it is apparently an equilibrium process. NEM reactions with cysteine and cysteine-containing peptides and proteins are known, from the interconversion of the diastereomeric adducts of cysteine and the loss of NEM label from red blood cells,[18] to be reversible.[19] These observations explain the fact that excess NEM is required to drive MT-NEM reaction (4) to completion.

NEM is less specific for thiols than is ESSE and some care must be exercised in the choice of reaction conditions. N-Terminal amino groups and to a lesser extent amino side chains such as lysine may also react with NEM.[15,20] The pH must be kept in the neutral to acidic range because at higher pH values base hydrolysis occurs at rates comparable to the thiolate reaction. In Tris buffer the NEM chromophore is lost at a rate comparable to the rate of its reactions with MT.[21] MOPS has proven to be a satisfactory buffer for NEM-MT reactions.[21]

To investigate reaction (4), the following typical stock solutions have been employed: NEM, 10 m*M* in MOPS buffer (which must be prepared freshly to avoid partial hydrolysis during storage); M_7-MT, $1-5 \times 10^{-4}$ in MOPS buffer; MOPS buffer, 10 m*M*, pH 7.4, containing 100 m*M* KCl.

To initiate the reaction appropriate amounts of reagents and buffer are added to a cuvette to give the following final concentrations of reactants: MT, 20 μM zinc = 57 μM thiolate; NEM, 200 μM–4 m*M*. The reactions are observed at 300 nm. The absorbance decreases to a minimum over 24 hr. A blank containing only NEM is monitored simultaneously and later subtracted from the reaction data prior to analysis.

[18] E. Buetler, S. K. Srivastava, and C. West, *Biochem. Biophys. Res. Commun.* **38,** 341 (1970).

[19] D. G. Smyth, O. O. Blumenfeld, and W. Konigsberg, *Biochem. J.* **91,** 589 (1964).

[20] C. F. Brewer and J. P. Riehm, *Anal. Biochem.* **18,** 248 (1967).

[21] L. He, M.S. Thesis, University of Wisconsin, Milwaukee, Wisconsin (1990).

Reactions with cis-Platin [$(H_3N)_2PtCl_2$]

The antitumor drug *cis*-platin [diamminodichloroplatinum(II)] reacts with MT *in vivo* and *in vitro*.[22] Studies of its reactions with biological binding sites are complicated by the reversible hydrolysis of the chloride ions to form the acidic aquo complexes, which undergo further equilibrium dissociation.[23]

$$\begin{array}{ccccc} (H_3N)_2PtCl_2 & \rightleftharpoons & (H_3N)_2Pt(OH_2)Cl & \rightleftharpoons & (H_3N)_2Pt(OH_2)_2 \\ & & -H^+ \Big\Downarrow\Big\Uparrow +H^+ & & -H^+ \Big\Downarrow\Big\Uparrow +H^+ \\ & & (H_3N)_2Pt(OH)Cl & & (H_3N)_2Pt(OH)(OH_2) \end{array} \quad (5)$$

The reactivity of these and additional complexes that can form *in situ* vary significantly. As a result, the pH and chloride content of the reaction must be carefully defined and controlled, if one is to obtain meaningful and reproducible results from kinetic studies.[24] The chloride/aquo equilibration is slow and 48 hr should be allowed for platinum stock solutions to equilibrate. The reaction conditions below illustrate one of many possible sets of conditions that may be interesting to study; in this case the high chloride ion content is designed to push the equilibria to the left, maintaining all of the platinum in the dichloro form. The reactions can be followed by the development of the Pt–S chromophore at 284 nm.

Typical stock solutions for these reactions are as follow: buffer, 50 m*M* phosphate and 500 m*M* NaCl, pH 7.4; 5 m*M* *cis*-dichlorodiamme platinum(II) (*cis*-DDP) in this buffer; and Zn_7-MT, 50–100 μM in zinc in the same buffer. Actual reaction concentrations are $[Zn]_{MT} = 20\ \mu M$ and platinum = 100 μM–5 m*M*. The reaction is initiated by adding *cis*-DDP to a solution of MT. The control solution should contain the protein and buffer. Reactions are monitored at 284 nm for at least two half-lives, then set aside for 24–72 hr for a final measurement of A_∞. These reactions are biphasic with an initial induction period before significant absorbance changes occur at 284 nm.

Metallochromic Dye Reagents for Monitoring Sulfhydryl Reactions

The reaction of the metal-bound sulfhydryl groups with various alkylating agents and certain metal ions can be usefully probed using metal-

[22] A. J. Kraker, J. Schmidt, S. Krezoski, and D. H. Petering, *Biochem. Biophys. Res. Commun.* **130,** 786 (1985).

[23] R. B. Martin, *ACS Symp. Ser.* **209,** 231 (1983).

[24] A. Pattanaik, Ph.D. Thesis, University of Wisconsin, Milwaukee, Wisconsin (1990).

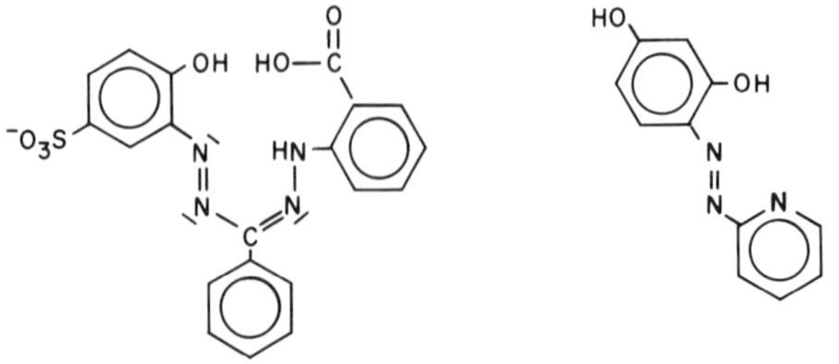

ZINCON PYRIDYLAZORESORCINOL

FIG. 2. Metallochromic dyes: zincon and pyridylazoresorcinol.

lochromic dyes such as pyridylazoresorcinol or zincon, shown in Fig. 2.[21,25] As the sulfhydryls are "modified" by such reagents, their metal-binding capacities are decreased or destroyed, leading to release of the thiolate-bound metal ions. The subsequent reaction of the metal ions, usually Zn^{2+} or Cd^{2+}, with the dye can be easily monitored at a visible wavelength characteristic of the metal–dye complex and usually with much greater sensitivity than is available by monitoring the 254- or 215-nm metal–thiolate charge-transfer bands.

When the conditions listed below are met, the rate of formation of the complex ML_n^{2+} will be the same as the rate of displacement of M^{2+} from MT:

$$X + M_7\text{-MT} \xrightarrow{\text{slow}} X_{20}\text{-MT} + 7M^{2+} \tag{6}$$

$$M^{2+} + nL \xrightarrow{\text{fast}} ML_n^{2+} \tag{7}$$

To be a suitable monitor for Zn^{2+} or Cd^{2+} displacement, a metallochromic dye must (1) be thermodynamically incapable of extracting the MT-bound metal ions or do so at a rate at least one to two orders of magnitude slower than the reaction to be studied, (2) complex with the displaced ions more rapidly than the sulfhydryl reaction occurs (complexation is usually quite rapid, occurring within the mixing time for conventional kinetic studies), (3) not react directly with the alkylating agent or other metal ion, and (4) must not participate in forming the activated complex for the reaction.

The following apparent binding constants at pH 7.4 have been calculated from literature data[25]:

$$Zn^{2+} + 2PAR \rightleftharpoons Zn(PAR)_2 \qquad \log K_{app} = 14.3 \text{ or } 12.6 \tag{8}$$

[25] C. F. Shaw III, J. E. Laib, M. M. Savas, and D. H. Petering, *Inorg. Chem.* **29**, 403 (1990).

$$Cd^{2+} + 2PAR \rightleftharpoons Cd(PAR)_2 \qquad \log K_{app} = 10.6 \tag{9}$$

$$Zn^{2+} + ZI \rightleftharpoons Zn(ZI) \qquad \log K_{app} = 4.9 \tag{10}$$

where PAR is pyridylazoresorcinol and ZI is zincon. From thermodynamic considerations, ZI is unable to extract Zn^{2+} and PAR is unable to extract Cd^{2+} from MT in the absence of other reagents. Thus, ZI is the reagent of choice for studying reactions in which Zn^{2+} is displaced and PAR for reactions in which Cd^{2+} is displaced. PAR reacts to remove one Zn^{2+} from Zn_7-MT with a rate constant, $k_f = 2.1 \times 10^{-3}$ sec^{-1}, and the remainder more slowly at a rate dependent on the PAR concentration, $k_{1s} + k_{2s}$ [PAR], where $k_{1s} = 8.5 \times 10^{-6}$ and $k_{2s} = 3.8 \times 10^{-2}$. As a result. PAR should be used to study reactions of zinc-containing thioneins only if they are considerably more rapid than the direct PAR-MT reaction.

Gold Sodium Thiomalate Reactions Monitored with Zincon and Pyridylazoresorcinol

The metallochromic dye concept was developed as method to follow the reactions of MT with gold sodium thiomalate (AuSTm), a gold-based antiarthritic drug.[26] AuSTm is a water-soluble oligomer in which the thiomalate ligand forms thiolate bridges between adjacent Au(I) ions. Since this agent also contains metal–sulfur linkages, the changes in UV spectra of the reaction mixture are expected to be small. In contrast, monitoring the formation of metal complexes of ZI and PAR produces reproducibly large absorbance changes that are easily monitored. AuSTm-MT reactions are complicated by the formation of different products, depending on whether gold or MT is the limiting reagent.[26] When gold is in excess ($Au/M^{2+} > 20$) it binds to a single MT cysteine residue retaining thiomalate to complete its usual two-coordinate structure; when MT is in excess ($Au/M^{2+} < 1$), gold binds to two cysteines and loses the thiomalate ligand. The metallochromic dyes can be used in either case, although the slow extraction of Zn^{2+} by PAR is a possible complication if it is used to probe the reactions of zinc-containing MTs.[25]

For these reactions, stock solutions are as follow: Tris buffer, 5 mM Tris-HCl containing 100 mM $NaClO_4$; PAR or ZI, 500 μM in Tris buffer; AuSTm, 500 μM in Tris buffer; M_7-MT, 20–50 μM in metal in Tris buffer. By adding appropriate aliquots of the reagents and buffer to a cuvette, the following final concentrations of reactants are obtained: PAR or ZI, 20 μM; AuSTm, 2 μM–1 mM; $[M^{2+}]_{MT} = 2$ μM. At these concentrations, ZI does not react with Cd^{2+} and, therefore, monitors exclusively

[26] J. E. Laib, D. H. Petering, C. F. Shaw III, M. K. Eidsness, R. C. Elder, and J. S. Garvey, *Biochemistry* **24**, 1977 (1985).

the displacement of Zn^{2+}. PAR reacts with both zinc and cadmium ions as they are displaced.

The reactions of metallothionein with AuSTm are rather rapid and biphasic.[26] Each phase is first order in protein and zero order in AuSTm over several orders of magnitude of gold concentrations. The most surprising finding is that they are independent of whether the MT-bound metal ions are Cd^{2+}, Zn^{2+}, or even traces of Cu^{+}.

Reactions with N-Ethylmaleimide Monitored with Zincon

The reactions of NEM with various M-MTs [metallothioneins with a particular metal (M) bound and stoichiometry unspecified] can also be monitored using metallochromic dyes.[21] The spectral properties of ZI and PAR are suitable for use in the presence of NEM. The slow extraction of Zn^{2+} by PAR, however, occurs at a rate comparable to the reaction of NEM with MT, thus precluding its use with Zn_7-MT, but not Cd_7-MT. ZI is the dye of choice for monitoring Zn_7-MT reactions with MT.

Reactions in three buffers (Tris, phosphate, and MOPS) were investigated. Phosphate interferes with development of the Zn–ZI chromophore; Tris accelerates the hydrolysis of NEM. The use of MOPS minimizes these interferences. The pH range is also limited at the low end (pH 5–6) by dissociation of Zn^{2+} from MT and at the upper end (pH 8 or above) by base hydrolysis of NEM.

Stock solutions for these reactions include NEM, 10 m*M* in MOPS buffer and prepared freshly to avoid partial hydrolysis during storage; ZI, 1 m*M* in MOPS buffer; Zn_7-MT, 115 μM in Zn^{2+} in MOPS buffer; MOPS, 10 m*M* MOPS and 150 m*M* KCl, pH 7.4. For the reaction, the NEM and ZI are typically present in pseudo first-order excess with Zn_7-MT (or other M-MTs) as the limiting reagent. Typical concentrations are $[Zn]_{MT} = 20\ \mu M$; ZI = 500 μM; NEM = 50–5000 μM. The blank should contain ZI, NEM, and buffer at the same concentrations. ZI, Zn_7-MT, and MOPS buffer are added to a cuvette. The final reagent to be added, which initiates the reaction, is the NEM.

The reaction is followed at the wavelength of the Zn–ZI complex, 620 nm. The reactions require a time of up to 2 hr to go to completion or equilibrium, depending on the NEM concentration. The plots of A_{620} vs time show a well-defined induction phase characteristic of a two-step mechanism. Plots of $\ln(A_\infty - A_t)$ vs time yield a biphasic curve characteristic of such a mechanism. One cannot assign the first and second steps to the fast or slow components, unless there is an independent method to measure the concentration of the intermediate or to measure the rate of the first step alone.

[47] Determination of Metals in Metallothionein Preparations by Atomic Absorption Spectroscopy

By MARTA CZUPRYN and KENNETH H. FALCHUK

Introduction

The detection and quantitation of metals in metallothionein are prerequisites to all subsequent experimental work with this protein. Zinc, copper, and cadmium are the metals most frequently associated with metallothionein, although biosynthesis of the protein can be induced by a wide variety of other d^{10} metal ions.[1,2] The measurement of these metals in metallothionein preparations in various stages of purification can be carried out using either flame (FAAS) or more sensitive electrothermal atomic absorption spectrometry (EAAS). To decide which of these methods is more suitable for a given determination it is necessary to consider such factors as the expected concentration of the metal, the amount of sample available, and the nature of the matrix. The FAAS technique is less susceptible to various interference effects, but requires considerably larger volumes and is ~100-fold less sensitive than EAAS. Usually cells and tissues from which metallothionein is isolated are available in sufficient quantities for metal analysis by FAAS. However, during later steps of metallothionein purification, sample volume and metal concentration can become limiting factors, making EAAS the method of choice. Although there are a number of potential inteference effects in metal analysis by EAAS, most of them can be avoided by appropriate sample preparation and chemical modifications of the sample matrix.

Chemical and Matrix Interferences

The EAAS technique is highly sensitive to chemical and matrix interference effects. Chemical interference occurs when the metal ion of interest combines with some other component(s) of the sample matrix to form a compound of high boiling point and low volatility. As a result, the number of neutral metallic atoms generated during atomization under standard conditions is reduced and there is a decrease in signal peak height.

Matrix interferences arise when the physical and chemical characteristics of the sample differ considerably from those of the reference solution, causing a change in the metal atomization rate. This type of interference is most commonly observed in samples containing large amounts of organic

[1] J. H. R. Kägi and A. Schäffer, *Biochemistry* **27,** 8509 (1988).
[2] M. Webb, *Experientia, Suppl.* **52,** 109 (1987).

METHODS IN ENZYMOLOGY, VOL. 205

materials and high concentrations of salts or acid. It can result in either a decrease or an increase in the atomization signal. In many cases, the interference effects can be eliminated by matrix modification.

Matrix Modification Techniques

Chemical modifications of the matrix include techniques that require pretreatment of the sample before it is dispensed into the furnace, and those that modify the sample directly in the furnace of the spectrometer.

Pretreatment of Sample before Analysis

Many of the samples containing metallothionein, for which metal content is to be determined, require extensive pretreatment due to the nature of the sample (solid tissues, intact cells, etc.), the complexity of the matrix (large amounts of organic material), or both. Such pretreatment involves sample oxidation by dry ashing or acid digestion, procedures that destroy organic matrices and thus minimize background absorption. If necessary, analyzed metals can then be separated from the bulk matrix components by classical separation techniques such as ion-exchange and solvent extraction.[3-5] These pretreatments, however, may cause sample contamination and loss of volatile elements and, if possible, should be avoided and matrix interferences evaluated by the method of standard additions.

Before either procedure can be performed, the sample must be freed of any external metal contaminants. Mammalian tissues and tissue culture cells should be rinsed with a metal-free isotonic salt solution. Plant cells, whose cell wall is sturdy, can be rinsed with metal-free water.

Dry Ashing. An aliquot of the material (usually 0.7 – 1 g) is placed in an acid-cleaned platinum crucible and dried to a constant weight in an electric oven at 106° (16 – 24 hr). The crucible is then transferred to a muffle furnace and heated at 500° for 12 – 24 hr until the sample is reduced to a white ash. The ash is taken up in a minimal amount of concentrated metal-free HCl and transferred to a suitably sized volumetric flask (usually 5 – 10 ml). The crucible is rinsed several times with 1% (v/v) metal-free HCl, and the rinses are quantitatively transferred to the volumetric flask. The sample is then brought to volume with H_2O or dilute metal-free acid. In general, the dry ashing procedure is simple and has a low probability of metal contamination.

[3] L. Murthy, E. E. Menden, P. M. Eller, and H. G. Petering, *Anal. Chem.* **53,** 365 (1973).
[4] F. W. Sunderman, Jr., *Hum. Pathol.* **4,** 549 (1973).
[5] R. Shapiro and M. T. Martin, this series, Vol. 158, p. 344.

Digestion. Wet digestion in the presence of oxidizing agents, such as concentrated H_2SO_4, HNO_3, $HClO_4$, or their combination constitutes another means of decomposing the organic components of biological samples. It is usually faster than dry ashing. Moreover, volatilization losses of metals are minimized because of the lower temperatures that are employed (100–200°).

The following procedure can be used for digestion of both animal and plant material.

Place the sample (~ 1 g) in a 50-ml acid-cleaned Erlenmeyer flask, add 15 ml nitric acid and 2.5 ml 72% (w/v) perchloric acid, and heat on a hot plate. After white perchloric acid fumes appear, swirl the flask occasionally, and continue heating at 180° to decrease the volume to 3–5 ml. Do not allow to dry. Dilute the resultant solution with 10–15 ml of water.

Modifications of Sample Matrix

In many cases chemical and matrix interferences can be overcome by modifying the matrix directly in the furnace, before atomization occurs.

For each type of sample (e.g., chromatographic fractions containing metallothionein) the extent of matrix interference must be established prior to analysis. This can be done by evaluating the effect of sample on the absorption signal of the standard (method of standard additions). Briefly, standard curves are generated in the presence and absence of the sample. The presence of interferences is indicated by obtaining nonparallel lines. These interfering effects frequently can be eliminated by sample dilution (provided the metal concentration is sufficiently high) or by chemical additions and change of atomization conditions.

Chemical modification of the sample facilitates removal of the matrix prior to atomization, either by reducing the volatility of the analyzed element or by increasing the volatility of the matrix. It can often be performed directly in the furnace of the atomic absorption spectrometer.

Modifications of the sample that reduce volatility of the analyte are required for elements more volatile than the matrix, such as cadmium and mercury. The addition of 50 m*M* $(NH_4)_2HPO_4$ to the samples containing cadmium and 1% (v/v) $(NH_4)_2S$ to mercury solutions obviates this problem by allowing the use of higher charring temperatures and volatilization of matrix components before the atomization of the element.[6]

Elimination of matrix interference by increasing its volatility is performed for samples containing sodium chloride. The presence of sodium chloride in the sample results in high background absorption in nearly all

[6] Analytical Methods for Atomic Absorption Spectroscopy Using the HGA Graphite Furnace. The Manual by Perkin-Elmer, 1977.

TABLE I
STANDARD CONDITIONS FOR METAL DETERMINATION BY EAAS

Metal	Matrix[a]	Sensitivity[b]
Ag	A	8
Au	B	30
Bi	C	60
Cd	D	4.5
Co	A	60
Cu	A	50
Fe	A	30
Hg	E	450
Ni	A	150
Pb	A	25
Sb	A	130
Sn	A	40
Zn	A	2

[a] A, 0.2% (v/v) HNO_3; B, 0.2% (v/v) HNO_3 + 0.2% (v/v) HCl; C, 1% (v/v) HNO_3; D, 0.2% (v/v) HNO_3 + 50 m*M* $(NH_4)_2\ HPO_4$; E, 1% (v/v) HNO_3 + 1% (v/v) $(NH_4)_2S$.

[b] The weight of the element (pg) that will produce a signal of 0.0044 absorbance unit (1% *A*).

elemental determinations, due to a relatively high temperature required for its volatilization.

In this case, the matrix volatilization temperature can be decreased by adding ammonium nitrate, which leads to formation of volatile sodium nitrate and ammonium chloride.[6]

Effect of Organic Solvents

Techniques used to purify metallothionein and resolve its various isoforms include high-performance liquid chromatography (HPLC) separations in the presence of organic solvents.[7-9] Such samples can be analyzed by EAAS, but the precision is degraded. This is due to "wetting" of the graphite tube by organic solvents, resulting in variable distribution of the

[7] S. Klauser, J. H. R. Kägi, and K. Y. Wilson, *Biochem. J.* **209,** 71 (1983).

[8] M. P. Richards and N. C. Steele, *J. Chromatogr.* **402,** 243 (1987).

[9] R. K. Mehra and D. R. Winge, *Arch. Biochem. Biophys.* **265,** 381 (1988).

sample. Precision degradation can be minimized by decreasing the organic solutions volume to 20 μl or less.

Preparation of Reference Solutions

Working standard solutions for EAAS are prepared by serial dilutions of certified atomic absorption standards (available from Fisher Scientific, Pittsburgh, PA). Solutions containing less than 1 μg metal/ml are not stable and should be prepared daily. Water used for dilutions should always be checked for contamination by the metal of interest. Acidification of the samples to 0.1–0.2% is advisable to maintain solution stability. Table I lists the recommended conditions[6] for determination of metals that can potentially bind to metallothionein.

[48] Amino Acid Analysis of Metallothionein

By YUTAKA KOJIMA and PETER E. HUNZIKER

Introduction

Metallothioneins (MTs) have been so designated because of their extremely high metal and sulfur content.[1] The sulfur content is accounted for by the preponderance of cysteine, which usually comprises about one third of all amino acid residues. All cysteine residues serve as ligands for metal binding.[2,3]

Metallothioneins are subdivided into classes on the basis of their structural characteristics[2-4]:

Class I: Polypeptides with locations of cysteine closely related to those in the first isolated equine renal MT

Class II: Polypeptides with locations of cysteine only distantly related to those in equine renal MT, such as yeast MT

Class III: Atypical, nontranslationally synthesized metal-thiolate polypeptides, such as cadystin, phytometallothionein, phytochelatin, and homophytochelatin

[1] J. H. R. Kägi and B. L. Vallee, *J. Biol. Chem.* **235**, 3460 (1960).

[2] J. H. R. Kägi, M. Vašák, K. Lerch, D. E. O. Gilg, P. Hunziker, W. R. Bernhard, and M. Good, *Environ. Health Perspect.* **54**, 93 (1984).

[3] J. H. R. Kägi and Y. Kojima, *Experientia Suppl.* **52** (Metallothionein II), 25 (1987).

[4] Y. Kojima, this volume [2].

The unique aspect of the class I mammalian MTs is their high content of cysteine (approximately 33% of the residues). In addition to cysteine, most MTs also contain a relatively large proportion of serine (approximately 14% of the residues) and basic amino acids, especially lysine and, occasionally, arginine (lysine and arginine being approximately 13% of the residues). The *Neurospora* protein in class I shows a similar abundance of cysteine (28% of the residues) and serine (28% of the residues). All contain a single methionine. Histidine is absent from mammalian MT but single residues are present in certain avian, reptilian, and amphibian forms. Also typical of all class I MTs is the complete lack of aromatic amino acids. The remaining amino acids are unexceptional.

Class II MTs contain less cysteine. Phenylalanine, tyrosine, and histidine have been found in some of them.

Class III MTs are unique because of their restricted amino acid composition that, in many cases, includes only cysteine, glutamic acid, and glycine. The highest amounts of sulfur (more than 13%; cf. class I MTs, 11%) occur in some class III MTs. Freshly isolated MT from cadmium-exposed maize roots contains both cysteine and cystine. In some forms glutamic acid is the most abundant component. Class III MTs lacking glycine or containing β-alanine instead of glycine as well as forms containing appreciable amounts of serine have also been found.

Procedures

The amino acid analysis of MT is carried out as follows: (1) modification of the cysteine residues, (2) acid hydrolysis, and (3) column chromatography of hydrolysate.

Modification of Cysteine Residues

The modification of the cysteine residues in MT is covered by another chapter in this volume.[5] The most convenient modification of MT for subsequent amino acid analysis is oxidation of the cysteine residues by performic acid. Since tyrosine is destroyed under these conditions, class II MTs must be analyzed in two separate steps: (1) analysis for tyrosine using a nonoxidized sample (cysteine cannot be quantified) and (2) analysis for cysteine following performic acid oxidation.

The following protocol for performic acid oxidation represents the slightly modified method described by Hirs.[6]

Preparation of Performic Acid. Distilled formic acid (950 μl) is mixed

[5] P. E. Hunziker, this volume [45].

[6] C. H. W. Hirs, this series, Vol. 11, p. 197.

with 50 μl of 30% H_2O_2 (Perhydrol, Fluka, Switzerland) and incubated at room temperature for 2 hr. Only freshly prepared reagent should be used.

Oxidation of Salt-Free Samples. Depending on the analysis system employed, appropriate amounts of MT are dried in a hydrolysis tube, dissolved in 25–50 μl distilled formic acid, chilled, and mixed with 50–100 μl of cold performic acid (the final ratio of formic acid to performic acid is 1:2). After incubation for 3 hr on ice 25–50 μl of water is added and the sample is dried in a Speed Vac concentrator (Savant, Hicksville, NY) or freeze-dried and hydrolyzed.

Acid Hydrolysis

Oxidized or alkylated salt-free MT is hydrolyzed for 22 hr at 110° *in vacuo* or under argon in 6 *M* HCl. Recoveries of threonine and serine may be calculated by extrapolation to zero time after hydrolysis for 16, 40, and 72 hr.

Amino Acid Analysis of Acid Hydrolysate

Normally, acid hydrolysates of MT are applied to a commercially available amino acid analyzer. With a Durrum D-500 amino acid analyzer the optimal amount of hydrolyzed protein for analysis is approximately 10 or 0.5 μg using detection with ninhydrin or o-phthalaldehyde reagent, respectively. With analyzers from other manufacturers (model 420A, Applied Biosystems, Inc., Foster City, CA; or Amino Quant, Hewlett-Packard) accurate results are obtained using approximately 50 ng of sample.

[49] Amino Acid Sequence Determination

By PETER E. HUNZIKER

Introduction

Free cysteine residues are usually not detectable as phenylthiohydantoin (PTH) derivatives obtained by Edman degradation. Because this amino acid represents up to 33% of all amino acid residues in metallothioneins (MTs), it must be chemically modified for the unambiguous determination of the primary structure of this protein. In addition to the variety of cysteine derivatives employed in peptide mapping[1,2] and sequence

[1] W. R. Berhnard, M. Vašák, and J. H. R. Kägi, *Biochemistry* **25**, 1975 (1986).

analysis,[3-7] good results were obtained in our laboratory using the S-methylated derivative of MT.[8] Complete primary structure of MT is achieved by N-terminal sequence analysis of the modified protein and by sequence analysis of peptides derived from a number of proteolytic digests.

In this chapter the chemical and enzymatic fragmentations usually necessary for complete determination of the amino acid sequence of this protein are described.

Material and Methods

Sequencing Equipment

Gas-phase sequencer, model 470A/900A, equipped with an online PTH analyzer, model 120A (Applied Biosystems, Inc., Foster City, CA)
Pulsed-liquid phase sequencer, model 477A, equipped with an online PTH analyzer, model 120A (Applied Biosystems, Inc.)

Chemicals

Chemicals for sequence analysis: All are purchased from Applied Biosystems, Inc.
Proteolytic enzymes ("sequencing grade"): All are supplied by Boehringer (Mannheim, Germany) and reconstituted according to instructions
Trifluoroacetic acid (TFA) for high-performance liquid chromatography (HPLC) is obtained from Pierce (Rockford, IL)
All other chemicals: All are of analytical or HPLC grade and are supplied by Fluka (Bucks, Switzerland) or by Merck

Cysteine Modification. Purified apoMTs are modified by S-methylation prior to peptide mapping and sequence analysis.[8,9]

Acid Cleavage. This procedure is necessary only for vertebrate MTs containing an acetylated N terminus. Up to 1 nmol of S-methylated MT is

---

[2] K. Münger and K. Lerch, *Inorg. Chim. Acta* **151**, 11 (1988).
[3] T. Kojima, C. Berger, and J. H. R. Kägi, *in* "Metallothionein" (J. H. R. Kägi and M. Nordberg, eds.), p. 153. Birkhäuser, Basel, 1979.
[4] K. Münger, U. A. Germann, M. Beltramini, D. Niedermann, G. Baitella, J. H. R. Kägi, and K. Lerch, *J. Biol. Chem.* **260**, 10032 (1985).
[5] I.-Y. Huang, A. Yoshida, H. Tsunoo, and H. Nakajima, *J. Biol. Chem.* **252**, 8217 (1977).
[6] I.-Y. Huang, M. Kimura, A. Hata, H. Tsunoo, and A. Yoshida, *J. Biochem.* **89**, 1839 (1981).
[7] D. R. Winge, K. B. Nielson, R. D. Zeikus, and W. R. Gray, *J. Biol. Chem.* **259**, 11419 (1984).
[8] P. E. Hunziker, this volume [45].
[9] P. E. Hunziker, this volume [53].

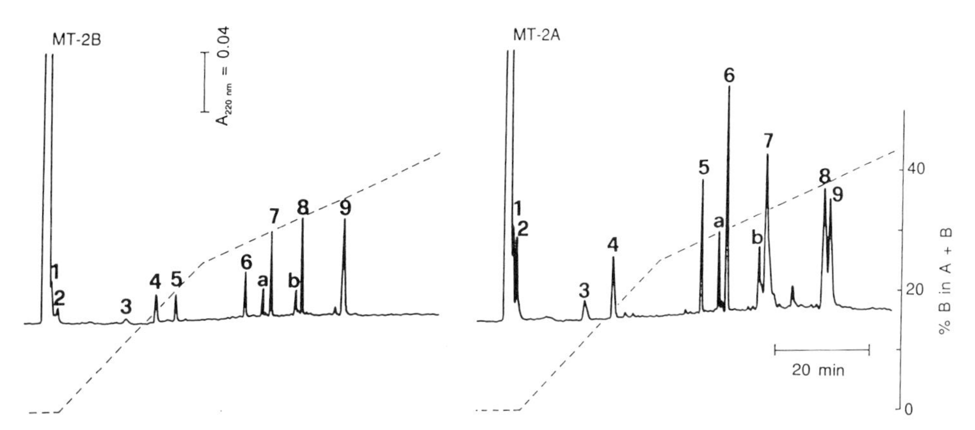

FIG. 1. Tryptic peptide mapping of S-methylated rabbit MT-2A and MT-2B. Samples were digested for 30 min at 37° in 0.2 *M N*-methylmorpholine/TFA, pH 8.3. Chromatography was performed on an Aquapore RP-300 column (4.6 mm i.d. × 250 mm; Applied Biosystems, Inc., San Jose, CA) at a flow rate of 0.6 ml/min. The gradient (dashed line) was formed between solvent A (0.1% TFA) and solvent B (same as A, containing 60% acetonitrile); the amounts injected were MT-2A, 6.8 nmol; MT-2B, 3.7 nmol.

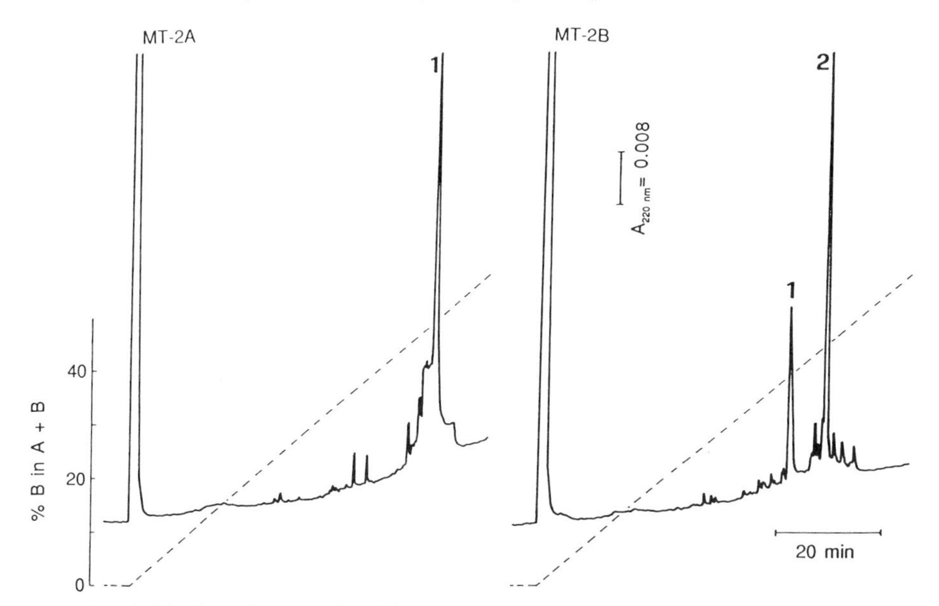

FIG. 2. Glu-C peptide mapping of S-methylated rabbit MT-2A and MT-2B. Samples were digested for 2 hr at 37° in 25 m*M* sodium phosphate, pH 7.8. Chromatography was performed on a Vydac C_4 column (4.6 mm i.d. × 250 mm, The Separations Group, Hesperia, CA) at a flow rate of 0.6 ml/min, using a gradient (dashed line) between solvent A (0.1% TFA) and solvent B (same as A, containing 60% acetonitrile). The amounts injected were MT-2A, 2.9 nmol; MT-2B, 2.9 nmol.

dissolved in 100 μl of 0.1% TFA and transferred to a 2-ml Reacti-Vial (Pierce). The sample is incubated for 20 min at 110° in an oven, cooled on ice, and sequenced or stored at −20° if immediate sequencing is not possible.

Tryptic Digestion. Up to 10 nmol of S-methylated MT is dissolved in 50 μl of 0.2 *M* *N*-methylmorpholine–trifluoroacetate, pH 8.3. Trypsin (2 μl) dissolved in 1 m*M* HCl (100 μg/ml) is added. The sample is incubated for 30 min at 37° and tryptic peptides are separated by reversed-phase high-performance liquid chromatography (HPLC)(Fig. 1).

Glu-C Digestion. Up to 10 nmol of S-methylated MT is dissolved in 100 μl of 25 m*M* sodium phosphate buffer, pH 7.8 Three microliters of Glu-C dissolved in water (500 μg/ml) is added. Digestion is performed at 37° for 2 hr and the resulting peptides are resolved by reversed-phase HPLC (Fig. 2).

Asp-N Digestion. Up to 10 nmol of S-methylated MT is dissolved in 80 μl 50 m*M* sodium phosphate, pH 8.0. Twenty microliters of Asp-N dis-

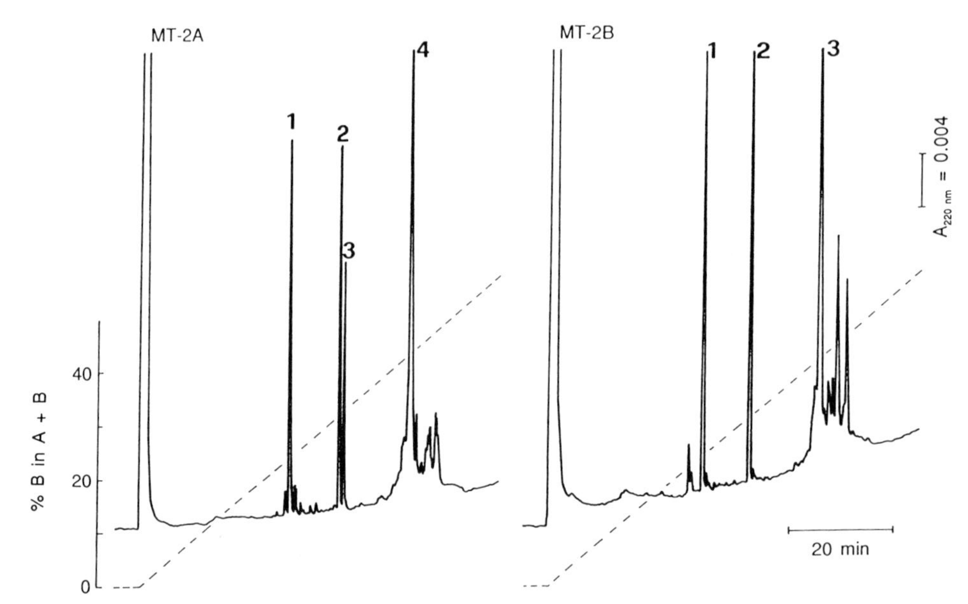

FIG. 3. Asp-N peptide mapping of S-methylated rabbit MT-2A and MT-2B. Samples were digested for 2 hr at 37° in 50 m*M* sodium phosphate, pH 8. Chromatography was performed on a Vydac C_4 column (4.6 mm i.d. × 250 mm, The Separations Group) at a flow rate of 0.6 ml/min, using a gradient (dashed line) between solvent A (0.1% TFA) and solvent B (same as A, containing 60% acetonitrile). The amounts injected were MT-2A, 2.9 nmol; MT-2B, 2.6 nmol.

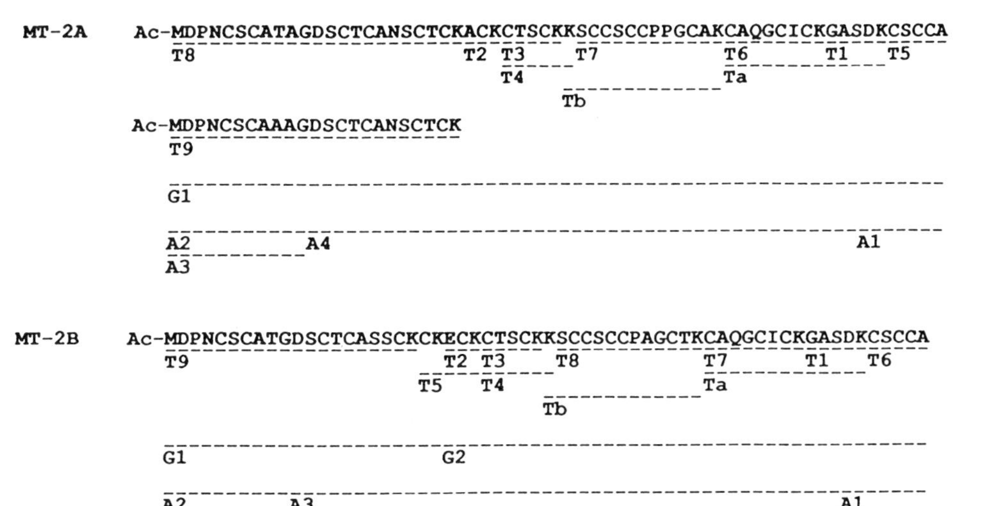

FIG. 4. Amino acid sequences of rabbit liver MT-2A and MT-2B. T, Peptides obtained by tryptic digestion; G, peptides obtained by Glu-C digestion; A, peptides obtained by Asp-N digestion. The numbering corresponds to the numbering given in Figs. 1–3.

solved in 10 m*M* Tris/HCl, pH 7.5 (40 μg/ml), is added. The sample is incubated at 37° for 2 hr and the resulting peptides are recovered by reversed-phase HPLC (Fig. 3).

Sequence Analysis. Sequence analyses have been performed according to the manufacturer's recommendations. The alignment of proteolytic peptides is illustrated in Fig. 4.

Comments

On the Applied Biosystems, Inc. model 120A PTH-amino acid analyzer, PTH-*S*-methylcysteine elutes as a single peak between PTH-tyrosine and PTH-proline. Coelution of PTH-*S*-methylcysteine with PTH-arginine can be prevented by altering the ionic strength of buffer A according to the separation instructions given by Applied Biosystems, Inc. PTH-*S*-methylcysteine obtained from Pierce is dissolved in acetonitrile and added to the PTH-amino acid standard supplied by Applied Biosystems, Inc., to give a final concentration of 1 pmol/μl.

Direct N-terminal sequence analysis of vertebrate MTs is impossible due to the acetylation of their N-terminal methionine residue. However, the protein can be deblocked by cleaving the acid-labile Asp-Pro bond, which is conserved in all known mammalian MTs. Using the acid-cleaved protein, reliable N-terminal sequencing is routinely performed up to residue 31. Because of the increased background of *S*-methylcysteine, due to unspecific hydrolysis of the protein during the cleavage step of the Edman degradation, reliable sequence information from residues 32 to 52 is only occasionally obtained. For vertebrate MTs the N-terminal Ac-Met-Asp, as released by acid cleavage of the intact protein or of the N-terminal proteolytic peptides, is analyzed by amino acid analysis following the separation of the hydrolyzed samples by HPLC.

[50] Differential Modification of Metallothionein with Iodoacetamide

By WERNER R. BERNHARD

Introduction

The partition of the seven d^{10} metal (M) ions Zn(II) and/or Cd(II) of mammalian metallothioneins (MT) into an amino-terminal $M(II)_3$-Cys_9 and a carboxyl-terminal $M(II)_4$-Cys_{11} cluster has been established both by nuclear magnetic resonance (NMR) and X-ray crystallographic studies.[1-3] The apparent average association constants for Zn(II) and Cd(II) (at pH 8.6) are approximately 10^{17} and 10^{21}, respectively.[4] However, the stability of the complexes in the two clusters is not uniform,[4] thereby predicating an ordered pathway in which the metal ions are bound when they encounter the apoprotein.

A direct approach for mapping the pathway of Cd(II) binding to apo-MT and for assessing the relative affinity of Cd(II) to the tetradentate thiolate sites is provided by differential modification of cysteine in metal-free apo-MT and in partially filled and fully reconstituted forms of Cd-MT using the alkylating agent iodoacetamide. This method relies on the known protective effect of metal coordination on the reactivity of the cysteine side chains of MT with this modifying reagent.[5,6]

[1] J. D. Otvos and I. M. Armitage, *Proc. Natl. Acad. Sci. U.S.A.* **77**, 7094 (1980).
[2] M. H. Frey, G. Wagner, M. Vasak, O. W. Sörensen, D. Neuhaus, R. R. Ernst, E. Wörgötter, J. H. R. Kägi, and K. Wüthrich, *J. Am. Chem. Soc.* **107**, 6847 (1985).
[3] W. F. Furey, A. H. Robbins, L. L. Clancy, D. R. Winge, B. C. Wang, and C. D. Stout, *Science* **231**, 704 (1986).
[4] M. Vašák and J. H. R. Kägi, *Met. Ions Biol. Syst.* **15**, 213 (1983).
[5] J. H. R. Kägi and B. L. Vallee, *J. Biol Chem.* **236**, 2435 (1961).
[6] T.-Y. Li, D. T. Minkel, C. F. Shaw III, and D. H. Petering, *Biochem. J.* **193**, 441 (1981).

Differential modification of MT involves a two-step exposure of the protein mercapto groups to an appropriate labeling reagent. In the first step the protein is incubated with a substoichiometric amount of the radioactive labeled form of the reagent, leading to partial derivatization. Subsequently, after removal of the protecting Cd(II) by ethylenediaminetetraacetic acid (EDTA) and denaturation of the protein, a large excess of unlabeled reagent is added to attain quantitative modification of the remaining unmodified mercapto groups. The reactivity of each cysteine is assessed on the basis of the specific radioactivity of its alkylated derivative determined after fragmentation of the protein and isolation of the peptides containing such residues.[7]

Materials and Methods

Chemicals and Analytical Procedures

^{14}C-labeled iodoacetamide is purchased from Amersham Radiochemical Center, Ltd. (England) and dissolved in water. Its specific radioactivity is 53 mCi/mmol and before use it is diluted to 2 mCi/mmol with unlabeled iodoacetamide. Solutions of iodoacetamide are made immediately before use. All other chemicals are of reagent grade or better. The nitrogen to purge the glovebox is delivered by the house nitrogen line.

Automated amino acid analysis is done using postcolumn derivatization of the eluted amino acids with ninhydrin for quantification. Radioactivity is determined by scintillation counting. High-performance liquid chromatography (HPLC) separations are done on a C_{18} reversed-phase column (Spheri 10, 10 μm; Brownlee Laboratories) using HPLC-grade chemicals and solvents.

The concentration of Cd(II) is determined by flame atomic absorption spectrometry (Instrumentation Laboratory, IL 157). The concentration of apo-MT is estimated by its absorption at 220 nm in 0.1 *M* HCl based on an extinction coefficient of 48,200 $M^{-1}cm^{-1}$. Protein concentration is determined subsequently by automated amino acid analysis of the derivatives obtained by oxidation with performic acid or alkylation with iodoacetamide. The calculations are based on the combined recovery of aspartate, proline, alanine, isoleucine, and lysine referred to the known amino acid composition.[8]

[7] K. Münger, U. A. Germann, M. Beltramini, D. Niedermann, G. Baitella-Eberle, J. H. R. Kägi, and K. Lerch, *J. Biol. Chem.* **260**, 1032 (1985).

[8] M. Kimura, N. Otaki, and M. Imano, *Experientia, Suppl.* **34**, 163 (1979).

Isolation of Metallothionein and Preparation of Apometallothionein

Rabbit liver MT-2 is prepared from liver of rabbits injected subcutaneously 15 times with 1 mg of $CdCl_2$/kg of body weight at 2- and 3-day intervals.[8] The protein is purified by a procedure adapted from Kägi *et al.*[9] and Kimura *et al.*[8] The purity is assessed by amino acid analysis and atomic absorption spectrometry. The native protein contains a total of 7 mole Zn(II) plus Cd(II)/mol of protein; copper is present in trace amounts only.

For the preparation of apo-MT, native MT is dissolved in 1.5 ml of 20 m*M* Tris-HCl, pH 8.6. To release the bound metal ions quantitatively from the protein an equal volume of 1 *M* HCl is added. After 3 min of incubation at room temperature the metal ions are removed by gel filtration on Sephadex G-25 equilibrated with 0.01 *M* HCl (column size: 2 cm in diameter, 40 cm in height). All fractions absorbing at 220 nm are pooled, diluted with an equal volume of water, and lyophilized.

Reconstitution of Cadmium-Metallothionein

All operations are done under oxygen-free conditions in a nitrogen-purged glovebox at room temperature. Prior to use, all solutions are degassed on a vacuum line by submitting them to three thaw, freeze, and pump steps. To prepare Cd-MTs of a given metal-to-protein ratio, 500 ($\approx$ 3 mg) of apo-MT-2 is dissolved in 1 ml 0.1 *M* HCl, and 35 μl of a 0.01 *M* HCl solution containing the appropriate amount of $CdCl_2$ is added. After a 10-min incubation 2 ml of 0.2 *M* Tris base is added slowly under vigorous stirring (Vortex mixer) to adjust the pH to 8.6. Before further use the solution containing the reconstituted Cd-MT is allowed to equilibrate for 1 hr.

Differential Modification

Under oxygen-free conditions 500 nmol of ^{14}C-labeled iodoacetamide (2 mCi/mmol, in H_2O) is added to the reconstituted sample of Cd-MT (500 nmol). After 10 min the excess of ^{14}C-labeled iodoacetamide is quenched by the addition of 5 μmol of dithioerythritol (DTE). After 20 min 2 vol of a saturated solution of guanidine hydrochloride containing 40 m*M* EDTA in 20 m*M* of Tris-HCl, pH 8.6, is added to remove the metal ions and to denature the partially modified MT. With a 10-fold excess (200 μmol) of unlabeled iodoacetamide over all mecapto groups, the remaining cysteines of the protein are carboxyamidomethylated. After 4 hr the mix-

[9] J. H. R. Kägi, S. R. Himmelhoch, P. D. Whanger, J. L. Bethune, and B. L. Vallee, *J. Biol. Chem.* **249**, 3537 (1974).

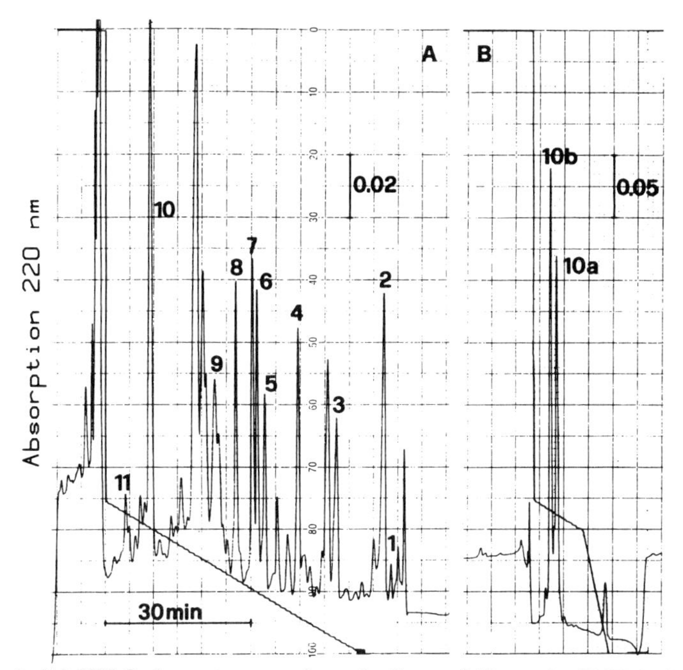

FIG. 1. (A) HPLC chromatogram of tryptic digest of 50 nmol of MT S-alkylated by iodoacetamide. Buffer A was 0.1% TFA; buffer B was 0.1% TFA/60% acetonitrile. Flow rate, 1 ml/min. (B) Rechromatography of tryptic peptides 10a and 10b under the same conditions using a less steep gradient. (From Ref. 10.)

ture is desalted on a Sephadex G-25 column equilibrated with 20 m*M* $(NH_4)HCO_3$. The fractions containing the labeled protein are pooled and lyophilized.

Peptide Mapping

The lyophilized samples are redissolved in 2 ml of 100 m*M* $(NH_4)HCO_3$ and digested with a total of 150 μg trypsin (dissolved in 1 m*M* HCl) added in 5 steps at 0, 1, 2, 4, and 6 hr. Two hours after the last addition of trypsin the samples are lyophilized. Subsequently, the digestion mixture is dissolved in 1.5 ml of 0.1% (v/v) trifluoroacetic acid (TFA) (buffer A), and approximately 200 μg of the protein digest is chromatographed by reversed-phase HPLC at a flow rate of 1 ml/min (Fig. 1A).[10] After 5 min of isocratic elution, a linear gradient ranging from 0 to 25% of the volume fraction of buffer B (0.1% TFA/60% acetonitrile) is applied for

[10] W. R. Bernhard, M. Vašák, and J. H. R. Kägi, *Biochemistry* **25**, 1975 (1986).

50 min. The peptide fractions are collected and evaporated to dryness.[7] Subsequently the peptides are dissolved in 500 μl of water. One peak (No. 10) is dissolved in 500 μl of buffer A and rechromatographed in the same buffer system on the same reversed-phase column, using a 5-min linear gradient extending from 0 to 20% of buffer B and a 10-min linear gradient extending from a 20 to 25% volume fraction of buffer B (Fig. 1B). The collected fractions are dried as before.

Identification of Peptides and Determination of Specific Radioactivity

The resolved peptides are identified on the basis of their amino acid composition with known segments of the published partial sequence of rabbit liver MT-2 (Table I). The atypical cleavage observed after Cys-5 and

TABLE I
TRYPTIC PEPTIDES OF RABBIT LIVER MT-2[a]

HPLC peak number	Sequence numbers: Residues	Sequence numbers: Ranking of Cys	Amino acid sequence of peptide[b]
1	21–22	6	CK
2	23–25	7	ACK
	52–56		GASDK[c]
3	26–30	8, 9	CTSCK[d]
4	57–61	18–20	CSCCA[d]
5	26–30	8, 9	CTSCK
6	57–61	18–20	CSCCA
7	1–5	1	MDPNC
8	6–15	2–5	SCAADG(SCTC)
9	32–43	10–14	SCCSCCPAGCTK
10a	44–51	15–17	CAQGCICK
10b	1–15	1–4	MDPNCSCAADG(SCTC)
11	1–20	1–5	MDPNCSCAADG(SCTCATSCK)

[a] Modified with iodoacetamide and separated by reversed-phase HPLC on an RP-18 column (Fig. 1).

[b] The partial sequence of the subform of rabbit liver MT-2 compatible with these peptides is N-Ac-MDPNC/SCAAD[10]G(SCTC/ATSCK)[20]CK/ACK/CTSCK[30]/K/SCCSCCPAGC[40]TK/CAQGCIC[50]K/GASDK/CSCC[60]A,[8] where (/) indicates the point of tryptic cleavage. Single-letter abbreviations for amino acid residues: A, alanine; R, arginine; N, asparagine; D, aspartic acid; C, cysteine; Q, glutamine; E, glutamic acid; G, glycine; H, histidine; I, isoleucine; L, leucine; K, lysine; M, methionine; F, phenylalanine; P, proline; S, serine; T, threonine; W, tryptophan; Y, tyrosine; V, valine.

[c] This peptide coeluted with peptide 23–25.

[d] Peptides of the same composition eluted as HPLC peaks 5 and 6, respectively, implying the occurrence of unidentified chemical heterogeneity. (From Ref. 10.)

Cys-15 is attributed to the use of relatively large amounts of trypsin and the long incubation times required for the digestion of carboxyamidomethylated MT. The specific radioactivity of the peptides is assessed by liquid scintillation counting and quantitative amino acid analysis. In peptides containing more than one cysteine only an average value for the specific radioactivity of each cysteine could be determined. The relative standard deviation evaluated from 36 analyses is about 10%.

Results and Interpretation

Figure 2 displays the variation in the specific radioactivity of the modified cysteine along the polypeptide chain of forms of MT with different cadmium-to-protein ratios. To allow for comparison, the results are expressed in percentage of the total radioactivity incorporated into each form, i.e., in percentage relative radioactivity. The data show that in the absence of bound metal (Fig. 2A) the label is distributed equally over all cysteines, indicating that in apo-MT all cysteines are equally reactive toward iodoacetamide. Distinct changes in the labeling profile are noticable when two or more equivalents of Cd(II) are incorporated. Starting with Cd_4-MT (Fig. 2E), there is a definite lowering of the specific radioactivity over the entire carboxyl-terminal portion of the chain. In the metal-saturated form, Cd_7-MT (Fig. 2H), there is a more than threefold difference in the radioactive labeling between the amino-and carboxyl-terminal halves. The break between the two portions coincides with the boundary of the metal–thiolate cluster domains in the protein.[11] Reflecting unequal rates of alkylation, the variations in the extent of isotopic labeling have been attributed to unequal masking of the nucleophilic cysteine side chains by their binding to Cd(II). As a consequence, differences in the kinetics of metal complex formation and dissociation are important determinants of the observed rates. Thus, the lower extent of labeling of the cysteine in the carboxyl-terminal $Cd(II)_4$-Cys_{11} cluster is consistent with the relatively tighter binding of the metal ions in this cluster.

On the assumption that iodoacetamide reacts mainly with dissociated cysteine ligands and that the alkylation reaction is the rate limiting step, it was calculated that the threefold lower rate of labeling of all carboxyl-terminal cysteines may arise from an about 30-fold higher association constant for Cd(II) in the $Cd(II)_4$-Cys_{11} cluster as compared to the amino-terminal $Cd(II)_3$-Cys_9 cluster.[10] This is also consistent with the greater kinetic stability of the $Cd(II)_4$-Cys_{11} cluster established by others.[12,13]

[11] D. R. Winge and K.-A. Miklossy, *J. Biol. Chem.* **257**, 3471 (1982).

[12] K. B. Nielson and D. R. Winge, *J. Biol. Chem.* **258**, 13063 (1983).

[13] J. D. Otvos, H. R. Engeseth, D. G. Nettesheim, and C. R. Hilt, *Experientia, Suppl.* **52**, 171 (1987).

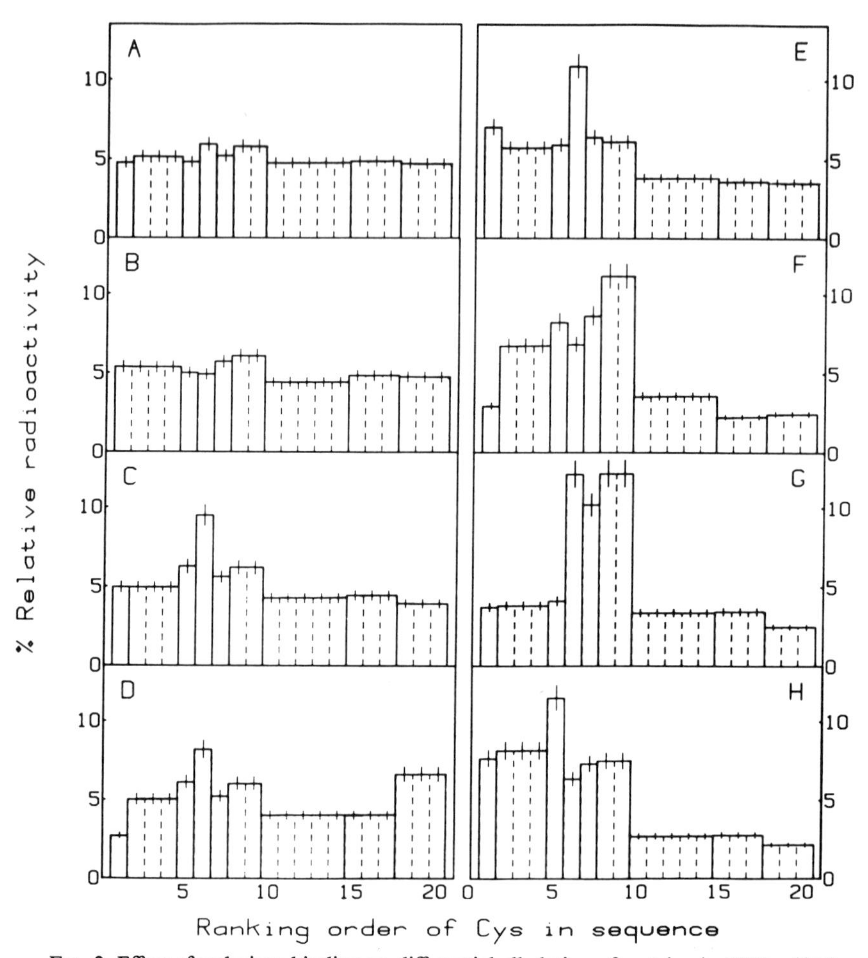

FIG. 2. Effect of cadmium binding on differential alkylation of cysteine in MT by ^{14}C-labeled iodoacetamide. The bars indicate the radioactivity of the individual carboxamidomethyl-Cys along the polypeptide chain (1–20) in percentage of total radioactivity incorporated. The length of the vertical lines indicates the relative standard deviation. In tryptic peptides containing two or more carboxamidomethyl-Cys, the same average radioactivity is assigned to each carboxamidomethyl-Cys.[10] The cadmium content (moles of cadmium per mole of MT) of the samples was as follows: (A) none, (B) 1, (C) 2, (D) 3, (E) 4, (F) 5, (G) 6, and (H) 7. (From Ref. 10.)

The changes in the labeling profiles observed in MT of differing cadmium-to-protein ratios provide information on the order in which the 20 cysteines are complexed with the metal. Thus, the nearly uniform lowering of the profile in the carboxyl-terminal half observed when 4 Eq of Cd(II) is bound (Fig. 2E) indicates the preferential formation of the $Cd(II)_4$-Cys_{11}

cluster. The differences in profile among the cysteines of the amino-terminal half of Cd_5-MT (Fig. 2F) and Cd_6-MT (Fig. 2G) suggest that binding of Cd(II) to cysteines in the $Cd(II)_3$-Cys_9 cluster is not uniform and that the least firmly bound Cd(II) is coordinated by the cysteine located near the interdomain junction.

Assessment of the Method

The two-step differential modification procedure introduced for probing the environment of amino groups in proteins[14] and for mapping contact surfaces in supramolecular structures[15] is very useful for resolving topographically the pathway of metal complexation and cluster formation in MT. The chief advantage of this method over the conventional one-step chemical modification is that the initial labeling step employs only a relatively small quantity of radioactive reagent, thereby minimizing the incidence of multiple modifications of the same molecule, leading to a bias in the course of the reaction. In the case of MT the absence of such effects was ascertained by comparing modification profiles obtained with 0.1, 0.25, and 0.45 mol of radioactive iodoacetamide/mol of cysteine in the initial labeling step. The subsequent alkylation of the remaining cysteines by a large excess of unlabeled reagent ensures chemical homogeneity of the product.

The use of iodoacetamide in the differential alkylation of MT is indicated by its high selectivity for mercapto groups at conditions where the metal is bound to the protein. Its small size and its lack of electric charge assures us that its reactivity is minimally influenced by steric hindrance and charges within the vicinity of the cysteine.

[14] H. Kaplan, K. J. Stevenson, and B. S. Hartley, *Biochem. J.* **124**, 289 (1971).
[15] H. R. Bosshard, *Methods Biochem. Anal.* **25**, 273 (1979).

[51] Lysine Modification of Metallothionein by Carbamylation and Guanidination

By JIN ZENG

Introduction

The conservation of lysine in class I metallothioneins (MTs) implies a structural or functional role for this basic amino acid residue.[1] From their juxtaposition to cysteine in the amino acid sequence it was suggested that

[1] J. H. R. Kägi and Y. Kojima, *Experientia, Suppl.* **52**, 25 (1987).

they might serve in neutralizing the negative charges of the metal–thiolate complexes. Indeed, both ^1H nuclear magnetic resonance (NMR) and pH titration studies of MT support an electrostatic interaction of the protonated side chains of lysine with the threefold negatively charged metal–thiolate clusters.[2,3] To probe these and other functions, a number of chemical modification studies of lysine residues have been undertaken. Arylation by trinitrobenzenesulfonic acid (TNBS) has led to the complete modification of all lysine residues in MTs.[3] However, both the reagent and the trinitrophenyl derivative of the ε-amino groups of the lysine residues also react with mercapto groups abundant in this protein, resulting in undesired secondary changes. None of these complications is encountered with the classical carbamylation and guanidination reaction here described.

Materials and Methods

Metallothioneins

Native rabbit $(Zn,Cd)_7$-MT-1 and human Zn_7-MT-2 are prepared by a combination of well-established procedures.[4,5] The homogeneously substituted rabbit Cd_7-MT-2 derivative is made by reconstitution from the metal-stripped form, i.e., apo-MT-2, and the appropriate amount of Cd(II) as described by Vašák and Kägi.[6]

Carbamylation

Reagents

Potassium cyanate (Fluka, Ronkonkoma, NY)
Sodium borate buffer, pH 9.2 (0.5 *M*)
Tris-HCl, pH 8.6 (10 m*M*)

Recrystallization of potassium cyanate from water–ethanol before use is strongly recommended.[7] With a slight modification carbamylation is carried out as previously described.[7,8] Lyophilized MT (2 mg) is dissolved in 0.2 ml of 0.5 *M* sodium borate buffer, pH 9.2, to a final concentration of

[2] M. Vašák, C. E. McClelland, H. A. O. Hill, and J. H. R. Kägi, *Experientia* **41**, 30 (1985).
[3] J. Pande, M. Vašák, and J. H. R. Kägi, *Biochemistry* **24**, 6717 (1985).
[4] J. H. R. Kägi, S. R. Himmelhoch, P. D. Whanger, J. L. Bethune, and B. L. Vallee, *J. Biol. Chem.* **249**, 3537 (1974).
[5] R. H. O. Bühler and J. H. R. Kägi, *FEBS Lett.* **39**, 229 (1974).
[6] M. Vašák and J. H. R. Kägi, *Met. Ions Biol. Syst.* **15**, 213 (1983).
[7] G. R. Stark, this series, Vol. XI, p. 590.
[8] S. Rimon and G. E. Perlmann, *J. Biol. Chem.* **243**, 3566 (1968).

10 mg/ml. The reaction is started by adding solid potassium cyanate to a final concentration of 1 *M*. The reaction mixture is incubated at 37° for 24 hr and then dialyzed thoroughly against a 500-fold volume of 10 m*M* Tris-HCl buffer, pH 8.6, or distilled water at 4°, using a membrane (Spectrum Medical Industries, Inc., Los Angeles, CA) with an M_r 3500 cutoff.

Guanidination

Reagents

O-Methylisourea hydrogen sulfate (Sigma, St. Louis, Mo)
Tris-HCl, pH 8.6 (10 m*M*)

Guanidination is performed by the method of Bregman *et al.*[9] *O*-Methylisourea is prepared from *o*-methylisourea hydrogen sulfate as described.[10] The reaction is started by dissolving 4 mg lyophilized MT in 0.8 ml of 1 *M* *o*-methylisourea, pH 11. The reaction mixture is stored at 4° for 6 days and then dialyzed thoroughly against an approximately 500-fold volume of 10 m*M* Tris-HCl, pH 8.6, or distilled water at 4°, using a membrane (Spectrum Medical Industries, Inc.) with an M_r 3500 cutoff.

Estimation of Extent of Modification

To determine the degree of carbamylation or guanidination, a combination of three methods—amino acid analysis, ninhydrin reaction, and polyacrylamide gel electrophoresis (PAGE)—is employed.

1. Amino acid analysis: The carbamylated or guanidinated derivative (about 10 nmol MT) is subjected to amino acid analysis after 6 *N* HCl hydrolysis for 20 hr at 110°. The appearance of homocitrulline or homoarginine indicates the desired modification.[11,12] Homoarginine is resistant to acid hydrolysis and may thus serve as a determinant of the extent of guanidination. The amount of homoarginine in the hydrolysate is calculated, using the color value of arginine. By contrast, approximately 30% of the homocitrulline is reverted to lysine during acid hydrolysis, thus providing no accurate measure of the degree of carbamylation.

2. Ninhydrin reaction: The ninhydrin test is used to monitor the disappearance of free amino groups in carbamylated or guanidinated MTs. Native MTs with free amino groups of lysine react with ninhydrin to give a red color, whereas the fully carbamylated or guanidinated derivatives do

[9] M. D. Bregman, D. Trivedi, and V. J. Hruby, *J. Biol. Chem.* **255**, 11725 (1980).
[10] J. R. Kimmel, this series, Vol. XI, p. 584.
[11] G. R. Stark, this series, Vol. XI, p. 594.
[12] J. R. Kimmel, this series, Vol. XI, p. 589.

not. Solutions containing approximately 10 mg/ml of either the derivative or of MT as reference in distilled water (Tris gives a false positive reaction!) are spotted onto a thin-layer chromatography plate and dried. After spraying the plate with a 5% ninhydrin solution in ethanol, the color is developed by heating the plate at 110° for 15 min.

3. Polyacrylamide gel electrophoresis: The extent of carbamylation is most conveniently assessed by PAGE. At different time intervals during carbamylation, 6-μl samples are withdrawn into a 50-μl volume of distilled water or of 10 mM Tris-HCl, pH 8.6, and dialyzed against a 500-fold volume of 10 mM Tris-HCl, pH 8.6, using a membrane (Spectrum Medical Industries, Inc.) with an M_r 3500 cutoff. Following lyophilization, the

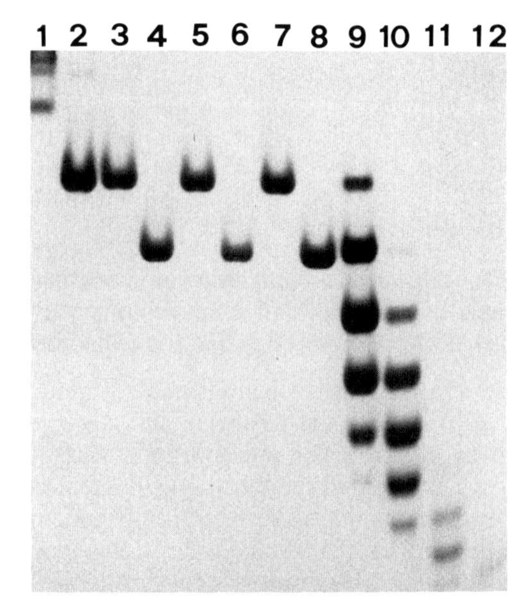

FIG. 1. Polyacrylamide gel electrophoresis of native mammalian MTs and of carbamylated human MT-1 at different degrees of modification. Lanes 1 and 2, horse kidney MT-1A (variant Arg-39) and MT-1B; lanes 3 and 4, rabbit liver MT-1 and MT-2; lanes 5 and 6, rat liver MT-1 and MT-2; lanes 7 and 8, human liver MT-1 and MT-2. The differences in mobility of the native MT isoforms denote the known charge differences.[1] Between pH 7 and 9.5 MT-1 isoforms carry two negative charges, MT-2 isoforms three negative charges. Horse kidney MT-1A, variant Arg-39 (lane 1) has a single negative charge.[13] Lanes 9 to 12 display human liver MT-1 reacted with potassium cyanate for 0.5, 1.5, 7, and 24 hr, respectively. It is clearly seen that the protein undergoes carbamylation progressively until all eight lysines are modified. Note also that with progressive modification (lanes 11 and 12) the bands are much less intensely stained by Coomassie Blue. Apparently, the dye binds less effectively when the protein loses its positive charges.

samples are redissolved in 20 μl of sample buffer containing 6.25 m*M* Tris/50 m*M* glycine, pH 8.4, 10% (v/v) glycerol, 0.05% (w/v) Bromphenol Blue, and electrophoresis is carried out on a 10% (w/v) gel (acrylamide:bis-acrylamide, 37.5:1), using a vertical system with 50 m*M* Tris/400 m*M* glycine electrophoresis buffer, pH 8.4, at 140 V. The electrophoresis is stopped before the marker dye Bromphenol Blue reaches the bottom of the gel. The protein bands are visualized by Coomassie Brilliant Blue staining. An illustration of the effects of partial carbamylation on the electrophoretic mobility of human liver MT-1 is given in Fig. 1.

Discussion

Cyanate is a relatively nonspecific reagent and might thus be expected to react not only with the amino group of lysine but also with the mercapto and carboxyl groups of MT.[7] However, at the alkaline conditions used, modification of carboxyl groups is precluded and, owing to their binding to the metal ions, the side chains of the cysteines are also unavailable to carbamylation. Thus, both cyanate and *o*-methylisourea react specifically only with lysine. Allowing sufficient time for the reaction, MTs from different sources are fully carbamylated and guanidinated. Carbamylation of lysine leads to the loss of positive charges from MT, whereas guanidination does not alter the charge. Like native MT, the carbamylated derivatives of $(Zn,Cd)_7$-MT-2, Zn_7-MT-1, Cd_7-MT-2, and the guanidinated derivative of Cd_7-MT-2 are stable at neutral and alkaline pH. They also tolerate lyophilization and can be stored in the dry state at −20° indefinitely. By contrast, guanidinated Zn_7-MT-1 is unstable in solution, yielding an insoluble aggregate on standing. The stable derivatives are homogeneous on gel filtration. In the electronic absorption, circular dichroism, and ^{113}Cd NMR spectra, both the carbamylated and guanidinated derivatives display features similar to those of unmodified MT, indicating a comparable organization of the metal–thiolate clusters. The use of PAGE allows for precise monitoring of the extent of carbamylation achieved. Thus, illustrating the linear dependence of the electrophoretic mobility on the number of charges, partial carbamylation of MT results in a ladder of electrophoretic bands (Fig. 1). Each of the Coomassie Blue-stained bands corresponds to forms of carbamylated MT containing the same number of carbamylated lysines. The enumeration of the bands confirms that full carbamylation of human MT-1 (lane 12) raises the negative net charge by eight units, corresponding to the loss of the positive charges of all eight lysines in the polypeptide chain of this isoform.[1]

[13] Y. Kojima, C. Berger, and J. H. R. Kägi, *Experientia, Suppl.* **34**, 153 (1979).

[52] Limited Proteolysis of Metallothioneins

By DENNIS R. WINGE

Introduction

Limited proteolysis has been widely used as a probe of protein structure. The technique has utility in dissection of proteins of high molecular weight into smaller fragments more amenable to study. Structural and functional domains of many proteins can be delineated by limited cleavage with proteases. The susceptibility of peptide segments to proteolysis correlates with the flexibility and solvent exposure of structural elements. The segments of proteins most susceptible to proteolysis are frequently exposed loops or interdomain hinges. The correlation of *in vitro* proteolysis and flexible structures is supported by numerous observations: (1) unfolded polypeptides are more effective substrates for proteases *in vitro* than folded molecules,[1-3] (2) proteins with increased rates of hydrogen exchange are better substrates for proteases,[4] (3) the activation of trypsinogen involves proteolysis of an amino-terminal peptide that is not well ordered in the crystal structure,[5] (4) binding of cofactors imparts stability to the tertiary fold of proteins and reduces the proteolytic degradation rate,[6] and (5) the abundance of prolyl residues in proteolytically susceptible interdomain hinge segments may preclude stable secondary structures.[7-10] Although this correlation is not applicable for *in vivo* proteolysis, the *in vitro* results justify the use of limited proteolysis as a structural probe.

Application of Limited Proteolysis to Metallothioneins

Limited proteolysis is an appropriate technique to apply to the study of metallothionein as the stabilization energy of the tertiary fold is dependent on metal ions. This observation has been found to be true for mammalian,

[1] K. Linderstrøm-Lang, *Cold Spring Harbor Symp. Quant. Biol.* **14,** 117 (1950).
[2] B. Hagihara, T. Nakayama, H. Matsubara, and K. Okunuki, *J. Biochem. (Tokyo)* **43,** 469 (1956).
[3] W. F. Harrington, P. H. von Hippel, and E. Mihalyi, *Biochim. Biophys. Acta* **32,** 303 (1959).
[4] J. J. Rosa and F. M. Richards, *J. Mol. Biol.* **145,** 835 (1981).
[5] R. Huber, *Trends Biochem. Sci.* **4,** 271 (1979).
[6] S. T. Perry, K. L. Lee, and F. T. Kenney, *Arch. Biochem. Biophys.* **195,** 362 (1979).
[7] K. E. Mostov, M. Friedlander, and G. Blobel, *Nature (London)* **308,** 37 (1984).
[8] B. Erni, *FEMS Microbiol. Rev.* **63,** 13 (1989).
[9] F. L. Texter, S. E. Radford, E. D. Laue, R. N. Perham, J. S. Miles, and J. R. Guest, *Biochemistry* **27,** 289 (1988).

fungal, and invertebrate metallothioneins, but detailed structural information is available only on the mammalian molecule. The structure of the mammalian Cd,Zn-metallothionein consists of two domains linked by a short peptide hinge segment (see [58, 59] in this volume). Each domain enfolds a polynuclear metal–thiolate cluster. The 20 cysteinyl thiolates in mammalian metallothionein are ligands for 7 divalent metal ions or 12 monovalent copper or silver ions. Nuclear magnetic resonance (NMR) studies using ^{113}Cd(II)-metallothionein established the existence of two metal : thiolate clusters with a distribution of three and four Cd(II) ions in the two polynuclear clusters.[11]

The structure of each domain is dominated by relatively long peptide segments linked by reverse turns.[12] Analysis of proton–proton couplings observed by nuclear Overhauser enhancement spectroscopy (NOESY) revealed no NOEs suggestive of α helices or β strands in either domain.[13] The numerous reverse turns consist of two 3_{10} helices and several half-turns generated from typical type II turns. The secondary structure appears constrained by metal coordination. Only minimal interdomain NOEs were observed by NMR spectroscopy and no hydrogen bonds were deduced from the crystal structure,[14] suggesting that the domains are structurally independent.

The metal-free protein appears to be devoid of secondary structure. The chemical shifts and intensities of proton resonances from various amino acids in the apomolecule are comparable with those parameters in a mixture of the constituent amino acids.[15] In addition, the fast exchange rate of amide protons in the apoprotein is consistent with a flexible, open structure approaching a random coil configuration. Addition of Zn(II) or Cd(II) ions to the apomolecule effects a structural change, as evidenced by the change in chemical shifts of certain residues in the proton NMR spectrum.[15] Cu(I) binding to metallothionein leads to formation of a metal:protein complex distinct from the Cd-metallothionein complex. The Cu(I)-binding stoichiometry is higher and extended X-ray absorption fine structure (EXAFS) analyses predict a trigonal geometry for Cu(I):

[10] R. J. Poljak, L. M. Amzel, H. P. Avey, B. L. Chen, R. P. Phizackerley, and F. Saul, *Proc. Natl. Acad. Sci. U.S.A.* **71,** 3440 (1974).

[11] J. D. Otvos and I. M. Armitage, *Proc. Natl. Acad. Sci. U.S.A.* **77,** 7094 (1980).

[12] W. Braun, G. Wagner, E. Wörgötter, M. Vašák, J. H. R. Kägi, and K. Wüthrich, *J. Mol. Biol.* **187,** 125 (1986).

[13] G. Wagner, D. Neuhaus, E. Wörgötter, M. Vašák, J. H. R. Kägi, and K. Wüthrich, *J. Mol. Biol.* **187,** 131 (1986).

[14] C. D. Stout, personal communication (1990).

[15] M. Vašák, A. Galdes, H. A. O. Hill, J. H. R. Kägi, I. Bremner, and B. W. Young, *Biochemistry* **19,** 416 (1980).

thiolate coordination in contrast to the tetrahedral coordination of Cd(II) or Zn(II) ions (see [55] in this volume).

The initial use of proteolysis as a structural probe for metallothionein was the demonstration of separate domains and the assignment of the peptides enfolding the three- and four-metal clusters within the primary structure.[16,17] Native metallothionein induced in rats injected with cadmium salts contains Cd(II) and Zn(II) in a 5:2 ratio.[16] Incubation of the protein with 1 m*M* ethylenediaminetetraacetic acid (EDTA) for 1.5 hr at pH 7.8 followed by gel filtration resulted in the selective depletion of the bound Zn(II) ions.[16] Incubation of zinc-depleted metallothionein with various proteases resulted in a discrete cleavage product whose identity was established by Edman degradation as the carboxyl-terminal α domain. The α fragment contained four bound Cd(II) ions that exhibited similar NMR chemical shifts to the four Cd(II) ions known to exist in one polynuclear cluster.[17] The presence of the four-metal center in the carboxyl-terminal α domain of metallothionein implied that the three-metal center was present in the β domain of the protein.

Incubation of the protein with EDTA appears to destabilize the structure of the β cluster. Zn(II) ions associate with metallothionein with a K_{ass} of 10^4 less than the binding affinity of Cd(II) ions,[18] so EDTA treatment can yield selective removal of Zn ions. The Zn(II) ions are not randomly distributed among the seven metal-binding sites; rather, the Zn(II) ions bind predominantly within the three-metal cluster.[11] The presence of Zn(II) ions in the β domain results in the selective lability of the β domain toward EDTA. The destabilization of the β domains is expected to yield a random coil-like configuration that is unstable in the presence of proteases. The proteolytic trimming of the zinc-depleted protein is restricted to the β peptide segment, yielding the intact Cd-α domain fragment. The short interdomain hinge segment must be sufficiently hindered sterically in the native protein, as only insignificant quantities of the α domain were produced by prolonged incubation of native metallothionein with proteases.

Reconstitution of apometallothionein with metal ions was found to impart protection against proteolysis. This observation permitted the technique of limited proteolysis to be of utility for two purposes: (1) production of fragments of metallothionein corresponding to each domain, and (2) monitoring the stoichiometry of metal binding to metallothionein.

First, domain peptides can be obtained by proteolysis of the protein reconstituted with metal ions at suboptimal stoichiometries. Cd-α frag-

[16] D. R. Winge and K. A. Miklossy, *J. Biol. Chem.* **257,** 3471 (1982).

[17] Y. Boulanger, I. M. Armitage, K. A. Miklossy, and D. R. Winge, *J. Biol. Chem.* **257,** 13717 (1982).

[18] J. H. R. Kägi and A. Schaffer, *Biochemistry* **27,** 8509 (1988).

ments are readily produced by reconstitution of apometallothionein with 4 mol equivalents Cd(II) followed by limited proteolysis at pH 7.5 with either subtilisin or proteinase K.[19] Cd(II) ions bind to metallothionein in a cooperative manner at pH 7.5, resulting in a cadmium-saturated α domain.[20] The ^{113}Cd NMR spectrum of Cd_4-metallothionein exhibits the characteristic spectrum of $^{113}Cd(II)$-α signals.[20] It is preferable to perform the reconstitution procedure below pH 8, as metal binding is clearly cooperative only below this pH.[20] Metal ion binding is random along the polypeptide chain above pH 8 in samples containing less than 3 mol equivalents Cd(II).[20-22] At Cd(II)-to-protein ratios between 3 and 5, the Cd(II) ions bind preferentially within the α cluster.[22] Therefore, the Cd-α domain may be prepared also above pH 8 by proteolysis of a Cd_4-metallothionein sample.

Likewise, the β domain fragment can be prepared by reconstitution of mammalian apometallothionein with 6 mol equivalents Cu(I) followed by proteolysis.[23] Unlike Cd(II) and Zn(II) ions, Cu(I) ions bind initially within the β domain in an all-or-nothing manner.[23] Incubation of the Cu_6-MT molecule with proteinase K or subtilisin leads to degradation of the peptide segment corresponding to the α domain. At stoichiometries below 6 mol equivalents Cu-β domain fragments are isolated by proteolysis at pH 7.5, but the yield of the domain fragment increases as the Cu(I) stoichiometry is increased to 6 mol equivalents. The metal-domain fragments can be readily purified by gel-filtration or ion-exchange chromatography.

Methods for Preparation of Metallothionein Domain Fragments

Apometallothionein Preparation

Metallothionein can be purified from animals injected subcutaneously with cadmium or zinc salts as described.[24] If metallothionein is purchased commercially, caution must be taken as the usual protein obtained from Sigma Chemical Company (St. Louis, Mo) is a mixture of sequence variants (isoforms). Before preparation of the apoprotein, it is important to ensure that the sample is homogeneous. Purity of metallothionein is best assessed by reversed-phase high-performance liquid chromatography (HPLC) using a C_4, C_8, or C_{18} bonded column equilibrated in 0.1% (v/v)

[19] K. B. Nielson and D. R. Winge, *J. Biol. Chem.* **258,** 13063 (1983).
[20] M. Good, R. Hollenstein, P. J. Sadler, and M. Vašák, *Biochemistry* **27,** 7163 (1988).
[21] M. Vašák and J. H. R. Kägi, *Proc. Natl. Acad. Sci. U.S.A.* **78,** 6709 (1981).
[22] W. R. Bernhard, M. Vašák, and J. H. R. Kägi, *Biochemistry* **25,** 1975 (1986).
[23] K. B. Nielson and D. R. Winge, *J. Biol. Chem.* **259,** 4941 (1984).
[24] D. R. Winge and M. Brouwer, *Environ. Health Perspect.* **65,** 211 (1986).

trifluoroacetic acid. A gradient of 0 to 60% acetonitrile in 0.1% trifluoroacetic acid is effective. A single peak eluting in the gradient between 36 and 38% acetonitrile is observed for pure mammalian Cd,Zn-metallothionein on a C_{18} reversed-phase column. Low-pH conditions dissociate metal ions from the protein, so the peak eluting in the gradient represents apometallothionein. Apometallothionein is conveniently prepared by acidification of the protein to pH 2 with HCl followed by gel filtration on Sephadex G-25 equilibrated in 0.02 *N* HCl. We routinely use a gel bed of dimensions 2.5 × 20 cm and limit the volume applied to 8 ml. This is a convenient volume for up to 25 mg metallothionein. The metal-free polypeptide elutes in the exclusion column volume. Cd(II) and Zn(II) ions elute in the internal column volume. The apometallothionein prepared from Cd,Zn-metallothionein maintains the cysteinyl sulfur atoms in the reduced thiolate state, so no reduction step is required. The fractions of apometallothionein are pooled and stored at −70° in vials sealed anaerobically. Multiple freeze-thawing cycles should be avoided.

Quantitation of Apometallothionein

The concentration of the apomolecule is quantified by use of thiol reagents [5,5′-dithiobis(2-nitrobenzoic acid) (DTNB) or dithiodipyridine] or amino acid analysis. Quantitative amino acid analysis is the method of choice for determination of the polypeptide concentration. After hydrolysis of the sample in 5.7 *N* HCl at 110° for 24 hr, the amino acids are resolved by standard chromatographic techniques described in [48] in this volume. The concentration is readily obtained by division of the molar concentration of several amino acids by their mole percentage, which yields the molar concentration of the polypeptide chain. If only a 24-hr hydrolysis is used, the calculation of polypeptide concentration avoids serine due to partial dehydration and cysteine due to oxidation.

Quantitation of cysteine residues by thiol titrations is also effective in the determination of protein concentration. Division of the total thiol concentration by 20 (the number of cysteinyl groups in the mammalian polypeptide) yields the concentration of the polypeptide chain. The implicit assumptions are that no oxidation of thiols occurred during sample preparation and that no reductants are present in column buffers. Analyses of protein concentration by amino acid analysis and thiol concentration provide unequivocal data on the concentration of the molecule as well as the extent of sample reduction. The use of Ellman's reagent [5,5′-dithiobis(2-nitrobenzoic acid)] has been described in detail in Volume 25 of this series. Dithiodipyridine is also effective in quantifying thiol groups.[25] In the

[25] D. R. Grasseti and J. F. Murray, *Arch. Biochem. Biophys.* **119**, 41 (1967).

dithiodipyridine protocol varying aliquots of apometallothionein are added to 0.1 *M* sodium acetate, pH 4, to a final volume of 0.98 ml. To this volume is added 0.02 ml of a stock solution of 2,2′-dithiodipyridine (7 mg dissolved in 2 ml 1 *N* HCl diluted finally to 8 ml with 0.1 *M* sodium acetate, pH 4). After a 1-hr incubation at room temperature, the absorbances at 343 nm are recorded for the different dilutions. The molar extinction coefficient for 2-thiopyridine is 7.06. The assay is quantitative below 70 nmol thiol. The low pH of the assay minimizes deleterious side reactions.

The concentration for apometallothionein can be conveniently estimated from the extinction coefficient for protein from rabbit liver at pH 1. The coefficient is 7.9 mg cm^{-1} ml^{-1} at 220 nm.[26] The assumption in using the extinction coefficient is that the protein is depleted of metal ions because metal binding contributes absorbance at 220 nm due to metal:thiol charge transfer transitions.

Metal Reconstitution

Reconstitution of apometallothionein is achieved by mixing the apopolypeptide at pH 2 with aliquots of metal ion stock solutions prepared at pH 2. Stock solutions of Cu(I) are prepared anaerobically as either Cu(I)-acetonitrile[27] or as $(CuCl_2)^-$ by dissolving cuprous chloride in 0.1 *N* HCl containing 4% NaCl by weight. Standardized metal solutions are available commercially and can be used for certain metals. Stock solutions containing nitric acid should be avoided due to potential oxidation of cysteinyl residues. In the preparation of α domain fragments the apopolypeptide (ranging from 1 to 20 mg) is mixed with 4 mol equivalents Cd(II), whereas the preparation of β domain fragments consists of using 5.5 mol equivalents Cu(I). The metal–protein mixtures are neutralized with buffer (potassium phosphate and HEPES–acetate are effective). The final pH is dictated by the protease to be employed. Buffers at pH 7.0 and 7.5 are preferred for samples incubated with proteinase K and subtilisin, respectively. The neutralized metalloprotein sample is incubated with protease at a 30:1 ratio of MT:protease for a period of 2–3 hr at 37°. The reconstitution procedure is performed routinely under anaerobic conditions, although the need for anaerobiosis is more critical for Cu(I)-metallothionein samples compared to Cd(II)-metallothionein. On completion of the incubation period, the sample is gel filtered on Sephadex G-50 (2.5 × 90 cm) in nitrogen-purged buffer at neutral pH to resolve the metallo-domain fragment from the digested peptides. The domain fragment is identified by

[26] R. H. O. Buhler and J. H. R. Kägi, *Experientia, Suppl.* **34,** 211 (1979).

[27] P. Hemmerich and C. Sigwart, *Experientia* **19,** 488 (1963).

either metal analysis or absorption measurements at 250 nm to detect the metal–thiolate charge-transfer transition.

Analysis of Product

The domain fragments are identified by amino acid analysis and/or sequence analysis by Edman degradation. The two domains of mammalian metallothionein have unique amino acid compositions, so amino acid analysis provides a clear indication of sample identity. The lysine content of either domain fragment can be variable depending on the extent of proteolysis of the Lys_{30}-Lys_{31} interdomain link. Sequence analysis by Edman chemistry provides an unequivocal verification of the α domain fragment and the nature of the proteolytic trimming. Edman degradation of the β fragment is precluded by the acetylation of the amino-terminal methionine. Sequence analysis is possible after CNBr digestion of the sample to remove the terminal methionine (see [49] in this volume). The products as isolated contain bound metal ions, so if the metal-free domain peptide is desired, an additional step is required to deplete the sample of metal. Metal depletion is readily accomplished in the Cd-α sample by acidification to pH 1, as described in the section, Apometallothionein Preparation. Copper depletion of β samples is accomplished by treatment with 0.2 *M* KCN at neutral pH for 1–2 hr. Copper cyanide is separated from the peptide by chromatography on Sephadex G-25 equilibrated and eluted with 10 m*M* Tris-HCl, pH 7.4. Alternative procedures for depleting β samples of bound copper ions include acidification to pH 0.5 followed by gel filtration at pH 1.5 with the sample loaded between a sandwich of 1 *N* HCl, or use of tetrathiolmolybdate[28] for Cu(I) chelation. The peptide peak in the elution profile is located by monitoring absorbances at 214 nm or reactivity with fluorescamine.[29] The peptide-containing fractions must be reduced prior to use, as the process of copper depletion results in oxidation of peptide thiols. Reduction is achieved by lyophilizing the copper-free β peptide followed by resuspension of the powder in 6 *M* guanidinium chloride containing 50 m*M* dithiothreitol and 0.1 *M* Tris-HCl, pH 8.6. Dithiothreitol (DTT) is preferred over 2-mercaptoethanol by virtue of the more negative redox potential and lower p*K*. The p*K* of DTT is 8.3.[30] After incubation for 4–12 hr at 23°, the sample is acidified to pH 2 followed by desalting on Sephadex equilibrated with 0.01 *N* HCl. The extent of reduction is monitored by reaction with Ellman's reagent (DTNB) or dithiodipyridine.

[28] K. H. Summer, personal communication.
[29] P. Bohlen, S. Stein, and S. Udenfriend, *Arch. Biochem. Biophys.* **163,** 390 (1974).
[30] P. C. Jocelyn, this series, Vol. 143, p. 246.

The yield of α peptide from the described procedure is approximately 6 mg α peptide/20 mg of starting apometallothionein. The yield for β peptide is less than 4 mg, as the procedure requires more steps.

Utility of Procedure

This procedure is of general use for any multidomained metallothionein. Metallothioneins from most vertebrate and invertebrate species appear to exist as multidomain molecules. Utility is based on the assumptions that a thermodynamic difference exists in metal binding to the two domains and that the domains are structurally independent. The proteolytic procedure is not unique as a method for generating domain peptide. Solid-phase peptide synthesis has been successfully employed to produce quantities of each domain peptide. Solid-phase synthesis has the advantage that modified sequences can be generated.

Limited proteolysis has proven useful in delineating the Cu(I)-binding domain of the Cu(I)-specific transcriptional activation protein from *Saccharomyces cerevisiae.* The transcriptional activator, designated *ACE1* or *CUP2*, appears to consist of two domains; the amino-terminal segment is the DNA-binding domain.[31] Specific DNA binding is dependent on Cu(I) binding to the amino-terminal domain. Cu(I) binding partially stabilizes the amino-terminal half of the protein toward limited proteolysis by trypsin. Formation of a Cu(I)-protein:DNA complex imparts greater resistance toward proteolytic digestion.

Methods for Use of Proteases in Determining Metal-Binding Stoichiometry

Proteolysis can be used in assessing the metal ion-binding stoichiometry in metallothionein samples. In this procedure apometallothionein is titrated with increasing quantities of a given metal ion according to the reconstitution procedure described above. The neutralizing buffer is kept at or below pH 7.5 to ensure cooperative metal ion binding. The neutralized samples are incubated with either proteinase K or subtilisin as described. After 2 hr, aliquots of each sample are reacted with fluorescamine to quantify amino groups. Each peptide bond cleaved generates an additional α-amino group capable of reaction with fluorescamine. The addition of increasing mole quantities of metal ion yields increasing concentrations of protein resistant to proteolytic digestion. Total protection is achieved at the point when each molecule is saturated in each cluster. At that point the fluorescence of the sample is minimal and is equivalent to that of a sample

[31] P. Furst, S. Hu, R. Hackett, and D. Hamer, *Cell (Cambridge, Mass.)* **55,** 705 (1988).

of metallothionein incubated in the absence of protease. The break point in the curve of fluorescence vs metal concentration is the binding stoichiometry. The assumptions in this assay are as follows: (1) binding is cooperative or all-or-nothing, (2) the metal ions used do not inhibit the protease, (3) the apometallothionein is fully reduced, and (4) a buffer not containing a primary or secondary amine is used.

The assay has proven effective for mammalian and yeast metallothioneins. The effectiveness of the assay is dependent on a significant difference in the fluorescamine reactivity of apometallothionein incubated in the presence and absence of protease. Mammalian metallothioneins exhibit a difference of five- to eightfold, depending on the isoform used. The individual domain peptides of mammalian metallothionein and yeast metallothionein have differences closer to threefold.

Detailed Protocol

Apometallothionein in 0.02 *N* HCl: 30–40 μg
Metal stock solution in 0.01 *N* HCl: appropriate quantity
Neutralizing buffer and protease: 0.5 *M* potassium phosphate, pH 7.5 for subtilisin incubation, pH 7 for proteinase K incubation
Protease stock solutions: 0.1 mg/ml in deionized water
Final volume: 200–400 μl
Incubation time: 1.5 hr at 37°

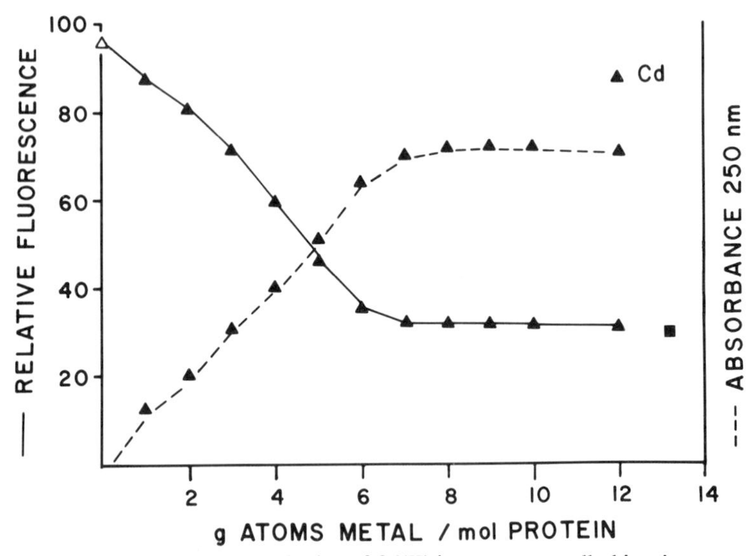

FIG. 1. Assay for the binding of Cd(II) ions to apometallothionein.

Take aliquots equivalent to 15 μg of apometallothionein and dilute with 1 ml of 0.2 *M* sodium borate, pH 9. Add 0.2 ml of a solution of fluorescamine (Sigma, St. Louis, MO) containing 25 mg dissolved in 50 ml of acetone. The emission of each sample is recorded at 475 nm after excitation at 390 nm. Appropriate controls include samples incubated in the presence and absence of protease. A typical assay for the binding of Cd(II) ions to apometallothionein from rat is shown in Fig. 1. The relative fluorescence of apometallothionein titrated with increasing quantities of Cd(II) is shown in filled triangles along the solid line. The filled square reflects the fluorescamine reactivity of the apoprotein in the absence of proteolysis. Also shown in Fig. 1 is the rise in the absorbance at 250 nm of each sample (dashed line). Maximal absorbance is observed at 7 mol equivalents, which is the maximal Cd(II)-binding stoichiometry of mammalian metallothionein. This is also the equivalency of maximal protection of the polypeptide against proteolysis.

Section VI

Physicochemical Characterization of Metallothioneins

[53] Metal Removal from Mammalian Metallothioneins

By PETER E. HUNZIKER

Introduction

Metal removal from metallothionein (MT) is a central step in the structural analysis of this protein. While zinc and cadmium are readily removed at low pH, copper remains partially bound and can be removed from MT only by using chelating agents. In this chapter, the methods for the preparation of apo-MT that have been successfully used for the primary structure analysis of mammalian MTs are described.

Removal of Zinc and Cadmium

Apometallothionein is usually prepared by gel filtration at pH 2.[1] For human MT isolated by gel filtration or preparative high-performance liquid chromatography (HPLC)[2] 10 μl of trifluoroacetic acid (TFA)/ml of holo-MT-containing fractions is added. The acidified sample is mixed vigorously and immediately desalted on a Sephadex G-25 column equilibrated with 0.1% TFA. If not used immediately, the recovered apo-MT fractions are stored at $-20°$.

For small sample amounts, apo-MT is prepared by HPLC. The pH of the sample is decreased by adding 10 μl of TFA/ml of sample. Chromatography is performed on an Aquapore RP-300 column (4.6×250 mm or 2.1×250 mm; Applied Biosystems, Inc., San Jose, CA) at a flow-rate of 0.6 or 0.2 ml/min, respectively. Apo-metallothioneins are eluted using a gradient formed between solvent A (0.1% TFA) and solvent B (same as solvent A, containing 60% acetonitrile). The usually employed gradient is as follows: 25% solvent B for 5 min; 25 to 50% solvent B within 50 min; apo-MT fractions as monitored by absorbance at 220 or 215 nm are frozen at $-20°$ as soon as possible.

Removal of Copper

In contrast to zinc and cadmium, copper is not removed completely from MT at low pH. Thus, apo-MT is prepared using the chelating agent diethyldithiocarbamate (DTC).[3,4] The pH of a Cu-MT-containing solution

[1] M. Văsák, this volume [54].
[2] P. E. Hunziker, this volume [27].
[3] A. G. Morell and I. H. Scheinberg, *Science* **127,** 588 (1958).
[4] P. E. Hunziker and I. Sternlieb, *Eur. J. Clinical Invest.* (in press).

as isolated by gel filtration on a G-50 or G-75 column is adjusted to pH 5.0 by adding 30 μl of 3 *M* sodium acetate, pH 5.0/ml of sample (final concentration of acetate, 0.1 *M*). One milligram of solid DTC/ml of solution is added and the sample is incubated at room temperature for 1 hr. The colloidal Cu-DTC complex is removed by filtration through a 0.22-μm filter (Millipore, Bedford, MA). The slightly yellow filtrate is desalted on a Sephadex G-25 column in 0.1% TFA and the apo-MT-containing fractions are freeze-dried. The dried product is reduced as described elsewhere[5] and desalted on a Sephadex G-25 column in 0.1% TFA.

Comments

The procedure described for the removal of copper may also be used for the removal of zinc and cadmium. Another chelating agent, ammonium tetrathiomolybdate, has also been used.[6] However, this reagent failed to remove copper from human hepatic Cu-MT derived from patients with hepatic copper overload.[4]

[5] P. E. Hunziker, this volume [45].
[6] R. K. Mehra and I. Bremner, *Biochem. J.* **219**, 539 (1984).

[54] Metal Removal and Substitution in Vertebrate and Invertebrate Metallothioneins

By MILAN VAŠÁK

Introduction

In mammalian metallothioneins (MT) 20 cysteine thiolates serve as ligands to 7 divalent metal ions; whereas in crab MTs there are 18 cysteine thiolates and 6 divalent metal ions.[1] Besides both divalent metal ions, another metal often found in the native protein is copper(I), for which increased metal-to-protein ratios have been found.[2] Depending on the environmental factors or the type of metal induction, homometallic Zn- and/or Cd-MT forms are often unavailable. Therefore, the need for metal substitution of MTs is twofold: (1) to introduce a spectroscopic probe by isostructural substitution for otherwise invisible metal ion(s), Zn^{2+} and

[1] J. H. R. Kägi and Y. Kojima, *Experientia, Suppl.* **52**, 25 (1987).
[2] D. R. Winge, this volume [55].

Cd^{2+} by, e.g., Co^{2+}, Fe^{2+}, and $^{113}Cd^{2+}$, and (2) to generate homometallic Zn- and/or Cd-MT forms sometimes required for comparative purposes. Metal substitution in MT can be performed by two different approaches: either by the metal removal from the native protein followed by insertion of the new metal or by direct displacement of the first metal by another metal. The basis of the latter procedure is differences in the relative magnitude of the affinity constants for the binding sites in MT, which follows the order typical of thiolate model complexes, i.e., Hg(II) $\gg$ Cu(I), Ag(I), Bi(III) $\gg$ Cd(II) > Pb(II) > Zn(II) > Co(II) > Fe(II). Because the use of this approach in MT is rather limited, due to the presence of the metal–thiolate cluster structure, some intrinsic metal ions often remain trapped. In this chapter the substitution methods via the apoprotein are described.

As with all metal substitution studies of metalloproteins, procedures to control adventitious metal concentrations should be followed to ensure that reagents and apparatuses are "metal free." Generally, solutions should be treated to remove metal contaminants either by extraction with dithizone or by passing over a Chelex 100 column. All glassware, and when possible plastic labware, should be rinsed with 20% (v/v) HNO_3 and then with metal-free water prior to use.[3,4] To prepare the apoprotein the metal ions in native Zn- or Zn,Cd-MT are removed by a competition with hydrogen ions at low pH. The metal substitution by the metal of choice is usually performed by the pH adjustment of the apoprotein–metal mixture to neutral (7.5) or higher, sometimes required to achieve full metal loading, by Tris base. The weak metal-chelating properties of this organic amine prove to be useful in maintaining the solubility of the free metal ions at these pH values. The presence of a large number of easily oxidizable cysteine thiolates poses a major difficulty in the metal substitution of MTs. Consequently, the metal substitutions are performed anaerobically.

Preparation of Metal-Free Protein

Native Zn- or Zn,Cd-MT having a high state of purity and well-established metal stoichiometry should be used. Apometallothionein is prepared by gel-filtration chromatography (Sephadex G-50) at pH 2.[5] Typically 20 mg of the lyophilized protein is dissolved in 2 ml of the 20 m*M* Tris-HCl buffer at pH 8.6, containing 20 mg of dithiothreitol, and incubated for 1 hr at room temperature. The pH is then adjusted to 1.0 by the rapid addition

[3] B. Holmquist, this series, Vol. 158, p. 6.
[4] F. W. Wagner, this series, Vol. 158, p. 21.
[5] M. Vašák, E. Wörgötter, G. Wagner, J. H. R. Kägi, and K. Wüthrich, *J. Mol. Biol.* **196,** 711 (1987).

of 1 *M* HCl. The sample is centrifuged for 5 min using an Eppendorf centrifuge and applied to a Sephadex G-50 column (50 × 2 cm) equilibrated with 10 m*M* HCl, and eluted with the same solution. After elution the protein fractions of the major monomeric peak are collected. The apoprotein in 10 m*M* HCl can be concentrated either by ultrafiltration (ultrafiltration cell using YM-2 membrane; Amicon, Danvers, MA) or by lyophilization after adjusting the pH to 2.5 by dilution with metal-free water. The latter form is best suited for storage (−20°). Changes in the high-performance liquid chromatography (HPLC) profile of apo-MT as a function of its exposure to the low pH have been observed. Therefore, prolonged periods at low pH should be kept at minimum; at pH 1 a maximum of 2 hr, at pH 2 a maximum of 24 hr. The quality of apo-MT may be determined by measuring metal, which should be zero, and the cysteine thiolate-to-apo-MT ratio. Typical values for the thiol/apo-MT ratio are between 19 and 21 and between 17 and 19 for mammalian and crab MTs, respectively. Protein concentration is determined spectrophotometrically by measuring the absorbance at 220 nm in 0.1 *M* HCl using $\epsilon_{220} = 48{,}200\ M^{-1}\ \mathrm{cm}^{-1}$[6] for mammalian and $\epsilon_{220} = 50{,}000\ M^{-1}\ \mathrm{cm}^{-1}$[7] for crab MT, respectively. Both latter analytical methods are described in detail elsewhere in this volume.[8]

Preparation of Metal-Substituted Metallothionein

The procedures for the preparation of metal-substituted MTs via apoprotein described in this section were used to generate various metalloforms of mammalian and crab MT. In the latter case, however, only cadmium and cobalt derivatives have been prepared.[7]

Although a similar strategy applies to all metal derivatives, depending on the stability of the fully substituted protein to air, two basic approaches will be discussed. While in the first approach, which concerns mainly the diamagnetic Zn^{2+},Cd^{2+}, Pb^{2+}, Hg^{2+}, and Bi^{3+}, a simple saturation of all solutions by nitrogen or argon prior to use is usually sufficient, in the second approach, strictly oxygen-free conditions must be maintained throughout the preparation of the paramagnetic Co^{2+} and Fe^{2+} metal derivatives and the easily oxidizable Cu^{+}-MT form.

Due to the different affinity of the metal ions for the metal-binding sites or the ease of thiolate ligand oxidation, the metal ion can be sometimes nonspecifically bound or leave binding sites unoccupied. The nonspecifi-

[6] R. H. O. Bühler and J. H. R. Kägi, *Experientia, Suppl.* **34,** 211 (1979).
[7] J. Overnell, M. Good, and M. Vašák, *Eur. J. Biochem.* **172,** 171 (1988).
[8] M. Vašák, this volume [7].

cally and usually more weakly bond metal ions can be removed by treatment with a small portion of Chelex 100. This treatment has been successfully applied by us to the Co-, Zn-, Pb-, and Cd-MT derivatives.[9] Detailed studies on its effect on the stability of Zn- and Cd-MT have also been performed.[10] The extent of the metal substitution may be determined from the metal–protein stoichiometry. For native mammalian Zn- and Zn,Cd-MTs the stoichiometry of 7 mol of metal/mol of protein is well established.[1] Well-defined metal complexes with the same stoichiometry exist also with previously mentioned divalent metal ions and bismuth. The generation of well-defined metal complexes may also be followed spectrocopically, e.g., electronic absorption and circular dichroism, as a function of the metal-to-protein ratio.

Examples of Metal Substitution Procedures

Oxygen-Insensitive Metal Derivatives

All metal substitutions were carried out at room temperature.

Cd-Metallothionein.[5] Apometallothionein (20 mg) in 50 m*M* HCl (0.5 mg/ml) is placed in a round-bottom flask and 8 equivalents of a $CdCl_2$ solution at the same pH are added. Under these conditions no metal binding occurs. The solution is rendered oxygen free by gently agitating under a stream of nitrogen or argon (~ 20 min) and kept under a positive pressure of either gas. The solution is then titrated with 0.5 *M* Tris base (oxygen free) to pH 8.6 in ~ 30 sec with fast stirring. Immediately after the pH adjustment, a small portion of Chelex 100 slurry is added and the suspension is stirred for 5 min. The Chelex 100 is removed by centrifugation and the supernatant concentrated by ultrafiltration. The sample is applied to a Sephadex G-50 column (140 × 2.5 cm) equilibrated with 20 m*M* Tris-HCl/10 m*M* NaCl buffer at pH 8.6. The metal–protein ratio is checked on a small aliquot of the sample added to a 3-ml quartz cuvette containing 0.1 *M* HCl. The apoprotein and metal concentration are determined using a small aliquot of the sample as described.[8] Typically, the Cd/apo-MT ratio ranges from 6.5 to 7.5. The available sulfhydryl groups should also be quantified.[8] The basic method as outlined above must be modified for the following metals.

Zn-Metallothionein.[11] After adding 8 zinc equivalents to apo-MT the titration with 0.5 *M* Tris base must be performed slowly (5 min) to a final

[9] M. Vašák and J. H. R. Kägi, *Met. Ions Biol. Syst.* **15,** 213 (1983).
[10] W. Cai and M. J. Stillman, *Inorg. Chim. Acta.* **152,** 111 (1988).
[11] M. Vašák, unpublished (1986).

pH of 7.4 and no Chelex 100 is added. The excess metal is removed in a gel-filtration step as described above.

Pb-Metallothionein.[12] As might be expected, the difficulties arising from a low solubility of lead chloride required a modification of the basic method. Into 5 ml of apo-MT (0.2 mg/ml) in 0.1 *M* $HClO_4$ 8 equivalents of the $Pb(ClO_4)_2$ solution in 0.1 *M* $HClO_4$ is added. The pH is adjusted quickly to 7.6 by the addition of 0.5 *M* Tris base and then Chelex 100 is added. Gel-filtration chromatography check is performed on a small aliquot of the sample at pH 7.6. The Pb/apo-MT ratio must be determined in the absence of choloride ions (0.1 *M* $HClO_4$).

Hg- and Bi-Metallothionein.[12,13] Both Hg^{2+} and Bi^{3+} have a very high affinity toward thiolate ligands, even at pH 1. Thus, besides the well-defined complexes with the metal–protein ratios of 7, increased metal loading also exists. Exactly 7 equivalents of the 10 m*M* $HgCl_2$ or $BiCl_3$ solution in 0.01 *M* HCl or 1 *M* HCl, respectively, is under vigorous stirring, mixed into at least 5 ml of apo-MT solution in 0.01 *M* HCl (0.2 mg/ml). The acidic solutions are then slowly titrated by 0.5 *M* Tris base to pH 8.6 for Hg-MT and to pH 7.6 for Bi-MT. The metal–protein ratio is determined using quantitative amino acid analysis and carbon furnace atomic absorption spectroscopy.[14] The presence of monomeric species is confirmed by gel-filtration chromatography on a small aliquot of the sample (see above). For Hg_7-MT the metal saturation can also be followed spectrophotometrically while maximizing the absorbance at 280 nm at pH 8.6 (Fig. 1[15]). The pH of 8.6 is crucial for the participation of all cysteine thiolates in metal binding.[13]

Comments

All aforementioned metal derivatives of MT form well-defined monomeric complexes. However, only Cd- and Zn-MT can be concentrated by ultrafiltration. In our hands attempts to concentrate the Pb-, Hg-, and Bi-MT solutions above 1 mg/ml have failed.

Oxygen-Sensitive Metal Derivatives

The preparation of oxygen-sensitive Co(II)- and Fe(II)-MT should be carried out in the argon-purged glovebox and prior to use all solutions must be rendered oxygen-free by at least three freeze-vacuum-thaw cycles

[12] M. Good, M.Sc. Thesis, University of Zürich, Zürich (1983).

[13] W. Bernhard, Ph.D. Thesis, University of Zürich, Zürich (1987).

[14] M. Czupryn and K. H. Falchuk, this volume [47].

[15] W. Bernhard, M. Good, M. Vašák, and J. H. R. Kägi, *Inorg. Chim. Acta* **79** (B7), 154 (1983).

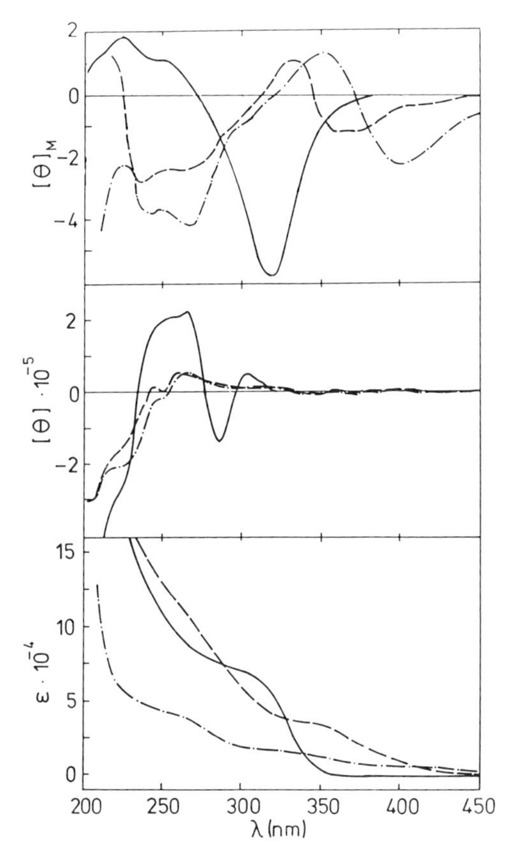

FIG. 1. Magnetic circular dichroism (MCD) (top), circular dichroism (CD) (middle), and absorption (bottom) spectra of $Hg(II)_7$-MT (—), $Pb(II)_7$-MT (– – –), and $Bi(III)_7$-MT (– · –). Units employed: ϵ (M^{-1} cm), $[\theta]_M$ (degrees cm^2 $dmol^{-1}$), and $[\Theta]_M$ (degrees cm^2 $dmol^{-1}$ G^{-1}). (Reproduced from Ref. 15.)

using a vacuum line. Metal substitutions can be performed only with the divalent oxidation state of these metal ions.

Preparation of Co(II) and Fe(II) Solutions.[11] Freshly prepared Co(II) solutions in H_2O from commercially available $CoCl_2 \cdot 6H_2O$ or $CoSO_4 \cdot H_2O$ salt of the highest purity must be used. As for the Fe(II) solution the commercially available Fe(II) salts contain varying amounts of the higher oxidation state, even the most stable $(NH_4)_2FeSO_4 \cdot 6H_2O$ form, and thus cannot be used without further reduction. Generally, 10 mg of $(NH_4)_2FeSO_4 \cdot 6H_2O$ is dissolved in 9 ml doubly-distilled water and 1 ml of 50 m*M* NH_4OH-NH_4Cl buffer, pH 9, is added. Iron is precipitated by the addition of Na_2S as black FeS. The reduced sample is heated for 15 min at 60°, cooled to room temperature, and centrifuged. The precipitate is washed with the same buffer, doubly-distilled water, and centrifuged. The precipitate is dissolved in 0.2 *M* H_2SO_4 and the resulting $FeSO_4$ solution used for the metal substitution. The metal concentration of both stock solutions is determined by flame atomic absorption spectroscopy.

Preparation of Co(II)- and Fe(II)-Metallothionein.[16,17] Degassed apo-MT (0.5 mg/ml) solution, 50 m*M* Co(II), or Fe(II) solution and 0.5 *M* Tris base are transferred into an argon-purged glovebox. Exactly 7 equivalents of either metal ions is mixed with the apo-MT solution and the pH titrated with Tris base to 8.6. The metal-to-protein ratios are determined by a combination of quantitative amino acid analysis[8] and flame atomic absorption spectroscopy.[14] Metal saturation can also be followed spectrophotometrically as described.[16,17] Both metal derivatives are extremely air sensitive. The green-colored Co(II)-MT derivative yields a light-brown product on exposure to air. Fe(II)-MT has a yellow color that, on exposure to air, immediately turns wine red and the protein slowly precipitates.

The preparation of the air-sensitive Cu(I)-substituted form of this protein is described elsewhere in this volume.[2]

Acknowledgments

The financial support of the Schweizerischer Nationalfonds throughout this work is gratefully acknowledged.

[16] M. Vašák and J. H. R. Kägi, *Proc. Natl. Acad. Sci. U.S.A.* **78**, 6709 (1981).
[17] M. Good and M. Vašák, *Biochemistry* **25**, 8353 (1986).

[55] Copper Coordination in Metallothionein

By DENNIS R. WINGE

Metallothionein (MT) exists naturally with zinc and/or copper ions bound to it. In the normal adult animal Zn-metallothionein is the predominant form of the protein, copper is usually observed as a mixed Zn,Cu-protein. Copper-metallothionein exists in animals at elevated concentrations during early development, in diseases affecting copper homeostasis, and in copper-treated animals.[1-3] The protein is normally localized within the cytoplasm but in copper-loaded tissues the protein is also found

[1] I. Bremner, *Experientia, Suppl.* **52**, 81 (1987).

within granules likely to be lysosomes.[3,4] Copper-metallothioneins from livers of fetuses and copper-loaded animals appear to be predominantly present within lysosomes in a particulate form requiring a reductant or sulfitolysis for solubilization.[5-7] Copper-metallothionein is also the predominant form of the protein in fungi.[8,9] Its concentration is regulated by the copper concentration in the growth medium.

The properties of the Cu-protein are not analogous to those of the well-studied Cd,Zn-metallothioneins. The purification of metallothioneins containing a high Cu:Zn mole ratio is complicated by the propensity of Cu-metallothioneins to oxidize and subsequently aggregate. The lability of the molecule to oxygen has hindered more detailed investigations. The physical properties of the Cu-protein led to the early mistaken identification of the molecule as distinct from metallothionein.[10] It is now clear that the copper-binding protein in animal cells is metallothionein. Metallothionein is presently defined structurally as a cysteine-rich polypeptide with multiple Cys-Xaa-Cys sequence motifs and as a protein capable of binding metal ions within metal–thiolate polynuclear clusters. The term metallothionein is used in diverse eukaryotic species without regard to primary structural homology. However, the name does not extend to small glutathione-related isopeptides involved in metal resistance in plants and certain fungi. These peptides of general structure $(\gamma\text{-EC})_n\text{G}$ will be designated by the structural formulas using the one-letter amino acid code (E, glutamic acid; C, cysteine; G, glycine). A full description of these peptides is found in [38]–[41] of this volume.

Much attention has been given to the nature of the copper-binding sites in metallothioneins. Studies have focused on the following areas: (1) the stoichiometry of copper ions within a cluster, (2) the valence state of the bound copper ions, and (3) the coordination geometry of the bound copper ions. Methodologies that are useful in addressing these issues will be reviewed in this chapter with the intent of providing a guide to potential approaches. References on specific procedures will be cited.

---

[2] I. Bremner, *J. Nutr.* **117,** 19 (1987).

[3] I. Sternlieb, *Gastroenterology* **78,** 1615 (1980).

[4] D. R. Winge and R. K. Mehra, *Int. Rev. Exp. Pathol.* **31,** 47 (1990).

[5] L. Ryden and H. R. Deutsch, *J. Biol. Chem.* **253,** 519 (1978).

[6] J. R. Riordan and V. Richards, *J. Biol. Chem.* **255,** 5380 (1980).

[7] K. Lerch, G. F. Johnson, P. S. Grushoff, and I. Sternlieb, *Arch. Biochem. Biophys.* **243,** 108 (1985).

[8] K. Lerch and M. Beltramini, *Chem. Scr.* **21,** 109 (1983).

[9] D. R. Winge, K. B. Nielson, W. R. Gray, and D. H. Hamer, *J. Biol. Chem.* **260,** 14464 (1985).

[10] D. R. Winge, R. Premakumar, R. D. Wiley, and K. V. Rajagopalan, *Arch. Biochem. Biophys.* **170,** 253 (1975).

Copper-Binding Stoichiometry

The copper-binding stoichiometry of metallothionein exceeds the binding stoichiometries of Cd(II), Zn(II), Ni(II), and Co(II) ions.[11] Cadmium-metallothioneins typically exhibit a Cys:Cd ratio of nearly 3, Cu-metallothioneins from animal and fungal species contain Cys:Cu ratios below 2. In the following cases detailed studies have defined the metal-binding stoichiometry of metallothioneins with various metal ions: mammalian MT β domain with 9 cysteinyl residues,[12] 3 Cd(II) or Zn(II) ions vs 6 Cu(I) ions; yeast MT with 12 cysteinyl residues,[9] 4 Cd(II) ions vs 7–8 CuI) ions; and *Neurospora crassa* MT with 7 cysteinyl residues,[8] 3 Cd(II) ions vs 6 Cu(I) ions.

An apparent exception to the Cys:Cu ratio being below 2 was the calf Cu,Zn-metallothionein reported by Briggs and Armitage.[13] This molecule was reported to contain three copper ions and four zinc ions. The demonstration of ^{113}Cd-exchangeable zinc ions in the α domain of the molecule implied that the three copper ions resided within the β cluster, yielding a Cys:Cu ratio of 3. It is unclear whether the calf sample was partially depleted of copper or whether metallothionein can exist with only three-Cu(I)ions within the β cluster. Calf liver metallothionein has now been purified with a metal stoichiometry of Cu_6,Zn_4.[14] This observation may imply that the Briggs and Armitage protein was partially copper depleted.

Binding stoichiometry is assessed by quantitation of the copper in the sample, typically by atomic absorption spectroscopy (see detailed procedures in Chapter 12, Vol. 158 of this series) and quantitation of the protein concentration. Protein concentration is best analyzed by quantitative amino acid analysis after hydrolysis in 5.7 *N* HCl (see [48] in this volume). Concentrations of amino acids are obtained directly, so determination of the protein concentration is dependent on the knowledge of quantity of each amino acid in the polypeptide chain. This information is available from primary sequence data (see [69] in this volume). Methods frequently used to quantify the concentration of Cd,Zn-metallothionein [including quantitation of thiolates with 5,5′-dithiobis(2-nitrobenzoic acid) or use of an extinction coefficient at 220 nm for peptide bond absorption at pH 2] are not effective with Cu-metallothioneins. Thiol reagents do not displace copper ions quantitatively and copper remains associated with most metallothioneins at pH 2. The ratio of copper to protein concentrations yields the Cu(I)-binding stoichiometry. This value can be misleading if anaerobic

[11] K. B. Nielson, C. L. Atkin, and D. R. Winge, *J. Biol. Chem.* **260,** 5342 (1985).
[12] K. B. Nielson and D. R. Winge, *J. Biol. Chem.* **259,** 4941 (1984).
[13] R. W. Briggs and I. M. Armitage, *J. Biol. Chem.* **257,** 1259 (1982).
[14] D. Petering, personal communication (1990).

conditions were not employed in the purification of Cu-metallothionein, because oxidation can deplete the molecule of bound copper ions. This has been a problem historically, as copper stoichiometry data have varied for certain proteins due to artifacts of isolation. Oxidation of thiolates catalyzed by copper ions can be minimized by purifying native samples with nitrogen-purged buffers at pH 5.5 or in buffers containing a reductant, e.g., 10-20 m*M* dithiothreitol.

Binding stoichiometry can be assessed also by *in vitro* reconstitution procedures (see [55] in this volume for detailed protocols). Ag(I) ions are isoelectronic to Cu(I) ions and can be used in place of Cu(I) ions for binding studies. Studies of reconstitution of apometallothioneins with increasing quantities of Cu(I) or Ag(I) have yielded data consistent with native binding stoichiometries.[11,15] The most useful procedures include titration studies with increasing quantities of Cu(I) followed by assays of luminescence, measurement of the Cu-thiolate, charge-transfer transition in the ultraviolet, or the susceptibility of the protein to proteolysis.[9,11,15]

In Vitro Protocols for Determination of Copper Stoichiometry

In vitro procedures for the determination of copper stoichiometry involve metal reconstitution protocols of apometallothionein. Apometallothionein is best prepared from Cd,Zn-metallothioneins, as metal depletion does not usually result in thiol oxidation. Preparation of apometallothionein is discussed in [53]-[55] in this volume. Depletion of metallothionein-bound copper is achieved by gel filtration at pH < 1 or incubation with KCN or tetrathiolmolybdate followed by chromatography at neutral pH. The use of acidification or KCN requires a subsequent reduction step to reduce oxidized cysteines (see [55] in this volume). Apometallothionein is reasonably stable near pH 2, but samples should be stored at -70° and stored on ice while in use. Multiple freeze-thawing is deleterious to apometallothionein samples, as is prolonged storage at room temperature.

Apometallothionein (30-40 μg) is mixed with increasing quantities of Cu(I) or Ag(I) stock solutions at pH 1.5-2. The apoprotein is preserved in 0.02 *N* HCl for Cu(I) reconstitutions or 0.1% trifluoroacetic acid for Ag(I) titrations. The mole ratio of metal:metallothionein is increased from 1 to a number equivalent to the cysteine content of the polypeptide. The mixture is neutralized to pH 7 with potassium phosphate or Tris-acetate such that the final volume is 1 ml. This process is carried out in an anaerobic bag or chamber and all solutions are deaerated prior to use. The Cu(I) samples are

[15] A. J. Zelazowski, Z. Gasyna, and M. J. Stillman, *J. Biol. Chem.* **264,** 17091 (1989).

scanned in anaerobic cuvettes in an ultraviolet (UV) spectrophotometer for absorbances in the ultraviolet and in a fluorimeter for emission measurements. Typical UV scans are from 220 to 350 nm and emissions scan from 450 to 700 nm after excitation at 300 nm. Copper-metallothioneins exhibit a shoulder near 260 nm in the absorption spectrum. The maximal binding stoichiometry is defined as the stoichiometry at which there is maximal shoulder absorbance. Addition of copper in excess of this stoichiometry enhances the absorbance in the ultraviolet but abolishes the shoulder feature. The physical basis of the absorbance shoulder is not clear. Maximal Cu(I)-binding stoichiometry as determined by emission measurements is defined as the Cu(I) stoichiometry at which emission is maximal. Care is required in the interpretation of emission data as emission is quenched by oxygen, so all samples must be anaerobic.

The use of partial proteolysis for determination of Cu(I) or Ag(I) binding stoichiometry is detailed in [52] in this volume.

The *in vitro* reconstitution studies have inherent limitations. Reconstitution studies require a fully reduced apometallothionein. This is verified by quantifying the protein by amino acid analysis and quantifying the thiol content by thiol titration. Copper reconstitutions require a stable form of Cu(I). This can be attained with Cu(I)-acetonitrile[16,17] or $[CuCl_2]^-$ prepared by anaerobically dissolving CuCl in 0.1*N* HCl containing 1 *M* NaCl.[12] Solutions of Cu-acetonitrile are stable for prolonged periods of time at $-20°$ whereas $[CuCl_2]^-$ must be used promptly. Acetonitrile suppresses disproportionation of Cu(I) and metal hydrolysis up to pH 7. Cu(I)-acetonitrile is prepared by dissolving Cu_20 in CH_3CN containing 2 *M* $HC1O_4$ at 100° and subsequently allowing the solution to slowly evaporate at room temperature to yield crystals.[16] A detailed account of the Cu(I) reconstitution procedure is contained in [55] in this volume. Cu(I)-metallothioneins are labile in air, so anaerobic procedures must be employed for reconstitutions and assays. Anaerobic bags are available from Instruments for Research and Industry (Cheltenham, PA) (I^2R) and anaerobic chambers are commercially available from several vendors. An important consideration of *in vitro* Cu(I) reconstitution studies is the propensity of excess Cu(I) ions to disrupt the cluster structure. Excess Cu(I) ions abolish the luminescence of native and reconstituted samples. This is especially a problem in using $[CuCl_2]^-$. The addition of stoichiometric amounts (relative to the cysteine content of metallothionein) of cysteine, glutathione, dithiothreitol or KCN will minimize the deleterious effects of excess Cu(I) on emission.[18]

[16] P. Hemmerich and C. Sigwart, *Experientia* **19,** 488 (1963).

[17] V. Vortisch, P. Kroneck, and P. Hemmerich, *J. Am. Chem. Soc.* **98,** 2821 (1976).

[18] J. Byrd, R. M. Berger, D. R. McMillan, C. Wright, D. Hamer, and D. R. Winge, *J. Biol. Chem.* **263,** 6688 (1988).

Valence State of Copper Ions Bound to Metallothionein

The absence of a significant electron paramagnetic resonance (EPR) signal and d-d transitions in the 600-nm spectral region in both mammalian and yeast Cu-metallothioneins were early indications that copper ions were present as cuprous ions or antiferromagnetically coupled cupric ions.[8,10,19,20] A copper EPR signal is characteristic of the unpaired electron in the d^9 state of cupric copper. Cuprous complexes or complexes of cupric ions in a configuration in which the unpaired electrons are spin-coupled are silent in the EPR spectrum. The initial assignment that bound copper ions were indeed cuprous was the observation that incubation of the protein at 100° anaerobically did not release any copper as cupric ions as detected by EPR spectroscopy.[10,19] Aerobic incubation at 100° released the total copper content as cupric ions. Likewise, the yeast Cu-metallothionein did not exhibit any EPR-detectable cupric copper at acid pH unless oxygen was present.[19] Native samples of yeast and mammalian Cu-metallothioneins were reported to contain an amount of EPR-detectable copper ranging from 1 to 12% of the total copper.[10,19,20] This amount of cupric copper was later found to be an artifact of isolation. Preparation of the protein under anaerobic conditions revealed the molecule to be EPR silent at both 19 and 173 K.[20] The lability of protein in air complicates the assignment of the valence state of copper. Integration of an EPR signal to quantify the percentage of cupric copper in a given sample is necessary in the interpretation of EPR data. EPR-silent copper was also reported for a Cu,Zn-metallothionein from calf liver.[13]

The presence of cuprous ions in metallothionein was confirmed by X-ray photoelectron spectroscopy.[21] A single homogeneous copper $2_{p_{3/2}}$ signal near 932 eV typically seen in Cu(I) complexes was observed for yeast Cu-metallothionein. No satellite signals characteristic of Cu(II) or spin-coupled binuclear cupric complexes were seen. Treatment of the protein with hydrogen peroxide yielded complexes with the characteristic Cu(II)-binding energies.[21]

More recently, spectroscopic techniques measuring luminescence, extended x-ray absorption fine structure (EXAFS), and X-ray absorption near edge structure (XANES) have typically been used to confirm the presence of Cu(I) in metallothioneins in diverse species. In 1981, Beltramini and Lerch[22] reported that the Cu-metallothionein from *Neurospora*

[19] H. Rupp, R. Cammack, H. J. Hartmann, and U. Weser, *Biochim. Biophys. Acta* **578,** 462 (1979).

[20] B. L. Geller and D. R. Winge, *Arch. Biochem. Biophys.* **213,** 109 (1982).

[21] U. Weser, H. J. Hartmann, A. Fretzdorff, and G. J. Strobel, *Biochim. Biophys. Acta* **493,** 465 (1977).

[22] M. Beltramini and K. Lerch, *FEBS Lett.* **127,** 201 (1981).

crassa exhibited orange luminescence when excited with ultraviolet radiation. The emission was maximal at 565 nm and the quantum yield was calculated to be 0.013 in comparison with the standard dye, rhodamine 6G. The emission was abolished by displacement of the copper ions by acid conditions or oxidation of the protein. Luminescence was property inherent in the native Cu(I)-protein. Similar luminescence with a large Stokes shift was reported with Cu(I) complexes with small thiolate ligands, including cysteine, glutathione, and 2-mercaptoethane sulfonate.[8,23] Cu(I) complexes with nitrogen and phosphorus ligands also are known to luminesce at wavelengths between 600 and 700 nm.[24,25] The emission in thiolate complexes is attributed to a triplet excited state of the Cu(I)–thiolate charge-transfer transition.[24,26] Paramagnetic ions quench emission, so luminescence is observed for Cu(I) and not Cu(II) complexes.[27] Luminescence can therefore be used as a probe for Cu(I).

The emission of small Cu–thiolate complexes was observed only in a glass at 77 K rather than emission of Cu-metallothionein in solution at room temperature. The fate of the excited state of these complexes is highly dependent on the interaction of the complex with the solvent. Solvent interaction facilitates the internal conversion of the excited state to ground state without emission of a photon.[27] This solvent quenching explains the lack of emission of Cu(I) chelates in solutions. The striking fact that emission is observed in Cu-metallothioneins implies that the polypeptide shields the luminophore from solvent interactions. The emission of Cu(I) complexes can also be quenched by oxygen, coordinating anions and other small molecules.[26,28] Quenching of fluorescence by small molecules is useful in assessing the solvent exposure of a fluor.[29] The rate of quenching Cu-metallothionein from *N. crassa* by oxygen was two orders of magnitude greater than that of acrylamide, implying a compact tertiary fold shielding the Cu(I)-thiolate cluster.[26]

Luminescence has been observed for Cu-metallothionein from diverse species. The protein from *Saccharomyces cerevisiae* exhibits a corrected

[23] J. H. Anglin, W. H. Batten, A. I. Raz, and R. M. Sayre, *Photochem. Photobiol.* **13,** 279 (1971).

[24] M. T. Buckner and D. R. McMillan, *J. Chem. Soc., Chem. Commun.,* 759 (1978).

[25] R. A. Rader, D. R. McMillan, M. T. Buckner, T. G. Matthews, D. J. Casadonte, R. K. Lengel, S. B. Wittaker, L. M. Darmon, and F. E. Lytle, *J. Am. Chem. Soc.* **103,** 5906 (1981).

[26] M. Beltramini, G. M. Giacometti, B. Salvato, G. Giacometti, K. Munger, and K. Lerch, *Biochem. J.* **260,** 189 (1989).

[27] F. E. Lytle, *Appl. Spectrosc.* **24,** 319 (1970).

[28] D. R. McMillan, J. R. Kirchoff, and K. V. Goodwin, *Coord. Chem. Rev.* **64,** 83 (1985).

[29] D. B. Calhoun, J. M. Vanderkooi, G. R. Holtom, and S. W. Englander, *Proteins* **1,** 109 (1986).

emission maximum at 609 nm and a quantum yield of 0.0058.[18] The lifetime of emission was determined to be 0.44 μsec, compared to 10 μsec for Cu-metallothionein from *N. crassa*. The emission of the *S. cerevisiae* Cu-metallothionein was diminished by exposure to oxygen but the decay occurred only slowly.[18] Related emission has been reported for Cu-metallothioneins purified from the livers of pig and calf, for Cu-metallothionein *in situ* from the liver of Bedlington terrier, and the Cu–(γ-EC)$_n$G peptide complex from *Schizosaccharomyces pombe*.[18,30–33] The emission of each mentioned complex implies Cu(I) coordination in a solvent-inaccessible coordination site.

The availability of synchrotron radiation has enhanced the utility of X-ray absorption spectroscopy. In the absorption of X rays by a given element, discontinuities exist in the rising absorption band within 50 eV of the energy peak.[34] The fine structure observed at absorption edges results from transitions of core electrons to discrete bound valence levels, although problems exist in assigning all the observed transitions.[34] The fine structure contains a wealth of information that is characteristic of the chemical environment of the X-ray-absorbing atom.[34] Edge measurements can reveal details about the oxidation state of a metal ion and the nature of the surrounding ligands.[34] Additional fine structure observed up to 1000 eV past the absorption edge results from interference between photoelectron waves propagating from the absorbing atom and waves back scattered by neighboring atoms. This extended fine structure designated EXAFS can provide information about the type, number, and distances of atoms in the vicinity of the absorbing atom.[34]

The dependency of the oxidation state of copper atoms on the X-ray absorption edge energies and intensities was defined in a study of a series of 20 Cu(I) and 40 Cu(II) model compounds.[35,36] Characteristic edge features assigned to $1s \rightarrow 4p$ transitions were observed between 8983 and 8984 eV and 8986 and 8987 eV for Cu(I) and Cu(II) complexes, respectively (Fig. 1). The appearance of a preedge peak below 8985 eV is indicative of Cu(I) in the sample.

Luminescence and analysis of the fingerprint region of the X-ray ab-

[30] R. K. Mehra and I. Bremner, *Biochem. J.* **219,** 539 (1984).

[31] M. J. Stillman, A. Y. C. Law, W. Cai, and A. Zelazowski, *Experientia, Suppl.* **52,** 203 (1987).

[32] I. Sternlieb, *Experientia, Suppl.* **52,** 647 (1987).

[33] R. N. Reese, R. K. Mehra, E. B. Tarbet, and D. R. Winge, *J. Biol. Chem.* **263,** 4186 (1988).

[34] S. P. Cramer and K. O. Hodgson, *Prog. Inorg. Chem.* **25,** 1 (1979).

[35] L. S. Kau, D. J. Spira-Solomon, J. E. Penner-Hahn, K. O. Hodgson, and E. I. Solomon, *J. Am. Chem. Soc.* **109,** 6433 (1987).

[36] E. I. Solomon, *Prog. Clin. Biol. Res.* **274,** 309 (1988).

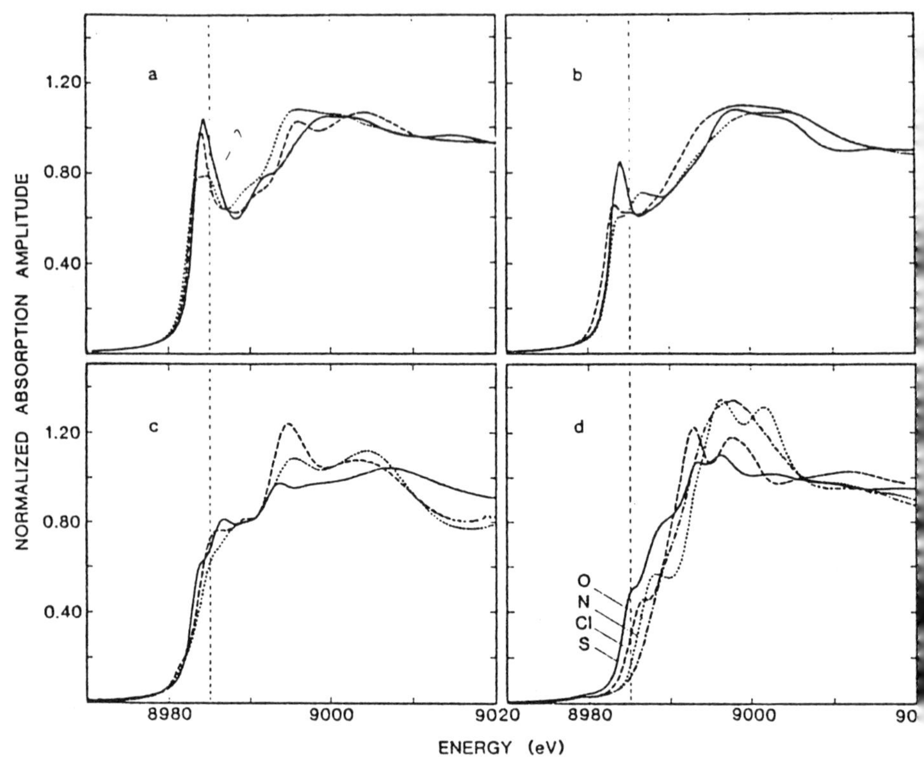

FIG. 1. Normalized X-ray absorption edge spectra of Cu(I) complexes (a,b,c) and Cu(II) complexes (d) compiled by Kau *et al.*[35] The multiple curves in each panel represent different copper complexes, the identity of which are listed in Ref. 35, *J. Am. Chem. Soc.* **109,** 6433 (1987). The vertical dashed line in each panel indicates the 8985.0-eV energy. The four curves in (d) represent Cu(II) tetragonal complexes with either S_4, Cl_4, or O_4 equatorial ligand sets. (Reprinted with permission of the American Chemical Society.)

sorption edge provide a clear indication of the valence state of copper in metallothionein. Copper-metallothioneins from all species studied to date and Cu–(γ-EC)$_n$G peptide complexes from fungi contain copper ions bound as Cu(I). Minor quantities of Cu(II) observed by EPR spectroscopy in certain Cu-metallothioneins are attributed to artifactual oxidation during sample preparation. If a given sample contained mixed valent copper ions, X-ray absorption edge spectral data can provide quantitative results on the amounts of Cu(I) and Cu(II) within a sample from calculations from difference edge spectra.[35] Cupric ions are not expected to be present in cysteine-rich polypeptides, as cupric ions are readily reduced by thiolates. One possible exception may be metallothioneins containing multiple histinyl residues in addition to multiple cysteines, such as metallothionein-1 from *Candida glabrata.*

Coordination Geometry of Cuprous Ions

Cuprous ions can form digonal, trigonal, and tetrahedral complexes. Many small Cu(I) complexes with thiolate ligands have been characterized crystallographically.[37] In structures of monometallic to polymetallic Cu(I)-thiolate complexes the most common binding geometry is digonal and trigonal.[37] A series of $[Cu_4(SR)_6]^{2-}$ complexes have been described with the cage consisting of doubly bridging thiolate ligands binding cuprous ions in a trigonal planar coordination.[38-40] The structures are stabilized primarily by the μ-bridging sulfurs.

X-Ray absorption edge spectroscopy is a powerful tool for assessment of coordination number. The combination of edge peak energies and peak shapes gives a clear picture of the coordination geometry. Kau *et al.*[35] demonstrated that digonal and trigonal Cu(I) complexes always exhibit X-ray absorption edge peaks at energies below 8984 eV (Fig. la and b). Three-coordinate Cu(I) complexes with one exception had a normalized absorption amplitude of 0.63 at 8984 eV (Fig. 1b). This is in contradistinction to the normalized peak amplitude of 0.99 for digonal Cu(I) complexes (Fig. 1a). Tetrahedral Cu(I) complexes exhibit an edge peak energy at 8984.7 with a peak amplitude of 0.49 (Fig. lc). The one "trigonal" Cu(I) complex that shows a peak amplitude in excess of 0.63 (shown in Fig. 1b, solid line) was later found to have a structure closer to linear two-coordinate Cu(I).

EXAFS spectroscopy has been employed more frequently than X-ray absorption edge spectroscopy for the determination of the Cu(I) coordination number in metallothionein. EXAFS spectroscopy is most accurate in the determination of distances and least accurate in the determination of coordination numbers. Prediction of coordination numbers by EXAFS data requires careful curve fitting of EXAFS Fourier transforms with model Cu(I) complexes. There is less ambiguity in XANES spectroscopy.

The initial EXAFS study was carried out on the Cu-metallothionein from *Saccharomyces cerevisiae.*[41] Tetrahedral copper coordination in a cubane structure was the initial conclusion. The study was flawed in that the protein was erroneously characterized as a Cu_4Cys_8-metallothionein. A

[37] I. G. Dance, *Polyhedron* **5,** 1037 (1986).
[38] D. Coucouvanis, C. N. Murphy, and S. K. Kanodia, *Inorg. Chem.* **19,** 2993 (1980).
[39] I. G. Dance, G. A. Bowmaker, G. R. Clark, and J. K. Seadon, *Polyhedron* **2,** 1031 (1983).
[40] I. G. Dance and J. C. Calabrese, *Inorg. Chim. Acta* **19,** L41 (1976).
[41] J. Bordas, M. H. J. Koch, H. J. Hartmann, and U. Weser, *FEBS Lett.* **140,** 19 (1982).

more careful analysis of the protein revealed the structure to be Cu_8Cys_{12}-metallothionein.[9] Reanalysis by X-ray analyses of the yeast Cu-metallothionein revealed a trigonal coordination of cuprous ions.[42] The yeast Cu-metallothionein exhibited X-ray absorption edge features analogous to trigonal Cu(I) complexes. A trigonal coordination Cu(I)–thiolate complex also gave the least error in the curve fitting of yeast Cu-metallothionein EXAFS data. The mean Cu–S distance of 2.23 Å in Cu-metallothionein is similar to the typical Cu–S distances of 2.24 to 2.27 Å for trigonal model complexes.[43-46] Cu–S distances of 2.16 to 2.17 Å are observed for digonal Cu(I) complexes and distances between 2.3 and 2.42 Å are observed for tetrahedral Cu(I) complexes.[43-46]

Trigonal Cu(I) coordination in mammalian metallothionein has been reported for rat Cu-β domain and calf liver Cu,Zn-metallothionein.[47,48] An average Cu–S distance of 2.25 Å was observed in each case, consistent with trigonal geometry. EXAFS amplitudes were also most consistent with three coordinate copper ions. More convincingly, the X-ray absorption edge structure of the Cu-β domain cluster had the characteristic features of three-coordinate Cu(I). The edge structure in Cu-metallothionein from *N. crassa* is also consistent with trigonal Cu(I) coordination.[49]

An EXAFS analysis of the lyophilized Cu-metallothionein from canine liver was reported to yield copper ions bound to four sulfur ligands at 2.27 Å.[50] The conclusion is questioned as the Cu–S distance is more typical of trigonal Cu(I) coordination and the reported X-ray absorption edge features are analogous to the edge features of model trigonal Cu(I) complexes shown in Fig. 1. The edge features of the canine Cu-metallothionein are analogous to those of the yeast Cu-metallothionein and the Cu-β domain cluster of rat metallothionein.[42,47]

The classification of Cu(I) complexes by X-ray absorption edge structures provides the most definitive assignment of coordination geometry in Cu-metallothioneins and therefore is the method of choice. EXAFS spectroscopy is ideal to determine Cu–S distances but the curve-fitting process of Fourier transforms introduces ambiguity in the assignment of coordina-

[42] G. N. George, J. Byrd, and D. R. Winge, *J. Biol. Chem.* **263,** 8199 (1988).
[43] I. G. Dance, *Aust. J. Chem.* **31,** 2195 (1979).
[44] E. H. Griffith, G. W. Hunt, and E. L. Amma, *J. Chem. Soc., Chem. Commun.*, 432 (1976).
[45] P. J. M. W. L. Birker and H. C. Freeman, *J. Am. Chem. Soc.* **99,** 6890 (1977).
[46] F. J. Hollander and D. Coucouvanis, *J. Am. Chem. Soc.* **96,** 5646 (1974).
[47] G. N. George, D. R. Winge, C. D. Stout, and S. P. Cramer, *J. Inorg. Biochem.* **27,** 213 (1986).
[48] I. I. Abrahams, I. Bremner, G. P. Diakun, C. D. Garner, S. S. Hasnain, I. Ross, and M. Vašák, *Biochem. J.* **236,** 585 (1986).
[49] T. A. Smith, K. Lerch, and K. O. Hodgson, *Inorg. Chem.* **25,** 4677 (1986).
[50] J. H. Freedman, L. Powers, and J. Peisach, *Biochemistry* **25,** 2342 (1986).

tion numbers. X-Ray absorption edge spectroscopy is also effective in the identification of Cu(I) complexes containing cuprous ions with different coordination numbers.[35] Copper-metallothioneins contain multiple Cu(I) ions within a cluster so XANES is well suited to provide a clear indication of heterogeneity in coordination number. The Cu-metallothioneins evaluated to date by XANES give no indication of coordination other than with trigonal geometry.

The summary of X-ray analysis of Cu-metallothioneins is that Cu(I) ions are bound in a trigonal geometry. The cysteinyl thiolates are likely to be μ-bridging sulfurs. This arrangement is distinct from Cd(II) or Zn(II) coordination in metallothionein, in which each metal ion is tetrahedrally coordination to terminal and μ-bridging sulfurs (see [58] in this volume).

Sample Preparation for Assessment of Copper Coordination

Luminescence measurements are routinely carried out with 3–30 μM quantities of Cu-metallothionein in dilute phosphate or Tris buffers. Measurements should be carried out at a fixed temperature as the emission is temperature dependent and in anaerobic cuvettes to minimize oxygen quenching. Procedural details for emission scanning can be found in [63] in this volume.

X-Ray absorption studies require 5–10 mM copper as Cu-metallothionein in a volume of 0.1–0.2 ml for packing of the cells. The sample should be prepared under anaerobic conditions to minimize oxidative destruction. Procedural details for X-ray absorption studies can be found in Vol. 117 of this series and Ref. 51.

[51] D. C. Koningsberger and R. Prins, eds., "X-Ray Absorption." Wiley, New York, 1988.

[56] *In vitro* Preparation and Characterization of Aurothioneins

By C. FRANK SHAW III, DAVID H. PETERING, JAMES E. LAIB, M. MERAL SAVAS, and KIM MELNICK

Gold(I) ions, like Cd(II), Pt(II), Hg(II), Cu(I), Zn(II), and Bi(III), bind to metallothionein (MT).[1,2] Depending on the ligation of gold, a variety of mixed metal- and all gold-thioneins can be formed. Their study is significant because gold(I) complexes are widely used as antiarthritic drugs. The

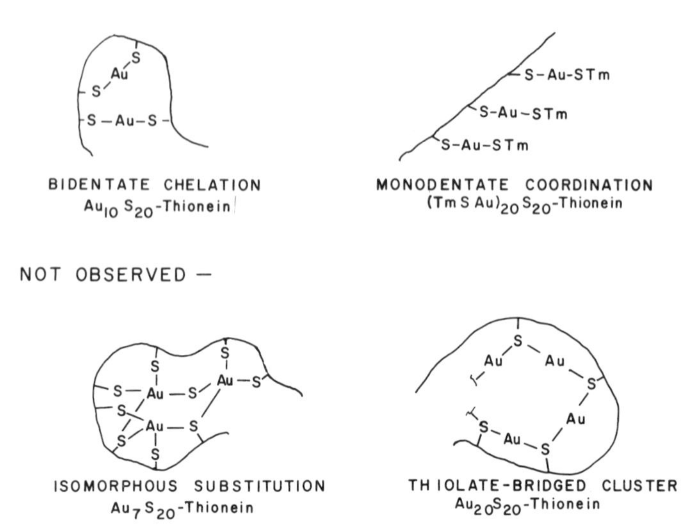

FIG. 1. Four plausible gold coordination environments for MT. Only the monodentate and bidentate modes have been experimentally observed to date.

gold, which is readily displaced from its carrier ligands, accumulates *inter alia* in the liver and kidneys of patients and animal models. When these tissues are examined after administering gold complexes to laboratory animals, a significant portion of the gold is found in the metallothionein fractions. Furthermore, cell culture studies demonstrate a protective role in which the ability to produce (or overproduce) metallothionein reduces the cytotoxicity of gold. Thus, it is of interest to understand which gold complexes react directly with metallothionein, the structures that result (gold coordination geometry), which initially bound metals can be displaced, and the rates at which gold displaces other metal ions.

Four plausible gold coordination environments have been identified: isomorphous displacement of Zn^{2+} or Cd^{2+}, monodentate coordination with retention of an "exogenous" ligand, terminal bidentate coordination, and gold–thiolate cluster formation[1,3] (see Fig. 1). Evidence for the monodentate and bidentate motifs has been obtained from the combination of

[1] C. F. Shaw III and M. M. Savas, *in* "Metal-Binding in Sulfur Containing Proteins" (M. Stillman, C. F. Shaw III, and K. Suzuki, eds.), in press. VCH Publishers, New York, 1900.

[2] A. Glennas, "Experimental Studies on Resistance to Certain Cytotoxic Effects of Three Gold Compounds, with Special Reference to Metallothionien." Dr. Med. Thesis, University of Oslo, 1986.

[3] J. E. Laib, C. F. Shaw III, D. H. Petering, M. K. Eidsness, R. C. Elder, and J. S. Garvey, *Biochemistry* **24,** 1977 (1985).

EXAFS/XANES spectroscopic data, metal exchange ratios, and retention of radiolabeled ligands.[3] Given the absence of any known tris- or tetrakisthiolatogold(I) complexes, the observation of isomorphous substitution seems unlikely. Although the thiolate cluster structure seems likely, there is, at present, no definitive evidence for its formation.

Preparation of Aurothioneins from Aurothiomalate

Reactions are carried out at 25° in aqueous solutions, adding small aliquots of concentrated aurothiomalate (AuSTm) solutions to metallothionein. Incubation for 1 hr should allow the bound metals and added gold to equilibrate completely. The mixture is then fractionated over Sephadex G-50 to resolve the resulting aurothioneins from unbound metal ions. The metal content of aurothioneins in which there is incomplete substitution will depend, obviously, on the metal content of the starting MT. Zinc ions are known to be more easily replaced than cadmium ions[3]; this is apparently a thermodynamic result, because the rates of displacement of Cd^{2+} and Zn^{2+} from homogeneous or mixed-metal M_7-MTs are identical.[3]

Au,Cd-Metallothionein

If Zn,Cd-MT reacts with more than 1.5 equivalents of gold per bound zinc, the zinc will be displaced preferentially, yielding Au,Cd-MT. It is convenient to work with Cd,Zn- or other M_7-MTs in a concentration range of 100 μM – 2 mM protein and add the desired quantity of AuSTm as a small aliquot of concentrated aqueous solution. In a typical case 57 μmol of AuSTm in 200μl was added to 120 μM MT containing 0.34 μmol Zn^{2+} and 1.33 μM Cd^{2+} in 2 ml of buffer. Although a wide range of buffers is suitable, it is best to avoid sodium salts, since Na^+ interferes with gold determination by flame atomic absorption spectroscopy (FAAS). The reaction mixture must be passed over Sephadex G-50 at the end of the incubation period. In our experience a column of 1.5 × 40 cm will suffice for reaction mixtures of this size. A trace of sodium ion added to the mixture will clearly identify the totally included (low-molecular-weight fractions) as the tubes are read for gold by FAAS. Relatively large fractions (ca. 3 ml) can be collected for a preparative reaction of this sort. After reading the tubes for gold, zinc, and cadmium, the metal content of the resulting aurothionein can be calculated by integrating the contents of the tubes bearing the MT peak. (Alternatively, just gold can be read on the individual fractions. Then after the MT-containing fractions are identified and pooled, the cadmium and zinc content of the combined sample can be measured.) The product isolated from the reaction mixture described above contained 0.001 μmol of zinc, 1.22 μmol cadmium, and 0.57 μmol of gold, confirming complete (>99%) displacement of the initially bound Zn.$^{2+}$ The total recoveries, including the displaced metals and unbound gold, were zinc (92%), cadmium (99%), and gold (104%).

Au,Zn,Cd-Metallothionein

If the procedure described above is followed using less than 1.5 equivalent bound Zn^{2+}, some zinc and the majority of the cadmium is retained, leading to a triply metalated protein.[3] In a typical case, treating 520 μM MT containing 7.5 μmol Zn^{2+} and 10.9 μmol Cd^{2+} with 7.5 μmol of AuSTm yielded a protein containing 2.7 μmol Zn^{2+}, 10.6 μmol Cd^{2+}, and 7.5 μmol AuSTm. In this case, to accommodate a slightly larger sample size, the mixture was eluted over a 2.5 × 60 cm Sephadex G-50 column.

$(TmSAu)_{20}$-Thionein

When a large excess of gold thiomalate is used, the displacement of zinc and cadmium is complete and the thiomalate ligand is retained in the coordination sphere of the gold, resulting in an unfolded structure with one gold bound to each MT sulfhydryl.[3] We have successfully employed ratios in the range of 175–200 equivalents of gold/protein to prepare an all gold-thionein. In a typical preparation, 640 μM MT (2 ml; 2.8 μmol Zn^{2+}, 6.2 μmol Cd^{2+}) were treated with 113 μmol of AuSTm. After gel-permeation chromatography the resulting thionein contained 24.0 μmol of gold. Less than 1% of the original zinc and less than 5% of the cadmium were retained in the MT.

Preparation of $(Et_3PAu)_xAu_y$-MT

Treatment of Zn,Cd-MT with substoichiometric amounts of chlorotriethyl phosphinegold(I) (Et_3PAuCl) led to displacement of zinc and incorporation of gold into the structure.[4] In a typical preparation 120 nmol of Et_3PAuCl (in 50 μl ethanol) was added to a solution of MT containing 170 nmol of Cd^{2+} and 86 nmol of Zn^{2+} in 1 ml of buffer. After 1 hr they were eluted over Sephadex G-50 at pH 8.6 in Tris buffer. Analysis of the fractions established that the zinc was extensively displaced, but very little cadmium (<3%) was lost. From the metal exchange ratios it was proposed that both monodentate coordination with retention of the Et_3P ligand and bidentate coordination of gold were occurring.[4]

[4] C. F. Shaw III and J. E. Laib, *Inorg. Chim. Acta* **123,** 197 (1986).

Similar reactions of the related antiarthritic drug auranofin (Et_3PAu-SATg, where AtgSH is 2,3,4,6-tetraacetyl-1-β-D-thioglucose) with apo-MT and M_7-MT were studied using triply labeled $Et_3PAuSATg$ with ^3H in the phosphine, ^{35}S, and ^{195}Au. It was demonstrated unambiguously that gold was binding with and without the phosphine ligand.[5] $Et_3PAuSATg$ does not, however, react with metal-saturated MT. This probably reflects the high affinity of the phosphine and gold ligands for gold(I) and the ability of free thiols (apo-MT) but not the metal-coordinated thiolates (M_7-MT) to compete for the ligands. The chloride of Et_3PAuCl and the thiomalate of AuSTm are, in contrast, relatively weak ligands for gold, and these complexes are sufficiently reactive to displace Zn^{2+} and Cd^{2+} from MT.

Large-Scale Preparations

Many physicochemical techniques require larger samples than those described above. In such cases, one can increase both the concentration of the metallothionein and the volume of solution in order to prepare a larger sample. We have scaled up both of the preparations described above to levels of 50 to 70 μmol of total cadmium and zinc in order to obtain sufficient aurothionein for the EXAFS studies previously reported.[3]

Ligand Tracer Studies

In the absence of spectroscopic techniques that are sufficiently facile and sensitive to probe the chemical environment of gold(I) thiolates, the use of radiotracer methods is the best method to follow the fate of the ligands present in the initial gold complexes. This approach has been applied to the reactions of AuSTm with MT[3] and $Et_3PAuSATg$ with MT and apo-MT.[5] After separation of the aurothionein from unreacted gold and displaced ligands, the activity of each ligand in the protein and low-molecular-weight fractions are determined. Using the specific activity of the ligand one can calculate the extent to which it is displaced from the gold and/or retained in the aurothionein. This approach was used with [^{35}S]thiomalate-labeled AuSTm to confirm the formation of $(TmSAu)_{20}$-Th with retention of the ligand and Au,Cd-thionein with loss of the ligand.[3]

Ecker et al.[5] used triply labeled auranofin ($Et_3PAuSATg$ with tetraacetyl[^{35}S]thioglucose; ^{195}Au, and [^3H]triethylphosphine) to study the stoichiomety of its reaction with apo-MT and M_7-MT.[5] Their results verified that auranofin does not react with M_7-MT and demonstrated that, when

[5] D. J. Ecker, J. C. Hempel, B. M. Sutton, R. Kirsch, and S. T. Crooke, *Inorg. Chem.* **25,** 3139 (1986).

apo-MT reacts with auranofin, 20% of the gold retains the triethylphosphine ligand and 80% binds with loss of both ligands.

Kinetic Monitoring of Aurothionein Formation

Although there are observable changes in the ultraviolet (UV) spectra as metallothionein reacts with gold complexes, the presence of gold–sulfur chromophores in the reactants and products renders any molecular interpretations ambiguous at best (D. H. Petering, W. Nelson, and C. F. Shaw III, unpublished observations, 1987). One can, in contrast, use metallochromic ligands that bind to the displaced cadmium or zinc ions to measure precisely the rates of displacement of these metals from MT.[6] The dyes zincon and pyridylazoresorcinol have been found to be suitable for monitoring displacement of zinc and cadmium, respectively.

Extensive studies of these processes have yielded a two-term rate law for the reaction of AuSTm with various MTs[6]:

$$\text{Rate} = k_{\text{fast}}[\text{MT}] + k_{\text{slow}}[\text{MT}]$$

Neither the choice of dye (zincon or pyridylazoresorcinol) nor its concentration nor the concentration of the AuSTm affected the reaction rates. The rate constants, $k_{\text{fast}} = 2.7 \times 10^{-2}$ and $k_{\text{slow}} = 6.9 \times 10^{-4}$, establish that the reaction is relatively rapid, although not instantaneous. A discussion of the underlying principles and detailed procedures for monitoring the kinetics with these dyes are discussed elsewhere in this volume.[7]

Further Characterization of Aurothioneins

To date EXAFS (extended X-ray fine-structure spectroscopy) and the related technique XANES (X-ray absorption near edge spectroscopy) have proven to be the most useful methods for determining the oxidation state and coordination environment of gold in aurothioneins.[3] Unfortunately, these techniques are limited by the need for access to X rays of relatively high energy (11.9 keV) and intensity, generally obtained only with synchrotron radiation sources. Because of these limitations, interested readers are referred to other sources for more detailed descriptions of the methods and their application to gold(I)–protein complexes.[8,9]

[6] C. F. Shaw III, J. E. Laib, M. M. Savas, and D. H. Petering, *Inorg. Chem.* **25**, 403 (1990).
[7] C. F. Shaw III, M. M. Savas, and D. H. Petering, this volume [46].
[8] R. C. Elder and M. K. Eidsness, *Chem. Rev.* **87**, 1027 (1987).
[9] R. C. Elder, M. K. Eidsness, M. J. Heeg, K. G. Tepperman, C. F. Shaw III, and N. Schaeffer, *ACS Symp. Ser.* **209**, 385 (1983).

Radioimmunoassay has been used to show that aurothioneins generated from horse kidney MT retain their cross-reactivity to antibodies generated from other mammalian MTs. Detailed procedures for radioimmunoassay (RIA) studies of MT were given in a previous chapter in this series.[10]

A potentially useful property of aurothioneins is luminescence.[11] An aurothionein prepared from "excess AuSTm" and apo-MT was examined at 77 K as a frozen solution. When excited with 260-, 280-, or 305-nm UV radiation, the sample was found to luminesce at 600 nm. The emission of the AuSTm starting material is "blue shifted" to ca. 525 nm.

The stability of aurothioneins can be addressed by the straightforward combination of chromatographic and analytical techniques used to measure the original stoichiometries of gold complexes. When radiolabeled $(TmSAu)_{20}$-MT was rechromatographed, there was a greater loss of radiolabel than of gold, suggesting a rearrangement from $(TmSAu)_{20}$-thionein to a species containing gold coordinated to two cysteine thiolates (K. Melnick, J. Laib, and C. F. Shaw III, unpublished data, 1991).

[10] J. S. Garvey, R. J. Vander Mallie, and C. C. Chang, this series, Vol. 84, p. 121.

[11] M. J. Stillman, A. J. Zelazowski, J. Szymanska, and Z. Gasyna, *Inorg. Chim. Acta* **161,** 275 (1989).

[57] Stability Constants and Related Equilibrium Properties of Metallothioneins

By DAVID H. PETERING and C. FRANK SHAW III

Introduction

The equilibrium metal-binding properties of metallothionein (MT) lie at the heart of its role as a protein involved in essential and toxic metal metabolism. When apometallothionein is synthesized in cells, a variety of metal ions is efficiently bound by the protein and sequestered away from other potential metal-binding sites.[1,2] The resulting molecules may be homogeneous metal- or mixed metal-metallothioneins (M-MT). In the latter structures, different metals may be segregated between the two do-

[1] D. R. Winge, R. Premakumar, and K. V. Rajagopalan, *Arch. Biochem. Biophys.* **188,** 466 (1978).

[2] A. J. Kraker, S. Krezoski, J. Schneider, D. T. Minkel, and D. H. Petering, *J. Biol. Chem.* **260,** 13710 (1985).

mains, leading to relatively homogeneous metal clusters, or may exist together in heterogenous metal clusters.[3,4]

Metal ion exchange studies have shown qualitatively that there are substantial differences in the equilibrium stability of various metal ions bound to metallothionein.[5] Furthermore, there are preferred distributions of metals in mixed-metal species, which seem to be based both on thermodynamic and kinetic characteristics of metallothionein.[6]

Apparent Stability Constants of Homogeneous Metal Metallothioneins

Methods of Measurement and Calculation

The general set of reactions that needs to be characterized is the following[1]:

$$M + MT \overset{K_1}{\rightleftharpoons} M\text{-}MT \quad (1)$$

$$M + M\text{-}MT \overset{K_2}{\rightleftharpoons} M_2\text{-}MT \quad (2)$$

$$\vdots$$

$$M + M_{n-1}\text{-}MT \overset{K_n}{\rightleftharpoons} M_n\text{-}MT \quad (3)$$

in which charges are omitted and $K_1, \ldots, K_n$ are the successive, pH-independent stability constants between the n metals (M) and the form of the protein (thiolates deprotonated) that binds to M. Since the sulfhydryl residues are deprotonated only above pH 9 and metal ions bind very strongly to MT under these conditions, it has not been possible directly to evaluate these constants in titrations of apometallothionein with metal ions at basic pH.[7] However, measurable equilibria are achieved under acidic conditions where M is zinc, cadmium, and possibly copper, and the sulfhydryl residues and at least some of the carboxyl groups of metallothionein are protonated [reactions (4) and (5), assuming that in each reaction the same number of protons (k) are displaced].

$$M + H_{nk}\text{-}MT \overset{K_{H,1}}{\rightleftharpoons} M\text{-}H_{k(n-1)}\text{-}MT + kH \quad (4)$$

$$\vdots$$

$$M + M_{n-1}\text{-}H_k\text{-}MT \overset{K_{H,n}}{\rightleftharpoons} M_n\text{-}MT + kH \quad (5)$$

[3] R. W. Briggs and I. M. Armitage, *J. Biol. Chem.* **257,** 1259 (1982).

[4] I. M. Armitage and J. D. Otvos, *in* "Biochemical Structure Determination by NMR" (A. A. Bother-By, J. D. Glickson, and B. D. Sykes, eds.), p. 65. Dekker, New York, 1982.

[5] J. D. Otvos, D. H. Petering, and C. F. Shaw III, *Comments Inorg. Chem.* **9,** 1 (1989).

[6] D. G. Nettesheim, H. R. Engeseth, and J. D. Otvos, *Biochemistry* **24,** 6744 (1985).

[7] M. Vašák and J. H. R. Kägi, *Met. Ions Biol. Syst.* **15,** Chap. 6 (1983).

Although $K_{H,1}$, . . . , $K_{H,n}$ can, in principle, be determined, the pK_a values of the associated thiolate ligands in metallothionein must also be known in order to convert these pH-dependent constants into K_1, . . . , K_n. At present they are undetermined. In addition, to identify these constants with those defined in reactions (1–3), it must also be assumed that the state of protonation of the carboxyl groups and, thus, the charge on the protein has no effect on these stability constants.

For practical calculations of the distributions of metal ions between metallothionein and competing ligands, it is the apparent stability constants $K_{app,1}$ to $K_{app,n}$ in reactions (6) and (7) that are of interest:

$$M_{all\ forms} + MT_{all\ forms} \overset{K_{app,1}}{\rightleftharpoons} M\text{-}MT_{all\ forms} \quad (6)$$

$$\vdots$$

$$M_{all\ forms} + M_{n-1}\text{-}MT_{all\ forms} \overset{K_{app,n}}{\rightleftharpoons} M_n\text{-}MT_{all\ forms} \quad (7)$$

These are defined as the constants for all forms of the species that exist at a given pH. Methods for evaluating apparent stability constants are described below.

Equilibria with Competing Ligands

One method to determine apparent stability constants for metal–ligand complexes (M–L) with large thermodynamic stability is to establish measurable equilibria between them and competing ligands (L′), which have known apparent stability constants with the metal [e.g., reaction (8)]:

$$M\text{-}L + L' \overset{K_c}{\rightleftharpoons} M\text{-}L' + L \quad (8)$$

Then, having determined K_c for reaction (8) and having calculated $K_{app,M\text{-}L'}$ for M-L′,

$$K_{app}(M\text{-}L) = K_{app}(M\text{-}L')/K_c \quad (9)$$

In the case of metallothionein, the possibility that there are different apparent stability constants for each bound metal ion makes the measurement of such constants potentially very difficult. However, studies to date indicate that there are only one or two kinetic classes of metals in zinc- or cadmium-metallothionein in ligand substitution reactions.[5] Each class behaves cooperatively and is either fully occupied or unoccupied with Zn^{2+} or Cd^{2+} in the presence of competing ligands.[5,8-10] In the case of the

[8] G. Bachowski, C. F. Shaw III, and D. H. Petering, submitted for publication.

[9] D. H. Petering, A. Pattanaik, P. Chen, S. Krezoski, and C. F. Shaw III, *in* "Metal Binding in Sulfur-Containing Proteins" (M. Stillman, C. F. Shaw III, and K. T. Suzuki, eds.), in press. VCH Publishers, New York.

[10] F. Vazquez and M. Vašák, *Biochem. J.* **253,** 611 (1988).

competing chelating agent, nitrilotriacetate (NTA), the two classes of bound metals have been identified with the two domain metal-binding clusters, C_α and C_β.[9] Therefore, a complicated problem of defining multiple equilibrium constants reduces to the measurement of two equilibrium constants. For example, considering Zn_7-MT, one writes for the β domain,

$$Zn_3\text{-}C_\beta + 3L' \overset{K_c}{\rightleftharpoons} 3Zn\text{-}L' + C_\beta \quad (10)$$

and

$$K_{app}(Zn_3\text{-}C_\beta) = (K_{app,Zn\text{-}L'})^3/K_c \quad (11)$$

In this formulation,

$$K_{app}(Zn_3\text{-}C_\beta) = K_{app,1}K_{app,2}K_{app,3} \quad (12)$$

in which the constants on the right-hand side of Eq. (12) are the stepwise constants for the addition of one, two, and three zinc ions to the cluster. Because of the cooperative nature of the release of metals from the cluster in the presence of a strong competing ligand, $K_{app,3} \gg K_{app,1}$ and $K_{app,2}$, the average apparent stability constant per zinc in the β domain is $K_{app}(Zn_3\text{-}C_\beta)^{1/3}$. In fact, within the error of the measurement, all of the zinc ions bound to homogeneous Zn_7-MT have the same apparent stability constant.[5,8]

To make the actual measurements, two competing ligands have been used with Zn_7-MT, nitrilotriacetate, and 3-ethoxy-2-oxobutyraldehyde bis(N^4-dimethylthiosemicarbazone) (H_2-$KTSM_2$).[8] The two ligands have the following properties used in these experiments: log K_{app}(Zn-NTA) = 8.11 at pH 7.4 and 25° in 0.1 *M* KCl. This is based on a pH-independent stability constant of 10.44 and a pK_a of 9.73 for the amine group of NTA.[11] The difference molar absorptivity for zinc bound to MT vs apoprotein is 13,000 $M^{-1}cm^{-1}$ at 215 nm, where the equilibrium measurements are made. The 1:1 complex of Zn^{2+} with H_2KTSM_2 has an absorbance maximum at 434 nm, which has a molar absorptivity of 15,500 $M^{-1}cm^{-1}$. Its log apparent stability constant at pH 7.4 is 9.73.[12] Each ligand reacts with biphasic kinetics with zinc metallothionein, revealing two kinetic classes of reactive metals, each of which approaches a measurable equilibrium with the competing ligand. Chapter [46] in this volume describes how these reactions are carried out.

[11] L. G. Sillén and A. E. Martell, "Stability Constants of Metal-Ion Complexes," Suppl. No. 1, Special Publ. No. 25, The Chemical Society, Burlington House, London, 1971.

[12] D. H. Petering, *Biochem. Pharmacol.* **23,** 567 (1974).

To illustrate the calculations used to determine $K_{app}(Zn_3\text{-}C_\beta)$, Fig. 1 shows an idealized semilog plot of absorbance changes of Zn_7-MT after reaction with NTA to establish an equilibrium mixture. The biphasic curve yields values for the equilibrium absorbance change at 215 nm in each component of the overall reaction, $(A_0 - A_\infty)_I$ and $(A_0 - A_\infty)_{II}$, as well as the total change in absorbance, $(A_0 - A_\infty)_t$, for the reaction that goes to completion. Allocating 4/7 of the total ΔA to step I and 3/7 to step II,[9] one can immediately calculate the fraction of zinc in each part of the reaction that has become bound to NTA. From these numbers and the actual concentrations of starting materials, K_c can be determined and the constants for zinc bound to each kinetic class of binding site evaluated. In the case of H_2KTSM_2, identification of domains with kinetic classes has not been done, so one must arbitrarily assign numbers of zincs to each class, about equal numbers to each. One can also simply analyze the equilibrium between Zn-MT and these chelating agents in terms of the overall equilibrium absorbance changes and ignore the fact that two classes of zinc exist in the protein. Then, the derived apparent stability constant represents an average for the seven metal ions.

It is important to use a series of concentrations of competing ligand,

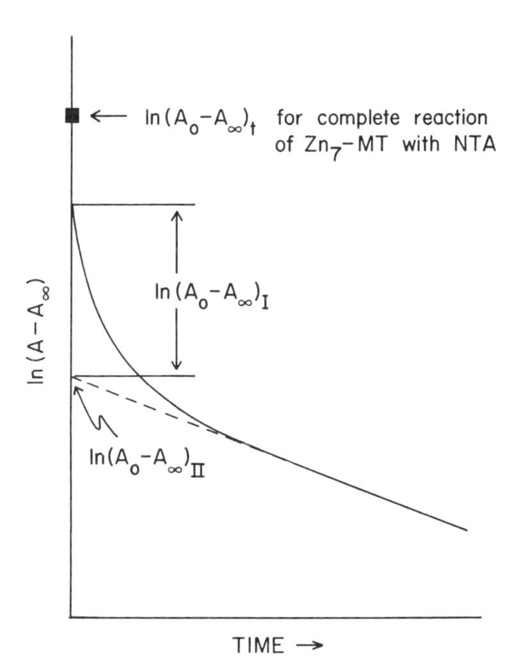

FIG. 1. Semilog plot of kinetics of reaction of a pseudo first-order excess of NTA with Zn_7-MT.

which removes increasing fractions of the total metal bound in metallothionein, to demonstrate that the derived apparent stability constants are independent of the amount of metal extracted from the protein. This validates the equilibrium model used in the calculation, particularly the all-or-nothing (cooperative) character of the binding of multiple metal ions to metallothionein, in the presence of high-affinity competing ligands.

A possible complication in this method arises if competing ligands also bind to metallothionein. In that case reaction equation 10 must be expanded to include this equilibrium process.

$$Zn_3\text{-}C_\beta + L' \xrightleftharpoons{K_{adduct}} L'\text{-}Zn_3\text{-}C_\beta \tag{13}$$

$$L'\text{-}Zn_3\text{-}C_\beta + 2L' \rightleftharpoons 3Zn\text{-}L' + C_\beta \tag{14}$$

Then, if one can only measure the sum of $[Zn_3\text{-}C_\beta]$ and $[L'\text{-}Zn_3\text{-}C_\beta]$, the expression for the equilibrium constant will include a term containing K_{adduct}.

$$K_{Zn_3-C_\beta} = K_{app}(Zn\text{-}L')^3(1 + K_{adduct}[L'])/K_c \tag{15}$$

pH-Dependent Equilibria

Starting from reactions (4) and (5), and following the approach of Kägi and Vallee, one assumes that each metal ion binds independently with the same stability constant and that the vacant metal-binding site (H_nS) is fully protonated with n hydrogen ions[12,13]:

$$M\text{-}S + nH \xrightleftharpoons{K_H} M + H_nS \tag{16}$$

The latter is reasonable in that, under acidic conditions, the sulfhydryl ligands for the metals will be protonated. It does not consider, however, the possibility that the protonation state of other residues such as carboxyl groups may affect K_H.

Following reaction (2), Zn- or Cd-MT have been titrated with hydrogen ion.[12,13] To establish that reversible equilibrium conditions exist during these titrations, it must be shown that the same curves of absorbance changes *versus* pH are obtained when the proteins, after titration with acid, are then brought back to neutral pH with base. In addition, because Cu(I) released from MT can readily be oxidized, the titration of Cu(I)-MT must be done anaerobically.

[13] J. H. R. Kägi and B. L. Vallee, *J. Biol. Chem.* **236,** 2435 (1961).

To determine K_H, the following equations have been used.

$$K_H = [M][H_nS]/[M\text{-}S][H]^n \tag{17}$$

$$\log[M\text{-}S]/[M][H_nS] = n\text{pH} - \log K_H \tag{18}$$

In general, to follow the loss or gain of metal ion from the protein, changes in absorbance (A) at 220 or 254 nm for zinc and cadmium metallothionein species, respectively, have been used as measures of metal–thiolate bonding in the proteins. This method assumes that each equivalent of metal binds to the protein with the same molar absorbance.

$$[M\text{-}S] = \frac{A - A_f}{A_i - A_f} C_M \tag{19}$$

$$[M] = [H_nS] = \frac{A_i - A}{A_i - A_f} C_M \tag{20}$$

In Eqs. (19) and (20), A is the absorbance at a given wavelength for one of the metallothioneins and at a given pH during the titration; A_i and A_f are the initial and final absorbances at the extremes of the titration where no further changes in absorbance occur. The total metal initially bound to the protein at neutral pH is given by C_M.

A plot of the left-hand side of Eq. (18), where the concentrations are calculated from Eqs. 19 and 20, versus pH should yield a straight line if the assumptions of the model are correct. The slope will be n and the intercept, K_H. If H_nS is an adequate representation of all forms of the ligand-binding site, then K_{app}, the apparent stability constant at a defined pH at least one pH unit below the average pK_a, 8.9, of the sulfhydryl groups of apometallothionein,[12] is given by

$$K_{app} = 1/K_H[H]^n \tag{21}$$

Reactions (4) and (5) are written in a general form in which as many stability constants may exist as there are metal ions in fully saturated metallothionein. According to published work, only one constant K_H for all of the zinc or cadmium ions is actually derived from the experimental results.[7,13] One explanation is that each metal ion binds independently with the same stability constant. That idea is inconsistent with the presence of two metal clusters in metallothionein, which are knit together by bridging thiolate residues. Still, titrations of apo-MT by Co^{2+} or Fe^{2+} indicate that at least with those metal ions the first three or four metal ions do bind relatively independently to the protein.[14,15]

[14] M. Vašák and J. H. R. Kägi, *Proc. Natl. Acad. Sci. U.S.A.* **78,** 6709 (1981).
[15] M. Good and M. Vašák, *Biochemistry* **25,** 8353 (1986).

In an alternative, cooperative metal-binding model, clusters are either completely filled with metals or they are vacant. There are, for example, no unsaturated species such as Zn_1-C_β or Zn_2-C_β. Both the zinc and cadmium proteins seem to exhibit this property, at least in the presence of strong competing ligands.[5,9,10] If the metals in each cluster bind cooperatively, then each metal ion has the same average stability constant that one sees if each metal ion binds independently with equal affinity to the protein.

Stability Constants in Heterogeneous Metal Metallothioneins

Metallothionein species isolated from organisms include Cd,Zn-MT and Cu,Zn-MT.[16,17] The Cd,Zn-protein exists as a set of molecules that includes mixed metal clusters.[4,6] The stability constants of the metals in such clusters have not been determined. If the metal ions are segregated between the two clusters, the approach taken in the Introduction above, can be used to determine apparent stability constants per metal ion in each domain. For example, with a Zn_4,Cu_6-MT isolated from bovine calf liver, zinc appears to be largely confined to the α domain. As a result, NTA reacts exclusively with zinc in the α domain to establish an equilibrium analogous to that in reaction (10).[9]

Titrations of Apometallothionein with Metal Ions

Metal ion titrations of metallothionein reflect in simple or complex ways the equilibrium binding properties of the protein. Cadmium, zinc, and cuprous ions are thought to bind cooperatively to metallothionein, forming cluster species even at small metal ion-to-protein ratios[18,19]; Co^{2+} and Fe^{2+} apparently bind initially at noninteracting sites and then cluster as the metal ion-to-protein stoichiometry increases beyond four.[14,15] Focusing on the Zn-, Cd-, and Cu-metallothionein species mentioned above, the progress of titration of apoprotein has been followed in several ways.

Chromatography after Proteolytic Digestion

In this method, apometallothionein is titrated with 1 to *n* g-atom equivalents of metal ion, where *n* is 7 for cadmium and zinc and 12 for

[16] B. Ujjani, G. Krakower, G. Bachowski, S. Krezoski, C. F. Shaw III, and D. H. Petering, *Biochem. J.* **233,** 99 (1986).

[17] D. H. Petering, J. Loftsgaarden, J. Schneider, and B. Fowler, *Environ. Health Perspect.* **54,** 73 (1984).

[18] K. B. Nielson and D. R. Winge, *J. Biol. Chem.* **258,** 13063 (1983).

[19] K. B. Nielson, C. L. Atkin, and D. R. Winge, *J. Biol. Chem.* **260,** 5342 (1985).

copper.[18,19] After the addition of metal, subtilisin is added to digest accessible peptide bonds. Such work has shown that fully occupied clusters, as well has holoprotein, are resistant to proteolysis. Apoprotein and possibly those domains only partially occupied by metal, are readily degraded. Then the resultant mixture is either chromatographed over Sephadex G-25 or electrophoresed through polyacrylamide to isolate metalloprotein species.[18,20] Results with this method of product definition indicate that clusters form cooperatively. That is, at all ratios of metal to protein, metal-saturated clusters or holoprotein are observed.[18,20] A cautionary note must be added. The use of subtilisin makes this a nonequilibrium system, since one of the reactants is being enzymatically degraded. Thus, the result may reflect the results of two paired reactions:

$$\text{MT} + n\text{M} \rightleftharpoons \text{M}_n\text{-MT} \tag{22}$$

$$\text{M}_n\text{-MT} + \text{subtilisin} \rightarrow \underset{\text{(modified)}}{\text{M}_n\text{-MT}'} + \text{amino acids} \tag{23}$$

The second reaction may drive the metal distribution toward particular arrangements that are resistant to proteolytic degradation.

Circular Dichroism Spectroscopy

There are large, characteristic changes in the circular dichroism spectrum of apometallothionein on the binding of Cd^{2+}.[21] These have been attributed to the formation of the Cd_4 cluster in the α domain. The three-metal cluster does not have a distinctive spectral signature. Thus, one can only monitor the formation of α domain clusters with this technique during the titration of apometallothionein with cadmium ion. An attractive feature of this method is that one can use very small concentrations of protein (ca. 10 μM).

Nuclear Magnetic Resonance Spectroscopy

In principle, the best method for following the formation of species of Cd-metallothionein is to do ^{113}Cd nuclear magnetic resonance (NMR) spectroscopy on the titration mixture.[4,6] Resonance assignments for members of the four- and three-metal clusters have been made so that one should be able to determine at what state of the titration clusters form in preference to other possible intermediates not seen in Cd_7-MT. To date such studies have not been published, although unpublished results have

[20] M. J. Stillman and A. J. Zelazowski, *Biochem. J.* **262,** 181 (1989).

[21] M. J. Stillman, W. Cai, and A. J. Zelazowski, *J. Biol. Chem.* **262,** 4538 (1987).

been mentioned. Minimum protein concentrations needed are about 1 m*M*.

Titrations Producing Heterogeneous Metal Metallothioneins

Mixed metal Cd,Zn-MT is typically found in cells after exposure to metallothionein. Similarly, Cu,Zn-MT is frequently observed in tissues. Titration experiments as summarized in reaction (24) have been done to explore the mechanism of formation of mixed metal species,

$$Zn_7\text{-MT} + nM \rightleftharpoons M_n,Zn_{7-n}\text{-MT} + nZn \tag{24}$$

Circular dichroism and ^{113}Cd nuclear magnetic resonance spectroscopy have been used to observe the detailed progress of this type of titration.[6,21] When M is cadmium, the competing metal ion distributes among the two clusters without forming homogeneous Cd_4 or Cd_3 clusters until n is greater than 4. In contrast, when Zn_7-MT is titrated with Cd_7-MT,

$$mZn_7\text{-MT} + nCd_7\text{-MT} \rightleftharpoons (m+n)Zn_{(7m-7n)/(m+n)},Cd_{(7n-7m)/(m+n)}\text{-MT} \tag{25}$$

the entire set of titration spectra are different and there is clear NMR evidence of Cd_4 cluster formation at values of n smaller than 4.

Studies by Stillman *et al.* on the effect of temperature on the products of reaction (24) indicate that this titration leads to a kinetically stable species distribution at 25° that is converted to another distribution favoring localization of cadmium in the α domain as the temperature is raised above 50°.[21] The new products are themselves stable on reduction of the temperature, suggesting that at higher temperatures rearrangement occurs to yield the thermodynamic product mixture. The product distribution bears some similarity to that generated in reaction 23 in that more Cd_4 cluster is present at low values of n.

Acknowledgments

The authors acknowledge the support of NIH Grant ES-04026.

[58] X-Ray Structure of Metallothionein

By A. H. ROBBINS and C. D. STOUT

Introduction

The structure of metallothionein (MT) in solution has been studied in detail by two-dimensional $^{113}Cd-^{1}H$ NMR methods. In this chapter we report a refined crystal structure for Cd_5,Zn_2-MT that is in good overall agreement with the NMR structure. A detailed comparison of the solution and crystal structures will be presented elsewhere. A detailed analysis of the crystal structure and the structure determination is also presented elsewhere.[1] This chapter focuses on problems unique to MT in solving the crystal structure and obtaining crystals.

Structure Determination

Crystals

Single crystals of Cd_5,Zn_2-MT (isoform 2) were grown as previously described.[2] To our knowledge these are the only crystals ever obtained of any MT that diffract to a high resolution. The crystallization procedure itself requires a precise ratio of sodium formate and potassium phosphate salts. The reader is referred to Ref. 2 for the details of the procedure. A possible explanation for the requirement for both salts is suggested by the structure itself (see below). The protein as isolated from rat liver has a metal composition of 5 mol Cd and 2 mol Zn/mol of MT. Samples from dissolved crystals have the same metal composition.[2] The amino acid sequence consists of 61 residues,[2a] including 20 cysteines, and contains *N*-acetylmethionine at the N terminus.[3] The crystals are tetragonal with unit cell constants $a = b = 30.9$ Å, $c = 120.4$ Å, and one molecule per asymmetric unit.

[1] A. H. Robbins, D. E. McRee, M. Williamson, S. A. Collett, N. H. Xuong, W. F. Furey, B. C. Wang, and C. D. Stout, *J. Mol. Biol.* in press (1991).

[2] K. A. Melis, D. C. Carter, and C. D. Stout, *J. Biol. Chem.* **258,** 6255 (1983).

[2a] Single-letter abbreviations are used for amino acid residues: A, alanine; R, arginine; N, asparagine; D, aspartic acid; C, cysteine; Q, glutamine; E, glutamic acid; G, glycine; H, histidine; I, isoleucine; L, leucine; K, lysine; M, methionine; F, phenylalanine; P, proline; S, serine; T, threonine; W, tryptophan; Y, tyrosine; V, valine.

[3] D. R. Winge, K. B. Nielson, R. D. Zeikus, and W. R. Gray, *J. Biol. Chem.* **259,** 11419 (1984).

Space Group

Diffraction patterns indicate the space group of the MT crystals to be $P4_12_12$ or its enantiomer, $P4_32_12$.[2] The space group derived in an earlier analysis was $P4_12_12$.[4] The present analysis shows that the correct space group is $P4_32_12$. The error in space group assignment was caused by an error in the positions of the cadmium sites used to calculate phases. The location of the cadmium sites is complicated by a pseudo centrosymmetric constellation of the heavy atoms due to packing of pairs of MT molecules about a twofold axis. The pseudo symmetry diminishes the magnitude of the anomalous differences and introduces ambiguity into space group assignment based on the cadmium positions alone. Therefore, the error in space group assignment in the MT structure originally reported was due to incorrect location of the cadmium sites, an error that was masked by pseudo symmetry. The following discussion deals only with aspects of the crystallographic analysis unique to this problem. Standard crystallographic procedures used in obtaining the refined structure are described in Ref. 1.

Direct Methods

The 2.5-Å resolution data were used for direct methods calculations with the program Multan.[5] Calculations were done in $P4_32_12$ using both the averaged $|F|$ values for the protein (2337 centric and acentric reflections) and the $|\Delta F| = \|F^+| - |F^-\|$ values (1522 acentric reflections) in an effort to independently locate the cadmium positions. A variety of indicators and test calculations, as well as the NMR structure, suggested that the previously determined cadmium positions were incorrect. The application of direct methods was predicated by our previous experiments[4], and by success with heavy atom derivative anomalous and isomorphous differences.[6] The application of Multan to the protein $|F|$ data was to our knowledge unprecedented, but was attempted after failure to locate the cadmium positions from the $|\Delta F|$ data. Also, consideration of the fact that the five cadmium positions represent 38% of the total scattering by the protein suggested that application of direct methods to the $|F|$ data was justified.

Both $|F|$ and $|\Delta F|$ values were normalized by assuming the contents of the asymmetric unit to be five carbon atoms. The E values derived in each way are listed in Table I. The E distribution based on $|\Delta F|$ is intermediate between centrosymmetric and noncentrosymmetric, indicating the large influence of the cadmium to the total scattering. The phase sets generated

[4] W. F. Furey, A. H. Robbins, L. L. Clancy, D. R. Winge, B. C. Wang, and C. D. Stout, *Science* **231,** 704 (1986).

[5] G. Germain, P. Main, and M. M. Woolfson, *Acta Crystallogr., Sect. A* **27,** 368 (1971).

[6] A. K. Mukherjee, J. R. Helliwell, and P. Main, *Acta Crystallogr., Sect. A* **45,** 715 (1989).

TABLE I
NORMALIZED STRUCTURE FACTORS FOR METALLOTHIONEIN 2.5-Å DATA FROM MULTIWIRE AREA DIFFRACTOMETER

	Experimental		Theoretical	
Parameter	From $\|F\|$	From $\|\Delta F\|$[a]	Centrosymmetric	Noncentrosymmetric
Number of reflections	2337	1552		
$\langle E^2 \rangle$	1.000	1.000	1.000	1.000
$\langle \|E^2 - 1\| \rangle$	0.8178	0.9798	0.9680	0.7360
$\langle \|E\| \rangle$	0.8643	0.7906	0.7980	0.8860
E limit	Percentages (%) of total Es greater than E limit			
0.5	72.2	60.8	61.7	77.9
0.6	64.5	54.8	54.9	69.8
0.7	55.3	48.7	48.4	61.3
0.8	47.5	41.4	42.4	52.7
0.9	40.6	36.1	36.8	44.5
1.0	33.9	31.0	31.7	36.8
1.1	27.9	26.1	27.1	29.8
1.2	22.2	21.9	23.0	23.7
1.3	17.6	18.1	19.4	18.5
1.4	13.8	15.4	16.1	14.1
1.5	11.4	13.0	13.4	10.5
1.6	8.7	10.6	11.0	7.7
1.7	7.0	8.9	8.9	5.6
1.8	5.6	7.6	7.2	3.9
1.9	4.6	6.2	5.7	2.7
2.0	3.1	5.0	4.6	1.8
2.1	2.2	4.1	3.6	1.2
2.2	1.7	3.4	2.8	0.8
2.3	1.3	2.8	2.1	0.5
2.4	0.9	2.0	1.6	0.3
2.5	0.6	1.8	1.2	0.2

[a] $\Delta F = \|\|F^+\| - \|F^-\|\|$.

by Multan for the 300 largest E values were evaluated by examining the E maps. All of the maps based on the $|\Delta F|$ data had a single very large peak on the diagonal twofold axis. For any given solution the combined figure of merit was either negative, or the residual was negative. The E map based on the $|F|$ data for the solution with the highest absolute figure of merit (0.395) had five high peaks, indicating the cadmium positions, and two lower peaks for the zinc positions (Fig. 1). The combined figure of merit for the correct solution was 2.20. The E map for the only solution with a higher combined figure of merit, 2.61, contained randomly scattered, uniformly weighted peaks.

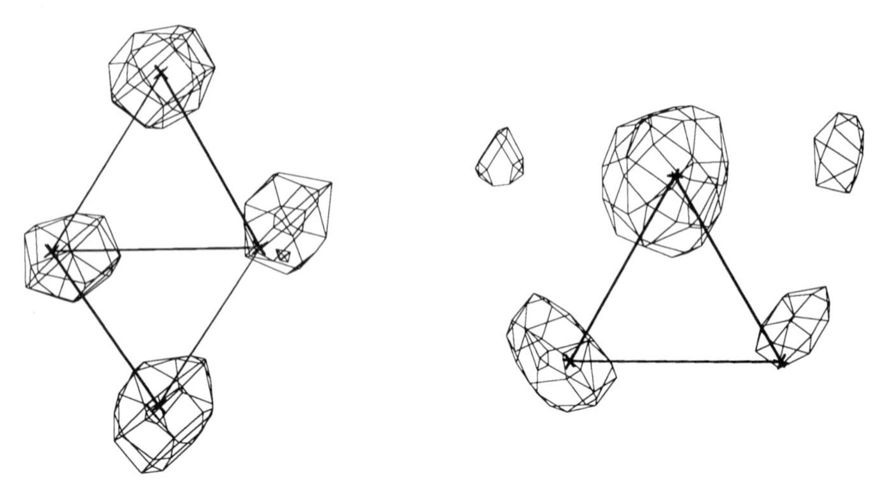

FIG. 1. Two regions of the E map in $P4_32_12$ showing the peaks for all seven metal sites in Cd_5,Zn_2-MT. Coordinates for the metals from the refined structure are shown connected by virtual bonds. Left-hand side, Cd_4 cluster; right-hand side, $CdZn_2$ cluster. Contoured at 0.30 the maximum density.

When the calculations were repeated in $P4_12_12$ using the $|F|$ data, the E map for the solution with second highest combined figure of merit also contained five high peaks for the cadmium sites (at inverted positions from the $P4_32_12$ solution). However, the $P4_12_12$ map had less contrast on the cadmium peaks and no clear peaks for the zinc sites. The E maps in $P4_12_12$ based on $|\Delta F|$ data were the same as those in $P4_32_12$.

Phase Calculations

The five cadmium positions from the direct methods solutions were refined against the 25% largest $|\Delta F|$ values.[7] The 2.5-Å acentric data were used, occupancy and *xyz* were refined with an overall B factor of 15.0 $Å^2$, and the R value at convergence was 0.31. These positions were used to calculate phases in both $P4_12_12$ and $P4_32_12$ by iterative single-wavelength, sincle anomalous scatterer (ISAS) methods.[8] The starting figure of merit of 2518 acentric reflections to 2.0-Å resolution was 0.37. Phases calculated based on the cadmium partial structure for 2518 acentric and 1800 centric reflections were combined with the ISAS acentric phases using phase probability profiles derived from the Sim weights (program Pmodel, B.-C. Wang, personal communication, 1989). After phase combination the fig-

[7] W. A. Hendrickson and M. M. Teeter, *Nature (London)* **290,** 107 (1981).

ure of merit for 4318 reflections to 2.0 Å was 0.55. The R factor for the cadmium partial structure was 0.504.

The combined 2.0-Å ISAS/partial structure phases were improved by solvent flattening methods assuming a solvent content of 40%.[8] Five cycles of refinement were done for acentric reflections, followed by extension to centric reflections for five cycles; the process was repeated for two solvent masks. The resulting values of figure of merit and map inversion R value were 0.77 and 0.207, respectively, in $P4_12_12$. In $P4_32_12$ these values were 0.79 and 0.183, indicating this to be the correct space group.

The assignment of space group as $P4_32_12$ was confirmed by the electron density maps. In $P4_32_12$ the map showed clear density for segments of the polypeptide chain, carbonyl bulges, all 7 metal sites, and peaks for the sulfurs of all 20 cysteines. In contrast, the $P4_12_12$ map had breaks in the density for the protein, lower solvent contrast, a peak for only one of the zinc sites, and missing density for some of the cysteine sulfurs, especially at the cadmium site in the β-domain cluster.

Tungsten Derivative

We previously reported that the compound $(NH_4)_2WS_4$, when soaked into MT crystals, stains them yellow and introduces significant isomorphous intensity differences[4] (samples of the compound were provided by E. Stiefel, Exxon Research Laboratories, Annandale, NJ). In addition, assay of soaked crystals showed that >0.7 mol of tungsten was bound per mole of protein. A 2.3-Å oscillation camera data set had been collected but we were unable to solve the difference Patterson map.[4] The film WS_4^{2-} derivative data was scaled to the multiwire area diffractometer (MAD) native data set (R_{merge} 0.12). An isomorphous difference Fourier map calculated with the refined native phases revealed three WS_4^{2-} sites, all near $z = \frac{1}{2}$, with one site on the diagonal twofold axis. The sites were used to calculate iterative single isomorphous replacement (ISIR) phases.[8] The electron density map based on the 2.3-Å WS_4^{2-} ISIR phases clearly confirmed the positions of the metal sites in both clusters as the highest peaks in the map.[1] Therefore the WS_4^{2-} derivative independently supports the results of the new direct methods analysis on the protein $|F|$ data.

The WS_4^{2-} ions bind between pairs of MT molecules related by twofold symmetry. A network of interactions is formed in the solvent channel at $z = \frac{1}{2}$. The first WS_4^{2-} site lies symmetrically between the N_ζ atoms of twofold related K_{30} residues, 3.95 Å from each. This distance suggests ionic interaction of the lysine side chains and the WS_4^{2-} ion. The second WS_4^{2-} site lies in a hydrophobic pocket formed by P^{38}, G^{40}, and A^{53} on sym-

[8] B.-C. Wang, this series, Vol. 115, p. 90.

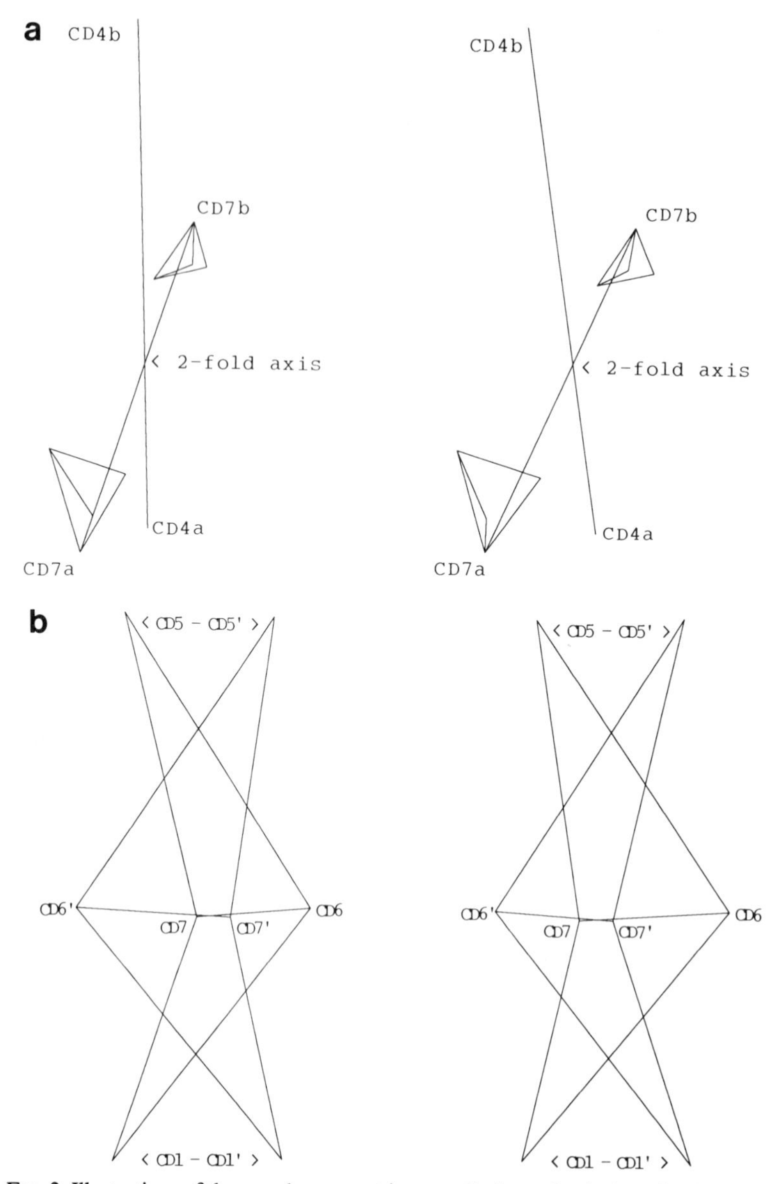

FIG. 2. Illustrations of the pseudo symmetric constellations of cadmium sites in the crystal structure. Only Cd–Cd virtual bonds are shown for the Cd_4 clusters. (a) Vectors connecting pairs of Cd-4 and Cd-7 sites in twofold related molecules at (x,y,z) (molecule a) and $(1 - y, 1 - x, \frac{1}{2} - z)$ (molecule b, see also Table II). The two vectors intersect on the diagonal twofold axis within 0.15 Å of each other, creating a center of symmetry. (b) Interpenetration of the Cd_4 clusters resulting from application of the pseudo inversion center created by the intersection of vectors between twofold related sites [see (a)]. The symmetrical arrangement of the two clusters is in agreement with the observed centrosymmetric distribution of E values derived from the anomalous scattering data (Table I).

metry-related molecules. This site is also 4.3 Å from C_β of K^{51} (side-chain density weak) and 6.1 Å from N_ζ of K^{30}. The third site also lies in a hydrophobic pocket formed by methyl groups from symmetry-related side chains of *N*-Ac-M^1, A^8, T^9, and V^{39}, and is 4.9 Å from N_ζ of K^{30}. The carbon-to-WS_4^{2-} contacts for the second and third sites are in the range 3.9–5.1 Å.

Pseudo Symmetry of Cadmium Positions

Figure 2a shows the disposition of 10 cadmium sites in two asymmetric units about the diagonal twofold axis at $z = \frac{1}{4}$. This arrangement results from crystal packing of the two-domain MT molecules. As a consequence of this packing a pseudo center of symmetry is created at 0.574, 0.426, 0.250. Vectors connecting twofold related Cd-4 and Cd-7 sites of the α and β domain clusters intersect within 0.15 Å (Fig. 2b).

To illustrate how the four-metal clusters are also related by the pseudo center of symmetry, the inversion operator was applied to one cluster to superpose the clusters (Fig. 2b). Figure 2b shows a symmetrical interpenetration of the clusters. The pairs of cadmium sites deviate in the superposition by 0.29, 1.70, 1.97, and 3.06 Å at Cd-7, Cd-5, Cd-1, and Cd-6, respectively. Considering Cd–Cd virtual bonds only, the four-metal cluster has internal symmetry, including a twofold axis normal to the Cd-6–Cd-7 virtual bond (Fig. 2b). If the clusters are superposed more perfectly such that pairs of Cd-6 and Cd-7 sites are aligned exactly, then the Cd-1 and Cd-5 sites deviate by 4.07 and 2.40 Å. Or, if the clusters are aligned on the Cd-1 and Cd-5 sites with twofold axes parallel, then the Cd-6 and Cd-7 sites are separated by 2.90 and 3.90 Å.

The point is that actual alignment of the clusters due to pseudo symmetry in the crystal is nearly as perfect as alignments resulting from exact superpositions of the clusters. In other words the cluster twofold axes are nearly parallel to the crystallographic twofold axis. Therefore, the orientation and internal symmetry of the four-metal cluster are such that the pseudo inversion center arising from crystal packing also relates individual sites within the four-metal clusters. Thus, the entire cadmium constellation is pseudo centrosymmetric.

Because cadmium dominates the anomalous scattering, the pseudo symmetry is reflected in the data. The distribution of E values derived from $|\Delta F|$ is almost ideally centrosymmetric (Table I). The centrosymmetric constellation diminishes the magnitude of the anomalous differences as well. The actual ratio of $\langle|\Delta F|\rangle / \langle|F|\rangle$ in the scaled data set is 0.04 whereas the expected value of this ratio for five Cd in this protein is 0.21.[7]

The presence of pseudo symmetry in the cadmium structure may be the reason the previous direct methods analysis based on film ΔF data yielded an incorrect solution.[4] In that case, the cadmium peaks were unequally weighted, and the model had to be translated to find an R minimum. Moreover, previous direct methods solutions also yielded E maps with a very large peak on the twofold axis. A similar phenomenon was observed with the new $|\Delta F|$ data.[1]

The pseudo symmetry may have prevented solution of the cadmium sites by other methods and at the same time introduced ambiguity in discriminating a correct solution. An R-factor search using all the $|\Delta F|$ data to 6.0 Å was performed with one site as a probe for the center of gravity of the four-metal cluster, using the program Hercules (D. E. McRee, unpublished results, 1989). The program had been successfully tested using anomalous difference and isomorphous replacement data from known protein structures. The R minimum occurred at a position translated ~6 Å on z from the originally reported solution.[4] (An ambiguity of z coordinate was also observed previously.) Repetition of the R-factor search procedure led to a five-site model for which the correlation coefficient (CC) for observed vs calculated Bijvoet difference Patterson maps was 0.76. The previously reported value of CC (6.0 Å) was 0.75[4] and yet both solutions are wrong. At 3.0-Å resolution, CC (Ref. 4) was 0.47 and CC (Hercules) was 0.59. For the correct solution CC (6.0 Å) is 0.60 and CC (3.0 Å) is 0.61. Therefore it appears that there are multiple incorrect solutions for the cadmium sites in the structure with reasonable values of CC and $R(\Delta F)$, especially at low resolution.

Rigid-body search calculations using Xplor,[9a] the NMR model for the four-metal cluster, and the $|F|$ data also failed to yield the correct solution. Again, the R minimum placed the cluster on the twofold axis under a variety of computational conditions. While the NMR and crystal coordinates for the clusters are very similar, this search may have suffered from omission of the fifth cadmium site. Molecular replacement calculations using the NMR model for the entire α domain failed to yield a consistent solution. This approach suffered from omission of the β domain, as the relative orientation of the two domains is not known from the NMR data.[9] Moreover, the R factor for the NMR α and β domains models, after fitting both the electron density map and rigid body refinement, is only 0.47. In addition to these limitations no other useful heavy atom derivative besides $(NH4)_2WS_4$ was found after searching over 100 compounds and/or conditions in the course of 7 years.

[9] P. Schultze, E. Wörgötter, W. Braun, G. Wagner, M. Vašák, J. H. R. Kägi, and K. Wüthrich, *J. Mol. Biol.* **203,** 251 (1988).

[9a] A. T. Brünger, M. Kardlus, and G. A. Petsko, *Acta Crystallogr., Sect. A* **45,** 50 (1989).

Results and Discussion

Domain Structure

The MT structure is illustrated in Figs. 3 and 4. There are no contacts between the α and β domains, except from K^{31} in the linker region to residues 19 and 21. The overall chain fold of both domains and the metal coordination of the 7 metal sites by the 20 cysteines is the same as in the solution structure of rat liver Cd_7-MT determined by NMR methods.[9] The structures of the isolated metal clusters, $Cd_4(S_\gamma)_{11}$ in the α domain and $CdZn_2(S_\gamma)_9$ in the β domain, are shown in Fig. 5. The metal sites are labeled to correspond to the NMR spectral assignments and numbering.[9] The NMR sites II and III correspond to Zn-1 and Zn-2 (Fig. 5).

Crystal Packing

The MT crystals are tetragonal, space group $P4_32_12$, $a = b = 30.9$ Å, $c = 120.4$ Å, and one molecule per asymmetric unit with a solvent content

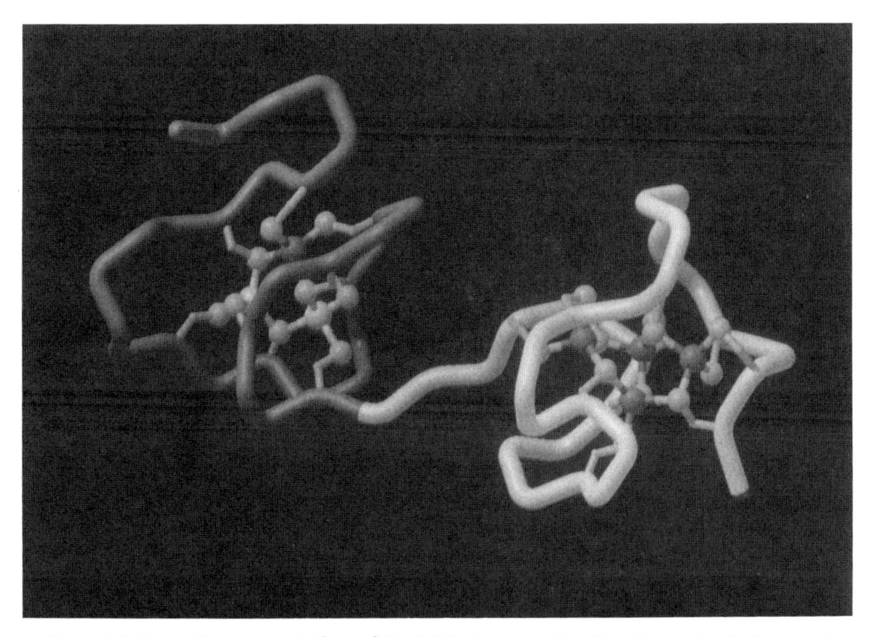

FIG. 3. Schematic representation of the MT structure showing the main chain, cysteine side chains, and metal sites. The N terminus and β domain are on the left, the C terminus and α domain are at the right.

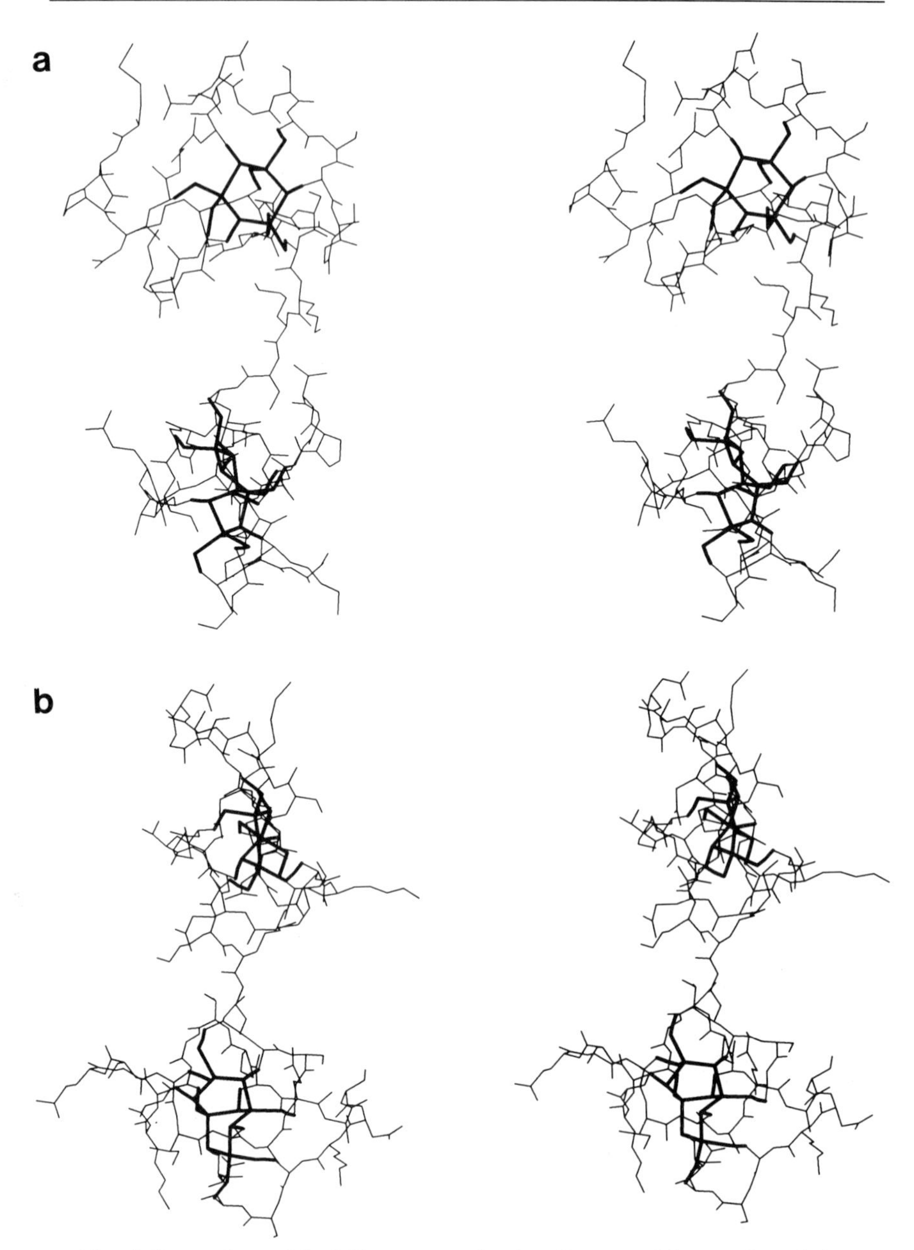

FIG. 4. Stereo views of the MT structure, showing all atoms of the protein and clusters. $C_\beta-S_\gamma$ and S_γ-metal bonds are highlighted. (a) View normal to the three-metal cluster of the β domain (top). The linker region, residues 30–32, is in the center of the figure, and the α domain is at the bottom. (b) View orthogonal to that in (a), emphasizing that the short

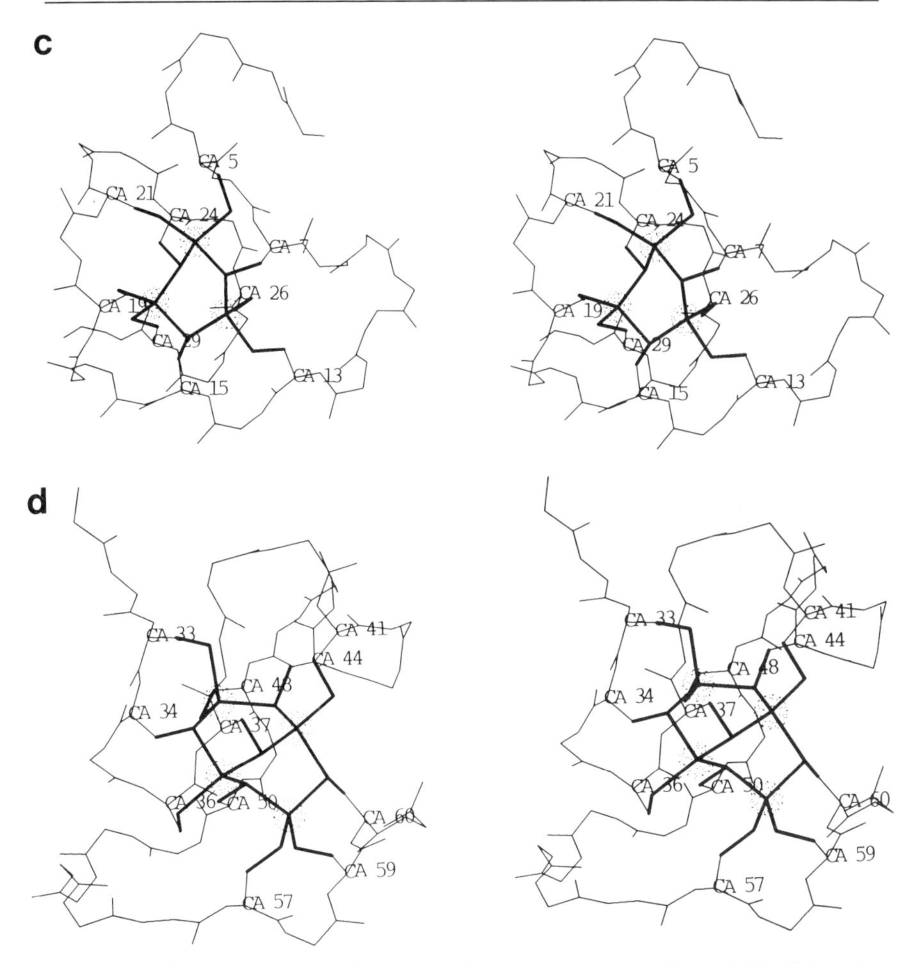

dimensions of the α and β domains are roughly normal to each other. (c) The β domain, showing the arrangement of the nine cysteine ligands around the three metal sites. Side-chain atoms of the noncysteine residues are omitted for clarity. (d) The α domain showing the arrangement of the 11 cysteine ligands around the 4 cadmium sites. Side-chain atoms of the noncysteine residues are omitted for clarity.

of 54%. Figure 6a shows the packing of eight MT molecules in one unit cell. The unique intermolecular contacts and symmetry operators that generate the packing are given in Table II. It can be seen that the MT molecules associate into two groups of four, leaving a solvent channel at $z = \frac{1}{2}$. All three WS_4^{2-} sites occur in this channel. Six of the eleven unique intermolecular contacts result from close association of pairs of molecules

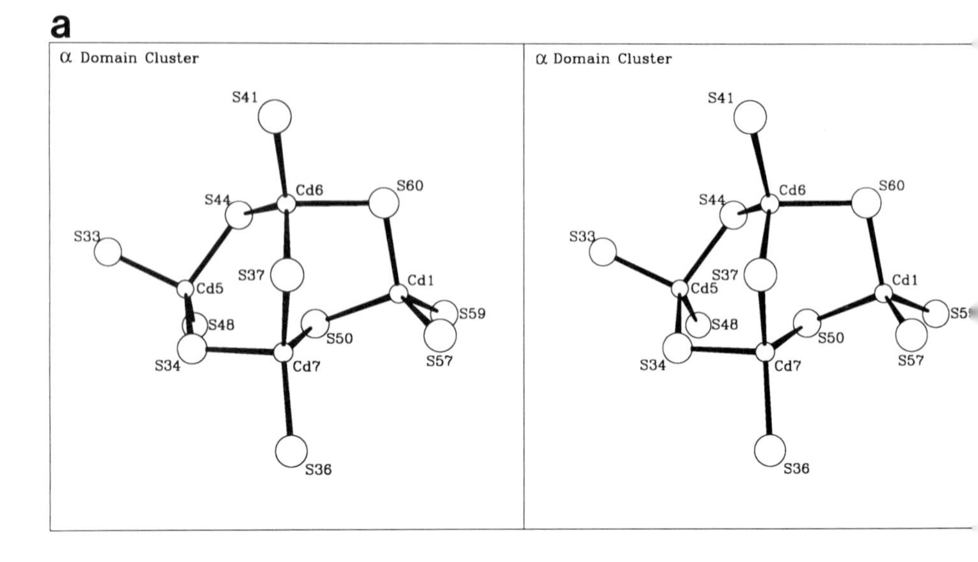

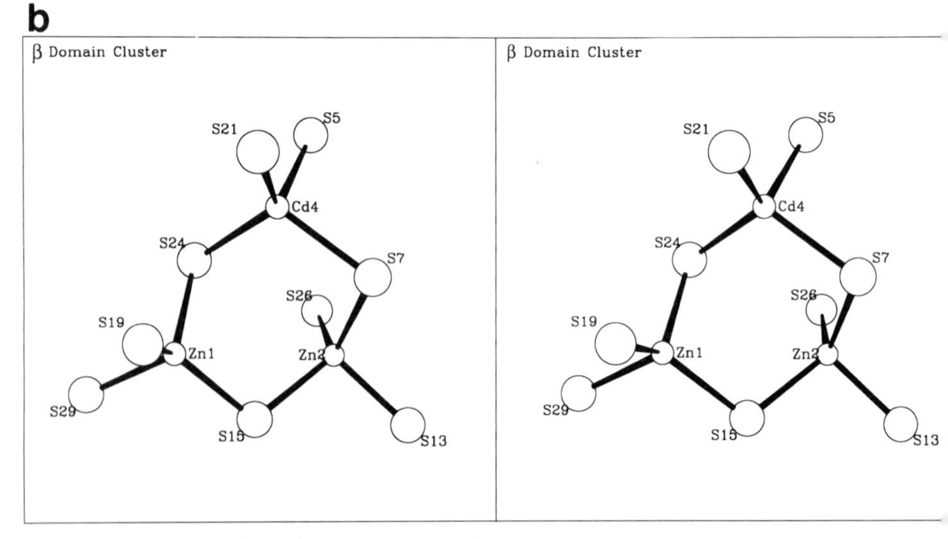

FIG. 5. Stereo views of the metal clusters in MT. S_γ atoms are labeled by residue number. (a) $Cd_4(S_\gamma)_{11}$ cluster. (b) $CdZn_2(S_\gamma)_9$ cluster.

about the diagonal 2-fold axis at $z = \frac{1}{4}$ (symmetry operators 1 and 6, Table II). These pairs are highlighted in Fig. 6b. It is this packing of MT molecules in pairs that creates the pseudo centrosymmetric constellation of cadmium sites in four adjacent metal clusters.

Crystallization of MT requires the correct concentration ratio of NaH-

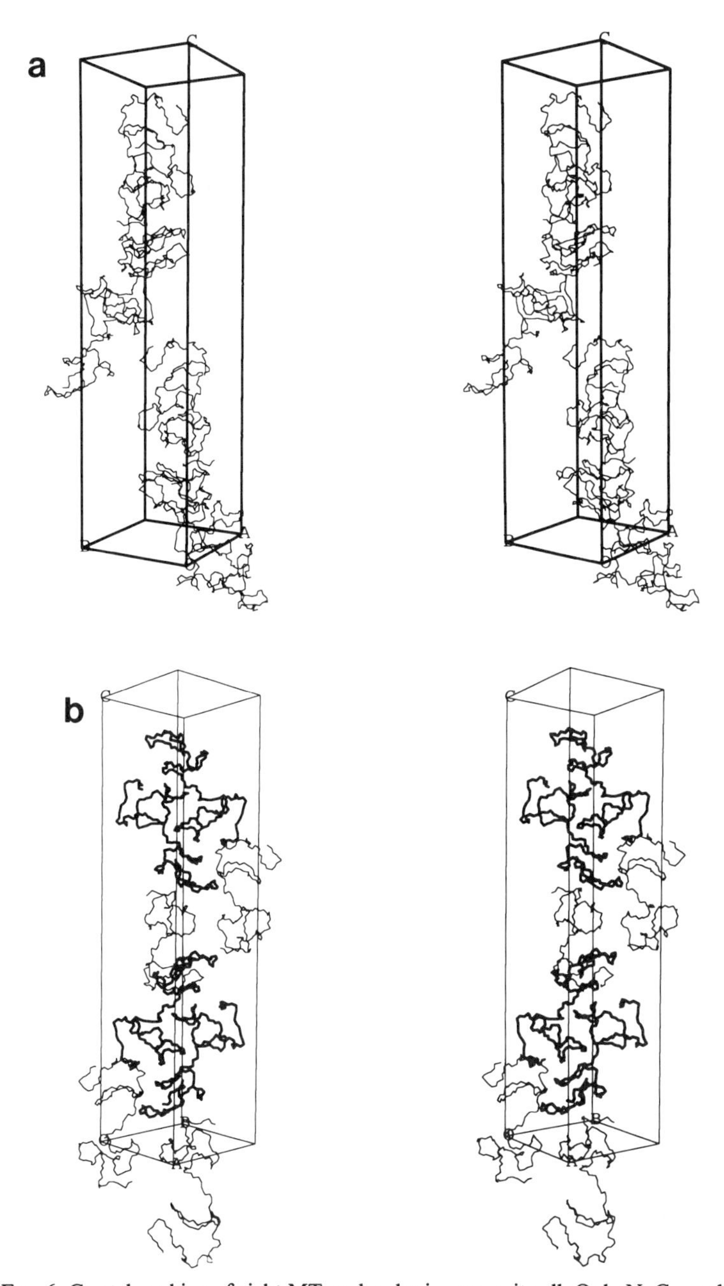

FIG. 6. Crystal packing of eight MT molecules in one unit cell. Only N, C_α, and C atoms of the main chains are shown. Symmetry operators are listed in Table II. (a) View approximately parallel to the diagonal twofold axes at $Z = 0, \frac{1}{2}$, emphasizing the packing of molecules into two groups of four and the solvent channel at $Z = \frac{1}{2}$. (b) View approximately parallel to the diagonal twofold axes at $Z = \frac{1}{4}, \frac{3}{4}$, emphasizing the packing of molecules in pairs about the twofold axes. Molecules 1 and 6 ($Z = \frac{1}{4}$) and 2 and 5 ($Z = \frac{3}{4}$) are shown in heavy lines. Molecules 1 and 6 are the reference pair for discussion of the MT dimer (see text).

TABLE II
INTERMOLECULAR CONTACTS OF CRYSTALLIZATION EXCLUDING WATER MOLECULES[a]

Molecule A		Molecule B			
Residue	Atom	Residue	Atom	Distance (Å)	Symmetry operators[b]
T^{9}	O	K^{30}	N	3.08	6/3
D^{10}	Cβ	K^{30}	O	3.17	6/3
S^{14}	O	K^{43}	Nζ	2.83	3/1
C^{15}	O	K^{43}	Nζ	2.73	3/1
S^{18}	Oγ	K^{43}	Nζ	2.82	3/1
C^{21}	O	Q^{46}	Nε	2.76	1/6[c]
K^{22}	O	Q^{46}	Nε	2.82	1/6[c]
Q^{23}	Nε	G^{47}	O	2.91	1/6[c]
C^{24}	N	Q^{46}	Oε	2.84	1/6[c]
K^{30}	Nζ	G^{40}	O	2.67	1/6[c]
K^{31}	O	S^{45}	Oγ	3.02	1/6[c]

[a]Contacts < 3.2 Å; hydrogen bond donor-H . . . acceptor angle 180 ± 40°. Na^+ contacts listed in Table III.
[b]Symmetry operators for crystal packing contacts:

Number	Symmetry	Translation a	Translation b	Translation c
1	x, y, z	0	0	0
2	$-x, -y, \frac{1}{2}+z$	+1	+1	0
3	$\frac{1}{2}-y, \frac{1}{2}+x, \frac{3}{4}+z$	0	−1	−1
4	$\frac{1}{2}+y, \frac{1}{2}-x, \frac{1}{4}+z$	0	+1	0
5	$y, x, -z$	0	0	+1
6	$-y, -x, \frac{1}{2}-z$	+1	+1	0
7	$\frac{1}{2}-x, \frac{1}{2}+y, \frac{3}{4}-z$	0	0	0
8	$\frac{1}{2}+x, \frac{1}{2}-y, \frac{1}{4}-z$	0	0	0

[c] Diagonal twofold axis at $z = \frac{1}{4}$.

COO^- and KH_2PO_4/K_2HPO_4 at pH 7.0.[2] It is not surprising then that both ions are found trapped in the crystal lattice. The electron density of what appears to be a phosphate ion is shown in Fig. 7a. The density is not strong, but its shape suggests a tetrahedral ion. An H_2O molecule refined at this site has $B = 44$ Å^2. The putative $H_2PO_4^-/HPO_4^{2-}$ ion makes contacts to C^{19} and K^{31} (Table III and Fig. 7a). K^{31} hydrogen bonds to C^{19} and C^{21} are the only interdomain contacts in the structure.[1] The fact that phosphate bridges this interaction between domains may explain the absolute requirement for phosphate in crystallization.

TABLE III
CONTACTS OF IONS OF CRYSTALLIZATION

Atom	Contacting atom[a]	Distance (Å)
Phosphate		
PO1	C^{19} O	3.00
PO1	K^{31} NZ	2.95
PO4	C^{19} O	2.60
Sodium		
Na^+	C^{29} O	2.16
Na^+	$A^{42'}$ O	2.26
Na^+	$S^{45'}$ OG	2.73
Na^+	O^{70}	2.14
Na^+	O^{71}	2.18
Na^+	O^{72}	1.98

[a] $A^{42'}$ and $S^{45'}$ are on a symmetry-related molecule at $1-y$, $1-x$, $\frac{1}{2}-z$. O^{70}, O^{71}, O^{72} are modeled as H_2O molecules in the refinement.

A second ion of crystallization is trapped at the interface of twofold related pairs of MT molecules. The electron density is shown in Fig. 7b. This site has contacts to carbonyls and a serine of twofold related proteins and three contacts to H_2O molecules (Table III). The overall geometry is octahedral. Because the crystallization medium contains 2 M $NaHCOO^-$ the density has been assumed to represent Na^+. An Na^+ refined at this site has $B = 25.4$ Å^2. Twofold association of MT molecules in the crystal appears to require mutual interactions to trapped, aquated Na^+. Without this interaction, the isolated domains may be too flexible to crystallize.

Two of the H_2O molecules associated with Na^+ (O^{71}, O^{72}) participate in a network of hydrogen bonds at the interface of twofold related proteins. This interaction surface includes residues 21–24, 30, 31, and 42 from one molecule with residues 27, 40, and 45–47 from the other in a head-to-tail packing of the domains (α/β' and β/α') (Fig. 6b). Of 12 residues involved in 7 unique contacts across the 2-fold axis (Table II), five of the residues and five of the contacts also involve direct interactions with Na^+, O-71, or O-72. The fact that Na^+ and phosphate are both involved in interactions that would inhibit motion of the α and β domains about the linker region may explain why a precise ratio of the two salts is required to obtain crystals.[2]

Metallothionein Dimers

Metal exchange occurs between pairs of MT molecules such that mixing of Cd_7-MT and Zn_7-MT leads to an equilibrium state of native

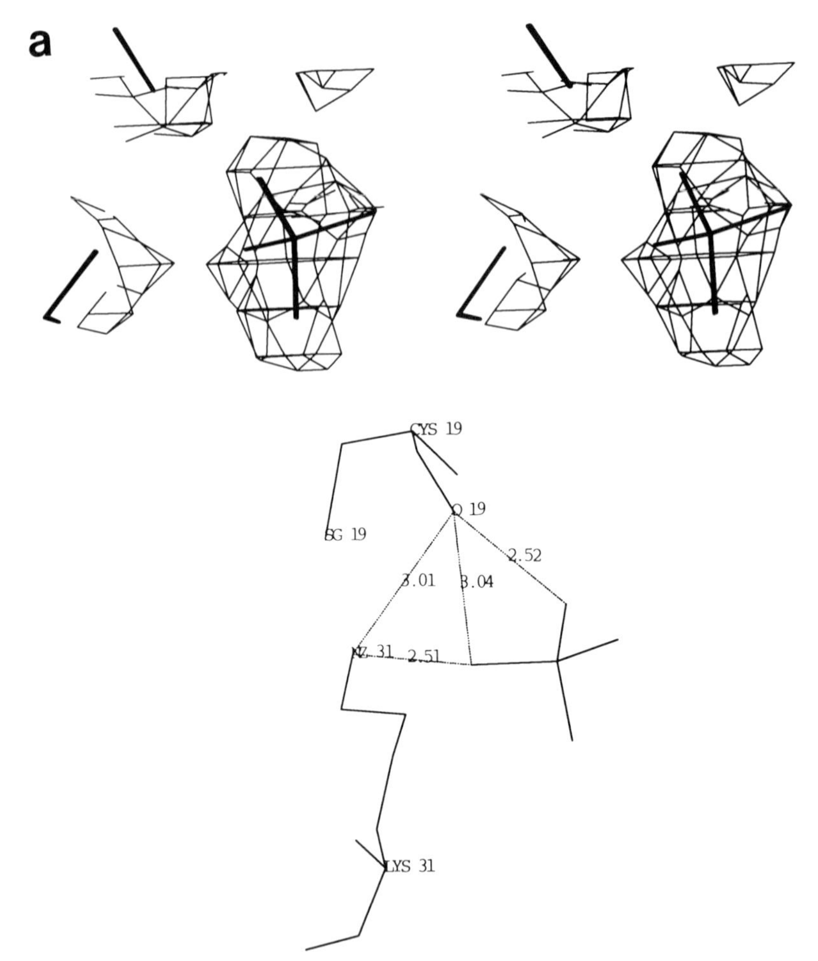

FIG. 7. Details of the interactions for two ions of crystallization. For (a) and (b) the electron density for the ion in the final 1.8-Å $2|F_o| - |F_c|$ map is shown in stereo above the schematic. The schematics and density figures show the ions in the same orientation. Contacts to the ions are listed in Table III. (a) Phosphate ion that assists in the bridging of the α and β domains by K^{31}. $N_\zeta(31)$ also has contacts to C^{21} and $S_\gamma(19)$ (Ref. 1). (b) Cation, modeled as Na^+, which has octahedral contacts at the interface of twofold-related MT molecules. Residues "545" and "542" are S^{45} and A^{42} on molecule 6 in the packing (Table II), and O^{70}, O^{71}, O^{72} are water molecules.

Cd_5,Zn_2-MT.[10] Metal exchange is therefore not only intramolecular but also intermolecular. Moreover, dimers are formed when Zn_7-MT is treated with free Cd^{2+}.[11] The presence of both monomers and dimers, and rapid exchange of cadmium and zinc among sites,[10,11] is apparent as heterogene-

[10] D. G. Nettesheim, H. R. Engeseth, and J. D. Otvos, *Biochemistry* **24,** 6744 (1985).

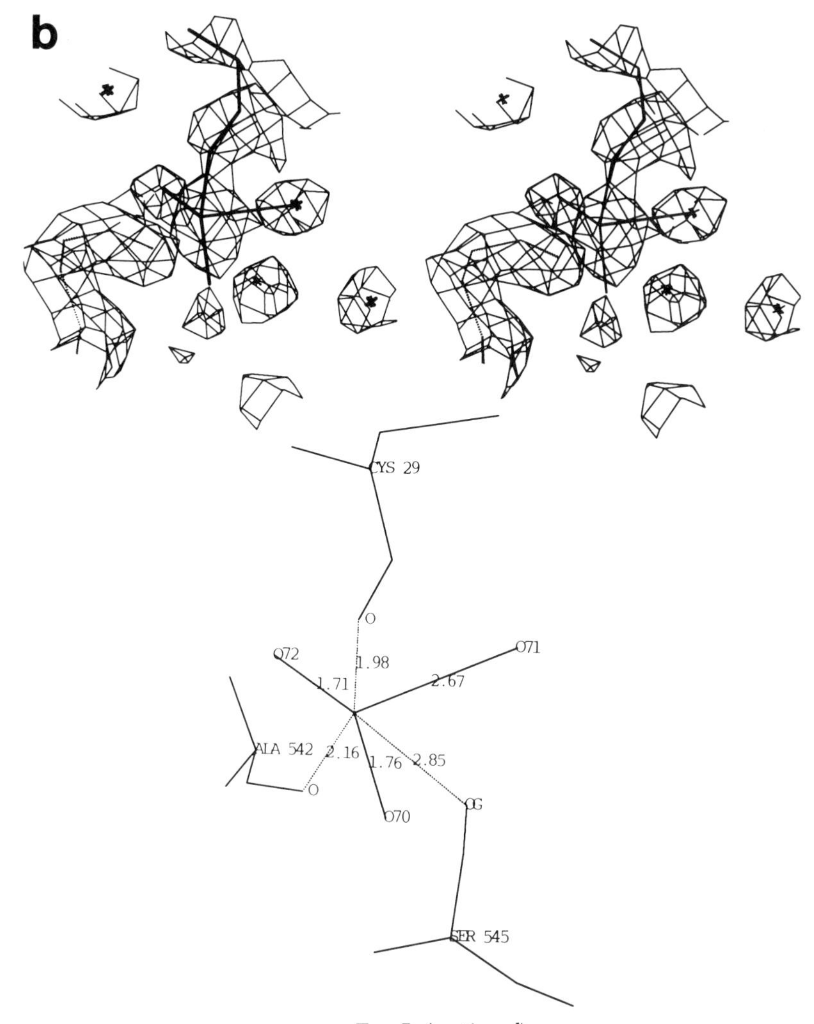

FIG. 7. *(continued)*

ity in NMR spectra of reconstituted MTs.[12,13] The implication is that MT dimers are capable of direct intermolecular metal exchange.

Two aspects of the crystal structure of Cd_5,Zn_2-MT support the idea that MT exists as a dimer in solution. First, the crystal packing contains

[11] J. D. Otvos, H. R. Engeseth, and J. Wehrli, *Biochemistry* **24,** 6735 (1985).

[12] M. Vašák, E. Wörgötter, G. Wagner, J. H. R. Kägi, and K. Wüthrich, *J. Mol. Biol.* **196,** 711 (1987).

[13] M. Vašák, G. E. Hawkes, J. K. Nicholson, and P. J. Sadler, *Biochemistry* **24,** 740 (1985).

intimately associated pairs of molecules related by the twofold axes (Fig. 6b). Second, a cation of crystallization is trapped between the twofold related MT molecules (Fig. 7b). Together these interactions would stabilize a MT dimer in solution. The putative dimer would then have a total of 6 bonds to 2 cations (Table III), 10 intermolecular hydrogen bonds (Table II), and a number of hydrogen bonds to trapped water molecules. Thus, the crystallization of MT may proceed from dimers rather than monomers. Intuitively, a dimer would crystallize more readily than a monomer because motion of the α and β domains about the linker could be restricted by head-to-tail packing of α/β' and β/α' (Fig. 7b).

The cysteines involved in contacts of the putative dimer, C^{21}, C^{24}, C^{29}, lie along one face of the three-metal cluster containing the Zn-1 site (Fig. 5b). Conceivably, the cation of crystallization may represent a site of divalent metal ion coordination during metal exchange. The shortest cation-to-metal cluster distance is Na^+–Zn-1, 6.09 Å; all other such distances are greater than 8.2 Å and the shortest intermolecular metal–metal distance about the twofold is Zn-1–Cd-6′, 14.25 Å. Therefore, both the local intramolecular environment of Zn-1[1] and its position in the putative dimer support the notion of rapid exchange at this site.

Acknowledgments

We are indebted to D. R. Winge for samples of the protein, E. Stiefel for samples of $(NH_4)_2WS_4$, and M. Pique and G. Gippert for computer graphics. We thank D. R. Winge, J. D. Otvos, and I. M. Armitage for discussions. This research was supported by NIH Grant GM-36535.

[59] Determination of the Three-Dimensional Structure of Metallothioneins by Nuclear Magnetic Resonance Spectroscopy in Solution

By KURT WÜTHRICH

Introduction

A metallothionein was among the first proteins for which the three-dimensional (3D) structure was determined by nuclear magnetic resonance (NMR) spectroscopy.[1] The course of the structure determination empha-

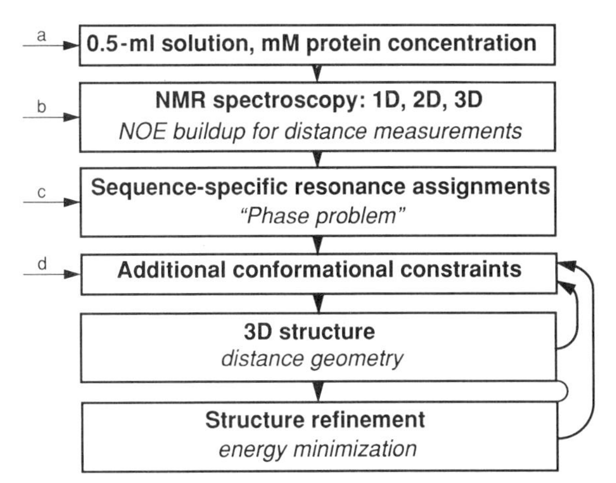

FIG. 1. Diagram outlining the course of a protein structure determination by NMR. The arrows (a to d, on the left) indicate the steps where special experiments have been used in studies with metallothioneins. (a) Isotope labeling with ^{113}Cd and ^{112}Cd, (b) homonuclear ^{113}Cd NMR; [^{113}Cd, ^{1}H]-COSY; ^{1}H NMR with X- and 2X-filter (X = ^{113}Cd); (c) identification of Cys spin systems from $J(^{113}\text{Cd}, ^{1}\text{H})$; and (d) Cd^{2+}–Cys coordinative bonds.

sized again that metallothioneins (MT) are a very special class of proteins, and that specialized experimental techniques are required for their characterization. This chapter presents a discussion of NMR experiments that were initially tailored for studies of metallothionein structures, but which may be of interest also for studies of other classes of compounds. We begin with a survey of the NMR method for protein structure determination as it is at present generally applied for polypeptides with up to approximately 150 amino acid residues.[2,3] In the second part, modifications of the method applicable for studies of metallothioneins are described in more detail. Finally, the results on metallothionein structures obtained by NMR in solution are briefly reviewed and critically evaluated.

Protein Structure Determination by NMR

Figure 1 presents an overview of the NMR method for protein structure determination in solution.[1-3] In the protein solutions used for the NMR data collection, the solvent is usually water, and the ionic strength, the pH, and the temperature should be adjusted so as to ensure near-physiological

[1] K. Wüthrich, *Acc. Chem. Res.* **22,** 36 (1989).
[2] K. Wüthrich, "NMR of Proteins and Nucleic Acids." Wiley, New York, 1986.
[3] K. Wüthrich, *Science* **243,** 45 (1989).

conditions. The protein concentration should be at least 1 m*M*, ideally 3 to 6 m*M* (or, with small proteins, even higher). Although this concentration is high relative to that of most proteins in their physiological milieu, it is not so far from the *total* protein concentration in many body fluids. The volume of a typical NMR sample is 0.5 ml, so that 10 to 20 mg of a protein with the size of a mammalian metallothionein should be available for a structure determination.

One-dimensional NMR experiments (and possibly circular dichroism studies) are used to define the conditions of pH, temperature, and ionic strength where the native, folded form of the protein is preserved. A set of conditions well within the stability range of the protein is then selected for the acquisition of the NMR data needed to prepare the input for the structure determination, usually with two- (2D) or three-dimensional (3D) experiments (Fig. 2). Here, we come to the first of *three fundamental elements* that constitute the foundations of the NMR method for protein structure determination,[1] i.e., the identification of an NMR parameter that can be directly related to the three-dimensional molecular structure. Nuclear Overhauser effects (NOE) can serve this purpose, provided that one takes into account the different spin physics in small and large molecules in solution and uses *NOE buildup experiments* to eliminate derogatory effects from spin diffusion.[2] NOEs can thus be used for measurements of short *through-space proton–proton distances* in the range of up to approximately 5.0 Å. Additional NMR experiments are recorded for establishing *through-bond* relations between hydrogen atoms that are part of the same amino acid residue. Frequently used experiments are correlation spectroscopy (COSY) and total correlation spectroscopy (TOCSY), and there is a large array of related experiments for use in special situations.[2,4]

The third step in Fig. 1 relates to the second fundamental element of the method, i.e., techniques for obtaining sequence-specific resonance assignments. By the fact that polypeptide chains in proteins generally contain multiple units of each amino acid type, spectral assignments are nontrivial. The problem was solved with the introduction of the *sequential assignment strategy*.[1,2] The importance of the resonance assignments is illustrated with Fig. 3. In the absence of sequence-specific resonance assignments each peak in the NOESY spectrum of Fig. 2 merely indicates the presence of two closely spaced hydrogen atoms, each of which may be located anywhere along the primary structure (top of Fig. 3). This type of information is clearly not suitable as a basis for 3D structure determination. However, once resonance assignments have been obtained, each cross-peak identifies

[4] R. R. Ernst, G. Bodenhausen, and A. Wokaun, "Principles of Nuclear Magnetic Resonance in One and Two Dimensions." Oxford Univ. Press (Clarendon), Oxford, 1987.

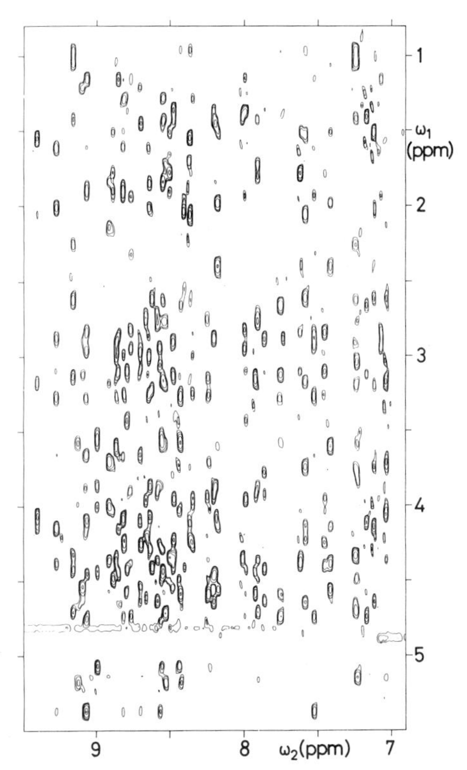

FIG. 2. Two-dimensional nuclear Overhauser enhancement spectroscopy (NOESY) of rabbit MT-2a. A contour plot of the spectral region ($\omega_1 = 0.8-5.4$ ppm, $\omega_2 = 6.9-9.5$ ppm) is shown, which contains the cross-peaks between amide protons and aliphatic protons. The spectrum was recorded with a 10 m*M* solution of [$^{112}Cd_7$]MT-2a in H_2O with a mixing time of 250 msec. The pH was 7.0, with 20 m*M* [$^2H_{11}$]Tris-HCl and 20 m*M* KCl, and the temperature was 14°. 160 scans were accumulated per t_1 value and 280 t_1 values were measured. To achieve an optimum signal-to-noise ratio, the time-domain data were truncated to 1024 points along t_2 and 130 points along t_1 before Fourier transformation. The horizontal ridge at $\omega_1 = 4.92$ ppm is a residual perturbation from the strong signal of the solvent H_2O. [Reproduced from A. Arseniev, P. Schultze, E. Wörgötter, W. Braun, G. Wagner, M. Vašák, J. H. R. Kägi, and K. Wüthrich, *J. Mol. Biol.* **201,** 637 (1988).]

an upper limit on the distance between two distinct locations along the polypeptide chain (bottom of Fig. 3), which is the input needed for a structure determination.[2] Thus, in its impact on the NMR structure determination method the sequential assignment strategy for obtaining sequence-specific NMR assignments can be compared to the use of isomorphous heavy atom derivatives for solving the phase problem in protein crystallography.[1]

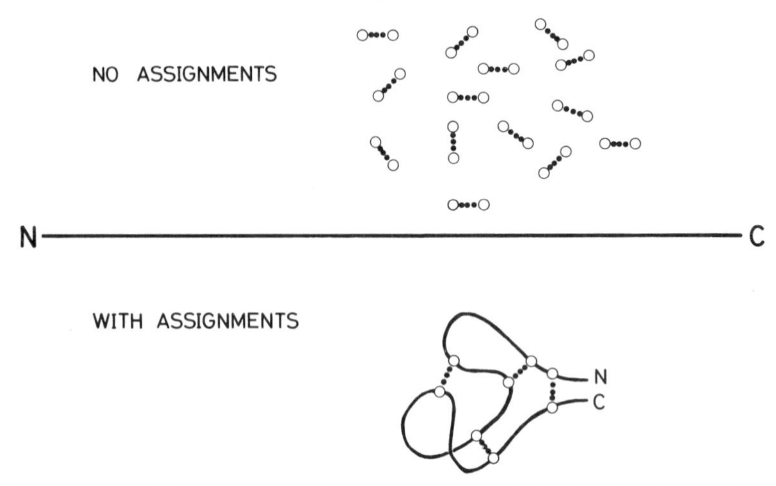

FIG. 3. Scheme illustrating the information content of $^1H-^1H$ NOEs in a polypeptide chain (represented by the horizontal line in the center) with and without sequence-specific resonance assignments. Open circles represent hydrogen atoms of the polypeptide, and dotted lines the short $^1H-^1H$ distances manifested by the NOEs (see text). (Reproduced from Ref. 2.)

Obtaining "complete sequence-specific resonance assignments" means also that the chemical shifts of all hydrogen atoms in the protein are determined. To achieve this goal only a small percentage of all cross-peaks in the NOESY spectra (Fig. 2) is needed. In contrast, the maximum possible number of conformational constraints must be collected as input for the calculation of the complete three-dimensional protein structure. Therefore, the individual chemical shifts are used to further assign all, or nearly all, remaining NOESY cross-peaks that were not needed for obtaining the resonance assignments. This laborious procedure can nowadays be largely automated. In the input for the structure calculations the resulting list of NOE distance constraints can be further supplemented by measurements of scalar spin–spin coupling constants, and by the identification of hydrogen bonds in regular secondary structures identified prior to the structure calculations.[2]

The last two steps of Fig. 1 relate to mathematical techniques for structure determination from NMR data, which is the third essential element of the structure determination method.[2,3,5] In present practice most structure calculation protocols include either a structural interpretation of the NMR data with a variable target function algorithm, which may be

[5] W. Braun, *Q. Rev. Biophys.* **19**, 115 (1987).

supplemented by molecular mechanics energy minimization, or an initial analysis with metric matrix distance geometry followed by molecular dynamics calculations.[2,5] As is indicated by the arrows in the lower right of Fig. 1, a structure determination usually goes through several cycles of data collection and structure calculations. One reason for this is that ^1H NOESY spectra of proteins often contain groups of two or several cross-peaks with identical chemical shifts along one of the two frequency axes ω_1 or ω_2 (Fig. 2). Although this initially prevents unambiguous assignments for these cross-peaks, most of the ambiguities can usually be resolved by reference to preliminary structures calculated without using these NOEs in the input. Multiple cycles of data collection and structure calculations include also checks of the protein structure against the experimental NOESY spectra.[6] Two different presentations of metallothionein structures determined by NMR are afforded by Figs. 4[7] and 5.[8]

Figure 4 leads the way to a brief discussion on the assessment of the quality of a protein structure determination by NMR.[2,3] Using the program DISMAN,[9] each of the 10 conformers in Fig. 4 was independently calculated from the same input of experimental conformational constraints but with different, randomly generated starting structures (although only the polypeptide backbone is shown in Fig. 4, the calculations were done with the complete structure, including the amino acid side chains and the metal ions). Statistical analyses of the residual constraint violations[2,3,5] showed that each of these conformers represents a good fit of the experimental data. To further decide whether the NMR data were sufficient to define a unique three-dimensional structure, the individual "good" distance geometry solutions were compared. A high-quality structure determination generates a tight bundle of conformers representing the solution structure of the protein, which corresponds to a small value for the average of the pairwise root-mean-square deviations (RMSD). Inspection of Fig. 4 shows that all parts of the polypeptide backbone are not equally well defined. In particular, the N-terminal segment 1–12 shows a significantly larger dispersion among the 10 conformers than the rest of the polypeptide chain. The observation that the local RMSDs among the group of distance geometry conformers used to represent the solution structure varies along the polypeptide chain is not unique for metallothioneins but has been observed

[6] A. Arseniev, P. Schultze, E. Wörgötter, W. Braun, G. Wagner, M. Vašák, J. H. R. Kägi, and K. Wüthrich, *J. Mol. Biol.* **201,** 637 (1988).

[7] B. A. Messerle, A. Schäffer, M. Vašák, J. H. R. Kägi, and K. Wüthrich, *J. Mol. Biol.* **214,** 765 (1990).

[8] P. Schultze, E. Wörgötter, W. Braun, G. Wagner, M. Vašák, J. H. R. Kägi, and K. Wüthrich, *J. Mol. Biol.* **203,** 251 (1988).

[9] W. Braun and N. Gō, *J. Mol. Biol.* **186,** 611 (1985).

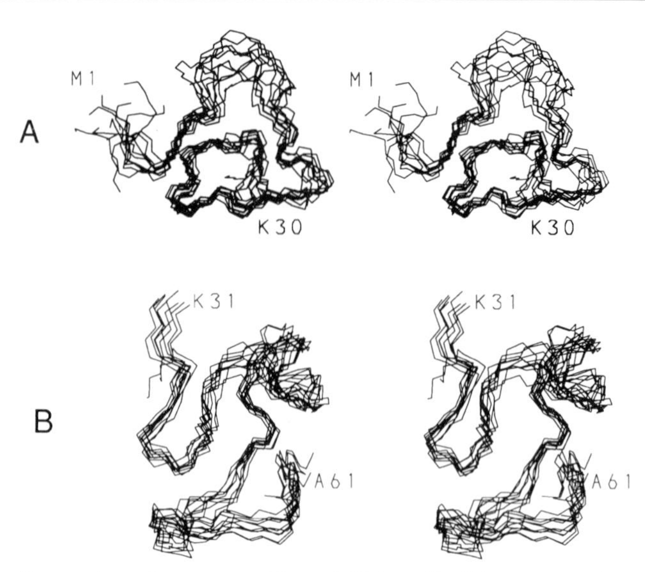

FIG. 4. Stereo view of a superposition of 10 "good" distance geometry structures of human MT-2, i.e., those with the lowest residual error functions, which were selected to represent the solution structure. The superposition is for minimum pairwise root-mean-square deviation (RMSD) between the best structure and each of the other nine conformers. Only the polypeptide backbone is shown. The two domains present in MT-2 are shown separately, since their relative spatial arrangement was not defined by the NMR data. (A) β Domain from residues 1–30, (B) α domain from residues 31–61. (Reproduced from Ref. 7.)

quite generally for globular proteins.[3] Figure 5 shows a more complete representation of a single conformer of rat MT-2 (taken from a corresponding group of distance geometry solutions as that in Fig. 4), which includes the metal ions and the cysteine side chains involved in the coordinative bonds with the metal ions.

Special Experimental Techniques for Metallothionein Structure Determinations by NMR

Modifications of the NMR method for protein structure determination described in the preceding section are needed primarily because of the quite unique amino acid compositions, amino acid sequences, and high contents of metal ions in metallothioneins. The principal additional experiments used are listed in the caption to Fig. 1 and are described in some detail in this section.

An important difference relative to typical globular proteins lies in the

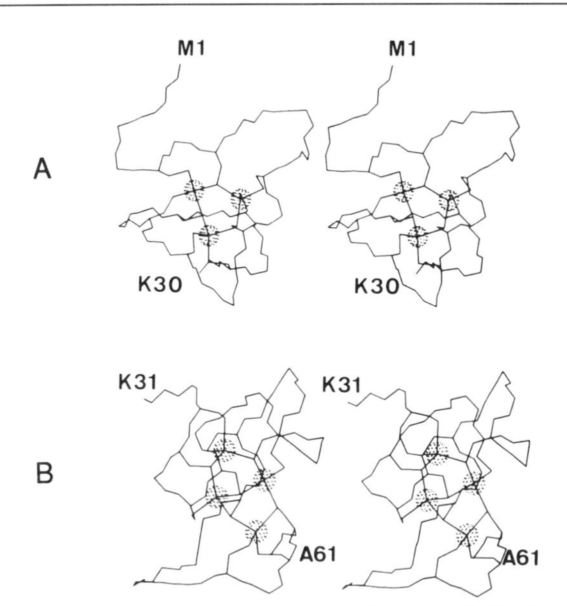

FIG. 5. Stereo view of the polypeptide backbone, the cysteinyl side chains, and the metal ions in rat MT-2. (A) β domain with residues 1–30 and the nine cysteinyl residues 5, 7, 13, 15, 19, 21, 24, 26, and 29. The cadmium ions are represented as dotted spheres of radius 0.9 Å. (B) α domain with residues 31–61 and the 11 cysteinyl residues 33, 34, 36, 37, 41, 44, 48, 50, 57, 59, and 60. (Reproduced from Ref. 8.)

fact that the core of the two domains in metallothionein consists of the metal–sulfur clusters (Fig. 5). In globular proteins, where this molecular region is built up of hydrophobic amino acid side chains, a large percentage of the $^{1}H-^{1}H$ NOEs used as input for the structure calculations arises from this hydrophobic core. Metallothioneins therefore yield comparatively small numbers of NOE distance constraints, which would hardly be sufficient to warrant a complete structure determination. However, if the protein contains the NMR-active isotope ^{113}Cd, the $^{1}H-^{1}H$ distance constraints derived from the NOEs can be supplemented by constraints corresponding to the coordinative bonds between the metal ions and the polypeptide chain. These can be identified using the special techniques (a)–(d) indicated in Fig. 1.

Isotope Labeling of Metallothioneins with ^{113}Cd and ^{112}Cd

In their natural state the mammalian-type metallothioneins contain primarily Zn^{2+} in the metal-binding sites, which may, however, also be occupied with Cd^{2+} or Hg^{2+}. For NMR studies we used Cd_7-MT, which

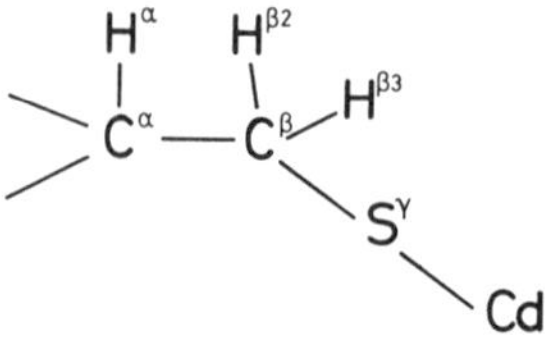

FIG. 6. Coordinative bond between cysteine and Cd^{2+} in metallothionein.

was obtained from the Zn_n,Cd_{7-n}-MT isolated from liver tissue either by displacement of the Zn^{2+} in the biosynthetic material by the addition of an excess of Cd^{2+}, or by reconstitution of the apoprotein with Cd^{2+}. At natural isotope abundance, cadmium contains 74.8% of ^{112}Cd with spin I equal to 0, and 12.9% of ^{111}Cd and 12.3% of ^{113}Cd, which both have a spin I of 1/2. The presence of ^{111}Cd or ^{113}Cd in the coordinative bonds with cysteine side chains in metallothionein (Fig. 6) is manifested by additional fine structure in the ^1H NMR spectra (Fig. 7). As a consequence, ^1H NMR spectra of Cd_7-MT with natural isotope abundance are exceedingly complex, because each ^1H resonance of cysteine contains a singlet from the protein molecules containing ^{112}Cd and two doublets from the protein molecules containing

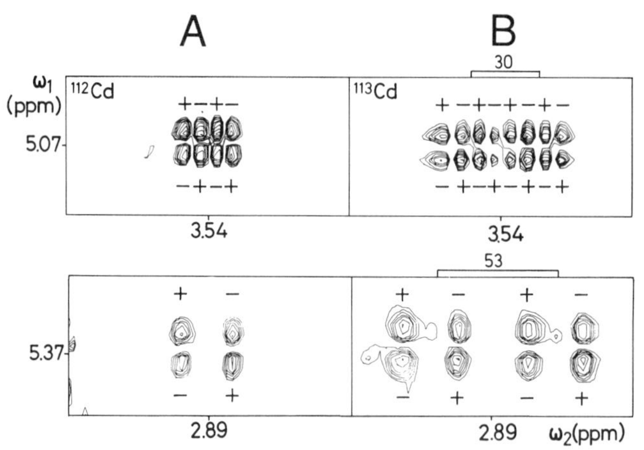

FIG. 7. Comparison of corresponding cross-peaks in homonuclear ^1H correlation spectroscopy (COSY) with $^{112}Cd_7$-MT-2a (A) and $^{113}Cd_7$-MT-2a (B) from rabbit liver. Two $C_\alpha H$-$C_\beta H$ cross-peaks of two cysteinyl residues, of which each is bound to a different single Cd^{2+} ion, are shown. The chemical shifts are indicated on the left and below the cross-peaks. Positive and negative fine-structure components are identified by the corresponding signs, and the size of the $^1H-^{113}Cd$ scalar coupling is indicated at the top of the $^{113}Cd_7$-MT-2a spectra. [Reproduced from D. Neuhaus, G. Wagner, M. Vašák, J. H. R. Kägi, and K. Wüthrich, *Eur. J. Biochem.* **143,** 659 (1984).]

^{111}Cd, or ^{113}Cd, respectively. For this reason $^{112}Cd_7$-MT as well as $^{113}Cd_7$-MT were prepared for the structure determinations. $^{113}Cd_7$-MT is needed to identify the cysteine spin systems, and subsequently the metal-binding sites in the polypeptide chain (see the following section). $^{112}Cd_7$-MT provides clean ^{1}H NMR spectra for the collection of NOE distance constraints [Fig. 2; in $^{113}Cd_7$-MT the NOESY cross-peaks with cysteine side-chain protons would be complicated by ^{113}Cd – ^{1}H scalar couplings (Fig. 7)].

Because the isolation and purification of metallothioneins is described elsewhere in this volume,[10] the following is limited to a brief account of the preparation of rat MT-2 for the NMR studies. Rabbit $^{113}Cd_7$-MT-2a and $^{112}Cd_7$-MT-2a, and human $^{113}Cd_7$-MT-2 and $^{112}Cd_7$-MT-2 were obtained from the Zn^{2+}-containing proteins either by metal displacement or by reconstitution of the apoprotein.[7,10–12]

Rat liver MT-2 was induced in male Sprague-Dawley rats by three subcutaneous injections of $CdCl_2$ (3 mg Cd/kg body weight) on subsequent days. The MT-2 isoform was purified as described elsewhere.[10] The resulting biosynthetic rat Cd,Zn-MT-2 usually contained five (± 0.1) Cd^{2+} ions and two (± 0.1) Zn^{2+} ions per protein molecule, and was homogeneous by high-performance liquid chromatography criteria. With the use of ^{113}Cd for the injections we obtained biosynthetic isotope-labeled $^{113}Cd_5$,Zn_2-MT-2, which was used for the purpose of comparisons with the $^{113}Cd_7$-MT-2 preparations obtained by *in vitro* modification.[13] However, biosynthetic $^{113}Cd_7$-MT-2 could not be obtained.

For the preparation of Cd_7-MT-2 by reconstitution, metal-free apo-MT-2 was prepared by gel filtration of a solution of the metalloprotein at low pH: A neutral solution of biosynthetic Cd,Zn-MT-2 was adjusted to pH 1.0 and subsequently passed over a Sephadex G-25 column equilibrated with 0.01 *M* HCl. For the reconstitution with $^{113}CdCl_2$ or $^{112}CdCl_2$, all solutions were degassed on a vacuum line before use, and all steps of the procedure were performed in a nitrogen-purged glovebox. Apo-MT-2 at 0.5 mg/ml in 0.1 M HCl was mixed with 7.5 Eq of an aqueous solution of either $^{113}CdCl_2$ or $^{112}CdCl_2$. The mixture was then adjusted to pH 8.5 with 0.5 *M* Tris base. Excess Cd^{2+} was removed by addition of a small amount of Chelex 100. In a subsequent centrifugation step, the resin was removed

[10] M. Vašák, this volume, [5]–[7].

[11] D. Neuhaus, G. Wagner, M. Vašák, J. H. R. Kägi, and K. Wüthrich, *Eur. J. Biochem.* **143,** 659 (1984).

[12] D. Neuhaus, G. Wagner, M. Vašák, J. H. R. Kägi, and K. Wüthrich, *Eur. J. Biochem.* **151,** 257 (1985).

[13] M. Vašák, E. Wörgötter, G. Wagner, J. H. R. Kägi, and K. Wüthrich, *J. Mol. Biol.* **196,** 711 (1987).

and the supernatant concentrated on an Amicon ultrafiltration apparatus using a YM2 membrane (Amicon, Danvers, MA).

As an alternative, rat $^{113}Cd_7$-MT-2 was obtained by Zn^{2+} displacement from the biosynthetic ^{113}Cd,Zn-MT-2, which was exposed to a 20-fold molar excess of $^{113}CdCl_2$ in 50 m*M* Tris-HCl at pH 8.6. The excess free metal ions were removed as described for the material reconstituted from apo-MT-2.

Before the NMR measurements, the MT-2 samples were passed over a Sephadex G-50 column to ensure size homogeneity of the preparations. The presence of a single peak in the high-pressure liquid chromatography profile was taken as an additional criterion for sample homogeneity. The NMR samples contained approximately 0.01 *M* of the protein in ^2H_2O or in a mixed solvent of 90% H_2O/10%^2H_2O, 0.02 *M* [$^2H_{11}$]Tris-HCl, and 0.05 *M* KCl at pH 7.0. With a sample volume of 0.5 ml, approximately 35 mg of MT-2 were thus used for each NMR sample.

^{113}Cd NMR

The discussion of ^{113}Cd NMR with metallothioneins[14] is limited to a comment on its use as an analytical tool. In our experience the recording of conventional one-dimensional ^{113}Cd NMR spectra to check on the homogeneity of the MT preparations is quite indispensable. As an illustration the spectrum (A) in Fig. 8 shows the seven lines of a homogeneous preparation of rat $^{113}Cd_7$-MT-2, and spectrum (B) corresponds to the aforementioned biosynthetic rat $^{113}Cd_5$,Zn_2-MT-2, which clearly contains at least three different molecular species. The ^{113}Cd NMR spectrum provides further information on the line widths of the ^{113}Cd resonances (for example, in Fig. 8A, line 3 is significantly broader than the other lines). Knowledge of the ^{113}Cd NMR line widths is helpful as a basis for optimizing the conditions of pH, ionic strength, and temperature for heteronuclear ^{113}Cd–^{1}H NMR experiments, and is also needed for the analysis of the heteronuclear experiments (see below). ^{113}Cd NMR measurements such as those in Fig. 8 are straightforward, but they depend on the availability of hardware that may not necessarily be part of the equipment in a laboratory specializing in protein structure determination by NMR. Ideally, the spectra are recorded either with a broadband heteronuclear probehead, or with a ^{13}C probehead tuned for ^{113}Cd. Alternatively, either of the corresponding reverse detection probes may be used, which, however, inevitably have reduced sensitivity for direct ^{113}Cd detection.

Of course, the use of ^{113}Cd NMR for checks on the homogeneity of MT preparations is complemented by similar checks using ^{1}H NMR.[6,12,13]

[14] I. M. Armitage and J. D. Otvos, *in* "Biological Magnetic Resonance" (L. J. Berliner and J. Reuben, eds.), Vol. 4, pp. 79–144. Plenum, New York, 1982.

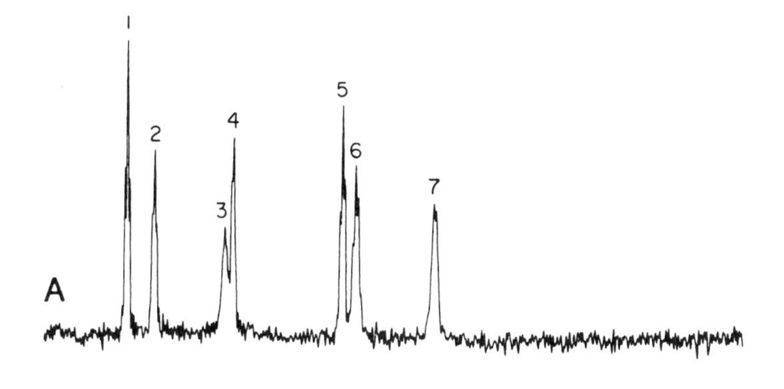

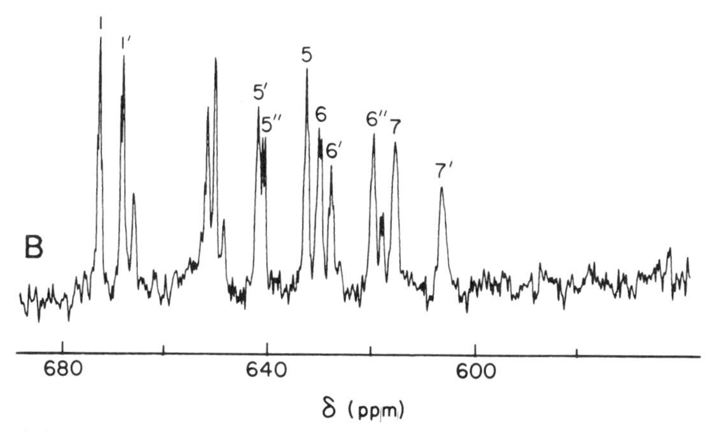

FIG. 8. ^{1}H-Decoupled 80-MHz one-dimensional ^{113}Cd NMR spectra at 25° of (A) reconstituted rat $^{113}Cd_7$-MT-2, and (B) biosynthetic rat $^{113}Cd_5,Zn_2$-MT-2. The ^{113}Cd signals of (A) are numbered according to decreasing chemical shift. The signals 1, 5, 6, and 7 of (B) have exactly the same chemical shifts as the corresponding signals in (A). The extra signals 1′, 5′, 6′, 7′, 5″, and 6″ indicate the presence of at least two additional, different molecular species. The chemical shifts in parts per million (ppm) are relative to 0.1 M $^{113}Cd(ClO_4)_2$. (Reproduced from Ref. 13.)

NMR Experiments Exploiting ^{113}Cd–^{1}H Scalar Couplings for Identification of Cysteinyl Spin Systems

As illustrated here with rabbit MT-2a (Fig. 9), metallothioneins have quite unusual amino acid compositions, so that in spite of the small molecular size the ^{1}H NMR spectral analysis is rather difficult. In rabbit MT-2a only 12 of the common amino acids are represented and even these

20C	9A	8S	7K
4G	3D	3T	3P
2N	1I	1M	1Q

62 Residues
12 Different amino acids
33 AMX spin systems ($C^{\alpha}H$-$C^{\beta}H_2$)

FIG. 9. Amino acid composition in rabbit MT-2a.

are unequally distributed (Fig. 9). Because the ^1H chemical shifts are primarily determined by the covalent structure, the spectral distinction of different residues of the same amino acid depends on the conformation-dependent dispersion of the shifts in globular proteins.[2] As a consequence the probability of residual degeneracy of two or more resonance lines in metallothioneins is considerably higher than in other proteins of similar size. This situation is all the more pronounced because mammalian metallothioneins do not contain any aromatic residues, which are known to cause particularly large conformation-dependent shifts.[2] In our work with metallothioneins these problems were solved by comparison of the ^1H NMR spectra of ^{113}Cd-MT and ^{112}Cd-MT (Fig. 7). Since only the resonances of the metal-bound cysteine residues are influenced by spin–spin coupling with ^{113}Cd, the comparison of the homonuclear ^1H correlation spectra (COSY) recorded with ^{112}Cd-MT (Fig. 7A) and ^{113}Cd-MT (Fig. 7B), respectively, enabled the distinction of the cysteine spin systems from the other $C_{\alpha}H-C_{\beta}H_2$ spin systems (Fig. 9).[2] On this basis nearly all the remaining spin systems could then be identified prior to the sequential assignments.[12] The latter, as well as the strategies for obtaining sequence-specific ^1H NMR assignments are the same for metallothioneins as for other proteins.[2]

In principle, with the use of ^{113}Cd broadband decoupling for obtaining ^1H NMR spectra corresponding to those recorded with ^{112}Cd-MT, the analysis of metallothionein ^1H NMR could be performed with a single sample of $^{113}Cd_7$-MT. Furthermore, X-filter[15] and $2X$-filter[16] ^1H NMR experiments ($X = {}^{113}Cd$) are available that enable direct, separate observation of subspectra containing only the ^1H resonances of cysteine side chains bound to either one or two $^{113}Cd^{2+}$ ions, respectively. For further details on these filter techniques the reader is referred to the original literature[15,16] and a recent review on heteronuclear filters.[17] Overall, for a high-quality NMR structure determination of a metallothionein the prepa-

[15] E. Wörgötter, G. Wagner, and K. Wüthrich, *J. Am. Chem Soc.* **108,** 6162 (1986).
[16] E. Wörgötter, G. Wagner, M. Vašák, J. H. R. Kägi, and K. Wüthrich, *J. Am. Chem. Soc.* **110,** 2388 (1988).
[17] G. Otting and K. Wüthrich, *Q. Rev. Biophys.* **23,** 39 (1990).

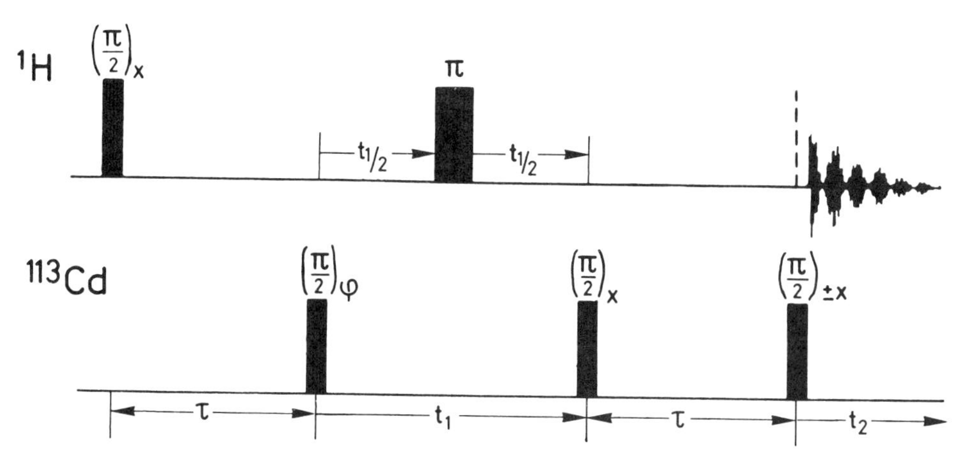

FIG. 10. Experimental scheme for correlating ^{113}Cd spins with directly J-coupled protons ($C_\beta H$) or coupling partners of the latter ($C_\alpha H$). The broken line in the ^{1}H channel represents either a z filter, a $(\pi/2)_y$ pulse, or a two-quantum filter (see text). The phase φ is cycled for the desired order p of ^{113}Cd coherence ($\varphi = k\pi/p$, $k = 0, 1, \ldots, 2p - 1$, with alternated addition and subtraction of the free induction decays). Time-proportional phase incrementation in steps of $\pi/2p$ along with t_1 can be added to distinguish positive and negative frequencies in ω_1. (Reproduced from Ref. 18.)

ration of both $^{112}Cd_7$-MT and $^{113}Cd_7$-MT appears advisable, and the X-filter and $2X$-filter experiments with $^{113}Cd_7$-MT can still provide supplementary, helpful information.

Heteronuclear ^{113}Cd–^{1}H Correlation Spectroscopy for Identification of Cys–Cd^{2+} Coordinative Bonds

The heteronuclear ^{113}Cd–^{1}H correlation experiment, ^{113}Cd,^{1}H-COSY, is a key experiment in metallothionein structure determinations, and it is therefore described in some detail. The basic experiment consists of four pulses, a delay τ, the evolution time t_1, and the observation time t_2 (Fig. 10).[18] It involves creation of in-phase proton magnetization by the initial ^{1}H($\pi/2$) pulse, which evolves during τ into coherence in antiphase with respect to the heteronuclear coupling. A ^{113}Cd($\pi/2$) pulse then transforms the proton coherence into heteronuclear two-spin coherence, which precesses during the evolution period, t_1. The ^{1}H(π) pulse in the middle of t_1 refocuses the heteronuclear J interactions and ensures that only the ^{113}Cd chemical shifts and the homonuclear couplings lead to modulation as a

[18] M. H. Frey, G. Wagner, M. Vašák, O. W. Sørensen, D. Neuhaus, E. Wörgötter, J. H. R. Kägi, R. R. Ernst, and K. Wüthrich, *J. Am. Chem. Soc.* **107,** 6847 (1985).

function of t_1. At the end of the evolution period a second $^{113}Cd(\pi/2)$ pulse transfers the two-spin coherence back to antiphase proton coherence, which then refocuses to in-phase coherence during the second τ delay. For the experiments with $^{113}Cd_7$-MT this basic experimental scheme was supplemented with purging procedures. A first purging process results from the insertion of a phase-alternated $^{113}Cd(\pi/2)$ pulse prior to acquisition (Fig. 10). Additional purging steps can be inserted at the position of the dotted line in the ^1H channel (see caption to Fig. 10). For example, with the use of a z filter, $(\pi/2)_x - \tau_z - (\pi/2)_\xi$, at this position the $C_\beta H$ resonances that are coupled to ^{113}Cd can be observed as in-phase absorptive signals (Fig. 11), which is preferable for measurements of $J(^{113}Cd,^1H)$. As seen in Fig. 11, imperfections in the z filter may lead to relayed cross-peaks between ^{113}Cd and $C_\alpha H$ of cysteine. Actually, because of the crowding of signals in the $C_\beta H$ region (Fig. 11), it is in practice often desirable to favor the $^{113}Cd-C_\alpha H$ relay peaks at the expense of the $^{113}Cd-C_\beta H$ correlations. This can be done by placing a single $(\pi/2)_y$ relay transfer pulse at the position of the dotted line in Fig. 10. When compared to Fig. 11, the $^{113}Cd-C_\alpha H$ cross-peaks are then strongly enhanced.[18] Alternatively, if simultaneous correlation of ^{113}Cd shifts with both $C_\alpha H$ and $C_\beta H$ is desired, a proton two-quantum filter, $(\pi/2)_x\,(\pi/2)_\Psi$, may be placed at the position of the dotted line. The two-quantum filter distributes the magnetization equally between the two positions and also acts as a phase-purging procedure. Finally, it must be emphasized that the intensity of the cross-peaks in such experiments depends also critically on the length of the delay τ (Fig. 10). In order to obtain a complete set of connectivities, it is therefore advisable to record several experiments with different τ values in the range from approximately 5–50 ms[13].

Provided that sequence-specific ^1H NMR assignments had been obtained for the polypeptide chain,[2,19] the $^{113}Cd,^1H$-COSY experiments provide a unique identification of the coordinative bonds between the metal ions and the polypeptide chain. From the presentation of these data in Fig. 12 it is readily apparent that the rat MT-2 molecular architecture consists of two domains with, respectively, three and four metal ions. The data visualized in Fig. 12 are an important supplement to the NOE distance constraints in the input for metallothionein structure calculations. Table I describes the format and the numerical values of the parameters used to represent the coordinative Cd-S bonds in the input for distance geometry calculations with the variable target function method.[2,5,6] Using the contents of Table I in combination with the NOE distance constraints and the

[19] G. Wagner, D. Neuhaus, E. Wörgötter, M. Vašák, J. H. R. Kägi, and K. Wüthrich, *Eur. J. Biochem.* **157**, 275 (1986).

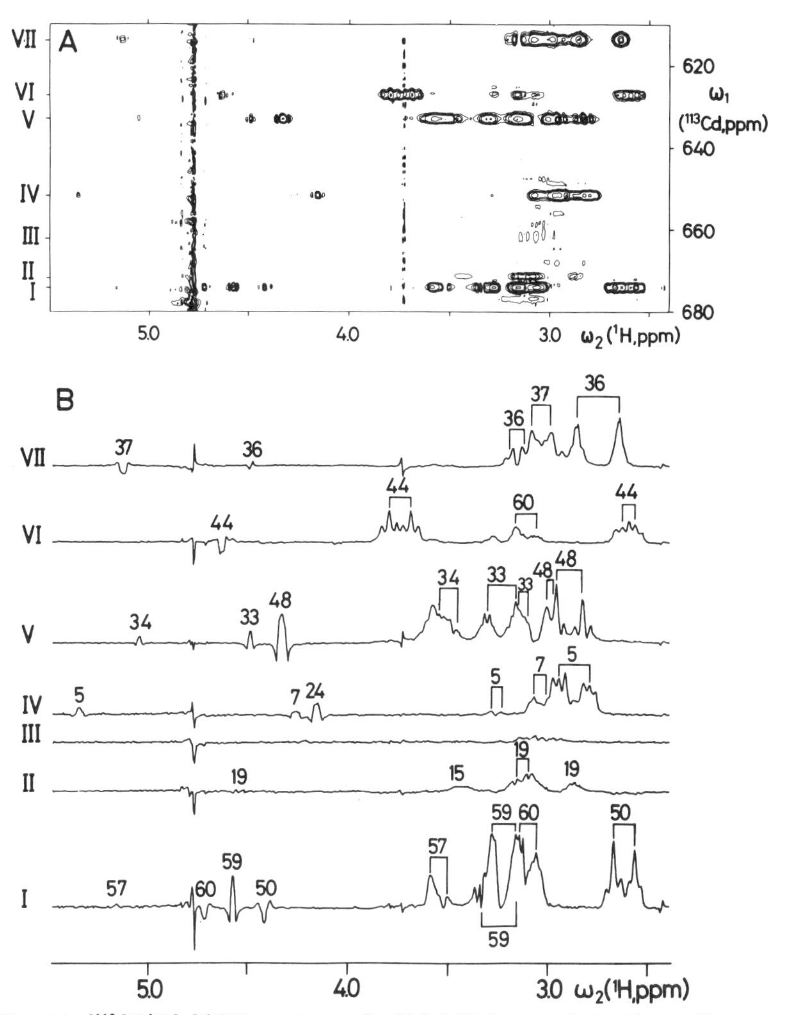

FIG. 11. [^{113}Cd,^{1}H]-COSY spectrum of rabbit MT-2a recorded with a z filter at the position of the broken line in Fig. 10. The delay τ was 8 msec, the z filter delay was fixed at 2 msec., $T = 20°$. (A) Contour plot representation. The ^{1}H spectrum is along ω_2; the ^{113}Cd spectrum is along ω_1. The locations of the seven ^{113}Cd resonances are indicated with roman numerals on the left-hand side of the spectrum. (B) Cross-sections along ω_2 at the ω_1 positions of the ^{113}Cd resonances indicated on the left. $C_\beta H$ resonances of cysteine appear between 2.5 and 4.0 ppm, $C_\alpha H$ resonances between 4.0 and 5.5 ppm. The cysteine signals are identified with the sequence positions. The resonance splittings due to (^{1}H, ^{113}Cd)-couplings are indicated with solid brackets. Any further splittings of the signals are due to ^{1}H-^{1}H scalar couplings. The spurious signal at 4.78 ppm is due to the residual water resonance. (Reproduced from Ref. 2.)

TABLE I
^{113}Cd–Cys CONNECTIVITIES IN INPUT FOR DISTANCE GEOMETRY CALCULATIONS WITH VARIABLE TARGET FUNCTION METHOD[a]

	Three-metal cluster[b,c]				Four-metal cluster[b,c]			
Cys	Cd(II)	Cd(III)	Cd(IV)	Cys	Cd(I)	Cd(V)	Cd(VI)	Cd(VII)
5			C	33		C		
7		+	C	34		+		C
13		C		36				C
15	+	C		37			+	C
19	C			41			C	
21			C	44		C	+	
24	C		+	48		C		
26		C		50	C			+
29	C			57	C			
				59	C			
				60	+		C	

[a] Formally, a pseudoatom, Cd_i^{δ}, representing Cd^{2+}, was attached to S^{γ} of *each* Cys residue. Since the 20 Cd^{δ} pseudoatoms must represent a total of 7 metal ions, an upper limit of 0.1 Å is imposed on the distance between any two Cd^{δ} pseudoatoms that correspond to the same Cd^{2+} ion. For bridging cysteines the coordination to the second metal ion is enforced by imposing a distance of 2.6 Å on the metal–sulfur bond. Reproduced from Ref. 6.

[b] C indicates that the Cd^{2+} ion identified by the roman numeral at the head of the column is covalently attached as a pseudoatom to the sulfur atom of the cysteinyl residue indicated on the left. A + indicates that the cysteinyl residue listed on the left is covalently bound as a bridging cysteine to a second Cd^{2+} ion to which it is formally not attached as a pseudoatom.

[c] Regular tetrahedral coordination of all Cd^{2+} and S centers is assumed and the following distance constraints are used: $d(Cd_i^{\delta} - Cd_j^{\delta}) < 0.1$ Å between all pairs of Cys having a C in the same column; $d(S_i^{\gamma} - S_j^{\gamma}) = 4.2$ Å between all pairs of Cys having C or + in the same column; $d(S_i^{\gamma} - Cd_j^{\delta}) = 2.6$ Å, $d(C_i^{\beta} - Cd_j^{\delta}) = 3.3$ Å, and $d(Cd_i^{\delta} - Cd_j^{\delta}) = 4.2$ Å between bridging cysteines i, marked by +, and all cysteines j having a C or a + in the same column.

dihedral angle constraints obtained from measurements of $^1H-^1H$ coupling constants as the input, MT structures such as those shown in Figs. 4 and 5 resulted from the distance geometry calculations.

Concluding Remarks

The techniques of sample preparation and NMR measurements described in this chapter are a reliable basis for the determination of 3D metallothionein structures in solution. Two experimental results represent

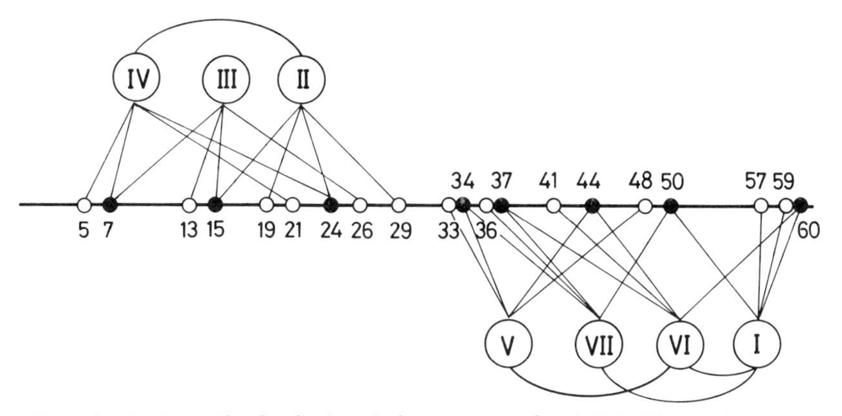

FIG. 12. Metal coordination in the solution structure of rat MT-2. The numbers near the horizontal line indicate the sequence positions of the 20 cysteinyl residues. The positions of bridging cysteinyl residues are identified by filled circles, those of singly bound cysteines by open circles. The seven Cd^{2+} ions are identified with roman numerals as in Fig. 11. The metal coordination was obtained as the direct result of the $^{113}Cd,^{1}H$-COSY measurements (same as Fig. 11) and the sequence-specific ^{1}H NMR assignments. (Reproduced from Ref. 8.)

direct support for this conclusion. First, independent structure determinations of the three homologous proteins rabbit MT-2a,[5] rat MT-2,[8] and human MT-2[7] yielded identical molecular conformations within the precision of the method. Second, a new crystallographic structure determination showed that the crystal structure of rat MT-2 published in 1986,[20] which was significantly different from the NMR structure in solution,[8] needed revision,[21,22] and that the rat MT-2 crystal structure is actually essentially identical to the corresponding solution structure determined by NMR. The foregoing text alluded to the fact that because of the presence of the metal–sulfur cluster in the core of the two metallothionein domains (Fig. 5), a much smaller number of $^{1}H-^{1}H$ NOEs was observed than in other proteins of similar size. In spite of the additional information on the coordinative polypeptide–metal bonds (Table I), this results in less precisely defined structures than the highest quality structure determinations that can otherwise be obtained by NMR. In more quantitative terms, in the metallothionein structures determined so far[6-8] the average of the pairwise

[20] W. F. Furey, A. H. Robbins, L. L. Clancy, D. R. Winge, B. C. Wang, and C. D. Stout, *Science* **231,** 704 (1986).

[21] C. D. Stout, D. E. Mc Ree, A. H. Robbins, S. A. Collett, M. Williamson, and X. H. Xuong, *Abstr. Int. Chem. Congr. Pacific Basin Soc.,* Honolulu, HW, U.S.A., December 18–22, 1989.

[22] A. H. Robbins and C. D. Stout, this volume [58].

RMSDs calculated for the polypeptide backbone atoms between the conformers selected to represent the solution conformation (Fig. 4) is of the order of 2.0 Å for the three-metal domain and 1.6 Å for the four-metal domain. For reference, the corresponding value for the protein Tendamistat (HOE 467, Hoechst, Frankfurt, Germany) was 0.8 Å.[3]

Compared to NMR structure determinations for proteins at large,[2] the essential additional experiments with metallothioneins all concern the use of the NMR spectral properties of the isotope ^{113}Cd. Applications of these techniques, which constitute the major message of this chapter, are not limited to metallothioneins. Rather, one can foresee that ^{113}Cd might quite generally serve as a NMR probe in structural studies of Zn^{2+} proteins. To cite just one example, DNA-binding proteins containing Zn^{2+} are currently in the forefront of the interest of molecular biologists. Corresponding experiments to those described here have been reported for a ^{113}Cd-substituted single zinc finger peptide,[23] where a tetrahedral coordination of the metal ion with three cysteine ligands and one histidine ligand was observed. In another DNA-binding protein, Gal4, the Zn^{2+} ions were also replaced by $^{113}Cd^{2+}$, and a ^{113}Cd-thiolate cluster comprising two ^{113} Cd ions and six cysteines was found, which contains two bridging cysteines (see Figs. 5 and 12).[24]

Acknowledgments

The author's research in the field of this chapter was supported by the Schweizerischer Nationalfonds and by special grants of the ETH Zürich. I owe a great debt to the many colleagues who participated in this work and whose names appear in the references. I would like to mention in particular the longstanding, most enjoyable collaboration with Prof. J. H. R. Kägi and Dr. M. Vašák. I thank Mr. R. Marani for the careful processing of the manuscript.

[23] M. F. Summers, T. L. South, B. Kim, and D. R. Hare, *Biochemistry* **29**, 329 (1990).
[24] T. Pan and J. E. Coleman, *Proc. Natl. Acad. Sci. U.S.A.* **87**, 2077 (1990).

[60] Paramagnetic Resonance of Metallothionein

By MILAN VAŠÁK

A common approach in investigating the structure and/or function of zinc(II)-containing metalloenzymes has been to substitute the native metal ion for a paramagnetic cobalt(II) ion[1] or for a cadmium(II) ion[2] and to use

[1] K. F. Geoghegan, B. Holmquist, C. A. Spilburg, and B. L. Vallee, *Biochemistry* **22**, 1847 (1983).

their spectral properties to probe the structure and the action of the protein. Both metal substitution approaches have proven to be indispensable in the elucidation of the structure and organization of the metal-binding sites in metallothioneins (MTs) from various sources.

Electron Paramagnetic Resonance Studies of Co(II)-Metallothionein

Electron paramagnetic resonance (EPR) is the method of choice when studying molecules with unpaired electrons. Native MT usually contains d^{10} Zn(II) and/or Cd(II) metal ions and is therefore diamagnetic. In order to probe the structure of the metal-binding sites, diamagnetic metal ions may be replaced by paramagnetic cobalt(II). Because the chemistry of cobalt(II) and zinc(II) is similar, the former ion usually accommodates the same coordination geometry as the latter. Cobalt(II) is always paramagnetic and as a d^7 metal ion it can be low spin ($S = \frac{1}{2}$) or high spin ($S = \frac{3}{2}$). When substituted for zinc(II) in proteins, cobalt(II) generally has a lower symmetry and is almost invariably in the high-spin state. The application of EPR spectroscopy to metalloproteins has been reviewed elsewhere[3,4]; however, in order to discuss its application to the study of Co-MT, the relevant EPR properties are briefly summarized.

The usefulness of EPR spectroscopy in biochemical research can be broadly divided into two categories. The first of these exploits the specific features present in the EPR spectrum to obtain structural and physical information characteristic of the paramagnet under investigation and requires a reasonable understanding of the EPR phenomenon. The second and more simple approach, applied also to Co-MT, exploits the ability of EPR to provide quantitative information on the concentration of the paramagnetic centers available in a given sample. EPR spectra of high-spin cobalt(II) are usually recorded below 20 K due to its high-spin relaxation rate. In the case of $Co(II)_7$-MT no EPR signal could be detected above 10 K. The primary spectroscopic parameter that is most readily extracted from an EPR spectrum is the g value. It is defined as

$$g = h\nu/\beta H = 714.44\ \nu\ (\text{GHz})/H\ (\text{gauss})$$

where h is Planck's constant; β, the electron Bohr magneton; and ν, the microwave frequency. In the studies on Co-MT, a microwave frequency of

[2] I. M. Armitage and J. D. Otvos, *Biol. Magn. Reson.* **4,** 79 (1982).

[3] G. Palmer, *in* "Methods for Determining Metal-Ion Environments in Protein: Structure and Function of Metalloproteins" (D. W. Darnal and R. G. Wilkins, eds.), Vol. 2, p. 153. Elsevier, New York, 1980.

[4] H. M. Swartz and S. M. Swartz, *Methods Biochem. Anal.* **29,** 207 (1983).

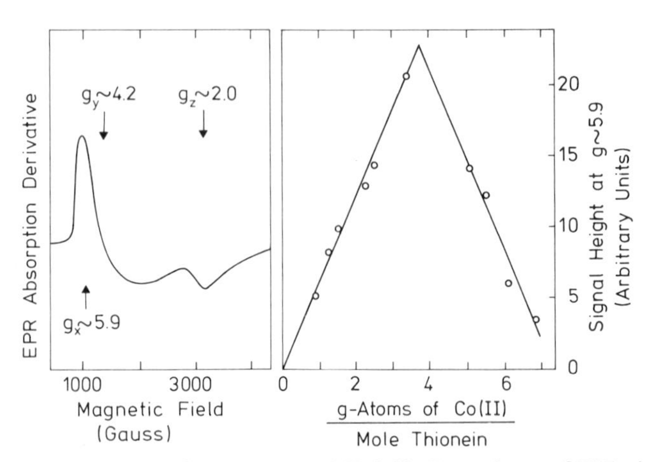

FIG. 1. EPR spectrum of Co(II)-MT at 4 K (left). Dependency of EPR signal size on Co(II)-to-protein ratio (right). As a representative measure of the magnitude of the high-spin Co(II) EPR spectrum, the amplitude at $g_x \sim 5.9$ is plotted [Reproduced from M. Vašák and J. H. R. Kägi, *Proc. Natl. Acad. Sci. U.S.A.* **78,** 6709 (1981).]

9.2 GHz, also known as *X*-band, was used. The position of the specific spectral feature(s) (see below) in terms of the applied magnetic field *H* is used when calculating the *g* value(s). Since the first-derivative EPR spectra of Co-MT were obtained in frozen solution as opposed to crystals, all possible orientations of the protein molecules are present. In general, as a consequence of the *g*-tensor anisotropy, the magnetic field required for resonance varies with the direction of the molecule relative to the magnetic field, giving rise to a spectrum that is the envelope of the individual spectra arising from each individual orientation. The orientation dependence of the *g* factor has a profound effect on the line shape of an EPR spectrum. Thus, the occurrence of one, two, or three *g* values in a spectrum makes it possible to differentiate among three basic types of the line shape, i.e., isotropic, axial, and rhombic.[3,4] In the case of Co-MT three principal *g* values can be determined (Fig. 1), indicative of a rhombically distorted high-spin Co(II) complex. The positions of the *g* values are clear when compared with an idealized first-derivative spectrum with rhombic symmetry (Fig. 2); the principal *g* values refer then to the positions of the maximum, cross-over, and minimum in the spectrum. It should be noted that the obtained *g* values are by no means a proof of the tetrahedral symmetry of the metal-binding site(s)[5] and thus additional spectroscopic

[5] A. Bencini, I. Bertini, G. Canti, D. Gatteschi, and C. Luchinat, *J. Inorg. Biochem.* **14,** 81 (1981).

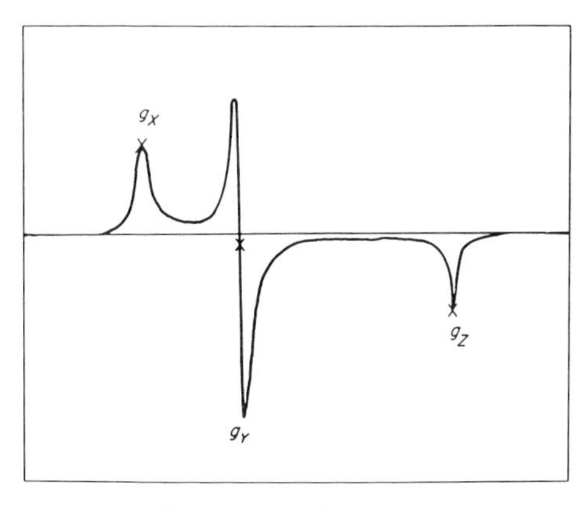

FIG. 2. Idealized first-derivative EPR spectrum with rhombic symmetry.

data must be obtained. In the case of Co-MT the spectral features of the *d–d* bands in the electronic absorption and magnetic circular dichroism (MCD) spectra (Fig. 3) proved to be indispensable in delineating the tetrahedral tetrathiolate–metal coordination.[6] Both latter methods and their application to MT are discussed in more detail elsewhere in this volume.[7] Based on the quantitation of the EPR-detectable Co(II) ions (see below), in the well-defined Co_7-MT sample (Fig. 1, right) only 0.5 Co(II) equivalents would be present. The dramatic decrease of the overall EPR intensity is due to the presence of the cluster structure, i.e., spin–spin interactions. In the metal–thiolate cluster structure of MT the paramagnets are linked by bridging thiolate ligands that transmit the interaction between two paramagnets, causing a spin pairing (antiferromagnetic coupling). As a consequence of spin–spin interactions, the four-metal cluster is diamagnetic ($S = 0$) and the three-metal cluster shows a residual paramagnetism ($S = \frac{3}{2}$).[8,9] Thus, the observed EPR signals of Co_7-MT originate from the three-metal cluster. Inspection of Fig. 1 (right) reveals a linear increase in the magnitude of the EPR signal as a function of the Co(II)/apo-MT ratio up to approximately four Co(II) equivalents, i.e., no antiferromagnetic coupling occurs, followed by an intensity drop, due to the generation of a cluster form. Consequently, the very weak paramagnetism in the EPR spectrum of the Co_7-MT sample provides evidence for a cluster structure in MT. This

[6] M. Vašák, J. H. R. Kägi, B. Holmquist, and B. L. Vallee, *Biochemistry* **20,** 6659 (1981).
[7] A. Schäffer, this volume [61].
[8] M. Vašák and J. H. R. Kägi, *Proc. Natl. Acad. Sci. U.S.A.* **78,** 6709 (1981).
[9] B. A. Dixon and M. K. Johnson, *Recl. Trav. Chim. Pays-Bas* **106,** 188 (1987).

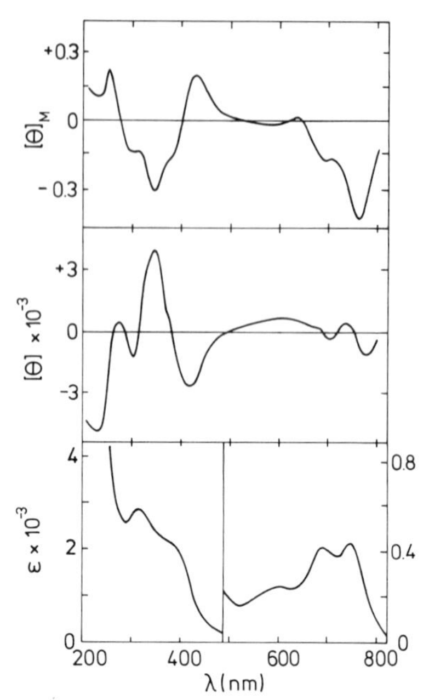

FIG. 3. Magnetic circular dichroism (MCD) (top), circular dichroism (CD) (middle), and absorption spectra (bottom) of Co(II)-MT (at 298 K). (Reproduced from Ref. 6.)

evidence is ambiguous, however, without additional data on the metal/protein stoichiometry[10] and the absorption spectrum (Fig. 3, bottom). In this instance the absence of well-resolved $d-d$ bands usually suggests sample oxidation. Note also that the d^6 Co(III) ion favors octahedral coordination and, as such, in almost all complexes is spin paired, i.e., diamagnetic. Overall, EPR spectroscopy of Co-MTs proved useful in establishing the presence of a metal cluster structure. The rather high sensitivity of this method can be of advantage when only small amounts of the protein (a few milligrams) are available.

Another paramagnetic metallothionein derivative is that containing Fe(II) metal ions.[11] Although a broad, so-called "$g = 10$," EPR signal appears to low field,[12] due to variations in spin-state populations and transition probability among various Fe(II) complexes, no reliable standards for the signal quantitation are available.[13]

[10] M. Vašák, this volume [7].

[11] M. Good and M. Vašák, *Biochemistry* **25,** 8353 (1986).

[12] M. T. Werth and M. K. Johnson, *Biochemistry* **29,** 3982 (1989).

[13] M. T. Werth, D. M. Kurtz, Jr., B. D. Howes, and B. H. Huynh, *Inorg. Chem.* **28,** 1357 (1989).

Experimental Procedures

Sample Preparation

The preparation of the extremely oxygen-sensitive Co-MT derivative is described in detail elsewhere in this volume.[14] Typically 0.2 ml of the Co-MT samples (0.5 – 1 m*M*) is placed in 3-mm quartz EPR tubes sealed under reduced argon pressure and stored in liquid nitrogen.

General Procedures

The determination of the Co(II) concentration relies on the comparison of the signal intensity between an unknown sample and a reference sample. Before such a comparison is made, the absence of any signal saturation must be confirmed. This phenomenon is due to nonequilibrium distribution in the two spin states between which the EPR transition is occurring, brought about by too-high incident microwave power. This effect may result in the signal loss due to saturation. Saturation behavior can be assessed by examining the signal intensity as a function of the square root of the applied microwave power. In the absence of saturation a linear increase in signal intensity will be obtained. Spin – spin interactions often enhance the electron spin relaxation rate. Thus, if titration studies are attempted, the absence of saturation must also be confirmed on the mononuclear Co-MT complex. Quantitation is performed by running the sample and a standard under identical conditions of temperature, modulation amplitude, and microwave power. As a standard, a Co(II) salt dissolved in 12 *M* HCl is used. Metal concentration is determined by atomic absorption spectroscopy. The tetrahedrally coordinated $CoCl_4^{2-}$ complex in 12 *M* HCl shows an EPR signal very similar to that of Co(II)-MT. The EPR spectra, recorded as the first derivative (see above), are double integrated and the areas compared. The details of this procedure have been described.[4]

^1H Nuclear Magnetic Resonance of Co(II)-Metallothionein

Nuclear magnetic resonance (NMR) can also be used to investigate paramagnetic centers in metalloproteins. Its application to paramagnetic metalloproteins and the underlying theory have been reviewed in this series.[15] The ^1H NMR spectrum of 1 m*M* rabbit liver Co_7-MT-2 shown in

[14] M. Vašák, this volume [54].

[15] I. Bertini, L. Banci, and C. Luchinat, this series, Vol. 177, p. 246.

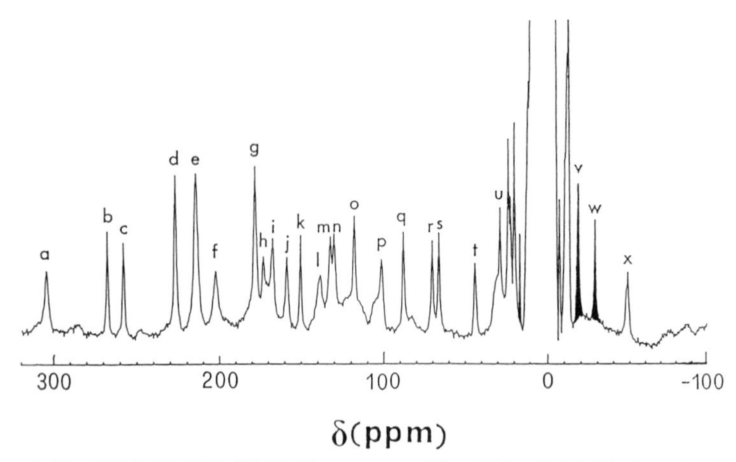

FIG. 4. The 298 K 90-MHz ^1H NMR spectrum of Co_7-MT, pH 8.5. The lettering denotes the isotropically shifted resonances. The filled signals disappear in D_2O. (Reproduced from Ref. 16.)

Fig. 4[16] reveals more than 30 relatively narrow resonances outside of the diamagnetic region (between 0 and 10 ppm range for Cd_7-MT-2a)[17] spread over a 400-ppm range.[16] It should be noted that although low-temperature EPR studies (4 K) revealed the Co_4 cluster to be diamagnetic, at room temperature a certain amount of paramagnetism is present. The signals outside of the diamagnetic region are so-called isotropically shifted resonances that are highly sensitive to subtle structural differences. The isotropic shift is the sum of contact and dipolar contributions. The application of ^1H NMR to Co-MT provided an insight into the details of the pathway of cluster formation and into the effect of an exchange-coupling within the Co_4 cluster on the isotropically shifted resonances.

The pathway of cluster formation may be examined using an approach similar to that taken in the EPR studies (see above). Samples of apo-MT are titrated with Co(II) ions and the ^1H NMR spectra are recorded.[16] In parallel, the room-temperature bulk magnetic susceptibilities at each titration step are obtained by the method of Evans.[18] This ^1H NMR method relies on the difference in chemical shift (measured in hertz) between the protons of the inert NMR standard DSS (4,4-dimethyl-4-silapentane 1-sulfonate) added to the Co-MT sample and the same amount of DSS in a capillary (external standard), dissolved in the same buffer. The diamagnetic contribution is corrected for by similar measurements using the apo-

[16] I. Bertini, C. Luchinat, L. Messori, and M. Vašák, *J. Am. Chem. Soc.* **111,** 7296 (1989).

[17] K. Wüthrich, this volume [59].

[18] D. F. Evans, *J. Chem. Soc.*, 2003 (1959).

MT form. Measurements of the magnetic susceptibility are important in assessing the nature of metal binding for the first three Co(II) equivalents, because no isotropically shifted ^1H NMR signals are detected. Based on these studies the first three Co(II) equivalents are bound into individual sites and are subjected to a chemical exchange. The ^1H NMR spectrum of Co_7-MT-2 (Fig. 4) develops already with five Co(II) equivalents and additional metal binding just increases its intensity. The majority of these signals originate from the Co_4 cluster. This conclusion is based on a comparison of the ^1H NMR spectrum of Co_7-MT-2 with that of Co_3,Cd_4-MT-2[19], in which homometallic cluster occupation occurs. The latter sample, whose preparation is described below, is characterized by absorption and ^{113}Cd NMR spectra. In this instance, the absorption spectrum confirms cobalt binding and the ^{113}Cd NMR spectrum the selective metal filling of the four-metal cluster with Cd(II) ions. The absorption spectrum is similar but not identical to that of Co_7-MT (Fig. 3, top). The ^{113}Cd NMR method and the characteristic ^{113}Cd chemical shift of the Cd_4 cluster resonances have been described elsewhere.[2] Since the overtitration of the four-metal cluster results in the formation of a corresponding population of Cd_7-MT-2, i.e., no mixed three-metal cluster(s) are being formed, the isotropically shifted resonances represent the "fingerprint" of the Co_3 cluster.

The studies of the effect of an exchange coupling within the Co_4 cluster on the chemical shift position of the isotropically shifted resonances of rabbit liver Co_7-MT-2 rely on their assignments to a theoretical quantum-mechanical model developed especially for the Co_4 cluster.[20] In this model advantage was taken of the basic structural features revealed in the solution NMR model of rabbit liver Cd_7-MT-2a.[21] As a consequence of the distance limitation of the paramagnetic effect, the isotropically shifted resonances originate predominantly from β-CH_2 groups of the 11 cysteine residues and from liable protons. The latter resonances are assigned based on their absence when the ^1H NMR spectrum of Co_7-MT-2 is recorded in D_2O.[16] Assignments to specific terminal and bridging β-CH_2 groups of the 11 cysteines rely on a comparison of the temperature dependence of the individual signals in the range between 278 and 323 K with that predicted by the quantum-mechanical model.[20] In the next step the geminal protons of each β-CH_2 group are assigned based on nuclear Overhauser effect (NOE) experiments using a more concentrated sample (4 m*M*). These

[19] M. Good and M. Vašák, *Recl. Trav. Chim. Pays-Bas* **106,** 184 (1987).

[20] I. Bertini, C. Luchinat, L. Messori, and M. Vašák, *J. Am. Chem. Soc.* **111,** 7300 (1989).

[21] A. Arseniev, P. Schultze, E. Wörgötter, W. Braun, G. Wagner, M. Vašák, J. H. R. Kägi, and K. Wüthrich, *J. Mol. Biol.* **201,** 637 (1988).

studies also afford supplementary spatial constraints.[22] For further details on the assignment and the technical details the reader is referred to the original literature[16,20] and a review.[15]

Experimental Procedures

The preparation of the Co-MT derivative is described in this volume.[14] However, the preparation of Co-MT samples at the concentration of 1 or 4 m*M* requires some changes in the procedure. While the reconstitution at the 1 m*M* concentration in a 0.5-ml volume, needed in the NMR measurements, is still manageable if the sample is stirred vigorously during the pH adjustment, this procedure is inadequate for higher concentrations. The latter problem can be overcome by using a lower starting protein concentration (~0.6 m*M*) followed by sample concentration using a 5-ml Amicon (Danvers, MA) ultrafiltration apparatus placed in an argon-purged glovebox. Prior to the sample concentration the Amicon apparatus and the YM2 membrane are rendered oxygen free by passing through several volumes of degassed water. The Co_3,Cd_4-MT-2 derivative is prepared by adding four Cd(II) equivalents to apo-MT followed by a pH adjustment to 7.5 with 0.5 *M* Tris base as described.[14] Subsequently, three Co(II) equivalents are added and the pH raised to 8.5. The H_2O/D_2O exchange is performed via a freeze-drying step on the concentrated sample using a vacuum line. The NMR measurements are run in 5-mm NMR tubes sealed under reduced argon pressure. For details of the NMR measurements, e.g., pulse sequence and required hardware, the reader is referred to the original literature[16] and a review.[15]

Acknowledgments

The financial support of the Schweizerischer Nationalfond Grant No. 31-25′655.88 is gratefully acknowledged.

[22] I. Bertini, C. Luchinat, L. Messori, and M. Vašák, *Inorg. Chem.* in press (1991).

[61] Absorption, Circular Dichroism, and Magnetic Circular Dichroism Spectroscopy of Metallothionein

By ANDREAS SCHÄFFER

Introduction

Despite gradually reaching maturity, electronic spectroscopy still plays a major role in the description of structural details of metalloproteins. More sophisticated methods, such as solution two-dimensional nuclear magnetic resonance (2D-NMR) or single-crystal X-ray crystallography, may provide unsurpassed accuracy and resolution in structure determinations, but they have inherent disadvantages when compared with ultraviolet (UV), circular dichroism (CD) or magnetic circular dichroism (MCD) spectroscopy. For example, the dissolved state and relatively low concentration of solute needed for spectroscopic studies can be considered as more nearly approaching physiological conditions. Furthermore, the ease and speed of performance both by conventional approaches and, particularly, in combination with stopped flow techniques, make electronic spectroscopy uniquely applicable to kinetic studies.

In this context, spectral studies on metallothionein (MT) have been concerned with the dynamics of the binding, exchange, or release of the cluster-organized metal ions.[1,2] Further applications of electronic spectroscopy to metallothionein (MT) include the determination of dissociation constants by spectrophotometric titrations,[3] secondary structure analysis by far-UV CD spectroscopy,[4] and, in favorable cases, evaluation of the origin and nature of electronic transitions by the combination of UV, CD, and MCD.[5,6] It was demonstrated that it is possible to predict the location, sign, and amplitude of the CD profile for the lowest energy absorption band of Cd_7-MT.[7]

The organization of the metal ions of MT into two independent clusters itself is the cause for highly specific absorptive, chiroptical, and MCD

[1] J. K. Nicholson, P. J. Sadler, and M. Vašák, *Experientia, Suppl.* **52**, 191 (1987).
[2] M. J. Stillmann, A. Y. C. Law, W. Cai, and A. J. Zelazowski, *Experientia, Suppl.* **52**, 203 (1987).
[3] J. H. R. Kägi and B. L. Vallee, *J. Biol. Chem.* **236**, 2435 (1961).
[4] J. Pande, C. Pande, D. Gilg, M. Vašák, R. Callender, and J. H. R. Kägi, *Biochemistry* **25** 5526 (1986).
[5] H. Willner, M. Vašák, and J. H. R. Kägi, *Biochemistry* **26**, 6287 (1987).
[6] H. Willner, Ph.D. Thesis, University of Zürich, Zürich (1987).
[7] H. Willner, W. R. Bernhard, and J. H. R. Kägi, *in* "Metal Binding in Sulfur-Containing Proteins" (M. Stillmann, C. F. Shaw III, and K. T. Suzuki, eds.), in press. VCH Publishers, New York, 1991.

features that have been exhaustively studied in both diamagnetic and paramagnetic MT derivatives.[8] For the latter, several investigators have reported on the mode of metal binding, selective cluster formation, and metal coordination geometries.[9-11]

Determination of the temperature and magnetic field dependence of MCD spectra, i.e., establishment of magnetization curves, allows the magnetic properties of individual metal centers in proteins to be probed.[12] The application of this technique, e.g., to the recently synthesized Fe_7-MT,[11] would represent an exciting and informative area of research.

Because this chapter is devoted mainly to practical aspects of UV, CD, and MCD spectroscopy, no theoretical introduction will be given. Detailed discussions of that matter can be found in excellent textbooks or reviews.[13,14]

Equipment

It is advantageous to obtain spectroscopic measurements with computer-interfaced instruments. First of all, instrument operation and balanced adjustment of experimental parameters can be easily set up in convenient menus organized for the user's needs. The online inspection of the developing spectra and the arbitrary adjustment of scale or selection of time intervals will provide high efficiency in optimizing experimental conditions. The main advantage, however, is an increased signal-to-noise ratio by repeated accumulation of spectra and subsequent averaging. In addition, the documentation and storage of data either for personal use or for incorporation into databanks is greatly simplified with results in digital format.

A further reduction in signal-to-noise ratio by a factor of $2^{1/2}$ can be obtained by mathematical utilization of the fact that CD properties of molecules, due to their chirality and the induction of chiroptical features in matter by an external magnetic field H, are additive. Thus, CD measurements in the presence of H aligned parallel to the direction of light (CD + MCD) followed by a scan with an antiparallel magnetic field (CD − MCD)

[8] J. H. R. Kägi and Y. Kojima, *Experientia, Suppl.* **52,** 25 (1987).

[9] M. Vašák and J. H. R. Kägi, *Met. Ions Biol. Syst.* **15,** 213 (1983).

[10] M. Good and M. Vašák, *Biochemistry* **25,** 3328 (1986).

[11] M. Good and M. Vašák, *Biochemistry* **25,** 8353 (1986).

[12] A. J. Thomson, D. G. Eglington, B. C. Hill, and C. Greenwood, *Biochem. J.* **207,** 167 (1982).

[13] A. B. P. Lever, "Inorganic Electronic Spectroscopy." Elsevier, Amsterdam, 1984.

[14] A. J. Thomson, *in* "Perspectives in Modern Chemical Spectroscopy" (D. L. Andrews, ed.), p. 243. Springer-Verlag, New York, 1990.

will allow the separation of the individual components by simple addition (2 CD) and subtraction (2 MCD) of both spectra.

As for the instruments, the optical system is the most significant quality criterion rather than the sometimes convenient but not necessarily essential facilities like automatic derivatization of spectra and simultaneous readout at multiple wavelengths. Although endowed with impressive scan speeds, diode array detector spectrometers do not have the optical specifications of traditional units equipped with a monochromator and photomultiplier. The calibration of CD and MCD spectrometers has been described.[15,16]

The magnetic field for MCD spectroscopy should be homogeneous, orientationally reversible, and its strength should reach a minimum of about 1.5 T. Because Zeemann splitting is proportional to the external field strength, higher values are desirable but at the expense of experimental effort in cooling the electromagnet with liquid helium.

For measurements in the far-UV at wavelengths below ca. 210 nm the optical assembly and the sample chamber of the spectrometers must be efficiently purged with nitrogen in order to reduce the concentration of absorbing oxygen and gaseous H_2O. For reduction of interfering background absorption, especially in this energy region, cells should be of short path length balanced by a corresponding increase in protein concentration. Assembled cells down to a path length of 0.01 cm are commercially available. Below, special cell windows with distance rings fixed by cell clamps may be used. The same solution and cell can be used for UV, CD, and MCD spectroscopy.

The quality of a titration experiment depends primarily on the reliability of the dispenser equipment used. Adjustable pipettes with disposable tips will provide an accuracy of 1% for volumes between ca. 0.5 and 1 ml but considerably lower precision in the microliter range. More accurate and reproducible results are obtained with glass microsyringes of the Hamilton type, in particular when controlled by a computer-interfaced motor drive.[6]

Most spectroscopic work on metallothionein requires strict anaerobic conditions. This is most efficiently achieved by the preparation of sample solutions in inert gas boxes that have been purged preferably with argon or high-grade nitrogen. For convenience, the box should be equipped with a pH meter, a magnetic stirrer, and a vortex, in addition to the essential

[15] P. H. Schippers and H. P. J. M. Dekkers, *Anal. Chem.* **53,** 778 (1981).

[16] B. L. Vallee and B. Holmquist, *in* "Methods for Determining Metal Ion Environments in Proteins: Structure and Function of Metalloproteins" (D. W. Darnall and R. G. Wilkins, eds.), p. 27. Elsevier North-Holland, New York, 1980.

bench working tools. The optical cells are filled with the test solutions inside the box, sealed with a stopper or a gas-tight septum, and are then ready for spectroscopic measurements.

Performance

Spectroscopy with Apometallothionein

The preparations for spectroscopic studies on MT include in most cases the production and processing of the metal-free form. Detailed information on this topic is provided in [54] in this volume. In regard to the subject presented here only a short survey of the essential steps is outlined.

Metallothionein lyophilized from dilute buffer is dissolved in H_2O, acidified under vigorous vortexing with a few drops of 6 *M* HCl, cooled on ice, and separated from the dissociated metal ions by Sephadex G-25 chromatography with 10 m*M* HCl as eluent. It is good practice to check the protein preparations beforehand for bound copper or mercury, because those derivatives will not release their metal under such conditions. The Sephadex pool is diluted to a protein concentration of ca. 0.5 mg/ml and to an HCl concentration of ca. 2 m*M* and lyophilized. For use in metal titration studies a stock concentration of ca. 1 m*M* apo-MT in 10 m*M* HCl is prepared. The protein concentration can be evaluated from its absorbance in 0.01 *M* HCl at 220 nm ($\epsilon = 48190\ M^{-1}\ cm^{-1}$) and should be confirmed by amino acid analyses. Finally, it should be checked that all 20 cysteines exist in the free SH form essential for metal binding. If a corresponding assay with DTNB [5,5′-dithiobis(2-nitrobenzoic acid)] proves a lower value, the protein must be reduced with DTE (dithioerythritol) at neutral pH and reprocessed via Sephadex G-25, 0.01 *M* HCl.

Spectroscopy with Metal Derivatives

The preparation of well-defined metal derivatives of MT requires the exclusion of oxygen during the reconstitution process. To an N_2-purged solution of ca. 0.5 – 1 mg/ml apo-MT in 0.01 *M* HCl is added the following: 8 Eq of metal (e.g., zinc or cadmium, $\cong$ 10 – 100 μM), 1 *M* Tris base (N_2 purged) up to a pH of ca. 8.2, and about 1 ml cation-exchange resin suspension (Chelex, 0.4 mEq/ml). The filtered solution is concentrated via ultrafiltration and passed over a Sephadex G-50 column equilibrated with 10 m*M* Tris-HCl, pH 8.6. The protein pool is concentrated by ultrafiltration, checked for its metal content by atomic absorption spectroscopy

TABLE I
EXTINCTION COEFFICIENTS OF MT DERIVATIVES[a]

Species	ϵ_{220}	ϵ_{250}	ϵ_{300}	ϵ_{350}	ϵ_{400}
Apo-MT					
pH 1.8	4.8	—	—	—	—
pH 8.1	7.6	1.3	—	—	—
pH 12	14.9	4.2	—	—	—
Zn_7-MT	15.9	1.5	—	—	—
Cd_7-MT	15.5	10.5	—	—	—
Hg_7-MT	15.8	10.0	7.2	0.3	—
Fe_7-MT			2.7	1.9	0.3
Co_7-MT			2.1	2.0	1.6

[a] Based on protein concentration. Data given as $10^4\ M^{-1}\ cm^{-1}$.

(AAS), and lyophilized. Mixed metal derivatives are prepared correspondingly by adding stoichiometric amounts to the apoprotein.[17]

Measurements

pH titrations of apo-MT and stepwise reconstitution with metal ions must be performed in an inert gas box (see above). For use in the box, solvents must be repeatedly degassed in the frozen state by means of a high vacuum line and subsequently thawed. The still evacuated flasks are then transferred to the box and ventilated inside.

Independent of the type of experiment planned, it is necessary to maintain reasonable optical densities of the test solutions; especially for CD and MCD spectroscopy the OD should never exceed 1.0–1.2; ideal absorbancies are in the range of 0.8. Spanning a wavelength region of 190–300 nm, which is sufficient for MT in most cases except for mercury or paramagnetic species, requires the use of two or more cuvettes with different path lengths and the optional postexperimental combination by mathematical methods. As a guide for considerations with regard to the concentration of samples or to cell path lengths, Table I lists some extinction coefficients of different MT derivatives at certain wavelengths.

For pH studies of apo-MT, defined amounts of gradually diluted NaOH or Tris base are added to the protein solution whose concentration is maintained constant for the individual experiments by appropriate addition of water, until a defined pH is reached that is measured directly inside

[17] M. Good, Ph.D. Thesis, University of Zürich, Zürich (1987).

the inert gas box. Similarly, for metal titration studies, stoichiometric quantities of metal are added stepwise to the apo-MT solution. For zinc and cadmium titrations, the apo-MT solution is adjusted to a pH above 8 prior to metal addition while with mercury derivatives the metal must be added to the acidic protein solution, which is subsequently pH adjusted. Each solution representing a defined titration step is separately investigated by UV, CD, and MCD spectroscopy and should finally be analyzed for metal concentration.

Data Representation

Relevant expressions and dimensions of data from UV, CD, and MCD spectroscopy will briefly be summarized. The molecular ellipticity $[\Theta]$ (in degrees cm^2 $dmol^{-1}$) is calculated from the experimentally observed ellipticity Θ (in degrees) by

$$[\Theta] = (\Theta)(100)(c^{-1}\, d^{-1})$$

with the concentration of the solute c (in molar units) and the cell path length d (in centimeters). $[\Theta]$ is converted to the difference in extinction coefficients of left and right circularly polarized light when interacting with optically active matter, $\Delta\epsilon = \epsilon_L - \epsilon_R$ (in M^{-1} cm^{-1}), by the equation

$$[\Theta] = (3300)(\Delta\epsilon).$$

Magnetic ellipticities Θ_M (in degrees G^{-1}) and molecular magnetic ellipticities $[\Theta]_M$ (in degrees cm^2 $dmol^{-1}$ G^{-1}) are calculated from the observed induced ellipticities in the presence of an external magnetic field by dividing them by the field strength in Gauss.

Mean residue ellipticities are obtained by dividing the corresponding data by the number of protein amino acids.

Results

Possible applications of UV, CD, and MCD spectroscopy to metallothionein are exemplified by some representative results.

Secondary Structure Analysis

A CD spectrum in the far-UV of apo-MT, Zn_7-MT, and Cd_7-MT is presented in Fig. 1.[4] As an instruction for experimental conditions, 0.114 mM apo-MT was measured in 0.1- and 0.02-cm cells for the low- and high-energy regions of the spectrum, respectively.

In the case of metal derivatives of MT, the ellipticities arising from $n-\pi^*$ and $\pi-\pi^*$ transitions of the amide group are overlapped by metal

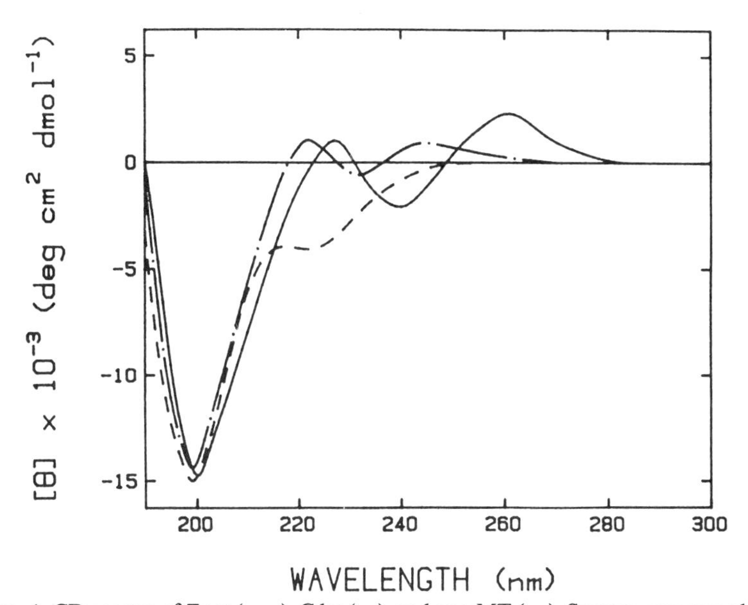

FIG. 1. CD spectra of Zn_7^- (– · –), Cd_7^- (—), and apo-MT (--). Spectra were recorded in 25 mM Tris-HCl buffer at pH 8.6 for the metal derivatives and in 0.02 M perchloric acid at pH 1.7 for apo-MT. Data are expressed as mean residue ellipticities. [Reprinted with permission from J. Pande, C. Pande, D. Gilg, M. Vašák, R. Callender, and J. H. R. Kägi, *Biochemistry* **25,** 5526 (1986). Copyright (1986) American Chemical Society.]

ligand charge transfer bands, which impedes structural calculations from CD spectra. However, the analysis of the apo-MT spectrum according to the method of Provencher and Glöckner[18] yields a contribution of 6% α helix, 18% β sheet, 21% β turn, and 55% unordered structure. In contrast to the mainly unordered structure of apo-MT, metal binding affects polypeptide chain folding by inducing numerous secondary structural elements, such as type II β turns, half-turns, and 3_{10} helical segments in holometallothionein, as determined by different spectroscopic studies.

Diamagnetic Derivatives

In Fig. 2 a spectral titration of 8 μM apo-MT with cadmium at pH 8.4 is shown.[7] The absorption profile typical for tetrahedral Cd(II)–S complexes reaches a maximum intensity at seven equivalents of metal per mole of protein. After the first three metals are bound the spectra shift toward the red by up to 5 nm with further metal addition. This has been inter-

[18] S. W. Provencher and J. Glöckner, *Biochemistry* **20,** 33 (1981).

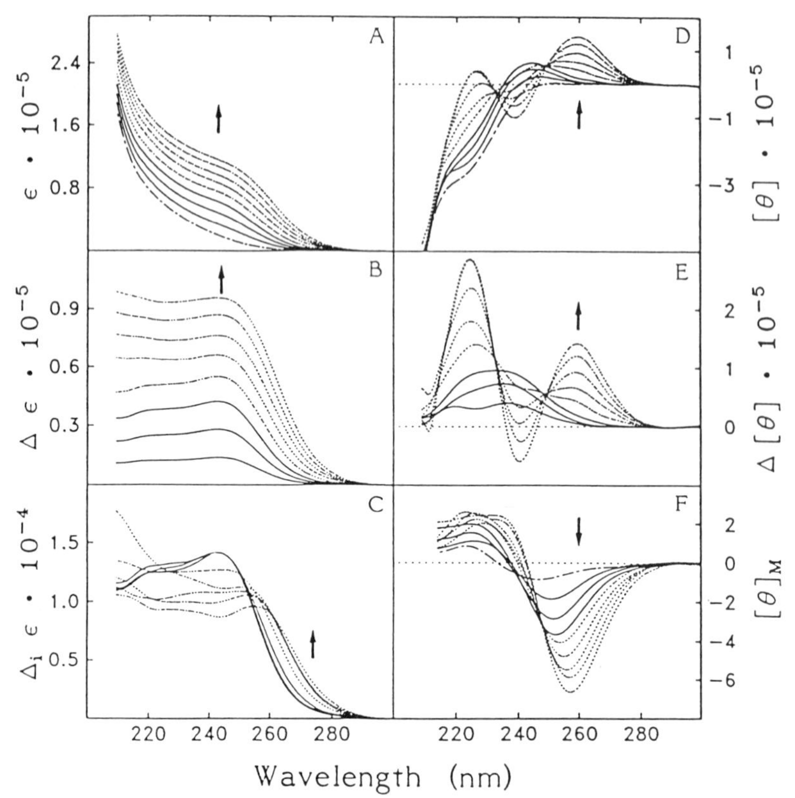

FIG. 2. UV, CD, and MCD spectra of apo-MT, 8 μM, in 10 mM Tris-HCl, pH 8.4, to which $CdCl_2$ is added in eight steps of 0.88 equivalents each. The arrows relate to increasing metal concentration. In A, D, and F the first spectrum corresponds to apo-MT, the last refers to the addition of seven cadmium equivalents. (A) Absorption spectra, (B) difference absorption spectra by subtracting the apo-MT spectrum at pH 8.4 from the remaining, (C) increment difference absorption spectra obtained by subtracting pairs of successive spectra of (A), (D) CD spectra, (E) difference CD spectra, and (F) MCD spectra. Dimensions: ϵ (M^{-1} cm^{-1}), [Θ] (degrees cm^2 $dmol^{-1}$), $[\Theta]_M$ (degrees cm^2 $dmol^{-1}$ G^{-1}). All units refer to protein concentration. [*Source:* H. Willner, W. Bernhard, and J. H. R. Kägi, *in* "Metal Binding in Sulfur-Containing Proteins" (M. Stillmann, C. F. Shaw III, and K. T. Suzuki, eds.), in press. VCH Publishers, New York, 1991.]

preted to indicate the transformation of an initially formed isolated $Cd(II)S_4$ complex at low metal-to-protein ratios, to cluster-organized metal ions[5] by analogy to results from experiments performed with Hg(II), Co(II), and Fe(II).[17,19] The biclustered structure implies the existence of 12

[19] W. Bernhard, Ph.D. Thesis, University of Zürich, Zürich (1986).

terminal and 8 bridging thiolates after all 7 metals are added. The shift of the first absorption band to lower energy can be described on the basis of an empirical concept of Jorgensen,[20] which relates the frequency $\tilde{\nu}$ (μm^{-1}) of this band to the difference in optical electronegativity of the ligands (X_L) and of the metals (X_M): $\tilde{\nu} = 3.0(X_L - X_M)$. Accordingly, the red shift of the first absorption band on cluster formation indicates a decrease in X_L for bridging thiolates compared to that for terminal sulfur ligands.

The MCD spectra undergo an analogous energy shift during stepwise metal incorporation and reveal characteristic positive Faraday A term signals expected for chromophores with high symmetry. Most conspicuously, cluster formation induces a distinct transition in the CD spectra from monophasic to multiphasic envelopes, which result from coupling of transition dipoles located and directionally fixed at the lone pair orbitals of the bridging thiolates within the clusters.[5]

Paramagnetic Metallothionein

Figure 3 contains a corresponding set of spectroscopic data resulting from a titration of apo-MT with Co(II) ions.[17] For the charge transfer region, analogous considerations as in the case of diamagnetic derivatives pertain to the initially formed Co(II)S_4 complexes and the mode of subsequent clustering with a pronounced energy shift of the first CT band from 350 to 390 nm. Similarly, protein saturation with seven metals is reflected by the plateau in absorptivity in both the charge transfer and the *d–d* region.

The use of paramagnetic metals for reconstitution provides additional insight into the structural details that can be extracted from inspection of the visible spectra.[21] The position and shape of the ν_2 band in the near-infrared region remains unchanged during titration, whereas the bands of the ν_3 transition preserve their profile only up to four Co(II) ions added. Changes on further addition reflect distortions of cobalt sites going from isolated to clustered complexes, which as judged from the MCD and absorption spectra still can be pictured as tetrahedral in character.

The sensitivity of the method is demonstrated by investigations of the Co(II) reconstituted α fragment of MT.[10] The first three metals bind to form a cluster resembling that of the N-terminal domain of holo-Co(II)$_7$-MT and the fourth appears to bind to an additional site that has a different coordination geometry.

[20] C. K. Jorgensen, *Prog. Inorg. Chem.* **12,** 101 (1970).

[21] M. Vašák and J. H. R. Kägi, *Proc. Natl. Acad. Sci. U.S.A.* **78,** 6709 (1981).

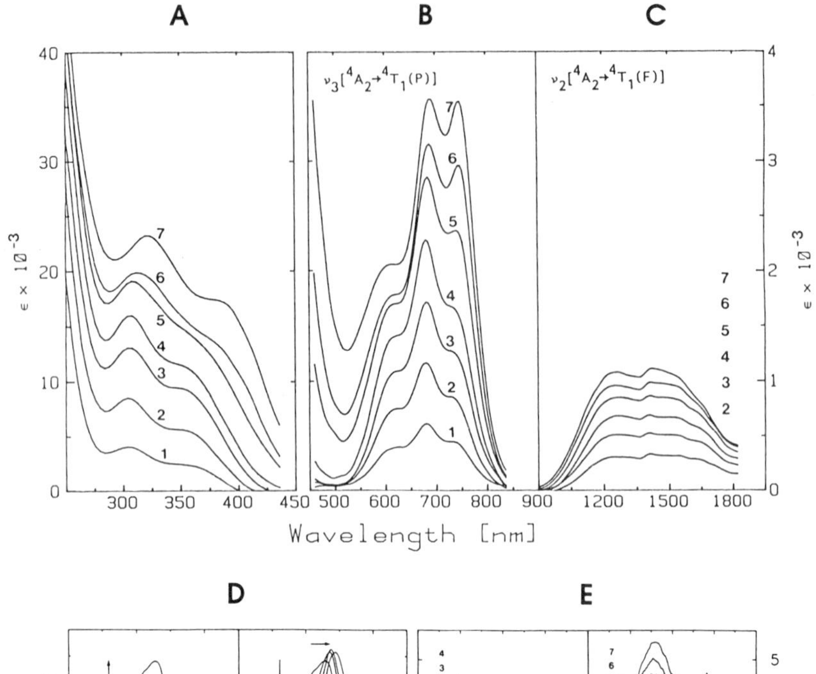

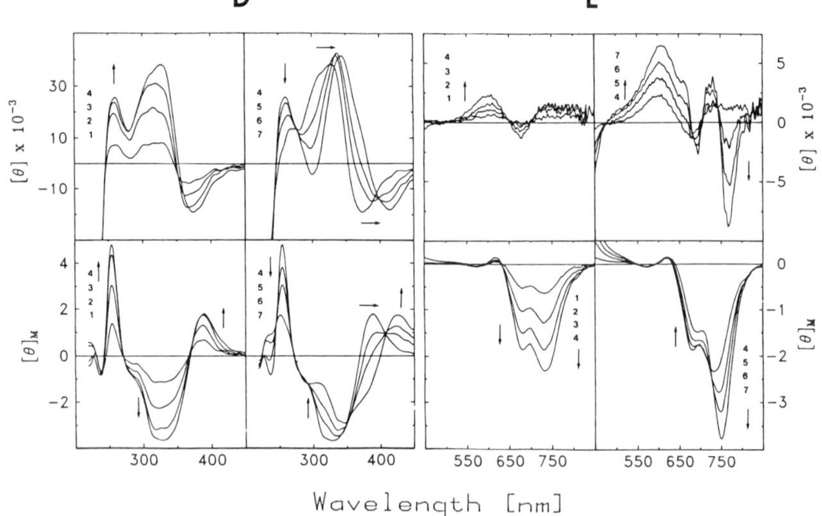

FIG. 3. Electronic absorption spectra of Co(II)-MT in the charge transfer (A), visible (B), and near-infrared (C) region (IR: in D_2O) as a function of the Co(II)-to-protein ratio. The numbers indicate the moles of Co(II) per mole of apo-MT. Corresponding CD and MCD spectra in the charge transfer (D) and the $d-d$ region (E). Arrows indicate the development of bands with increasing Co(II) concentration. Conditions: 0.17 m*M* Co(II)-MT in 160 m*M* Tris-HCl, pH 8.3. Dimensions: ϵ (M^{-1} cm^{-1}), [Θ] (degrees cm^2 $dmol^{-1}$), $[\Theta]_M$ (degrees cm^2 $dmol^{-1}$ G^{-1}). All units refer to protein concentration. [*Source:* M. Good, Ph.D. Thesis, University of Zürich, Zürich (1987).]

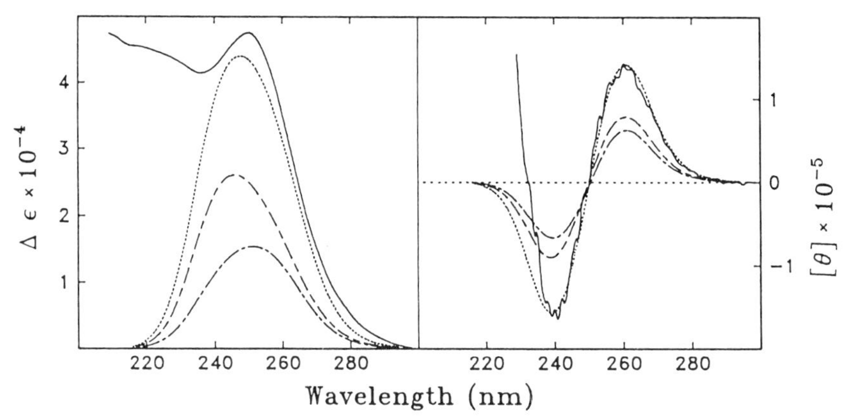

FIG. 4. Comparison of calculated and observed UV absorption (left) and CD spectra (right) of bridging thiolates in Cd_7-MT. The spectra of the three-metal cluster (– · –) and the four-metal cluster (–––) were separately calculated. The dotted lines (· · ·) represent the sums of the calculated spectra. The experimental spectra (—) are difference spectra obtained by subtracting the spectra of Cd_3-MT from those of Cd_7-MT. Units refer to protein concentration. [*Source:* H. Willner, W. Bernhard, and J. H. R. Kägi, *in* "Metal Binding in Sulfur-Containing Proteins" (M. Stillmann, C. F. Shaw III, and K. T. Suzuki, eds.), in press. VCH Publishers, New York, 1991.]

Calculation of Spectra and Comparison with Theory

The UV absorption and CD properties of metallothioneins result mainly from interactions of sulfur chromophores within the metal clusters and, in the far UV, from electronic excitations of the amide groups. By restriction to a certain transition, knowledge of its nature, and information about the localization and mutual distances of the involved oscillators it becomes feasible to predict accurately the location, sign, and amplitude of the related CD signal. For metallothionein, such calculations have been performed on the basis of the formalism of DeVoe,[22,23] which deduces by classical means optical properties of aggregates from those of the corresponding monomers.[7]

The lowest energy absorption band of Cd_7-MT as resolved by Gaussian analysis (249 nm) is thought to arise from a ligand-to-metal charge transfer (LMCT) excitation of a sulfur *p* electron to an acceptor molecular orbital (MO) with a strong linear combination of atomic orbitals (LCAO) component from the sulfur ligand. This interpretation is supported by the finding that the extinction coefficient of sulfur in thiolates as well as in metal

[22] H. DeVoe, *J. Chem. Phys.* **41,** 393 (1964).
[23] H. DeVoe, *J. Chem. Phys.* **43,** 3199 (1965).

complexes either in terminal or bridging form remains unchanged ($\epsilon_{240} = 5900\ M^{-1}\ cm^{-1}$ per sulfur). It also entails that the oscillator associated with this transition is sterically fixed in the direction of the lone pair sulfur orbitals whose asymmetry is the result of the folding of the peptide chain. By calculating the pairwise electrostatic interactions between the linearly polarized electric transition dipole moments of the bridging thiolate ligands from the oscillator strength of the transition and the geometric parameters extracted from a 2D-NMR structure analysis of Cd_7-MT,[24] the calculation of absorption and CD spectra of Cd_7-MT reveals a high degree of coincidence with the experimental data (Fig. 4). Thus, the established relation of CD properties and the chiral configuration of bridging sulfur provides—in the converse direction—structural insight into clustered metal complexes from spectroscopic measurements.

[24] A. Arseniev, P. Schultze, E. Wörgötter, W. Braun, G. Wagner, M. Vašák, J. H. R. Kägi, and K. Wüthrich, *J. Mol. Biol.* **201,** 637 (1988).

[62] Luminescence Spectroscopy of Metallothioneins

By MARTIN J. STILLMAN and ZBIGNIEW GASYNA

Introduction

Although metallothionein (MT) can bind many different metals both *in vivo* and *in vitro,* detailed structural information about the metal-binding properties is available for only a limited number of metals. This is partly because the inherent "chromophoric silence" of nd^{10} metals like Cd(II), Zn(II), Cu(I), Ag(I), and Hg(II) is a major obstacle to finding a suitable spectroscopic method to monitor metal binding.[1,2] Luminescence spectroscopy is well known to be able to provide quantitative information about the electronic configuration and interstate mixing within the chromophore. For metalloproteins, luminescence properties provide information about the way that the orbitals on the metal and coordinating amino acid groups mix in the excited states. This mixing directly depends on the coordination geometry and bonding structures, such as cluster formation. Luminescence data can be obtained from both steady state measurements of luminescence intensity, either fluorescence or phosphorescence, and dynamic measurements of excited state lifetimes.

[1] J. H. R. Kägi and Y. Kojima, *Experientia, Suppl.* **52,** 52 (1987).
[2] C. F. Shaw, M. J. Stillman, and K. T. Suzuki, *in* "Metallothionein. Synthesis, structure and properties of metallothionein, phytochelatins and metal-thiolate complexes" (M. J. Stillman, C. F. Shaw III, and K. T. Suzuki, eds.), p. 1. VCH Publishers, New York, 1991.

Yellow luminescence from metallothionein was first reported by Lerch's group for the copper-containing metallothionein from *Neurospora crassa*.[3-6] Since then, studies by our group have shown that copper-containing mammalian metallothioneins are generally luminescent, and that the luminescent intensity reaches a maximum when 12 coppers are bound to the protein.[7-11] Similar results are obtained for Ag-MT, Au-MT, and Pt-MT, but the intensity is often very low, so that reasonable quality emission spectra must be recorded from frozen solutions (in our case, at 77 K).[12-14] For each of these metals, the luminescence intensity is detected above 500 nm and is directly dependent on metal coordination. The emission process in these metallothionein species may be quite complicated, because, for example, for Cu-MT, the absorption maxima are near 280 nm, whereas the emission maximum is detected near 600 nm.[1,2,8]

In this chapter, we will describe methods for using emission spectroscopy through the application of luminescence techniques for the analysis of metal binding in metallothioneins, using examples from our laboratory.

Use of Luminescence Spectroscopy for Metallothioneins

Luminescence from a metallothionein is not limited to a single chromophore, and with absorption of 280- to 300-nm photons, many different excited states can occur in metallothionein. Absorption in this region may be into peptide chain, charge transfer, metallic, or ligand field states. In the *absence* of a single, highly luminescent aromatic amino acid chromophore in the metallothionein peptide chain, the emission spectrum recorded following excitation near 300 nm of dilute solutions of the protein at room temperature, will comprise Raman bands, weak phosphorescence from groups in the peptide chain, and bands related to the presence of the metal.

[3] M. Beltramini and K. Lerch, *FEBS Lett.* **127,** 201 (1981).

[4] M. Beltramini and K. Lerch, *Biochemistry* **22,** 2043 (1983).

[5] M. Beltramini, K. Munger, U. A. Germann, and K. Lerch, *Experientia, Suppl.* **52,** 237 (1987).

[6] K. Munger and K. Lerch, *Biochemistry,* **24,** 6751 (1985).

[7] M. J. Stillman and J. A. Szymanska, *Biophys. Chem.* **19,** 163 (1984).

[8] Z. Gasyna, A. J. Zelazowski, A. R. Green, E. Ough, and M. J. Stillman, *Inorg. Chim. Acta* **153,** 115 (1988).

[9] M. J. Stillman, Z. Gasyna, and A. J. Zelazowski, *FEBS Lett.* **257,** 283 (1989).

[10] A. Y. C. Law, J. A. Szymanska, and M. J. Stillman, *Inorg. Chim. Acta* **79,** 112 (1983).

[11] J. A. Szymanska, A. J. Zelazowski, and M. J. Stillman, *Biochem. Biophys. Res. Commun.* **115,** 167 (1983).

[12] M. J. Stillman, A. J. Zelazowski, and Z. Gasyna, *FEBS Lett.* **240,** 159 (1988).

[13] A. J. Zelazowski, Z. Gasyna, and M. J. Stillman, *J. Biol. Chem.* **264,** 17091 (1989).

[14] M. J. Stillman, A. J. Zelazowski, J. Szymanska, and Z. Gasyna, *Inorg. Chim. Acta* **161,** 275 (1989).

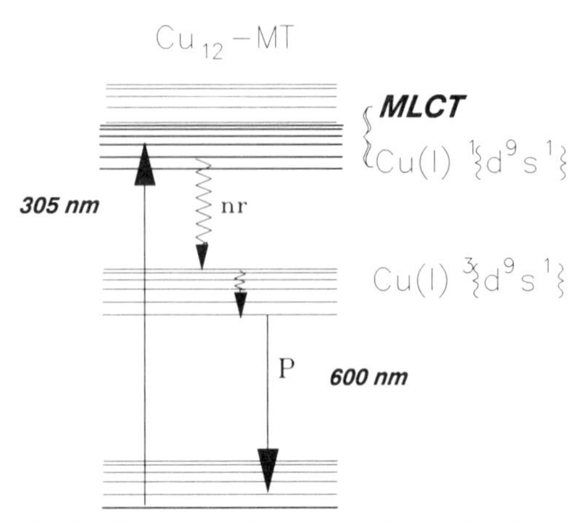

FIG. 1. The Jablonski diagram showing proposed radiative (P) and nonradiative (nr) transitions following absorption of a photon by Cu_{12}-MT. (Adapted from a scheme in Ref. 2.)

In metallothioneins, the charge-transfer states (both thiolate to metal, and metal to thiolate in character) that give rise to absorption to the red of 220 nm in Cd-MT (250 nm), Zn-MT (228 nm), and Hg-MT (280 nm) are expected to be nonluminescent, singlet states.

It is useful to review briefly the processes that can occur following absorption of a photon by a chromophore in which a ground state electron is promoted into an excited state. Figure 1 shows the relationship between absorption, luminescence, and nonradiative pathways, in terms of the familiar Jablonski diagram, drawn for proposed states in Cu_{12}-MT.[2] Singlet-state emission, fluorescence, will be characterized by short lifetimes (of the order of nanoseconds). If the potential surfaces are the same for the ground and excited states, then the emission band will lie close to the absorption band, appearing as a "mirror" image to lower energy. If the two potential surfaces are quite different, then the absorption transition will end on a vibrational level well above zero, which will result in a large Stokes shift moving the band center of the emission to much lower energy. In addition, if the electron crosses to the triplet states, then phosphorescence will be observed, which will be characterized by long lifetimes (microseconds to seconds).

Excited states are quenched to varying extents by nonradiative pathways to the ground state, in which case the emission intensity will be diminished. For a molecule as large as metallothionein, quenching of metal-related excited states by the solvent will play an important role if the solvent can gain access to the metal-binding site.

Figure 1 shows proposed energies of both metal-to-ligand charge transfer (MLCT) and intrametallic states for Cu-MT.[2] Luminescence is observed in the 500- to 650-nm region for Cu-MT, Ag-MT, Au-MT, and Pt-MT.[12-14] It is proposed here that only triplet luminescence is taking place in the case of metals like Cu(I) and that absorption between 280 and 350 nm takes place into overlapping MLCT and Cu(I) $^1[d^9s^1]$ excited states.

Quantitative estimates of excited state properties can be obtained by measuring the emission spectrum at a variety of temperatures, and following excitation at a range of wavelengths.[2] For metallothioneins, the lack of absorption to the red of 350 nm for most metallated species, and the strong peptide-based absorbance to the blue of 215 nm, reduces the usable excitation range to 230–350 nm. For measurements below 0°, either antifreeze or glassing agents must be added to the solution. The molar absorptivity of a chromophore controls the efficiency in absorbing photons of a specified energy, but the intensity of the luminescence depends on a number of interrelated factors. Although increasing the concentration of a solute increases the fraction of incident light absorbed, self-absorption of the emitted light can then become a problem.

The quantum yield of the luminescence is an important quantity in making comparisons between the type of molecular orbitals involved in the excited state and the efficiency of the nonradiative modes. For the metal-binding sites in metallothionein, where we envisage a tightly wound peptide chain covering the metal coordination sites, changes in the quantum yield can be used to probe the exposure of the binding sites to solvent molecules.

Similarly, changes in the excited state lifetime, τ, which is defined in $I_t = I_0 e^{-t/\tau}$, where I_0 is the intensity at time 0 (the time of the initial excitation pulse) and I_t is the intensity at time t, indicates how the various competing mechanisms deactivate the luminescent excited state. When more than one pathway is involved in the decay of the excited state, two or more lifetimes may be needed to fit the observed decay trace of the intensity. The DECAYFIT program of Gasyna and Stillman[15] illustrates techniques that can be used to obtain lifetime data for multiple pathways.

Luminescence from Metal Complexes that Model Metallothionein Binding

Cu(I) and Ag(I) inorganic complexes readily emit in the solid state or as frozen solutions.[16-21] Henary and Zink[20] report that a cubic, four-metal cluster of Cu(I) emits near 662 nm following excitation at 488 nm, with a

[15] Z. Gasyna and M. J. Stillman, *Photochemistry* **38,** 83 (1987).

[16] P. A. Breddels, Ph.D. Thesis, University of Holland, Utrecht (1983).

[17] R. Tamilarasan, S. Ropartz, and D. R. McMillin, *Inorg. Chem.* **27,** 4082 (1988).

lifetime of ca. 7.3 μsec at 13 K. The emission is assigned to a metal-centered, $3d^9 4s^1$ excited state that is strongly modified by copper–copper interactions of the Cu–X–Cu cluster.[20] Copper(I)–imidazole complexes emit near 580 nm at 77 K, following absorption into the lowest energy band at 280 nm.[22] These spectra most closely resemble those of Cu(I)-MT and Ag(I)-MT. With phenanthroline ligands, broad emission near 700 nm (lifetime ca. 80 nsec) is stimulated by absorption, assigned as MLCT, near 475 nm.[17]

In a comprehensive description of the luminescent properties of Cu(I) complexes, Breddels[16] describes the emission from a number of powdered samples of triphenylphosphine and phenanthroline complexes, in which the Cu(I) is tetrahedrally coordinated. He reports that the emission is generally broad and featureless for these compounds. The maxima of the lowest energy absorption bands, which are assigned[16] as metal-to-ligand charge transfer, lie near 400 nm.

Methods for Protein Preparation and Luminescence Measurements

Materials

Generally we use Zn-MT as the starting point in our titrations.[13,23,24] Zn-MT-2 can be prepared by isolating the protein from the livers of rabbits injected eight times with a solution of $ZnSO_4$, at a dose of 20 mg Zn/kg body weight, over a 2-week period.[13,23,24] Apo-MT-2 can be prepared from the Zn-MT-2 by passing the protein through a Sephadex G-25 column, which was previously equilibrated with 0.01 *M* HCl.

Metallothionein α fragment can be prepared from rabbit liver apo-MT as described previously by Winge and Miklossy.[25] In this preparation, 4 mol Eq of cadmium is added to a solution of the apo-MT and the pH of the solution is brought up to 8. The protein is then partially digested with the enzyme subtilisin (Sigma, St. Louis, MO). Apo-α-MT is isolated from Cd_4-α-MT by stripping off the cadmium on a Sephadex G-25 column, previously equilibrated with 0.01 *M* HCl. The β fragment is prepared from

[18] K. L. Stevenson, J. L. Braun, D. D. Davis, K. S. Kurtz, and R. I. Sparks, *Inorg. Chem.* **27,** 3472 (1988).
[19] N. P. Rath, E. M. Holt, and K. Tanimura, *Inorg. Chem.* **24,** 3934 (1985).
[20] M. Henary and J. I. Zink, *J. Am. Chem. Soc.* **111,** 7407 (1989).
[21] R. E. Gamache, R. A. Rader, and D. R. McMillin, *J. Am. Chem. Soc.* **107,** 1141 (1985).
[22] T. N. Sorrell and A. S. Borovik, *J. Am. Chem. Soc.* **108,** 2479 (1986).
[23] M. J. Stillman, W. Cai, and A. J. Zelazowski, *J. Biol. Chem.* **262,** 4538 (1987).
[24] A. J. Zelazowski, J. A. Szymanska, and H. Witas, *Prep. Biochem.* **10,** 495 (1980).
[25] D. R. Winge and K. A. Miklossy, *J. Biol. Chem.* **257,** 3471 (1982).

apo-MT-1 after substitution with 6 mol Eq of Cu^+ and digestion as described previously.[13,26] In this procedure, apo-β fragment is prepared from the Cu_6-β fragment by incubating the protein fragment with 5 *M* excess of KCN in HCl at pH 0.3 for 1 hr. The β fragment is then separated on a Sephadex G-25 column that has been equilibrated with 0.01 *M* HCl.[13,23] Protein concentrations can be estimated from measurements of SH groups using 5,5′-dithiobis(nitrobenzoic acid) in 6 *M* guanidine hydrochloride.[27] Calculations of the total protein concentration are based on the assumption that there are 20 SH groups in the whole protein, with a distribution of 11 and 9 in the α and β fragments, respectively. The total SH plus RSSR concentration can be estimated using the method of Cavallini *et al.*[28]

Luminescence Measurements

Emission spectra can be measured on most commercial spectrometers that have a wavelength region extending to 700 nm. In our laboratory we have used Perkin-Elmer (Norwalk, CT) MPF-4 and Photon Technology, Inc. (Santa Clara, CA) LS1 spectrometers. The spectra are processed using the programs Spectra Manager[29] and Plot3D,[30] and replotted on an HP 7550A plotter. The luminescence decay measurements and excited state lifetime data must be measured on instruments capable of detecting nanosecond–microsecond decays in the 500- to 700-nm region. In our laboratory we have used a PRA model 3000 fluorometer for the nanosecond region, which is based on the technique of time-correlated single-photon counting, and the lifetime capability of the LS1 spectrometer for both nanosecond and microsecond measurements. Fluorescence lifetimes and calculated decay curves were obtained using the DECAYFIT program[15] and proprietary Photon Technology, Inc., software.

Spectral Data for Copper-Metallothionein

Mammalian metallothioneins bind up to 12 mol Eq of Cu(I).[1,2] Yeast metallothioneins also contain copper(I).[31] Spectral data have been reported by Weser's group[32] for mammalian metallothioneins, by Lerch's group for

[26] K. B. Nielson and D. R. Winge, *J. Biol. Chem.* **257,** 4941 (1984).

[27] W. Birchmeier and P. Christen, *FEBS Lett.* **18,** 208 (1971).

[28] D. Cavallini, M. T. Graziani, and S. Dupre, *Nature* (*London*) **212,** 294 (1966).

[29] W. R. Browett and M. J. Stillman, *Chem.* **11,** 73 (1987).

[30] Z. Gasyna and M. J. Stillman, unpublished program (1990).

[31] C. F. Wright, K. McKenney, D. H. Hamer, J. Byrd, and D. R. Winge, *J. Biol. Chem.* **262,** 12912 (1987).

[32] H. Rupp and U. Weser, *Biochem. Biophys. Acta* **533,** 209 (1978).

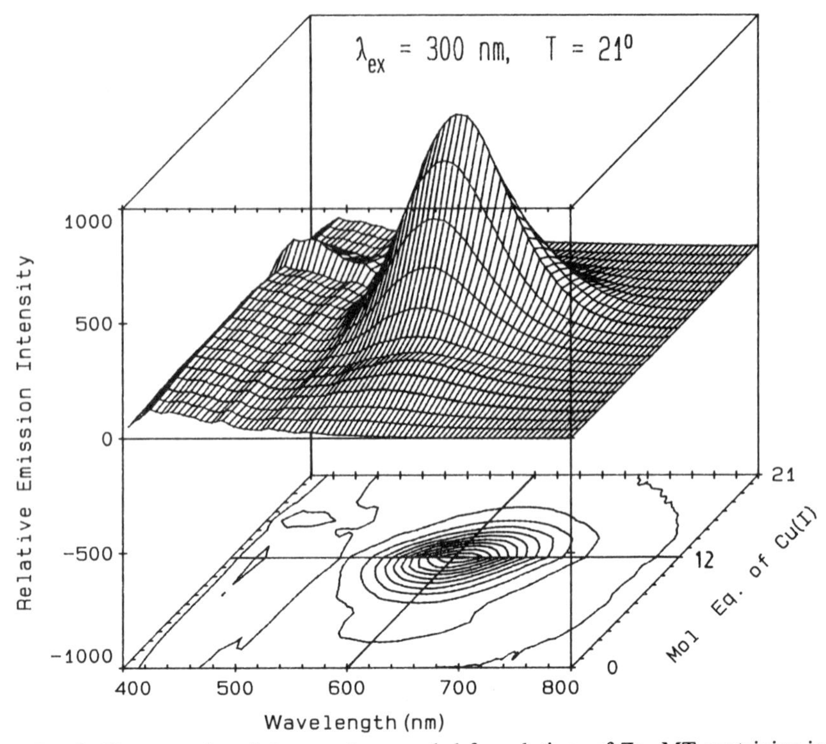

FIG. 2. Uncorrected emission spectra recorded for solutions of Zn_7-MT containing increasing mole equivalents of Cu^+. The excitation wavelength was 300 nm. The spectra are plotted against moles of Cu(I) added. (Reproduced with permission from Ref. 36.)

Neurospora crassa,[3-6] and by Winge's group for yeast metallothioneins.[33] Our laboratory has examined the competitive binding of Cu(I) in rabbit liver apo-MT, Zn_7-MT,[8,34] and also the binding of Cu(I) to isolated α fragment,[35] using both circular dichroic (CD) and emission spectroscopy to probe for cluster formation.

Methods

Aliquots of Zn_7-MT-2 in aqueous solutions were titrated with Cu^+, in 30% (v/v) aqueous solutions of CH_3CN, to form the mixed metal Cu,Zn-

[33] J. Byrd, R. M. Berger, D. R. McMillin, C. F. Wright, D. Hamer, and D. R. Winge, *J. Biol. Chem.* **263,** 6688 (1988).

[34] M. J. Stillman, A. Y. C. Law, W. Cai, and A. J. Zelazowski, *Experientia, Suppl.* **52,** 203 (1987).

[35] A. J. Zelazowski, J. A. Szymanska, A. Y. C. Law, and M. J. Stillman, *J. Biol. Chem.* **259,** 12960 (1984).

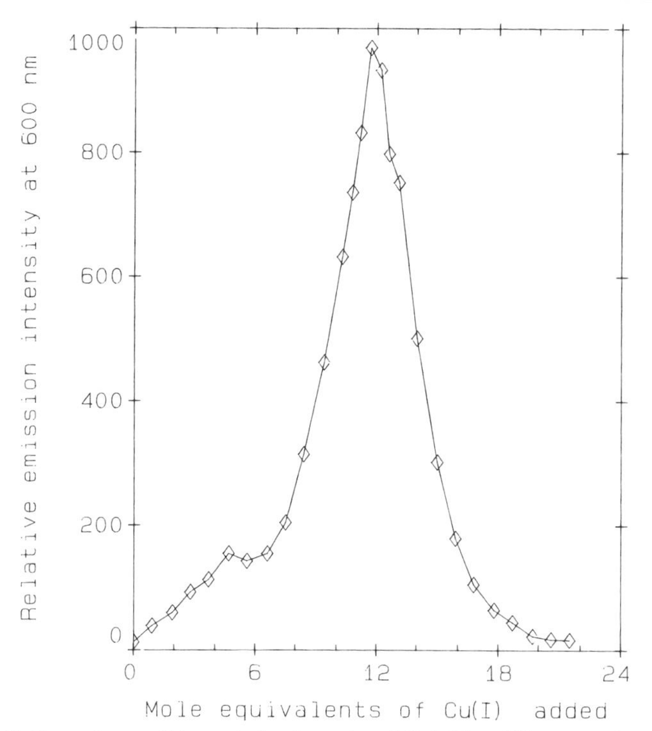

FIG. 3. Dependence of the emission intensity of Cu-MT at 600 nm on the molar ratio of Cu(I) added to Zn-MT, following excitation at 300 nm. (Reproduced with permission from Ref. 36.)

MT with decreasing amounts of Zn^{2+}, up to Cu_{20}-MT, at which point no Zn^{2+} was present. Apometallothionein was titrated with Cu^{+} in the same way.

Spectral Data

Figure 2[36] shows the emission spectra, following excitation at 300 nm, recorded from solutions of Zn-MT-2 as up to 20 mol Eq of Cu^{+} is added.[8,34,36] The emission intensity maximum at about 600 nm rises as Cu^{+} is added up to 11–14 mol Eq, then falls rapidly as up to 20 mol Eq Cu^{+} is added. The peak intensity occurs when Cu_{12}-MT is formed (Fig. 3[8,34,36]). The emission spectrum provides evidence that only Cu_{12}-MT involves a tight structure because, unlike the case with Ag_{18}-MT (see below), the emission intensity for Cu_{20}-MT is very much weaker than that for Cu_{12}-MT, which strongly suggests that the Cu–thiolate binding sites are much more exposed to solvent deactivation.[2]

[36] A. R. Green, Z. Gasyna, A. J. Zelazowski, and M. J. Stillman, submitted for publication.

The emission intensity from Cu-MT is strongly dependent on the nature of the solvent and the temperature of the measurement. With less polar solvents and at lower temperatures, the emission intensity increases considerably. In a frozen solution of Cu-MT, the peak emission intensity is found at 535 nm for Cu_6-MT (77 K).[36] At low temperatures (77 K) the emission spectrum is also very much dependent on the stoichiometric ratio of Cu:MT. As with the room-temperature results, the emission intensity from frozen solutions is also clearly dependent on the hydrophobicity introduced by cluster formation within the protein.

Lifetime Data

The luminescence decay from $Cu(I)_n$-MT in liquid solution is not a one-exponential function, but involves at least three decay components. These component lifetimes at 298 K have values of 20 nsec, 700 nsec, and 3.5 μsec, with intensity contributions to the observed decay trace that are dependent on the stoichiometric ratio of Cu(I):MT (Fig. 4[8]). The luminescence lifetime is strongly dependent on temperature; at 77 K the lifetime increases to over 100 μsec.

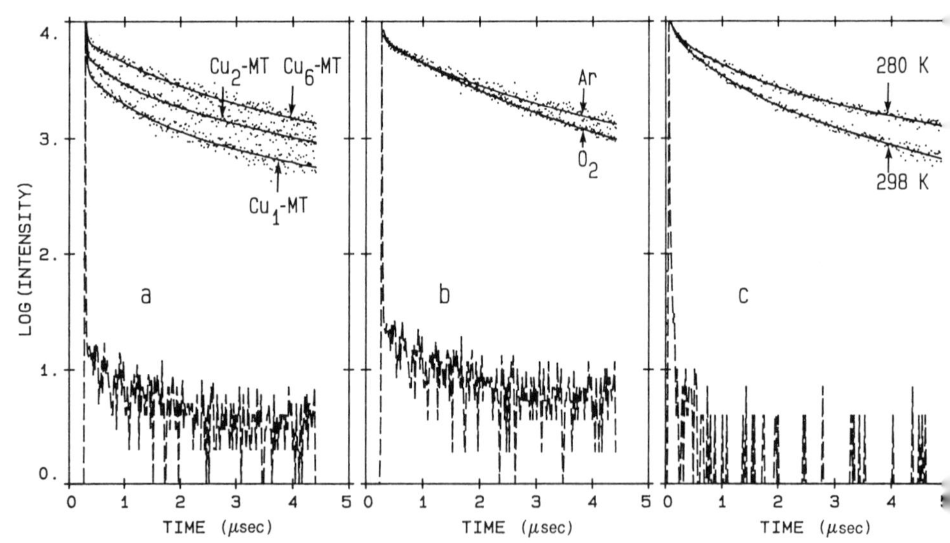

FIG. 4. Luminescence decay curves from Cu_n-MT in solution. (a) Solutions containing 1, 2, and 6 mol Eq Cu(I) and rabbit liver Zn-MT-2, following excitation at 310 nm at 298 K. (b) Effect of O_2 on the lifetime of Cu_6-MT-2. (c) Effect of a change in temperature (298–280 K) on the lifetime of Cu_{12}-MT-2. The experimental points (dots) were fitted with a three-exponential function (solid line). (Reproduced with permission from Ref. 14.)

Silver Metallothionein

Silver binds to apo-MT[13] and Zn-MT[12] in a series of steps that can be followed by absorption, circular dichroism, and emission spectroscopies. Silver also binds to the isolated α and β fragments.[13] Both CD[13] and emission[12-14] spectra recorded during titrations of the protein change substantially as the molar ratio of silver to protein increases. When the solution is frozen at 77 K, Ag-MT complexes exhibit luminescence properties that are dependent on the Ag:MT molar ratio. Most significant is the appearance of a species near 16 mol Eq of Ag(I), a complex that could resemble the Hg_{18}-MT species.[37]

Methods

Aliquots of Ag^+ are added to individual aqueous solutions of Zn_7-MT-2 or apo-MT-2 at pH 3.8 under argon. Portions of these solutions are frozen in liquid nitrogen and added to a quartz tube for emission measurements.

Spectral Data

The emission from Ag(I)-MT exhibits a large Stokes shift: with excitation at 300 nm the emission spectrum has a maximum near 550 nm. Ag^+ binding to Zn_7-MT-2 occurs in a series of stages that can be distinguished by changes in the emission spectrum. Both the position of the luminescence band maximum in the 460- to 600-nm region, and the emission intensity, are strongly dependent on the silver-to-protein stoichiometric ratio, as can be seen in Fig. 5.[13] The change in the emission spectrum when Ag^+ binds to apo-MT-2 resembles that observed for the reaction of Ag^+ with Zn_7-MT-2,[12] although the different steps are much better resolved. Analysis of the spectral data suggests that Ag^+ binds in a domain-specific mechanism to apo-MT-2. Both of the spectral techniques used to study the stoichiometry of silver binding to metallothionein, luminescence, and CD spectroscopies[13] are in complete agreement as to the stoichiometric ratios at which major species form. Clearly, monitoring the luminescence enhancement is a sensitive method for following Ag(I) binding to metallothioneins in general.

Lifetime Data

The emission lifetime of these Ag_n-MT complexes at 77 K is relatively long (longer than 0.1 msec).[8] This suggests that the electronic transition is spin forbidden, probably from an excited triplet to a singlet ground state,

[37] W. Cai and M. J. Stillman, *J. Am. Chem. Soc.* **110,** 7872 (1988).

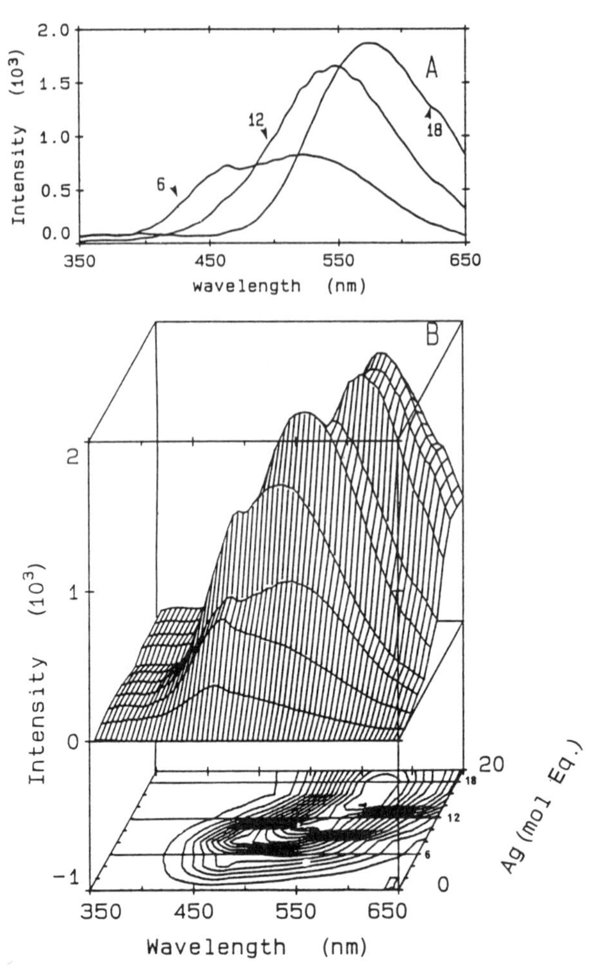

FIG. 5. (A) Emission recorded at 77 K for Ag_6-MT(6), Ag_{12}-MT(12), and Ag_{18}-MT(18) following excitation at 300 nm. (B) Dependence of the emission spectrum of Ag_n-MT for $n = 0-20$, following excitation at 300 nm at 77 K. (Reproduced with permission from Ref. 13.)

and can be described as phosphorescence. Thus the luminescence characteristics are very similar to those of Cu-MT.

Other Luminescent Metallothioneins

Metals such as gold and platinum have an important physiological chemistry that involves MT. We find that metals in addition to Cu(I) and Ag(I) also luminesce when bound to metallothionein.[14] Thus luminescence is observed from Au(I) and Pt(II),[14] and is expected for metals like Pd(II), as well as other nd^{10} and nd^8 complexes. For certain symmetries of the nd^8 metal complexes, especially, in dimeric or cluster forms, the $nd \rightarrow (n+1)p$ excitations produce low-energy luminescent states, which have properties reminescent of the nd^{10} systems.

Methods

Au(I)-MT-1 is prepared by incubation of the apo-MT-1 with an excess of sodium aurothiomalate (Na_2AuTM) in 0.02 *M* phosphate buffer at pH 8.0 under nitrogen.[14] Pt(II)-MT-1, *cis*-Pt(II)-MT-1, and *trans*-Pt(II)-MT-1 are prepared by incubation of the apo-MT-1 in 0.02 *M* phosphate buffer at pH 8.0 under nitrogen, with an excess of potassium tetrachloroplatinate, *cis*-dichlorodiammineplatinum, and *trans*-dichlorodiammineplatinum, respectively.[14] Portions of these solutions are frozen in liquid nitrogen and added to a quartz tube for emission measurements.

Spectral Data

Figure 6[14] shows the absorption, uncorrected excitation, and corrected excitation spectra for Au(I)-MT-1 formed by addition of an excess of Au^+ to apo-MT-1. The inner filter effect of the strong absorbance below 310 nm results in the distortion of the uncorrected excitation spectrum. When the absorbance at each wavelength is factored into the excitation

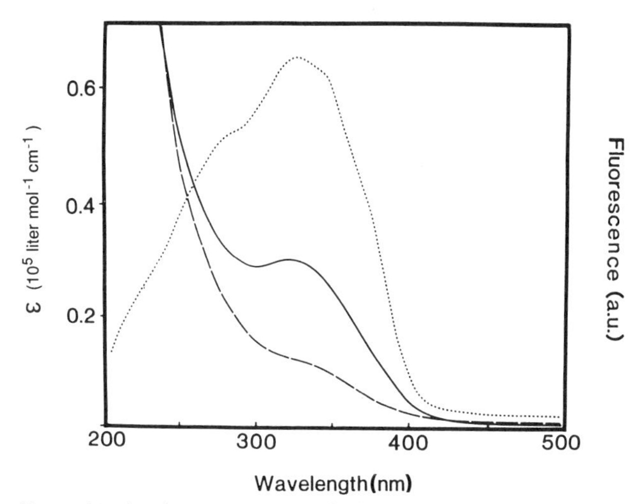

FIG. 6. Absorption (---), uncorrected excitation (· · ·), and corrected excitation (—) spectra of Au(I)-MT-1, recorded at 77 K by monitoring the emission intensity at 600 nm. (Reproduced with permission from Ref. 14.)

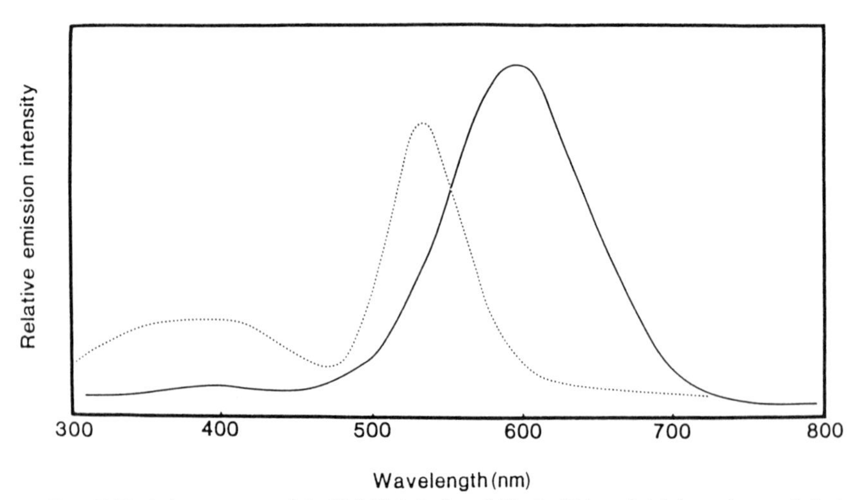

FIG. 7. Emission spectra of Au(I)-MT-1 (—) and Na_2Au(thiomalate) (· · ·) recorded at 77 K following excitation at either 260, 280, or 305 nm. (Reproduced with permission from Ref. 14.)

intensity, the corrected excitation spectrum now resembles the observed absorption spectrum. The emission spectrum of Au(I)-MT (Fig. 7),[14] exhibits a maximum at 600 nm, which is independent of the exciting wavelength. The extended Stokes shift is similar to that observed for other nd^{10} metal-MTs.

Luminescent Metallothionein Measured Directly from Tissue

Since large fractions of heavy metals, such as copper, gold, and platinum, can be expected to be bound to metallothionein in mammalian organs, the direct measurement of luminescence intensity from metallothionein located in tissue is a new and useful method in probing metallothionein binding *in vivo*. In the case of copper, about 50–80% of all copper is bound to MT in the liver from a patient with Wilson's disease.[38] Similar quantities are reported in mammalian liver if MT is stimulated by the injection of copper salts.[39] We have reported that emission spectral data similar to those observed from Cu-MT *in vitro* are observed from Cu-MT located in rat liver tissue.[9]

[38] N. O. Nartley, J. V. Frei, and M. G. Cherian, *Lab. Invest.* **57**, 397 (1987).

[39] D. R. Winge, R. Premakumar, R. D. Wiley, and K. V. Rajagopalan, *Arch. Biochem. Biophys.* **170**, 253 (1975).

Methods

Copper-metallothionein synthesis is induced in female rats that have been injected with aqueous $CuCl_2$ solution, seven times, every second day, over a 2-week period.[9] Copper in the liver is determined using atomic absorption spectrophotometric (AAS) methods, after digestion of the tissue in the mixture of HNO_3 and $HClO_4$. Metallothionein concentrations are estimated using the mercury-binding method[40] and the concentration of mercury is determined by AAS methods. The protein concentration method is calibrated by adding a standard solution of metallothionein that contains six Cu(I) ions per molecule of MT. A value of 58 ± 10 μg of copper and 134 ± 30 μg of MT/g of wet liver tissue is found in the liver from rats treated with 3 mg copper, which can be compared to the value of 2.0 ± 0.4 μg of copper and 41 ± 1 μg of MT/g of wet liver tissue determined for the control rat.[9] A gel-chromatography analysis shows that an estimated 65% of the total copper is bound to MT. The liver is sliced and transferred into a quartz Dewar filled with liquid nitrogen for the luminescence measurements.[9]

Spectral Data

Figure 8[9] shows the uncorrected emission spectra from liver isolated from a rat that had been treated only with saline to that from a rat treated with a solution containing 2 mg Cu/kg body weight. The spectra were obtained as a function of excitation wavelength in the range of 280–350 nm. The new intense luminescence with a maximum near 605 nm grows when the rat is treated with copper, whereas insignificant luminescence is observed in this region from liver of the control rats that was not exposed to copper. Enhanced emission intensity has also been observed from liver tissue isolated from a patient with Wilson's disease as compared to a normal liver (we gratefully acknowledge the assistance of Prof. M. G. Cherian, Department of Pathology, University of Western Ontario, in providing this tissue sample).

The emission spectrum of Cu-MT, measured directly from liver tissue,[9] exhibits two useful properties: (1) the large Stokes shift results in the luminescence band being located in the 600-nm region, well clear of other cellular emission, and (2) the long luminescence lifetime is a distinctive parameter of Cu(I) luminescence. Such properties are an advantage in developing a method for using the Cu-MT emitting chromophore as an

[40] J. K. Piotrowski, J. A. Szymanska, E. M. Mogilnicka, and A. J. Zelazowski, *Experientia, Suppl.* **34,** 363 (1979).

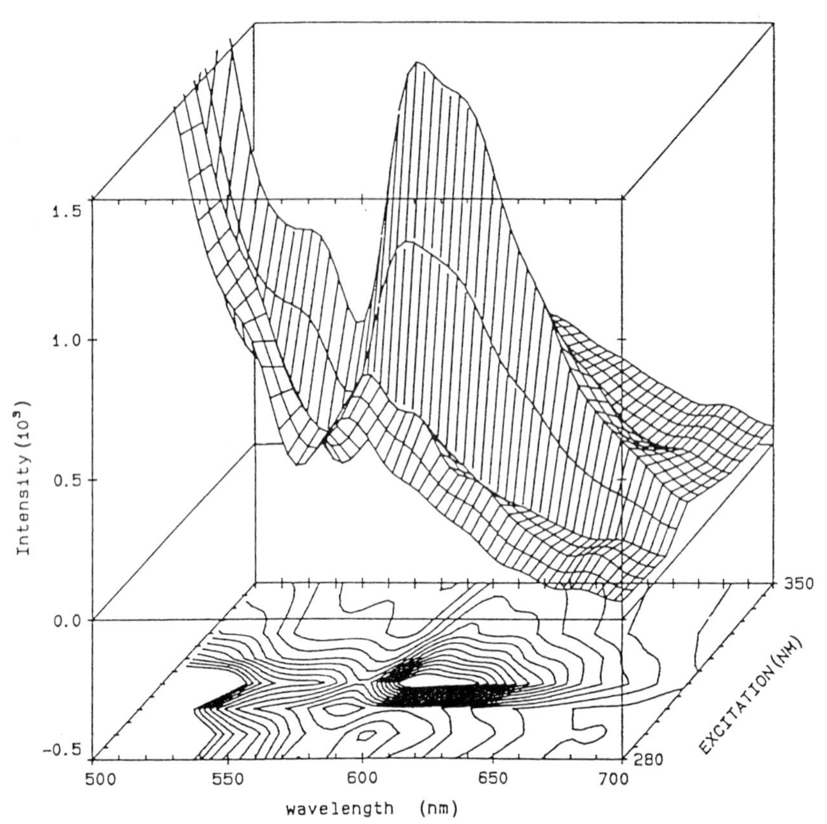

FIG. 8. Uncorrected emission spectra from whole-liver tissue isolated from a copper-treated (2 mg/kg body wt) rat. Excitations were in the range of 280–350 nm at 77 K. (Reproduced with permission from Ref. 9.)

internal imaging label, or as a specific staining method for copper in tissue.[9]

Analysis of Spectral Data

The emission in Cu-MT and Ag-MT can be attributed to $nd^9(n+1)s^1 \rightarrow nd^{10}$ intrametal(I) transitions, where $n = 3$ for copper and 4 for silver, as indicated in Fig. 1.[2] It appears from the form of the emission intensity dependence on the stoichiometric ratio of copper or silver, that the various S–M–S clusters in $M_n - MT$ (M = copper or silver) exhibit emission spectra that are characteristic of the individual clusters. Taking Ag-MT as the example, we find that there are three bands, with maxima at

465, 520, and 575 nm, that can be tentatively assigned to clusters present in the Ag_6-MT, Ag_{12}-MT, and Ag_{18}-MT species.[13] The formation of the Ag–S–Ag bonds in the clusters leads to the delocalization of the metal electrons, which modifies the observed luminescence. The luminescence of such species is expected to be shifted to lower energy. The luminescence enhancement on binding of Ag^+, as well as the other nd^{10} ions, to MT is due to the increased hydrophobicity of the binding region that occurs with the extrusion of water molecules.

Acknowledgments

We gratefully acknowledge funding from NSERC of Canada. We also thank Anna Rae Green, Edward Ough, and Dr. Andrzej Zelazowski for their collaboration with studies on the luminescence of metallothioneins. M.J.S. is a member of the Centre for Chemical Physics and the Photochemistry Unit at the University of Western Ontario.

Section VII

Induction of Metallothioneins

[63] Induction of Metallothionein in Rats

By FRANK O. BRADY

Introduction

Induction of metallothionein (MT) in animals was first postulated in 1964 by Piscator[1], who observed increased quantities of Cd-MT in rabbit liver after exposure to cadmium. Subsequent studies documented the ability of cadmium, zinc, copper, and mercury to induce MT greatly in liver and to a lesser extent in kidney.[2,3] Induction of MT proceeds until 90–100% of the exogenous metal has been bound by newly synthesized MT. Because of its ability to bind toxic metals, such as cadmium and mercury, a great deal of effort was focused in the 1970s and early 1980s on the potential role of MT in detoxification of and protection against heavy metals. After the first Zurich meeting on MT in 1978,[2] a few research groups became very interested in the induction of rat hepatic Zn_7-MT[4-6] by specific hormones and by a variety of chemical and physical stresses. This activity was prompted by an interest in elucidating the physiological function of the normally occurring all-zinc metallothionein rather than the artificially occurring Cd_4Zn_3-MT. Two early reports on the induction of rat hepatic ZnMT by a variety of physical and chemical stresses[7] and following bacterial infection[8] were seminal in attracting interest to the characterization of the hormonal processes involved in the *in vivo* induction of rat hepatic zinc-metallothionein.

Currently, a wide spectrum of agents are known to be *in vivo* inducers of rat hepatic metallothionein (Table I). Metals, catecholamines, adenosine, and phorbol esters are strong inducers. Glucocorticoids and polypeptides (glucagon, angiotensin II, interleukin 1, and interferon) are modest inducers. These communicate with the promoters of *MT* genes via direct effects (metals and glucocorticoids) or via the adenylate cyclase (epinephrine, adenosine, glucagon) or phospholipase C (norepinephrine, angiotensin II, phorbol esters) cascades to stimulate transcription and synthesis of

[1] M. Piscator, *Nord. Hyg. Tidskr.* **45,** 76 (1964).
[2] J. H. R. Kägi and M. Nordberg, *Experientia, Suppl.* **34** (1979).
[3] J. H. R. Kägi and Y. Kojima, *Experientia, Suppl.* **52** (1987).
[4] F. O. Brady, *Trends Biochem. Sci.* **7,** 143 (1982).
[5] R. J. Cousins, *Physiol. Rev.* **65,** 238 (1985).
[6] M. Webb and K. Cain, *Biochem. Pharmacol.* **31,** 137 (1982).
[7] S. H. Oh, J. T. Deagen, P. D. Whanger, and P. H. Weswig, *Am. J. Physiol.* **234,** E282 (1978).
[8] P. Z. Sobocinski, W. J. Canterbury, C. A. Mapes, and R. E. Dinterman, *Am. J. Physiol.* **234,** E399 (1978).

TABLE I
KNOWN INDUCERS OF HEPATIC METALLOTHIONEIN

Pathway	Agent	*In vivo* effect[a]
Direct	Metals (Zn, Cu, Cd, etc.)	+++
Glucocorticoid receptor	Dexamethasone	++
Adenylate cyclase	Epinephrine (β)	+++
	Glucagon	+ to ++
	Adenosine	+++
Phospholipase C	Norepinephrine (α_1)	+++
	Angiotensin II	+
	Phorbol ester (TPA)	++
?	Interleukin 1	+
	Interferon	+
Acute phase response	Turpentine (interleukin 6?)	+++
?	Physical and chemical stresses	+ to +++

[a] +++, $>$10-fold; ++, 5- to 10-fold; +, $<$5-fold.

MT mRNA and protein. The actions of the other agents are not well delineated, but the acute-phase response induction of MT may involve the direct action of interleukin 6 on the liver.

This chapter describes the protocols used in our laboratory for induction and analysis of Zn_7-MT in rats, particularly in liver; a brief discussion on metal induction of MTs with other metal contents (cadmium, copper, mercury) is also included.

Protocols for *in Vivo* Induction of Metallothionein

Metals

Zinc, copper, and cadmium are the metals that routinely are used to induce rat MT. Stock solutions of metals are made in sterile 0.01 *N* HCl. Dilutions for injection are made in 1.0 ml of sterile normal saline. Concentrations of stock solutions are 10 mg Zn/ml as $ZnSO_4$ or $ZnCl_2$, 10 mg Cd/ml as $CdCl_2$, and 10 mg Cu/ml as $CuSO_4$. Injection dosages are well below the LD_{50} values: zinc (5 mg/kg body weight), cadmium (0.25 mg/kg body weight), and copper (0.5 mg/kg body weight). Injections are administered intraperitoneally. Maximum levels of hepatic MT occur 18–24 hr after administration. Induction with zinc results in increased hepatic levels (20- to 50-fold) of Zn_7-MT. Induction with cadmium results in elevated levels of Cd_4,Zn_3-MT. Induction with copper results in elevated levels of MT containing comparable amounts of zinc and copper. Repeated injections with metals on a daily basis can result in further increases in MT in liver. However, kidney and testicular damage does occur with repeated injections of cadmium. Induction of MT in kidney occurs also, but it reaches more modest levels than those seen in liver.

Other metals have been administered to rats, resulting in increased levels of hepatic and renal MT. Not all metals that induce MT end up complexed with it, reflecting complex processes in which hepatic and serum zinc levels are altered and the inducing metal has low affinity for MT.

One metal that has an interesting *in vivo* behavior with respect to MT induction is mercury. Administration of Hg^{2+} (2 mg/kg body weight as $HgCl_2$) leads to a significant induction of hepatic Zn_7-MT with very little, if any, mercury attached. Rather, most of the mercury appears in kidney with a significant amount attached to MT. Apparently, kidney preferentially accumulates mercuric ions as compared with liver, unlike the comparable uptakes of other metals by the two tissues.

Glucocorticoids

Naturally occurring glucocorticoids, such as cortisol and corticosterone, are poorly soluble in water solutions. Rather, they must be administered intraperitoneally in an oil suspension. Corticosterone (40 mg/kg body weight) is injected in a suspension of peanut oil or olive oil (1 ml), containing 5% (v/v) ethanol. At least an 18-gauge needle must be used in order to allow adequate flow of the suspension.

Dexamethasone (stock solution, 2 mg/ml; dosage, 20 mg/kg body weight) is a water-soluble glucocorticoid that is much easier to use. Induction of hepatic MT by glucocorticoids is more modest (fivefold) than induction by metals. Dexamethasone appears to be more potent than natural glucocorticoids and longer lasting for hepatic MT induction.

Catecholamines

Rat liver possesses α_1- and β_2-adrenoceptors, the former communicating with the phospholipase C cascade and the latter with the adenylate cyclase cascade. Norepinephrine and epinephrine are the naturally occurring catecholamines that prefer to bind to α- and β-receptors, respectively. However, this is only a preference, not an exclusivity; epinephrine can stimulate α-adrenoceptors and norepinephrine can stimulate β-receptors. We have used multiple injections of norepinephrine, epinephrine, or isoproterenol (β-agonist) to stimulate induction of hepatic Zn_7-MT.[9]

Catecholamine stock solutions were prepared at 1 mg/ml in sterile 0.01

[9] F. O. Brady and B. Helvig, *Am. J. Physiol.* **247,** E318 (1984).

N HCl. Dilutions were made in 1.0 ml sterile saline immediately before use. Norepinephrine and epinephrine were administered at 20 μg/kg body weight and isoproterenol at 10 μg/kg body weight every 2 hr, and the animals were sacrificed at 11 hr. Inductions of MT were 4- to 12-fold, with the best inducer being isoproterenol. This is probably because it persists longer in the blood than does epinephrine or norepinephrine.

To confirm adrenoceptor-mediated induction of MT, blockers were used to inhibit the inductions. Propranolol (β-blocker) and phentolamine (α-blocker) (10 mg/ml in sterile saline; dosage 10 mg/kg body weight) can be used separately, but they are most effective when used together. Blockers were effective when administered at -1 and $+5$ hr relative to catecholamine administration beginning at 0 hr.

Phorbol Esters

12-*O*-Tetradecanoylphorbol 13-acetate (TPA), also called phorbol myristate acetate (PMA), is a tumor promoter and a direct activator of protein kinase C (PK-C). Since the phospholipase C cascade results in the activation of PK-C by diacylglycerol, use of TPA to activate PK-C directly is a way of circumventing the phospholipase C pathway, preventing the increase of cytosolic calcium by inositol 1,4,5-trisphosphate. TPA is a potent *in vivo* inducer of Zn_7-MT[10] with an effective dose for 50% of maximal induction at a concentration of 26.5 nmol/kg body weight. Maximal induction (seven- to eightfold) occurs at about 100 nmol/kg body weight and 18–25 hr after a single injection.

Stock solutions of TPA are made in dimethyl sulfoxide (DMSO) at 1 mg/ml. Dilutions in 1.0 ml of sterile saline are made immediately before injection. TPA solutions are light sensitive, and TPA hydrolyzes in water solutions. To confirm direct effects of TPA, an inactive phorbol ester, such as 4β-phorbol, can be used at identical dosages. 4β-Phorbol does not induce MT, nor does the vehicle, DMSO.

Purinergic Agonists

Purinergic agonists cause activation of either adenylate cyclase (ATP, ADP) or phospholipase C (adenosine, AMP) via P_1 or P_2 receptors, respectively. Rat liver possesses the A_2 subtype of the P_1 receptor and the P_{2y} subtype of the P_2 receptor. Adenosine (100 μmol/kg body weight) is a modest inducer (three to fourfold) of hepatic Zn_7-MT. ATP (10 μmol/kg body weight) was not effective. There are available several good synthetic

[10] F. O. Brady, B. S. Helvig, A. E. Funk, and S. H. Garrett, *Experientia, Suppl.* **52,** 555 (1987).

agonist analogs of adenosine. We[11] have used these at the following dosages: 2-chloroadenosine (100 μmol/kg body weight) (7.3-fold induction, 11 hr); 5-*N*-ethylcarboxamide adenosine (NECA) (10 μmol/kg body weight) (5.5-fold, 11 hr); and 5′-chloro-5′-deoxyadenosine (50 μmol/kg) (2.1-fold, one injection; 3.9-fold, three injections; 11 hr). Single injections of adenosine or single/multiple injections of ATP failed to induce hepatic MT, probably due to their rapid uptake and metabolism *in vivo.*

Stock solutions of all these compounds were made in sterile saline at either 100 or 10μM. Dilutions were made in sterile saline immediately before injections.

Methylxanthines are blockers of adenosine receptors. We have used caffeine and theophylline (100 μmol/kg body weight) to block induction by 2-chloroadenosine (25 μmol/kg body weight, $ED_{0.5}$), which confirms the purinergic effect *in vivo* on MT induction. These blockers are not very soluble. Solutions (100 μM) in sterile saline must be heated briefly in a boiling water bath before dilution and injection.

Polypeptide Hormones

Glucagon[12,13] and angiotensin II[13] are modest inducers of rat hepatic MT. Glucagon communicates via the adenylate cyclase cascade and angiotensin II via the phospholipase C cascade.

Glucagon is administered as a suspension in sterile saline (5 mg/ml stock; 1 mg/kg body weight). Angiotensin II is administered as a solution in sterile saline (5 mg/ml stock; 1 mg/kg body weight). We have used these either intraperitoneally or intravenously. Intravenous injections were done through the tail vein under light ether anesthesia using a total volume of 0.25 ml after cleansing with 70% (v/v) ethanol. Single injections gave two- to threefold inductions at 11 hr, while multiple injections (0,4, and 8 hr) gave four- to sixfold inductions for glucagon and fourfold inductions for angiotensin II. Another peptide hormone, Arg-vasopressin, did not induce hepatic MT *in vivo.*

Acute-Phase Response

A variety of organic chemical insults will induce hepatic MT. This is probably due to the cytokines that are liberated during the acute phase response.[14] Of current interest is the role of hepatocyte-stimulating factor,

[11] F. O. Brady, S. H. Garrett, and K. Arizono, *in* "Metal Ion Homeostasis: Molecular Biology and Chemistry" (D. Hamer and D. Winge eds.), p. 81. Alan R. Liss, New York, 1989.

[12] K. R. Etzel and R. J. Cousins, *Proc. Soc. Exp. Biol. Med.* **167,** 233 (1981).

[13] B. S. Helvig and F. O. Brady, *Life Sci.* **35,** 2513 (1984).

[14] H. Baumann, G. P. Jahreis, D. N. Sauder, and A. Koj, *J. Biol. Chem.* **259,** 7331 (1984).

or interleukin 6. This cytokine has multiple functions on many cell types, but it appears to be responsible for the stimulation of synthesis in liver of the acute-phase proteins. Metallothionein is inducible by agents that cause the acute-phase response.[11] The historical way of inducing the acute-phase response is to inject a rat subcutaneously with turpentine (3.0 ml/kg body weight). We have done this, observing a five- to sixfold induction of hepatic Zn-MT. It was maximal at 18 hr, a time when serum α_2-macroglobulin was just beginning to rise and before serum ceruloplasmin began to increase. Hypozincemia occurred at 11–39 hr. The specific effects of cytokines, such as interleukin 6, on MT induction deserve further study, particularly in cell culture.

Analysis of Metallothioneins after *in Vivo* Inductions

Numerous methods of analysis of MT are available, and they have been covered in detail elsewhere in this volume.[15] In our studies with MT *in vivo* there is usually sufficient protein available to use the Sephadex G-75/atomic absorption spectroscopy method. This is a convenient method and yields information about the various metals that can be present in MT.[16] This method is satisfactory for analysis of tissues that are large and/or rich in MT. When the amount of MT is limiting, we use a metal replacement technique with radioactive ^{109}Cd or ^{203}Hg, followed by Sephadex G-75 superfine column chromatography.[17] A sample of tissue as small as 0.2 g can be analyzed by this method.

Sample Preparation

Tissue is removed from a rat after decapitation and is frozen at −20° until further use. Individual tissues are weighed, partially thawed, chopped with a pair of scissors, and homogenized with a Teflon stirrer–glass tube apparatus, using 4 vol of ice-cold 0.01 *M* Tris-HCl, pH 8.0, 0.25 *M* sucrose, 1.0 m*M* NaN_3. The homogenate is centrifuged at greater than 140,000 *g* for 45 min at 4° in a Beckman (Fullerton, CA) ultracentrifuge. The supernatant is carefully decanted, its volume is measured, and a 4.0-ml aliquot is subjected to molecular sieving chromatography on Sephadex G-75 fine (1.5 × 90 cm, Econoline; Bio-Rad, Richmond, CA) at 4°. Three-milliliter aliquots are collected, and the profile of metal content is assessed with atomic absorption spectroscopy (AAS). The MT peak is located at V_e/V_0 of 2.0–2.5. The amount of metal in MT is quantitated

[15] This volume [9]–[24].

[16] F. O. Brady and M. Webb, *J. Biol. Chem.* **256,** 3931 (1981).

[17] S. H. Garrett, Ph.D. Thesis, University of South Dakota, Vermillion, South Dakota (1989).

and expressed as micrograms metal or microgram-atoms metal per gram of tissue wet weight. This method is useful in studies in which various metals may be associated with MT because each metal can be quantitated, using AAS. If only the amount of MT protein is desired, several of the other methods cited in this volume may be used.

For very small samples (< 1.0 g) or for tissues with low levels of MT the radioactive metal displacement method is used. A sample of tissue is placed in a 1.5-ml microfuge tube on ice. Two volumes of ice-cold 0.1 *M* NaCl, 0.01 *M* Tris-HCl, pH 8.6, 1 m*M* NaN_3 is added, and the tissue is disrupted by brief sonication or grinding with a Teflon pestle. After centrifugation at 16,000 *g* in a microfuge the supernatant was used for analysis of MT. An aliquot (0.2 ml) is mixed with dithiothreitol to make the solution 1 m*M* and with 50 μg of phenol red, as a marker. The sample is doped with ^{109}Cd (12 μCi/μg Cd) to 1.5 times as much cadmium as there is zinc in the sample (measured by AAS). After incubation on ice for 15 min the sample is applied to a Sephadex G-75 Superfine column (0.8 $\times$ 29 cm, Econoline; Bio-Rad) and run in a cold room. Elution buffer is the same as sample buffer. Fractions (0.5 ml) are collected and counted on a γ counter. Three radioactive peaks are obtained, the second being due to MT. This peak is calculated, and the amount of MT is reported as picomoles of cadmium in MT per gram tissue or per milligram protein in the sample. Because of gel binding of some free metal and its subsequent slow leakage, we discard the gel after each column run and repour with fresh Sephadex. When using this technique, the method should be calibrated initially with authentic, purified MT to verify that all of the zinc in MT is being displaced by the added cadmium. If the tissue being examined already contains some cadmium bound to MT, ^{203}Hg can be used as the displacing metal instead. Remember, though, that the stoichiometry of cadmium and zinc binding to MT is 7, while mercury is 10–12. Neither of these metals is useful for quantitative displacement of endogenous metals from MT that contains significant amounts of copper.

^{109}Cd has a half-life of 470 days, with a γ-ray emission energy of 88 keV. It is obtainable from New England Nuclear (Boston)/Du Pont as cadmium chloride in a 0.5 *M* HCl solution. ^{203}Hg has a half-life of 47 days, with a γ-ray emission energy of 279 keV. It also emits β particles with an energy of 210 keV. It is obtainable from New England Nuclear/Du Pont as mercuric nitrate in 0.5 *M* HNO_3 or as mercuric chloride in 0.5 *M* HCl.

Pitfalls with in Vivo Inductions of Metallothionein

The major problem in doing *in vivo* inductions of MT is the condition of the animals when the experiment is begun. Hepatic MT is inducible by so many things that it is important to know what types of environmental activities will raise the baseline level of MT for a group of animals. We routinely house a new group of animals for 5–7 days in a controlled temperature-lighting room, feeding laboratory chow and tap water *ad libitum*. This allows them to adjust hormonally to their new environment before an experiment is begun. We do injections in the same room in which the animals are housed. Sacrifice of the animals is done after transport to a separate room. Starvation of rats induces MT, even just overnight. Female rats have a higher basal level of MT than do males. Juvenile rats (< 100 g) have slightly higher basal levels than adults. Neonatal rats (0–21 days) have very elevated levels of hepatic MT. Any animals that are exposed to organic chemicals, such as occurs in an animal house when pesticides are used to eliminate bugs, may have a transient rise in basal levels of MT.

Caution should always be used in doing *in vivo* inductions of MT to ensure that basal levels of MT are in the range of 0.5–1.5 μg zinc in MT/g liver (7.7–23 ng atoms zinc in MT/g liver) (6–18 μg MT protein/g liver). If

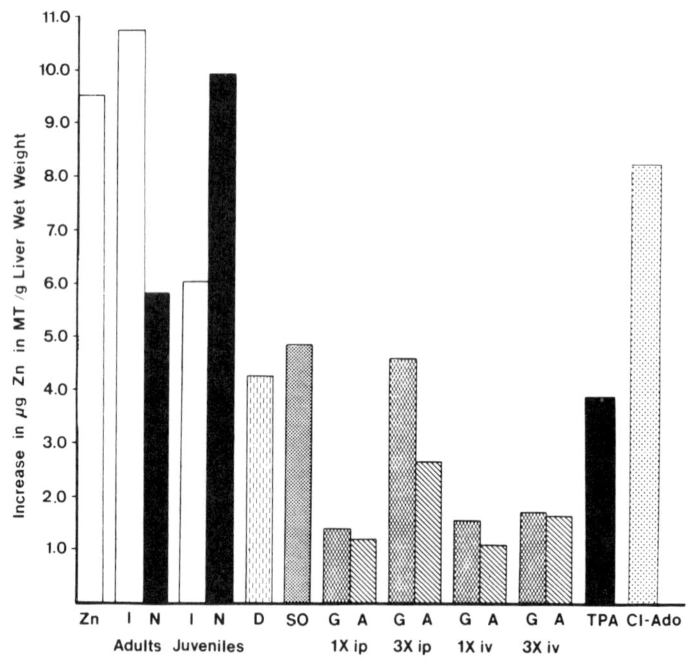

FIG. 1. Increases in rat hepatic Zn_7-MT 11 hr after various inductions. Zn, Zinc; I, isoproterenol; N, norepinephrine; D, dexamethasone; SO, sham operation; G, glucagon; A, angiotensin II; TPA, phorbol ester; Cl-Ado, 2-chloroadenosine.

they are higher, then the environmental source of the induction of MT should be sought out and eliminated.

Relative Levels of Metallothionein Inductions

A composite of the inductions of hepatic MT that we have performed is shown in Fig. 1.[12] The increase in micrograms zinc in MT per gram liver wet weight above basal levels 11 hr after various treatments is shown. Zinc, isoproterenol (adults), norepinephrine (juveniles), and 2-chloroadenosine are strong inducers. Dexamethasone, sham operation, glucagon (three injections, intraperitoneal), and TPA are moderate inducers, and glucagon (one injection, intraperitoneal or intravenous; three injections, intravenous) and angiotensin II are weak inducers.

Zn_7-MT induction in liver appears to be a fundamental response to stresses on an animal. The induction by metals, hormones, and other agents remains an interesting area of research. The elucidation of response elements in the promoters of *MT* genes has confirmed the direct interaction of various cascade signaling systems with the elevation of gene transcription.

Acknowledgments

This work was supported by USPHS Grants ES-01288, ES-02879, and DK41459, a grant from the Dakota Aerie of the Fraternal Order of Eagles, and a grant from the Parson's Fund of the University of South Dakota School of Medicine.

[64] Induction of Metallothionein in Primary Rat Hepatocyte Cultures

By CURTIS D. KLAASSEN and JIE LIU

Introduction

Metallothionein (MT) is induced *in vivo* by metals, steroids, various organic chemicals, and acute stimuli. The mechanism of MT induction can be either a direct genomic event or an indirect, integrated physiological response to changes in neural, hormonal, or hemodynamic homeostasis. Because of this complexity, classification of various MT inducers as direct or indirect inducers may be confounded when studied in intact animals. Use of *in vitro* cell systems has facilitated the observation and elucidation of the mechanisms of MT induction. *In vitro* cell systems for MT induction have included nonmammalian cell cultures, embryonic cells in culture, various cell lines, primary cell cultures, and human peripheral blood cells. Some of the representative cell types used for MT study are shown in Table I.

Many of the *in vitro* studies on MT induction have used various malignant cell lines, which are easy to maintain and survive many passages. The *MT* gene can be amplified by growing the cells in media containing metals (such as cadmium and zinc). Eventually high constitutive MT levels can be achieved and passed from generation to generation. These cells are thus resistant to the toxic effect of metals. The amplified *MT* genes are also responsive to other MT inducers. However, these cells are transformed cells, and they do not always respond to MT inducers in a manner similar to intact animals. Primary cultures of adult rat hepatocytes represent a normal, differentiated cell type that resembles, in many respects, the *in vivo* metabolic and morphologic characteristics of liver parenchymal cells.[1-3] The induction of MT in primary cultured hepatocytes by metals and steroids is very similar to that observed in the intact animals.[4,5]

In this section, we will focus on the induction of MT in rat primary hepatocyte cultures, and describe cell isolation and culture, sample preparation, and analysis of MT at the protein level. Most of the culture and assay procedures are also suitable for the study of MT induction using other types of cell cultures.

Isolation of Rat Hepatocytes

Hepatocytes are isolated by a two-stage single-pass perfusion method[6] as modified by Bissell and Guzelian.[1] The procedures are as follows:

1. Rats are anesthetized with pentobarbital (65–75 mg/kg, *ip*), and the abdomen is cleaned with Betadine followed by 70% (v/v) ethanol. The abdomen is opened and the small intestine moved to one side to visualize the portal vein. Two ligatures are loosely tied around the portal vein, one proximal and one distal to the liver, and an 18-gauge catheter is inserted into the portal vein between the two ligatures. The needle guide is re-

[1] D. M. Bissell and P. S. Guzelian, *Ann. N.Y. Acad. Sci.* **349,** 85 (1980).
[2] T. Nakamura and A. Ichihara, *Cell Struct. Funct.* **10,** 1 (1985).
[3] C. Guguen-Guillouzo, P. Gripon, Y. Vandenberghe, F. Lamballe, D. Ratanasavanh, and A. Guillouzo, *Xenobiotica* **18,** 773 (1988).
[4] W. M. Bracken and C. D. Klaassen, *Toxicol. Appl. Pharmacol.* **87,** 381 (1987).
[5] W. M. Bracken and C. D. Klaassen, *J. Toxicol. Environ. Health* **22,** 163 (1987).
[6] M. N. Berry and D. S. Friend, *J. Cell Biol.* **43,** 506 (1969).

TABLE I
INDUCTION OF METALLOTHIONEIN IN REPRESENTATIVE CELLULAR SYSTEMS[a]

Cell systems	Inducers	Index	Refs.
Nonmammalian cells			
Drosophila cell lines	Cd	Pr	*b*
Turkey embryos	Zn	Pr	*c*
Rainbow trout hepatocytes	Cd, Zn, dex	Pr	*d*
Various cell lines			
CHO	Cd, Zn, IFN	mRNA	*e*
	Gold compounds	mRNA, Pr	*f*
Mouse hepatoma (1A)	Metals, etc.	mRNA	*g*
Rat hepatoma (H4IIE)	Butyrate, Cd, dex	mRNA	*h*
Embryonal carcinoma	Butyrate, Cd, Zn	mRNA	*i*
Hepatoma (G2)	Il-1, IFN	mRNA	*j*
TRL 1215 rat liver	Butyrate, Zn, Cd	Pr	*k*
FRSK (epidermal keratinocyte)	Vitamin D_3, etc.	mRNA, Pr	*l*
Primary hepatocyte cultures			
Rat	Steroids	Pr	*m*
	Metals, etc.	Pr	*n*
	cAMP, dex	mRNA	*o*
	IL-6, IL-1	mRNA, Pr	*p*
Mouse	Zn	mRNA	*q*
Human cell lines and blood cells			
HeLa cell lines	Cd, Zn	Pr	*r*
	Cd	mRNA, Pr	*s*
Human trophoblasts	Cd, Zn	Pr	*t*
Human skin fibroblasts	Carcinogens	mRNA	*u*
Human hepatic Chang cell	Metals, dex	Pr	*v*
	Macrophage factor	Pr	*w*
Human blood			
Leukocytes	Cd	mRNA, Pr	*x*
T and B cells	Cd	Pr	*y*

[a] Pr, protein; Cd, cadmium; Zn, zinc; dex, dexamethasone; IFN, interferon; IL-1, interleukin 1.

[b] A. Debec, R. Mokdad, and M. Wegnez, *Biochem. Biophys. Res. Commun.* **127,** 143 (1985); [c] M. P. Richards, *J. Pediatr. Gastroenterol. Nutr.* **3**, 128 (1984); [d] S. J. Hyllner, T. Andersson, C. Haux, and P. Olsson, *J. Cell. Physiol.* **139,** 24 (1989); [e] S. Morris and P. C. Huang, *Mol. Cell. Biol.* **7**, 600 (1987); [f] T. R. Butt, E. J. Sternberg, C. K. Mirabelli, and S. T. Crooke, *Mol. Pharmacol.* **29**, 204 (1986); [g] D. M. Durnam and R. D. Palmiter, *Mol. Cell. Biol.* **4**, 484 (1984); [h] B. W. Birren and H. R. Herschman, *Nucleic Acids Res.* **14**, 853 (1986); [i] G. K. Andrews and E. D. Adamson, *Nucleic Acids Res.* **15**, 5461 (1987); [j] M. Karin, R. J. Imbra, A. Heguy, and G. Wong, *Mol. Cell. Biol.* **5**, 2866 (1985); [k] M. P. Waalkes and M. J. Wilson, *Toxicol. Lett.* **32**, 289 (1986); [l] M. Karasawa, J. Hosoi, H. Hashiba, K. Nose, C. Tohyama, E. Abe, T. Suda, and T. Kuroki, *Proc. Natl. Acad. Sci. U.S.A.* **84**, 8810 (1987); [m] W. M. Bracken and C. D. Klaassen, *Toxicol. Appl. Pharmacol.* **87**, 381 (1987); [n] W. M. Bracken and C. D. Klaassen, *J. Toxicol. Environ. Health* **22**, 163 (1987); [o] V. L. Nebes, D. DeFranco, and S. M. Morris, Jr., *Biochem. Biophys. Res. Commun.* **166,** 133 (1990); [p] J. J. Schroeder and R. J. Cousins, *Proc. Natl. Acad. Sci. U.S.A.* **87**, 3137 (1990); [q] J. E. Piletz and H. R. Herschman, *J. Nutr.* **117,** 183 (1987); [r] C. J. Rudd and H. R. Herschman, *Toxicol. Appl. Pharmacol.* **47**, 273 (1979); [s] S. Koizumi, T. Sone, N. Otaki, and M. Kimura, *Biochem. J.* **227**, 879 (1985); [t] L. D. Lehman and A. M. Poisner, *J. Toxicol. Environ. Health* **14**, 419 (1984); [u] P. Angel, A. Pöting, U. Mallick, H. J. Rahmsdorf, M. Schorpp, and P. Herrlich, *Mol. Cell. Biol.* **6**, 1760 (1986); [v] S. Kobayashi, T. Okada, and M. Kimura, *Chem. Biol. Interact.* **55**, 347 (1985); [w] Y. Iijima, T. Fukushima, and F. Kosaka, *Biochem. Biophys. Res. Commun.* **164,** 114 (1989); [x] C. B. Harley, C. R. Menon, R. A. Rachubinski, and E. Nieboer, *Biochem. J.* **262,** 873 (1989); [y] H. Yamada, S. Minoshima, S. Koizumi, M. Kimura, and N. Shimizu, *Chem.–Biol. Interact.* **70**, 117 (1989).

moved, the cannula is secured in place by tightening the ligatures, connected to the perfusion tubing, and the perfusion started. The thoracic cavity is then opened, vena cava and aorta clamped using a hemostat, and the inferior vena cava cut for the perfusate to exit. In this manner, a single-pass retrograde perfusion is established.

2. An oxygenated perfusion buffer (137 m*M* NaCl, 5.4 m*M* KCl, 4.2 m*M* $NaHCO_3$, 0.4 m*M* KH_2PO_4, 0.3 m*M* Na_2HPO_4, 5 m*M* glucose, 0.5 m*M* EGTA and 25 m*M* Tris-HCl, pH 7.4) is perfused *via* the portal vein for 12–15 min (18 ml/min) at 37°. The liver turns pale when the perfusion begins and the blood is cleared from the tissue. One needs to be careful that the surface of the liver does not dry out and that no air bubbles in the perfusion line enter the liver, as this will create an air embolus and block further perfusion.

3. After the perfusion buffer (250 ml) has been used, the liver is perfused with 200 ml oxygenated medium containing 0.05% collagenase II (Worthington Biochemical Co., Freehold, NJ) at a flow rate of 18 ml/min for the first 2 min, 14 ml/min for another 5 min, and 10 ml/min for the remaining media. The liver is enlarged and soft at the end of the enzymatic perfusion.

4. The liver is carefully removed to a sterile 100-ml beaker and any attached tissue removed. The liver is gently minced with a pair of scissors, and 60 ml of medium is added to suspend the cells. The cell suspension is filtered through four layers of sterile gauze into a sterile 100-ml beaker.

5. The cell suspension is divided into two 50-ml sterile plastic centrifuge tubes. Parenchymal cells are separated from nonparenchymal cells and debris by low-speed centrifugation (50 *g*, 1 min), the cell pellet is washed three times with medium, and the final pellet resuspended in 30 ml of medium.

The cell yield is approximately 150×10^6/rat (200–300 g body weight). The purity of parenchymal cells is more than 95% and cell viability is usually more than 85% as determined by Trypan Blue dye exclusion.

Primary Culture of Hepatocytes

1. Collagen coating of culture dishes: Sixty microliters of collagen (Collagen Corp., Palo Alto, CA) is spread over 60-mm plastic dishes and allowed to dry overnight. The dishes are washed with media, and ready for use. Collagen coating of dishes is essential for hepatocyte attachment.

2. Culture media: A variety of standard media (*e.g.*, L-15, 199, Ham's F10, minimum essential media, Basal media Eagle) have been used successfully for rat hepatocytes in primary culture. Medium is usually supple-

mented with insulin (30 nM), ascorbic acid (0.3 mM), antibiotics (penicillin G, 100 U/ml, streptomycin, 100 μg/ml, and Fungizone, 0.25 μg/ml) and fetal calf serum (10%) for monolayer growth. For establishing cultures under serum-free conditions, modified Waymouth's medium 752/1 is often used.[1] In addition, 0.5 mM methionine and 0.5 mM cysteine are included as sulfhydryl sources for MT synthesis.[7]

3. Approximately $2-3 \times 10^6$ cells in a final volume of 3.0 ml is placed in each 60-mm collagen-coated dish, and the dishes are then incubated at 37° (5% CO_2 and 95% humidity). After a 1-hr incubation, medium and unattached cells are removed and fresh medium added. The culture medium is changed every 24 hr thereafter to provide fresh nutrient for cell viability.

Metallothionein Induction

Hepatocytes are usually left to equilibrate on the monolayer for 20–24 hr before treatment. This period allows the hepatocytes to recover from the isolation stress and to form a monolayer. At this time, the medium is changed, and various inducers may be added (1% volume of media). The final concentration of commonly used inducers is as follows: cadmium, 0.1–5.0 μM; zinc, 20–100 μM; dexamethasone, 0.1–10 μM, as indicated in Fig. 1.

Sample Preparation for Metallothionein Protein Analysis

For the quantitation of MT protein, hepatocytes are usually incubated with inducers for 24–72 hr to achieve the maximum MT protein synthesis, as indicated in Fig. 2. Medium is removed and the dishes are rinsed three times with isotonic saline. Two milliliters of 10 mM Tris-HCl (pH 7.4) is added to each dish to harvest the cells. The cells are scraped with a rubber policeman and transferred to a fresh tube. Cells are then disrupted by sonication (W-10 cell disruptor, Heat Systems-Ultrasonics, Inc., Plainview, NY), and the suspension is aliquoted and stored at −70° for later analysis.

Quantitation of Metallothionein Protein

Metallothionein is quantitated by the cadmium-hemoglobin radioassay.[8] A 100-μl aliquot of sonicated cell suspension is mixed with 100 μl $^{109}CdCl_2$ solution (1.0 μCi/2.0 μg Cd/ml) in a 400-μl polyethylene micro-

[7] A. F. Stein, W. M. Bracken, and C. D. Klaassen, *Toxicol. Appl. Pharmacol*, **84**, 276 (1987).

[8] D. L. Eaton and B. F. Toal, *Toxicol. Appl. Pharmacol.* **66**, 134 (1982).

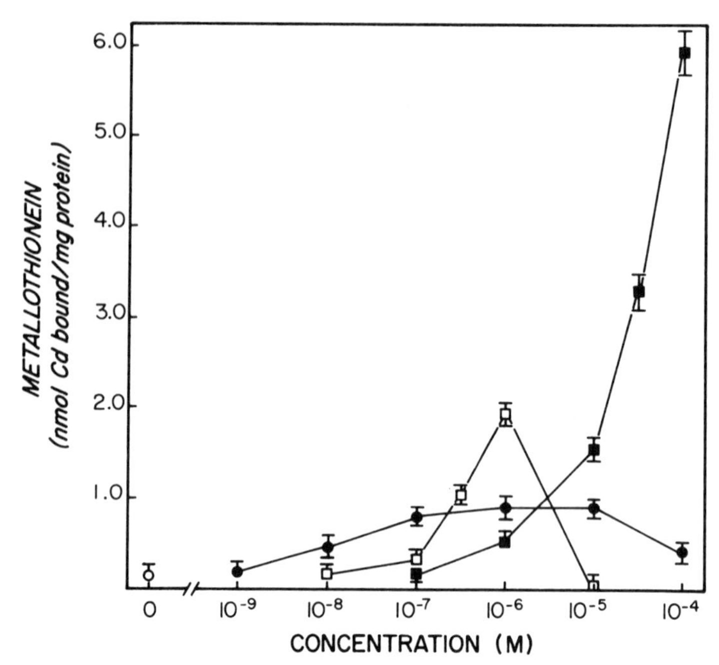

FIG. 1. Metallothionein content of hepatocyte cultures following a 24-hr zinc (■), cadmium (□), or dexamethasone (Dex; ●) treatment. Each value represents the mean ± SE for four experiments. (○), Control. Reproduced from Bracken and Klaassen[4] by permission of the publisher.

centrifuge tube. After adding 50 μl of 2% (v/v) hemoglobin, the tube is mixed well and heated at 100° for 1 min. The tube is put into ice water to cool and centrifuged (10,000 *g*, 3 min) using a microfuge E (Beckman Instruments, Inc., Palo Alto, CA). The sample is then mixed with another 50 μl of 2% hemoglobin, heated, and centrifuged again. An aliquot of 100 μl of supernatant is removed and quantitated for ^{109}Cd. A tube without sample is used as sample blank, and a tube without hemoglobin is used as total blank. All the solutions are prepared in 10 m*M* Tris-HCl buffer, pH 7.4. The following formula is used for calculation of total MT: micrograms MT per milliliter cell suspension = (cpm in sample − cpm in blank) × 17.8/(cpm in total − cpm in blank). Although a number of assays have been used (*e.g.*, Sephadex G-75 gel filtration, ^{203}Hg binding, [^{35}S]cysteine incorporation, and radioimmunoassay), the cadmium-hemoglobin assay is a convenient and accurate assay for MT quantitation.[9]

[9] H. H. Dieter, L. Muller, J. Abel, and H. Summer, *Toxicol. Appl. Pharmacol.* **85**, 380 (1985).

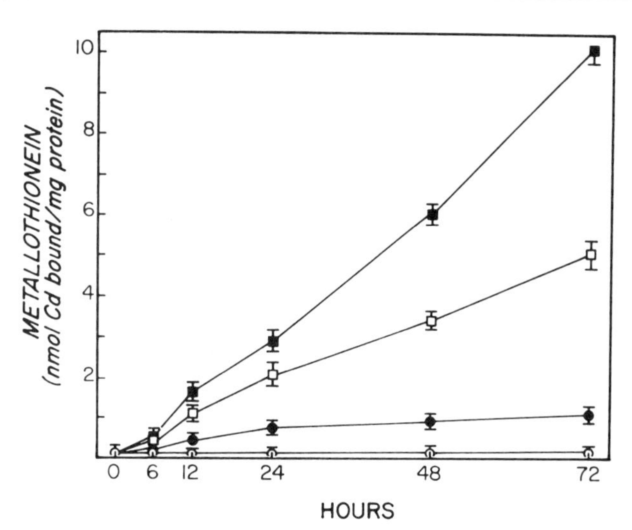

FIG. 2. Time course for MT induction by zinc (30 μM; ■), cadmium (1 μM; □), or dexamethasone (1.0 μM; ●). Hepatocytes were exposed to the inducers for up to 72 hr. Each value represents the mean ± SE for four determinations. (○), Control. Reproduced from Bracken and Klaassen[4] by permission of the publisher.

Quantitation of Total Cellular Protein

A 50-μl aliquot of sonicated cell suspension is used for total protein quantitation by the method of Bradford,[10] using bovine serum albumin as standard. Protein concentration per milliliter is used to normalize the MT data (μg MT/mg protein). One may also use total cellular DNA for this purpose.

Evaluation of Cytotoxicity

Cytotoxicity can be assessed by measuring intracellular potassium content.[5] Sonicated cell suspensions are centrifuged (10,000 *g*, 3 min) and a 100-μl aliquot of the supernatant is added to 10 ml of 15 mEq lithium, and potassium content of the sample can be quantitated by flame photometry. One may also use extracellular activities of lactate dehydrogenase and aspartate aminotransferase for this purpose.

Precautions and Comments

1. Cell yield and cell viability vary from different lots of collagenase. If the viability of isolated hepatocytes is less than 85%, purification with

[10] M. M. Bradford, *Anal. Biochem.* **72,** 248 (1976).

Percoll is recommended.[11] Isotonic Percoll is mixed with an equal volume of final cell suspension ($5-10 \times 10^6$ cells/ml). After centrifugation at 70 *g* for 5 min, damaged cells will separate to the top. The viability of purified cells is usually 95%.

2. The matrix for hepatocyte attachment is necessary. In addition to commonly used collagen, Matrigel, fibronectin, and specially modified surface plastic culture dishes are all suitable for this purpose.

3. All the surgical and cell culture procedures and treatments should be performed under sterile conditions.

4. The concentration of inducers added should not retard cell growth, inhibit cellular protein synthesis, or cause cell death. In this regard, intracellular potassium content is a sensitive index for cytotoxicity, and used to monitor the appropriate concentration of inducer.[5]

5. The components of the culture medium may play an important role for some MT inducers. For example, IL-6 induction of MT in cultured hepatocytes is dependent on the media containing glucocorticoid hormones and zinc.[12]

6. *In vitro* studies should be complemented by *in vivo* studies. Many inducers of MT, such as cadmium, zinc, and dexamethazone, increase MT both *in vivo* and *in vitro.* However, some inducers of MT, such as ethanol and urethane, appear to induce MT synthesis only in intact animals.[5] Therefore, negative results of MT induction *in vitro* do not exclude the possibility that a compound can induce MT *in vivo.*

The induction of MT by common inducers in rat primary hepatocyte cultures is reproducible.

Acknowledgment

Supported by USPHS Grant ES-01142. The authors acknowledge the suggestions of W. M. Bracken of Abbott Laboratories.

[11] B. L. Kreamer, J. L. Staecker, N. Sawada, G. L. Sattler, M. T. S. Hsia, and H. C. Pitot, *In Vitro Cell. Dev. Biol.* **22,** 201 (1986).

[12] J. J. Schroeder and R. J. Cousins, *Proc. Natl. Acad. Sci. U.S.A.* **87,** 3137 (1990).

[65] Metallothionein and Zinc Metabolism in Hepatocytes

By JOSEPH J. SCHROEDER and ROBERT J. COUSINS

Introduction

Monolayer cultures of rat liver parenchymal cells provide a novel and convenient system to study hepatic zinc metabolism at the cellular level. Preparations of fresh cells show similar glycolysis, gluconeogenesis, urea synthesis, ketogenesis, and cholesterol and fatty acid synthesis to the liver *in vivo*.[1] The phenotypic stability of these cells and the changes that occur during culture have been described.[2]

Kinetic analyses of zinc uptake by isolated rat liver parenchymal cells define two intracellular zinc pools.[3] One pool is small and binds zinc relatively weakly. The other pool accounts for the bulk of cellular zinc and binds zinc more tightly. Although the smaller, labile-zinc pool accounts for net zinc accumulation, both intracellular pools respond to hormonal stimuli.

Zinc metabolism in hepatocytes is intimately linked to metallothionein. Not only does metallothionein bind zinc taken up from culture medium, but metallothionein expression is regulated by the zinc content of the medium.[4-6] In addition, glucocorticoids and stress-related mediators that augment zinc uptake and accumulation in hepatocytes also stimulate metallothionein expression.[5,6] Changes in zinc flux into intracellular pools are directly related to the metallothionein content of hepatocytes.[6,7] Metallothionein may provide zinc for metalloenzymes or for "zinc finger" motifs of DNA-binding transcription factors or may act to buffer intracellular zinc concentrations for these or other functions.[7-9] Metallothionein may also play a cytoprotective role as a radical scavenger[10] or as a zinc donor for membrane stabilization.[11]

[1] N. W. Cornell, *Fed. Proc.* **44,** 2446 (1985).
[2] D. M. Bissell and P. S. Guzelian, *Ann. N.Y. Acad. Sci.* **349,** 85 (1980).
[3] S. E. Pattison and R. J. Cousins, *Am. J. Physiol.* **250,** E677 (1986).
[4] M.L. Failla and R. J. Cousins, *Biochim. Biophys. Acta* **538,** 435 (1978).
[5] M. L. Failla and R. J. Cousins, *Biochim. Biophys. Acta* **543,** 293 (1978).
[6] J. J. Schroeder and R. J. Cousins, *Proc. Natl. Acad. Sci. U.S.A.* **87,** 3137 (1990).
[7] M. A. Dunn, T. L. Blalock, and R. J. Cousins, *Proc. Soc. Exp. Biol. Med.* **185,** 107 (1987).
[8] R. J. Cousins and A. S. Leinart, *FASEB J.* **2,** 2884 (1988).
[9] J. D. Otvos, D. H. Petering, and C. F. Shaw, *Comments Inorg. Chem.* **9,** 1 (1989).
[10] P. J. Thornalley and M. Vašák, *Biochim. Biophys. Acta* **827,** 36 (1985).
[11] J. P. Thomas, G. J. Bachowski, and A. W. Girotti, *Biochim. Biophys. Acta* **884,** 448 (1986).

This chapter will describe methodology for isolating and maintaining primary hepatocyte monolayers for metallothionein and zinc metabolism studies. Attention will be focused on quantitating zinc concentration and kinetics, metallothionein expression, and zinc metalloenzyme activity in these cells.

Methods

Isolation and Maintenance of Rat Hepatocytes

Parenchymal cells can be prepared from livers of a variety of marine, avian, rodent, livestock, and primate species.[1] We routinely isolate cells from male Sprague-Dawley strain rats weighing 150–250 g. The animals are housed in stainless steel, suspended cages in a room with a 12-hr light–dark cycle (0700 to 1900 hr and 1900 to 0700 hr, respectively) and are fed a standard commercial diet (Purina rodent chow; Ralston Purina, St. Louis, MO) and distilled water *ad libitum.*

Parenchymal cells are isolated by a modification of the collagenase perfusion technique, which can be completed in 90 min.[12] All buffers used in the technique are perfused at 37°. Rats are anesthetized by intraperitoneal injection of sodium pentobarbital (65 mg/kg body weight) and surgery is initiated between 0900 and 1000 hr to reduce variability between cell preparations due to circadian rhythms. Following cannulation of the portal vein, cell wash buffer [0.14 *M* NaCl, 7 m*M* KCl, 5 m*M* glucose, 10 m*M* *N*-2-hydroxyethylpiperazine-N′2-ethane-sulfonic acid (HEPES), pH 7.4] supplemented with 0.5 m*M* ethyleneglycol-bis (β-aminoethylether)-N, N′-tetraacetic acid (EGTA) is perfused through the liver for 30 sec at a rate of 10 ml/min. Simultaneously, the subhepatic vena cava is severed distal to a ligation. The thoracic vena cava is then severed and the perfusion rate increased to 40 ml/min for 4 min. During this period, the liver is excised and placed on nylon screening. Subsequently, residual EGTA is washed from the liver by perfusion with 50 ml of cell wash buffer. Then intercellular junctions are dissolved by recirculating 60 ml of enzyme perfusion buffer (65 m*M* NaCl, 7 m*M* KCl, 5 m*M* $CaCl_2$, 5 m*M* glucose, 100 units collagenase/ml, 0.1 *M* HEPES, pH 7.6) through the liver at a rate of 40 ml/min for 15 min.

A crude cell suspension is obtained by gently raking the dispersed liver with a plastic spatula in ice-cold modified Waymouth's MB 752/1 medium (pH 7.4) containing 15 m*M* HEPES, 10 m*M* Tris(hydroxymethyl)methyl-2-aminoethanesulfonic acid (TES), 30 m*M* $NaHCO_3$, 0.5 m*M* sodium

[12] M. N. Berry and D. S. Friend, *J. Cell Biol.* **43,** 506 (1969).

pyruvate, streptomycin sulfate (0.1 mg/ml), penicillin G (100 units/ml), gentamicin sulfate (50 μg/ml), alanine (0.41 μ*M*), and serine (0.53 μ*M*). The cell suspension is filtered through cheesecloth and nylon mesh to remove aggregated and undispersed cells. Cells are pelleted by centrifugation (100 *g*, 4 min, 4°), resuspended in fresh Waymouth's medium, and pelleted again. This pelleting procedure is repeated two more times (50 *g*, 3 min, 4°). The parenchymal cells are counted with a hemocytometer and viability is assessed by Trypan Blue exclusion. The isolation procedure routinely yields 3×10^8 cells from one liver with viability greater than 85%.

Parenchymal cells are purified by a selective attachment procedure.[13] Cells are suspended (10^6 cells/ml) in modified Waymouth's medium supplemented with fetal bovine serum [FBS, 10%(v/v)] and insulin (1 μg/ml).[3,4] Aliquots (3 ml) are inoculated into 60-mm polystyrene culture dishes, previously coated with collagen. Smaller dishes of 35 mm (in trays) or larger dishes (100 or 150 mm) can be used with proportional adjustments in volumes of medium and preparative materials. The dishes are incubated at 37° in a humidified 95% air/5% CO_2 atmosphere, for a 3-hr selective attachment period. Unattached cells and debris are washed free with a 3-ml aliquot of cell wash buffer and are removed by aspiration.

Attached cells are cultured with fresh, modified Waymouth's MB 752/1 medium supplemented with bovine serum albumin (BSA 2 mg/ml) and appropriate treatments.[3] BSA is included in the medium because it is the major plasma zinc-binding protein,[14] it functions as a zinc transport protein in portal plasma,[15] and it produces saturable zinc uptake kinetics in hepatocytes.[3] For experimental culture periods lasting longer than 24 hr, culture media are renewed every 24 hr.

After culture periods, cells are washed twice with 3-ml aliquots of cell wash buffer. Then an aliquot of the appropriate assay or extraction buffer is added. The cells are dislodged from culture dishes with a rubber policeman and transferred to a test tube with a Pasteur pipette. Protein concentrations are determined by the method of Lowry *et al.*,[16] using BSA as a standard.

Collagen Coating of Culture Dishes

Originally it was believed that culture dishes must be collagenized with rat tail collagen.[2] Subsequent studies have shown that calf skin collagen can

[13] B. A. Laishes and G. M. Williams, *In Vitro* **12,** 521 (1976).
[14] R. J. Cousins, *Physiol. Rev.* **65,** 238 (1985).
[15] K. T. Smith, M. L. Failla, and R. J. Cousins, *Biochem. J.* **184,** 627 (1979).
[16] O. H. Lowry, N. J. Rosebrough, A. L. Farr, and R. J. Randall, *J. Biol. Chem.* **193,** 265 (1951).

be used.[6] We prepare a collagen stock solution by adding 200 mg of calf skin type III collagen (Sigma, St. Louis, MO) to 50 ml of 0.5% (v/v) acetic acid. The collagen is dissolved by stirring for 24 hr at 4°. This solution can be stored at 4°. An aliquot of the collagen stock is diluted to 20 mg/liter with sterile, distilled H_2O. Then 3 ml of diluted collagen is added to each 60-mm dish so that the bottom is covered. The next day the solution is removed by aspiration and the dishes are allowed to dry before they are used.

Sera and Media Zinc

The extracellular zinc concentration is an important variable in zinc metabolism studies and must be monitored by atomic absorption spectrophotometry (AAS). Endogenous levels of zinc in culture media and sera vary. The zinc concentration of Waymouth's and L-15 media range from 0.7 to 1.2 μM and that of FBS used in the attachment medium is often as high as 45 μM. Higher levels of zinc are attained by adding a soluble zinc salt. We use a 0.01 M $ZnSO_4$ stock, which can be sterilized by filtration [0.20-μm Millipore (Bedford, MA) filter] and stored at 4°.

Zinc can be extracted from media and sera via ion exchange using Chelex 100 resin (Bio-Rad Laboratories, Richmond, CA).[17] The chelating resin is prepared by adding 200 g to 300 ml NH_4OH and mixing for 1 hr. Then NH_4OH is removed by filtration (#1; Whatman, Milford, NJ) and the resin is washed sequentially with 1.5 liters of distilled water, 500 ml of 0.1 N HCl, and 1 liter of distilled water. The filter-dried resin is mixed with serum or medium in a ratio of 1:4 (w/v). After mixing, the resin is removed by centrifugation (2000 g, 10 min, 4°), the pH is adjusted to 7.4, and the serum or medium is sterilized by filtration (0.45-μm Millipore filter). When zinc is extracted from culture medium, cations such as calcium and magnesium must be replenished to ensure cell attachment to dishes and to prevent confounding nutrient deficiencies. Mineral levels in untreated serum and medium, chelated serum and medium, and calcium- and magnesium-repleted medium are shown in Table I.

Hormone Additions to Medium

Glucocorticoid hormones are important mediators of zinc metabolism in hepatocytes.[3-6] The synthetic glucocorticoid dexamethasone maximally induces metallothionein expression and zinc accumulation at a concentration of 0.1 μM.[5] We routinely use a fresh 1 mg/ml dexamethasone sodium

[17] A. Flynn, *Nutr. Res.* **5,** 487 (1985).

TABLE I
COMPOSITION OF CULTURE MEDIA AND SERA[a]

Medium or serum	Zinc (μM)	Magnesium (mM)	Calcium (mM)
Fetal bovine serum	43.0	—	—
Chelated fetal bovine serum[b]	4.0	—	—
Waymouth's medium	0.7	2.9	0.8
Chelated Waymouth's medium[b]	0.3	0.3	0.2
Repleted Waymouth's medium[c]	0.3	2.3	0.8

[a] Reproduced from Schroeder and Cousins[17a] with kind permission of the American Institute of Nutrition.
[b] Serum and medium were treated with a chelating resin to remove divalent cations as described in Methods.
[c] Medium was treated with a chelating resin and then magnesium and calcium were restored to their initial levels by adding $CaCl_2$ and $MgCl_2$, respectively.

phosphate stock. This form of dexamethasone is soluble in culture medium.

A variety of stress-related mediators affect hepatic metallothionein expression and zinc metabolism *in vivo*.[7,8] False negative results may be obtained *in vitro* if polypeptide hormones are applied to hepatocyte cultures immediately after attachment, because receptors may be damaged during the isolation procedure.[18] Usually this problem can be circumvented by maintaining the cells in BSA-supplemented medium overnight before adding medium containing polypeptide hormones. However, in some cases damaged receptors may not be replaced.

Cell Zinc Concentration and Kinetics

For measurement of cellular zinc content, hepatocytes from one or two dishes are washed twice with ice-cold ethylenediaminetetraacetic acid (EDTA) buffer (10 mM EDTA, 150 mM NaCl, 10 mM HEPES, pH 7.4) prior to two washes with ice-cold cell wash buffer to remove nonspecifically bound zinc. Then hepatocytes are harvested in 1 ml of 0.2% sodium dodecyl sulfate in 0.2 N NaOH. This solution can be stored for up to 1 week and should be kept in a plastic bottle, because it leaches zinc from glass. Hepatocytes are allowed to solubilize overnight at room temperature in covered tubes. Zinc is measured by atomic absorption spectrophotometry.

[17a] J. J. Schroeder and R. J. Cousins, *J. Nutr.* **121,** 844 (1991).
[18] A. Ichihara, T. Nakamura, K. Tanaka, Y. Tomita, K. Aoyama, S. Kato, and H. Shinno, *Ann. N.Y. Acad. Sci.* **349,** 77 (1980).

Hepatocyte monolayer cultures are advantageous for kinetic studies because they lack many of the diffusional barriers associated with intact liver and liver slice systems. Medium containing ^{65}Zn used to follow zinc fluxes can be rapidly removed from hepatocyte monolayers by aspiration. This allows reproducible pulse times on the order of 10 sec. For zinc uptake and efflux experiments we routinely use 10–200 nCi ^{65}Zn/dish. The hepatocytes are harvested similar to those used for total cellular zinc measurements, using EDTA buffer to remove nonspecifically bound ^{65}Zn. The remaining ^{65}Zn is measured by liquid scintillation.

Figure 1 shows the total cellular zinc concentration and ^{65}Zn uptake exchange of hepatocytes over time. Total cellular zinc does not increase dramatically over 20 hr of culture with L-15 medium containing 16 μM zinc. In comparison, a major uptake of ^{65}Zn is observed.

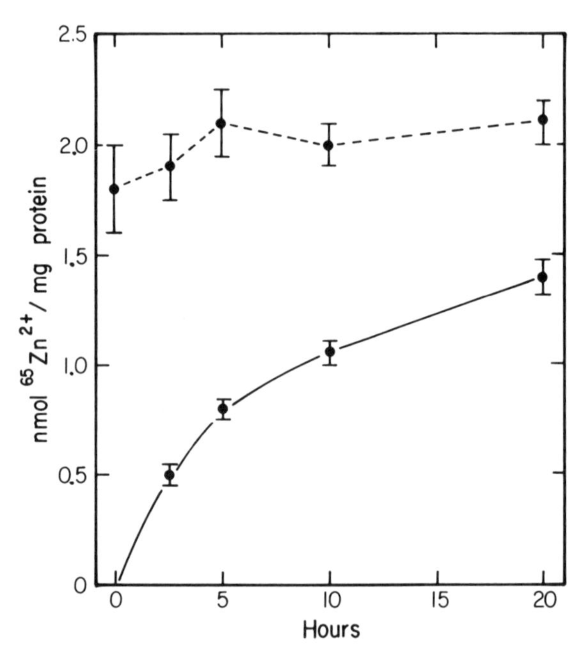

FIG. 1. Time course for zinc uptake by hepatocytes. Hepatocytes were incubated in L-15 medium [2 mg/ml bovine serum albumin (BSA), 16 μM Zn^{2+}] for up to 20 hr. Total cellular zinc (dashed line) was measured by atomic absorption spectrophotometry. $^{65}Zn^{2+}$ uptake (solid line) was determined by incubation in L-15 medium, which contained ^{65}Zn added at 0 hr. Each point represents the mean ± SEM from a representative experiment that was repeated with three separate hepatocyte preparations. (Reproduced from Pattison and Cousins[3] with kind permission of The American Physiological Society.)

Zinc Metalloenzymes

Most assays for zinc metalloenzymes can be scaled down to measure activity in hepatocytes. Often cells can be harvested from one or two dishes directly in the assay buffer. Since these samples are relatively dilute, precautions should be taken to prevent degradation.

Metallothionein mRNA

To measure levels of metallothionein (MT) mRNA, total RNA is extracted from hepatocytes using the method of Chomczynski and Sacchi.[19] Cells from three to five dishes are harvested in 1 ml of guanidinium isothiocyanate, protein is removed via extraction using a phenol:chloroform:isoamyl alcohol mixture, and RNA is precipitated with ethanol. RNA is dissolved in sterile, distilled, deionized water and the concentration of each sample is calculated using A_{260}.

We have conducted dot-blot and Northern blot hybridization analyses using 60-mer oligonucleotide probes specific for *MT-1* and *MT-2* genes and corresponding to bases 16–76 from the 5′ terminus.[20] The probes are 5′-end labeled with [γ-^{32}P]ATP (DuPont/NEN Research Products, Boston, MA) using T4 polynucleotide kinase (Bethesda Research Laboratories, Gaithersburg, MD) and purified by chromatography (Sephadex G-50, Sigma) before hybridization. The specific activity of each probe is routinely 3.0 μCi/pmol, as measured by Cerenkov counting.

Nitrocellulose or nylon membranes are used for both Northern blots and dot blots. For Northern blots, total RNA is electrophoresed in a 1.1% agarose gel and then transferred to the membrane. Dot-blot analyses are used to quantitate metallothionein mRNA. After hybridization, the ^{32}P content of each dot is measured by liquid scintillation. Then molecules of metallothionein mRNA per cell are calculated using an RNA/DNA ratio of 4.0, 6.4 pg of DNA/cell, and an assumed hybridization efficiency of 100%.[21]

Northern blot analysis of total RNA from hepatocytes provides strong qualitative evidence regarding the effects of mediators on metallothionein gene expression. The autoradiogram in Fig. 2 shows that increasing the level of zinc in the culture medium from 1 to 16 μM increases metallothionein mRNA. At each level of zinc, the addition of interleukin 6 alone

[19] P. Chomczynski and N. Sacchi, *Anal. Biochem.* **162,** 156 (1987).

[20] R. D. Anderson, S. J. Taplitz, B. W. Birren, G. Bristol, and H. R. Herschman, *Experientia, Suppl.* **52,** 373 (1987).

[21] R. J. Cousins, M. A. Dunn, A. S. Leinart, K. C. Yedinak, and R. A. DiSilvestro, *Am. J. Physiol.* **251,** E688 (1986).

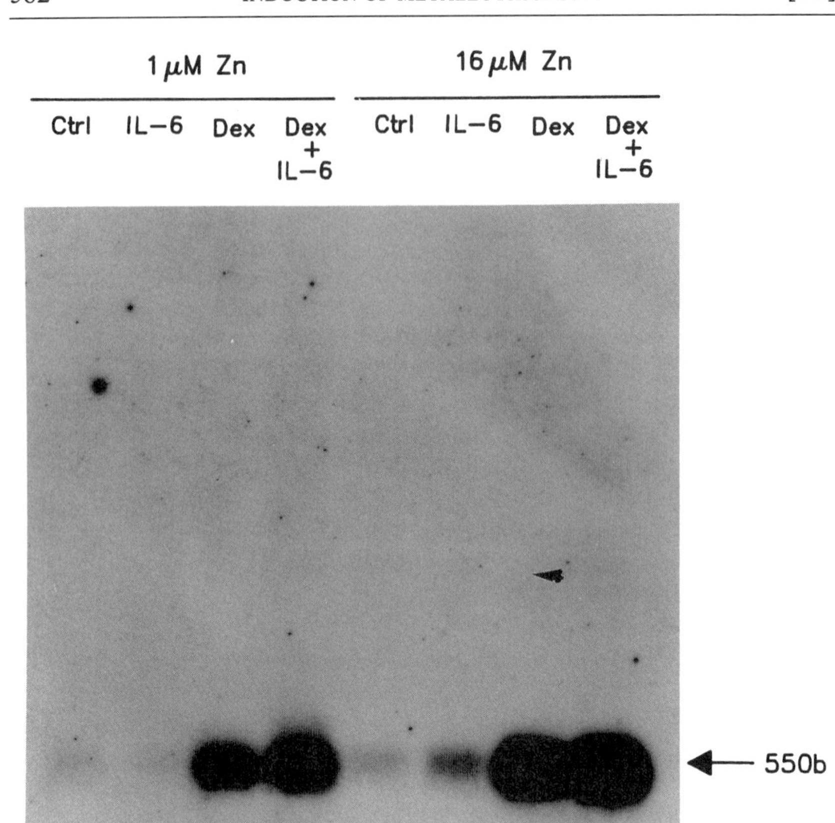

FIG. 2. Northern blot illustrating effects of zinc, dexamethasone, and interleukin 6 combinations on MT mRNA in rat hepatocyte monolayer cultures. Hepatocyte cultures (24 hr) were incubated with Waymouth's medium containing BSA (2 mg/ml) and the treatment combinations indicated. Concentrations of dexamethasone and interleukin 6 were 1 μM and 10 hepatocyte-stimulating factor units/ml, respectively. After 24 hr the hepatocytes were harvested and total RNA was extracted for Northern blot analysis. The 550-b band represents MT mRNA. (Reproduced from Schroeder and Cousins[6] with kind permission of the National Academy of Sciences.)

has little or no effect, while the addition of dexamethasone dramatically increases metallothionein mRNA. However, the addition of interleukin 6 and dexamethasone together increases metallothionein mRNA above that of the corresponding dexamethasone control. The dexamethasone dependency for interleukin 6 regulation of metallothionein expression in hepatocyte monolayer cultures may reflect a need for a basal level of glucocorti-

TABLE II
DEPENDENCE OF INTERLEUKIN 6 STIMULATION OF METALLOTHIONEIN EXPRESSION AND CELLULAR ZINC ACCUMULATION ON LEVELS OF EXTRACELLULAR ZINC AND GLUCOCORTICOIDS[a]

Treatments	MT mRNA (molecules/cell)	MT (ng/mg protein)	Cell Zn (nmol/mg protein)
1 μM Zn	60[a]	66[a]	1.8[ab]
+IL-6	57[a]	69[a]	1.7[a]
+Dex	275[b]	305[a]	2.2[c]
+Dex + IL-6	495[c]	1415[b]	2.6[d]
16 μM Zn	115[ab]	179[a]	2.0[bc]
+IL-6	115[ab]	202[a]	1.9[ab]
+Dex	576[c]	933[b]	3.4[e]
+Dex + IL-6	1289[d]	3964[c]	5.3[f]
Pooled SEM:	54	196	0.1
n:	3	3	4

[a] Hepatocyte cultures were incubated for 24 hr in Waymouth's medium containing BSA (2 mg/ml) as indicated. Concentrations of dexamethasone and interleukin 6 were 1 μM and 10 U HSF (hepatocyte-stimulating factor)/ml, respectively. After 24 hr the hepatocytes were harvested and metallothionein mRNA (MT mRNA), metallothionein (MT) protein, and cell zinc concentrations were measured. Each value represents the mean ± SEM ($n = 3$ or 4). Values with differing superscript letters are significantly different ($p \leq 0.05$). (Reproduced from Schroeder and Cousins[6] with kind permission of the National Academy of Sciences.)

coids, which normally bathe the liver *in vivo,* to facilitate expression of some liver functions.

Analysis of dot-blot or Northern blot autoradiograms by densitometry or visually does not provide an accurate estimation of metallothionein mRNA levels because of technical limitations associated with autoradiography. Alternatively, a more accurate estimation can be achieved by measuring ^{32}P in dot blots via liquid scintillation and expressing metallothionein mRNA levels as molecules per cell. This is illustrated by comparing metallothionein mRNA levels in Fig. 2 with values quantitated from dot-blot analysis of the same RNA shown in Table II. The small apparent increase over dexamethasone controls induced by interleukin 6 and dexamethasone together, visible in the autoradiogram (Fig. 2), actually represents a doubling in relative levels of metallothionein mRNA (Table II).

Metallothionein

Metallothionein in hepatocytes can be quantitated by a variety of methods described in this volume. We have employed a modification of the cadmium-binding assay,[22] which uses a high specific activity cadmium solution (0.2 μg Cd and 0.5 μCi ^{109}Cd/ml of 10 mM Tris-HCl buffer, pH 7.4). Cells are harvested in 10 mM Tris buffer (pH 7.4) and homogenized using a Polytron (Brinkmann, Westbury, NY) with a P-10 generator. Hepatocytes from two or three dishes are required to measure basal metallothionein levels, while one dish usually suffices to measure the metallothionein concentration in induced cells.

Metallothionein protein and cellular zinc concentrations measured by methods described in this chapter are shown in Table II. The effects of glucocorticoids and cytokines on metallothionein mRNA are reflected in metallothionein and cell zinc. Assuming each molecule of metallothionein binds seven atoms of zinc, then the increases in cellular zinc in hepatocytes cultured with added zinc or dexamethasone are accounted for by the increases in cellular metallothionein.[6] In contrast, the increases in cellular zinc in hepatocytes cultured with dexamethasone and interleukin 6 together are less than expected when compared to the corresponding increases in metallothionein. Therefore, the addition of dexamethasone and interleukin 6 together may cause a change in the intracellular zinc distribution such that the portion of zinc not associated with metallothionein is reduced.

Acknowledgments

Methods and data presented in this chapter were supported by NIH Grant DK 31651, USDA Fellowship (J.J.S.), and Boston Family Funds.

[22] D. L. Eaton and B. F. Toal, *Toxicol. Appl. Pharmacol.* **66**, 134 (1982).

[66] Metallothionein and Copper Metabolism in Liver

By IAN BREMNER

Introduction

Porter[1] and co-workers were first to demonstrate that metallothionein (MT) is a major copper-binding protein in liver when they reported that the insoluble copper protein, neonatal hepatic mitochondrocuprein, yielded on sulfitolysis a polypeptide resembling MT in molecular weight and amino acid composition. Since then, copper-containing forms of MT

[1] H. Porter, *Biochem. Biophys. Res. Commun.* **56**, 661 (1974).

have been isolated and characterized from the hepatic cytosol of the pig[2] and fetal calf.[3] In addition, aggregated forms of copper-MT have been obtained from the particulate fractions of the livers from the human fetus,[4] Bedlington terriers,[5,6] and pigs[7] by extraction with alkaline or mercaptoethanol-containing solutions. A high proportion of the hepatic copper may be bound to those cytosolic and particulate forms of MT, indicating that it may have a storage or detoxification function. Thus 80% of the hepatic copper is bound to MT in copper-loaded pigs, provided they are of normal zinc status.[7] However, in zinc-deficient pigs and in copper-loaded rats and sheep little of the hepatic copper is associated with MT,[7] suggesting that the accumulation of copper-MT is controlled by factors other than liver copper content.

Induction of Metallothionein Synthesis

Nevertheless the fact that liver MT concentrations do often increase in line with liver copper concentrations suggests that copper is able to induce MT synthesis. This was confirmed by the isolation and characterization of copper-MT from the liver of copper-injected rats.[8,9] Since synthesis was accompanied by increased incorporation of [^{35}S]cysteine into the protein[10] and increased production of MT mRNA,[11,12] inducibility of gene expression appears to occur at the level of transcription. Although it was suspected at one stage that zinc and not copper was the primary inducer of MT synthesis in copper-injected rats, this is now known not to be the case, as MT mRNA levels increase prior to the associated increase in hepatic uptake of zinc.[11] Indeed it appears that the increase in liver zinc content and in zinc binding to MT in copper-injected rats may be an inflammatory response.[13] Copper is less effective than zinc at increasing MT mRNA

[2] I. Bremner and B. W. Young, *Biochem. J.* **155,** 631 (1976).

[3] K. Munger, U. A. Germann, M. Beltramini, D. Niedermann, G. Baitella-Eberle, J. H. R. Kägi, and K. Lerch, *J. Biol. Chem.* **260,** 10032 (1985).

[4] J. R. Riordan and V. Richards, *J. Biol. Chem.* **255,** 5380 (1980).

[5] G. F. Johnson, A. G. Morell, R. J. Stockert, and I. Sternlieb, *Hepatology* **1,** 243 (1981).

[6] K. Lerch, G. F. Johnson, P. S. Grushoff, and I. Sternlieb, *Arch. Biochem. Biophys.* **243,** 108 (1985).

[7] R. K. Mehra and I. Bremner, *Biochem. J.* **219,** 539 (1984).

[8] I. Bremner and B. W. Young, *Biochem. J.* **157,** 517 (1976).

[9] D. R. Winge, B. L. Geller, and J. Garvey, *Arch. Biochem. Biophys.* **208,** 160 (1981).

[10] I. Bremner, W. G. Hoekstra, N. T. Davies, and B. W. Young, *Biochem. J.* **174,** 883 (1978).

[11] S. A. Wake and J. F. B. Mercer, *Biochem. J.* **228,** 425 (1985).

[12] D. M. Durnam and R. D. Palmiter, *J. Biol. Chem.* **256,** 5712 (1981).

[13] J. C. Fleet, K. A. Golemboski, R. R. Dietert, G. K. Andrews, and C. C. McCormick, *Am. J. Physiol.* **258,** G926 (1990).

levels in mouse liver at low dose levels but there are no differences in efficacy at higher dose levels.[12,14] The induction of hepatic *MT* gene expression is therefore not linear with copper dose in the mouse.

The relative ineffectiveness of low doses of copper in inducing hepatic *MT* gene transcription may partially explain why dietary copper supplementation of rats does not stimulate copper MT synthesis in the liver.[15,16] Only when liver copper concentrations exceed a high threshold value do copper MT levels increase. The fact that only a small proportion of the copper in the liver of copper-loaded sheep is associated with MT[7] is consistent with the report that copper is a poor inducer of *MT* genes in sheep.[17] Transcriptional regulation of MT synthesis by copper appears to be determined by cis-acting DNA regulatory sequences and by trans-acting binding factors.[18] Specific DNA sequences in the 5′-flanking region are responsible for metal responsiveness and are termed the metal regulatory elements. Copper, like cadmium and zinc, probably binds to the trans-acting proteins and so facilitates their binding to the metal regulatory elements in the promoter region of the *MT* gene. Although copper is less effective than zinc in inducing *MT* gene transcription, the turnover of the mRNA in copper-treated hepatoblastoma cells is slower than that in zinc-treated cells.[19] Concentrations of mRNA in copper-treated cells remained at an elevated plateau level for 15 hr, whereas they declined after 6 hr in zinc-treated cells.

Because copper is a poor inducer of liver MT synthesis, the binding of copper to MT may sometimes depend on prior synthesis of the protein by other agents, such as zinc, glucocorticoids, or "stress" factors. Binding of hepatic copper to MT in the rat is greatly enhanced if zinc is administered before or after copper loading.[20] There are also marked differences between species in their ability to incorporate liver copper into MT, with the dog and pig being particularly effective in this regard.[5,7] It is interesting that these species also show abnormalities in the copper-binding site on albumin, which could affect the kinetics of hepatic copper uptake.

Turnover of Copper-Metallothionein

Although excessive copper loading leads to the accumulation of aggregated forms of copper MT in lysosomes and other organelles in the liver,

[14] V. C. Culotta and D. H. Hamer, *Mol. Cell. Biol.* **9,** 1376 (1989).
[15] I. Bremner, R. K. Mehra, J. N. Morrison, and A. M. Wood, *Biochem. J.* **235,** 735 (1986).
[16] T. L. Blalock, M. A. Dunn, and R. J. Cousins, *J. Nutr.* **118,** 222 (1988).
[17] M. G. Peterson and J. F. Mercer, *Eur. J. Biochem.* **174,** 425 (1974).
[18] D. H. Hamer, *Annu. Rev. Biochem.* **55,** 913 (1986).
[19] C. Sadhu and L. Gedamu, *Mol. Cell. Biol.* **9,** 5738 (1989).
[20] F. A. Day, M. Panemangalore, and F. O. Brady, *Proc. Soc. Exp. Biol. Med.* **168,** 306 (1981).

the binding of injected radioactive copper tracers to hepatic MT is rapid and transient. In rats injected with ^{64}Cu or ^{67}Cu, the isotope rapidly associates with cytosolic MT but is transferred within a few hours to superoxide dismutase and other hepatic proteins, as though MT is involved in hepatic uptake and intracellular distribution mechanisms.[21] However, if pharmacological amounts of copper are injected, the binding of copper to liver MT does not occur until after induction of MT synthesis. The clearance of copper from the MT pool is also much slower, because the half-life of copper-induced MT in rat liver, as measured by pulse-labeling techniques, is about 17 hr in rats of normal zinc status and 12 hr in those that are zinc deficient.[10] This is similar to the half-life of zinc-MT and very much less than that of cadmium-MT and of other cytosolic proteins. The half-life of copper-MT in cultured Chang liver cells is also very much less than that of zinc- or cadmium-MT.[22] Unfortunately, no studies have been made of the turnover of copper-MT under steady state conditions *in vivo.*

The disappearance of copper-MT from the liver probably results mainly from degradation of the protein, although some secretion into plasma and bile may also occur.[15,23] It has commonly been assumed that such degradation occurs within lysosomes, because this organelle plays a general role in protein degradation. However, it now appears that degradation of proteins with a half-life of less than 50 hr is usually achieved by cytoplasmic and not lysosomal enzymes.[24] Nevertheless, Chen and Failla[25] obtained evidence using chloroquine and a neutral protease inhibitor that degradation of zinc-MT in cultured hepatocytes occurs within both lysosomal and nonlysosomal compartments. Unlike zinc- and cadmium-MT, copper-MT is resistant to degradation with lysosomal enzymes *in vitro,*[26] which could explain the accumulation of copper-MT in the lysosomes of copper-loaded livers from animals such as Bedlington terriers[5] but is difficult to reconcile with the short half-life of copper-induced MT *in vivo.*[10]

The rapid turnover of cytosolic copper-MT could involve selective removal of the associated zinc from the protein, which causes major conformational changes and increased susceptibility to neutral proteases *in vitro.*[27] Metal removal could result from chelation with endogenous ligands or reaction with oxygen free radicals.[28]

[21] N. Marceau and N. Aspin, *Biochem. Biophys. Acta* **293,** 338 (1973).
[22] S. Kobayashi, M. Imano, and M. Kimura, *Chem.-Biol. Interact.* **52,** 319 (1985).
[23] M. Sato and I. Bremner, *Biochem. J.* **223,** 475 (1984).
[24] S. W. Rogers and M. Rechsteiner, *J. Biol. Chem.* **263,** 19843 (1988).
[25] M. L. Chen and M. L. Failla, *Proc. Soc. Exp. Biol. Med.* **191,** 130 (1989).
[26] R. K. Mehra and I. Bremner, *Biochem. J.* **227,** 903 (1985).
[27] K. B. Nielson and D. R. Winge, *J. Biol. Chem.* **259,** 4941 (1984).
[28] J. R. Arthur, I. Bremner, P. C. Morrice, and C. F. Mills, *Free Radical Res. Commun.* **4,** 15 (1987).

The disappearance of copper-MT from the liver is also associated with its secretion into plasma and bile. The bile is a major excretory route for copper and plays an important role in maintaining copper homeostasis. Several copper-binding components in the bile of copper-loaded rats react with antibodies to metallothionein and have been tentatively identified as degradation and aggregation products of MT. The biliary excretion of MT-1 and these immunoreactive components is closely related to liver MT-1 content but their contribution to the total biliary copper output or to the turnover of hepatic copper-MT is relatively minor.[15,23] Secretion of copper-MT into plasma is also related to hepatic copper-MT content but only makes a minor contribution to plasma copper binding. Parenterally administered copper-MT is rapidly cleared from the plasma with an apparent half-life of only a few minutes. If the half-life of endogenous copper-MT is similar, only about 10% of the turnover of hepatic copper-MT can be accounted for by secretion into plasma.[15] There is therefore still considerable uncertainty as to the mechanisms involved in the turnover of copper-MT in the liver.

Intracellular Distribution of Copper-Metallothionein

As stated above, copper-MT occurs not only in the cytosol but also in the particulate fractions of the liver. Subcellular fractionation studies have indicated that the protein is distributed among several organelles but most attention has been paid to its occurrence in the lysosomes.[5] However, interpretation of the results of subcellular fractionations can be difficult, because copper loading often affects the density of the organelles. Copper-loaded lysosomes containing insoluble aggregates of copper-MT may sediment, for example, with the nuclear fraction. A better indication of intracellular MT distribution may be obtained by immunocytochemical techniques. These have consistently shown the presence of MT in both the cytoplasm and the nucleus of the liver from copper-loaded rats,[29,30] neonates,[31] and human subjects with Wilson's disease and other diseases with liver copper retention.[32,33] Certain aspects of the distribution of MT in copper-injected rats are consistent with a role in the excretion of the metal.[29] Striking changes in intracellular MT distribution occur during the

[29] L. M. Williams, H. Cunningham, A. Ghaffar, G. I. Riddoch, and I. Bremner, *Toxicology* **55,** 307 (1989).

[30] W. E. Evering, S. Haywood, M. E. Elmes, B. Jasani, and J. Trafford, *J. Pathol.* **160,** 305 (1990).

[31] D. M. Templeton, D. Banerjee, and M. G. Cherian, *Can. J. Biochem. Cell Biol.* **63,** 16 (1985).

[32] N. O. Nartey, J. V. Frei, and M. G. Cherian, *Lab. Invest.* **57,** 397 (1987).

[33] M. E. Elmes, J. P. Clarkson, N. J. Mahy, and B. Jasani, *J. Pathol.* **158,** 131 (1989).

fetal and neonatal period. In developing rats there is a gradual increase in levels of cytoplasmic MT during gestation. At day 20, however, MT is detectable in the nucleus and remains so for several days after birth, whereafter the protein is once again found only in the cytoplasm.[31] The presence of MT in the nucleus may be indicative of a unique mechanism regulating MT synthesis.

Surprisingly, the occurrence of MT in the lysosomes has never been confirmed by immunocytochemical techniques. This may reflect blocking of the epitopes in the aggregated copper-MT molecules in the lysosomes,[34] although fractionation of fetal deer liver (which has an extremely high liver copper content) before and after selective disruption of lysosomal membranes indicated that MT was in fact absent from lysosomes and occurred mainly in the nucleus.[35]

Functions of Metallothionein in the Liver

Radiocopper that is taken up by hepatocytes rapidly becomes associated with MT and other cytosolic proteins before its transfer to superoxide dismutase and mitochondrial and nuclear fractions.[21,36] However, the amount of copper that is bound to MT depends on copper status and on species, with mouse hepatocytes incorporating less ^{64}Cu into MT than rat hepatocytes.[36] These results could reflect isotopic exchange of the copper or exchange reactions between copper and zinc on MT; the kinetics of uptake are certainly inconsistent with copper-induced synthesis of MT. It seems unlikely that MT plays an obligatory role in copper uptake, as other cytosolic proteins were also involved in the initial binding of copper.[36] Moreover, yeast cells from which the related thionein gene has been removed continue to grow and divide in standard media without any loss of copper–enzyme activities, indicating that MT is not required for the uptake and distribution of copper in these cells.[37] The transfer of copper from MT to hepatic cuproenzymes could take place through the involvement of an intermediary transport molecule. Alternatively there could, in theory, be direct transfer of copper from MT to the apoenzymes, which would imply that MT regulates cuproenzyme activity by supplying metal to the catalytic sites. Such transfer has been reported *in vitro* between MT and superoxide dismutase but only under conditions where oxidation of the

[34] A. R. Janssens, F. T. Bosman, D. J. Ruiter, and C. J. A. Van den Hamer, *Liver* **4,** 139 (1984).

[35] M. Leighton, Ph.D. Thesis, University of London, London (1987).

[36] F. A. Palida, A. Mas, L. Arola, K. Bethin, P. A. Lonergan, and M. J. Ettinger, *Biochem. J.* **268,** 359 (1990).

[37] D. J. Thiele, M. J. Walling, and D. H. Hamer, *Science* **231,** 854 (1986).

copper to Cu(II) or of the thiol ligands has first occurred.[38] However, there is no evidence that direct transfer occurs *in vivo.*

It has been suggested that binding of copper to MT favors its excretion into bile, since much of the copper in pig liver is bound to MT[7] and there is a direct relationship between liver and bile copper concentrations.[39] In contrast, in sheep liver only a small proportion of the copper is associated with MT[7] and biliary copper excretion does not increase as liver copper content increases. However, treatments that increase hepatic MT content in the rat are generally associated with a decrease in biliary copper excretion[40] and only a small proportion of hepatic copper-MT is directly secreted into bile.[15,23]

Conversely the occurrence of high concentrations of copper MT in fetal livers[4,31] probably reflects the immaturity of biliary copper excretion mechanisms rather than the deliberate laying down of copper stores during fetal development to satisfy copper needs during the postnatal period. Metallothionein can almost certainly act as a copper storage depot but it seems more likely that its primary role in copper metabolism is as part of a cellular detoxification mechanism. Binding of copper to the thiol ligands of MT would prevent the participation of copper ions in oxidative attack on membranes, nucleic acids, and enzymes. The finding that the expression of monkey MT cDNAs in a yeast cell line restores its ability to grow and survive in a high copper medium strongly supports this view.[37] The uptake of copper-MT and its aggregation in organelles such as lysosomes would also limit the toxicity of the metal. There is circumstantial evidence that the differences between species in their susceptibility to copper toxicity are related to their ability to incorporate hepatic copper into MT.[7] Species that have a low tolerance of copper tend to have low copper-MT concentrations in the liver. However, it has also been suggested that overproduction of copper MT could be a contributory factor to the cell death seen in Wilson's disease and copper-poisoned rats.[30,33]

Conclusion

A schematic representation of the involvement of MT in hepatic copper metabolism is presented in Fig. 1.[41] As copper is absorbed from the intestine and is transported to the liver as copper-albumin or copper-histi-

[38] B. L. Geller and D. R. Winge, *Arch. Biochem. Biophys.* **213,** 109 (1982).

[39] M. Skalicky, A. Kment, I. Halder, and J. Leibetseder, *in* "Trace Element Metabolism in Man and Animals" (M. Kirchgessner, ed.), p. 163. Arbeitskreis für Tierernährungsforschung, Freising-Weihenstephan, 1978.

[40] C. D. Klaassen, *Am. J. Physiol.* **234,** E47 (1978).

[41] M. P. Richards, *J. Nutr.* **119,** 1062 (1989).

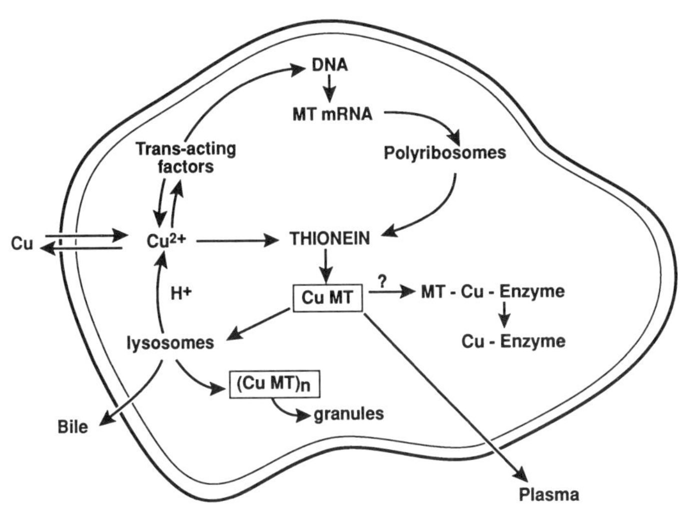

FIG. 1. A general model describing the function of MT as an autoregulated copper buffer in the cellular metabolism of copper. (See Ref. 41.)

dine complexes, it binds to thionein, either by occupying vacant sites or by displacement of zinc. If no binding sites are available on MT, the copper ions interact with the trans-acting factors that bind to the *MT* genes and promote the synthesis of MT mRNA and MT protein. This then chelates the excess copper ions. In this way, free copper ions are removed from the system and, at the same time, the stimulus for more thionein synthesis is removed. The fate of the copper after its binding to MT probably depends on the needs of the animal. The copper-MT may be degraded and the released copper used in copper-dependent processes or excreted. A proportion of the copper-MT may also be secreted into the blood or bile. However, if copper balance is positive and there is a large influx of copper into the liver, increasing amounts of the copper MT may be taken up into lysosomes and other organelles where it polymerizes to form insoluble aggregates. These may eventually be secreted into the bile.

Synthesis of MT therefore forms part of a complex but closely regulated homeostatic system designed to prevent the accumulation of free copper ions in the cell without compromising the supply of copper for essential metabolic processes.

[67] Biological Indicators of Cadmium Exposure

By JUDITH A. GLAVEN, ROBIN E. GANDLEY, and BRUCE A. FOWLER

Introduction

Cellular exposure to cadmium is known to elicit a number of responses in addition to metallothionein induction. A mechanistic understanding of these processes may permit development of biological indicators of cadmium exposure. This approach may be applied to both mammalian and nonmammalian tissues. This chapter will examine a number of these methods and attempt to evaluate their utility for monitoring how cells respond to this metal at the organelle and molecular levels of biological organization. Ideally, such indicators should provide early, sensitive, and readily measurable responses that are specific for cadmium.

In order for a potential indicator to be accepted as an indicator of effect, it must ultimately show a mechanistic correlation with the biological activity of this metal in living tissues. As discussed below, there are a number of potential candidate indicators of exposure and effect for cadmium that have each shown some promise for monitoring the actions of cadmium in cells.

Indicators of Cadmium Exposure

Total Tissue Burden

In addition to cadmium induction of metallothionein (MT), monitoring of total tissue cadmium burdens via either atomic absorption spectroscopy[1,2] or *in vivo* neutron activation analyses[3,4] may provide useful data on cadmium exposure. The atomic absorption spectroscopy methods involve sampling and digestion of tissues prior to analysis and hence are of limited value for monitoring living organisms. These analyses are frequently conducted on autopsy samples of target tissues such as the kidney[5] or for

[1] R. Lauwerys, H. Roels, M. Regniers, J. P. Buchet, A. Bernard, and A. Goret, *Environ. Res.* **20,** 375 (1979).

[2] R. Lauwerys, J. P. Buchet, and H. Roehls, *Inst. Arch. Occup. Environ. Health* **26,** 275 (1976).

[3] K. J. Ellis, W. D. Morgan, I. Zanai, S. Yasumura, D. Vartsky, and S. H. Cohn, *J. Toxicol. Environ. Health* **7,** 691 (1981).

[4] H. A. Roels, R. R. Lauwerys, J. P. Buchet, A. Bernrd, D. R. Chettle, T. C. Harvey, and I. K. Al-Haddad, *Environ. Res.* **26,** 217 (1981).

[5] C. G. Elinder, T. Kjellstrom, L. Friberg, B. Lind, and L. Linnman, *Arch. Environ. Health* **31,** 292 (1976).

nonmammalian aquatic organisms such as mussels,[6] where field sampling methods may be employed to assess cadmium exposure of populations. *In vivo* neutron activation analyses of renal cadmium concentrations in exposed workers is another potentially useful methodology[3,4] but further studies are needed to validate this approach with regard to direct wet chemical analyses, because the technique appears to be sensitive to geometric considerations.

Mineral Concretions

X-Ray microanalysis and atomic absorption spectroscopy are potentially useful monitoring tools for detecting cadmium in mineral concretions of marine mollusks.[7-11] In these species, the concretions represent a larger intracellular storage compartment than the metallothionein pool[7,8] and have the potential advantage of being extensively excreted from the kidney when cadmium exposure exceeds the tissue capacity to store this metal.[8,11] This response may thus also represent an indicator of cadmium toxicity. Disadvantages of monitoring concretions are that exposure to other metals such as copper will also cause this phenomenon,[11] so that there is a lack of specificity for cadmium in this response if other metals are also present.

Stress Proteins

Prokaryotic and eukaryotic cells respond to heat shock and various other forms of stress, including metal exposure, by synthesizing a discrete set of proteins.[12-16] These molecules were first observed in cells exposed to conditions of thermal stress and as a result have been termed heat-shock proteins (hsps).[12,13] Subsequently, it has been found that various forms of

[6] E. D. Goldberg, M. Koide, V. Hodge, A. R. Fleagal, and J. H. Martin, *Estuarine Coastal Shelf Sci.* **16,** 69 (1983).
[7] N. G. Carmichael, K. S. Squibb, and B. A. Fowler, *J. Fish. Res. Bd. Can.* **36,** 1149 (1979).
[8] N. G. Carmichael and B. A. Fowler, *Mar. Biol.* **65,** 35 (1981).
[9] S. G. George and B. J. S. Pirie, *Biochem. Biophys. Acta* **580,** 234 (1979).
[10] S. G. George, B. J. S. Pirie, and T. L. Combs, *J. Exp. Mar. Biol. Ecol.* **42,** 143 (1980).
[11] B. A. Fowler and E. Gould, *Mar. Biol.* **96,** 207 (1988).
[12] A. Tissieres, H. K. Mitchell, and V. M. Tracey, *J. Mol. Biol.* **84,** 389 (1974).
[13] M. Ashburner and J. J. Bonner, *Cell. (Cambridge, Mass.)* **17,** 241 (1979).
[14] S. Lindquist, *Annu. Rev. Biochem.* **55,** 1151 (1986).
[15] W. J. Welch, J. I. Garrels, and J. R. Fermaisco, *in* "Heat Shock from Bacteria to Man" (M. J. Schlesinger, M. Ashburner, and A. Tissieres, eds.), p. 257. Cold Spring Harbor Laboratory, Cold Spring Harbor, New York, 1982.
[16] W. J. Levinson, H. Opperman, and J. Jackson, *Biochem. Biophys. Acta* **606,** 170 (1980).

stress, including drugs,[17] amino acid analogs[17,18] and metals,[16] induce the expression of these molecules, which are now more commonly termed the stress proteins. This response is highly conserved and presumed to be universal, serving a general protective function in the stressed cell.

Principal stress proteins have molecular masses ranging from 28,000–174,000 Da. Of these, those between 70,000 and 90,000 are seen in response to most situations of stress; the other stress proteins appear to be induced more selectively. The occurrence of specific protein responses to specific conditions of stress may have great utility as biological indicators of exposure, and in elucidating the mechanisms of toxicity at the subcellular level.

Specific patterns of protein synthesis can be detected through metabolic labeling of proteins and standard techniques of two-dimensional gel electrophoresis.[19-21] The technique consists of an isoelectric separation in the first dimension and separation based on molecular weight in the second dimension. With the development of computerized digitizing image analysis systems, autoradiographs of the gels may be evaluated in a more quantitative manner.

Because of the environmental significance of cadmium there is an interest in identifying early indicators of exposure and delineating their relationships to mechanisms of action. Levinson *et al.* described the induction of four proteins in chicken and human cell lines in response to several transition series metals, including cadmium.[16] Along with the induction of hsp70, several low-molecular-mass proteins between 25,000 and 35,000 were induced. In a study examining the effect of sublethal acute doses of cadmium on *Drosophila* cells, it was shown that some of the induced proteins were the same as those seen in heat-shocked cells. However, cadmium also stimulated the synthesis of several other unique low-molecular-weight proteins.[20] These proteins are thus candidates as biomarkers of exposure to this element. There are several other examples of stress proteins detected in response to cadmium. Induced gene expression in two fish cell lines exposed to cadmium has been investigated by Price-Haughy and Gedamu[22]; they described the induction of metallothionein (MT), a 14,000 metal-inducible protein, and several stress proteins with molecular masses between 28,000 and 29,000 in response to cadmium.

A number of studies have been focused on the biological function of those molecules in an effort to better understand their roles in cell survival

[17] L. E. Hightower, *J. Cell. Physiol.* **102,** 407 (1980).
[18] P. M. Kelley and M. J. Schlesinger, *Cell (Cambridge, Mass.)* **15,** 1277 (1978).
[19] P. H. O'Farrell, *J. Biol. Chem.* **250,** 4007 (1975).
[20] A. Courageon, C. Maisonhaute, and M. Best-Belpomme, *Exp. Cell Res.* **153,** 515 (1984).
[21] Y. Aoki, M. M. Lipsky, and B. A. Fowler, *Toxicol. Appl. Pharmacol.* **106,** 462 (1990).
[22] J. Price-Haughey and L. Gedamu, *Experientia, Suppl.* **52,** 465 (1987).

during chemical or physical stress. Riabowol *et al.* demonstrated the importance of stress proteins to the cell while examining the function of the heat shock protein with molecular mass 70,000 (hsp70) in rat fibroblasts.[23] Cells injected with monoclonal antibodies raised against hsp70 were unable to withstand heat-shock treatment. This work shows that the protein has an important protective function in the stressed cell. A stress protein seen in response to cadmium with a relative molecular mass of 34,000 has been identified as heme oxygenase.[24] Specific antibodies for human heme oxygenase were used to identify the protein by an immunoblot technique.[25] These data suggest that the rate-limiting enzyme in the oxidative metabolism of heme[26] may be of importance during early response to sublethal injury.

The direction of future research in this area will be to identify molecular triggers for stress response following exposure to cadmium, and to relate specific stress proteins to adaptation/protection in the stressed cell. Research that may provide further information in this direction includes DNA footprinting techniques to examine the trigger for the induction of a specific stress protein by identifying DNA-binding proteins,[27] and various immunological techniques to determine the intracellular location of specific stress proteins during and after the stress.[15,28] As specific stress proteins induced in response to cadmium exposure are identified and examined from the perspective of the role they play in the protection of the cell, a more basic understanding of the mechanisms of cadmium biological activity will emerge.

Metallothionein Gene-Binding Factors

The significance of metallothionein (MT) as an indicator of cadmium exposure is discussed later in this chapter. Briefly, we would like to describe cellular factors involved in the regulation of MT, because they may serve as indicators of cadmium exposure prior to MT induction. Investigation into the structure and function of genes encoding for MT have identified trans-acting nuclear factors (NF) involved in the regulation of MT expression.[29,30] Cadmium induces the formation of complexes of NF with metal regulatory elements (MRE) of the *MT* gene. Initially it was hoped that exposure to cadmium would cause detectable, diagnostic increases in the

[23] K. T. Riabowol, L. A. Mizzen, and W. J. Welch, *Science* **242,** 433 (1988).
[24] S. Taketani, H. Kohno, T. Yoshinga, and R. Tounaga, *FEBS Lett.* **245,** 173 (1989).
[25] H. Towbin, T. Stahelin, and J. Gordon, *Proc. Natl. Acad. Sci. U.S.A.* **76,** 4350 (1979).
[26] M. D. Maines, *FASEB J.* **2,** 2557 (1988).
[27] D. J. Galas and A. Schmitz, *Nucleic Acids Res.* **5,** 3157 (1978).
[28] W. J. Welch and L. A. Mizzen, *J. Cell Biol.* **106,** 117 (1988).
[29] C. Seguin, B. K. Felber, A. D. Carter, and D. Hamer, *Nature (London)* **312,** 781 (1984).
[30] C. Seguin and D. Hamer, *Science* **235,** 1383 (1987).

cellular concentration of NF; unfortunately, however, comparisons of cellular concentrations of MRE-binding NF in cells grown in the presence and absence of cadmium showed that cadmium does not act by increasing the synthesis of these proteins.[30] Instead, cadmium may act on NFs directly to increase their affinity for the MRE of the gene, or cadmium may act by altering protein–protein interactions between the NF and some as yet unidentified "coactivator" protein.[30] If this is the case it would be necessary to detect the cadmium-induced change in the NF, or the induction of the "coactivator."

The use of DNA footprinting[30,27] to reveal MREs of the *MT* gene protected from exonuclease activity due to bound NF, and synthetic oligonucleotides from the MRE sequence to detect DNA-binding proteins immobilized on blots,[32,33] has resulted in the description of two *MT* gene-binding NFs with molecular masses of 74,000 and 108,000.[31,32] One of these, the 74K molecule, has been purified by affinity chromatography on trout MRE[31]; further characterization of this protein and its interaction with regulatory regions of the mouse *MT* gene are awaited.

Proteinuria

Cadmium-induced tubular proteinuria can be used as an early indicator of cadmium exposure and effect. Alterations in renal function produced by cadmium can be due to alterations in tubular reabsorption or glomerular filtration. A distinctive low-molecular-weight proteinuria has been associated with elevated cadmium exposure and is characterized by an increase in low-molecular-weight urinary proteins and a decrease in high-molecular-weight proteins, suggesting damage to the renal proximal tubule.[34] Alternatively, another possible effect of cadmium exposure is a decrease in glomerular permeability to larger proteins.[35,36] Studies by Squibb *et al.*[37] provided a mechanistic basis for development of cadmium-induced proteinuria and its use as an early indicator of cadmium exposure following parenteral administration of cadmium-metallothionein (Cd-MT) to rats. Since Cd-MT is the major form of cadmium in the circula-

[31] J. Imbert, M. Zafarullah, V. C. Culotta, L. Gedamu, and D. Hamer, *Mol. Cell. Biol.* **9,** 5315 (1989).

[32] C. Seguin and J. Prevost, *Nucleic Acids Res.* **22,** 1052 (1988).

[33] W. K. Miskimins, M. P. Roberts, A. McClelland, and F. H. Ruddle, *Proc. Natl. Acad. Sci. U.S.A.* **82,** 6741 (1985).

[34] P. L. Goering, K. S. Squibb, and B. A. Fowler, *Trace Subst. Environ. Health* **19,** 22 (1985).

[35] A. Bernard, H. Roels, G. Hubermont, J. P. Buchet, P. L. Masson, and R. Lauwerys, *Int. Arch. Occup. Environ. Health* **38,** 19 (1976).

[36] A. Bernard, J. P. Buchet, H. Roels, P. Masson, and R. Lauwerys, *Eur. J. Clin. Invest.* **9,** 11 (1979).

[37] K. S. Squibb, J. B. Pritchard, and B. A. Fowler, *J. Pharmacol. Exp. Ther.* **229,** 311 (1984).

tion, this model mimicks the actual *in vivo* condition. Within 8 hr of Cd-MT administration, an increase in low-molecular-weight proteins was detected in the urine, via sodium dodecyl sulfate (SDS) gel electrophoresis,[37] that was temporally correlated to altered lysosome structure and function in renal proximal tubule cells. This study demonstrated the utility of proteinuria as a biological indicator and the relationship of the proteinuria to pathophysiological changes in target cell populations.

Specific Urinary Proteins. The proteinuria induced by cadmium may not increase the overall urine protein content significantly until the cadmium exposure is quite extensive. Clinical diagnosis of proteinuria is based on detection of an increase in total protein in the urine. The methods used for early detection of cadmium-induced proteinuria must discriminate between high- and low-molecular-weight proteins in the urine. Immunological assays for β_2-microglobulin, retinol-binding protein, and urinary metallothionein are used to detect low-molecular-weight proteinuria.

β_2-Microglobulin. β_2-Microglobulin is used to monitor tubule reabsorption capabilities. There are several problems with using this protein to detect proteinuria. First, there is an increase in β_2-microglobulin in the urine of adults, beginning at the age of 50 years.[38] Second, β_2-microglobulin is unstable at urine pH < 5.5. The most common assay for β_2-microglobulin is a radioimmunoassay based on latex particle agglutination. Samples are chromatographed on Sephadex G-75 and fractions are pooled and concentrated, using ultrafiltration. The sample is then analyzed via immunoassay using an antiserum and immunoglobulin G-coated latex particles.[39] The samples can be read via an automated latex immunoassay system.[40] Another common protein assay used for cadmium exposure is an electrophoretic technique using silver staining with sodium dodecyl sulfate-polyacrylamide gel.[41] A detection level of 0.8 mg protein/liter urine can be achieved with this technique.

Retinol-Binding Protein. Retinol-binding protein (RBP) differs from β_2-microglobulin in that it is stable in urine of pH 4.5 and higher, and its degradation is slower.[42] Like β_2-microglobulin, RBP is present at elevated levels in older human subjects due to an overall loss of renal function. Enzyme-linked immunosorbent assay (ELISA) can also be used to detect urinary retinol-binding protein.[43] It has a range of detection of 25–250 μg

[38] K. Nomiyama, M. Yotoriyama, and H. Nomiyama, *Arch. Environ. Contam. Toxicol.* **12,** 147 (1983).

[39] C. Viau, A. Bernard, and R. Lauwerys, *J. Appl. Toxicol.* **6,** 185 (1986).

[40] A. Bernard and R. Lauwerys, *Clin. Chem.* **29,** 1007 (1983).

[41] K. Nomiyama, H. Nomiyama, M. Yototiyama, and K. Matsui, *Ind. Health* **20,** 11 (1982).

[42] A. M. Bernard, D. Moreau, and R. Lauwerys, *Clin. Chim. Acta* **126,** 1 (1982).

[43] M. D. Topping, H. W. Forster, C. Dolman, C. M. Luczynska, and A. M. Bernard, *Clin. Chem.* **32,** 1863 (1986).

protein/liter urine. This method compares favorably to the commercially available test kits for β_2-microglobulin.[44] The assays described above can be utilized in both the clinical or laboratory setting.

Metallothionein. A third protein that may be used as an indicator of cadmium exposure is metallothionein (MT).[42] A significant correlation between urinary cadmium and urinary MT levels has been seen as a more accurate indicator than either β_2-microglobulin or retinol-binding protein because MT binds the cadmium directly and plays a central role in the biology of this metal *in vivo*.[45] Preliminary studies have considered several routes of exposure to cadmium (oral and injection), while trying to determine the utility of urinary MT as a biological indicator of cadmium exposure. Studies considering both routes have found favorable results for the use of urinary MT as a biomonitoring tool.[46,47] The assays recommended for MT are radioimmunoassay (RIA) or an enzyme-linked immunosorbent assay (ELISA). Monoclonal and polyclonal antibodies are used for both assays. The RIA method has been found to be slightly more accurate, while the ELISA is faster. The RIA method is also recommended for detection of low levels of MT.[48] This method is a double-antibody assay using radiolabeled MT. Competition between labeled and unlabeled MT is measured via scintillation counting to determine a curve that is compared to a standard to determine the amount of MT present in a sample.[49] The ELISA is a similar procedure that is not as accurate but can be useful for faster results.[50]

Another possible biological indicator of cadmium exposure is *N*-acetyl-β-D-glucosaminidase (NAG). Work to determine the usefulness of NAG is still in preliminary stages. NAG, like MT, is more closely related to the cadmium exposure than to the damage done by the exposure. In this sense, these indicators may be more specific to cadmium than β_2-microglobulin and retinol-binding protein levels which may be caused by any agent causing renal tubular damage.[51,52]

Problems with the use of proteinuria as a biological indicator of cadmium exposure are the possibility of delayed development and uncertain

[44] A. M. Bernard, D. Moreau, and R. Lauwerys, *Clin. Chem.* **28,** 1167 (1982).

[45] R. R. Lauwerys, A. Bernard, H. A. Roels, J. P. Buchet, and C. Viau, *Environ. Health Perspect.* **54,** 147 (1984).

[46] Z. A. Shaikh, K. M. Harnett, S. A. Perlin, and P. C. Huang, *Experientia* **45,** 146 (1989).

[47] M. Sato, Y. Nagai, and I. Bermner, *Toxicology* **6,** 23 (1989).

[48] M. P. Waalkes, J. S. Garvey, and C. D. Klaassen, *Toxicol. Appl. Pharmacol.* **79,** 524 (1985).

[49] J. S. Garvey, R. J. Vander Mallie, and C. C. Chang, this series, Vol. 84, p. 121.

[50] D. G. Thomas, H. J. Linton, and J. S. Garvey, *J. Immunol. Methods* **89,** 239 (1986).

[51] T. Kawand, C. Tohyama, and S. Suzuki, *Int. Arch. Occup. Environ. Health* **62,** 95 (1990).

[52] C. Tohyama, Y. Mitane, E. Kobayashi, N. Sugihira, A. Nakano, and H. Saito, *J. Appl. Toxicol.* **8,** 15 (1988).

relationship to other manifestations of renal disease. Proteinuria is an early sign of cadmium exposure; however, at subchronic levels it is far more difficult to initially detect due to other signs of renal dysfunction that may be occurring concomitantly. The physiological mechanism of cadmium toxicity in the sense of renal damage seems to be progressive. Many epidemiological studies at a variety of exposure levels have noted an increase in the degree of proteinuria[53] during followup evaluations. This problem appears to be more a complication of progressive cadmium toxicity than a problem with the use of proteinuria as an indicator. In the future this problem may be bypassed if assays for MT and NAG prove to be as accurate and early at indicating cadmium exposure as they appear at present. By more fully understanding the mechanism of toxicity of cadmium, biological indicators of exposure will be much easier to develop.

[53] H. A. Roels, R. R. Lauwerys, J. P. Buchet, A. M. Bernard, A. Vos, and M. Oversteyns, *Br. J. Ind. Med.* **46,** 755 (1989).

Section VIII

Chapter Related to Section IV: Isolation and Purification of Metallothioneins

[68] Cadmium-Binding Peptide Complexes from *Schizosaccharomyces pombe*

By DONALD J. PLOCKE

Introduction

The synthesis of small, cysteine-rich metal-binding peptides by the fission yeast *Schizosaccharomyces pombe,* and by plants of a wide variety of species when these are grown in the presence of cadmium and certain other metal ions, is well documented.[1] These peptides, which have the structure (γ-Glu-Cys)$_n$Gly, or (γ-EC)$_n$ G, where $n = 2$ to 11, have been termed cadystins,[2] phytochelatins,[3] γ-glutamyl metal-binding peptides,[4] and class III metallothioneins[5] by different investigators, and the specific properties of these peptides in plants and in *S. pombe* are discussed elsewhere in this volume.[6-9] In 1989 Grill *et al.*[10] reported the discovery and purification of an enzyme from *Silene cucubalus* that catalyzes the synthesis of these peptides. The enzyme, γ-Glu-Cys-dipeptidyl transpeptidase (phytochelatin synthase), was found to be constitutively present in *Silene cucubalus* and several other plant species; synthesis of phytochelatins from glutathione as the initial substrate was effected by activation of the enzyme when free metal ions such as cadmium, copper, lead, zinc, or mercury were present in the growth medium. Although γ-glutamyl peptides are produced by the fission yeast *Schizosaccharomyces pombe,*[2,11] there is no evidence for their production by the budding yeast *Saccharomyces cerevisiae.*[12] The latter responds specifically to excess copper in the growth medium by the

[1] N. J. Robinson, *in* "Heavy Metal Tolerances in Plants: Evolutionary Aspects" (A. J. Shaw, ed.), p. 195. CRC Press, Boca Raton, Florida, 1990.
[2] N. Kondo, M. Isobe, K. Imai, and T. Goto, *Agric. Biol. Chem.* **49,** 71 (1985).
[3] E. Grill, E.-L. Winnacker, and M. H. Zenk, *Science* **230,** 674 (1985).
[4] R. N. Reese, R. K. Mehra, E. B. Tarbet, and D. R. Winge, *J. Biol. Chem.* **263,** 4186 (1988).
[5] B. A. Fowler, C. E. Hildebrand, Y. Kojima, and M. Webb, *Experientia, Suppl.* **52,** 19 (1987).
[6] W. E. Rauser, this volume [38].
[7] E. Grill, E.-L. Winnacker, and M. H. Zenk, this volume [39].
[8] N. Mutoh and Y. Hayashi, this volume [40].
[9] Y. Hayashi, M. Isobe, N. Mutoh, C. W. Nakagawa, and M. Kawabata, this volume [41].
[10] E. Grill, S. Löffler, E.-L. Winnacker, and M. H. Zenk, *Proc. Natl. Acad. Sci. U.S.A.* **86,** 6838 (1989).
[11] A. Murasugi, C. Wada, and Y. Hayashi, *Biochem. Biophys. Res. Commun.* **103,** 1021 (1981).
[12] E. Grill, E.-L. Winnacker, and M. H. Zenk, *FEBS Lett.* **197,** 115 (1986).

production of a class II metallothionein that has been the object of numerous studies.[13]

Although there is ample evidence for the formation in plants and *S. pombe* of clusters of phytochelatin molecules complexed with metal ions such as cadmium and copper, and also for the presence of inorganic sulfide in some of these complexes,[8,14] little has been reported on the stoichiometry of these complexes. Peptide heterogeneity within a cluster has been suggested, but the precise composition of the clusters has remained obscure, and the separation methods reported have produced peptide clusters of uncertain stoichiometry and apparent inhomogeneity.[4]

This chapter describes methods for the production of γ-glutamyl peptides by *S. pombe* when cadmium is added to the growth medium, and a procedure for the separation of discrete clusters or complexes of these peptides having a definable composition with respect to the number and species of peptides within a cluster, as well as the content of cadmium and inorganic sulfide in a given cluster.

Growth and Harvesting of *Schizosaccharomyces pombe*

The cell strain used for the work described here is *S. pombe* L 972 (h^-), which is maintained by the National Collection of Yeast Cultures (Food Research Institute, Norwich NR4 7UA, England), as well as the American Type Culture Collection (Rockville, MD). The growth medium contained 2% Bacto-peptone (Difco, Detroit, MI), 2% D-glucose, and 1% yeast extract (Difco). The cells were grown in 2-liter Erlenmeyer flasks at 30° on a large rotary shaker at 110 cpm. The flasks were modified with four indentations near the base of each flask to provide for maximum turbulence of the growth medium during shaking and hence for vigorous aeration of the cells, which is necessary for obtaining a maximum yield of cells. Typically, 10 ml of an inoculum of midlogarithmic phase cells was added to 800 ml of culture medium in a 2-liter growth flask. Growth may conveniently be monitored by turbidimetry measurements at 660 nm.[12] $CdCl_2$ was added to give a concentration in the medium of 1.0 m*M* when the cells approached the end of exponential growth (an OD_{660} of 7.8 under our conditions, which corresponds to approximately 8×10^7 cells/ml). The cells were exposed to cadmium for 48 hr, during which time vigorous aeration of the cells was maintained. After 48 hr, the optical density of the cell suspension increased to 11.9, corresponding to a cell concentration of

[13] D. R. Winge, K. B. Nielson, W. R. Gray, and D. H. Hamer, *J. Biol. Chem.* **260,** 14464 (1985).

[14] A. Murasugi, C. Wada, and Y. Hayashi, *J. Biochem. (Tokyo)* **93,** 661 (1983).

approximately 1.2×10^8 cells/ml. The cells were harvested by centrifugation, washed with H_2O, and resuspended in a small volume of 0.1 *M* Tris-HCl buffer (pH 8.0). Typically, about 60 g wet weight of cells was obtained from 10 liters of growth medium, and this amount of cells was resuspended in 30 ml of buffer.[15]

Isolation of Cadmium-Containing Peptide Complexes

Preparation of Cell Extract

Cell disruption was accomplished by vigorous agitation of the cell suspension (see above) with acid-washed glass beads (diameter, 0.5 mm; 1:1 ratio by volume of cell suspension to glass beads) in a reciprocal mechanical agitator.[16] The cells were shaken for a total of 5 min, using 10 cycles of 30 sec each of shaking followed by 1-min intervals without shaking (to avoid excessive heating). Examination of the cells with a light microscope after this procedure revealed that the extent of breakage was greater than 95%. After removal of the glass beads by filtration through several layers of cheesecloth, a crude cell extract was obtained by centrifugation of the disrupted cells at 12,000 *g* for 30 min, removal of the supernatant, resuspension of the precipitate in buffer, recentrifugation, and combination of the two supernatants. The combined supernatants were lyophilized and the lyophilate was dissolved in 60 ml of 0.01 *M* Tris-HCl buffer (pH 8.0), centrifuged at 23,000 *g* (30 min), and the supernatant was centrifuged at 40,000 *g* (60 min) to obtain a final high-speed cell extract.

DEAE-Sephadex Chromatography

The supernatant from the 40,000 *g* spin was applied to DEAE-Sephadex A-25 (Pharmacia Fine Chemicals, Inc., Piscataway, NJ) in a column of dimensions 2.5×60 cm, and elution performed first with 400 ml of 0.01 *M* Tris-HCl buffer (pH 8.0), and then with 600 ml of 0.5 *M* NaCl in the same buffer. Fractions (16 ml each) from the elution with 0.5 *M* NaCl were collected and analyzed for absorbance at 254 nm as well as for cadmium content, the latter being determined by atomic absorption spectroscopy. A single major peak of cadmium concentration was obtained, which coincided approximately but not exactly with the peak of A_{254}

[15] The growth conditions given here are essentially those described by Grill *et al.* (Ref. 12). However, the final yield of cells obtained using the conditions of vigorous aeration described in the text would appear to be significantly greater than that reported by Grill.

[16] The commercially available Bead Beater (BioSpec Products, Bartlesville, OK) or other similar device may also be used.

absorbance. The fractions containing the highest cadmium content (approximately 90 ml) were pooled, and then the pooled fractions were desalted on Sephadex G-10 (Pharmacia), followed by lyophilization of the cadmium-containing material.

Sephadex G-50 Chromatography

The lyophilate was dissolved in 4.0 ml of 0.01 *M* Tris-HCl buffer (pH 8.0) containing 0.1 *M* NaCl, applied to a Sephadex G-50 (Pharmacia) column (2.5 × 100 cm), and eluted with the same buffer using a flow rate of 20 ml/hr. The results of this fractionation are shown in Fig. 1; the cadmium-containing fractions appear in the middle peak. The later eluting (lower molecular weight) fractions in this middle peak were pooled separately (fraction B) from the fractions eluting earlier (fraction A). Fraction B (the lower molecular weight material) was then desalted on Sephadex G-10, and the cadmium-containing fractions (approximately 20 ml) pooled for further purification.

DE-52 Chromatography

The material from fraction B was applied to a DE-52 (Whatman, Clifton, NJ; preswollen microgranular DEAE) column (2.5 × 40 cm) that had been previously equilibrated with 0.05 *M* Tris-HCl buffer (pH 7.6) and elution of the cadmium-containing material achieved using a gradient of NaCl (0–0.5 *M*) and a flow rate of 20 ml/hr with fractions of approximately 5.0 ml being collected. As shown in Fig. 2, DEAE fractionation yielded three peaks of absorbance at 254 nm, which were in close conformity with the peaks of cadmium-containing material.

Characterization of Cadmium-Binding Clusters

Spectral analyses as well as analyses for cadmium by atomic absorption spectroscopy (Model IL 157 atomic absorption spectrometer; Instrumentation Laboratory) amino acid composition (500 automated amino acid analyzer; Durrum),[17] sulfhydryl content,[18] and inorganic sulfide[19] were performed on peak fractions (Fig. 2). These fractions were also analyzed by high-performance liquid chromatography (HPLC) after acidification to determine the peptide composition using the conditions employed by Grill *et al.*,[12] with subsequent analysis of the HPLC-fractionated material for sulfhydryl and amino acid content. In all instances, amino acid analysis

[17] Dried samples were treated for 3 hr at 0° with performic acid prior to hydrolysis to convert cysteine to cysteic acid.

[18] G. L. Ellman, *Arch. Biochem. Biophys.* **82,** 70 (1959).

[19] T. E. King and R. O. Morris, this series, Vol. 10, p. 635.

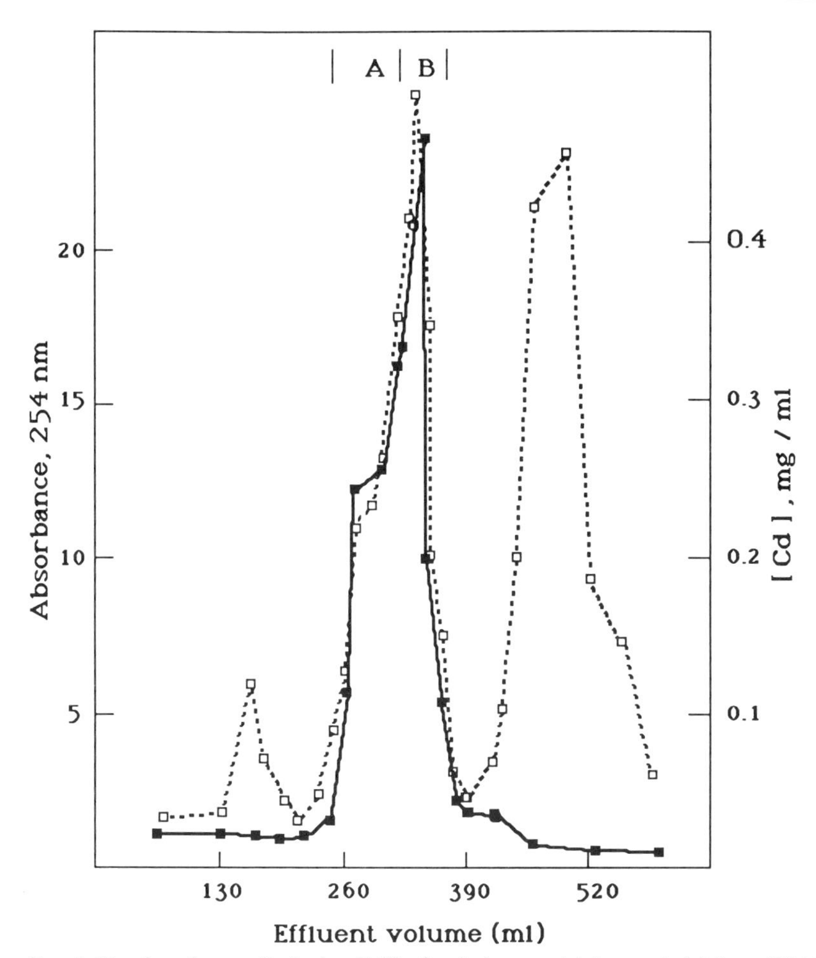

FIG. 1. Fractionation on Sephadex G-50 of cadmium-containing material from DEAE-Sephadex A-25. The column (2.5 × 100 cm) was equilibrated with 0.01 *M* Tris-HCl buffer, pH 8.0, containing 0.1 *M* NaCl. Rate, 20 ml/hr; 6.5-ml fractions were collected. (□), A_{254}; (■), [Cd] in milligrams per milliliter.

revealed the presence of glutamic acid, cysteine (as cysteic acid), and glycine as the only constituents in the fractions analyzed. From these measurements it was possible to propose stoichiometries for three distinct clusters as resolved by DEAE chromatography. The results, which have also been reported elsewhere,[20] are summarized in Table I; fractions I, II, and III refer to the DEAE column fractions so designated in Fig. 2. Fraction I consists of a cluster of apparently lower molecular weight than the

[20] D. J. Plocke and J. H. R. Kägi, *Biophys. J.* **55,** 520a (1989).

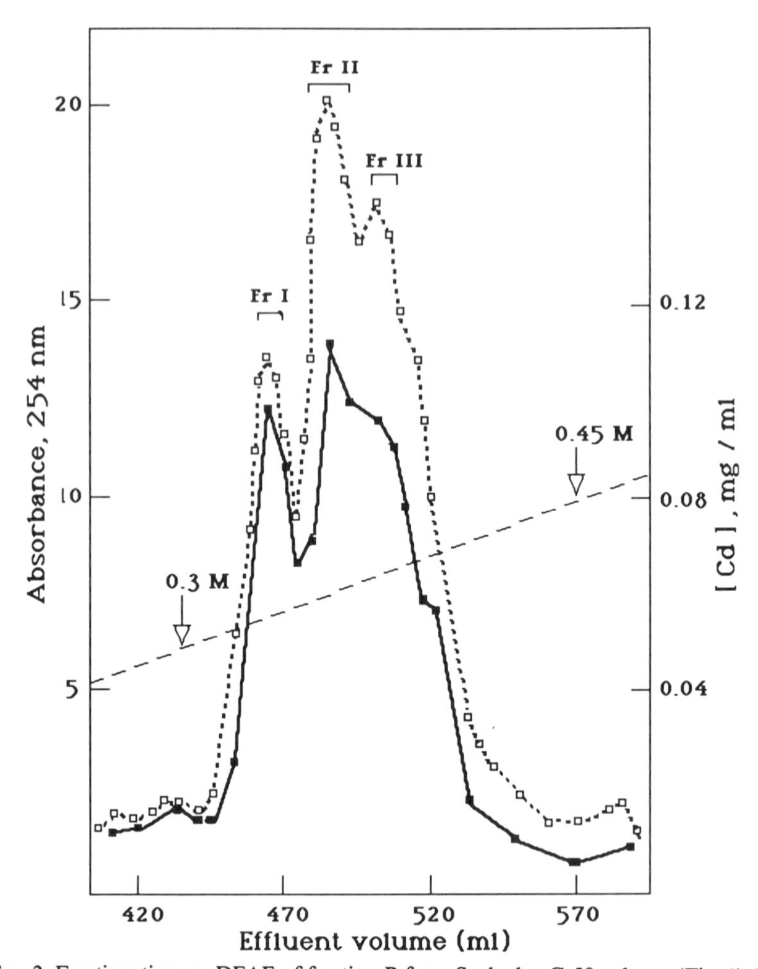

FIG. 2. Fractionation on DEAE of fraction B from Sephadex G-50 column (Fig. 1). The column (2.5 × 40 cm) was equilibrated with 0.05 *M* Tris-HCl buffer, pH 7.6. Gradient elution was begun by allowing 0.7 *M* NaCl in this buffer to flow into a mixing chamber containing 400 ml of the buffer with no NaCl. Rate, 20 ml/hr; 5.0-ml fractions were collected. The arrows indicate the NaCl concentrations at the corresponding elution volumes as measured by conductivity. (□), A_{254}; (■), [Cd] in milligrams per milliliter.

material in fractions II and III, which appear to be composed of clusters of approximately the same molecular weight. These two clusters (II and III) have similar absorption spectra, each displaying a maximum at 265 nm; however, the CD (circular dichroism) spectra of these two clusters are strikingly different. The CD spectrum of fraction III is similar to CD

TABLE I
STOICHIOMETRY OF CADMIUM-BINDING PEPTIDE COMPLEXES[a]

Parameter	Fraction I	Fraction II	Fraction III
Component			
[Cd](×10³)	0.883	1.026	0.848
[SH](×10³)	2.36	2.94	2.04
[S^{2-}](×10³)	0.00	0.174	0.149
[Cys](×10³)	1.89	1.94	1.95
Stoichiometry (HPLC)			
$n = 1$	2	—	—
$n = 2$	2	6	2
$n = 3$	—	2	3
$n = 4$	—	—	1
Molecular weight (from stoichiometry)	1918	5490	5219

[a] Fractions I, II, and III refer to fractions so labeled in Fig. 2.

spectra previously reported for cadmium-binding peptide complexes in *S. pombe*[21] and various plants,[3,22,23] with a positive band at 279 nm and two negative bands at 259 and 234 nm. The CD spectrum of fraction II, however, displays a single negative band at 272 nm and a positive band at 255 nm. The MCD (magnetic circular dichroism) spectra of these two fractions are, however, nearly identical, with minima at 275 nm and maxima at 255 nm. The CD and MCD spectral bands of fraction I are extremely weak in comparison to the spectral bands of fractions II and II. The significance of these differences and the interpretation of the CD and MCD spectra in structural terms will be discussed in detail elsewhere.[24]

Comments

Growth and Production of Phytochelatins

The amounts of cadmium-binding peptides produced, and the relative proportions of peptides of different lengths, apparently depend on the amount of free cadmium in the growth medium as well as the time of contact of the cells with the free metal ion. While the conditions for maximal production by *S. pombe in vivo* of peptides of given lengths have

[21] A. Murasugi, C. Wada, and Y. Hayashi, *Biochem. Biophys. Res. Commun.* **103,** 1021 (1981).
[22] W. E. Rauser, *Experientia, Suppl.* **52,** 301 (1987).
[23] M. Fujita and T. Kawanishi, *Plant Cell Physiol.* **27,** 1317 (1986).
[24] D. J. Plocke and J. H. R. Kägi, in preparation.

not been established, Grill *et al.*[12] have compared the amounts of peptides of different lengths produced on the addition of 1 m*M* Cd^{2+} at two different stages of growth. The predominant peptide produced on addition of Cd^{2+} to early logarithmic cells was found to be $(\gamma\text{-EC})_2G$, whereas the predominant peptide produced on addition of Cd^{2+} to stationary phase cells was $(\gamma\text{-EC})_3G$. The conditions used in the procedure described above (addition of 1 m*M* Cd^{2+} to late logarithmic phase cells) result in the production of $(\gamma\text{-EC})_2G$, $(\gamma\text{-EC})_3G$, and $(\gamma\text{-EC})_4G$ (see Table I), as well as peptides of greater length in clusters of undetermined stoichiometry (fraction A in Fig. 1).

Isolation Procedure

The isolation procedure described here utilizes as a final step a salt-gradient DEAE fractionation of the lower molecular weight material from Sephadex G-50 chromatography (fraction B in Fig. 1). This protocol results in the separation of three distinct peptide clusters, including two clusters of approximately the same molecular weight but of different stoichiometry and markedly different CD spectra. For the most part, the methods that have been reported previously for the isolation of these peptides from plants or *S. pombe* have either relied primarily on gel filtration, or, where DEAE fractionation has been used, it has been employed at an early stage in purification on material that was probably quite inhomogeneous in size. It seems likely that separation of the three clusters described here was made possible by the removal of the higher molecular weight material (fraction A in Fig. 1), which to date has not been fractionated successfully by DEAE chromatography or any other means.

Acknowledgment

The work described in this chapter was performed by the author in the laboratory of J. H. R. Kägi at the Biochemistry Institute of the University of Zürich.

Section IX

Overview

[69] Overview of Metallothionein

By JEREMIAS H. R. KÄGI

Introduction

Metallothioneins (MTs) were discovered in 1957 when Margoshes and Vallee searched for a tissue component responsible for the natural accumulation of cadmium in mammalian kidney.[1] Metallothioneins are still the only biological compounds known to contain this metal. However, as documented already in the earliest reports, cadmium is but one of several optional metallic components, the others being most commonly zinc and copper.[2-4] In fact, in subsequent studies zinc[5] and/or copper[6] were often found to be the sole constituents, indicating at an early stage that the major role of the MTs is to be sought in the metabolism and disposition of these biologically essential metals. The MTs differ from most other metalloproteins by their much larger metal content, their unusual bioinorganic structure, and their remarkable kinetic lability, properties now believed to enable them to function in rapid metal transfer and metalloregulatory processes. In this chapter, the conspicuous features of these compounds and the principal results of structural and functional studies of mammalian forms are summarized.

Definition

Mammalian MTs have a molecular weight of 6000–7000, usually containing 61 amino acid residues, among them 20 cysteines and binding a total of 7 equivalents of bivalent metal ions.[7] All cysteines occur in the reduced form and are coordinated to the metal ions through mercaptide bonds, giving rise to spectroscopic features characteristic of metal–thiolate clusters. In view of these unique chemical characteristics, the phenomenological definition was adopted that any polypeptide resembling mammalian MT in several of these features can be named an MT.[8] Accordingly,

[1] M. Margoshes and B. L. Vallee, *J. Am. Chem. Soc.* **79,** 4813 (1957).
[2] J. H. R. Kägi and B. L. Vallee, *J. Biol. Chem.* **235,** 3460 (1960).
[3] J. H. R. Kägi and B. L. Vallee, *J. Biol. Chem.* **236,** 2435 (1961).
[4] P. Pulido, J. H. R. Kägi, and B. L. Vallee, *Biochemistry* **5,** 1768 (1966).
[5] R. H. O. Bühler and J. H. R. Kägi, *FEBS Lett.* **39,** 229 (1974).
[6] Cited in K. Lerch, *Met. Ions Biol. Syst.* **13,** 299 (1981).
[7] Y. Kojima, C. Berger, B. L. Vallee, and J. H. R. Kägi, *Proc. Natl. Acad. Sci. U.S.A.* **73,** 3413 (1976).
[8] Cited in M. Nordberg and Y. Kojima, *Metallothionein Exp. Suppl.* **34,** 41 (1979).

TABLE I

FACTORS THAT INDUCE METALLOTHIONEIN SYNTHESIS IN CULTURED CELLS OR *in Vivo*

- Metal ions[a]
  - Cd, Zn, Cu, Hg, Au, Ag, Co, Ni, Bi
- Hormones and second messengers
  - Glucocorticoids[b]
  - Progesterone[b]
  - Estrogen[b]
  - Catecholamines[c]
  - Glucagon[c]
  - Angiotensin II[c]
  - Arg-Vasopressin[c]
  - Adenosine[c]
  - cAMP[d]
  - Diacylglycerol[e]
  - Ca[f]
- Growth factors
  - Serum factors[e]
  - Insulin[e]
  - IGF-1[e]
  - EGF[e]
- Inflammatory agents and cytokines
  - Lipopolysaccharide (LPS)[g]
  - Carrageenan[b]
  - Dextran[b]
  - Endotoxin[b]
  - Interleukin-1[g]
  - Interleukin-6[h]
  - Interferon-α[i]
  - Interferon-γ[g]
  - Tumor necrosis factor[g]
- Tumor promoters and oncogenes
  - Phorbol esters[e,j]
  - *ras*[k]
- Vitamins
  - Ascorbic acid[l]
  - Retinoate
  - 1α,25-Dihydroxyvitamin D_3[m]
- Antibiotics
  - Streptozotocin[b]
  - Cycloheximide[e]
  - Mitomycin[j]
- Cytotoxic agents
  - Hydrocarbons[n]
  - Ethanol[b]
  - Isopropanol[b]
  - Formaldehyde[o]
  - Fatty acids[n]
  - Butyrate[p]
  - Chloroform[b]
  - Carbon tetrachloride[b]
  - Bromobenzene[q]
  - Iodoacetate[a]
  - Urethane[q]
  - Ethionine[q]
  - Di(2-ethylhexyl)phthalate[q]
  - α-Mercapto-β-(2-furyl)acrylate[r]
  - 6-Mercaptopurine[s]
  - Diethyldithiocarbamate[t]
  - Penicillamine[t]
  - 2,3-Dimercaptopropanol[t]
  - 2,3-Dimercaptosuccinate[t]
  - EDTA[t]
  - 5-Azacytidine[q]
  - Acetaminophen[q]
  - Indomethacin[q]
- Stress-producing conditions
  - Starvation[b]
  - Infection[b]
  - Inflammation[b]
  - Laparotomy[b]
  - Physical stress[b]
  - X-irradiation[b]
  - High O_2 tension[b]
  - Ultraviolet radiation[j]

[a] Cited in R. D. Palmiter, *Experientia, Suppl.* **52**, 63 (1987).

[b] Cited in I. Bremner, *Experientia, Suppl.* **52**, 81 (1987).

[c] Cited in F. O. Brady, S. H. Garrett, and K. Arizono, *in* "Metal Ion Homeostasis: Molecular Biology and Chemistry," p. 81 (D. H. Hamer and D. R. Winge, eds.). Alan R. Liss, New York, 1989.

[d] V. L. Nebes, D. DeFranco, and S. M. Morris, *Biochem. J.* **255,** 741 (1988).

[e] Cited in R. J. Imbra and M. Karin, *Mol. Cell. Biol.* **7**, 1358 (1987).

[f] T. Miyahara, S. Nemoto, T. Kaji, H. Yamada, M. Takeuchi, M.-A. Mori, and H. Kozuka, *Toxicol. Lett.* **31**, 113 (1986).

such compounds have now been identified not only in vertebrates but also in invertebrates, plants, eukaryotic microorganisms, and in some prokaryotes.[8-10] On structural grounds, they are subdivided into three classes.[11] Class I MTs are defined to include polypeptides related in primary structure to mammalian MTs, while those of class II display none or only very distant evolutional relationships to mammalian forms. Class III MTs are polypeptides composed of atypical γ-glutamylcysteinyl units.[12]

Biological Function

The most conspicuous biological feature of the MTs is their inducibility by a variety of agents and conditions (Table I). Thus, their biosynthesis is greatly enhanced both *in vivo* and in cultured cells by metal ions and by certain hormones, cytokines, growth factors, tumor promoters, and many other chemicals. A massive accretion of MT is also observed in the livers of animals submitted to physical and chemical stress. The regulation of biosynthesis occurs mainly at the level of transcription initiation where a variety of cis- and trans-acting elements are involved.[9,13,14]

[9] Cited in D. H. Hamer, *Annu. Rev. Biochem.* **55,** 913 (1986).
[10] Cited in J. H. R. Kägi and Y. Kojima, *Metallothionein II Exp. Suppl.* **52,** 25 (1987).
[11] Y. Kojima, this volume [2].
[12] Cited in W. E. Rauser, *Annu. Rev. Biochem.* **59,** 61 (1990).
[13] Cited in R. D. Palmiter, *Metallothionein II Exp. Suppl.* **52,** 63 (1987).
[14] Cited in R. Chiu, R. Imbra, M. Imagawa, and M. Karin, *in* "Essential and Toxic Trace Elements in Human Health and Disease" (A. S. Prasad, ed.), p. 393. Alan R. Liss, New York, 1988.

---

[g] S. K. De, M. T. McMaster, and G. K. Andrews, *J. Biol. Chem.* **265,** 15267 (1990).
[h] J. J. Schroeder and R. J. Cousins, *Proc. Natl. Acad. Sci. U.S.A.* **87,** 3137 (1990).
[i] S. Morris and P. C. Huang, *Mol. Cell. Biol.* **7,** 600 (1987).
[j] P. Angel, A. Pöting, U. Mallick, H. J. Rahmsdorf, M. Schorpp, and P. Herrlich, *Mol. Cell. Biol.* **6,** 1760 (1986).
[k] C. J. Schmidt and D. H. Hamer, *Proc. Natl. Acad. Sci. U.S.A.* **83,** 3346 (1986).
[l] S. Onosaka, D. Kawakami, K.-S. Min, K. Oo-Ishi, and K. Tanaka, *Toxicology* **43,** 251 (1987).
[m] M. Karasawa, J. Hosoi, H. Hashiba, K. Nose, C. Tohyama, E. Abe, T. Suda, and T. Kuroki, *Proc. Natl. Acad. Sci. U.S.A.* **84,** 8810 (1987).
[n] S. Onosaka, Y. Ochi, K.-S. Min, Y. Fujita, and K. Tanaka, *Eisei Kagaku* **34,** 440 (1988).
[o] P. L. Goering, *Toxicol. Appl. Pharmacol.* **98,** 325 (1989).
[p] G. K. Andrews and E. D. Adamson, *Nucleic Acids Res.* **15,** 5461 (1987).
[q] Cited in M. P. Waalkes and J. M. Ward, *Toxicol. Appl. Pharmacol.* **100,** 217 (1989).
[r] E. Giroux and P. J. Lachmann, *J. Biol. Chem.* **259,** 3658 (1984).
[s] K. Amemiya, L. S. Hurley, and C. L. Keen, *Biol. Trace Elem. Res.* **11,** 161 (1986).
[t] Cited in P. L. Goering, S. K. Tandon, and C. D. Klaassen, *Toxicol. Appl. Pharmacol.* **80,** 467 (1985).

The large number of factors stimulating the biosynthesis of MT makes it difficult to pinpoint a specific biological role. In fact, more than three decades after their discovery, their functional significance is still a topic of discussion. The hypothesis that MTs mainly serve an unspecific protective function, limiting the intracellular concentrations of reactive heavy metal ions and shielding cellular structures from the harmful influences of toxic metals such as cadmium, mercury, platinum, bismuth, silver, and gold, has long been favored. However, the conservation of the structure of MT in evolution (see below), their ubiquitous occurrence, the redundancy of genes (see below), and the programmed synthesis of MTs in development, regeneration, and reproduction are weighty arguments for suspecting MT to also serve other and perhaps more specific metal-related cellular roles.[15,16]

That MT provides a reservoir for supplying zinc and copper in the biosynthesis of metalloenzymes and metalloproteins is supported by studies demonstrating that these metals are transferred *in vitro* from MT to the appropriate apoforms.[17,18] In animals where zinc is the major and often the sole metallic constituent[19] it has also been postulated that by controlling the flow of zinc, MT may serve a metalloregulatory role in zinc-dependent processes in replication, transcription, and translation.[20,21] This suggestion is sustained by the observation that the apoform of MT, apo-MT or thionein, readily removes zinc from zinc finger transcription factors such as Sp1[22] and *Xenopus* TFIIIA,[23] thereby abrogating their transcription activation competence. Hence, it appears that the changeable ratio of apo-MT to zinc-containing MT governs the direction of the flow of zinc. Thus, MT may serve as a zinc-dispensing and collecting system that both protects the cellular constituents against fluctuations in zinc supply and modulates the action of zinc-dependent processes fundamental to cell activation in proliferation and differentiation.

Primary Structure

A representative selection of primary structures from the three classes of MT is given in Fig. 1. Class I and II MTs are single-chain proteins. All

[15] Cited in M. Webb, *Metallothionein II Exp. Suppl.* **52,** 483 (1987).

[16] Cited in J. H. R. Kägi and A. Schäffer, *Biochemistry* **27,** 8509 (1988).

[17] F. O. Brady, *Trends Biochem. Sci.* **7,** 143 (1982).

[18] M. Beltramini and K. Lerch, *FEBS Lett.* **142,** 219 (1982).

[19] P. E. Hunziker and J. H. R. Kägi, *in* "Metalloproteins, Part 2: Metal Proteins with Non-redox Roles" (P. M. Harrison, ed.), p. 149. Macmillan, Houndmills, Basingstoke, 1985.

[20] B. L. Vallee, *Metallothionein Exp. Suppl.* **34,** 19 (1979).

[21] M. Karin, *Cell* **41,** 9 (1985).

mammalian forms contain 61 to 62 amino acid residues. Larger chains with 72 and 74 residues are found in mollusks[24] and in a nematode,[25] respectively. Shorter chains occur in insects[26] and certain fungi,[27,28] the shortest one, with 25 residues, in *Neurospora crassa*.[6] Class III MTs, which thus far have been found only in plants and certain microorganisms, are of variable length with usually less than 20 residues. Their complexes with cadmium form aggregates composed of two or more chains.[29] Class III MTs containing cadmium often include inorganic sulfide.[30,10] In some microorganisms class III MTs are components of CdS crystallites.[31]

Following the early studies on equine MT,[7] amino acid sequences of over 50 MTs have now been documented.[10] As evident from Fig. 1, the most conspicuous structural motif of all forms is the recurrence of Cys-X-Cys tripeptide sequences, where X stands for an amino acid residue other than cysteine. The hallmark of the class I MTs is the very close correspondence in the alignment of the cysteines along the chain. In the 37 mammalian MTs now known, 56% of all residues are conserved in evolution, among them all 20 cysteines and nearly all lysines and arginines. Most amino acid substitutions are conservative, both with respect to the chemical and the space-filling properties of the residues. Interestingly, aromatic amino acid residues are totally absent and there are also only a very few bulky aliphatic ones.

Both class I and II MTs display genetic polymorphism. Mammalian tissues usually contain two major fractions, MT-1 and MT-2, differing at neutral pH by a single negative charge. In lagomorphs, ungulates and primates there are also subforms within these fractions, separable by high performance liquid chromatography (HPLC)[32] and specified by lower-case letters, i.e., MT-1a, MT-1b, etc.[11] The most complex polymorphism is found in the human, where at least 10 iso-MT genes are expressed, some of

[22] J. Zeng, R. Heuchel, W. Schaffner, and J. H. R. Kägi, *FEBS Lett.* **279,** 310 (1991).

[23] J. Zeng and J. H. R. Kägi, *Experientia* **47,** A38 (1991).

[24] E. A. Mackay, Ph.D. thesis, University of Aberdeen, Scotland, 1989.

[25] M. Imagawa, T. Onozawa, K. Okumura, S. Osada, T. Nishihara, and M. Kondo, *Biochem. J.* **268,** 237 (1990).

[26] P. Silar, L. Théodore, R. Mokdad, N. E. Erraïss, A. Cadic, and M. Wegnez, *J. Mol. Biol.* **215,** 217 (1990).

[27] D. R. Winge, K. B. Nielson, W. R. Gray, and D. H. Hamer, *J. Biol. Chem.* **260,** 14464 (1985).

[28] R. K. Mehra, J. R. Garey, and D. R. Winge, *J. Biol. Chem.* **265,** 6369 (1990).

[29] D. J. Plocke, this volume [68].

[30] Y. Hayashi, C. W. Nakagawa, and A. Murasugi, *Environ. Health Perspect.* **65,** 13 (1986).

[31] C. T. Dameron, R. N. Reese, R. K. Mehra, A. R. Kortan, P. J. Carroll, M. L. Steigerwald, L. E. Brus, and D. R. Winge, *Nature (London)* **338,** 596 (1989).

[32] P. E. Hunziker, this volume [27].

Class I[a]	1[b] 20 40 60	Ref.
Human (MT-2)[c]	MDP NCSCAAGDSCTCAGSCKCKECKCTSCKKSCCSCCPVGCAKCAQGCICKGASD KCSCCA	f
Pigeon (MT-2)	MDPQDCTCAAGDSCSCAGSCKCKNCRCQSCRKSCCSCCPASCSNCAKGCVCKEPSSSKCSCCH	g
Trout (tMT-B)	MDP CECSKTGSCNCGGSCKCSNCACTSCKKSCCPCCPSDCSKCASGCVCKGKTC DTSCCQ	h
Crab (MT-2)	PDP C C NDKCDCKEGECKTGCKCTSCRCPPCEQCSSGC KCANKEDCRKTCSKPCSCCP	i
N. crassa	GDCGCSGASSCNCGSGCSCSNCGSK	j
Class II	1 20 40 60	
Sea urchin (Mtb)	MPDVKCVCCKEGNECACTGQDCCTIGKCCKDGTCCGKCSNAACKTCADGCTCGSGCSCTEGNCPC	k
Nematode (CeMT-II)	VCKCDCKNQNCSCNTGTKDCDCSDAKCCEQYCCPTASEKKCCKSGCAGGCKCANCECAQAAH	l
Yeast	QNEGHECQCQCGSCKNNEQCQKSCSCPTGCNSDDKCPCGNKSEETKKSCCSGK	m
Cyanobacterium	TSTTLVKCACEPCLCNVDPSKAIDRNGLYYCCEACADGHTGGSKGCGHTGCNC	n
Class III	1 10	
S. pombe (cadystin A)[d]	eCeCeCG[e]	o
R. canina (phytochelatin, PC_7)	eCeCeCeCeCeCeCG	p
P. vulgaris (homophytochelatin, h-PC_6)	eCeCeCeCeCeC-β-alanine	q

them tissue specifically.[33] The abundance of so many iso-MTs in mammals largely precludes the construction of an evolutional pedigree. Nonetheless, in class I MTs the increasing differences in amino acid sequence noticeable with increasing taxonomic distance is consistent with divergent evolution from an ancient precursor gene.[34] The concurrence of Cys-X-Cys and Cys-Cys sequences in otherwise unrelated class I and class II MTs is believed to be the result of convergent evolution guided by the requirements of metal complexation.

Metal-Binding Sites

The abundance of cysteines and their arrangement in the chain predisposes MT for the chelation of "soft" metal ions. In the fully occupied forms all cysteines participate as thiolate ligands in metal binding.[7] The details of coordination have been elucidated by the application of a large variety of spectroscopic methods (Table II). For many of these studies, homogeneously substituted derivatives had to be prepared in which the naturally bound and predominantly spectroscopically "silent" metal ions, Zn(II), Cd(II), and Cu(I), were replaced by spectroscopically "active" metal ions or isotopes suitable for the chosen spectroscopic method. In particular, replacement studies with Co(II) proved useful in establishing the mode

[33] P. E. Hunziker and J. H. R. Kägi, *in* "Essential and Toxic Trace Elements in Human Health and Disease" (A. S. Prasad, ed.), p. 349. Alan R. Liss, New York, 1988.

[34] J. H. R. Kägi, M. Vašák, K. Lerch, D. E. O. Gilg, P. Hunziker, W. R. Bernhard, and M. Good, *Environ. Health Perspect.* **54,** 93 (1984).

FIG. 1. Amino acid sequences of representative forms of MT. Key to superscript letters: [a] open positions denote deletions introduced for optimal alignment of class I MTs; [b] numeration refers to the sequence determined for human MT-2; [c] specified subform of MT is indicated in parentheses; [d] also designated "phytochelatin PC_3" (Ref. *p*); [e] e, glutamic acid residue linked by γ-glutamyl bond; [f] M. M. Kissling and J. H. R. Kägi, *FEBS Lett.* **82,** 247 (1977); [g] L. Lin, W. C. Lin, and P. C. Huang, *Biochem. Biophys. Acta* **1037,** 248 (1990); [h] K. Bonham, M. Zafarullah, and L. Gedamu, *DNA* **6,** 519 (1987); [i] K. Lerch, D. Ammer, and R. W. Olafson, *J. Biol. Chem.* **257,** 2420 (1982); [j] K. Lerch, *Experientia, Suppl.* **34,** 173 (1979); [k] M. Nemer, D. G. Wilkinson, E. C. Travaglini, E. J. Sternberg, and T. R. Butt, *Proc. Natl. Acad. Sci. U.S.A.* **82,** 4992 (1985); [l] M. Imagawa, T. Onozawa, K. Okumura, S. Osada, T. Nishihara, and M. Kondo, *Biochem. J.* **268,** 237 (1990); [m] D. R. Winge, K. B. Nielson, W. R. Gray, and D. H. Hamer, *J. Biol. Chem.* **260,** 14464 (1985); [n] R. W. Olafson, W. D. McCubbin, and C. M. Kay, *Biochem. J.* **251,** 691 (1988); [o] N. Kondo, K. Imai, M. Isobe, T. Goto, A. Murasugi, C. Wada-Nakagawa, and Y. Hayashi, *Tetrahedron Lett.* **25,** 3869 (1984); [p] E. Grill, E.-L. Winnacker, and M. H. Zenk, *FEBS Lett.* **197,** 115 (1986); [q] E. Grill, W. Gekeler, E.-L. Winnacker, and H. H. Zenk, *FEBS Lett.* **205,** 47 (1986).

TABLE II
SYNOPSIS OF SPECTROSCOPIC METHODS APPLIED TO THE STUDY OF METAL-BINDING SITES IN METALLOTHIONEIN[a]

Objective of study	Spectroscopic method	Metal derivative examined
Ligand identification	Electronic absorption spectroscopy	Cd, Zn, Hg, Co
	X-ray photoelectron spectroscopy	Cd, Zn, Hg, Cu
Elucidation of coordination geometry	Electronic absorption spectroscopy	Co, Fe
	Electron paramagnetic resonance (EPR)	Co
	Magnetic circular dichroism	Co
	^{113}Cd nuclear magnetic resonance (NMR)	^{113}Cd
	Perturbed angular correlation of γ ray (PAC)	^{111m}Cd
	Mössbauer spectroscopy	^{57}Fe
	Extended X-ray absorption fine structure (EXAFS)	Zn, Cu
Demonstration of metal-thiolate clusters	^{113}Cd nuclear magnetic resonance (NMR)	^{113}Cd
	Electron paramagnetic resonance (EPR)	Co
	Mössbauer spectroscopy	^{57}Fe
	Electronic absorption spectroscopy	Zn, Cd, Hg, Co, Fe
	Circular dichroism	Zn, Cd, Hg, Co, Fe
	Magnetic circular dichroism	Zn, Cd, Hg, Co, Fe
	Luminescence	Cu, Ag, Au

[a] For citations see M. Vašák and J. H. R. Kägi, "Metal Ions in Biological Systems" (H. Sigel, ed.), Vol. 15, p. 213. Marcel Dekker, New York, 1983; J. H. R. Kägi and Y. Kojima, *Metallothionein II Exp. Suppl.* **52,** 25 (1987); as well as appropriate chapters in this volume.

of metal coordination.[35-37] They documented that Co(II) and by inference also Zn(II) and Cd(II) are bound to four thiolate groups of cysteine side chains in tetrahedral MeS_4 geometry.

Metal-Thiolate Clusters

To reconcile the existence of these tetrathiolate complexes with the overall stoichiometry of 7 bivalent metal ions per 20 cysteines, one must postulate that they are joined to oligonuclear structures or clusters in which 8 of the cysteine side chains as bridging ligands are shared by adjacent metal ions. The first conclusive evidence for the existence of such metal–thiolate clusters and their topological organization has come from ^{113}Cd nuclear magnetic resonance (NMR) studies of an exclusively ^{113}Cd-containing form of MT, i.e., $^{113}Cd_7$-MT, which revealed ^{113}Cd–^{113}Cd scalar

[35] M. Vašák, *J. Am. Chem. Soc.* **102,** 3953 (1980).
[36] M. Vašák and J. H. R. Kägi, *Proc. Natl. Acad. Sci. U.S.A.* **78,** 6709 (1981).
[37] Cited in M. Vašák and J. H. R. Kägi, *in* "Metal Ions in Biological Systems," Vol. 15 (H. Sigel, ed.), p. 213. Marcel Dekker, New York, 1983.

coupling via the bridging thiolate ligands.[38] Analogous indications of metal–metal interaction via the sulfur bridges were also provided by the observation of antiferromagnetic coupling of the metal centers in electron spin resonance and magnetic susceptibility measurements of the Co(II)-substituted derivative[36] as well as by Mössbauer studies of the ^{57}Fe(II)-substituted form.[39] The clustered arrangement of the metals also manifests itself in the electronic absorption spectrum and in the natural and magnetic chiroptical properties of MT. Thus, the metal-specific charge-transfer transitions of forms containing 7 equivalents of Cd(II), Hg(II), Co(II), and Fe(II) are red shifted with respect to the spectral position of the corresponding mononuclear tetrahedral tetrathiolate complexes[10] and are associated with elaborate multiphasic profiles in the optical rotatory dispersion[40,4] and circular dichroism spectra.[37,41] The latter features originate from excitonic coupling of like transitions of neighboring bridging thiolate ligands within the cluster.[42]

The ^{113}Cd–^{113}Cd NMR decoupling experiments revealed that in mammalian and crustacean MTs the ^{113}Cd resonances are partitioned into two separate linkage groups. In the mammalian forms the 7 ^{113}Cd ions were inferred from such data to be arranged in two distinct clusters, one with 3 ^{113}Cd ions and 9 cysteines, $^{113}Cd(II)_3Cys_9$, consistent with a cyclohexane-like structure, and the other with 4 ^{113}Cd ions and 11 Cys, $^{113}Cd(II)_4Cys_{11}$, consistent with a bicyclo[3.3.1]nonane-like structure.[38] This partition was also supported by independent protein-chemical studies that revealed a two-domain structure of MT and which established that the three-metal cluster and the four-metal cluster are associated with the amino- and carboxyl-terminal domains, respectively.[43]

Spatial Structure

The two-cluster, two-domain model derived from spectroscopic and chemical studies is fully corroborated by the spatial structures derived from two-dimensional NMR spectroscopic analysis in aqueous solution. The essential step in this structure determination was the unambiguous allocation of the 28 coordination bonds connecting the 7 metal ions with the 20

[38] J. D. Otvos and I. M. Armitage, *Proc. Natl. Acad. Sci. U.S.A.* **77,** 7094 (1980).

[39] X. Ding, E. Bill, M. Good, A. X. Trautwein, and M. Vašák, *Eur. J. Biochem.* **171,** 711 (1988).

[40] D. D. Ulmer, J. H. R. Kägi, and B. L. Vallee, *Biochem. Biophys. Res. Commun.* **8,** 327 (1962).

[41] H. Willner, M. Vašák, and J. H. R. Kägi, *Biochemistry* **26,** 6287 (1987).

[42] H. Willner, W. R. Bernhard, and J. H. R. Kägi, *in* "Metal Binding in Sulfur-Containing Proteins" (M. Stillman, C. F. Shaw, and K. T. Suzuki, eds.). VCH Publishers, New York, 1991.

[43] D. R. Winge and K.-A. Miklossy, *J. Biol. Chem.* **257,** 3471 (1982).

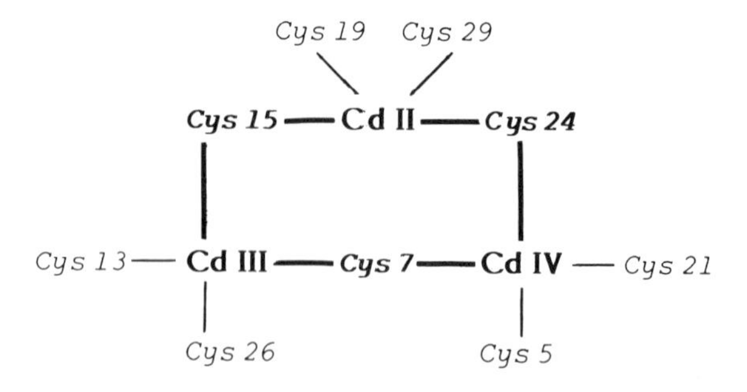

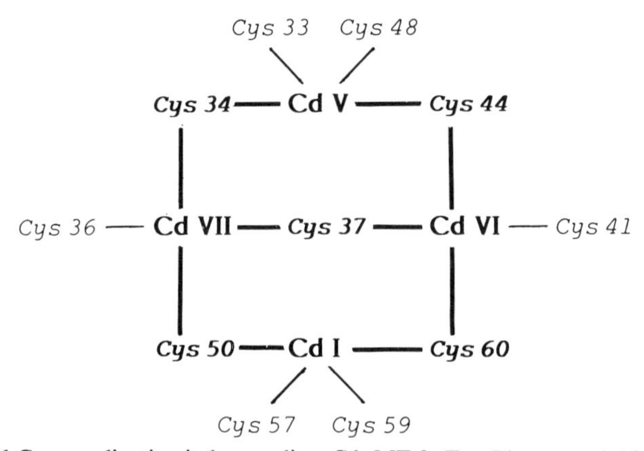

FIG. 2. Cd-Cys coordination in human liver Cd_2-MT-2. *Top:* Three-metal thiolate cluster. *Bottom:* Four-metal thiolate cluster. The roman numerals specifying the metal sites refer to the corresponding ^{113}Cd NMR resonance positions.[38] The arabic numerals refer to the residue position in the sequence (see Fig. 1). The thick- and thin-lettered residues denote the bridging and terminal cysteines, respectively.

cysteines by 2-dimensional heteronuclear ^{1}H^{113}Cd NMR correlation spectroscopy of $^{113}Cd_7$-MT.[44] This method, developed for the study of MT, allows the direct observation of the coupling of each of the seven ^{113}Cd resonances differentiated in the ^{113}Cd NMR spectrum with the ^{1}H resonances of the appropriate cysteines, which beforehand had been identified in the ^{1}H spectrum by two-dimensional homonuclear ^{1}H NMR correlation and nuclear Overhauser enhancement spectroscopy (COSY and NOESY) measurements.[45] The networks of Cd-Cys bonds established by this analy-

[44] M. H. Frey, G. Wagner, M. Vašák, O. W. Sørensen, D. Neuhaus, E. Wörgötter, J. H. R. Kägi, R. R. Ernst, and K. Wüthrich, *J. Am. Chem. Soc.* **107,** 6847 (1985).

[45] K. Wüthrich, this volume [59].

sis in ^{113}Cd-substituted human MT-2[46] are displayed in Fig. 2. The arrangements show that of the 20 cysteines coordinated to the metal ions, Cys-7, Cys-15, and Cys-24 are the bridging ligands of the three-metal cluster and residues Cys-34, Cys-37, Cys-44, Cys-50, and Cys-60 those of the four-metal cluster, forming the predicted cyclohexane-like and the bicyclo[3.3.1]nonane-like ring structures, respectively. Exactly the same topological arrangement of Cys-Cd bonds is found in rabbit liver MT-2a[47] and rat liver MT-2.[48]

From the interresidue distances given by the geometric requirements of the established 42 intramolecular Cys-Cd-Cys cross-links and from long-range through-space ^{1}H–^{1}H distance constraints provided by 2-dimensional homonuclear ^{1}H NMR NOESY measurements, the most probable spatial fold of the polypeptide chain of human liver $^{113}Cd_7$-MT-2 was deduced using a distance geometry algorithm developed for tertiary protein structures.[49] A stereo view of this computed structure is shown in Fig. 3. For the sake of clarity, the model is simplified to show only the course of the polypeptide backbone, the orientation of the cysteine side chains, and the positions of the seven ^{113}Cd(II). It documents that MT is composed of two about equally sized globular domains. Their structures, which in reality are connected through the conserved Lys_{30} - Lys_{31} segment, are drawn separately, because their mutual orientation is not sufficiently defined by the available 2-dimensional NMR data. Each domain contains in its interior the appropriate metal–thiolate cluster as a "mineral core" wrapped by two large helical turns of the polypeptide chain. In the amino-terminal domain the chain fold is right handed; in the carboxyl-terminal one it is left handed.

Figure 4 shows that the spatial structure of human MT-2 is closely similar to those determined previously for rabbit MT-2a and rat MT-2. The agreement in the course of the polypeptide backbone is the more remarkable, as two of them (rabbit MT-2a and rat MT-2) differ in more than 25% of their noncysteine amino acid residues.[10] It attests that the organization of the metal–thiolate clusters and the conformation of the polypeptide chain enfolding them are dictated by the conserved arrangement of the cysteines in the chains (see above). The spatial structures established by 2-dimensional NMR spectroscopy are also in excellent

[46] B. A. Messerle, A. Schäffer, M. Vašák, J. H. R. Kägi, and K. Wüthrich, *J. Mol. Biol.* **214,** 765 (1990).

[47] A. Arseniev, P. Schultze, E. Wörgötter, W. Braun, G. Wagner, M. Vašák, J. H. R. Kägi, and K. Wüthrich, *J. Mol. Biol.* **201,** 637 (1988).

[48] P. Schultze, E. Wörgötter, W. Braun, G. Wagner, M. Vašák, J. H. R. Kägi, and K. Wüthrich, *J. Mol. Biol.* **203,** 251 (1988).

[49] W. Braun and N. Gõ, *J. Mol. Biol.* **186,** 611 (1985).

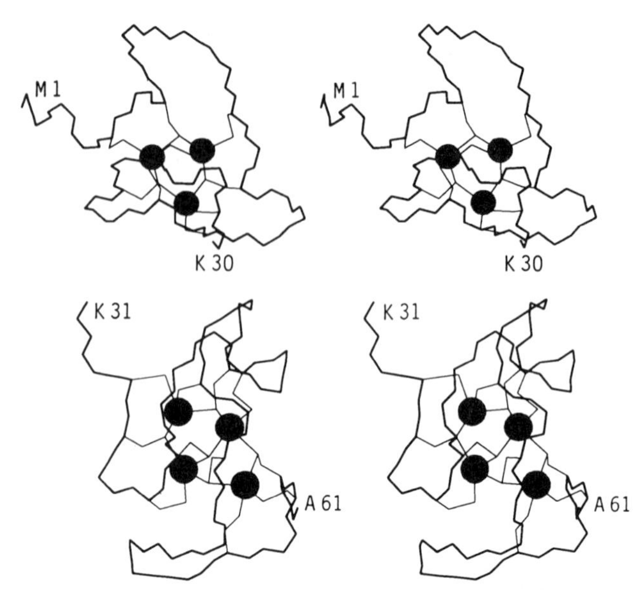

FIG. 3. Two-dimensional NMR solution structure of amino-terminal domain (*top*) and carboxyl-terminal domain (*bottom*) of human liver Cd_7-MT-2. Stereo view of the polypeptide backbone (thick line), cysteine side chains (thin lines), and metal positions (filled spheres) as determined by 2D ^{113}Cd ^1H heteronuclear and 2D $^1H^1H$ homonuclear NMR correlation spectroscopy. The separately drawn domains are in reality connected through the K_{30} - K_{31} segment (see text). The figure is adapted from Ref. 46.

agreement with the redetermined crystal structure of native rat liver MT-2.[50] Thus, as with most other globular proteins, there are only minor structural differences between the dissolved and the crystalline state.

Dynamic Aspects

Notwithstanding their well-defined equilibrium structures ascertained by 2-dimensional NMR and by crystallographic studies, the MTs are flexible molecules undergoing dynamic fluctuations within the clusters and the polypeptide chain. This lack of rigidity is disclosed by the quality of the ^{113}Cd NMR resonances[51,52] and other spectroscopic features[53] as well as by

[50] A. H. Robbins and C. D. Stout, this volume [58].

[51] M. Vašák, G. E. Hawkes, J. K. Nicholson, and P. J. Sadler, *Biochemistry* **24,** 740, 1985.

[52] J. Otvos, S.-m. Chen, and X. Liu, *in* "Metal Ion Homeostasis: Molecular Biology and Chemistry" (D. H. Hamer and D. R. Winge, eds.), p. 197. Alan R. Liss, New York, 1989.

[53] M. Vašák, *Environ. Health Perspect.* **65,** 193 (1986).

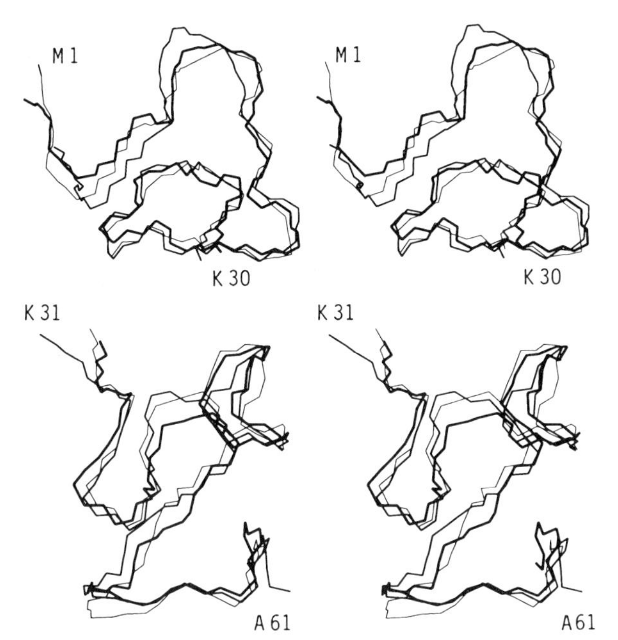

FIG. 4. Stereo view of a superposition of the 2D NMR structures of the polypeptide backbone of human liver MT-2 (thickest line, Ref. 46), rabbit MT-2a (medium thickness line, Ref. 47), and rat MT-2 (thinnest line, Ref. 48). *Top:* Amino-terminal domain. *Bottom:* Carboxyl-terminal domain. (The figure is adapted from Ref. 46.)

the free accessibility of most peptide hydrogens to D_2O,[54,55] and by the ready reactivity of the buried cysteine side chains with alkylating agents.[56] These features also reveal substantial differences between the two domains. Thus, in the mammalian forms, the amino-terminal domain is more loosely structured than the carboxyl-terminal domain.[55,56] The high structural flexibility of the MTs has its likely physical basis in the kinetic lability of the many coordination bonds to group IIB metal ions [Me(II)] in the clusters.[57,58] Thus, while thermodynamically stable,[37] the Cys-Me(II)-Cys cross-links are to be visualized as undergoing a continuous

[54] D. D. Ulmer and B. L. Vallee, *Adv. Chem. Ser.* **100,** 187 (1971).

[55] B. A. Messerle, M. Bos, A. Schäffer, M. Vašák, J. H. R. Kägi, and K. Wüthrich, *J. Mol. Biol.* **214,** 781 (1990).

[56] W. R. Bernhard, M. Vašák, and J. H. R. Kägi, *Biochemistry* **25,** 1975 (1986).

[57] M. Eigen, *in* "Coordination Chemistry. Plenary Lectures Presented at the Seventh International Conference on Coordination Chemistry, Stockholm, 1962," p. 97. Butterworths, London, 1963.

[58] G. K. Carson and P. A. W. Dean, *Inorg. Chim. Acta* **66,** 157 (1982).

breaking and reforming of their noncovalent bonds. A patent manifestation of this "fluxional" state is the facile exchange of the metal ions within the cluster,[59] with metal ions in solution,[60] and, surprisingly, also with metal ions in clusters of other MT molecules.[60] The latter reaction is unique to MT and proceeds with a rate of about 2 sec^{-1} for the three-metal cluster[52] and a rate of about 1×10^{-3} sec^{-1} for the four-metal cluster,[61] allowing for a fast dispersion of labeled metal ions among all MT molecules in the solution. As established by ^{111}Cd NMR saturation transfer experiments[52] and radioisotope redistribution studies with ^{109}Cd, ^{65}Zn, or ^{203}Hg between MT isoforms,[61] these metal exchange reactions occur by direct transfer between the clusters and thus implicate transient pairwise associations of MT molecules and coordinated intermolecular ligand exchange processes.[62] We believe that this capacity of the metal–thiolate clusters to facilitate intermolecular metal ion transfer is fundamental to the biological roles of MT. It may represent the mechanistic basis for the chaperoning function that MT was suggested to have in channeling and regulating the flow of essential metals, in particular of zinc, to and from their many sites of action.[52,63]

[59] J. D. Otvos, H. R. Engeseth, D. G. Nettesheim, and C. R. Hilt, *Metallothionein II Exp. Suppl.* **52,** 171 (1987).

[60] D. G. Nettesheim, H. R. Engeseth, and J. D. Otvos, *Biochemistry* **24,** 6744 (1985).

[61] R. F. Schmid, Diploma thesis, University of Zürich, Switzerland, 1991.

[62] R. Schmid, J. Zeng, and A. Schäffer, *Experientia* **46,** A36 (1990).

[63] B. L. Vallee, *Metallothionein II Exp. Suppl.* **52,** 5 (1987).

Author Index

Numbers in parentheses are footnote reference numbers and indicate that an author's work is referred to although the name is not cited in the text.

A

B

D

H

L

M

O

P

Q

R

S

T

U

V

W

X

Y

Z

Subject Index

A

B

H

I

M

N

Q

R

S

T

U

V

W